D0931253

YOUR HEREDITY AND ENVIRONMENT

Also by Amram Scheinfeld

You and Heredity (1939)
Women and Men (1944)
The *New* You and Heredity (1950)
The Human Heredity Handbook (1956)
The Basic Facts of Human Heredity (1961)

FOR CHILDREN

Why You Are You (1959)

FICTION

Postscript to Wendy (1948)

YOUR HEREDITY AND ENVIRONMENT

YOUR HEREDITY AND ENVIRONMENT

by

AMRAM SCHEINFELD

Assisted in research and editing by
Herbert L. Cooper, M. D. and by
others herein mentioned

Illustrated by the Author

•

J. B. LIPPINCOTT COMPANY
Philadelphia and New York

SECOND PRINTING, WITH CORRECTIONS

Library of Congress catalog card number 64-14468

PRINTED IN THE UNITED STATES OF AMERICA

THE HUMAN CHROMOSOMES

Shown here are the author's *chromosomes in their doubled stage.** Your own *chromosomes, which carry everything you inherit and transmit genetically, would look just like these under the microscope when similarly prepared.*

A. The 46 chromosomes, in the doubled stage.

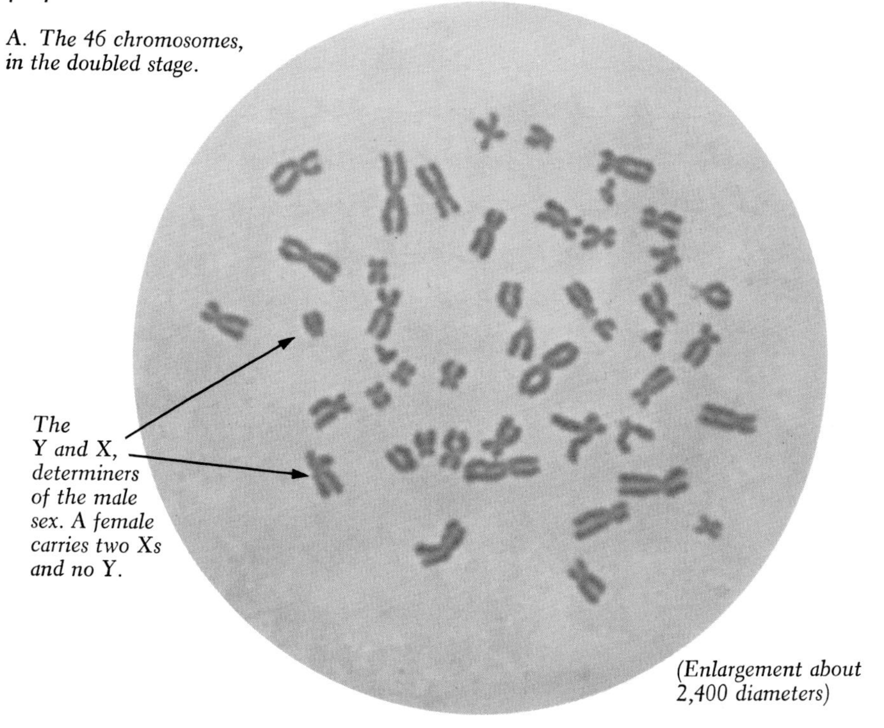

(Enlargement about 2,400 diameters)

B. Below, the chromosomes (as cut out from the above photographic print) arranged schematically in order of size and shape.

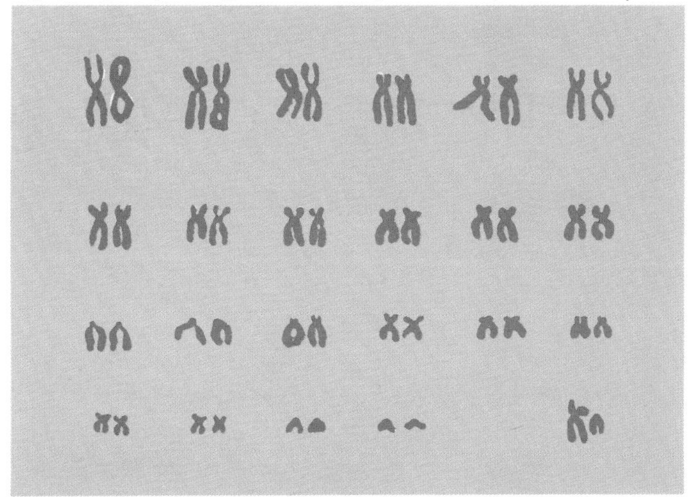

* The author's chromosomes were prepared from one of his white blood cells and photographed under supervision of Dr. Orlando J. Miller, Depart. of Obstetrics and Gynecology, Columbia University College of Physicians and Surgeons.

Also by Amram Scheinfeld

You and Heredity (1939)
Women and Men (1944)
The *New* You and Heredity (1950)
The Human Heredity Handbook (1956)
The Basic Facts of Human Heredity (1961)

FOR CHILDREN

Why You Are You (1959)

FICTION

Postscript to Wendy (1948)

TO DOROTHY
SYLVIA AND AARON
RUHAMAH AND LEO
AND THE MEMORY OF
ROSALIE AND EMMANUEL

Contents

Illustrations

Illustrations

Tables

Preface

It is now more than a quarter of a century since my first book on human genetics, *You and Heredity*, was published (officially, on September 2, 1939, in the very hours that World War II was breaking out); and it is fifteen years since that book's much-revised successor, *The New You and Heredity*, made its bow. In the periods that have followed, the world of mankind has undergone tremendous changes, and so have all of the sciences—human genetics in particular—which deal with the forces that shape human lives. Not least, my increasing personal identification with these sciences has added greatly to what I can report about them. All of which is the *raison d'être* for this new book and can explain why it goes so far beyond its predecessors in size, scope and depth.

It is hard to realize how much was not known about the subject three decades ago, when I first began concentrating on human heredity, and how little even of what was known had seeped through to the public. The overwhelming majority of persons, including many of the most educated, had never heard of genes or chromosomes, and knew next to nothing about how the mechanism of heredity really worked in human beings or about the distinctions between the effects of heredity and environment. Against the present background it may also seem incredible that attempts to popularize the facts about human heredity were then frustrated by the prudish reluctance of all but a few American newspapers and mass magazines (those for women, especially) to permit mention of details regarding human reproduction or even of references to sperms and eggs. (In 1940, an explanation I was scheduled to give, on one of the leading movie newsreels, about how twins were produced—to accompany film shots of the identical-twin family shown in this book at the end of Chapter 17—was canceled when the producer became fearful that the demonstration would "offend family audiences.")

Fortunately, today's general readers, students and, of course, professional persons are far more tolerant, knowledgeable and sophisticated than were those of a generation ago. Not only "genes" and "chromosomes" but also "mutations," the "Rh factor" and many other terms dealing with genetics have become household words; and almost everyone is aware of differences between various acquired and inherited characteristics. Nevertheless, it is painfully clear that grievous misapprehensions, areas of ignorance and ingrained prejudices regarding human heredity still persist and contribute to countless personal fears, family worries, national problems, ethnic and racial conflicts and human miseries of many kinds which could be greatly lessened if the facts were known. And more than ever, the increasing interdependence among human beings has made it necessary that these facts be broadcast and accepted if mankind is to progress—or even to survive.

Thus, I believe that modern human genetics is now prepared to aid significantly in enlightening people on many points of vital importance in their everyday lives and in their relationships to their fellow humans. When I speak of "modern" human genetics I have in mind that this science—to an extent relatively greater than almost any other science—has undergone a revolutionary growth and advancement in recent years. In fact, one could set the beginning of the new era in human genetics at the opening of the First International Congress of Human Genetics in Copenhagen, the week of August 1 to 6, 1956.* Presented there, among other new developments, was an exhibit I saw of striking photographs of human chromosomes which a young geneticist, J. H. Tijo,

* The lagging progress of the science until then was set forth in a paper I gave at the closing plenary session of that Congress, in which I reported: "Only about 1 in 10 of the accredited medical colleges [in the United States] and hardly 1 in 100 of all the other colleges and universities, offer as much as a single semester course in human genetics. In the country at large there probably are no more than a dozen full-time human geneticists, and no more than two or three annual graduates in this specialty. For the most part, those active in the field are part-time workers, mainly medical men, general geneticists and social scientists. . . . Except for Denmark, the situation may not be much different in most other countries." This situation was to begin improving almost immediately after the Congress, and to such an extent that at this writing the college courses and the number of workers in human genetics may well be five times what they were in 1956. (Paper cited: "The Public and Human Genetics," Amram Scheinfeld. *Proc. 1st Intl. Congr. Hum. Genetics, Acta Genet. et Statist. Med.*, 7:2, 1957. Also, revised, in *Eug. Qu.*, 5:3, Sept., 1958.)

and a colleague, A. Levan, had made by a new technique. For the first time, human chromosomes could be accurately counted and identified as to their different shapes, sizes and various distinguishing characteristics. Soon hundreds of other investigators were preparing, photographing and examining vast numbers of human chromosomes, and in quick order came amazing new findings regarding the chromosomal causes of many previously baffling abnormalities. In fact, it was now possible to *see* and diagnose many genetic traits of individuals under the microscope. Graphically illustrating this amazing advance is the frontispiece of this book, showing my own chromosomes. These may be compared with the different views of human chromosomes on the plate facing page 62, made by earlier techniques, and which had been the frontispiece of my 1950 book.

Further concrete evidence of the recent great strides in genetics (both general and human) will be found in the detailed discussions of a great many new subjects about which little or nothing could be said in my previous books: As examples, DNA and the "genetic code"; human organ transplants; the "sex-identity" tests; the "inactivated X" (Lyon hypothesis); the Klinefelter and Turner syndromes; many newly identified "errors of metabolism" and other "wayward gene" conditions; many new blood elements; atomic-fallout–induced mutations; Zinjanthropus and other of man's earliest forebears; and so on. Additional advances are evident in the greatly enhanced discussions of prenatal influences, mental diseases and defects, twins and twinning, longevity, behavior, sex life, human evolution, race differences, the "population explosion," birth control and genetic problems of parenthood. While certain areas still are clouded—intelligence, personality, talents, crime and sexual behavior among them—enough new light about these has come through to show that hereditary influences are at work in even the most complex human traits.

More important than any individual development may be the dramatic emergence of human genetics as a whole from the laboratories, the clinics, the technical journals, to become an active and meaningful force in modern life. At the same time, since no science functions in a vacuum, the course of human genetics has been much influenced by world events and changes in the cultural milieu. Thus, the atom bomb and its threats to the human germ plasm gave enormous impetus to genetic studies and brought forth lavish subsidies for carrying them out; and the launching of the U.S.S.R.'s first sputnik caused an additional stepping-up of scientific research in the United States, with human genetics sharing the benefits. The explosive outcroppings of many new nations in Africa and

Asia, the race conflicts in many countries, the political upheavals everywhere, have jolted the world into revised thinking about the genetic differences among human groups. The emergence of women into new areas of activity has forced a re-evaluation of the basic sex differences in aptitudes, while the steadily widening margin of female survival has focused new interest on the possible genetic causes of female biological superiority. The increased attention given to individual differences in intelligence, aptitudes and personality in education and industry has stimulated the search for genetic influences in these traits. Mushrooming crime and delinquency, alcoholism, drug addiction, homosexuality, sexual maladjustments and various other social problems, however much environment may be responsible, are also being re-examined from the standpoint of contributory genetic predispositions. The "population explosion" has spurred research into the problems of differential human reproduction and the possible effects on mankind's genetic makeup. And continuing scientific advance in every area has inspired speculation as to what may lie ahead in the control even of human heredity.

So human genetics, as I have viewed it and am seeking to present it, is far more than a limited scientific specialty dealing with the mechanism of human heredity—to be spoken of in terms of recessives, dominants and other genes, and of such and such inherited traits. Rather, I have regarded human genetics as an introspective, personal, all-embracing science which can reveal people in the round and help us to see why we as individuals and groups are what we are, and what our inborn capacities might enable us to be for the better. At the same time, in viewing matters from the readers' standpoints, I have tended to place more stress on what I think people generally want to know about, rather than on what may often seem most important to the scientist. Specifically, this book will be found to devote relatively much more space to the inheritance of normal physical traits (features, coloring, body form), mental and behavioral traits (intelligence, personality, behavior, sex life), cultural race differences and social problems—even where the genetic elements are indistinct—than to the mechanics of heredity and to the pathological traits with which the average technical book in the field is more largely concerned.

In short, it is hardly surprising that, with its broadened outlook and many new subjects presented, this book should differ greatly in many ways from its predecessors. Precisely how and why it differs can best be told by the following brief history.

You and Heredity, my first book, grew out of a scientific soil that by today's standards would be considered almost pathetically meager and malnourished.* Human genetics in 1939 was merely a pigeonhole in the science of general genetics (in itself barely thirty-five years old), which dealt mainly with plants, livestock and experimental creatures. There were no human genetics journals in English (although there was one in German), and the few journals that gave considerable attention to human heredity carried "Eugenics" in their titles and stressed theories that even then were repudiated by many geneticists. There was no organization of human geneticists—the American Society not having been established until 1948 (its journal was first published two years later). Nevertheless, by a diligent gleaning of the available literature, discussion with well-informed scientists and smatterings of original research, I found it possible to achieve and communicate a good understanding of whatever of importance was known about human heredity at the time. As it happened, the pioneers in the field had explored, hewed out the pathways and built well, for the basic concepts they established have stood up firmly through the years; and if there were many vague or unknown areas three decades ago, it was possible even then to dispel some of the most grievous myths and misconceptions about human heredity and to tell a good deal of what was true or likely to be true.

The knowledge of human heredity (and my own acquaintance with it) continued to grow, and eleven years later I produced a much revised and enlarged edition of my book, with the title *The New You and Heredity*.† This book was almost twice the size of the first one, had mostly new illustrations, and—reflecting advances in education and the increasing use of my book in high schools and colleges—was considerably upgraded in readership and bibliographical material. (In 1961 a special edition of the book was issued with a ten-page Addendum.)

With continuing advances in human genetics, still another revision was planned. Active work on that book was begun in 1958 for con-

* While the formal publication date of *You and Heredity* was September 2, 1939, a preliminary edition had come from the press in November, 1938, inasmuch as the original publication date had been set for the following February. Selection by the Book-of-the-Month Club caused a postponement, which also provided an opportunity for limited revisions. Publication in the United States was by the Frederick A. Stokes Company (later taken over by my present publishers, J. B. Lippincott Company), and simultaneously in Great Britain by Chatto & Windus. Subsequently, other editions of the book were published in Swedish, Danish, Spanish, Portuguese and Italian.

† In the meantime I had written another book, *Women and Men*, dealing with the interplay of heredity and environment in producing the differences between the sexes. That book was published in 1944, by Harcourt, Brace & Company; in England, by Chatto & Windus; and in various translations, including Danish, Swedish, Portuguese, Spanish, Italian and Japanese.

templated publication in 1960. But neither I nor my publishers could foresee the momentous developments in the science which lay ahead and which would sweep the project along by tidal waves of new findings. Many existing concepts began to give way to new ones; long-standing hypotheses crystallized into facts; innumerable previous assumptions underwent radical change; impressive new structures of evidence appeared on the horizon, exciting new vistas opened, and the whole landscape of human genetics took on a more intricate shape. The planned two-year job stretched out into five years, and eventually resulted in this largely new book. While limited parts of the previous book are retained —mostly in areas not greatly changed or augmented by new findings— one can regard this carry-over as contributing flavor and mellowness (as with aged spirits blended in with the new). In concrete terms, these comparisons can be made between my new book and its forebears:

In wordage (with its larger and more closely-set type pages) the new book is almost three times the length of the original *You and Heredity* and almost two-thirds again as large as *The New You and Heredity.* Of the contents, only about 10 per cent of that in the original book, and about 20 per cent of that in the second book, has been retained. Of my line illustrations, most are entirely new or completely redrawn, and the remainder have been completely reset; all of the photographs are new as compared with those in the original book, and only a few have been retained from the second one; and all of the text tables, almost all of the Appendix material and all of the Bibliography (except for a few items) are entirely new.

A wholly new title for the book was therefore in order. To explain the choice that was made, the words "heredity and environment" express the great emphasis in this book on the constant interaction between the two forces, while the "your" stresses the book's objective to relate the discussions wherever possible to the reader personally, and to those aspects of the subject which might interest and concern him most.

Achieving the aims of this book, it can be assumed, entailed many problems of selecting and evaluating the vast amount of material pouring forth; of balancing the discussions wherever there were moot points; of stating the facts with the thought not so much of always aiming for technical preciseness as of conveying to readers the most accurate impression—which might often mean, as in ballistics, aiming somewhat away from the target in order to come closest to hitting it. Intensifying all of these problems is the fact that the growing complexity of human genetics, and of the other sciences on which it draws, has made it wellnigh impossible for any one person to master, keep abreast of and completely evaluate the developments in all of the fields. So, if I was to be thought of as a "generalist," it was imperative that in preparing this book I would have to seek aid from a great many specialists. Fortunately, there were many of these among my friends, colleagues and professional acquaintances, and their responses to my calls were most generous.

In expressing my thanks, I must begin first with Dr. Herbert L. Cooper, one of the new generation of brilliant human geneticists whom I had the good fortune to enlist as an aide in research and editing. During most of the period he was helping me, he was at the New York University–Bellevue Medical College, concentrating mainly on human chromosome study while there.* It would be hard to overestimate the importance of the insights he provided into the latest developments in chromosomal and medical genetics, of necessary data he often supplied and of the aid he gave in checking various sections of the book.

Dr. F. Clarke Fraser, Professor of Medical Genetics of McGill University, and an esteemed friend of long standing, deserves another large share of my gratitude for his extensive help in checking all of the major medical genetics sections of the book (including the Wayward Gene Tables in the Appendix), as well as the Glossary. Additional direct aid by other friends and professional acquaintances in reading individual chapters or sections of the manuscript, or in connection with the illustrations, has been so important that I feel the acknowledgments warrant a special section following this Preface. In that section there also are references to the assistance received in preparing the illustrations. Here, however, I should like to make particular mention of the important contributions of my good friend Dr. Orlando J. Miller, and his assistants, in painstakingly preparing the frontispiece photograph and "karyotype" of my chromosomes.

More general, but no less important, have been the aid and stimulation I received through continuous contacts with my colleagues of the Columbia University Seminar on Genetics and the Evolution of Man. Under the spirited guidance of Professor Howard Levene, the frequent seminar gatherings (informal dinners, followed by scheduled talks and discussions) provided an opportunity for keeping abreast of important developments in the science and of becoming acquainted with notable visiting scientists who played important roles in these developments. I think back fondly to innumerable of these gatherings as among my most

* Dr. Cooper subsequently became affiliated with the National Institute of Dental Research of the National Institute of Health in Bethesda, Md., where he continues his chromosome and genetic studies.

valued and enjoyable experiences; and to all of my seminar colleagues, I extend thanks for the privilege of having been one of them.*

I also recall with gratitude some who aided me importantly with my previous books, and whose influence has carried over into sections retained in this new book. Among them are Dr. Morton D. Schweitzer (now Associate Professor of Epidemiology, Columbia University), who helped greatly with the research and editing for *You and Heredity;* Dr. J. B. S. Haldane, who, in editing a preliminary edition of that book for British publication, made many pertinent suggestions, which were incorporated in the subsequent American and other editions; and Dr. Julian Huxley, who also made important suggestions for the same book. Others who were constant sources of aid and inspiration with my previous books were two dear and now departed friends, Dr. Abraham Stone, a pioneer and great leader of the Planned Parenthood movement, and Professor Irving Lorge, with whose Institute of Psychological Research of Teachers College, Columbia University, I became affiliated. Undoubtedly, many others who should be mentioned have been overlooked, and where so, I hope that I will be forgiven for faulty memory and that their names will be considered included in the general thanks extended to the numerous estimable persons in all of the human sciences whose combined efforts have helped to make this book possible.

One other bouquet: No gratitude could be adequately expressed for the aid given by my devoted and capable wife, Dorothy, who, on countless occasions, stepped in to do emergency typing and correspondence, proofreading, editing and other chores, and who through the years patiently accepted the interference of "The Book" in vacations and social activities.

Not least, I am profoundly grateful to the science of human genetics itself and to the other related sciences for what affiliation with them, and my acquaintance with the dedicated persons engaged in them, has meant to me in so many important ways. The Preface of my 1950 book included a passage that I should like to repeat here:

> Thinking back to my earlier years, I can recall some of my confusion and uncertainty, and to a considerable degree of cynicism, regarding human makeup and "human nature," and the possibility of man-

* The seminar "regulars" have included, among others, Drs. Alexander G. Bearn, William C. Boyd, Herbert L. Cooper, Theodosius Dobzhansky, L. C. Dunn, Arthur Falek, I. Lester Firscheim, Franz J. Kallmann, Frances V. de George, Jerry Hirsch, Kurt Hirschhorn, Edward E. Hunt, Jr., James C. King, Howard Levene, Philip Levine, Max Levitan, Ashley Montagu, Richard H. Osborne, John Rainer, Richard Rosenfield, Diane Sank, Winnifred Seegers.

kind's becoming better. So it is heartening to say that the more science has made clear to me the amazing life processes, the greater has grown my awe and humility, and my respect for the human organism; and the more I have learned about people—what is inherent in them and what is not, how they develop and how they *can* develop—the stronger has grown my faith in them. At a time of continuing conflict and tragedy, of doubt and pessimism, the knowledge of what human beings have the power to become, rather than what they still are, can offer reassurance and hope. I, for one, have never been more convinced than I am now of the fundamental soundness, decency and goodness of the great majority of my fellow men; more certain that their genetic assets far outweigh their liabilities; or more confident that as people the world over really get to know and understand themselves, to make the most of their great endowments, and to give expression to their true impulses to work together for their common good, their lives will be increasingly fuller, worthier and happier.

The foregoing sentiments, I may add now, have been strengthened with the passing years and have been a never-ending source of comfort and optimism in a chaotic and often discouraging world.

Altogether, as I near the end of my work on *Your Heredity and Environment*, I can say, happily, that it has been as absorbing, rewarding and satisfying a project as any writer could wish. Assembling and organizing the material was in certain respects like doing a huge jigsaw puzzle while hundreds of the pieces were still in the process of being manufactured. It has been exciting to fit the pieces together, to see open areas filling up and becoming colored and clarified by new details, to find misty hypotheses turning into realities and, even where blank spaces remain, often to guess by the shapes what the forms of future findings would be.

But as much as anything else, the exploration into the nature of man, and writing about it, has been for me a continuously exciting adventure in human understanding. I hope this book has made it possible for the reader to share some of the excitement and the rewards of that adventure.

Amram Scheinfeld

New York City
November 10, 1964

Acknowledgments

THE profound thanks of the author are extended to the following distinguished persons for reading sections of the book in manuscript (the numbers in parentheses designating chapters), and offering important criticisms and suggestions, or rendering other aid. In extending thanks it should be clear that no responsibility is attributed to those mentioned for any errors of commission or omission which may be found in this book. (Where a name appears more than once, the identification of the individual is given only at first mention.)

Prenatal Life (6): Dr. Virginia Apgar, Director, Division of Congenital Malformations, The National Foundation, New York.

Dr. Lester W. Sontag, Director, Fels Research Institute, Yellow Springs, Ohio.

Human sperms and ovum (photographs): Dr. Landrum B. Shettles, Asst. Prof., Obstetrics and Gynecology, College of Physicians and Surgeons, Columbia University.

Human embryos (photographs): Dr. Mary Rawles, Curator of Collection, and Mr. Richard D. Grill, photographer, Carnegie Institute of Washington Dept. of Embryology (Baltimore).

Sex Determination (7): Dr. Orlando J. Miller, Assoc. Prof., Obstetrics and Gynecology, College of Physicians and Surgeons, Columbia University.

Genetic Mechanisms (8, 9, 19): Dr. H. Bentley Glass, Prof. of Biology, Johns Hopkins University.

"DNA" (charts and discussions, 19): Dr. Bruce M. Alberts, Graduate School of Biochemistry, Harvard University.

Coloring—Eyes, Hair, Skin (10, 11, 12): Drs. R. E. Billingham and Willys K. Silvers, Wistar Institute, Philadelphia, Pa.

Features, Body Form (13, 14, 15): Prof. Wilton M. Krogman, Director, Philadelphia Center for Research in Child Growth, University of Pennsylvania.

Twins and Supertwins (17, 18): Dr. Norma Ford Walker, Prof. of Zoology, University of Toronto, Canada.

Dr. Alan F. Guttmacher, former Chief, Obstetrics & Gynecology, Mt. Sinai Hospital, New York; President, Planned Parenthood–World Population.

Dr. Richard Osborne, Prof., Medical Genetics and Anthropology, University of Wisconsin.

All Medical Sections (19–29): Dr. F. Clarke Fraser, Prof., Medical Genetics, McGill University, Montreal, Canada.

Mental Diseases and Mental Defects (25, 26): Drs. Franz J. Kallmann, Arthur Falek, Lissy Jarvik, Lewis Hurst, Dept. of Psychiatry, College of Physicians and Surgeons, Columbia University.

Dr. Torsten Sjögren, Director, Dept. of Psychiatry, Karolinska Institute, Stockholm, Sweden.

Dr. George Jervis, Director, New York State Research Institute in Mental Retardation, Staten Island, N.Y.

Sex abnormalities (27): Dr. Orlando J. Miller.

Blood factors and diseases (28): Dr. Philip Levine, Director, Division of Immunohematology, Ortho Research Foundation, Raritan, N.J.

Dr. Richard E. Rosenfield, Director, Blood Bank, and Attending Hematologist, Mt. Sinai Hospital, N.Y.

Dr. Alexander S. Wiener, Senior Serologist, Office of Chief Medical Examiner, New York.

Dr. Alexander G. Bearn, The Rockefeller Institute, New York.

Longevity (29): Drs. Lissy Jarvik, Arthur Falek, Franz J. Kallmann. Edward Lew, Actuary & Statistician, and associates, Metropolitan Life Insurance Company, N.Y.

Special Medical Sections:

Cancer, Dr. George W. Woolley, Sloan-Kettering Institute, New York. Six Differences, Dr. I MacKay Murray, Assoc. Prof., College of Medicine, State University of New York, Brooklyn, N.Y.

Skin, Dr. George B. Kostant, Assoc. Clinical Prof. of Dermatology, New York University Postgraduate Medical School.

Eyes, Dr. Dan Gordon, Assoc. Prof. of Surgery (Ophthalmology), Cornell University Medical College.

Deafness, Dr. John D. Rainer, New York State Psychiatric Institute. Dr. Kenneth S. Brown, Human Genetics Branch, National Institute of Dental Research, Bethesda, Md.

Intelligence (31): Dr. David B. Wechsler, Chief Psychologist, Bellevue Psychiatric Hospital, New York; Prof. of Psychology, New York University.

Dr. Anne Anastasi, Prof. of Psychology, Fordham University, New York.

Gifted Ones, Genius (32, 33, 34): Dr. Anne Anastasi.

Dr. Anne Roe, Prof., Graduate School of Education, Harvard University.

Music talent: Dr. Raleigh M. Drake, Prof., Psychology, Kent State University, Kent, Ohio.

Chess prodigies, photo and data: Mr. Edward Lasker, New York.

Behavior, Personality (35, 36): Drs. John Fuller and J. Paul Scott, Jackson Memorial Laboratory, Bar Harbor, Me.

Dr. Abraham Maslow, Prof. of Psychology, Brandeis University.

Dr. Jerry Hirsch, Prof. of Psychology, University of Illinois.

Somatotypes (chart and discussion): Dr. William H. Sheldon, Clinical Prof., Medicine, University of Oregon Medical School.

Crime (38): Dr. Sheldon M. Glueck, Roscoe Pound Professor of Law, Harvard University.

Sexual Behavior (39, 40): Prof. Abraham Maslow.

Dr. Daniel G. Brown, Psychologist, Captain, U. S. Air Force, Forbes Base, Kansas.

Dr. Lena Levine, Assoc. Medical Director, Margaret Sanger Bureau, New York.

Evolution (41): Dr. Theodosius Dobzhansky, The Rockefeller Institute, New York.

Dr. George Gaylord Simpson, Museum of Comparative Zoology, Harvard University.

Radiation section: Dr. Howard Newcombe, Head, Biology Branch, Atomic Energy of Canada, Ltd., Ontario, Canada.

Race—Physical, Cultural (42, 43): Dr. Theodosius Dobzhansky.

Dr. W. W. Howells, Prof. of Anthropology, Harvard University.

Dr. Howard Levene, Prof., Mathematical Statistics & Biometrics, Columbia University.

Section on blood groups and race: Dr. Philip Levine.

Illustration, "Evolution of Man and Tools" (43): Dr. Harry L. Shapiro, Chairman, Dept. of Anthropology, American Museum of Natural History.

Ancestors, Relatives (44): (Legal aspects), Walter Frank, Attorney, New York.

Personal Problems of Heredity (45): (Artificial insemination, fertility, etc.), Dr. Sophia Kleegman, Clinical Prof., Obstetrics and Gynecology, New York University—Bellevue College of Medicine.

Dr. John MacLeod, Assoc. Prof., Anatomy, Cornell University Medical College.

Paternity blood tests: Dr. Philip Levine.

Appendix, Glossary and "Wayward Gene" Tables: Dr. F. Clarke Fraser.

PART I

From Egg to Adult: Normal Physical Traits

CHAPTER ONE

A Human Life Begins

STOP AND THINK about yourself:

In all the history of the world there was never anyone else exactly like you, and in all the infinity of time to come, there will never be another.

Whether or not you attach any importance to that fact, undoubtedly you have often wondered what made you what you are; what it was that you got from your parents and your ancestors and how much of you resulted from your own efforts or the effects of environment; and finally, what of yourself you could pass on to your children.

Until the dawn of the present century, all of this was a matter of theory and speculation. Then came the birth of the new science of genetics—the study of heredity—and as it has grown, matured and combined its findings with those of many other sciences, the facts about human inheritance and the answers to your questions have become increasingly clear and precise.

We can now open our story of heredity with the first instant of your own debut into the world.

A sperm and an egg: You, like every other human being and most other animals, began life as just that.

A single sperm enters a single egg, and a new individual is started on its way.

Leaving aside for the present the part played by the mother, we know that a father's role in his child's heredity is fixed at the moment of conception. Whatever the father passes on to his child must be contained within that single sperm.

But to find out exactly what that sperm contains has not been so simple a matter.

Consider, first, its size:

One hundred million sperms may be present in a single drop of seminal fluid. Three billion sperms—three thousand million, as many as fathered all the people in the world today—could be comfortably housed in a teaspoon!

The microscope had to be well perfected before a sperm could even be seen. Then, in the first flush of discovery, carried away by their desire to believe, and just as children and lovers imagine that they see a man in the moon, some scientists (circa 1700) reported excitedly that every sperm contained a tiny embryonic being. With professional gravity they gave it the name of "homunculus" (little man), and scientific papers appeared showing careful drawings of the little being in the sperm—although there was some dispute as to whether it had its arms folded or pressed against its side and whether its head had any features.

Presently, however, it became apparent that imagination had run away with scientific perspicacity. The head of the sperm (in which interest rightly centered, as the tail was merely a means for propelling it) proved to be a solid little mass that defied all attempts at detailed study. Many scientists thought it was hopeless to try to find out. Others concluded that even if the sperm head itself could never be dissected and its contents examined, they might still find out what it carried if they could learn what happened *after* it entered the egg. And in this they were right.

After years of painstaking study, we know at last that what a human sperm carries—the precious load that it fights so desperately to deliver—*is twenty-three minute threadlike bodies called chromosomes.*

When the sperm enters the egg and penetrates its substance, the head begins to unfold and to reveal itself as having been made up of the twenty-three chromosomes closely packed together. And we know beyond any doubt that *these chromosomes must comprise all the hereditary material contributed by the father.**

What of the egg? Although many thousands of times larger than the sperm, the egg is yet smaller than a period on this page, barely visible to the naked eye. (The weight of a human egg is estimated to be about one-millionth of a gram.) Under the microscope we see that it consists largely of foodstuffs with the exception of a tiny globule, or nucleus. What that contains we see when the sperm head enters the egg and releases its chromosomes. *Almost at the same time, the egg nucleus*

* Only a single sperm enters and fertilizes the egg. In some way as yet not entirely clear, the moment the first sperm in the race penetrates the egg's outer coating (perhaps previously softened by the seminal fluid), this coating tightens and toughens so that all other sperms are shut out.

FERTILIZATION OF A HUMAN EGG

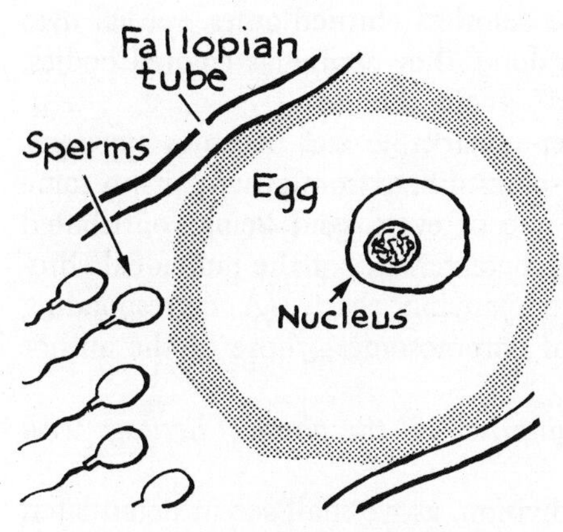

1. *A single sperm—one of millions moving up a Fallopian tube—enters the waiting egg.*

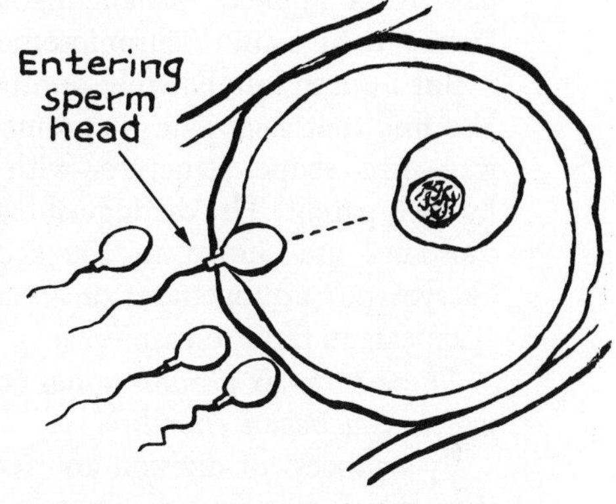

2. *The egg coating hardens instantly and shuts out all other sperms, while the entering sperm head moves toward the egg nucleus.*

3. *The sperm head and egg nucleus fuse, and both release their chromosomes—the father's 23, the mother's 23.*

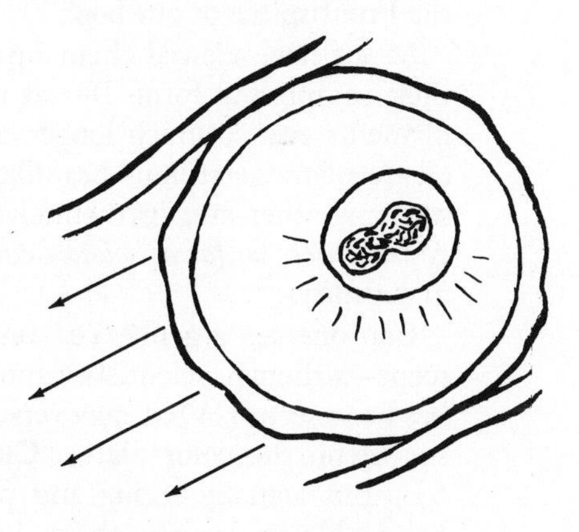

4. *The combined 46 chromosomes of the two parents now begin to work to start the development of the new baby.*

also releases its twenty-three similar chromosomes—the contribution of the mother to the child's heredity.

Thus, as the parents' two sets of hereditary factors come together, the new individual is started off with *forty-six* chromosomes.

In order to reveal the otherwise colorless chromosomes, special dyes have to be applied. When this is done, they appear as colored bodies. Hence their name "chromosomes" (color bodies).

But almost immediately another remarkable fact becomes apparent. We find that each pair of chromosomes differs from other pairs in some way—size, shape, structure—with one of every kind being contributed by each parent. The distinguishing characteristics of the individual chromosomes are shown in the accompanying chart. (A corresponding "karyotype" arrangement of actual chromosomes—those of the author —appears in the Frontispiece.)

These forty-six chromosomes comprised all the physical heritage with which you began your life.

By a process of division and redivision, as we shall see in detail later, the initial forty-six chromosomes are so multiplied that eventually every cell in the body contains a replica of every one of them. This is not mere theory. By simple and painless techniques now possible, an expert could take some of your own cells and show you the chromosomes in them looking exactly like those of the author, as pictured in the Frontispiece of our book.

As we have viewed them up to this point, the chromosomes are in their compressed form. But at certain times they may stretch out into filaments ever so much longer, and then we find that what they consist of are many gelatinous beadlike sections closely fitted together. These sections either are, in themselves, or contain the *genes*. And *it is the genes which, so far as science can now establish, are the ultimate factors in heredity.*

Can one see a gene? Yes, probably, by means of an electron microscope—although scientists cannot yet clearly identify what it is that has been seen. (What may very well be genes—and genes in action—are shown on the color plate, "Close-ups of Chromosomes," facing page 63.) But actually seeing and photographing genes is not essential to being able to describe them. In later chapters you will find many details about the chemical composition of genes and the ways in which the innumerable differences among genes are or can be produced.

But whatever we have yet to learn about genes, we already know beyond question that each gene has a definite function in the creation and development of every living person.

Altogether, of all the miraculous particles in the universe, one can

CLASSIFICATION OF HUMAN CHROMOSOMES

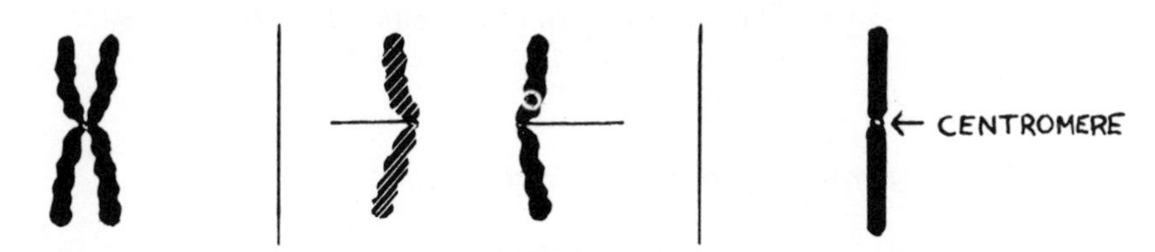

An actual chromosome in the doubled stage, as usually seen in photographs.

The single-form chromosome, as it would look when separated from its replica.

Same chromosome in a stylized drawing. Centromere *is junction point of chromosome.*

"IDIOGRAM"—DIAGRAMMATIC ARRANGEMENT OF INDIVIDUAL CHROMOSOMES

*Numbered and grouped according to relative sizes, shapes and other characteristics.**

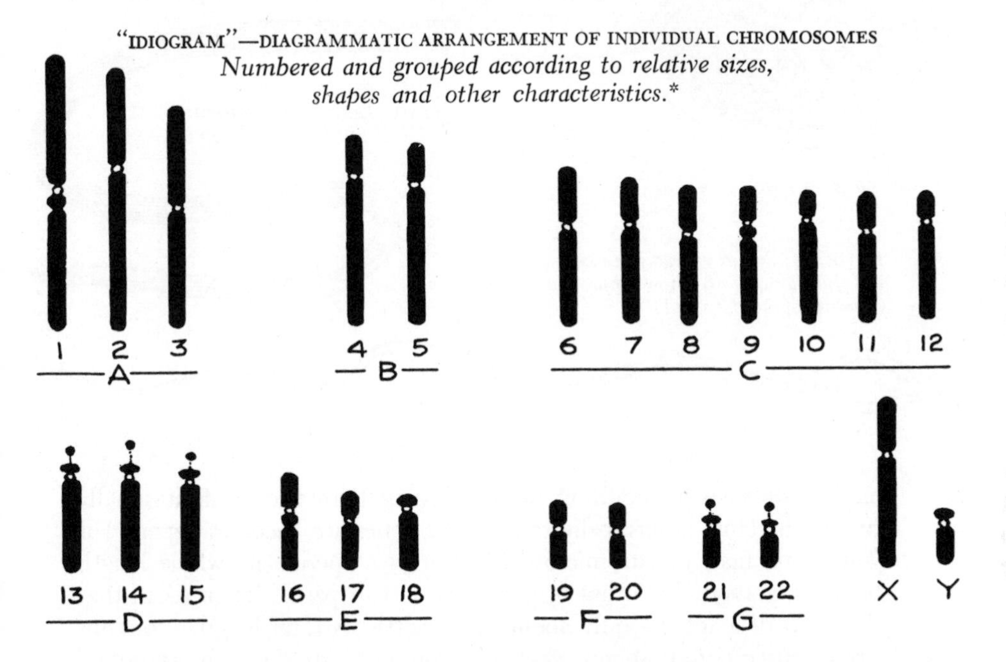

SPECIAL DETAILS. Satellite: *A small, pinched-off section of a chromosome, often seen as joined by a threadlike link to one end. Found with chromosomes 13, 14, 15, 21, 22.*

Constriction: *Pinching together of a chromosome on its long arm, near the centromere. Often seen in the* A, C *and* E *groups, as in* Nos. *1, 9, 16.*

Centricity: *Location of the* centromere *marking relative lengths of chromosome arms. A chromosome is called* "metacentric," *or approximately so, when centromere is close to mid-point, and the arms are not too disproportionate, as in* Nos. *1, 2, 3; 6, 7, 8, 11; 19, 20 and the* "X." "Sub-metacentric" *is applied to* Nos. *4, 5, 9, 10, 17, 18. (No. 16 can be metacentric or sub-metacentric.) When the centromere is far off center, and the arms are very different in length, the chromosome is called* "acrocentric," *as in* Nos. *13, 14, 15, 21, 22 and the* "Y."

Chart prepared with aid of Dr. Orlando J. Miller, College of Physicians and Surgeons, Columbia University.

* For a "*karyotype*," or graded arrangement of a full set of actual chromosomes, see the Frontispiece, showing chromosomes of the author.

THE HUMAN SPERM: A REMARKABLE MECHANISM

(A billion *like this can be produced by the average fertile man every week.*)

A. *Shown diagrammatically, as viewed from above:*

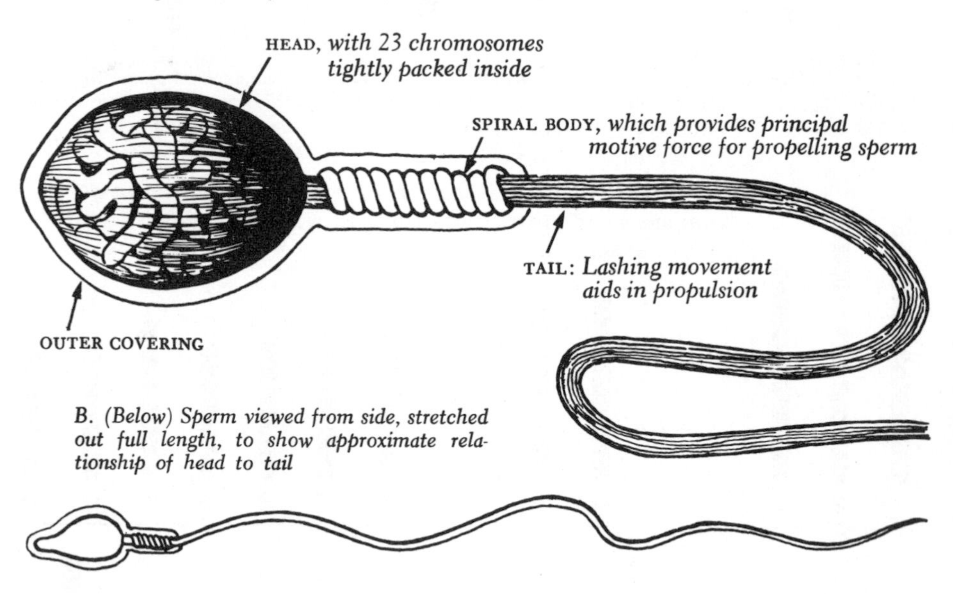

B. *(Below) Sperm viewed from side, stretched out full length, to show approximate relationship of head to tail*

hardly conceive of anything more amazing than these infinitesimally tiny units. How fantastically minute the genes are becomes apparent if you think, first, of the microscopic size of a sperm—its whole length being only about 1/25 that of a comma on this page.* Then recall that the head of a sperm—only about one-twelfth its total length—contains twenty-three chromosomes. And now consider that these chromosomes may contain, collectively, about thirty thousand genes (with hundreds, or a thousand or more of them in any one chromosome) and that *a single gene, in some cases, may be able to change the whole life of an individual!*

To grasp all this you must prepare yourself for a world in which minuteness is indeed carried to infinity. Contemplating the heavens,

* The average sperm is about 60 microns in length, from tip of head to end of tail, the head alone being about 5 microns long. The average unfertilized human egg measures about 130 microns (1/200 of an inch) in diameter. In weight a human egg is about one-millionth of a gram, a sperm about 35 trillionths of a gram.

you already may have adjusted yourself to the idea of an infinity of bigness. You can accept the fact that the sun is millions of miles away; that stars, mere specks of light, may be many times larger than the earth; that the light from a star which burned up and ceased to exist hundreds of thousands of years ago is reaching us only now; that there are billions of stars in outer space which our most powerful telescopes cannot yet reveal. This is the *infinity of bigness outside of you.*

Now turn to the world *inside* of you. Here there is an *infinity of smallness.* As we trace further and further inward we come to the last units of life that we can distinguish—the genes and the particles composing them. We may not be able to see all the way into your "inner space"—just as we are stopped in our exploration of outer space by the limitations of existing equipment. But, as with the unseen stars and planets, we can make many accurate deductions about our as yet un-

HOW A CHROMOSOME MIGHT CHANGE ITS SHAPE

(*Beads inside represent the genes.*)

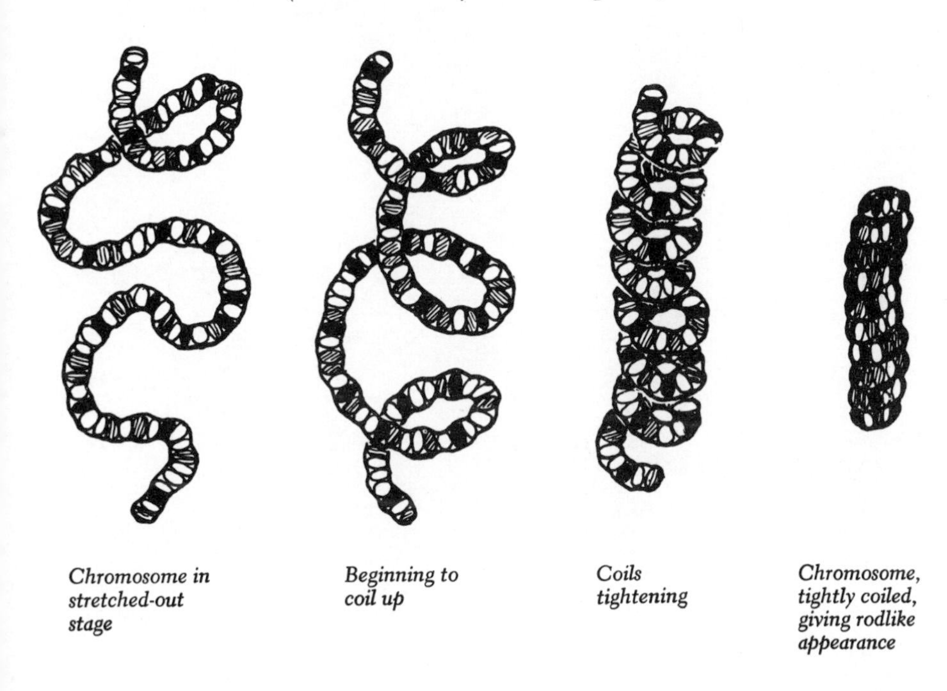

seen elements of heredity from what we already know of their effects and behavior.

You believe the astronomer when he tells you that, on October 26 in the year 2144, at thirty-four minutes and twelve seconds past twelve o'clock noon, there will be a total eclipse of the sun. You believe this because time and again such predictions have come true.

You must now likewise prepare to believe the geneticist when he tells you that a specific gene, whose presence can as yet only be inferred, will nevertheless at such and such a time do such and such things and create such and such effects—*under certain specified conditions.* The geneticist may have to make many more reservations than the astronomer, for unlike the celestial bodies, the genes are parts of *living* entities, whose actions are complicated by all the unpredictable experiences which can befall the human beings housing them. This becomes apparent, as we shall now see, from the moment of conception.

CHAPTER TWO

The Human Seeds

No less important than knowing what heredity is, is knowing what it *is not*.

Before we examine in detail the determiners of heredity—the chromosomes and their genes—let us first find out how the sperms, and the eggs which carry these hereditary factors, are produced in the parent. That in itself will clear away much of the deadwood of the past with its innumerable false theories, beliefs and superstitions about the life processes.

Not so long ago the most learned of scientists believed that whatever it was that the sperms or eggs contained, these were products of the individuals, in which were incorporated in some way *extracts* of themselves. That is to say, that each organ or part of a person's body contributed something to the sperm or egg. Darwin, who had gone along with that theory, called these somethings *gemmules*.

By the gemmule theory, all the characteristics of both parents could be transmitted to the child, to be blended in some mysterious way within the egg and reproduced during development. A child would therefore be the result of what its parents were at the time it was conceived. As the parents changed in their traits and physical makeup through life, so would their eggs or sperms, and the chromosomes within them, also change. That is what scientists used to believe and what a great many people still believe—erroneously.

Today we know beyond question that the hereditary properties in the human seeds do not change as the individual changes. For we have learned finally that the chromosomes which the sperms and eggs contain *are not new products of the individual,* and are most certainly not

THE HEREDITY PROCESS

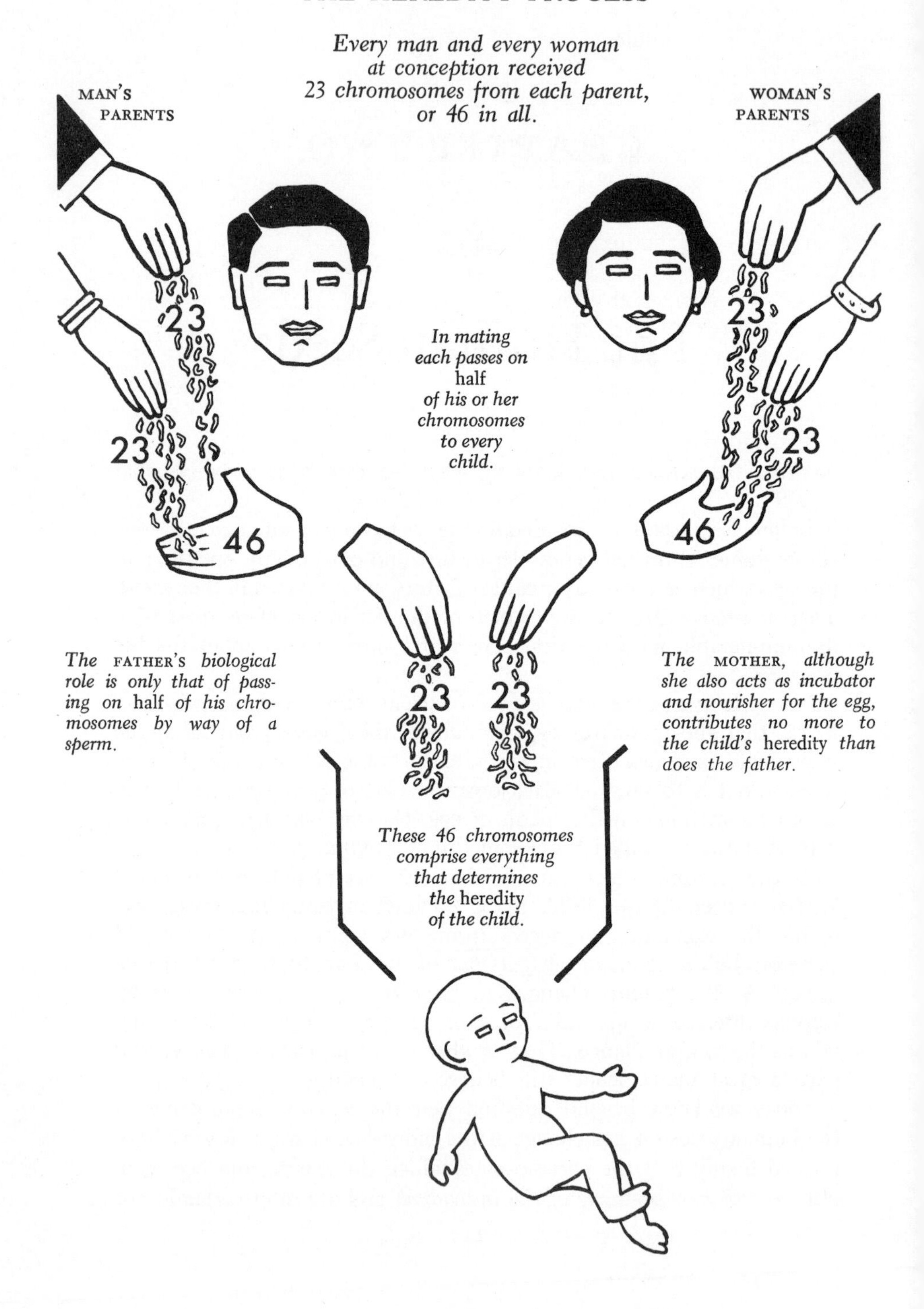

made up of "gemmules" or contributions from the various parts of the body.

As we have seen, a human being starts life as just a single cell containing forty-six chromosomes. That initial cell must be multiplied into billions of other cells to produce a fully developed baby, and this is accomplished by a process of division and redivision, with the material needed after that in the egg is exhausted coming from nourishment provided by the mother.

But the cells do not all remain the same, by any means. After the earliest stages, when they are still very limited in number, they begin specializing. Some give rise to muscle cells, some to skin, blood, brain, bone and other cells, to form different parts of the body. But a certain number of cells are reserved for another function.

These cells are the germ cells, dedicated to posterity. *It is from these cells that the sperms or eggs are derived.*

When a boy is born, he already has in his testes all the germ cells out of which sperms will eventually be produced. When he reaches puberty, a process is inaugurated that will continue throughout his life—or most of his life, at any rate. In the same way that billions of cells grew from one cell, millions of more germ cells are manufactured from time to time by division and redivision. Up to a certain point the process is the same as that previously explained—but just before the sperms themselves are to be formed, something different occurs. The chromosomes in the germ cell remain intact, and the cell divides in a special way so that each half gets *only twenty-three chromosomes, or one of every pair.*

The process of forming the sperms is illustrated in the accompanying chart (several stages are omitted for simplification). This should make clear how, from a parent germ cell with the regular quota of forty-six chromosomes, sperms are formed, each carrying only twenty-three chromosomes. The reason and necessity for this "reduction" division will be explained presently.

Before we go on, let us stop to answer a question which has undoubtedly caused concern to many a man:

Is it true that the number of sperms in a man is limited and that if he is wasteful with them in early life, the supply will run out later?

No, for as we have seen, the sperms are made out of germ cells thrown off without decreasing the "reserve" stock. Endless billions of sperms can continue to be discharged from a man's body (200 million to 500 million or more in a single ejaculation!) and the original quota of germ cells will be there to provide more—so long as the necessary machinery

HOW SPERMS ARE PRODUCED

*The process ("meiosis") by which only one member of every pair of the parent's chromosomes goes into a germ cell.**

1. *Father's sperm-forming cell with his 23 pairs of chromosomes. (Only four pairs shown in detail:* black, *representing those from his father,* white, *those from his mother.)*

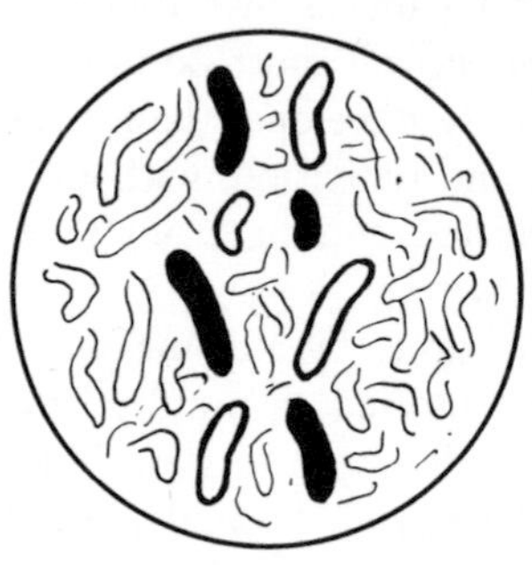

4. *Another chromosome division follows, but now only the already doubled chromosomes are pulled apart and each of the four cells produced gets* only one chromosome *of every pair.*

2. *(A) Chromosomes pair up and each doubles. (B) Doubled pairs line up at cell center. (C) Doubled members of each pair are separated and go to opposite sides of the cell.*

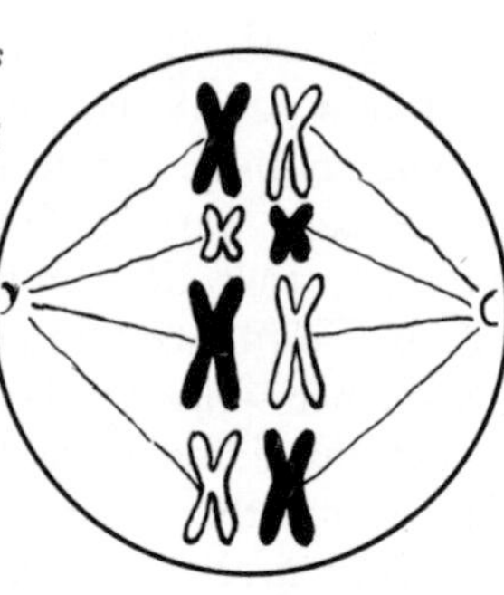

5. *The four cells, each with 23* single *chromosomes, separate completely, and each shapes into a head (with its chromosomes) and a tail.*

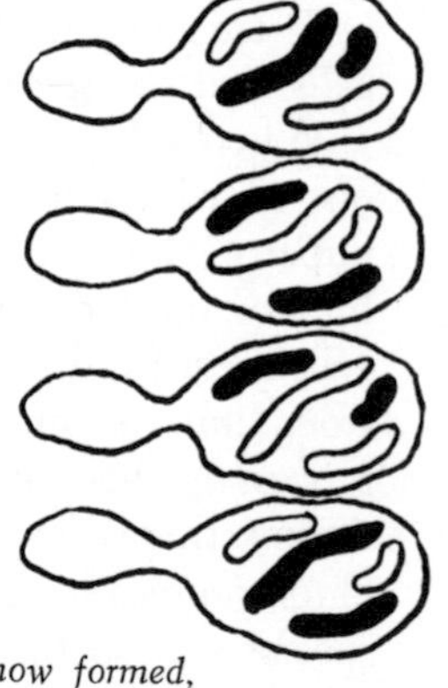

3. *Cell divides and two cells are formed, each cell with one doubled member of every pair. The "paternal" and "maternal" members of pairs are distributed at random.*

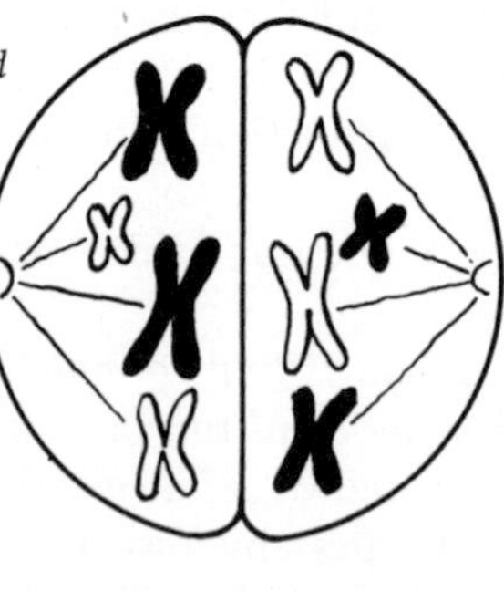

6. *Four sperms are now formed, the heads containing the tightly packed single chromosomes* (but not the same assortment in each sperm), *and the remaining cell contents squeezed out into long tails.*

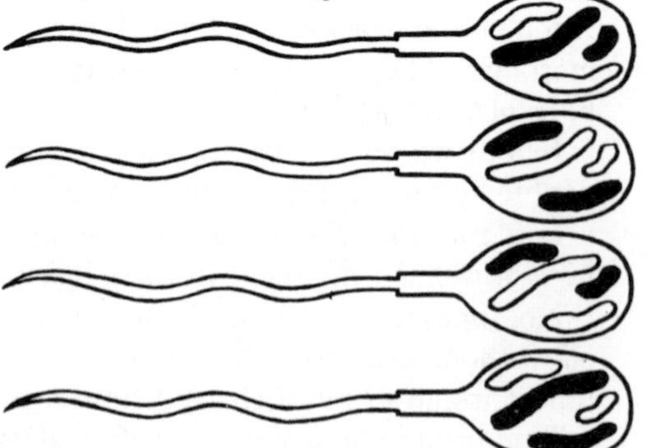

* The process whereby eggs are formed in the mother and each egg receives only half of her chromosomes—one of every pair—is basically like that in sperm formation, with this exception: When the equivalents of stages 4 and 5 above are reached, only one of the four parts is fashioned into an egg, receiving the bulk of the cell material (cytoplasm), whereas the other three parts presently disintegrate.

functions and the body can supply the material out of which to make them. However, dissipation to an extreme point which might injure or weaken the body—and similarly, disease, accident or old age—may curtail the production of sperms or greatly reduce the number of those which will be normal and fertile.

In the female, although the eggs are also manufactured out of germ cells, the process does not provide for an endless number, running into billions, as in the case of the sperms. The female, when she reaches puberty, will be required normally to mature only one egg a month, for a period of about thirty-five years. So, when a girl baby is born, the fundamental steps in the process have already been taken, and the germ cells have already been turned into eggs. In other words, her ovaries at birth contain tiny clusters of all the eggs, in rudimentary form, which will mature years later. The chromosomes which she will pass on to her future children are also already present within each egg. The maturing process will merely increase the size of the egg by loading it with a store of food material with which to start a new individual on its way.

Although we need not go into the complicated details of the egg-formation process, it is essential to point out that in the production of eggs there also is a "reduction" division, just as there is in the case of the sperms. This results in only half of the mother's chromosomes, like half of the father's, going to the child at conception; and with each parent contributing twenty-three single chromosomes, the child thus receives two of every type of chromosome, making the full quota of twenty-three *pairs*, or *forty-six* chromosomes in all, required for a normal human being.

If that reduction process hadn't taken place, each sperm or egg would carry 46 chromosomes; on uniting they would start off an individual with 92 chromosomes; the next generation would begin with 184, the next with 368, and so on and on to an absurd and impossible infinity. However, the reduction division, as will soon be seen, has much more than a mathematical significance.

One fact should be constantly kept in mind: The sperms or eggs receive chromosomes and genes which are almost invariably replicas of those which the parents themselves received from their parents. The rare exceptions are the cases where genes and/or chromosomes undergo *mutations*. These mutations, we shall learn in a later chapter, are changes in workings or arrangements which occur at infrequent intervals, either spontaneously (through "errors" in the genetic mechanisms) or through some *outside influence* (radiation or strong chemical effects).

Apart from the mutations, the hereditary factors remain exactly the same from generation to generation. *No nonhereditary trait in a person's body developed during his or her lifetime can cause the genes to change so that they will produce a similar trait in the child. And nothing that we ourselves or any scientist is as yet able to do can change the makeup of our germ cells in any specified direction.*

It is as if, when Nature creates an individual, she hands over to him billions of body cells to do with as he wishes and in addition, wrapped up separately, a small number of special germ cells whose contents are to be passed on to the next generation. And, because Nature apparently does not trust the individual, she sees to it that the hereditary factors in those germ cells are so sealed that he cannot tamper with them or alter them in the slightest degree to suit his purposes.

CHAPTER THREE

What We *Don't* Inherit

MEN SINCE THE WORLD BEGAN have taken comfort in the thought that they could pass on to their children not merely the material possessions they had acquired, but also the physical and mental attributes they had developed.

To both types of inheritance, as previously conceived, serious blows have been dealt within recent years. The passing on of worldly goods has been greatly limited by huge inheritance taxes in most countries. But this is as nothing compared with the "inheritance tax" levied by Nature on a man's mental or physical assets—the qualities of mind and body which he might wish to pass on to his children. For with respect to such biological heredity, all pre-existing notions have been shaken by the finding we have just dealt with:

The chromosomes in our germ cells are not affected by what we do with our body or brain cells.

What this means is that no change that we make in ourselves or that is made in us in our lifetimes, for better or for worse, can be passed on to our children through the process of biological heredity. Such changes—made in a person by what he does, or what happens to him—are called "acquired characteristics." Whether such characteristics could be passed on has provided one of the most bitter controversies in the study of heredity. It has been waged by means of thousands of experiments and is still being carried on by a stubborn few, including, chiefly, certain Soviet scientists who are motivated by reasons of their own. (We shall deal with them and their theories in a later chapter.) But now that the smoke of battle has cleared away, there remains no verified evidence that any acquired characteristic can be inherited in human beings.

Reluctantly we must abandon the belief that what we in one generation do to improve ourselves, physically and mentally, can be passed on through our germ plasm to the next generation. It may not be comforting to think that all such improvements will go to the grave with us. And yet the same conclusion holds for the defects developed in us, of the things we may do in our lifetimes to weaken or harm ourselves. If we cannot pass on the good, we likewise cannot pass on the bad.

Why we can't should now be obvious. Inasmuch as all that we transmit to our children, genetically, is the chromosomes and their genes, it is clear that in order to pass on any change in ourselves, every such change as it occurred would have to be communicated to the germ cells and accompanied by some corresponding change in every chromosome and gene concerned with the specific characteristic involved.

Just imagine that you had a life-sized, plastic statue of yourself and that inside of it was a small, hermetically sealed container filled with millions of microscopic replicas of this statue. Suppose now that you pulled out of shape and enlarged the *nose* of the big statue. Could that, by any means you could conceive, automatically enlarge all the noses on all the millions of little statues inside? Yet that is about what would have to happen if a change in any feature or characteristic of a parent were to be communicated to the germ cells, and thence to the child. It applies to the binding of feet by the Chinese, to circumcision among the Jews, to facial mutilation and distortion among savages, to all the artificial changes made by people on their bodies for generations which have not produced any effect on their offspring. And it applies to the *mind* as well.

Nature performs many seeming miracles in the process of heredity. But it would be too much to ask that every time you completed a study course or crammed for an examination every gene in your germ cells concerned with the mental mechanism would brighten up accordingly. Or that, with every hour you spent in a gymnasium, the genes concerned with the muscle-building processes would increase their vigor.

Thinking back to your father, you will see that what he was, or what he made of himself in his lifetime, might have little relation to the hereditary factors he passed on to you.

Remember, first, that your father gave you replicas of only *half* of his chromosomes—and which ones he gave you depended entirely on chance. It is possible that you didn't receive a single one of the chromosomes whose genes helped to produce your father's outstanding characteristics.

Aside from this fact, what your father was, or is, may not at all indicate what hereditary factors were in him. The genes, as we shall see presently, do not necessarily *determine* characteristics. What they determine are the *possibilities* for a person's development under given circumstances.

Thus, your father may have been a distinguished citizen or a derelict, a success or a failure, and yet this may provide no clear indication of what genes were in him. But whether or not the nature of his genes did reveal themselves through his characteristics, you can make only a guess as to which of them came to you, by studying unusual traits that your father and you have in common.

What of your mother? The situation with respect to her is different only in that she provided the initial "soil" in which the human seed that was to become you took root and flourished. Her contribution in the way of chromosomes—half of her own—was essentially the same as your father's. Later we will hear more about the possible effects of the prenatal environment provided by the mother. But that her role in heredity is no greater than the father's is clearly proved by the fact that children by and large resemble their mothers no more than their fathers.

You may already be thinking, "What about my children? How much of *myself* did I, or can I, pass on to them?"

Let us first see what you *can't* pass on.

You may have started life with genes that tended to make you a brilliant person, but sickness, poverty, hard luck or laziness kept you from getting an advanced education. Your children would be born with exactly the same mental equipment as if you had acquired a string of degrees from Yale to Oxford.

Suppose you are a woman who had been beautiful in girlhood, but as a result of an accident, illness, or aging had lost your looks. The children born to you thereafter would have not one whit less chance of being pretty or handsome than if you had developed into a movie queen.

Suppose you are a war veteran who was blinded, crippled or permanently invalided. If you had a child today, his heredity would be basically the same as that of any child you might have fathered in your fullest vigor when you marched off to service.

Suppose you are old.

The sperms of a man of ninety-five, if he still is capable of producing virile sperms—and there are records of men who were—would be the same in the quality of their genes as when he was sixteen. (Livestock breeders, acting on this principle, use prize bulls and stallions for breeding just as long as they can procreate.) And although the span of reproductive life in a woman is far shorter than in a man, the eggs of a woman of forty-five would be no different in their *genes* from those of her girlhood.*

Nevertheless, there may be considerable difference in the offspring born to parents under different conditions. But not because of *heredity*.

The older the mother—at least near forty—the more likely she is to have had or to have diseases or various bodily disorders, and so the *prenatal environment* she provides for a child she carries may tend to be less favorable than in her younger years. Moreover, as a woman ages, her egg-producing mechanism may more often go awry, increasing the chance that chromosome abnormalities or various deficiencies may occur before or during conception.

But with the mother as well as the father, it is the *external environment* provided by parents which may be most important with respect to children born at different stages of their lives, or under different circumstances. And this obviously refers not merely or even as much to the physical environment as to the psychological environment.

Take this situation: A young man is a teetotaler when he marries and fathers a son. Twenty years later he becomes a drunkard and then fathers another son. Will the second son be more likely to take to drink than the first? Quite definitely. Not because the genes passed on by the man to his second son could be any more "alcoholic," but because the first son had grown up under the influence of a sober father, while the much younger brother will grow up under the influence of a drunken one. There is no evidence whatsoever that drunkenness or other bad habits can be inherited directly as such. (Geneticists have

* With regard to the latest age at which a woman can bear children, motherhood has been reported for some women who were in their sixties. But claims of pregnancies at age seventy or over are questioned by medical authorities. (Although the Bible reports the birth of a child, Isaac, to Sarah when she was ninety and her husband Abraham one hundred, ages were probably not figured on the same basis then as they were later. For instance, Abraham was reported as dying at the age of one hundred seventy-five. [The phenomenal Biblical ages will be discussed in Chapter 29.] However, Sarah was apparently still so old that, after her previous barrenness, her pregnancy was considered amazing at the time; and when told she would bear him a son, ". . . Abraham fell upon his face, and laughed.")

kept generations of mice, rabbits, guinea pigs and poultry virtually stupefied with alcohol, without the slightest sign of any transmission of the behavioral effects to their offspring.) If children of drunkards are so frequently drunkards themselves, the major reason certainly would be, "through precept and example."

As often as not, similarities between child and parent (mother as well as father) which are ascribed to heredity are really the effects of similar influences and conditions to which they have been exposed. In fact, *so interrelated and so dependent on each other are the forces of environment and heredity in making us what we are that they cannot be considered apart, and at every point in this book will be discussed together.*

Thus where heredity falls short, environment may be able to compensate. And if you ask, "Can I pass on to my child any of the accomplishments or improvements I have made in myself?" the answer may be, "Yes! You can pass on a great deal—not by heredity, but by training and environment!"

The successful, educated, decent-living father can give his son a better start in life. The athletic father can, by example and training, help his son to develop a better physique. The healthy, intelligent, foresighted mother can greatly increase the likelihood that her child will have a favorable entry into the world and every chance thereafter to make the most of his or her inherited capacities. And just as easily, bad initial environment and bad parental influences can thwart or cancel the effects of good heredity.

So, more and more as we proceed in this book we will see that every individual, in makeup and behavior, is the end result of many forces, and that, at every stage of one's life, heredity and environment have interacted, and continue to interact, to make a person what he or she is, can be, or will be.

CHAPTER FOUR

Myths of Mating

Of the various myths about mating and parenthood, one that has been most ardently cherished by many loving couples is that about "putting themselves in the right state for the conception of a child." It is too bad to disillusion them; but what we have just learned should convince us of the unromantic fact that whether a child is conceived during a glamorous sojourn on sunny strands or in the depressing air of a dingy tenement, whether in the height of passion or when its parents are barely on speaking terms with one another, the hereditary factors transmitted to it will be not one whit different.

What, then, of a "love child"? Popular belief is that a child born out of wedlock is in some ways different from a legitimate child, that it is likely to be more delicate, more sensitive, developing to extremes—sometimes a genius, often a criminal. Weren't Leonardo da Vinci and Alexander Hamilton illegitimate? Wasn't Hitler? Yes, but it need hardly be stated that Nature never took and still does not take any note of marriage certificates.

An illegitimate child may well be different from a legitimate child—if its *environment* is different. Where society is relentless toward illegitimacy, the child born to an unmarried mother may come into the world and be reared under handicaps which may have continuing effects throughout its life, sometimes in peculiar directions. But wherever consideration is shown to unmarried mothers, where good care is given them during pregnancy, and where the child has anything like a normal start, the factor of illegitimacy has little importance. This is now being proved in countless cases where children born out of wedlock are being lovingly and intelligently reared by foster parents.

The *age of parents* is also believed to affect the nature of the child, but where it does, it is only through environmental influences. A "child of old age," born, let us say, when a mother is in her mid-forties and the father in his sixties, frequently appears to be frailer and sicklier than others. The explanation will be found not in any weakness in the parents' germ cells, but, first, in the less favorable intra-uterine environment provided by the older mother (as we noted in the preceding chapter); and, second, in the fact that such late births are so often unwanted and occur mostly where conditions for childbearing are bad. Following birth other factors enter. The child of old age, surrounded, as is usually the case, by much older brothers, sisters and their friends in addition to the older parents, is frequently pampered and spoiled, and quite understandably might become high-strung and precocious.

In children born to very young mothers (under seventeen), both the intra-uterine and postnatal conditions are also likely to be unfavorable. Immaturity in the mother may not only have physical ill effects both on her and on her child, but may be equally damaging psychologically owing to the mother's lack of social experience.

It should hardly be necessary now to dwell at length on certain other erroneous beliefs associated with mating and parenthood which, while prevalent largely among breeders of domestic animals, are also applied to human beings:

> *Telegony* is the theory that if a female is mated with two or more successive males, the influence of an earlier sire may carry over to offspring of a later father.
>
> Similarly, by *infection* is meant that a male mated first with an undesirable female (a blooded bull with a "scrub" cow) may communicate some of her characteristics to offspring of the next female with whom he is mated.
>
> Or there is the belief that continued mating together may cause a male and female to resemble each other; and that by *saturation* the oftener a female is mated with the same male, the more the successive offspring will resemble him.

All of these theories have been disproved both by innumerable experiments with lower animals and by observing the results of matings among human beings. With regard to the telegony theory, there have been many instances where a woman has been married first to a man of one race, then to a man of another, without the children of her second husband showing in the slightest any of the racial traits of her first husband. So, too, the infection theory has been disproved both by animal experiments and by observation of the progeny of men's first and second marriages.

However, regarding the beliefs about the mutual influences on a couple of continuous mating, or of the saturation imprints on their offspring, the explanation which should readily occur is that individuals who live together for a long time, whether lower animals or humans, may show common effects of the same environment, diet, habits and other living conditions. Husbands and wives sometimes may get to look alike in the same way that any persons living in the same environment may develop similarities of physique and appearance. One often hears it said of a couple, "You'd think they were brother and sister!" The same environmental influence also tends to increase the resemblance between children and parents.

The myths and superstitions associated with mating and parenthood could fill a book by themselves. Back of all of them lie sometimes coincidences, sometimes mistaken assumptions of paternity, and most often the cropping out of hidden factors (recessive genes)—which will be dealt with more fully further along.

One common question regarding mating may deserve special attention:

"Can there be such a thing as conflict between the chromosomes of one person and another—a genetic incompatibility that would seriously affect or prevent the birth of children?"

Yes, but only as applied to certain individuals, not to types or races. This is an important distinction. Later we shall see how in any two given persons there may be specific "dangerous" genes or other genetic factors in each, which when combined by their mating may offer serious threats to their prospective offspring. But one should not jump to the conclusion that this would single out persons of radically different surface types. For the fact is that *all* human beings, regardless of race, type, color, or any other classification in which they are placed, are, as members of the same species, sexually and genetically compatible with one another.

No such chromosome incompatibility exists between any two kinds of humans as there does between animals of different species—a cat and a dog, a chicken and a goose, or even a horse and donkey (as our illustration shows).* The tallest, blondest Nordic might mate with the

* While male mules are invariably sterile, rare instances have been reported of fertile female mules. However, suspicion about these cases has been aroused by a recent study of the chromosomes of two alleged fertile female mules, which showed that the animals were genetically typical *donkeys*, despite their physically large size. Also tending to confirm why true mules are infertile were findings that among the mule's 63 chromosomes, those received from the horse parent differ from those from the donkey parent in structure as well as in number (32 from the horse—

MATING CONFLICT AND COMPATIBILITY

A CAT *and a* DOG *cannot mate and have offspring together because they are of different species and their chromosomes differ in kind and number (cat, 38, or 19 pairs; dog, 76, or 38 pairs). Chromosomes in a germ cell of a cat would not work with those in a germ cell of a dog.*

A HORSE *and a* DONKEY, *while of different species, are related enough to be able to mate and produce* MULE *offspring. But these offspring are sterile because* HORSE *and* DONKEY CHROMOSOMES *are different.*

A HORSE *produces eggs or sperms each with 32 of certain kinds of chromosomes.*

A DONKEY *produces eggs or sperms with 31 of other kinds of chromosomes.*

While these chromosomes can work together somehow to produce MULE *offspring . . .*

The conflicting chromosomes cannot form in the MULE *fertile germ cells. (For possible exceptions, see Text.)*

BUT ALL VARIETIES OF HUMAN BEINGS ARE FERTILE WITH ONE ANOTHER

Because all humans are of the same species with the same kind of chromosomes. Thus an AFRICAN PYGMY . . .

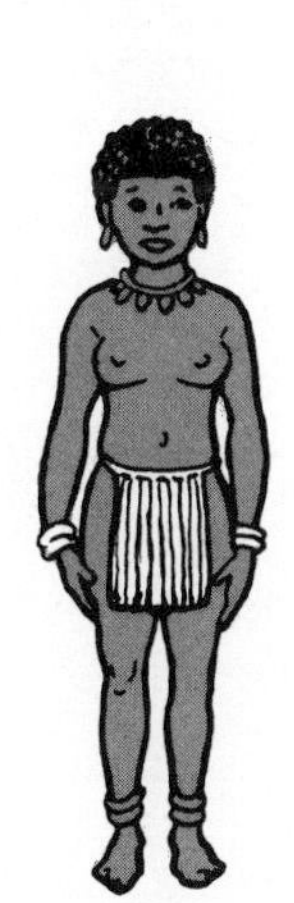

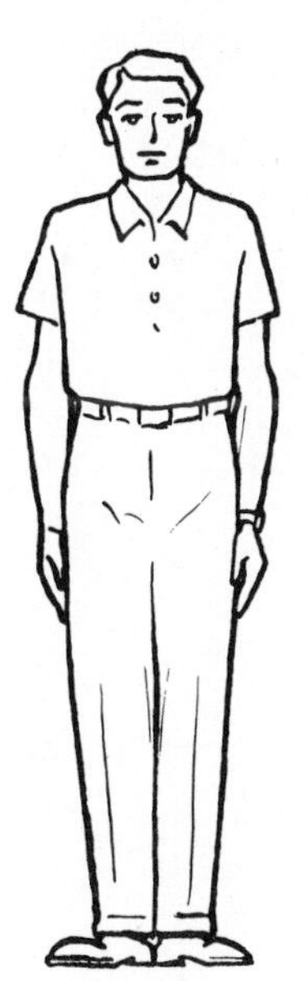

. . . and any CAUCASIAN *or person of another race could mate and produce a child perfectly normal in the eyes of Nature.*

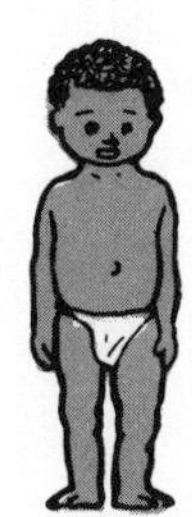

smallest, blackest Pygmy and produce children perfectly normal in the eyes of Nature. This, however, has certain qualifications, for if a tall, big-boned European mated with a Pygmy woman, the child might be too large for her to bear without danger to it or to herself. The same, however, would apply to the mating of any extremely large man with a small woman, even of the same race, where sometimes a Caesarean operation is required to deliver the child.

The theory has been advanced that in the mating of parents of radically different types, serious disharmonies may result in the bodily structure and features of their offspring. The evidence on that score is far from conclusive and until better established, must be placed among beliefs rather than the facts with which we are dealing.

As matters stand the fear of having a freak child because of parental differences in features or bodies need hardly worry any of you who read this. Unless, perhaps, you happen to be a large-headed circus giant mated to a pin-headed midget.

---

half of its 64; and 31 from the donkey—half of its 62), making for conflicts between the two sets of chromosomes when the mule tries to produce germ cells. Lions and tigers are two other related animals which, while genetically distinct, can sometimes be crossbred, producing a *tiglon*. Several of these have been in zoos, including one in New York. Also crossbred have been a zebra and a donkey, producing a *zebronkey*, and a chicken and a turkey, producing a *churk*.

CHAPTER FIVE

Your *You*-niqueness

WHAT WAS THE MOST extraordinary adventure in your life?

Whatever you might answer, you are almost certain to be wrong. For the most remarkable series of events that could possibly have befallen you took place before you were born. In fact, it was virtually a miracle that *you* were born at all!

Consider what had to happen:

First, *you*—that very special person who is *you* and no one else in this universe—could have been the child of only *two specific parents* out of all the untold billions past and present. Assuming that *you* had been ordered up in advance by some capricious Power, it was an amazing enough coincidence that your parents came together. But taking that for granted, what were the chances of their having had *you* as a child? In other words, *how many different kinds of children could any couple have, theoretically, if the number were unlimited?*

This is not an impossible question. It can be answered by calculating how many different combinations of chromosomes any two parents can produce in their eggs or sperms. For what every parent gives to a child are replicas of just half of his or her chromosomes—one representative of every pair taken at random. In that fact you will find the explanation of why *you* are different from your brothers and sisters, and why no two children (except "identical" twins) can ever be the same in their heredity.

Putting yourself in the role of parent, think for a moment of your fingers (thumbs excluded) as if they were four pairs of chromosomes, of which one set had come to you from your father, one set from your mother. (To distinguish between the two, we've made the paternal set

CHROMOSOME RECOMBINATIONS

Fingers representing 4 of the 23 chromosomes received from one's MOTHER

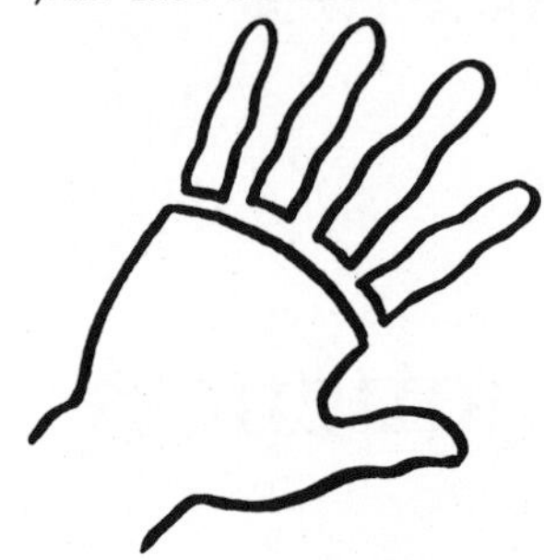

Fingers representing 4 of the 23 chromosomes received from one's FATHER

Combinations of chromosomes which could be produced with the four pairs above

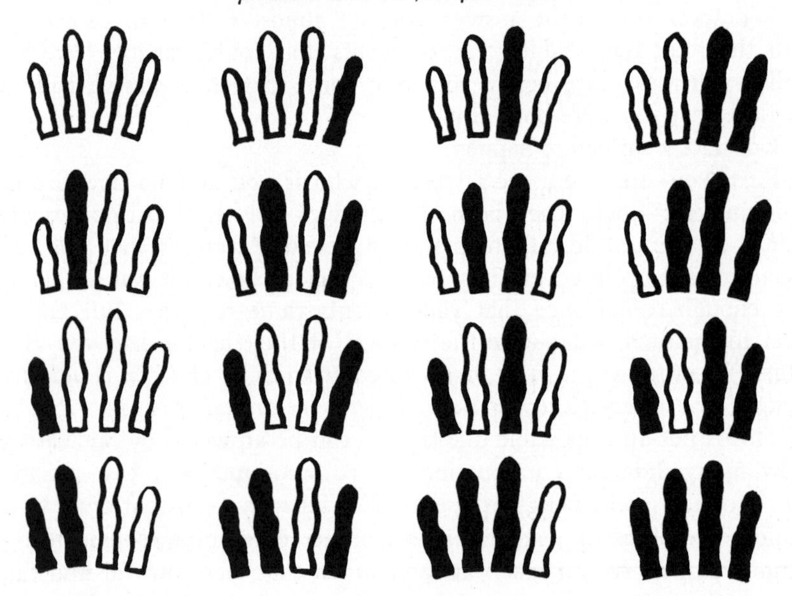

black in the diagram, the maternal set white.)

Now suppose that these "chromosomes" were detachable and that you had countless duplicates of them. If you could give a set of four to every child and it didn't make any difference whether a particular "chromosome" had come from your father or your mother, how many different combinations would be possible?

Sixteen (see diagram), in which every combination differs from any other in from one to four "chromosomes."

But this is with just *four pairs* involved. If now you put the thumb of each hand into play, representing a *fifth* pair of "chromosomes," you could produce twice the number of combinations, or thirty-two. In short, as you can readily see, with every added pair of factors the number of possible combinations is doubled. So in the case of the actual chromosomes, with *twenty-three* pairs involved—where one from each pair is taken at random—every parent can theoretically produce 8,388,608 combinations of hereditary packets, each different from any other in from one to *all twenty-three* chromosomes.

Whether we are dealing with the millions of sperms released by a male at one time, or the single egg matured by a woman, the chance of any specific combination occurring in either would be that once in 8,388,608 times.

But to produce a given individual, *both* a specific sperm and a specific egg must come together. So think now what had to happen for *you* to have been born.

At exactly the right instant, the one out of 8,388,608 sperms which represented the potential half of you had to meet the one specific egg which held the other potential half of you. That could happen in less than once in some 64 trillion times! If we add to this all the other factors involved, the chance of there having been or ever being another person exactly like you is virtually nil.

At this point you might say, with modesty or cynicism, "So what?" Well, perhaps it wasn't worth all the fuss, or perhaps it wouldn't have made any difference whether or not *you* were born. But it was on just such a miraculous coincidence—the meeting of a specific sperm with a specific egg at a specific time—that the birth of a Lincoln, or a Shakespeare, or an Edison, or any other individual in history, depended. And it is by the same infinitesimal sway of chance that a child of yours might perhaps be a genius or a numskull, a beauty or an ugly duckling!

However, the first great coincidence was only the beginning.

The lucky sperm, which has won out in the spectacular race against millions of others, enters the chosen egg which has been waiting in a Fallopian tube of the mother. Immediately, as we previously learned, the sperm and the nucleus in the egg each releases its quota of chromosomes, and thus the fertilized egg starts off on its career.

Already, from this first instant, the fertilized egg is an individual with all its inherent capacities mapped out—so far as the hereditary factors can decide. Will the baby have blue eyes or brown eyes? Dark hair or blond hair? Will it have six fingers or a tendency to diabetes? Will its brain be normally efficient? These and thousands of other characteristics are already largely predetermined by genes in the child's particular chromosomes.

But as yet the individual consists of only one cell, like the most elemental of living things—for example, an amoeba. To develop it into a full-fledged human being, trillions of cells will be required. How this multiplication is effected we have seen in a previous chapter: The chromosomes split in half and separate, then the cell divides, making two cells, each with exact replicas of the forty-six chromosomes that were in the original whole. Again the process is repeated, and the two cells become four. Again, and the four cells become eight. So it continues, and as you could figure out if you wished, the doubling process would have to be repeated only forty-five times to provide the trillions of cells which constitute a fully developed baby.

However, as the cells go on to specialize, some divide and multiply much more slowly than others. But regardless of how they multiply or what they turn into, every one of the cells to the very last will still carry in its nucleus descendants of each of the original forty-six chromosomes.

HUMAN EGG AND HUMAN SPERMS

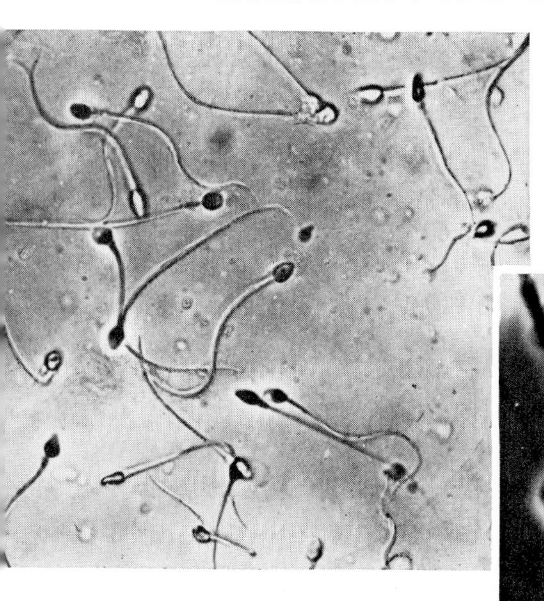

At left: Living human sperms in motion. From slide prepared for author by Dr. Abraham Stone; microphotographed by Joseph Weber. Magnification as shown about 500 times.

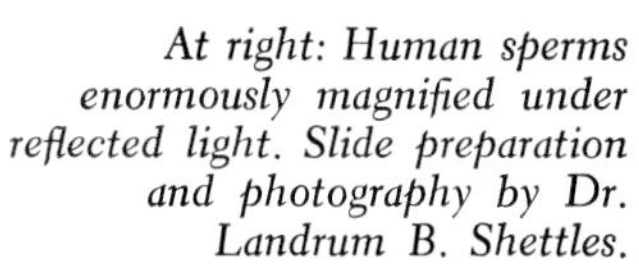

At right: Human sperms enormously magnified under reflected light. Slide preparation and photography by Dr. Landrum B. Shettles.

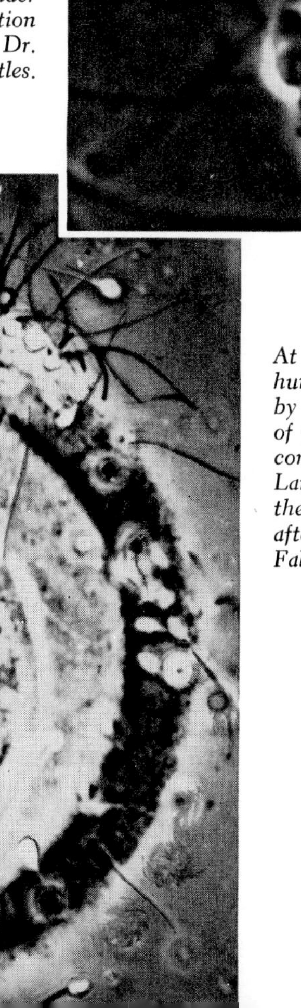

At left: Mature, living human ovum surrounded by sperms, as at moment of fertilization. Phase contrast photo by Dr. Landrum B. Shettles of the whole transparent egg, after its withdrawal from Fallopian tube.

HOW YOU LOOKED BEFORE YOU WERE BORN

The human embryo at successive stages dating from conception, shown in various magnifications, from almost actual size to about four times actual size.

All photographs from Carnegie Institution of Washington, Department of Embryology, Baltimore, Maryland. Magnifications as shown: 1. 3.8×; 2. 2.8×; 3. .7×; 4. 2.1×; 5. 1.4×; 6. 1.4×.

1. At 26 days.

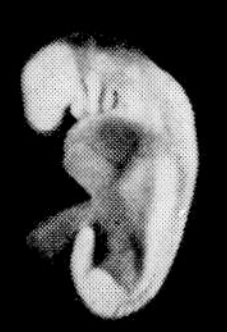

2. At 33 days.

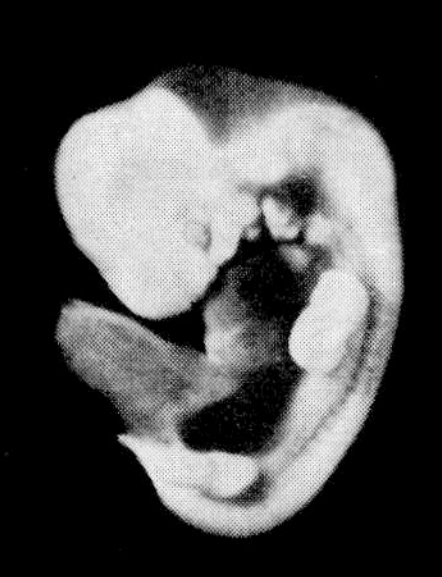

3. At 40 days. (Embryo shown in sac, within opened root-covered capsule which attaches to womb.)

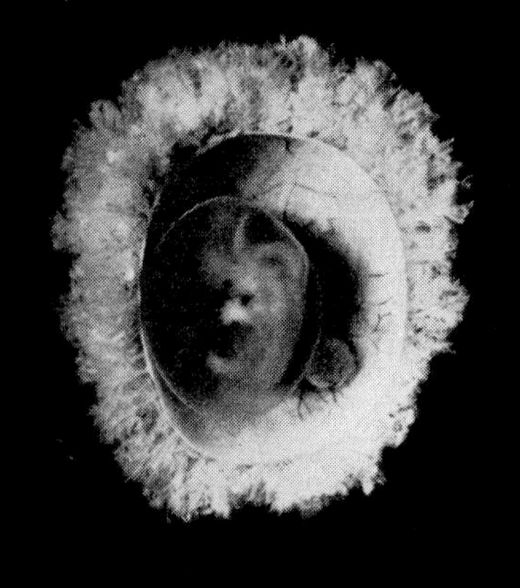

4. At 44 days.

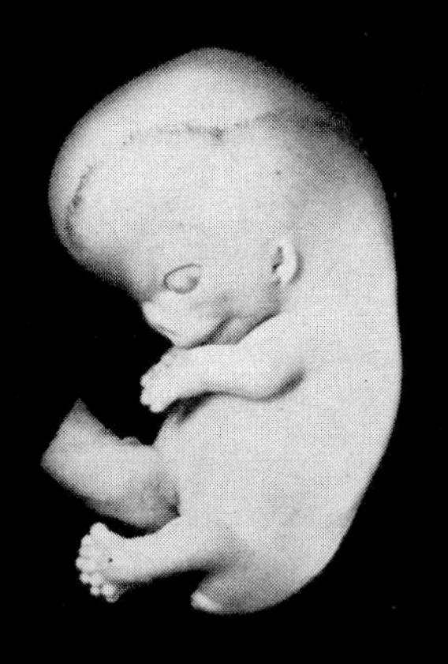

5. At 56 days (8th week).

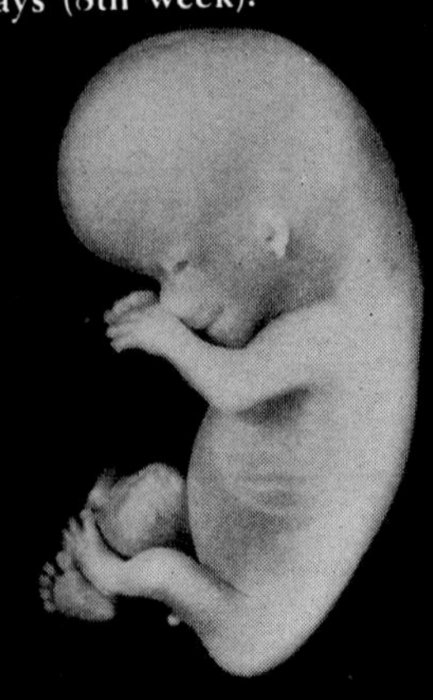

6. At 70+ days (10th-11th week).

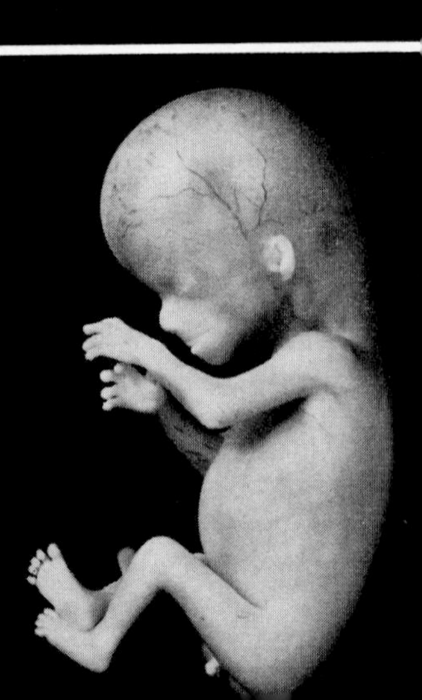

CHAPTER SIX

The Rocky Road to Birth

WE HAVE FOLLOWED the first stage in *your greatest adventure*—the remarkable coincidence by which you were conceived. But conception is a long way from birth, and no adventurer could face more hazards than does a human egg as it embarks on life.

In the first days after fertilization, while all the cell multiplication and activity have been going on inside the egg, it has been slowly making its way down a Fallopian tube toward the mother's uterus. Within a few more days the egg finds itself at the entrance of what must be—to this tiny droplet of substance, smaller than a period on this page—a vast, foreboding universe. If you can think of *yourself* at that stage, your life hung precariously in the balance. Innumerable adverse forces confronted you. At any moment you might be swept away to destruction. In short, the odds were all against your survival.

But to become impersonal once more, the immediate concern of the human egg at this stage is to take root somewhere. Already it has prepared itself for this by developing microscopic little tendrils from its outer surface, so that it somewhat resembles a tiny thistle. Thus it can attach itself to the mother's membrane, assuming (which is not always the case) that that membrane is receptive. If luck is with the egg, it is hospitably received—a hungry and thirsty little egg that has almost exhausted the store of food with which it started out. Immediately, with the maternal tissues cooperating, arrangements are begun for its food, oxygen and water supply through the development of a receiving surface, the *placenta*, which grows into the mother's membrane *but does not become part of it*. And so, about nine or ten days from the start of its existence—the most perilous days in any per-

CONCEPTION

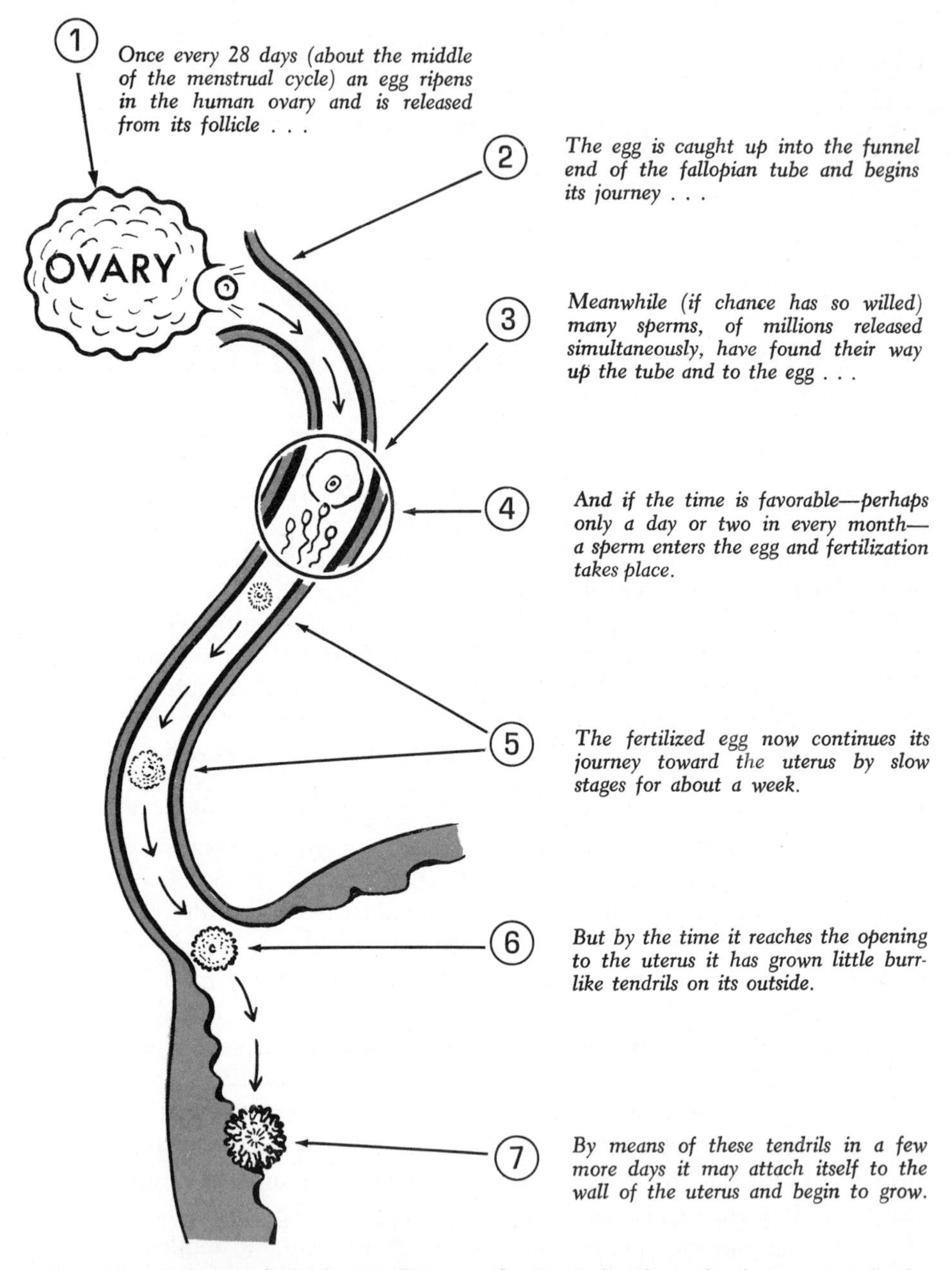

Note: The rhythm method of controlling reproduction is based on the facts presented—the method of estimating the time in a woman's menstrual cycle when an egg is likely to be released and avoiding sex relationships several days prior to and following that time. However, the method is not dependable because in few women does ovulation occur with regularity at predictable points in successive months. For further discussion of the rhythm method, see Chapter 31.

son's life—the new individual becomes what is really a parasite on its mother.

And now we may ask, how far can the mother, from this point on, affect the development and future of the child? The answers should do much to dispel many popular misconceptions.

Skipping some of the early stages, we presently find the embryo encased in a fluid-filled sac, attached to the placenta and connected with it by the umbilical cord, which acts as the conduit that brings in the food from the mother and carries out the waste products. But the umbilical cord is not, as is commonly supposed, a tube that goes directly into the mother's body. In fact, there is no direct connection anywhere, or at any time, between the mother and child. The child is from the earliest stage until birth as distinct an individual as if it were developing outside of the mother's body, like a chick within the eggshell.

There is a wall between the mother and child. On one side, the open ends of some of the mother's blood vessels empty into the wall. *But the mother's blood does not go into the child, nor do any mother and child have a single drop of blood in common* (although, sometimes, stray cells may be interchanged). For the food substances in the mother's blood are first broken down and strained out. And like moisture soaking through a blotter, it is these which are drawn into the placenta pressed hungrily on the other side of the wall and then conducted by the umbilical cord to the embryo.*

Not only is there no direct blood connection between mother and child, but there is, moreover, no direct nerve connection, and hence no such mental or psychological relationship as mothers have always liked to believe exists between them and the little one they are carrying. In the light of all this, another batch of myths, about prenatal influences and maternal impressions, about "strawberry marks" or other marks and deformities in the child resulting from the mother's having seen or done this or that, vanish into thin air. (We say "vanish" with reservations, for myths die hard, and even some of the most enlightened mothers still cherish a few of them.)

If the mother goes to concerts while she is carrying the child, that will not make it one whit more musical. Nor will thinking pure thoughts, reading elevating books or doing kind deeds during pregnancy improve the child's character. Nonetheless, inasmuch as the mother's nervous system and her physical functioning work closely together, any-

* "Embryo" is the term usually applied to the developing human individual in its initial stage, up to about the end of the eighth week. Thereafter, until birth, it is technically called a "fetus."

THE RELATIONSHIP OF MOTHER AND CHILD

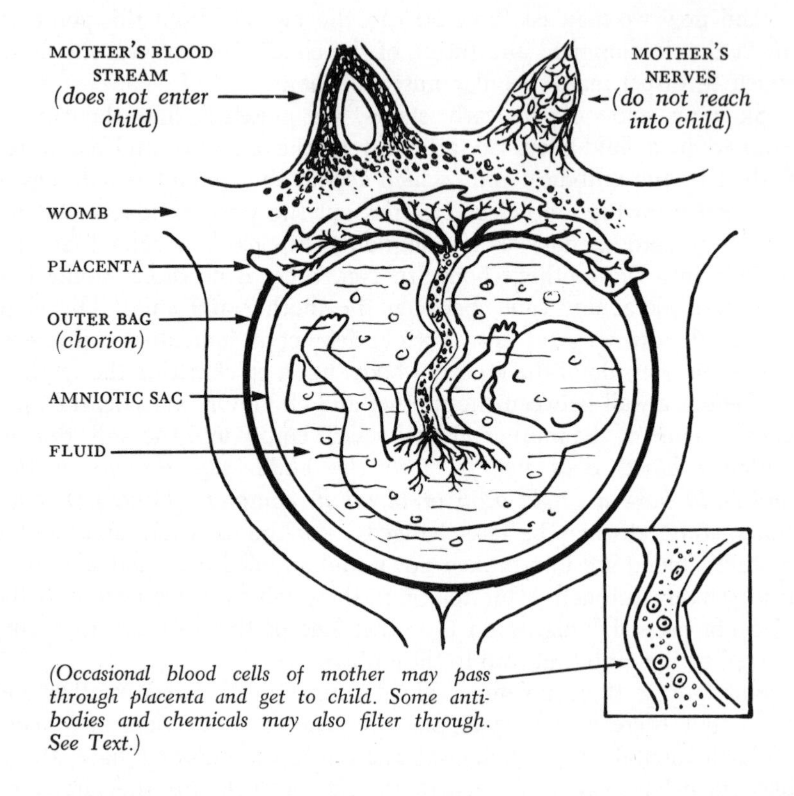

(Occasional blood cells of mother may pass through placenta and get to child. Some antibodies and chemicals may also filter through. See Text.)

thing that will contribute toward making her physical condition better will be to the child's advantage. By the same token, if the mother is unhappy and constantly upset, the child, too, may suffer.

We have evidence of this in studies by Dr. L. W. Sontag and others, showing that when a pregnant mother undergoes emotional upsets there is a marked increase in the body movements of the child within her. (The fetus can also hear and react to sounds.) Prolonged maternal stress, Dr. Sontag reported, may cause the baby to come into the world a "hyperactive, irritable, squirming, crying infant"—to all intents a "neurotic" person at birth. (The arrival of such a baby is often heralded by its having been a fretful fetus in the womb, with its unusual amount of activity felt by the mother.) However, such prenatally produced "neurosis" may have only temporary effects if the child is born into a subsequent relaxed atmosphere. But the child may continue to be neurotic if the mother continues to be.

Some of the greatest hazards to the embryonic child arise from the fact that the porous placental wall through which it gets its food *may also permit the passage through from the mother of many harmful substances.* Doctors had known for years that this might apply to a variety of drugs. But how far-reaching and drastic the effects of a drug taken by the mother during pregnancy can be on her unborn child was startlingly brought home to the world by the "thalidomide babies" of the early sixties. As readers may recall, several thousand of these babies—most of them with badly deformed or totally absent limbs and/or other abnormalities—were born in Germany and other countries (many surviving to this day), before the tragedies could be traced to their mothers' use of the supposedly safe new thalidomide sedative and sleeping tablet. While indeed found safe for the women themselves who took it, the drug, when seeping through to the fetal babies in their first few weeks, could act disastrously in impeding normal bodily development.* One constructive result of the tragedies was to increase precautions and alertness with respect to the use of almost any drug by pregnant women. (The possible relationship of heredity to thalidomide babies will be discussed in Chapter 24.)

Other sedatives and sleep-inducing drugs, and particularly the stronger narcotics, if used by pregnant women, had previously been established as dangerous to their fetal babies in varying degrees. If an expectant mother is an habitual user of morphine or opium, her child may actually come into the world a *drug addict.* Twelve such drug-addicted newborn infants arrived at New York's Bellevue hospital within a nine-month period in 1960. The babies had been getting drugs through their mother's system, and after birth, suffered the typical withdrawal symptoms of an addict, some dying of convulsions before treatment could prove effective. Such tragedies usually may be averted if a drug-addicted mother is kept without drugs for at least ten days before delivery, or, if the baby does exhibit withdrawal symptoms at birth, it is immediately given the standard treatments for drug addiction to cure it of its "habit."

The effects on a fetus of a pregnant mother's smoking or drinking

* In the United States the thalidomide tragedies were almost entirely averted because a wary government drug expert, Dr. Frances O. Kelsey, had held up the marketing of the sedative. However, some American women had received the drug from abroad or as samples from doctors, and a few who used it while pregnant did bear deformed babies—in one case deformed twins.

are less well established. Some studies have indicated that if the mother smokes to excess, the nicotine may get to the child and be harmful (Dr. Sontag having shown that the fetal heartbeat reacts to the mother's smoking just as does her own.) Premature births, or births of babies below average in weight, have been reported as considerably greater to mothers who are heavy smokers than to those who smoke sparingly or not at all. However, it is a myth that babies of heavily smoking mothers will be born yellow. As for drinking, alcoholic beverages taken in moderation during pregnancy are not likely to affect the fetal child. But there is evidence that heavy drinking may do so, perhaps making the baby more emotional for a while after birth.

There are many other ways in which the mother can adversely affect her child, as we shall see. Lest we sound too alarming, though, as we recount the hazards, it should be emphasized that the great majority of babies pass safely and happily along the road to birth. In fact, the more that medical science has learned about prenatal hazards—and the more the individual mother has profited by the knowledge—the safer the journey has become for the fetal traveler. This is reflected in the steady drop in the proportion of prenatal and birth casualties wherever conditions for childbearing have been improved. Moreover, the effects of many of the prenatal accidents can now be overcome or much reduced in gravity by treatment or operation after birth. Nevertheless, large numbers of babies continue to be lost or rendered permanently defective, and while in many cases the blame can be placed on the prenatal environment, a good many of the prenatal casualties result from something wrong in the baby because of defective hereditary factors. Only by knowing the different causes of possible mishaps before birth, and especially, those which are preventable, can we make much greater headway in reducing the prenatal casualty list.

One of the most important ways in which the fetal child can be helped is by seeing that the mother has a proper diet. If the food she provides is inadequate in various ways, the fetus, being a voracious and exacting little parasite, will suffer malnutrition. By this is meant much less an insufficiency of food (for the mother doesn't really have to "eat for two"), than deficiencies in quality in terms of essential elements needed by the child. Thus, lack of certain vitamins in the mother's diet (C, B-6 or B-12, D, K, or E), may lead to rickets, scurvy or other defects in the baby and may account for some of the stillbirths. There also is evidence that infants are more often aborted or born underweight in groups where mothers have deficient and improper diets than in groups of similar hereditary stock where pregnant mothers have adequate nutrition. (It is uncertain, however, whether any prenatal dietary deficiencies can cause congenital malformations in babies or can produce any other permanent and uncorrectible defects.)

Offering further dangers to the embryonic child are various diseases in the mother which can be transmitted through germs or viruses penetrating the placenta. *German measles* (*rubella*) is one of the virus conditions which, if contracted by a mother during the first months of pregnancy (particularly in the early weeks) may often result in such conditions in the child as congenital cataracts, deafness, heart defects, or any of a variety of malformations. *Asian flu* is another virus disease which, if present in the pregnant mother, may menace her baby. Among the serious germ diseases that can be passed on to the child prenatally are diphtheria, typhoid, tuberculosis and syphilis, while other infections, such as gonorrhea, may be acquired by the baby during the final hours of birth. Moreover, whether directly through germs, or indirectly through their effects on the mother's condition, if she has scarlet fever, cholera, malaria, smallpox, erysipelas or pneumonia during pregnancy, in many cases the child may be seriously harmed. (We should note here, prior to a detailed discussion in later chapters, that a child *cannot* inherit any germ disease, or start off with it at conception; for no sperm, and so far as we have record, no human egg, could carry a disease germ of any kind and still successfully go through the fertilization process.)

There may be further dangers to the fetal child if the pregnant mother has nephritis (kidney disease), certain heart conditions, diabetes or goiter. In the case of goiter, the thyroid deficiency in the mother which causes it may result in a child's being born with the mental defect called "cretinism" as well as with other abnormalities. Where the mother is diabetic, or is on the way to becoming so (prediabetic), her condition may cause a miscarriage or result in a baby who is abnormally large at birth and in danger of not surviving. (These latter threats may often be averted by medical treatment—insulin, vitamins, etc.—of the diabetic mother during pregnancy and/or a shortening of the gestation time, if necessary by a Caesarean operation.)

Most dramatic of the prenatal dangers is a possible *incompatibility between the blood of a mother and her fetal child,* resulting from inherited differences in some of their blood elements. (Many readers, especially those who have been parents, may already know about the *Rh* factor or, more recently, about the *A-B-O* group blood conflicts.) Despite there being no direct blood connection between mother and child (except for occasional cells as we said before), there still can be an interchange of *chemical substances* in their bloods; and if there is a clash between these substances, the child's blood cells may be seriously damaged, sometimes with death or severe injury resulting. In a later chapter we will give the full story of this amazing situation, how in former years it cost the lives of thousands of infants, and how discovery of the causes and precautionary measures have now greatly reduced the threats.

The possibility of transmission of chemical substances from the mother to the child has a favorable side: the fact that she can also transmit *immunities* which she has developed to various diseases through previous exposure or vaccination. For instance, a mother who has had polio shots can transmit enough of the immunization to the child, via the placenta, to protect it against the disease for months after it is born. There is every likelihood that a mother can also immunize her fetal child—or in time will be enabled to—against other diseases and conditions.

The *age* of the mother during pregnancy also has an important bearing on the prenatal environment provided for the baby. Many studies have shown that the risks of a baby's being miscarried, or being born defective, increase steadily as the mother approaches or passes age forty. In the case of congenital malformations, or some types of mental defect (notably mongolism), the incidences are much greater for babies born to mothers aged thirty-six and over than for those born to mothers in their twenties. Not only may the older mother's system have begun to run down in its chemical efficiency, but with added years she is more likely to have suffered diseases, accidents or other adverse experiences prejudicial to pregnancy. (One remarkable recent finding is that chromosome derangements, leading to abnormalities in a child, may also occur with much above average frequencies in the eggs of the older woman.) On the other hand, if a mother is too young—perhaps only fifteen or sixteen—her baby can suffer by the fact that her system may not yet have achieved the chemical efficiency needed for childbearing. Thus, immature mothers have a higher than average rate of stillbirths.

However, the mother's age cannot be considered without regard to other factors which have a bearing on her physical condition. If she is healthy and is well-cared for, the fact that she is in her late thirties will of itself seldom interfere with her having a fine baby. But one must keep in mind that in the modern world, women of the more favored groups have fewer children and tend to restrict their childbearing to the best maternal years, whereas the majority of women who have very large families and who bear children at both too young and too advanced an age are more often of the underprivileged groups. Thus, the too-old or too-young mother is also as a rule more likely to be one whose health and maternity conditions are below par. After a child is born, the same inferior maternal conditions may act further to interfere with its chances of a healthy start in life.

Also greatly affecting a baby's chance of survival—and again much influenced by the mother's age, health and childbearing conditions—is the question of whether or not it is carried through to full term and is born big enough to grapple with life. Premature babies run the highest risk of being defective and/or of not surviving. (They account for half of all infant deaths.) Another danger is that of *postmaturity*—of being carried too long. The mortality of babies carried twenty or more days beyond the average due date is about three times that of normal-term babies. One might note that both prematurity and postmaturity are greatest among babies born to older mothers or to those who have not had a child for many years previously. The prematurity rate is also much above average among babies born to very young mothers.

A new danger of recent decades is the harm that can come to a fetal child through *radiation*. Excessive radiation of the mother, particularly during the early period of pregnancy, may often cause malformations or other defects in the child. Following the atomic bombings of Hiroshima and Nagasaki, much above average numbers of defective children were born to mothers who had been pregnant at the time. Medical X rays, of the pregnant mother, if frequent and/or in heavy doses, may also sometimes menace the fetal child. (Where such X-raying is done today, it is usually only because of critical need for it.) It should be added that some amount of radiation which can affect a fetal child has always existed both in the air and in emanations from the earth.

So far, in discussing perils on the road to birth, we have put stress mainly on the environmental factors—the effects on the baby of influences coming from outside itself. But it may be pointed out that in many of the situations the environmental factors are not divorced from hereditary factors. For instance, the possible blood incompatibilities between mother and child derive from differences in their inherited blood substances. Again, where diabetes in the mother poses a threat to her baby, it can often be shown that the disease may well have arisen from an hereditary tendency (with the chances, though, that the child itself does not and will not have diabetes). Various other ailments, or chemical deficiencies in the mother, may have an hereditary basis. So, too, a fetal child's degree of resistance or susceptibility to various prenatal adversities, or its chances of developing a particular defect, may be much influenced by its own hereditary makeup. Thus, the food or care a pregnant mother receives is far from being the sole determinant of her expected baby's welfare. With the same diet, even if inadequate, one mother may be able to convert the substances needed for the child more efficiently than the other. With the same conditions, one baby may be able to run the gauntlet of birth hazards, and another not.

On the whole, however, it is remarkable how efficiently the overwhelming majority of mothers can carry through the requirements of pregnancy, even under adverse conditions. Further, the fetal baby is usually a most adjustable little creature. Floating in its fluid-filled bag, it is cushioned against many shocks to the mother and can ride along unharmed while she performs manual work, indulges in sports, or travels in bumpy vehicles or in airplanes. Only if the mother is suffering from the diseases mentioned or has a previous history of unsuccessful pregnancies, or her doctor sees other reasons for special precautions, need she worry that pursuit of her normal activities will endanger her fetal child. Indeed, under modern maternity conditions, it can be said that where accidents to babies do occur before or during birth, or when a baby is born defective, the mother is seldom responsible because of anything she herself did or did not do.

In later chapters we'll have much more to say about the hereditary hazards—very often unforeseen and unavoidable—which can menace babies. At this point, though, it has already been made apparent how from the very beginning of life both heredity and environment are at work to influence the fate of the individual. An egg might start off with *good heredity,* or the best of genes, but through bad prenatal environment it may never have a chance to develop properly. Or an egg might start off with *bad heredity* in the form of some seriously defective genes, which might under average conditions destine it to be killed off. But if the mother's condition is unusually good—in other words, if the environment is extremely favorable—the "weak" egg may develop through to birth, and the individual may survive. However, where conditions are equally bad, or equally good, the chances for the egg with the better heredity will always be brighter than for the egg starting out under a genetic handicap.

In the nine months before birth every human being faces the most severe test that he will ever undergo. With seeming ruthlessness Nature exercises her "Law of Selection," killing off the weakest individuals more relentlessly than ever the ancient Greeks ventured to do. In fact, so stringent is the initial ordeal that many experts believe the children who achieve birth represent only a *portion of the eggs fertilized.* In other words, in innumerable instances—perhaps a majority of the cases—women conceive and the egg is killed off without their even knowing it. What frequently are described as "false alarms" may have been actual conceptions. And more often than not, there would have been little reason to cry over the lost babies, for had they survived, they might have brought tragedy with them.

Undoubtedly, and unfortunately, many worthy individuals are also sacrificed in the initial weeding process. On the whole, however, most parents may have reason to be grateful for the rigid selection. Where birth is achieved, it can generally be taken as an indication that the individual is qualified to face life. From that point on, how he fares is up to his parents, to himself and to the environment created by society.

CHAPTER SEVEN

"Boy or Girl?"

NEXT TO BEING BORN, the most important single fact attending your coming into the world was whether you were to be a male or a female. As it was to your parents and as it is or will be to you as a prospective parent, the sex of the expected child is undoubtedly the first thing that is thought about. Going with this has always been speculation as to what it is that determines a child's sex, and whether and how the parents-to-be can influence the outcome.

The beliefs and theories on the subject have been endless: that the "boy or girl" decision is governed by the time or circumstances under which conception occurs or by the parents' physical or mental states; or that it depends on whether the mother's egg comes from the right ovary (for a boy) or the left one (for a girl); or that it is influenced by the moon and the stars; or that it is only *after* conception that the baby's sex is decided by what happens in the womb.

With so much attached to the outcome, many people have hardly been content to leave the matter of sex determination to chance. So as far back as there are records, and even to this day throughout the world, a fascinating variety of methods—potions, prayers, midwives' formulas, diets, drugs or quasi-medical treatments—have been and are being employed to influence the sex of a hoped for or expected child. Most often, undoubtedly, the objective has been or is a boy. But an ample list could also be compiled of the "what-to-dos" to make it a girl.

Alas then, whatever the methods employed, hocus-pocus or supposedly enlightened, all can now be equally dismissed with this disillusioning answer:

The sex of a child is fixed at the instant of conception; and it is not through the mother, but through the father, that sex is determined.

The moment that the father's sperm enters the mother's egg, the child is started on its way to being a boy or a girl. Subsequent events or influences may possibly affect the *degree* of maleness or femaleness, or thwart normal development; but from that first instant of conception, nothing now known within human power can change what is to be a girl into a boy, or vice versa.

The solution of the mystery of sex determination came about through this discovery:

That the only difference between the chromosomes carried by a boy and a girl, or by a man and by a woman, lies in just one of the pairs—in fact, in a single chromosome of this pair.

Of the twenty-three pairs of human chromosomes, twenty-two pairs are alike in both males and females. That is to say, any one of them could just as readily be in either sex. But when we come to the twenty-third pair, the sexes are not the same. For, as we see in the accompanying diagram, every woman has in her cells two of what we call the X chromosome. But a man has just one X—its mate being the much smaller Y. It is the presence of that "Mutt and Jeff" pair of chromosomes in the male (the XY combination), and the XX pair in the female, that sets the machinery of sex development in motion and results in all the genetic differences that there are between a man and a woman.

(The terms "X" and "Y" for the sex chromosomes are simply identifying letters: the X was so called—much in the way that the X ray was named—for want of a better identifying name. After another chromosome was found to be paired with the X in the process of sex determination, this second one was then called the Y. All of the other chromosomes, referred to as *autosomes*, are now identified by numbers ranging from 1 to 22, inclusive. See illustration, page 7.)

We have already seen how when human beings form eggs or sperms, each gets just *half* the respective parent's quota of chromosomes or one of every pair. When the female, then, with her two Xs, forms eggs and gives to each egg one chromosome of every pair, *every egg gets an* X.

But when the male forms sperms and his unevenly matched pair of sex chromosomes separate, *an X goes into one sperm, a Y into the other.*

Thus we have this important difference: With regard to the sex factor, the mother produces only one kind of egg, each containing an X. But the father produces *two* kinds of sperms, in *equal numbers;* which is to say that of the tens of millions of sperms released by a man at each ejaculation, half are X-bearers, half Y-bearers.

HOW SEX IS DETERMINED

This is what makes all the inherent differences there are between a human male and female.

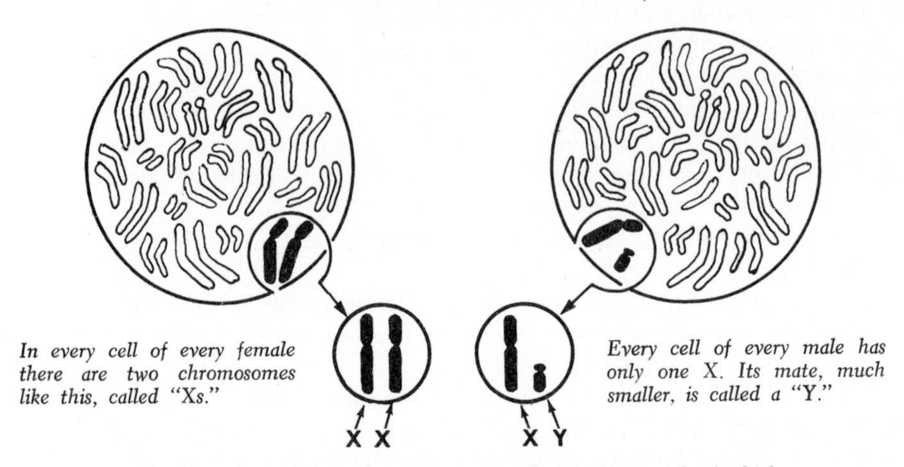

In every cell of every female there are two chromosomes like this, called "Xs."

Every cell of every male has only one X. Its mate, much smaller, is called a "Y."

For reproduction, a female forms eggs, a male sperms, to each of which they contribute only HALF *their quota of chromosomes, or just one of every pair.*

Since a female has TWO *Xs, each egg gets one X, so in this respect every egg is the same:*

But as the male has only ONE *X, he forms* TWO *kinds of sperms:*

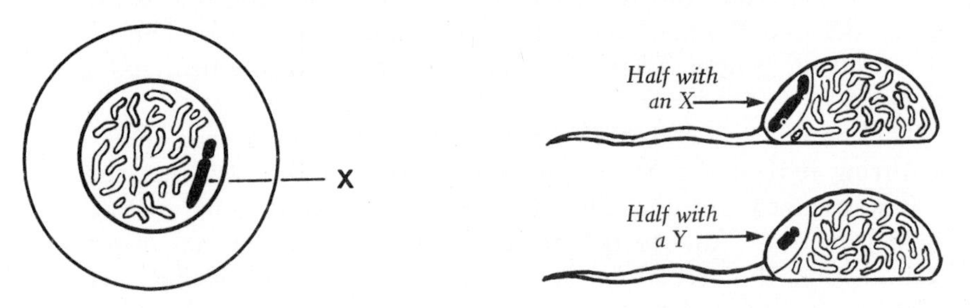

Thus: If an X-bearing sperm enters the egg, the result is an individual with TWO *Xs:*

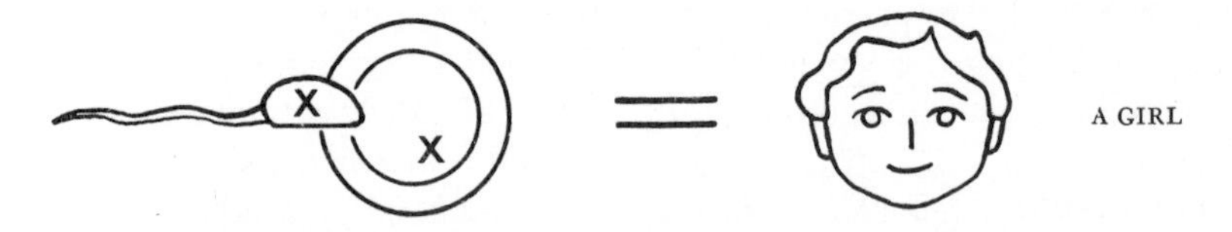

But if a Y-bearing sperm enters the egg, the result is an "XY" individual,:

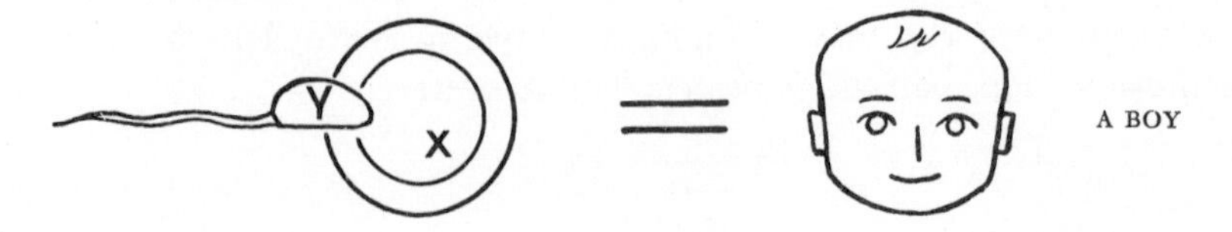

Science having established that only one sperm fertilizes an egg (as the egg coating hardens the instant any sperm enters, shutting out all the others), the result should be self-evident. If a sperm with an X gets to the egg first, it pairs up with the X already there, an XX individual is started on its way, and eventually a girl baby is produced. But should a Y-bearing sperm win the race, the result will be an XY individual, or a boy.

Here at last is the comparatively simple answer to what was long considered an unfathomable mystery!

Moreover, we see that sex, as it is determined through chromosomes, is an *inherited* characteristic (perhaps the most important of all inherited traits for the individual). But precisely how does the process work?

The key to the situation in human beings is held not by the X chromosomes, as was formerly thought, but by the Y. It had long been believed that the sex chromosomes in people work basically like those in the fruit flies and certain other experimental creatures. In these species the X chromosomes carry genes heavily slanted in the "femaleness" direction, whereas the "sex" genes distributed among the other chromosomes are slanted in the "maleness" direction. The Y may be a blank so far as sex genes are concerned (and, in fact, some species have no Y at all). At conception, then, there is a tug of war. If the fertilized egg has two Xs, the double extra load of "femaleness" genes can pull the individual over toward development as a female. If the egg has only one X, the excess of "maleness" genes in the other chromosomes pulls the individual over toward development as a male.*

The belief that the foregoing also applied to human beings was radically changed by the new techniques that made possible accurate identification of the sex chromosomes in persons with either normal or abnormal sex development. It was immediately apparent that the human Y chromosome was by no means an innocent bystander in the sex-determination process. On the contrary, far from being dwarfed by the much bigger X, the Y was found to be tremendously potent

* This principle of balance between the "sex" genes can also explain what used to be a puzzling fact with respect to poultry, birds and certain insects: In these species the sex-determining mechanism is reversed, with the XX combination producing a *male*, the XY combination producing a *female*. As we now know, in these species the arrangement of the "sex" genes is transposed, the X chromosomes being the ones that are overloaded with "maleness" genes, the other chromosomes carrying most of the "femaleness" genes, so that two Xs in the bird species pull the individual toward male development, whereas a single X permits it to be pulled toward female development.

and masterful—reminiscent of the famous line in the Gershwin song, "Li'l David was small, but oh my!" For, as now clearly established, it is actually the presence and strength of the Y that causes the XY individual to develop as a normal male, and the absence of the Y that permits the XX individual to develop as a normal female.

All of this has been proved by the exceptional cases (somewhat more common than was previously supposed) of human individuals who are deficient or abnormal in their sex chromosomes. In some situations an individual is started off at fertilization with only an X, having failed to receive, or losing on the way, the other sex chromosome—either the Y or a second X. Yet such an "XO" individual still develops as a female (though a defective one). In still other rare situations, a fertilized egg may start off with a Y and two Xs, or even three Xs; but not even the double or triple quota of "X" genes can prevent the little Y from causing the individual to become a male (although again, usually, a defective one).

The exceptional XO or XXY and XXXY individuals and others who are abnormal in sex development will be dealt with fully in a later chapter. At this point we need only keep in mind that in the overwhelming majority of conceptions the sex decision is immediately clear-cut, the two-X baby developing step by step into a girl, and in succeeding years, into a woman; the XY baby becoming a boy and then a man. Once the initial decision is made, the human female remains a female, the human male, a male. (Which, everything considered, is pretty fortunate, for otherwise wouldn't this be a pretty mixed-up world?)

The successive stages by which a fetus develops into either a boy or a girl are illustrated in the accompanying chart. It should be stressed that sex differentiation, though not outwardly noticeable until the fetus is several months old, actually gets under way at once. For as the cells multiply following conception, into every cell of the male goes the XY combination of sex chromosomes, distinguishing it from every cell of the female with an XX combination. Stemming from this difference, XY cells in the sex region of a fetus go on to produce the male sex glands, or testes, with an output of male-slanted hormones. By contrast, XX cells go on to produce the female sex glands, or ovaries, with an output of female-slanted hormones. As growth proceeds, other glands—also influenced by the XY and XX combinations—join with the two types of sex glands and their hormones to further enhance the physical development of the male in the masculine direction, and of the female in the feminine direction. This differentiation goes on throughout life, the developmental paths of the two sexes diverging more and more

HOW THE INTERNAL SEX MECHANISM DEVELOPS

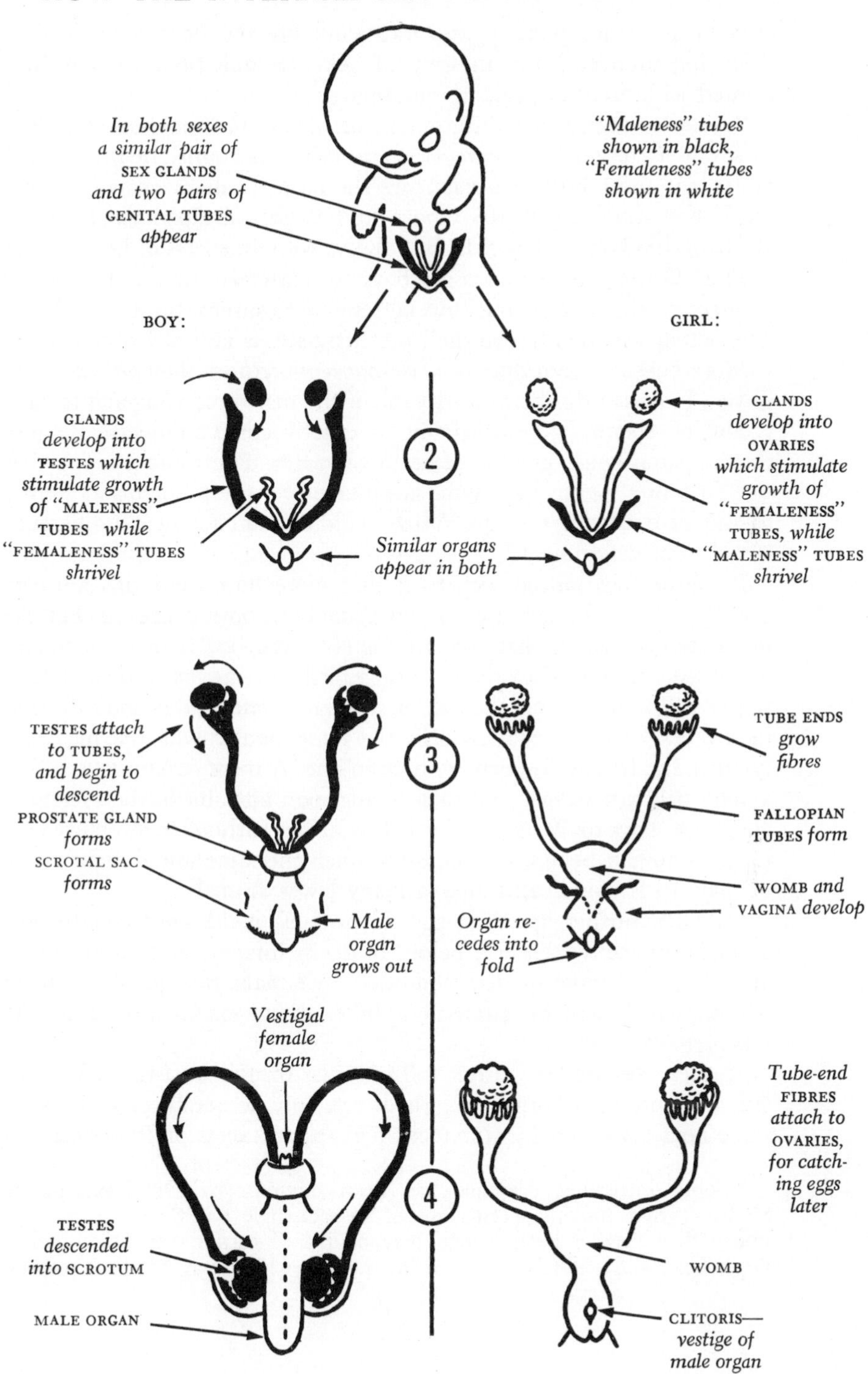

after birth, during puberty, and into adult life and beyond.

Having unraveled the mystery of sex determination, we are confronted with another puzzling question:

Year after year, when the millions of births are tabulated according to sex, it is found that *more boys are born than girls.* In the United States and most European countries the ratio for many years has been much the same: about 105.5 boys to 100 girls. Or perhaps 105.6, or 105.8, or 106 boys to 100 girls. But always with an excess of boys. Why?

The old theory was that more boys are born because boy fetuses are stronger on the average and thus better able to survive through to birth. The actual situation, as we shall presently see, is radically different.

More boys are born only because many more boys than girls are conceived. How can this be, you may ask, if the male-producing and female-producing sperms are turned out in exactly equal numbers? Because they apparently are not the same in character. The assumption is that the Y-bearing sperms have some advantage over the X sperms in getting to and/or fertilizing the egg. What would account for such a difference is still to be determined.*

If Nature has indeed seen to it that more boys than girls are conceived, it would be because a reserve supply of boys is needed. For the unquestioned fact is that proportionately more boy than girl embryos or fetuses succumb before or during birth, or are born with defects which will imperil their chances of survival. Some studies indicate that in the very first prenatal months the male fetuses who die may outnumber the females by two or three to one. A more certain fact is that among stillborn babies, and those dying soon after birth, the average is about 130 boys to 100 girls. Nor is this higher early male mortality confined to human beings, for one finds much the same initial higher ratio of male to female deaths among many lower animals.

In short, further upsetting age-old notions, *males—not females—are biologically the weaker sex,* before birth, in infancy, and always thereafter (at least under modern conditions: we make this qualification for reasons which will be given in a fuller discussion of the subject in Chapter 29).

If, then, despite the greater toll of males on the road to birth, there still are more boys born than girls, it can only be because a good many more boys are entered in the race—perhaps an excess of 30 per cent, or

* Some intriguing evidence has been offered (by Dr. Landrum B. Shettles) that human sperms are of two distinct types and head sizes: one with a smaller head, which may be the Y carrier, the other with a larger head and a different-shaped inner mass, which may be the X carrier.

even more, in the belief of many authorities. But this starting surplus is hardly sufficient, for by the time of birth the excess of males has been whittled down to 6 per cent, as previously noted; and with a continuing higher toll of males at all ages, a point is eventually reached where females outnumber males.

The explanations as to *why* males are less able to survive will be presented in later chapters (22 and 29). For the moment, knowing that males are less hardy than females, we might gather that where prenatal conditions are most unfavorable, the chances of sons being born are diminished; where prenatal conditions are at their best, the chances of a boy's achieving birth are greatest. Evidence to support this comes from many studies. Young, healthy mothers, with their first pregnancies, tend to produce a higher average ratio of boys than do older mothers who have had many previous pregnancies. A major reason is that the stillbirth and involuntary abortion rate—always with a marked excess of boys among the babies lost—increases with successive births, in part because of deterioration of the mother's general health as she grows older. Thus, in some groups the sex ratio has been reported as high as 120 boys to 100 girls, while the ratio among veteran mothers of forty or over has been reported as low as 90 boys to 100 girls.

It also has been found that among the more favored social and economic groups the birth ratio of boys to girls is considerably higher than in the underprivileged ranks. To a great extent this may be because inferior prenatal conditions are more prejudicial to the survival chances of boy fetuses. But a further, if not equally important fact, is that women of the less favored groups tend to have more children and to continue childbearing into the later years, which would significantly lower the ratio of boys. As the proportion of less favored mothers to whom this might apply, and those with large families, has decreased in recent decades, while at the same time childbearing conditions generally have improved, one could expect an increase in the ratio of boys born. This seems to have been so in England, where records of boy-girl birth ratios show an overall advance from a ratio of about 104 boys to 100 girls at the turn of the century, to more than 106 boys to 100 girls today.

In the United States, however, the available records on this point, which go back only to 1915, show no similarly marked upward trend in the sex ratio, despite great improvements in childbearing conditions. It seems obvious, then, that group variations in the boy-girl ratio cannot be accounted for solely by what is known about prenatal environments. This is most strikingly shown in comparisons between the boy-girl ratios of Negroes and Whites in the United States. For many years the

Negro ratio has averaged only about 102.5 boys born to 100 girls, in contrast with a White ratio of about 105.5 to 100.* A first thought might be, "If bad prenatal conditions tend to lower the sex ratio, the inferior conditions of the Negro could explain why so many fewer boys are born." This answer is clouded by the fact that increasing improvement in childbearing conditions among Negroes (relatively even greater than among Whites, since there was much further to go) has brought no change in the sex ratio. Further, the sex ratio among Negroes in New York City, with better conditions, is just as low as it is in the deep South, with inferior conditions.

There is reason to suspect, then, that we may be dealing with a true racial difference in the sex ratio at birth.† This suspicion is heightened by evidence from other sources. In Korea (where childbearing conditions are if anything inferior to those among Americans, whether White or Negro), the sex ratio has been reported as averaging 115 boys to 100 girls—the highest known ratio among world populations. In Japan, on the other hand, the ratio is about 106 to 100, almost exactly what it is among American Whites. Lower or higher sex ratios have been reported in other racial and ethnic groups (among the Turks, for instance, it has been given as 110 boys to 100 girls). While the facts about inborn human racial differences in the sex ratio at birth remain to be established, there is no doubt that such sex-ratio differences do exist among lower animals in various breeds of a given species. For instance, Jersey cattle produce an excess of male calves, Guernseys an excess of female calves; and in mice, fish and fruit flies, there are both high and low male-producing strains.

What, then of the possibility that apart from possible racial or ethnic differences within any human group, a tendency to bear either sons or

* In recent years the sex ratio given for Whites in the United States has dropped off somewhat, reaching down to 105.2 in 1962. Contributing to this in some degree may be the fact that the government statistics include among Whites the growing numbers of Puerto Ricans who have a heavy Negro admixture and thus a sex ratio lower than that of the general White population.

† Marked race differences definitely do exist in another area of human reproduction—*multiple births*—with Negroes having the highest rates, Japanese almost the lowest, and Whites intermediate. Twins will be dealt with in a later chapter. Pertinent here, however, is the fact that the sex ratio among twins and triplets is much lower than it is among the singly born. The apparent reason, in line with what has been noted previously, is that prenatal conditions are especially adverse when more than one baby is being carried and that this would be more harmful to boy than to girl twin fetuses.

daughters may run in certain families? The belief that this is so would seem borne out by very large families with all daughters and no sons; or, the reverse, of families with eleven or twelve sons in a row, and no daughters. And what of the *fifteen all-son* family which a West Virginia schoolmaster, Grover C. Jones, and his wife, brought to the New York World's Fair in 1940?

Remarkable as these one-sex families may be, skeptical scientists still do not think they need be other than matters of chance. That is to say, if there is an approximately even chance that a given baby will be a boy or a girl, the outcome in successive births is much as it is when tossing coins. With the odds 1 in 2 for the first baby being a boy, the odds for two in a row being boys (or being girls) are 1 in 4; for three of the same sex in succession (doubling the odds each time), 1 in 8; four in succession, 1 in 16; and so on, with the odds for eight children of the same sex in a row, 1 in 256; ten in a row, 1 in 1,024; twelve in a row, 1 in 4,096. Even the fifteen-son family mentioned could happen by chance once in every 32,768 fifteen-child families. Naturally, since such big families are unusual, those with all children of one sex are extreme rarities. Nevertheless, scientists who have checked carefully on enormous numbers of families of all sizes have found that the incidences of one-sex families of any given size conform pretty much to the expected odds.

The possibility remains that some families, or two given parents, might have an hereditary tendency to produce children of one sex, if not exclusively, at least with above average frequencies. This is based on various assumptions: That in some parents either the "maleness" genes or the "femaleness" genes might be especially potent (or, conversely, defective); or in some fathers the Y sperms might be more than ordinarily vigorous; or that some mothers may be genetically endowed to provide prenatal environments especially favorable for the conception and development of male fetuses, while with others it may be the reverse. These theories are still to be proved. Also inconclusive are reported findings that the "boy- or girl-prone" parents may be identified by various traits. For instance, some data have suggested that fathers who are physically stronger and more active tend to sire sons with greater frequency than do sedentary fathers engaged in mental work; or that when both parents are in robust health, there is a better chance of their having a boy than when both are ailing; or that when and if parental age affects the sex ratio of offspring, it is the father's age rather than the mother's that counts, with older fathers being less likely than younger ones to sire sons. Conflicting reports have been published on all the foregoing theories. Also currently in dispute is the question of whether sex ratios are influenced by the A-B-O blood groups—that is, by whether the parents and/or fetuses are of one blood type or another.

Various other possible influences on the sex ratio have been extensively studied. One long-standing popular belief which scientists have explored is that during and after wars more boys than usual are born, presumably because of Nature's plan to make up for the men killed. That the ratio of boys born does go up—although slightly—in wartime, is a fact. After World War I there was an advance of about 1 to 2 per cent in the ratio of male births in the principal war countries, the largest increase being in Germany where the ratio rose to 108.5 boys to 100 girls. During World War II there was only a small increase in the boy-to-girl birth ratio in the United States, with a high for Whites in 1946 of 106.3, and for Negroes of 103.3 boys to 100 girls. In England there was a slightly greater rise from 1942 to 1946, which brought the birth sex ratio to a point higher than it had been before. But there is nothing supernatural or mysterious about these wartime changes. They can be explained simply and chiefly by the fact that during and immediately following wars there is an increase in childbearing among the younger and first-time mothers, who normally would produce a higher ratio of boys.

Similarly, there is now no mystery as to why, in illegitimate births, there was, and often still is, a below-average ratio of boys. For this, too, there are obvious reasons. More often than not the conditions in illegitimate maternities are inferior, or the mother is immature, or in other ways inadequately equipped for childbearing, any of which factors discriminate against boy births. However, if the conditions for an unmarried mother are on a par with those for a married one, there will be no difference in the sex ratio. Thus, a study made by the author some years ago in five New York homes for unmarried mothers, where the best of care was provided, showed that among the more than three thousand babies born to these mothers the ratio was 108.6 boys to 100 girls.

There are other questions which may have occurred to you. Can the weather or the climate where you happen to live influence the chances of conceiving, or bearing, a son or daughter? Apparently not, for studies in the United States tend to show that the sex ratios average out to about the same in one section of the country as compared with another, despite marked differences in climate. Does the season of the year affect the chances of having a boy or a girl? Some investigators say "Yes, to some extent," but if this is true, it may only be because of differences in the times that conceptions occur, or are planned, by mothers of "high-boy ratio" groups as compared with those of "low boy-ratio" groups. One also hears that certain foods, vitamins or chemicals if eaten or taken by the mother, or even by the father, can influence sex determination. These claims are still to be substantiated.

All of which isn't to say that some means may not eventually be found which will give parents a measure of power with regard to having a child of the sex they desire. Such a possibility, if not a probability, may lie in any of several methods. First, if there are physical differences between the X-bearing and the Y-bearing sperms (as has already been indicated), in size, structure or movement, these sperms might be separated in advance, and then those of only one type could be introduced into the prospective mother's reproductive tract. Experiments in this direction have been made so far only with lower animal sperms. One method employed has been *electrophoresis,* with the thought that the X and Y sperms would respond differently to positive and negative electric charges (or may carry different electric charges); another method is *centrifuging,* the means used to separate out from a solution two substances of different densities. While complete separation of X and Y sperms has not been achieved so far by either method, experimenters have reported success with it in causing rabbits and cattle to produce much above average ratios of males, and rabbits to also produce marked excesses of females.

A second possibility explored is that the X and Y sperms may be affected differently by the chemical states of the female sex organs and tubes through which they must pass on their way to the egg. Thus, by modifying these chemical states in one direction or the other, fertilization would be achieved only by the sperms for the desired sex. One means by which that has been attempted, the much publicized *acid-alkali* treatment (an acid state for a girl, alkali for a boy), has failed to stand up after many experiments.

Whether by one of the foregoing methods or another, scientists believe strongly that human beings will presently have the means of controlling or greatly influencing the sex-determination process. This might be a godsend to parents who desperately want a child of the one sex. For society at large, though, it might open up a Pandora's box of new headaches and social complications, at least for a while. A first thought is that there would immediately be an inordinate rush to have boys. The general preference for a boy baby (at least, for the first child, or for a majority of boys) has long prevailed in many groups and in many parts of the world. The reasons are many and deep-rooted. Various

POSSIBLE FUTURE METHODS OF SEX CONTROL

DIFFERENCES MAY EXIST, CHEMICALLY OR STRUCTURALLY, IN ELECTRICAL REACTION OR IN SIZE OR SPEED, BETWEEN THE

FEMALE-PRODUCING SPERM

MALE-PRODUCING SPERM

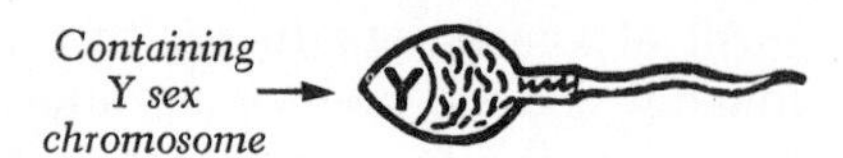

ACCORDINGLY, THESE METHODS MAY POSSIBLY LEAD TO PRODUCING OFFSPRING OF THE SEX DESIRED:

1 A. *Before introducing the father's sperms into the mother, it may be possible to separate the "male-producing" sperms from the "female-producing" sperms by "centrifuging" or by chemical or electrical action.*

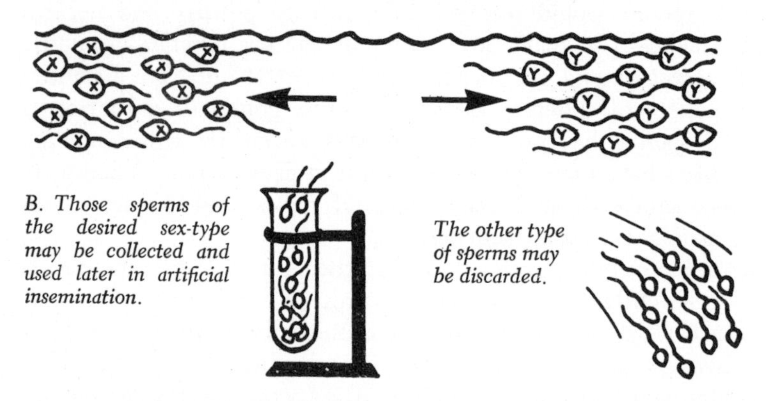

B. *Those sperms of the desired sex-type may be collected and used later in artificial insemination.*

The other type of sperms may be discarded.

2 *Some chemical solution may be found in which the sperms of one type can swim faster than the other, so that when injected into the mother, only one type of sperms will reach the egg.*

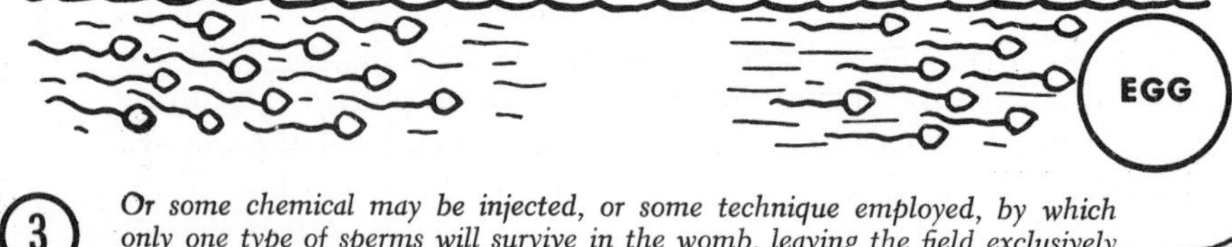

3 *Or some chemical may be injected, or some technique employed, by which only one type of sperms will survive in the womb, leaving the field exclusively to the other type.*

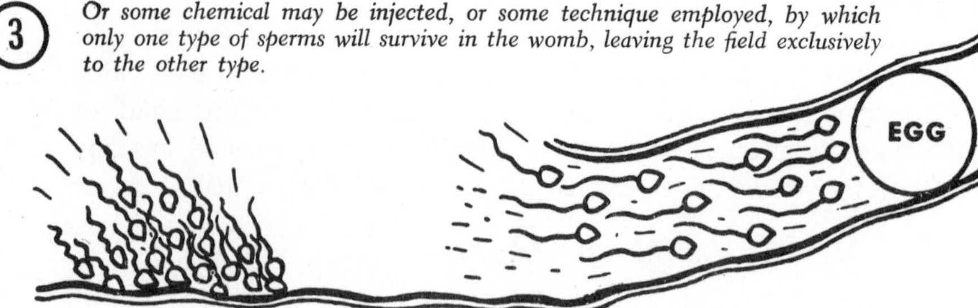

religions specify that a man must have a son to pray for his spirit if he is to have peace and happiness in the Hereafter. Kings and potentates want or must have male heirs for their thrones. (The ex-King of Egypt and the present Shah of Iran both secured divorces for the official reason that their wives had borne no sons. The third marriage of the Shah, in 1959, was blessed with a boy.) In our own country industrialists want sons to carry on their enterprises, and farmers want them as helpers who in time will be successors; many men are anxious to have sons to perpetuate their family names; and for the first-born, most parents prefer a boy, inasmuch as it works out better socially if a brother is older than a sister.

But any one-sided preference for sons has been steadily diminishing in the United States, and probably in other countries, as social changes have taken place. Where only one or two children are possible, many parents now say they favor having girls; and, in fact, adoption agencies report a much greater demand for girl babies than for boy babies. Looking at the question broadly, the law of supply and demand will probably work out with respect to sex control, as in other matters.

For the time being the question of a baby's sex remains largely a matter of chance for individual parents. The extent, degree and causes of variations in the sex ratio have been told at length, but one must always remember that this is speaking in terms of large averages. With regard to any given baby you are expecting, no matter how your general chances are slanted, Mr. Stork might still ignore the odds and cantankerously grab up whatever little bundle is handiest.

More immediate than sex control is the fact that expectant parents will at least be able to know the baby's sex well in advance of its birth. Various scientific methods of doing this are or ultimately will be possible. So far, unfortunately, the method that was shown to be most effective—that of withdrawing fluid from the amniotic sac and studying cells given off by the fetus—has proved to entail serious risk to the baby. Safer "sex-forecasting" methods will undoubtedly be forthcoming. If nothing else, parents will then be able to assuage their disappointment well in advance of the baby's coming and to have in readiness a blue layette or a pink layette, as required—plus a name for the baby.

CHAPTER EIGHT

Peas, Flies and People

NOT ONLY YOUR SEX, but a myriad of your other characteristics had been potentially determined or influenced by your genes at the instant of conception.

Before going into details about all of these other inherited traits, let us turn back the pages of genetic history to learn something of how the mysteries of the genes and chromosomes were unraveled.

Whatever we know about heredity is based largely on the original findings of two men: one, an obscure Austrian monk of the past century, Gregor Mendel, who cultivated garden peas; the other, an American scientist, Thomas Hunt Morgan, who until his death in 1945, cultivated fruit flies.

How, you might ask, could ordinary peas (much like those you get at any club luncheon with your chicken and candied sweets) and fruit flies (the tiny kind that buzz around bunches of bananas)—how could these have any bearing on what you or other humans are? They could, because, as we know now, the mechanism of heredity in peas and flies, as in all other living things, is basically the same as it is in man. This is one of the amazing facts that may be hardest to accept for those who think of human beings as altogether unique.

When in 1857 the Abbot Gregor Mendel, working in the garden of the monastery at Brno (in what is now Czechoslovakia), set out to clarify his mind about the heredity of peas, he himself did not dream that he was at the same time about to throw lasting light on the heredity of human beings. Mendel had a brilliant mind, but it was simple and direct. And this is why he succeeded where others failed. He resolved to confine his studies to his own little 30-by-7-foot garden patch, where he had plants with many different characteristics. Mendel also decided to concentrate on just one character at a time. So, as one instance, he set out to see what would happen when he mated peas of a pure red-flowering strain with those that habitually bore white flowers. Thorough in his methods, he bred together hundreds of such plants. And this was the result: *The offspring were all red-flowered.*

Had the influence of the *white* parent been completely lost? No, because when Mendel mated any two second-generation red-flowered plants together, three out of four of the offspring were *red*-flowered, but *one out of four was pure white like the white grandparent.* This proved that the white element had been carried along *hidden* in the preceding generation.

Further investigation showed that the third generation red-flowered plants were not all alike, even though they *looked* the same. In only one out of three cases were they "pure" red-flowered, like the grandparent, and when mated with each other would produce only red flowers. In the other instances the plants had both kinds of colors, red and white, like their immediate parents.

Mendel checked the results in planting after planting. Meanwhile, in different patches of his garden, he experimented with other matings—breeds of tall pea plants with breeds of short ones; plants having yellow seeds with those having green; and wrinkled seeds with smooth. For everything he kept exact figures, carefully tabulated, until finally the evidence pointed to these conclusions:

1. The inherited characteristics are produced by genes (although they were not called that by Mendel) which are passed along *unchanged* from one generation to another.*
2. In each individual these genes are found in pairs, and where the two genes in a pair are different in their effects, one gene's action often *dominates* over that of the other, so that the former gene might be referred to as a *dominant,* the latter as a *recessive.*
3. A recessive gene, although its effects may be masked by a dominant, will assert itself if coupled with a matching recessive.
4. When seeds are formed in any individual, the members of each pair of genes segregate out, with just one of every two matched genes going from each parent to each offspring.
5. When traits are separate and governed by different genes, the

* The word "gene" was first used in a publication by W. Johanssen in 1909. Mendel himself talked mostly about the segregation of "characters" and once used the term "elements" to refer to whatever it might be that made the germ cells different in respect to alternative traits.

genes for each trait are transmitted independently, unaffected by the genes for the other traits.

The foregoing principles, sometimes stated in other ways, are frequently referred to as the "Mendelian Laws." In the technical classification, three laws are given: (1) the Law of Segregation; (2) the Law of Independent Assortment; and (3) the Law of Dominance. They were embodied by Mendel in a paper which he read before his local scientific society in 1865 and which was published in 1866. But almost no attention was paid to it. The scientific world of the time was in a turmoil over Darwin's theory of evolution. The few who saw Abbot Mendel's paper ignored it. And so Mendel, bitterly disappointed that no one appreciated the scientific treasure he had unearthed, turned to other things and passed on in 1884, at the age of sixty-two.*

But recognition did come—sixteen years too late. In 1900, three biologists, almost at the same time (although they were working independently) chanced on Mendel's paper and quickly realized its importance.† Their reports set the world of biology feverishly experimenting to see whether the Mendelian findings applied to other living things—including man. Yes, in many cases Mendel's findings, or "laws" did seem to operate. But in other instances the results were either inconclusive or flatly contradictory. Much of the confusion started to clear up, fortunately, when one of the leading biologists, Thomas Hunt

* Even in Mendel's own community "not a soul believed that the experiments of the kindly cleric were anything more than a pastime, and his theories anything more than the maunderings of a charming putterer." This was reported some years ago by C. W. Eichling, Sr., of New Orleans, Louisiana, who as a youth, representing a French botanical firm, had visited Mendel in 1878. When asked about his work with peas, Mendel (Mr. Eichling said) deliberately changed the subject, as if it were a sore spot with him. Others have reported that biology was only one of Mendel's interests, which included playing chess, organizing fire brigades, running banks and fighting government taxes. During World War II, the Mendel "shrine" in the monastery at Brno was struck by a shell, and many of the priceless mementos kept there were destroyed. However, the little plot where he worked with his peas was not harmed.

† The three "discoverers" of Mendel were Hugo de Vries (Dutch), C. Correns (German), and Erich von Tschermak (Austrian). The claim has been made, however, that an American, Dr. L. H. Bailey, founder of the Cornell University Hortorium bearing his name, referred to Mendel's work in one of his papers in 1892, and that it was this reference which caused De Vries to look up Mendel's work. But while other scientists, too, may have taken some note of Mendel's publication well before 1900, the real implications were not seen or applied until that year by the three mentioned.

GREGOR MENDEL'S FINDINGS

As deduced from experiments such as this one with peas

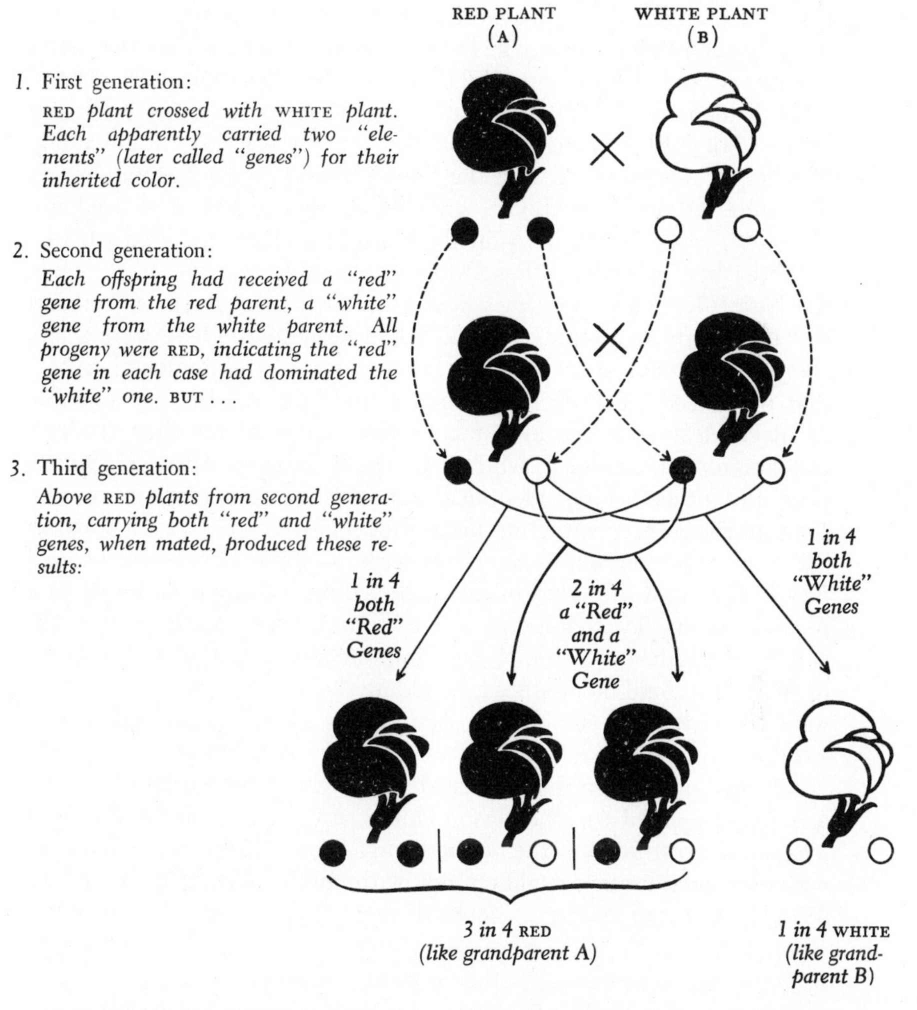

1. First generation:
   RED *plant crossed with* WHITE *plant. Each apparently carried two "elements" (later called "genes") for their inherited color.*

2. Second generation:
   Each offspring had received a "red" gene from the red parent, a "white" gene from the white parent. All progeny were RED, *indicating the "red" gene in each case had dominated the "white" one.* BUT . . .

3. Third generation:
   Above RED *plants from second generation, carrying both "red" and "white" genes, when mated, produced these results:*

. . . evidence that the genes had sorted out and recombined independently, and that the "white" genes had asserted themselves in the one-in-four cases where two came together with no "red" gene to dominate them.

Morgan, became intensely interested in the little fly named *Drosophila melanogaster*.

The drama which might be called "Man Meets Fly" began in 1907. Morgan, then at Columbia University, found the fruit fly an ideal subject for his experiments. For one thing, the Drosophila, *mère* and *père*, do not believe in birth control. At the age of twelve days they are ready to breed, and within another twelve days each female produces some three hundred offspring. Starting from scratch, within two years one can get sixty generations of flies, as many as there have been generations of mankind from the time of Christ. Moreover, the fly has many easily distinguished variations, and the cost of boarding it is trifling. The reward for all this is that the Drosophila has today become one of the most famous experimental animals in science and is assured immortality, even though individually it might prefer a speckled banana.

With the Drosophila, then, Morgan was able to prove that, while the basic Mendelian principles held firmly, the mechanism of heredity was not nearly so simple as Mendel had suggested. There were many complicated forms of gene operation and many environmental factors that influenced what the genes did or could do. All this he was able to make clear with the aid of a brilliant corps of his then student-collaborators: Hermann J. Muller, Calvin B. Bridges, Alfred H. Sturtevant and others. They identified hundreds of specific genes in the Drosophila, even constructing maps showing at exactly what points on the flies' chromosomes these genes were located. They actually bred flies with almost any kind of designated trait, just about as easily as a pharmacist would compound a prescription. And subsequently Dr. Muller showed how, with X rays, innumerable changes in the genes of these flies could be produced, heralding the dire radiation threats that were to come with the Atomic Age. (Both Drs. Morgan and Muller were in turn awarded Nobel prizes for their contributions.)

All this time, everything learned about the gene workings in flies was being applied to other living things, up the scale from the most elemental creatures to the highest organisms. There were Jennings with the paramecium, Goldschmidt with moths, Castle with rabbits, Wright with guinea pigs, Stockard with dogs, Crew with livestock, Cuénot and Little with mice, and Demerec, Beadle and Tatum with bacteria, bread molds and other microorganisms.

Nor was the study of human inheritance being neglected. True enough, it presented many special difficulties. Human beings could not be experimented with and cultivated like plants, nor bred in bottles like flies. They could not be interbred like mice, livestock and pedigreed dogs—parents with their offspring, brothers with sisters; and even if it were only a matter of observing human matings, one geneticist within his lifetime could follow through on only a few generations, compared with thousands of generations of fruit flies and countless generations of bacteria. Yet there were many compensating advantages in the study of people. They offered a far greater variety of traits for study than any lower animal species, and they lived in and interacted with far more varied environments; they could be much more intensively studied under different conditions; there were often detailed records of their family histories; and as individuals, they could reveal a great deal about themselves personally, by their speech and writing, as well as their behavior and performance. Not least, there was the special fascination of studying human inheritance, for in a sense, the geneticist was then also studying himself, his own family, and the other persons he knew.

Unfortunately, in the first decade or so of the century, many investigators were unaware of the complexities of human inheritance. Eagerly and rather naively, they hurried to apply Mendelian principles to every imaginable human trait, not merely of physical makeup, but of behavior. Even such traits as "decency" and "degeneracy," "nomadism" (the "wanderlust"), "temperament" ("excitability" or "placidity," "cheerfulness" or "depression"), and "alcoholism," were ascribed to simple dominant or recessive genes, with hardly a thought to possible environmental factors. Before long, though, more knowledgeable and perceptive researchers—Haldane, Hogben, Penrose, Snyder—began developing a true science of human genetics. Most important, it was increasingly recognized that human beings cannot be studied simply as biological or physical mechanisms. And more and more, as human geneticists went on with their researches, they began drawing in with them not merely medical men, physiologists, chemists and statisticians, but psychologists, anthropologists and sociologists.

Thus, it is through the continuing teamwork of thousands of brilliant scientists from many ranks that our present knowledge of human inheritance has been made possible. But as far as we may have come, and as far yet as we may go, it is always well to remember that it was by way of Mendel with his peas, and Morgan with his flies, that we have arrived at this understanding of heredity in people.

CHAPTER NINE

The Genes and Chromosomes

In this atomic and space age, much that would have seemed incredible a few decades before has become fact, and much that had been thought impossible has come to be regarded as probable.

Having grown quite accustomed to the idea of fantastically big effects emanating from fantastically small things, we may no longer think it so miraculous that a single gene, millions of times smaller than the smallest speck one can see with the naked eye, should be capable of causing the difference between a blond and a redhead, a dwarf and a six-footer, or, in some instances, a sane person and an insane one. Nor, as we penetrate ever farther into the infinite realms of outer space, is it surprising that science is also carrying us deeper and deeper into the realms of man's "inner space"—to the very core of our existence and into the minutest detail of the material of life itself.

Thus, with much new knowledge at hand, we can find ever clearer answers to these questions: What *is* a gene? And how, and by what processes, do genes do their work—not merely in the interval between conception and birth, but throughout one's life?

To begin, we must talk a bit about molecules and atoms, taking it for granted that you know that all substances in the universe—air, water, earth, vegetables, your cat, your hat, and you yourself—are composed of minute molecules, which in turn are made up of still tinier atoms. Genes, then, are also aggregates of atoms; but whether a particular gene is a complete molecule, or only a section of one of the molecules in the chain making up a chromosome, is not yet clear. We do know, though, that the molecule which houses or embodies a gene—any gene—is unique among all other types of molecules in that it is a *living thing*—

HUMAN CHROMOSOMES AT TWO STAGES

(Other than the stage shown in the Frontispiece.)

A. As they look in the stretched-out stage under microscope (after staining). Magnification about 1,500 times.

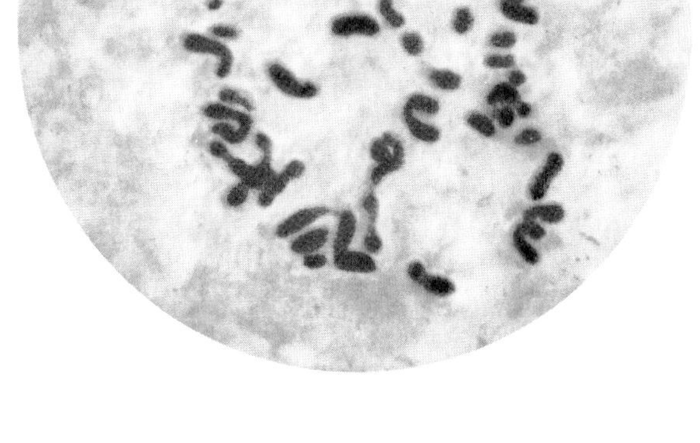

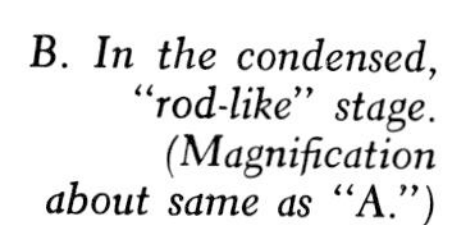

B. In the condensed, "rod-like" stage. (Magnification about same as "A.")

Above plates made from slides prepared by Dr. Jack Schultz and photographed under his direction for this book by Dr. George T. Rutkin and F. Carroll Beyer. (Colored prints processed by the author.)

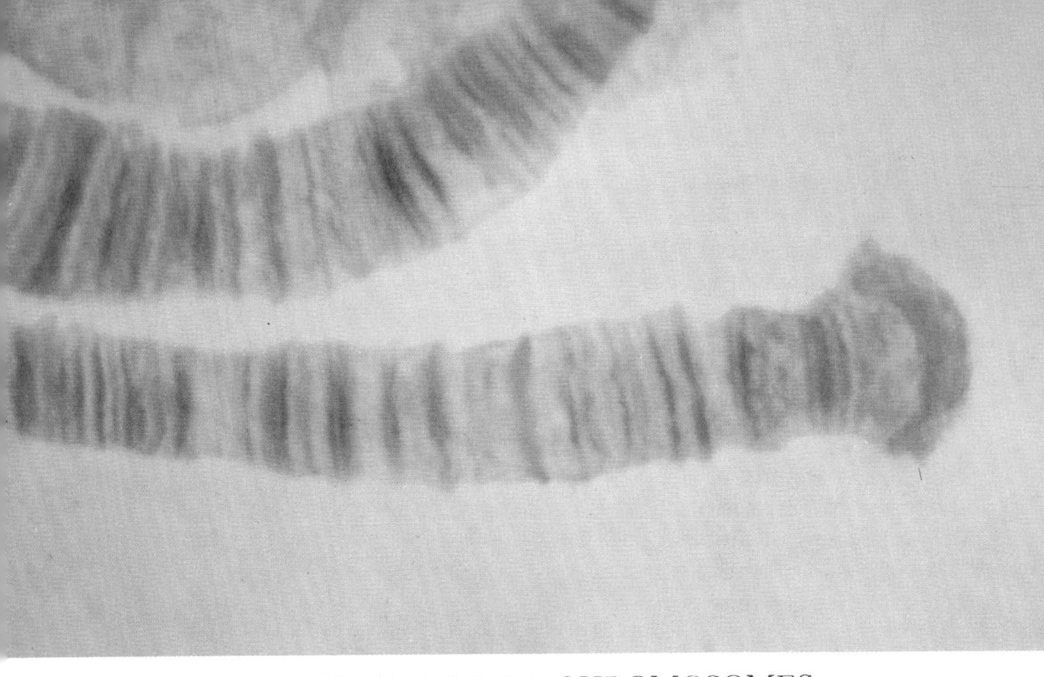

CLOSE-UPS OF CHROMOSOMES

Enormously magnified sections of giant chromosomes of the midge fly (Chironomus tentans). *Human chromosomes, while they cannot yet be so photographed, may look much like this.*

Above: *Sections of the midge-fly chromosomes showing bands which mark the locations of genes.*

Below: *Section of a midge-fly chromosome which has been chemically treated to produce puffs where genes are activated. (These puffs have been found to produce RNA, the carrier of the code messages of genes.)*

From slides prepared and photographs made by Drs. Wolfgang Beermann and Ulrich Clever, Max Planck Institute for Biology, Tübingen, West Germany.

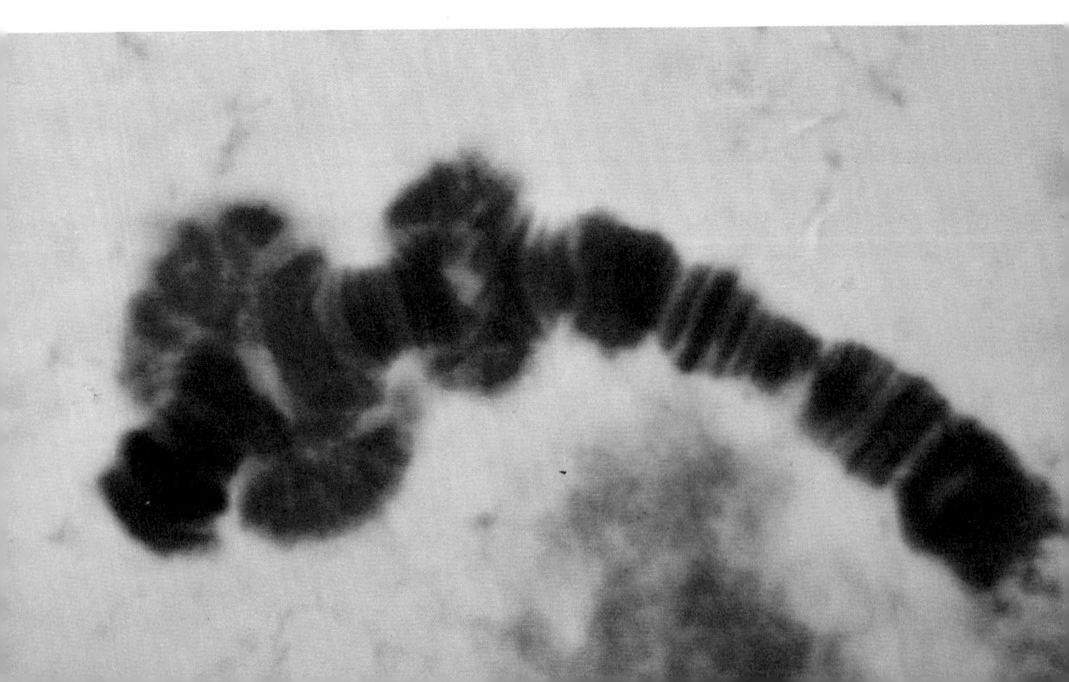

the ultimate unit of life, so far as has been established.

The chemical key to the life force in human beings and all other animals (and plants, too) is an amazing compound, *deoxyribonucleic acid* (DNA). Although known for some years previously to be an important constituent of genes, it was not until 1958 that three scientists, Drs. J. D. Watson of Harvard and F. H. C. Crick and Maurice Wilkins of Cambridge (all three were awarded Nobel prizes in 1962), were able to show, by brilliant X ray–diffraction detective work, just how the DNA molecule is constructed. In effect, it consists of two very long, coiled, mated strands of atoms which can be likened to a flexible, twisted rope ladder, with the short rungs in between made up of certain other groups of atoms. (For the technical-minded, details of the DNA "rope ladder" and the endless chemical variations which different arrangements of its rungs can make possible will be given in Chapter 19.)

Coming back, then, to the chromosomes and genes, we find that each chromosome is actually a string of DNA molecules; and each gene, as a section of a chromosome, is either a whole DNA molecule, or more likely, just a part of one. In any case, every gene has a dual role, first to perform specific duties in the cell housing it; and second, to join with other genes in reproducing itself and the entire chromosome whenever required. With regard to the first function, each gene acts according to a certain chemical code embodied in its DNA atomic arrangement. Somewhat as a telegraph operator taps out dot-and-dash messages according to the Morse code, a gene "taps out" its chemical messages into the cell housing it, and issues instructions for the formation of designated substances, the production of designated structures, and the carrying out of specific functions according to the most detailed genetic blueprints.

In some respects, what each gene does or causes to be done is accomplished through the workings of an *enzyme*—an organic catalyst, or substance which can produce a certain change in surrounding material without itself being affected. You yourself may often employ enzymes. You do so when you put yeast in dough to make it rise; or when, with a pellet of rennet, you turn milk into junket; or when you use as a meat tenderizer the enzyme *papain* derived from papaya juice. Home brewers in the old Prohibition days depended on enzyme action when they put a few raisins in their jug of mash. And readers engaged in various industries may be familiar with hundreds of catalytic substances (small pieces of platinum, for instance) used in given processes to bring about desired chemical changes. Thus, particular genes, too, produce their effects in the body cells by controlling the presence and activity of particular enzymes.

In creating an individual following conception, the genes work first upon the raw material in the egg, then upon the materials which are sent in by the mother, converting these into various products. The products, in turn, react again with the genes, leading to the formation of new products. So the process goes on, with specific materials being sorted out meanwhile to go into and to construct the millions of cells of the body at different locations.

But as the genes are *living* units, we cannot quite regard them as mere chemical substances. Considering the amazing things they do, we may well think of them as workers endowed with varied special capacities and personalities. No factory, no industrial organization, has so diversified an aggregation of workers and specialists as the genes in a single individual; and no army of workers can do more remarkable jobs. Architects, engineers, plumbers, decorators, chemists, artists, sculptors, doctors, dieticians, repairmen, masons, carpenters, common laborers—all these and many others will be found among the genes. In their linked-together forms (the chromosomes) we can think of them as "chain gangs"—twenty-three of these gangs of workers sent along by each parent to construct and operate a new human being—and each chain gang capable of reproducing itself endless times.

Turn back to the moment of your conception. The chain gangs contributed by your mother are packed together closely in a shell (the nucleus) suspended in the sea of nutrient material which constitutes the egg. Suddenly, into the sea, is plunged a similar shell (the sperm) filled with the chain gangs sent by your father. Almost immediately both shells open up and out come the thousands of gene workers, stirred to activity.

The first thing the tiny workers do is to fatten themselves on the sea of materials around them in the egg. They then reproduce, by a process made possible by another unique property of the DNA of which each gene is composed: the capacity to fashion an exact replica of every part of itself by drawing on the chemicals in the cell. In this process (referred to earlier in connection with chromosome-doubling) each gene is guided by its code to build an exact copy of itself. As the entire chromosome doubles in this way, the duplicates so formed pull apart. (See illustration.) All the other chromosomes having been similarly duplicated, the replicas of each chromosome move to opposite parts of the cell, and the cell itself divides. With continuing repetition of the process, the two cells become four, the four eight, the eight

MULTIPLICATION OF CHROMOSOMES AND CELLS

1. Egg after fertilization, with cell nucleus containing 46 chromosomes. (Only four shown for simplification.)

2. The individual chromosomes line up in center, and each doubles itself.

3. The duplicated chromosomes are pulled apart toward opposite sides of the nucleus.

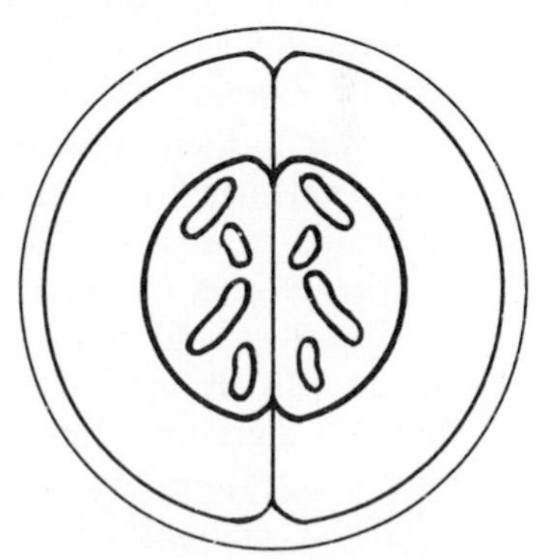

4. The nucleus divides to form two halves with matching full sets of the chromosomes.

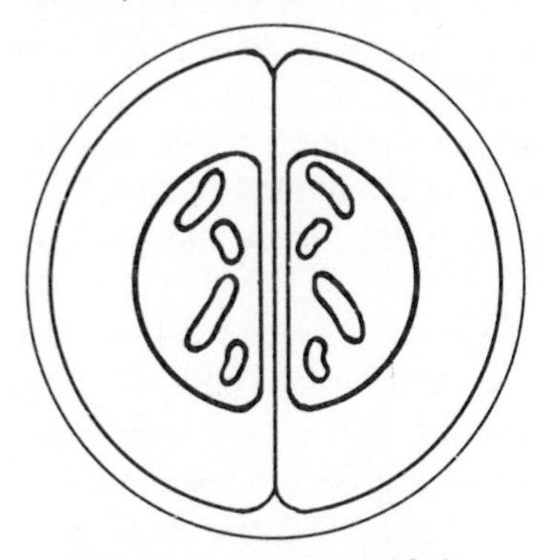

5. The whole inner egg with its contents also divides, to form two cells, each with duplicates of the original chromosomes in nuclei surrounded by foodstuff.

6. The process is repeated and the two cells become four, the four eight, the eight sixteen, etc., each with replicas of all the original chromosomes.

sixteen, and so on. But always, into every newly formed cell, goes *a replica of each original chain gang with each of its gene workers.*

However, the cells as they multiply to hundreds, thousands, then millions and billions, do not all remain the same. In the initial stages the genes have been doing mostly ordinary construction work or concerning themselves largely with producing basic body materials. But more and more, as the process of multiplying themselves and the body cells continues, the specialists get into action and begin constructing *different kinds of cells at different locations.*

The details of how this is done—such details as are known or surmised—fill thousands of pages in scientific treatises. To state this matter briefly, we can assume that on a set cue each gene steps out for its special task—fashioning this or that product, setting it in place, mixing and making a particular chemical compound—all the time working in cooperation with other genes.

Throughout one's lifetime the genes are in a constant ferment of activity, carrying on and directing one's life processes at every stage according to their prescribed genetic codes. The step-by-step process has been explained as a sequence of reactions, the gene workers being motivated to each step by the effects of the preceding one. By observing the process in lower experimental animals, we can see how first the broad general construction of the body is worked out; then how certain cells are marked off for the organs, certain ones for the respiratory and digestive systems, certain ones for the muscles, others for the skin and features.

The generalized cells now begin to develop into special ones. In those cells marked off for the circulatory system the rudiments of hearts, veins, and arteries begin to be formed (here is where the "plumber" genes step in to construct the great series of pumps and pipelines); from the generalized bone cells the skeleton begins to be shaped; from the skin cells, the rudiments of features, etc. With each stage the specialization is carried further along in the developing embryo. The amazing way in which the development of every human being closely parallels that of every other proves how infinitely exact and predetermined the genes can be in their workings.

Another remarkable aspect of the developmental process is this: Despite the growing differences in the various specialized cells, *every cell still receives and carries exact replicas of all the chromosomes with their genes.* Thus, the same genes which produced eye color in your eye cells will also be found in your big toe cells, and the same genes which directed the fashioning of your big toe will also be found in your eye cells—or in your ear and liver cells, for that matter. Probably, then, in addition to producing a special effect in certain cells, a given gene may or may not take part in other functions elsewhere.

("Some genes," as geneticist J. B. S. Haldane has put it, "are pretty lazy on the average. They only get an assignment of work if they happen to find themselves in a cell concerned with making pigments, forming bone or transmitting messages. But the majority seem to have some work to do in almost every kind of cell, even though it may be a more vital job in a gland cell than in a skin cell.")

To go on, we may recall that a person starts life with two chromosomes of every kind, which means also two genes of every kind (except with respect to the male's sex-chromosome genes). If, in terms of chain gangs, we designate the chromosomes by numbers, there would be two No. 1 chains, two No. 2 chains, two No. 3 chains, and so on through No. 22 up to the last, or twenty-third pair—where in the case of a girl, as noted in a previous chapter, there would be two X chains, but in the case of a boy, only one X chain, the other being a Y chain. With this latter exception, the corresponding chain gangs (1–1, 2–2, 3–3, etc.) would be exactly alike in the number of gene workers each contained and in the *type* of worker at each point in the chain.

If the third gene in the No. 1 chain contributed to you by your father was an "architect" gene, so would be the third gene in the No. 1 chain from your mother. The sixth gene in each No. 1 chain might be a "carpenter," the fifteenth a "decorator." All the way from the No. 1 chain to the X chain, the genes at each point on matching chromosomes in all human beings are the same in the *type* of work to which they are assigned. In other words, every individual starts life with one gene worker sent along for the job by the mother, and one by the father (save only, again, in the case of a son's Y chromosome).

But here is a vitally important point: The two paired gene workers delegated to the same job very often do not work at all in the same way. The two No. 1-chain "carpenter genes," one coming from your father, the other from your mother, might be as different in their characters and abilities as any two human carpenters. So, too, just as any two plumbers might differ (even though they belonged to the same union), or any two architects, artists or chemists might differ, any two paired genes might be radically different in what they do and how they do it. One gene can be highly efficient, the other can make a botch of the same job; one gene can do things the expected way, another may go off on the most peculiar tangents.

There are strong (dominant) genes and weak (recessive) genes; highly active genes and sluggish genes; constructive genes and destruc-

OUR GENE "WORKERS"

The CHROMOSOMES *may be thought of as chains of gene workers sent along by the parents to create the child. Each parent contributes 23 of these chains, consisting of many linked-together workers assigned to given tasks. The drawings represent short sections of two matching chromosomes.*

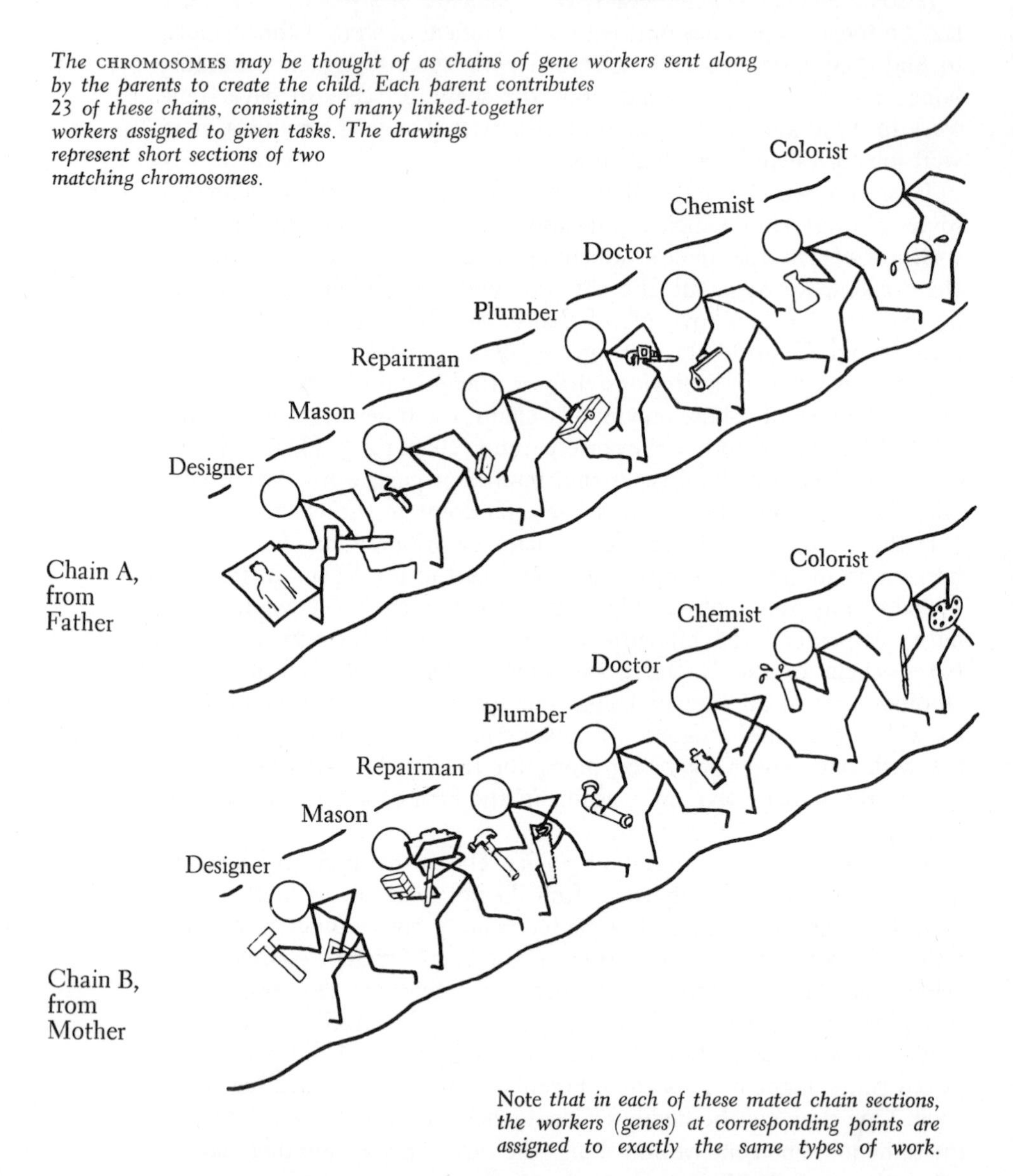

Note *that in each of these mated chain sections, the workers (genes) at corresponding points are assigned to exactly the same types of work.*

tive genes; superior genes and inferior genes; steady, dependable genes and temperamental genes; genes which will work one way with some companions, and in an entirely different way with other gene company. Indeed, if we endow them with personalities, genes individually have as many different characteristics as have the people they create. Ever present, moreover, are the many factors of the environment within the cell—its chemical composition, its nutrition and its state of health—as well as its location with respect to other cells, all of which may greatly affect the work of any given gene, just as a human artisan is affected by diet, weather, hygiene and working conditions.

A pertinent question here is: How did all the many differences in genes assigned to the same job come about? (Or, in fact, how were all the many genes for many different jobs themselves produced?) The answers were touched on briefly at the end of Chapter 2, and there will be more details on this involved subject later in our book. For the moment, we'll say merely that a change in a gene's workings (a mutation) can take place if anything happens to shake up the arrangements of atoms inside it, or to alter the nature or number of its atoms, as may result through some potent chemical or physical influences. Such mutations, at extremely rare intervals for any individual genes—but often enough among all the genes, collectively, over the long history of the human species—could easily account for many varieties of every gene.

Another possibility is that of changes occurring in the structures or numbers of chromosomes. For instance, during the process of egg or sperm formation, when two matching chromosomes twist about one another and then break away, the two may interchange sections (an effect called "crossing over"—see illustration).* Or one chromosome may lose a section carrying some of its genes (like a lopped-off chain),

* Facts about chromosome crossovers have been highly important in genetic research. First, as may be seen from the illustration, entwining and crossing over would tend to be most frequent near the centers of chromosomes, and decreasingly frequent toward the ends. Second, the further apart two given genes are on a chromosome, the greater the chance of their being separated by a crossover; and the closer together two given genes, the less often separation and relocation through crossovers would occur. Thus, by estimating the frequency of crossovers of particular genes, geneticists—experimenting first with drosophila—were able to construct "maps" showing precisely where on given chromosomes many genes were located. Subsequently, chromosome maps were made of other experimental creatures; and eventually, it is hoped, maps will also be made of human chromosomes. (So far this has been done in only a limited way for the human "X" chromosome.)

THE CHROMOSOME CROSSOVER PROCESS*

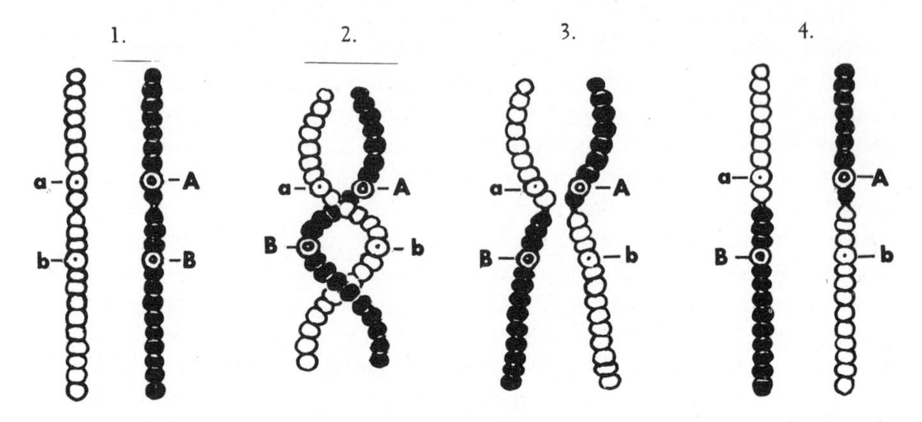

1. Paired chromosomes, received by individual from mother (white) and from father (black), showing locations of certain corresponding genes: recessives ("a" and "b") on maternal chromosome, and dominants ("A" and "B") on paternal chromosome.

2. During the process of sperm or egg formation, the paired chromosomes twine about each other.

3. In pulling apart and separating again, the chromosomes may exchange parts which have "crossed over" from one to the other. (Crossing over may occur between any sections of mated chromosomes, and sometimes there may be double crossovers. Shown here is only one crossover.)

4. The chromosomes as now formed and passed on to offspring consist in part of the original maternal chromosome, in part of the original paternal chromosome. Note that where, before, genes "a" and "b" were linked on one chromosome, "A" and "B" on the other, they have changed company, the "a" now being linked with "B," the "A" with "b."

* To be technically correct, chromosomes when they engage in crossing over are in the *doubled stage*. However, for simplicity and greater clarity, the process is illustrated here only with single chromosomes.

and these may hook onto another chromosome. Or in a given fertilized egg, one of the chromosomes may get left out entirely, while in another fertilized egg an extra chromosome may crowd in with the others.* If we think again of chromosomes as chains of workers, any radical changes in the arrangements or numbers of gene workers (as with workers in an industrial plant or large business organization) may

* A more detailed discussion of chromosomal accidents and abnormalities, with illustrations, appears in Chapter 20.

cause great changes in the behavior of the genes affected, and in what they produce.

There are many other complexities in the makeup and functioning of our genes and chromosomes which are highly meaningful for the specialist but which the general reader shouldn't have to worry about—at least, not at this stage in our book.

What is most important to know for the present is that with the countless variations and combinations among human genes and with the manifold possibilities offered by their interplay with different environments, a limitless variety of human beings can be produced. The best way of illustrating this is to tell what the genes do in specific instances with regard to our features, organs, minds, and all other individual characteristics.

CHAPTER TEN

Coloring: Eyes

ALTHOUGH THERE ARE NO STATISTICS on the subject, we dare say that millions of fathers since the world began, and a not inconsiderable number of mothers, noting that one of their children looked very different from themselves, have had the cold suspicion creep up that it wasn't actually their own. And a great many children, in turn, having noted how little they resembled others in their family, may have wondered if they had really had the same parents.

Sometimes the suspicions may have been justified. But in the great majority of such cases the doubts have been, or are, groundless, for today we know that the process of heredity can act in many peculiar ways, to produce not only resemblances, but marked *differences*, between parents and their offspring.

Among the traits which often are most puzzling and unexpected are the coloring of eyes, hair and skin. These warrant special attention because of the great *social* significance that can be attached to color differences, first with regard to the attitudes of whole peoples, and second with regard to individual beliefs and feelings.

Where a person's coloring is identified with a particular race, and there is hostility or conflict for whatever reason between this race and another, we can easily see why inherited color differences may have much meaning. But where race is not at issue, why are individuals still so much swayed by the tints of each other's faces? Why do you, personally, "go" more strongly for a member of the opposite sex with one kind of eye or hair color rather than another? Or why, if you are a prospective parent, are you so likely to be concerned about what the coloring of your baby will be? Some of the reasons may be sociological—bound up with general preferences or prejudices in certain groups or countries. But for many of the individual attitudes toward color, only a psychoanalyst could dig up the reasons.

What we can explain more easily is how human color differences are produced, and even more, what kind of coloring you usually can or cannot expect in a baby, if you happen to be awaiting one.

Being on the surface, color differences can quite easily be perceived and graded, and therefore offer us some of the simplest means of studying and analyzing the action of genes. In fact, eye color was the first normal trait in human beings for which genetic mechanisms were worked out.

The first point to clarify is that any color, as we know it, is not a substance, but an *effect* produced by the reflection of light on different materials. "Blue" eyes, for example, have no blue substance in them, but merely look blue to us (as we'll explain later). When we speak, then, of different genes "producing" different colors, what is actually meant is that they take part in producing different amounts or kinds of substances which give the various color effects.

Have the pigment substances any practical value? Yes. Among many lower animals coloration may make possible concealment from enemies or victims and may enhance natural attractiveness to mates. Among human beings coloring may also have mating importance, but only in a social way. *There is no natural attraction or repulsion between people because of their coloring,* so far as science can establish. The practical value of pigmentation in human beings, however, is mainly in ensuring protection from the sun. In the eyes, the pigment deposited in the otherwise translucent iris shades the retina within. The pigment in the skin protects the raw flesh underneath. Even the pigment in the hair affords some protection to the living cells in the hair roots and in the scalp.

Although we recognize many different kinds of coloring in human beings, all of the color effects, in the eyes, skin and hair, are produced by virtually the same pigments. In fact, one basic pigment, *melanin,* varying from black and dark brown to light yellow, accounts for most of our color differences, as determined by its particle size and shape, degree of concentration and opacity, and the way in which it is distributed. Supplementing melanin are a few other pigments which lead to special color effects.

In human eyes, specifically, the wide range of colors, from black to light blue, are all results of the way different "eye-color" genes produce and distribute melanin, plus one or two other pigments. It

THE HUMAN EYE

(*Key to Color Plate*)

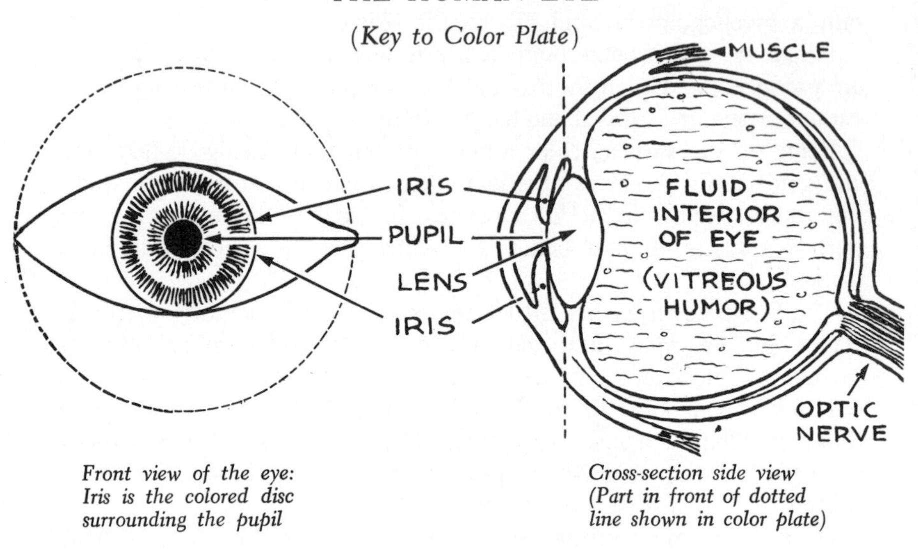

Front view of the eye: Iris is the colored disc surrounding the pupil

Cross-section side view (Part in front of dotted line shown in color plate)

is believed that the original "eye-color" genes in the human species may have been the highly active ones producing in all of the first human beings heavily pigmented eyes, dark brown or black; and that by subsequent mutations through the ages, weaker variations of these genes developed which now provide us with all the many lighter shades.

Eye color is determined in the cells of the *iris*, the small ring (annulus) around the pupil (or rather, one should say, the pupil is the hole in the iris). Without any pigment the iris would look something like a tiny, transparent doughnut. It has, however, two clearly defined parts, as if it had been slit in half and pasted together again. Thus we speak of the part facing out as the "front" of the iris and the other half as the "rear." And it is in the way that pigment is produced both in the front and in the rear of the iris—in some cases separately—that eye-color differences result.

While a number of genes participate in this eye-pigmenting process, it is a single key gene which usually determines the main outcome.

In *blue eyes* (of various types) the genes are weaklings which produce little or no pigment in the front of the iris (as you can see if you look at a light-blue-eyed person sideways), but manage to produce a certain amount in the rear. But this pigment itself, remember, is *not*

WHAT MAKES YOUR EYE COLOR

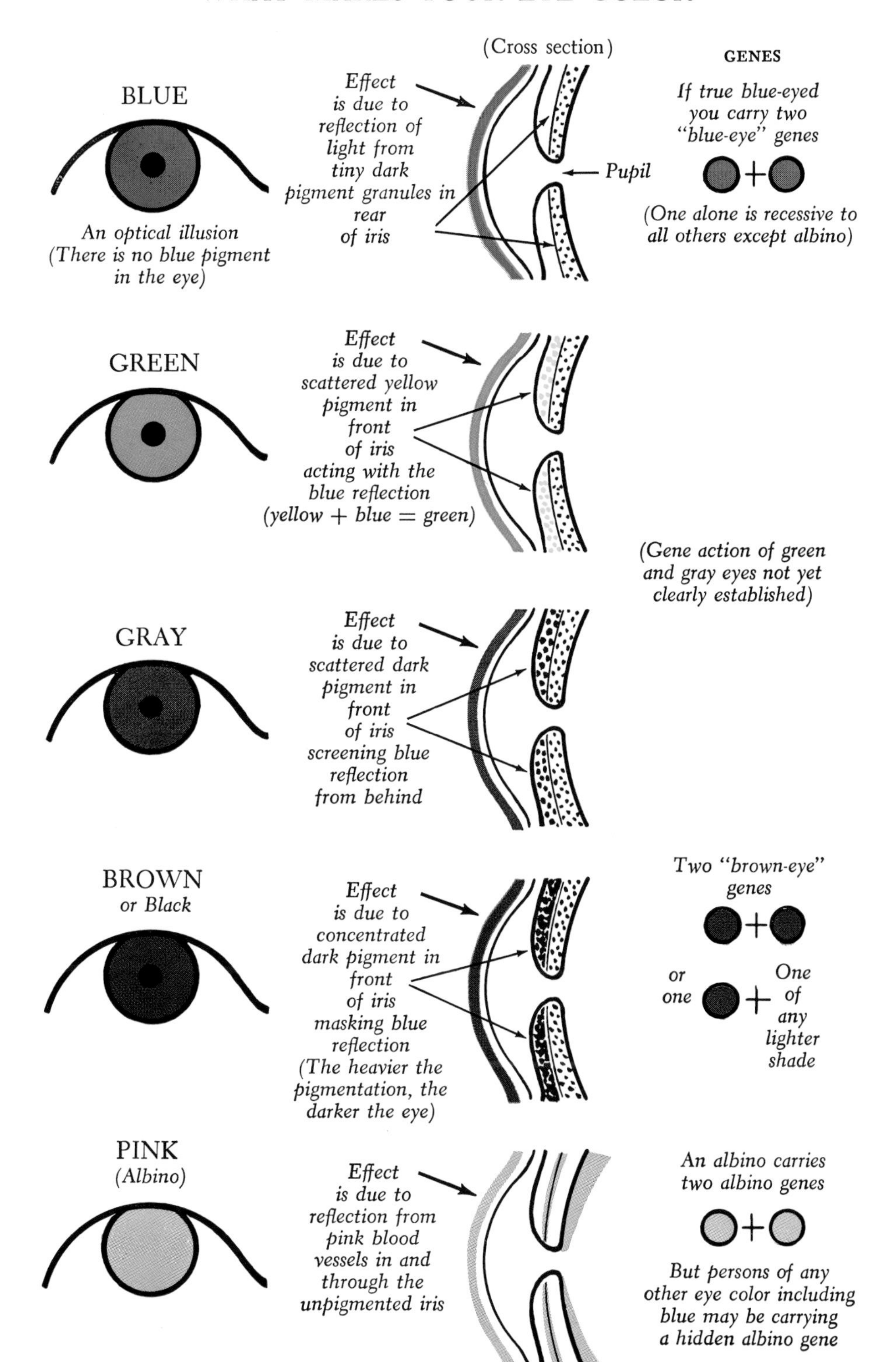

blue. What is present in the "blue" eyes is merely a scattering of the brownish melanin particles, which produce the *optical effect* of blue, through the refraction and dispersion of light rays. It is precisely in this way that dust particles make the sky look blue. The other eye colors are due mostly to the addition of pigment in the front of the iris.

In *gray eyes* the gene action produces a somewhat heavier concentration of the pigment in the rear of the iris, plus a scattering of dark pigment in the front, enough to give a gray or grayish-blue effect, depending on the depth of the pigmentation. (Color from muscle fibers behind may contribute to, or modify, the gray effects, as also the blue.)

In *green eyes* the rear of the iris has the same kind of pigmentation as in blue or gray eyes, but in addition there is a special gene which acts to produce a certain amount of dilute brown or yellow pigment in the front of the iris. Superimposed on the "blue" background, this produces the effect of green (just as you might make any blue surface look green by stippling it with yellow dots). With a little more pigment in the front and rear of the iris one may have grayish-yellow or grayish-brown eyes.

In *brown eyes* the key gene is active enough to fill up the front of the iris with pigment so that no reflection can be seen.

And, finally, in *black eyes* (or very dark brown) the gene is of the most vigorous type, which dictates the laying down of an intense deposit of pigment in the front of the iris, and perhaps also in the rear, making it completely opaque.

In addition to these main eye-color types, various other color effects can be produced by the patterns in which the pigment in the iris is distributed. (Altogether, if we tried to get down to the finer shadings, the problems of tracing eye-color inheritance would be very complex. In blue eyes alone it has been estimated there are nine color classes which can be inherited in forty-five possible combinations!)

Albino eyes should be dealt with separately because they are not due to any "eye-color" gene but to a defective "general" gene which interferes with all pigmentation processes. Thus, the true albino eyes have *no pigment whatsoever*, in either the front or rear of the iris. The *pink* effect is a result of the tiny blood vessels in the otherwise colorless iris and the reflection from other blood vessels behind.

Now let us see what happens when an individual receives one kind of "eye-color" gene from one parent, a different gene from the other parent.

In eye color, as in most other processes, some genes can do the same work singly quite as well as if there were two. The blackest, or darkest "eye-color" gene, generally works this way. That is to say, if a child should receive just one dark "eye-color" gene from one parent, no matter what other gene it received from the other parents—green, gray, blue or even albino—that child would have dark brown eyes.

This follows the principle of *dominance* and *recessiveness* which Mendel discovered. Just as the gene for red flower dominated that for white flower in his garden peas, in human beings the gene for black (or brown) eyes dominates that for blue eyes (or any other of the lighter shades).* But, you might ask, doesn't the "blue-eye" or other recessive gene do *anything* when coupled with the dominant "brown-eye" gene?

Possibly you may recall having gone to a party planning to sing, or to play the piano or exhibit some other accomplishment, and just as you were preparing to perform, someone else got up and did the same sort of thing, much more impressively than you could have done. If you were an ordinarily shy person, the chances are that you kept your performance to yourself for the rest of the evening.

That is about what happens when a little Blue-Eye gene arrives and finds a big domineering Brown-Eye gene on the scene. Little Blue-Eye may sit back with never a move out of it through all the long lifetime of the individual in which it finds itself. (Perhaps it does do some work, though if so, the effects are obscure.) But there is always "another time." Just as you might go to the next party and, in the absence of the menacing competitor, perform handsomely, so the Blue-Eye gene need not be permanently squelched. To a gene, the "next time" means the next individual to which it is sent—that is, some future child. Again little Blue-Eye gene goes forth hopefully (and if necessary again and again, generation after generation) until in some child it finds itself coupled, not with a bully Brown-Eye gene, but with a kindred Blue-Eye gene. And this time, glory be, the two Blue-Eye genes happily go to work, and the result is a blue-eyed baby!

In all mixed matings the "blue-eye" genes have a hard time of it, for they are also dominated in whole or part by the "light-brown-," "green-" and "gray-eye" genes. As for the other eye-color "contests," the general rule is that the genes for darker colors dominate those

* Some authorities believe that the "black-eye" or "dark-brown-eye" genes of Negroes and Mongolians may be of different types, or at least more intense in their workings, than the "black-eye" gene found among Whites. It is possible, then, that the former types of "dark-eye" genes may be more completely dominant over lighter "eye-color" genes in some cases than is the "dark-eye" gene of Whites.

for the lighter shades. When a "green-eye" gene and a "gray-eye" gene, both of apparently equal potency, get together, the result may be grayish-green.

Always to be kept in mind is the fact that eye colors are not simple effects, such as might be produced by painting little discs with flat colors from a palette or paintbox. In addition to the actual pigment, there are many structural details of the eye which contribute to the impression of color it gives, leading to many varieties of each basic eye color. Thus, the specialist may distinguish among numerous shades and types of blue eyes genetically determined through many combinations. Nevertheless, while the finer details of eye-color inheritance are very complex, and there still are many points to be worked out, the general facts are well established.

In short, you may make these guesses about the "eye-color" genes you are carrying:

If you have black or brown eyes:

1. Where both your parents, all your brothers and sisters and all your near relatives also have dark eyes, in all probability you carry *two* "black" (or "brown-eye") genes.

2. Where both your parents have dark eyes but one or more of your brothers and sisters or other near relatives have eyes of a lighter shade (gray, green, blue), you may be carrying, in addition to the "dark-eye" gene, a "hidden" gene for the lighter shade.

3. Where one of your parents has black or brown eyes and the other light-colored eyes, you definitely carry one "dark-eye" gene and one for a lighter shade. If the light-eyed parent has gray or green eyes, your "hidden" gene may be either a gray, green, or blue one. If the parent has blue eyes, then you almost certainly carry a hidden "blue-eye" gene.

If you have gray or green eyes:

Regardless of what eye colors your parents have, in all probability you carry no "dark," but only "light-color" genes, which may be gray, green or blue. If *one* parent has blue eyes, however, you may be quite sure you carry at least one hidden "blue-eye" gene.

If you have blue eyes:

Regardless of the eye colors of your parents, you are almost certain to be carrying *two* "blue-eye" genes of one type or another.

If you have albino eyes:

In all probability you are carrying two "albino" genes. (In rare instances in men there is a special sex-linked single gene which may prevent color production.)

To all the foregoing deductions there may be occasional exceptions. Once in a blue moon a freak gene in a person with blue eyes may rear up and dominate a freak "brown-eye" gene, in which case the usual eye-color forecasts would be upset. More often, though still with great infrequency, environmental factors may swerve a gene from its normal course. For example, one cannot always be positive that a person whose eyes are, or appear to be blue, really is carrying two "blue-eye" genes, there being a 1 to 2 per cent chance of error. (One possibility of this happening will be discussed presently under "iris-pattern differences.") Cases also occur from time to time where an individual receives at conception one, or even both genes for some darker eye color, and at some stage thereafter something happens to inhibit or modify the usual "eye-color" gene workings, so that blue or blue-appearing eyes result. Diseases (or dietary deficiencies) which upset the body chemistry and affect pigment production may be among these modifying influences. In another way, cataracts, or some other eye defects, may rob dark eyes of their color and make them look watery blue. In still other instances eye-pigment production may be speeded up abnormally, so that, despite there being genes for blue eyes, dark eyes may result.

Age is the most important modifier of eye color throughout life. Generally, eyes darken from birth to maturity and thereafter begin to lighten up a bit with aging. Among Whites (and in some cases among Negroes), most babies at birth have slate-blue eyes, which may in time either become more heavily pigmented, turning to brown, or in other cases may become clear blue, depending on the genes carried. As any mother knows, it may take a number of years before the true eye color of her child is revealed. Borderline blue and gray eyes may turn to green or brown, or brownish eyes may become green or gray, but "pure" blue or gray eyes usually remain fairly constant in color. In old age, brown eyes may again become slate blue.

There is an interesting sex difference in that the time and manner in which age affects eye color is different in females and males. The development toward physical maturity of girls is more rapid than that of boys (girls' full maturity being reached about two or three years earlier). Hence, as eyes tend to darken at puberty, girls and young women on the average may have somewhat darker eyes than males of the same ages. In addition, throughout life sex differences in body chemistry may affect eye-pigment production in varying ways.

Iris-pattern differences among individuals may also modify the coloring or appearance of eyes. Such patterns of pigment distribution, with the pigment in rings, in "clouds," in flecks or patches, in radial stripes, or even spread evenly over the whole iris, are inherited in various ways. So, too, may be differences in the thickness and surface of the iris (an iris that is full and smooth usually going with darker eyes, while the opposite type is found as a rule in lighter eyes).

Predictions can be given for some of the iris-pattern or iris-thickness genes, but it is doubtful whether the average reader would be much concerned with them. (Do you know what your own iris pattern is?) However, these patterns may be important in that if a child has a different inherited iris pattern from that of a given parent, even though the color genes are the same, their eyes may look somewhat different. Another unexpected situation may arise with some presumably blue-eyed persons whose irises on close inspection are found to be flecked with brown. Persons of this type may be carrying a modified "brown-eye" gene, and two such parents might have a brown-eyed child. There may be other rare genetic situations where apparently blue-eyed persons carry a brown-eye gene: allowances must always be made for such exceptions.

One peculiar phenomenon, which occurs about once in every six hundred persons, is that of unmatched eyes (*heterochromia iridis*) where the eyes of an individual are of different colors. Most commonly one eye is brown, the other blue. This may occur in several ways: A person may inherit one "brown-eye" gene and one "blue-eye" gene, which normally would make both eyes brown. But in the very earliest stage something may happen to the "brown-eye" gene in one of the rudimentary cells, leaving the field clear for a blue eye to develop on that side, whereas on the other side the "brown-eye" gene may be doing its work normally. Or, starting with two "blue-eye" genes, some disease may increase pigment production in one of the eyes, making it brown. Conversely, a person starting with two brown eyes may develop glaucoma, or a cataract in one of the eyes, which will make that eye bluish.

While most unmatched eyes are believed due to environmental factors, the condition may also be *inherited* as the result of a one-sided nerve defect which can be transmitted through a certain gene (dominant or recessive). On record are a number of families where unmatched eyes of the same type have appeared for several generations. The condition, by the way, is quite prevalent among many domestic animals, including, dogs, cats, rabbits and horses.

Although the range of human eye color is ordinarily confined to variations of brown, blue, green, and gray eyes, rare instances have been reported of persons with tortoise-shell eyes (mottled yellow and black) and also, of persons with ruby-colored eyes. We may eventually see eyes of other colors. One cause may be through spontaneous gene changes such as have occurred in our past and as geneticists see popping out from time to time in flies, mice and other laboratory creatures.

Another remote possibility is that humans who want new eye colors may not have to wait for their genes to oblige them. One noted geneticist, the late Professor H. S. Jennings, seriously suggested that new eye colors in people might conceivably be produced by means of chemicals. So the time may come when women may be able to change the color of their eyes just as they now change their hair color. When that day arrives and a man says to a girl, "Where did you get those big blue eyes?" she may reply, "At Antoine's, corner of De Peyster Avenue and Thirty-second Street!"

CHAPTER ELEVEN

Hair Color

GENTLEMEN, we used to be told, prefer blondes.

There must still be some truth to this—or at least, many women must think so—for how otherwise account for the large numbers of dark-haired damsels who go to beauty parlors for a bleaching?

Red hair, too, has a certain social significance, as the brisk demand for henna rinses and other hair-reddening alchemies bears out.

Which might lead to this question, assuming that you are a prospective parent:

"What are the chances of a child of yours being a natural blond or a redhead?"

As you already have gathered, looking at your own hair and that of your mate may not in itself provide the answers. You must try to ascertain what *genes* for hair color you both carry.

Pigmentation of the hair follows the same general principle as that of the eyes. Often, in fact, but not always, the two are related. Where the general pigmenting process of the body is very intense, as in "pure" Negroes, black hair and black eyes (and also dark skin) almost invariably go together; while among Mongolians the black hair is almost invariably accompanied by dark brown eyes. In most White stocks, too, one finds darker hair going as a rule with darker eyes, and lighter hair going with lighter eyes. But as evidence that the "hair-color" genes and "eye-color" genes may work independently, one frequently finds black hair with blue eyes (most often among the Irish), or *natural* blond hair with dark brown eyes.

One's hair color is determined chiefly by the way genes work to produce pigment in the hair cells, although, as in eyes, the basic actions of the "hair-color" genes may be modified or changed by a variety of factors, some inherent and some environmental. Also, as with eyes, the distinction between one hair color and another is not always clearly marked. You know what difficulty there often is in deciding whether a person's hair is blond or brown, or brown or red. (Among women, this hasn't been made any easier by the use of bleaches, dyes, and rinses.) So we are dealing here, mostly, with the quite definitely distinguishable shades of hair color.

In structure your hair is a tube (somewhat like the twig of a tree), with an inner section through which pigment granules and tiny air bubbles are distributed, and an outer cover which may also contain pigment granules. The same pigment found in eyes, melanin, is also the principal element in our hair coloring. And this is the way the "hair-color" genes work:

If the key "hair-color" gene acts to produce a heavy deposit of melanin in and among the hair cells, the result is black hair; a little less melanin, dark brown hair; still less, light brown; very dilute, blond hair. The shade of hair color is further influenced by the gradations and types of melanin produced, by the way the melanin granules are distributed in the hair cells and how the latter are constructed, and by the air content of these cells and their amount of natural oil or greasiness. But keep in mind that hair color, as you see it, is an *effect*, governed by the way the pigment reflects light. You need hardly be told that your hair looks different under different lights.

Red hair is due to a supplementary gene which produces a diffuse red pigment. The "red-hair" gene is often present with the key "melanin" gene. If this key gene is very active, making the hair black or dark brown, the effect of the "red" gene will be completely obscured. (It is claimed, however, that a hidden "red" gene may betray its presence in the black-haired persons by a special *glossiness* of the hair.) Where the "melanin" gene is weaker, the "red-hair" gene can manifest itself, and the result will be a reddish brown, or chestnut shade. If the "brown-hair" gene is an utter weakling, or if it is absent, distinctive red hair will be produced.

That the "red-hair" gene is an independent little cuss is shown by the fact that red hair may be present with almost any eye color, whereas brown or black hair usually goes with brown or black eyes, and blond hair usually goes with blue or gray eyes. However, red eyelashes seldom go with anything but very red head hair.

How the "red-hair" gene works in relation to a "blond" gene is not yet completely clear. Theoretically, it should dominate the blond, but

WHAT MAKES YOUR HAIR COLOR

DUE TO — **GENES YOU CARRY**

WHITE
(*Natural*)

No pigment granules among hair cells

(If white hair is not due to age or disease)

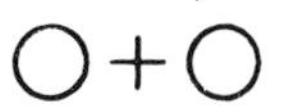

(But all hair color types may be carrying one hidden "White-Hair" Gene)

BLOND

Yellow effect produced by dilute pigment

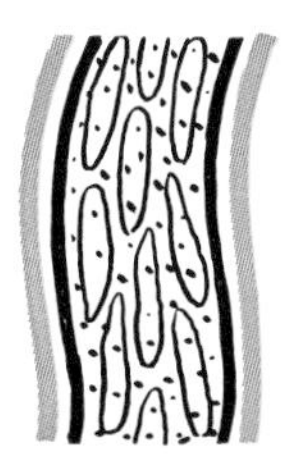

● + ●

RED

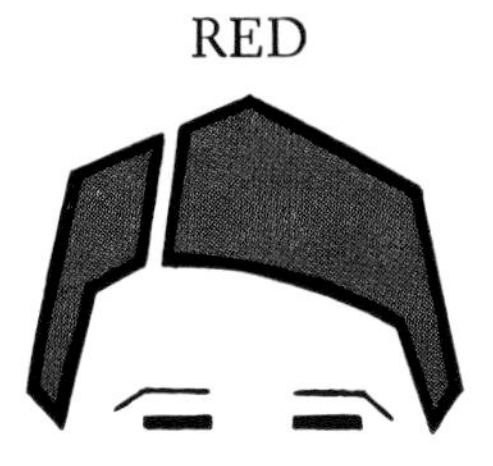

Effect produced by dissolved red pigment diffused with the scattered pigment granules

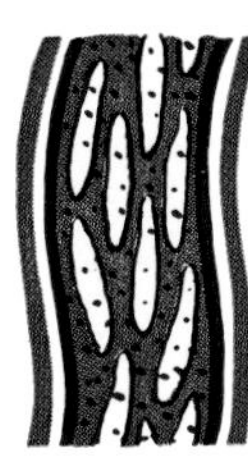

● + ?

(The "Red-Hair" Gene is a special one which shows its effect if not masked by very "Dark-Hair" Genes)

BROWN

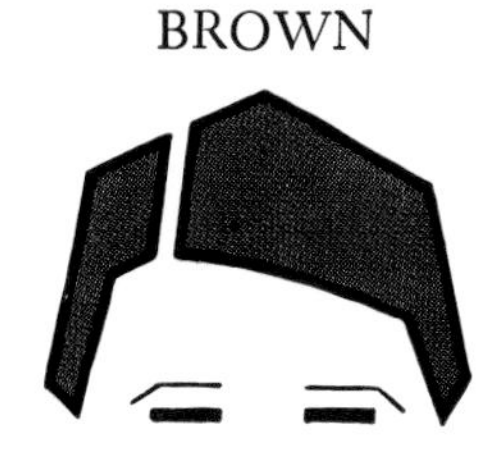

Effect produced by heavy deposit of pigment granules (The heavier the deposit, the darker the brown)

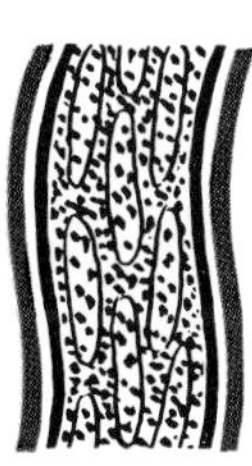

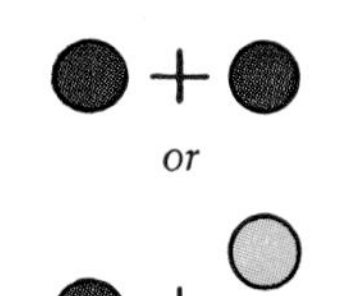

● + ● or ●

(Red gene may make hair reddish-brown)

BLACK

Intense deposit of pigment granules *

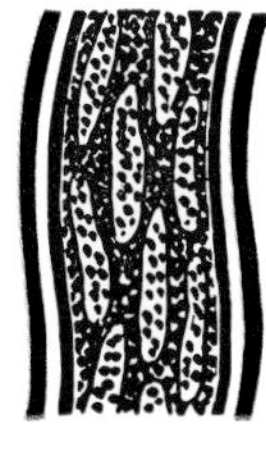

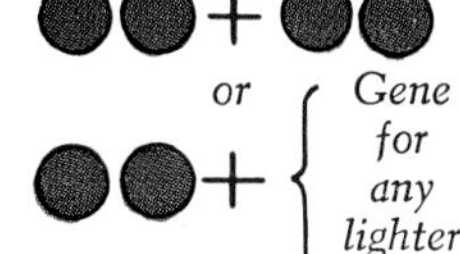

* The double-brown gene mechanism suggested here refers to the black hair among Whites. In Negroes and Mongolians the "black-hair" genes may be of a different, more intensified kind.

we have cases, nevertheless, where blond parents have a red-haired child. With rare exceptions, however, the "blond" gene is definitely recessive to those for all darker hair shades.

This leads to these general conclusions:

If you have dark hair, you are carrying either two "dark-hair" genes, or one dark and one for any other shade.*

If you have blond hair, you carry two "blond" genes.

If you have red hair, you are carrying either one or two "red" genes, supplementing "blond" or "brown" genes.

The basic "hair-color" genes are found among all peoples, although not by any means in the same proportions. Redheads are found even among Negroes and are quite frequent among the usually black-haired Latins. (The highest proportion of redheads—11 per cent—is among the Scottish Highlanders.) While blonds are also not uncommon among Latins and other black-haired peoples, we have no way of knowing to what extent the "blond" gene may have arisen among them by mutation, or to what extent it was introduced through interbreeding. The mutation theory seems to be the most plausible one in the case of blond Indians found among certain black-haired tribes, notably in Panama.

White hair can be due to various factors, genetic and otherwise. In its most striking form it is caused by the "albino" genes, which, as we have already seen, also rob the eye of color. White hair may also be due to extreme "blond" genes, or to some "inhibiting" genes or conditions which would interfere only with the hair-pigmenting process. White-haired persons of this type, quite common among Norwegians and Swedes, differ from albinos in that they are normal in eye and skin pigmentation. And finally, there is the white hair due to age and disease. In fact, in all hair colors there is the constant possibility that other factors may alter the effects of the key genes.

Age plays a much more important part in hair color than it does in eye color. The "hair-color" genes may be slow in expressing themselves, and also, the structure of the hair changes with age, which may affect its coloring. Light hair as a rule has a tendency to turn darker from

* The black hair of Negroes and Mongolians may be due to stronger genes, producing a more intense dark pigmentation than in the black hair of Whites. Thus, while a Negro or Mongolian "black-hair" gene completely dominates any "light-hair" gene, there is some indication that the "black-hair" gene of Whites is not always completely dominant over the "blond" gene and that in some cases, when these are paired, an in-between shade of hair color may result.

childhood on through maturity, as may also be true of red hair. (How often have mothers wailed as they have seen the golden or auburn locks of a young child turn later into an indefinite murky brown!)

Among Whites, babies are frequently born with a temporary growth of dark hair which in a few weeks is replaced by light hair. But if the dark hair remains, only rarely does it become lighter in color as a child grows up. Constant exposure to the sun, or bleaching by salt water, drugs, or some other artificial means, can of course easily lighten or change hair color, and climate can also be an influence. But regardless of surface changes, the pigment particles will still be there, so that under a microscope a scientific Sherlock Holmes could easily tell whether a blond was natural or artificial.

The hair-color change that comes with age is one of *decolorization.* Not merely the pigment, but the air content, oil content and structure of the hair are affected. The time at which hair pigmentation begins to slow down often seems to be governed by heredity. Where a parent has grayed prematurely, in many cases a child will begin to gray at about the same time. The exact cause of normal graying, though, has not yet been established, nor has any medically approved way been found of preventing it or of restoring hair color once it is lost. (Incidentally, when gray hair comes naturally, it has no necessary connection with health or physical fitness, and need not be related to aging in other ways.)

Nerve or gland disorders, various diseases, peculiarities of the body chemistry and other physiological factors have been shown to affect the hair-forming cells and the hair-pigmenting process. Thus, occasionally, persons during a long illness or as a result of some harrowing experience may have their hair *gradually* turn gray or white. But this is far from supporting the popular belief that a person's hair can turn white "overnight."* While a sudden nerve upset might cause the *new hair* to grow out white, a little study of the hair structure will show that no nervous shock could instantaneously knock out all the pigment particles in the hair already grown out. For lack of authenticated cases "the stories of hair turning white overnight" will have to be lumped with the myths about children being born with white hair because their mothers were

* A classic reference to this belief was in Byron's famous poem *The Prisoner of Chillon* (1816), Stanza 1:

> *My hair is gray, but not with years*
> *Nor grew it white*
> *In a single night,*
> *As men's have grown from sudden fears.*

frightened by white horses during pregnancy.

Another point of interest is that discrepancies in color or shade between the hair on the head and that elsewhere may be caused by localized conditions in different parts of the body. Some contributing factors may be differences in the various hair areas and the structure and nature of their hair cells, gland action, and degree of exposure to air and sunlight. Whatever the reasons may be, we sometimes find men with brown or black hair having red mustaches and beards, and fair or reddish pubic hair, while others with light head hair have dark pubic hair. (One of the strangest possibilities is that head hair may become *green* if continually exposed to copper—a phenomenon long noted among some copper-ore miners.)

We have dealt with environmental influences in hair coloring at some length because it is important for parents to consider them before drawing conclusions as to what "hair-color" genes they carry and can transmit to their children. However, in the great majority of cases a person's hair coloring during maturity offers some pretty accurate clues as to the genes he is carrying, for hair color is among the traits most definitely determined by heredity. This is unquestionably proved by the fact that all types of hair color are found among humans in many different environments; and also by the fact that among lower animals which are bred for their color—dogs, cats and other fur-bearing animals—a great variety of hair colors, down to the most minute shades, can be accurately produced through carefully controlled breeding which follows genetic principles.

If there was any point in breeding human beings in this way, and if we did so, we could also most certainly produce redheads, blonds and brunettes, of every shade, at will. A truly hair-raising prospect.

CHAPTER TWELVE

Skin Color

JUDY O'GRADY and the Colonel's lady, both being White, might very well be "sisters under their skins." But whether the Black man, Yellow man and White man are brothers under their skins is a question that has long agitated the world and continues to cause strife and bitterness.

If skin color itself were the only point at issue, modern scientists would say there is little or nothing to justify the prejudices associated with it, or even to warrant making clear-cut distinctions among peoples on the basis solely of their skin pigmentation. For the old notion that each race has a pigment peculiar to itself—that Negroes have special "black" pigments in their skins, Chinese "yellow" pigments, Indians "red" pigments which are not found in Whites—is all wrong.

What is now established is that *exactly the same skin-coloring elements are present in all human beings, of all races.* The only differences are in the degrees and amounts in which these elements are produced and blended, chiefly through inheritance.

Thus, in depicting flesh color of any kind—always a headache—artists may come close to the desired results by using mixtures of brown, red, yellow and blue. This, too, is the palette of colors and effects used by Nature in tinting human skins.

Whatever your own skin coloring may be, it is produced by a combination of these five pigments:

1, 2. Our old friend *melanin*—the same sturdy brownish pigment found in eyes and hair—and a junior relative, *melanoid,* which two are mainly responsible for the degree of darkness of your skin. (Melanin is present in granule form, melanoid in diffuse form.)

3. Possibly *carotene,* a yellowish or yellowish-red pigment, chem-

ically very much like one found in carrots (hence its name).

4, 5. Two blood pigments in the two forms of hemoglobin—the bright red one in the arteries and a darker kind in the veins—which, as they shine up through the skin, give it its pinkish flesh tint.

Finally, not involving a pigment, a bluish tinge is contributed to the general skin color by the deep, opaque underlying layers of the skin. Without this bluish effect, produced through light-scattering (in the same way that eyes or the sky are made to appear blue), "the normal skin would appear much like a piece of cellophane-wrapped, raw beefsteak" (to quote Drs. Edward A. Edwards and S. Quimby Duntley, active in the field of skin-color research).

In individual cases, the effects of skin pigmentation are also influenced by the structure of the skin, its thickness, translucency and oiliness, while the pigmentation process itself may be intensified or diminished by hormonal action, sunlight and other factors to be dealt with later. But all important differences in human skin colors, racial and individual, can be attributed to the way the basic pigments mentioned are produced and distributed through gene action.

The key genes in the whole process are those which govern the production of the varying amounts of melanin and melanoid. Differences in concentration of these brownish pigments alone can produce grades of color from yellow, in dilute form, through orange, orange-red, dark brown and almost black. The most active "melanin" and "melanoid" genes are found in Negroes, with decreasingly active genes being present, in order, in Mulattos, Hindus, Japanese and Whites. Where the darkest skins are produced, there is probably a special "intensifying" gene which causes extra pigmentation not only in the skin but in the eyes and hair. This may account for the fact that whereas in the lighter shades of eyes, hair and skin, colors of various kinds may go together, "black" skin is almost invariably linked with very dark eyes and hair.

For *carotene* there may quite likely be another gene. Inasmuch as this pigment is yellowish or yellowish-red, we might suspect (as a theory, for there is as yet no proof) that the most active "carotene" genes are in the Chinese and other Mongolians, and in their offshoot relatives, American Indians. (Incidentally, Indians don't have "red" skins: at best, they're bronze, or copper-colored.) However, many authorities believe there is actually no difference in carotene among the races, and that "yellow" skin is probably due to fine melanin particles, plus effects produced by the small blood vessels in the way (or depth) that they lie beneath the skin.

As for the blood pigments, these have no racial connections, the blood of all peoples being the same in coloring. (The exceptions are only in individual cases where some persons are more anemic than others.)

Thus, all the important differences in skin color among human beings—ranging all the way from the darkest Negroes to the whitest Whites—can be related mainly to the graded action of a few separate genes governing the production, in varying degrees, of melanin and melanoid. Further, with this simplified understanding of how much more limited are the actual differences in skin colors than we used to think, and how much overlapping there is among races and subgroups even in these differences, it becomes clear now why we can't make any accurate classification of the world's peoples merely on a color basis. (You yourself probably know Whites who are darker than many Negroes, or "yellower" than many Chinese, or "redder" than many Indians.) Indeed, modern anthropologists no longer speak of Black, Brown, Yellow, Red, or White races, but refer instead to Negroids, Mongolians, Australoids, and Caucasoids, and to various divisions of each race, basing their classifications on many physical traits in addition to skin color.

Also, the science of genetics has made a lot clearer what happens in matings between persons of markedly different complexions. The old notions were that skin colors were "blended" in the offspring of such matings, much as colors are blended by mixing different pigments. True, this seemed the case when a Negro mated with a White and they had Mulatto children of an in-between shade. But always baffling was the result of matings between one Mulatto and another, for instead of their offspring continuing to be all of the same blended shade, it was found that they could be of all different colors, ranging from the darkest black of any Negro grandparent, to the very light skin of the fairest White grandparent.

This mystery is ended now by proof that in matings between persons with different-colored skins, there is no mixing of colors, but merely a recombining of pigment-producing genes. Further, in accordance with the Mendelian laws, these genes remain separate and distinct in their workings, and only in their effects is there any blending. In subsequent matings the genes segregate out again and may form new combinations and produce new effects.

How this sorting out and recombining of the "skin-color" genes may work in Negro-White and Mulatto matings is shown in our accompanying Plate. To simplify the process, we have depicted the "Negro" genes and "White" genes as each of two kinds (which might refer, theoreti-

If a Negro mates with a White:

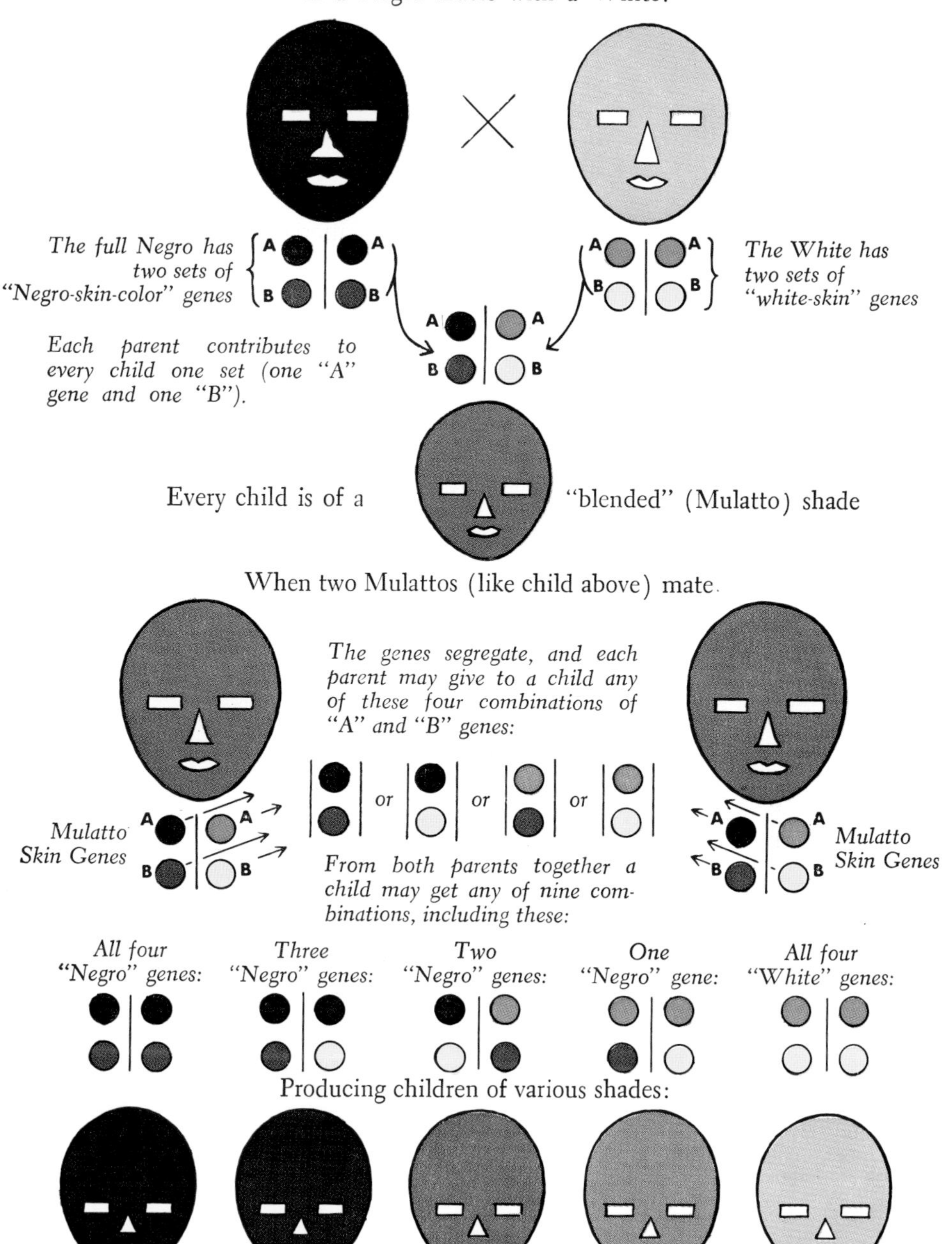

Note: Only two types of "skin color" genes are shown, but there probably are more.

cally, to their different "melanin" and "melanoid" genes) *although there are probably a number of other genes involved.* With these multiple genes we introduce a new genetic principle. In the case of eye and hair color, you will remember that a "dark" gene dominates a "light" one: Where a "brown-eye" gene is coupled with a "blue-eye" one, brown eyes result quite as well as if there are two "brown-eye" genes, and blue eyes result only with two "blue-eye" genes. With the multiple "skin-color" genes, however, *each gene* can assert itself and can add to the production of pigment. Individually, a "dark-skin" gene is stronger than a "light-skin" gene, but it is not nearly as strong as *two* "dark-skin" genes together. Thus, as you will see by our Plate, if at least two pairs of different "dark" genes are required to produce the full Negro color, one pair alone coupled with a "White" pair will produce the in-between Mulatto shade, and still other shades will result when three "Negro" genes are linked with one "White" or one "Negro" gene is linked with three "White" ones. Should additional types of genes be involved in producing Negro skin color as compared with White, the possible graded color combinations would be even more complex.

Anyway, we can see now why a truly black-skinned child can result only if *both* parents carry some "Negro skin-color" genes. This should dispose of a still widespread superstition that a woman with a little "hidden Negro blood," passing as White and married to a White man, might give birth to a "coal-black" baby. One such supposedly throw-back case, reported to geneticist Curt Stern, was thoroughly investigated by him and found—as always with such cases—to be groundless. Where a black baby does unexpectedly turn up, it can be taken for granted that (1) both parents have Negro ancestry, or (2) the parentage is doubtful. Conversely, it would be equally impossible for a Negro woman with "hidden White blood," mated to a genetically pure Negro, to give birth to a White child.

One important question with respect to Negroes in the United States is whether their average skin color hasn't been becoming steadily lighter through the infiltration of "White" genes. No accurate scientific studies have been made on this point, nor could they be made easily. For one thing, while a great many "White" genes were mixed into the Negro population in the past (some authorities estimating that not more than 10 per cent of the Negroes are free of them), there is little evidence of any increase in interbreeding in recent years; in fact it is more probable that there has been a decline. On the other hand, the lighter-skinned Negroes, who tend to be economically more favored, are reported as having a lower birth rate than the darker-skinned ones; and further decreasing the pool of "lighter" genes would be the fact that a certain number of those with the fewest "Negro" genes pass over into the White population each year. If we take together all these factors (and others to be touched upon later in the chapters on race), the answer might be that our Negro population is not at present turning genetically lighter to any great average extent.

But, you may ask, haven't we overlooked the possibility that living in a temperate climate for generations has made Negroes much lighter-skinned than their African ancestors? Before answering this specifically, let's go on to the general subject of environment in relation to skin color.

First, it can be said that pigment production in the skin is much more responsive to outward and inward influences than is eye coloring or hair coloring. (Perhaps you don't have to be reminded that you may look like a boiled lobster after your first day at the seashore and that at other times your mirror may have told you, you look positively black, or green, or yellow, or pasty white.) Most often, skin color is affected by sun and climate. One of the principal purposes of the skin (aside from its heat control and excretory functions) is to serve as a protective wrapping, and where the pigment has its special value is in shading the delicate blood vessels from strong sun or light rays. Because human beings may live in many different climates and under different conditions, the mechanism of skin-pigment production is so adjusted that it can be governed to some extent by the needs of the individual. Thus, under some conditions the pigment genes step up their production, and a darkening of the skin results; under other conditions they lazy along, and there is less pigmentation.

However, the way that different skins react to outer influences—for instance, the sun's rays—may also be governed by genetic differences, both in the way the skin is constructed and in the way the pigment genes work within it. Take the familiar examples of stenographers on their vacations. In brunet Gertie's case, the more she basks on a blazing beach, the more her genes will rally to increase pigmentation, with a beautiful deep tan resulting. In light-blond Flossie's case (as is usually true of very fair-skinned people, and often of redheads) the pigment genes may be unequal to the task. Unless she properly protects herself, her flesh may actually be *broiled*, sometimes with serious consequences.

There are White persons who through constant exposure to a hot sun can become almost as dark as Negroes. But a heavily tanned White man never has quite the same skin color as a "full-blooded" Negro, for even with the same degree of darkness, other factors contributing to their coloration would differ. Also to be stressed is that if a normally fair White man does become dark-skinned, it is only an *acquired* characteristic which can have no effect on his children. White families can live in the tropics for generations, and yet their children will continue to be born as fair-skinned as if their forebears had never strayed from Hoboken, New Jersey.

What of the differences in complexions between the very light-skinned Nordics of the cold climates and the swarthy southern Italians, or the very dark Arabs? These are only in part due to degrees of exposure to the sun, for one can find light-skinned Whites and swarthy Whites of native stocks in every climate. Most Mediterranean peoples have a considerable admixture of "dark-skin" pigment genes, whose effects would persist regardless of the climates in which they lived. (We need hardly point out that in the United States offspring of peoples of many stocks living side by side reflect in their skin colors all shades of their ancestral differences.)

Which brings us back again for a moment to Negroes. By now it should be apparent that it wasn't "the shadow'd livery of the burnish'd sun" which accounted for the skin color of their ancestors—regardless of how many generations they had lived in Africa—but certain genes which they happened to have to start with (and which may have become more concentrated through selective breeding). Nor has the lack of any such burnished sun made too much difference in the skin color of their descendants living in New York, or Minneapolis, although it is true that Negroes living in cool climates will, like Whites, be less heavily pigmented than those of their race in the tropics. We might also add that Negroes of different ancestral stocks have almost as many genetic variations in skin color as have Whites, and that they, too, react differently on exposure to hot sun. Not only do some Negroes sunburn more easily than others, but many Negroes run almost the same risk of sunstroke as do Whites.

Not too much is known about other racial aspects of skin-color inheritance, or the pigmenting results of crossbreedings between Chinese or Japanese and Whites, Indians and Whites, etc., because they have not been given detailed study, which in turn is largely because they do not have as much social significance or interest as do the facts relating to Negro skin color. However, the basic principles of skin-color inheritance that apply to Negro-White and Mulatto-Mulatto matings also hold for interracial matings of all kinds, and to a lesser extent for matings between persons of the same race with very different skin colors. For instance, where a Caucasian of swarthy stock (southern Italian, Arab, or East Indian) mates with one of the fair-skinned stock, the offspring would be of in-between shades; and in matings of Whites of mixed stock, the offspring may be of a variety of skin colors, ranging from dark- to light-skinned. In some instances, though, the "darker" genes appear to dominate the "lighter" ones.

Of special interest are the cases where a remote Mongol ancestor, or sometimes a Negro ancestor, may reveal himself in a White European or American infant through the *Mongolian spot*. This is a temporary bluish patch in the skin near the base of the spine which is present through heredity in almost all infants of Mongolian race (Chinese, Japanese, Eskimos, American Indians); in many infants of some Negro stocks; and occasionally among White infants of Armenian, Italian or other darker-skinned Caucasian stocks. It disappears by the end of the first year. The Mongolian spot appears to be produced by dominant genes in some cases and by recessive genes in others, but is not reported in blonds, indicating that highly active general "pigment" genes are needed to bring it out.

Freckles provide a more familiar example of spotting with an hereditary basis, a dominant gene being reported as the usual cause. Freckling is often transitory, appearing in childhood and disappearing in maturity. Frequently it is associated with red hair and white skin, indicating that where there is such a combination, the "freckling" gene may have some special opportunity to assert itself.

Albinism, which deprives eyes and hair of their natural coloring, also greatly reduces the pigmentation of the skin, most strikingly so in the case of Negro albinos (two of whom are shown in the photo opposite page 234). But this condition belongs with other skin oddities and abnormalities to be dealt with in Chapter 23.

In the general skin-pigmenting process, the genes do not assert themselves in full strength until after infancy. Characteristic pigmentation, however, begins during the embryonic stages and is quite well advanced at birth. Newborn Negro babies are usually of light chocolate or brick-red color but then darken steadily, so that by the end of the first month they are quite fully colored. Mulatto babies may look white for a time. but if both parents are Mulattos, their babies, as we've previously indicated, may be of almost any color at birth and thereafter. As for White babies, any mother knows they're anything but white at birth, their flaming redness being due to their very thin and sparsely pigmented skin allowing the actively charged blood vessels beneath to shine through.

Skin Color

Changes in skin color in later years may be due to various influences other than those already mentioned. Many diseases, especially the ones arising from glandular upsets, may affect the pigmenting processes, leading to either darkening or lightening effects. Addison's disease, for example, gives the sufferer a bronzed skin, jaundice produces a yellowish skin, and tuberculosis may give a person very white skin with very red cheeks. Pregnancy tends to darken the skin for a while, though normally a woman's skin is somewhat lighter in color than that of a man of the same stock; for female skins have somewhat less of the melanin pigment and somewhat more of the carotene, and they tan less deeply than do male skins. These sex differences are due largely to hormonal influences on the pigmenting process, as well as on the development of the blood vessels, which in individuals of either sex may modify skin color (over the body as a whole or in specific areas) all the way from infancy through old age.

But, as we must constantly keep in mind, whatever the varying states or environmental conditions of individuals may be, and whatever effects these may have on their skin pigmentation, they can in no way change the "skin-color" genes which persons transmit to their children.

CHAPTER THIRTEEN

The Features

IF YOU'RE AMONG THOSE who believe that "your face is your fortune," you might want to know how much of that "fortune" is inherited.

Making faces happens to be one of the most interesting jobs done by genes, though in the plastic details, such as nose, eye shapes, ears, and lips, examples of gene dominance and recessiveness are not quite as clear-cut as in our coloring processes. This isn't because the "sculptor" genes are necessarily less definite in their workings than the "color" genes, but because many more of them may be involved in producing any given feature, and many more things can happen to modify and alter what these genes set out to do. Added to all this is the difficulty of classification.

In eye color, for instance, "blue" is blue, applied to anyone the world over. We know quite clearly what is meant by "blue" when we say a child or a man of sixty has blue eyes, a girl or a boy has blue eyes. But in describing features or body forms we can use such terms as "large," "small," "broad," or "narrow," only relatively. A nose that would be large on a child would be small on an adult; a nose that would look handsome on a Romeo would be a whopper on his Juliet. Moreover, in appraising the features of persons of the same age, sex and stock, when you say this one has a "long" nose, that one a "small" nose, you might be surprised to find that the actual difference in length is perhaps no more than three-sixteenths or a quarter of an inch. (The famed "schnozzola" of the comedian Jimmy Durante may be no more than half an inch longer than average.) Similarly, "enormous" eyes, or a "wide" mouth, may represent only a fraction of an inch more in width than the average.

In general, most of the feature differences considered important among persons of the same family or stock are so small as to require very detailed and minute measurements if they are to be reduced to statistical terms. For this reason, geneticists and anthropologists who have studied feature inheritance have so far confined themselves largely to the very marked racial differences—as between Negroes and Whites, or Whites and Mongolians—and the results of crosses among them. However, these interracial studies have been checked sufficiently with observations of matings of people of similar stocks to lead to some fairly definite conclusions as to how features are inherited.

Let's start with the face as a whole, not only because it is the way you ordinarily see a person at first, but because each individual feature may be much influenced in its appearance by the sum total of all the features. This goes much further than mere looks, however. In the actual process of facial construction, during prenatal life and in all the developmental stages thereafter, the molding of the face in its entirety can affect the shaping of each separate feature, and each feature can individually affect its neighbors.

Have you ever modeled a head out of clay or soft wax, and then, for fun (or out of disappointment), squashed it? You've chuckled as you've seen how every feature is altered and how different are the effects if you push in the clay head from the sides, or from top to bottom. There are much the same results when you stand in front of the distorting mirrors in a fun house or look at your reflection in bowls or vases, some of which widen the face while others narrow and lengthen it to exaggerated degrees.

These examples illustrate a basic principle of facial construction. First, there are genes which govern the structure of the skull as a whole, as you can see most clearly in the case of dogs, ranging all the way from the pulled-out, lengthened faces of wolfhounds and dachshunds to the squashed-together faces of bulldogs and Pekingese. Humans, fortunately, have not been bred for any such extremes, but nonetheless there are abnormal cases, as in certain kinds of dwarfs, where a gene or two can radically change the structure of the whole face and affect every feature. There are many other cases, to be dealt with in later chapters, where a few abnormal genes, or glandular abnormalities in whole or in part hereditary, may shape the face as a unit in characteristic ways, or greatly distort individual features. (Certain pituitary disorders may transform the face, broadening the nostrils, thickening the lips and darkening the skin. Again, in the inherited blood disease *Cooley's anemia,* the facial changes produced in afflicted children of different families may cause them to look more like each other than like their respective sibs.) In normal persons, however, no gene is known which singly can do much to affect the features in unison, although a few genes together may act to broaden or lengthen the face.

As a general rule, wider heads and stronger jaws (which can be much influenced by chewing habits) tend to make for broader faces. Longer or deeper jaws contribute to facial lengthening, particularly in the case of men, who, on the average, have longer faces than women (and not only when the bills come in at the end of the month). Either in the sex differences or otherwise, hormonal influences may also have much to do with the broadening or lengthening of the face. Thus, so many factors are involved in face-shaping that it is quite impossible to speak of inheriting such or such a type of face as a whole.

For the most part, the construction of every feature individually is governed by special genes. While these usually work together with the genes for the neighboring features, they can also go off on tangents of their own. The ears and the nose, which grow away from the face, may be most independent in their development, whereas the mouth, as we shall see, is much influenced by its surroundings.

But the best evidence of how closely "feature" genes follow their basic blueprints is presented by something which we've taken too lightly for granted: the precision and symmetry with which opposite sides of the face and particularly paired features—eyes, ears, teeth, and nostrils—are constructed. If the genes did not influence the features in specific ways, down to the most minute details, or if environment were the major factor, one eye would be radically different from another, one ear from another ear. As it is, minor differences do exist between corresponding features and sides of the face, but except in some abnormal instances, these result from the slightly different conditions encountered by the duplicate genes as the body develops.

Now let's consider each of the features in detail.

THE NOSE. The "nose" genes are among those that can be most clearly analyzed. Some studies seem to indicate that there is one key gene producing the general shape of the nose, but most authorities agree that quite a number of genes are at work, each on a different part. That is, there may be separate genes for the bridge (its shape, height and length); the nostrils (breadth, shape and size of apertures); the root of the nose and its juncture with the upper lip; and the bulb, or point of the nose.

Often, it is true, the nose as an entire unit appears to be "inherited" from one parent. Where such resemblance occurs it may be assumed

DOMINANCE AND RECESSIVENESS IN FEATURES*

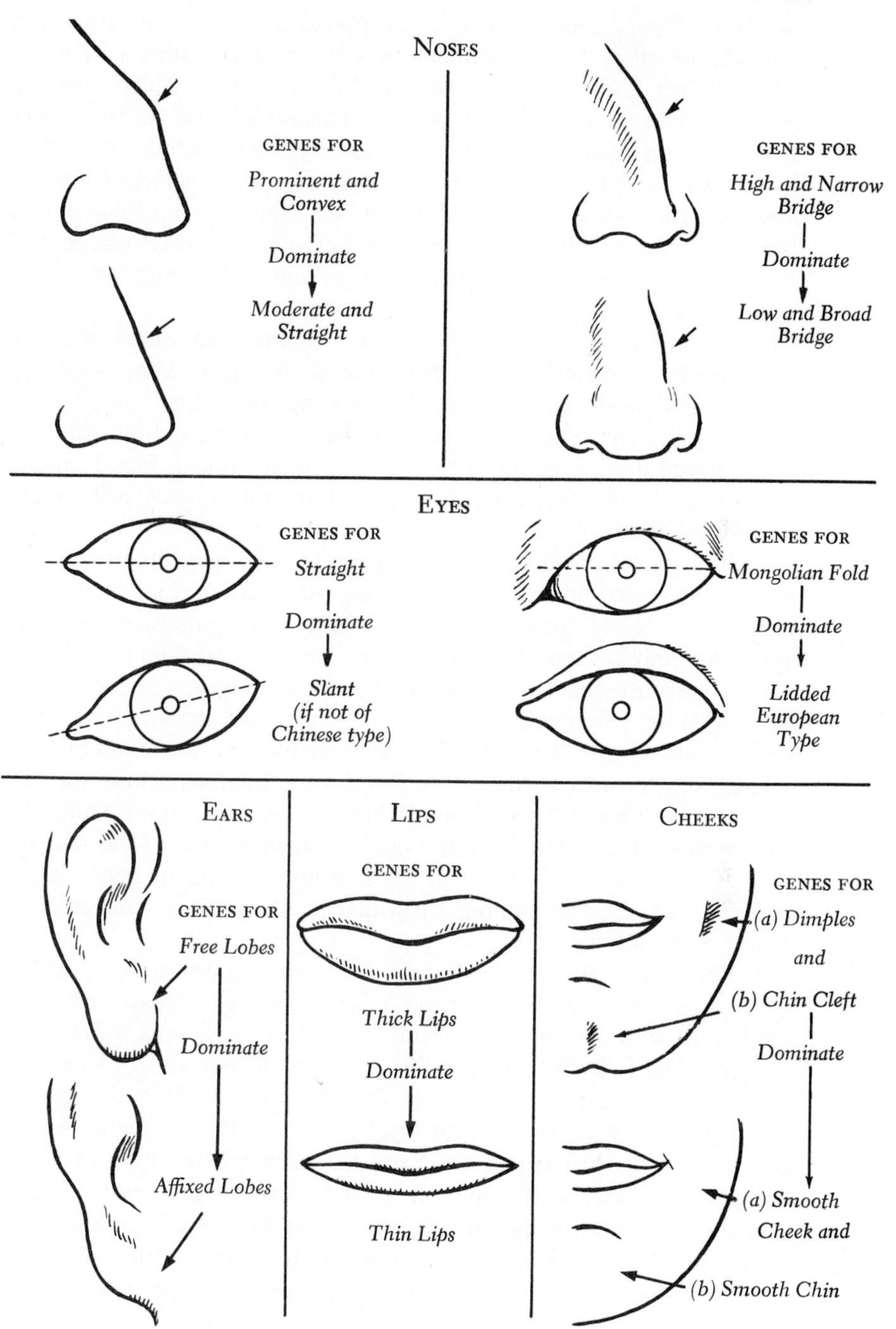

* It should be kept in mind that it is not one feature itself which dominates another but the genes for a given feature which dominate other genes.

that the different genes involved were passed over in combination and were almost all of them dominant over those of the other parent.

Equally often, on the other hand, a child has a nose which seems to be a cross between that of both parents. This would bear out the theory that several or many unit factors are involved. At any rate, it is clear that distinctive genes are at work, and that they sort out independently; otherwise, the nose of every child would be a blend of its parents' noses. Even in the most inbred peoples, however, noses of every shape and size appear, proving the Mendelian segregation and sorting out of the "nose" genes.

In the *bridge*—the most important part of the nose—shape and size are dependent on how far out from the skull and at what angle the "bridge" gene works until it stops. At the same time, other "nose" genes are acting to determine the *breadth* of the nose, ranging from the thin bridge found among many Whites to the broad bridge found among Negroes; and still other genes are at work on the nostrils and on the bulb of the nose.

What happens in children when parents with very different noses mate is suggested in our accompanying illustration. In general, the more active "nose" genes—those for larger, more prominent or more accentuated noses—tend to dominate the more conservative ones. Thus, in most cases the genes for prominent and convex noses, or for hooked and pug noses, dominate those for moderate and straight; the genes for high and narrow bridges dominate those for low and broad bridges; the genes for broad nostrils, of the Negro type, dominate those for narrow nostrils (although in White-White matings, the reverse may be true in some cases). The "potato nose" of Dutch families (and perhaps some Russian families) seems to result from a dominant gene, as apparently does also the bulbous tip which in other strains continues to enlarge with old age.

However, with the various "nose" genes sorting out independently, we can see how persons may have a large nose with small nostrils, a small nose with a wide bridge and large nostrils, or any other combination. The full effects of the "nose" genes do not assert themselves until maturity; in fact, these genes may keep on working throughout life to produce increases in both length and breadth. As many readers have learned to their regret, the pertest, daintiest little noses of childhood may blossom after adolescence into veritable caricatures. Moreover, during and after middle age, there may be a final spurt in nose development, or a putting on of "finishing touches," so that often, in later years, racial or familial characteristics become most apparent.

As a rule, the general growth of the individual affects nose form and size, with tall people tending to have longer noses than short people. Sex differences, too, play an important part, the male (androgenic) hormones acting to make the noses of males somewhat larger, relatively, than those of females.

Nose-shaping may also be affected by many environmental factors, including climate, diet, structural defects (deviated septums), diseases (sinusitis and blood-vessel disorders), alcoholism and, of course, accidents. Despite all this, however, the "nose" genes do manage to assert themselves quite strongly, and a knowledge of what genes the parents carry can lead to pretty fair predictions of what their children's noses will look like.

THE EYE. The form and shape of the eyes are governed chiefly by the shape of the individual eye socket and by the way the lids grow. An eye may look large either because the eyeball and the socket are themselves large or because the eyeball protrudes and pushes back the lids around it, perhaps as the result of some disease, such as goiter. In normal cases the gene for a wide eye dominates that for a narrow eye. (Apart from this, the eye, or eyeball, in females, tends to be relatively longer than in males, even when they have the same "eye" genes. Another oddity is that the right eyeball is usually larger than the left in males, though not in females.)

The "slant" or "almond" eye is often confused with the Mongolian eye found among Chinese, Japanese and Eskimos. In the slant eye the inner corner is rounded, the outer pointed and slightly higher. The Mongolian eye, however, is due to fat in the eyelid and a thick skin fold overlapping the inner corner of the eye, giving it its oblique appearance. (Other facial elements, including the cheekbone, eye socket and nose bridge, contribute to the appearance of the true Mongolian eye.) While the gene for a slant eye is recessive to that for a straight eye, the gene for the Mongolian eye fold usually is dominant.

Another hereditary eye condition, which, alas, may be unjustly confused with the effects of dissipation (or of disease), is a form of "baggy eyelids." While more often occurring in Mongolian individuals, it is also found among White persons.

Deep-set eyes seem to be a recessive trait, with multiple genes probably involved. Wide-apart or close-together eyes, where inherited, are determined not by "eye" genes but chiefly by nose-bridge width, or the width structure of the face, and the genes for these.

Unusually long eyelashes are inherited through a dominant gene (fortunately for the ladies). Where a mother has very long lashes, she can count on an average of one in every two children inheriting them.

THE EAR. Human ear shapes are so distinctive, as between one individual and another, that some authorities have suggested using ear prints of newborn babies to avoid mixups in hospitals and to aid in identification in later years. In ear shapes the inheritance of several characteristics has been noted.

The gene for a long ear seems to dominate that for short.

The gene for a wide ear dominates that for a narrow ear.

The gene for a full and free lobe usually dominates that for affixed (women with the latter type being at a disadvantage when they wish to wear earrings).

A minor, very common peculiarity in some ears, inherited as a qualified dominant, is *Darwin's tubercle* (or point), a small projection from the inner rim (helix) of the ear, varying in size, and found either in one ear or in both. (See Illustration.) It is named after Darwin because he had called attention to it as a possible vestigial trait from man's earliest ancestors.

Other hereditary ear peculiarities include *ear pits*, or tiny holes, sometimes in the ear lobes, making them look as if they'd been pierced (and once believed evidence of inheritance of an acquired trait), and sometimes appearing midway where the ear joins the face. These pits, inherited usually as dominants, may appear either in one ear or in both ears. A rarer oddity is the small *cup-shaped ear*, likewise inherited as a dominant.

Racial differences in ear form include the fact that long and narrow

"DARWIN'S TUBERCLE"

(Vestigial trait from man's earliest ancestors)

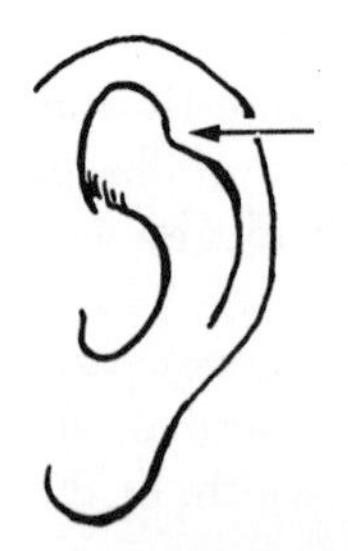

Point on ear flange when rolled and infolded

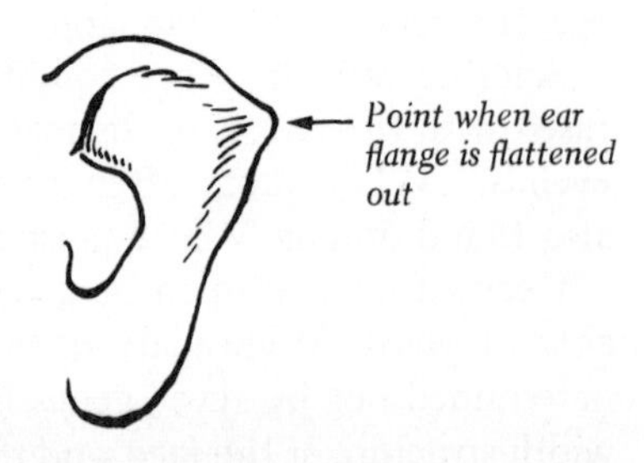

ears more often occur in Whites, and still longer and narrower ears in Mongolians, whereas Negroes tend to have short and wide ears, usually smaller than those of other racial groups.

THE MOUTH AND TEETH. That lovely mouth which poets rave about is a bugaboo for geneticists. Though the mouth may look to you like a fairly simple feature, in reality it is more complex, more dependent on other factors for its formation, and more subject to change than any other part of the face.

First, these are genes which govern the shaping of the general surface details of the mouth and lips, as can be gathered from comparisons between Negroes and Whites, or by noting characteristic mouth and lip forms running in many families. But the formation of the mouth as a whole is much influenced by the underlying structure of the teeth, jaws and palate, for all of which details there are separate gene combinations. When we add to this the fact that the way the mouth looks at any stage of life is also much governed by how one has used it and how the muscles controlling it have been "toned" or set, we can see the difficulty of working out any simple formulas for mouth-shape inheritance.

We know only what happens in a few cases where extremely different lips are involved, or where there are certain mouth abnormalities. In Negro-White crosses, the thick, broad lips appear to dominate the thin ones, but that there must be a number of genes participating is shown by various degrees of in-between lip shapes among offspring of Mulatto matings. (Note also that what often appear to be thick lips may be the result in part of the way the lips are turned outward, though this, too, is conditioned by gene action. Try the effect on yourself of pushing your lips outward, and upward and downward.) As for matings between one White and another, we do not yet know what happens when "big-mouth" genes run into "small-mouth" genes, or when "cupid's-bow" genes meet with other "lip" genes.

An abnormal condition, the protruding Hapsburg lip (named for its prevalence in the Spanish-Hapsburg royal family, where it has been carried along for hundreds of years) reveals itself clearly as dominant in inheritance. The narrow, undershot Hapsburg jaw also goes with the lip as a rule. Dimpled or cleft chins may be inherited either through dominant or recessive genes. (Cheek dimples may also be inherited, as a rule through dominant genes.) The receding chin is generally recessive in inheritance to the straight chin, and narrow or pointed chins seem to be recessive to the wide one. However, the appearance of the chin is much dependent on how the lower jaw is set against the upper jaw, and in this the whole structure of the head may be involved.

In *teeth*, genes are at work to produce many of the variations noted among individuals and groups in sizes and shapes of teeth as a whole and of particular teeth; in the time of eruption; and in abnormalities and defects (the latter will be dealt with in a later chapter).* Most striking are some of the racial differences in dentition. For example, in Mongolians there is a very high incidence of shovel-shaped incisors whereas in Whites the incisors are usually chisel-shaped. (For more on race differences in teeth, see Chapter 42, pages 592–593.)

Hair Form. The hair has been subjected in times past and present to more artificial changes than probably any other human feature. But of all hairdressers, most important are the "hairdresser" genes with which a person starts life. They determine whether one's hair is to be as straight as a poker, naturally curly, or so kinky "you can't do a thing with it." Whatever one may do to alter the surface effects of the genes' workings, we may see that under the microscope the different forms of hair are actually different in their basic construction.

In cross section, straight hair generally tends to be almost round, wavy hair more oval, curly hair still more so, and kinky hair a flattened oval. However, this refers to *averages*, for all forms of hair have been reported with all forms of cross sections. Even on the same head the forms of individual hairs may differ considerably. Thus, the cross-section shape of the hair would appear to be only one of the factors associated with the form it takes, another factor, perhaps, being the shape of the follicles out of which hair grows. (In curly hair the follicles tend to be curved; in straight hair, straight.) Yet as the late noted anthropologist Earnest A. Hooton suggested, the more circular cross-section form in hair may make for greater rigidity and straightness, while the flattened-oval cross section may permit of greater curliness or kinkiness. "I should doubt," he said, "that it would be possible to put a 'permanent wave' into really straight, coarse Mongolian hair, or to make permanently straight the flattened oval, twisted Negroid hair."

In the manner of growth, kinky hair is characterized by "bunching" (or "matting") in spirally twisted locks. In woolly hair, the extreme form of kinky, these spiral twists are very small and clumped together close to the scalp. This is most accentuated in the "pepper-corn" hair

* One special item: A space between the upper center teeth—much valued by some little boys, though hardly by adults—appears to be inherited as a dominant.

THE "HAIRDRESSER" GENES

They Account for These Characteristics:

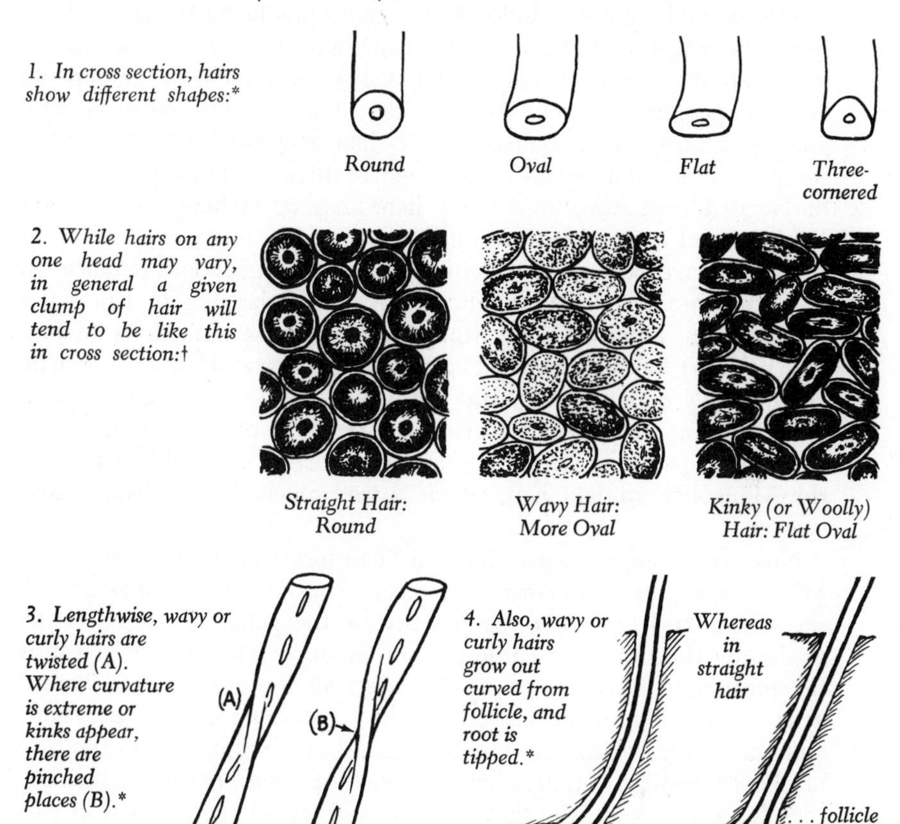

5. Where a different "hair" gene is received from each parent (particularly if parents are of similar racial stocks), the effect generally is as follows:

* Figures 1, 3 and 4 after sketches and data by Dr. Stanley M. Garn.
† Figure 2 after microphotographs by Dr. Morris Steggerda.

of some Negro stocks, where the bare skin may show between the tightly twisted clumps. (Added to the genes producing the form of the individual's hair, there may be other genes governing the way it grows.) Although the matted woolly or kinky hair occurs chiefly among the Negroid peoples, the gene for it seems to have arisen independently among Whites, for occasionally woolly hair is found in White families —some very fair-skinned and blond—where there can be no presumption of Negro blood. Also, some kinky hairs may occur here and there on the head and body of most Whites.

As will have been gathered, the most marked genetic differences in hair form are racial. While there may be overlapping in hair types among races, almost all the Mongolians (Chinese, American Indian, and Eskimo) have straight and coarse hair; the Negroid peoples, chiefly curly, woolly or kinky hair; and the Whites, the in-between forms of wavy, curly or *fine* straight hair. The latter type of straight hair, it is important to note, is genetically and in structure quite different from the Mongolian straight hair, which is very much thicker, heavier and coarser.

Now what happens when different "hair-form" genes come together? With respect to dominance and recessiveness their workings are on the whole fairly distinct, the potency of the genes appearing to be graded by the degree of curliness they produce. The "woolly" gene is the most potent, and seems to dominate all the others. The "kinky" gene, in turn, dominates the "curly"; the "curly" dominates the "wavy"; and the "wavy" dominates the "straight."* An exception is in the case of the Mongolian "straight-hair" gene, which may be dominant over that for woolly hair (the thickness and stiffness of the Mongolian hair being here probably the determining factor). There may also be other occasional exceptions in the interaction of the "hair-form" genes, due to various modifying factors. But in matings among Whites, it appears quite certain that the "straight-hair" gene is recessive to all the rest. As a rule, then, *two straight-haired parents can expect to have all straight-haired children.* Where the parents have other types of hair, the results for their children cannot be so easily predicted, unless we have a fair idea as to which two genes each is carrying.

Also, in attempting genetic forecasts, or in comparing hair of parents

* A sex difference has been reported in a number of studies, with findings that the male hair often tends to be wavier than the female hair in the same families (to the distress of many girls who view their brothers' hair with envy). This may lead to the fact that some apparently straight-haired parents may have boys with some degree of wavy hair.

and children, it should be kept in mind that hair form and structure undergo changes with aging. Wavy or curly hair may become straighter, and the diameter of the hair in the same individual may vary from very fine in childhood to its maximum thickness in adult life, then thinning again in old age. Apart from this there undoubtedly are genes governing degrees of hair thickness (as is evident in comparing the hair of Mongolians and Whites). But how these genes work in people of the same stock to produce finer or coarser hair hasn't yet been established.

Environmental factors have not been shown to have much influence on hair form. (We are excepting, of course, what beauty parlors or home treatments may do. But these cannot alter the way the new hair grows out.) Whether in the tropics or in Alaska, the hair of Negroes is equally kinky or woolly and the hair of Mongolian peoples is equally straight. Nor, among Whites, does curly hair, wavy hair or straight hair undergo any change in different climates. Also, contrary to popular notions, hair doesn't grow thicker or coarser if it is cut repeatedly, or, in the case of men's facial hair, if it is shaved. (Of special interest to the ladies, this also applies to hair on the legs.) Likewise, there is evidence that nutrition, sunlight or the covering of hair with clothing has little effect on forms of hair growth. In short, hair form must be set down as one of the traits most directly determined by heredity and least influenced by environment.

An interesting hair characteristic is the *whorl,* the manner in which the hair grows wheel-like around the point at the top center of the head. Some people tend to inherit a clockwise whorl, others a counterclockwise, though which dominates which seems to vary in different families and may also be influenced by chance factors in prenatal development. About 5 per cent of the people have double whorls. (What kind do you have?)

The distribution and arrangement of body and head hair—the form and density of beard and mustache, the shape and thickness of eyebrows, the growth of hair on arms, legs and chest—are determined or strongly influenced by genes. (This is indicated when we look at the characteristic patterns of hair growth in lower animals.) Hormonal influences, in turn, account in large measure for the sex differences in hair growth, for differences in amount of hair at various ages, and for some of the abnormalities in hair loss or excessive hairiness.

Where the hereditary differences in hair distribution and density appear most strongly is again in comparisons between racial groups. If you have never seen a Heap Big Indian Chief with a long, luxurious beard, there's a reason. Facial and body hair are sparsest in full-blooded American Indians and are also sparse in Chinese, Japanese and other Mongolians. While Negroes have more hair than the latter, they are not nearly as hairy, on the average, as Whites. The hairiest of all peoples are the Ainus, a primitive group who, though inhabiting the northern parts of Japan, are basically of White stock. (It is believed that they are the remnants of a White people who very long ago lived in Asia.) In women with a special tendency to hairiness, it may assert itself temporarily after pregnancy and permanently after the menopause.

The recent revival of beard-growing among young men in various groups may warrant giving some attention to this adornment. Easily observable is the fact that there are varied natural patterns with which beards grow around the mouth and on the chin, jaw and cheeks, pointing to the existence of "barber" genes delegated to this work. The size and luxuriance of the beard, and the speed with which it can be cultivated, are determined by hormonal activity interoperating with the genetic tendencies. Ordinarily, men with an abundant production of male hormones (androgens) will be most successful in cultivating heavy beards. But this is by no means proof of masculinity in other ways, for it will be recalled that American Indian, Chinese and Japanese males, no matter how masculine, can grow only scanty beards. As summed up by Dr. James B. Hamilton, an expert on hair growth, "An extensive beard owes far more to genetic predisposition . . . than it does to quantities of androgens in circulation." However, where the genetic tendencies are present, hormones do serve to trigger beard growth; and if the male hormone supply is interfered with (as in castration, or in damage to the sex glands) beard growth may be much curtailed.

Coalescent eyebrows—heavy ones growing together—are most common among Mediterranean peoples (Greeks, Turks, Armenians). There are also inherited racial differences in baldness (to be discussed in Chapter 22); but the individual genetic differences in hair distribution within each race and group have as yet been little studied. Among the few traits which have been given such attention is "*widow's peak*," found to run in families and indicated as due to a dominant gene. (Incidentally, this trait—a point formed by hair down the middle of the forehead—got its name from an old belief that women who had it were destined to early widowhood.)

Looks and Environment. In discussing the separate features, we have given some attention to the effects of environment on each. However, environmental changes or modifications in the structure of any given feature, barring accidents, are apt to be much less marked, and certainly much less sudden, than can be such changes in the appearance of the face as a whole, especially in its fleshy formation. Throughout life the face as a unit is constantly influenced by habits of eating, sleeping and speaking, by thinking, emotions and various eccentricities of behavior, and, of course, by disease and aging. Where these influences are consistently the same for members of families, or even for large groups of people, many of the similarities in looks which we think of as hereditary may actually be in large part environmental. It is only when individual differences in features develop in the very same environments that we can be pretty sure they are hereditary.

But it must also be kept in mind that the environmental effects on the "features" genes include not only outside influences but those *within* the individual as well—all the glandular and other bodily workings which can shift the courses taken by the genes. Where these internal environmental factors differ from the outer ones is that they may be *hereditary* in part or whole. This is most clearly shown in comparisons between the features of women and men. We know that the two sexes carry the same "feature" genes, and yet the differences in their internal environments (started off initially by the XX- and XY-chromosome mechanisms) are sufficient to cause their features to develop differently in a number of ways and degrees.

In general, the bony foundations and frameworks of male faces are more rugged, and thus their features are more pronounced than those of females, given the same genes. Males within a given ethnic group tend to have heavier brows, higher cheekbones, more prominent noses, squarer and heavier chins, somewhat larger teeth, and heavier facial muscles. And need we again mention beards and mustaches? All these distinctions arise mainly through different glandular influences in the two sexes. But it is important to remember that in individuals of the same sex any degree of difference in the workings of the glands which control or modify growth—especially before or during puberty—can similarly affect the formation of individual features or the face as a whole. Eunuchs, for instance (if castrated before puberty), will develop feminized features, and girls and women in whom the glandular balance is radically upset will tend to develop masculinized features (including, sometimes, beards and mustaches).

CHAPTER FOURTEEN

Your Size and Shape

As A HUMAN BEING, your size and shape are pretty closely determined by your genes (as are the sizes and shapes of other animals). Otherwise you might have developed with the dimensions of an elephant or a rabbit, or what not.

But we take the general proportions of humans for granted. What concerns us are the relatively small differences—the few inches more or less in height, the ten pounds either way in weight, which may cause you to be considered tall or short, plump or scrawny. Not to mention the little bulges here and there, or the lack of them, which may classify your figure as "gorgeous" or "terrible"—depending on the viewpoints of the people among whom you live. (For what rates as an enviable shape varies considerably according to group, place and period.)

If we grant that the little differences in human form may mean a lot, to what extent are they inherited? As in the case of the features our answers will have to take note of various environmental influences, and these we will find to be very great with respect to whatever genes there are for body shape and stature. In fact, many of the characteristics of figure or size in yourself or other people which you've thought of as inherited may be largely environmental. On the other hand, it has also become clear that hereditary factors, working directly or indirectly, account for most of the major differences.

Comparing individuals, we find that the greatest differences in body size and formation may arise through *internal* environments, chiefly glandular, which may be in part or in whole hereditary. The most striking examples, of course, are provided by comparisons between women and men. Here we have the fact that two physically distinct groups of humans are produced not through differences in specific "body-form" or "stature" genes, but through over-all developmental differences, set in motion by the sex-chromosome balances which cause the same genes to work in radically different ways. For one thing, with the same "stature" genes and under the same outward conditions, a man will grow to an average height of about one-sixteenth greater than a woman, or somewhat over three-quarters of an inch more to every foot of her size. (The present averages in the United States as a whole—with many diverse groups included—are, for men, about 5 feet 8½ inches [175 cm], for women 5 feet 4 inches [164 cm], with an increase for both in sight.) And with the same "shape" genes, we also know how different the results can be in the two sexes. On all of these points there will be more to say as we go along.

STATURE. In the matter of human sizes we are indeed speaking for the most part of relatively small differences, for as a species we present no such extremes in variability as are found among many other animals: a giant St. Bernard dog, for instance, alongside a pint-size Mexican Chihuahua.

To take the extremes in normal human breeds (abnormal cases will be touched on later), the very tallest humans, as a group, are the Watusis of the Ruanda-Urundi region of Central Africa, who average well over 6 feet in height, with many males in the 7-foot range; while the smallest are the pygmy Negrillos, averaging 4 feet 6 inches. (By a coincidence, one pygmoid group, of part pygmy ancestry—the Ba-Twas—has long been living in subjection to the Watusis.) While the differences in stature between these African tribes may be considered for the most part hereditary, it involves two isolated groups who for thousands of years may have been inbred among themselves.

Between and among large groups of humans—whole races or nationalities—where there has been a constant intermingling and dispersal of "stature" genes, there are no such marked contrasts. Although Americans may consider themselves very tall in comparison with the Chinese and Japanese, their average differences in height are no more than three inches; and it is not unusual to find many strapping Orientals. Among Europeans, the tallest, the Norwegians, Swedes and Montenegrins, may average over 5 feet 8 inches for males (about the same as Americans), and the "short" Italians, 5 feet 6 inches.

What would happen if we deliberately tried to breed humans for size we can only guess. Two such attempts of which we have record can hardly be called scientific. The Prussian king, Friedrich Wilhelm I, set out to produce a race of tall soldiers by marrying his towering grenadiers to tall women. His death stopped the experiment. Catherine de Médicis took the opposite tack by setting out to breed a race of midgets. She did promote quite a number of midget-matches, but these, unfortunately (or perhaps fortunately), proved sterile, as such matches usually do. However, if she had chosen for her experiment the type of larger dwarfs known as *achondroplastic*—the ones with stubby limbs and relatively large heads—she could have produced as many as she wanted. (The abnormal individuals, and also the "pituitary" giants who would be useless for breeding, will be dealt with in a later chapter.)

Much more significant are the informal experiments which people themselves have carried out in choosing mates near their own size. In one Canadian family which went in for tall matings—started off by a man 6 feet 6 inches who married, in succession, three wives each 6 feet or over—there were thirteen male descendants, all between 6 feet and 6 feet 4 inches in height and ten females all between 5 feet 8 inches and 6 feet, with one exception. In the opposite direction, several Canadian families of "shorties" have produced successions of sons ranging between 5 feet 2 and 5 feet 4 inches.

These facts, coupled with much other evidence, have led to the conclusion that, conditions being equal, there are definite genes for tall stature and short stature, and that they may work in fairly simple ways. But how these genes operate when they are put together is not yet too clear. The prevailing theory is that the "tallness" genes tend to be recessive, the "shortness" genes dominant. What this means, specifically, is that *a tall person probably does not carry "shortness" genes, whereas a short person may quite often be carrying hidden "tallness" genes.*

Whatever an individual's "stature" genes, their effects can be greatly stepped up or retarded by environment. And we've needed no scientific experiment to prove this. Right under our very eyes in the United States we have been observing, during the past few decades, one of the most remarkable biological phenomena in human history: an increase in stature as well as various other changes in bodily development, more marked than any which in previous epochs came only through many centuries of slow evolution. In little more than three or four decades average heights of males have gone up over 2 inches, with the largest gains occurring in many of the groups of foreign stock. In innumerable families over the country there are sons taller by 4 inches or more than their fathers or any male relatives of previous generations.

This upshooting of our youth has made six-footers commonplace, and even seven-footers a familiar sight in basketball games.* Moreover, as practical proof of the physical changes, the whole range of clothing sizes for children and adults has been markedly increased, with "tall girls" and "big men" shops springing up everywhere; higher desks are being installed in schools; and longer beds are offered for sale. The phenomenon is not confined to the United States, for in most European countries, and also in Japan and Hawaii, people have been growing bigger. When we look back into history, we find that the redoubtable ancient Greeks and the brave knights of old were veritable shrimps compared with their modern descendants.

What has caused this change? We know there can hardly have been any sudden wholesale mutations in our genes, so the most obvious reason for at least a good part of the increase must be the great improvements in diet, living conditions, hygiene and medical care. Some scientists also think that our growth genes may have been speeded up further by changes in world climatic conditions, and even by cosmic rays or other still unknown meteorological effects. (These refer only to natural phenomena, for it should be kept in mind that the growth changes began long before any atom-bomb radiation was unloosed.) In any event, environmental factors, whatever they may be, cannot obscure the importance of hereditary differences as well. For, as anthropologists point out, some of the tallest statures in the United States are found in some of the most backward regions, such as the Ozark mountain areas, where the poorest nutrition and the worst living conditions prevail. Again, some of the tallest basketball players and athletes are Negroes who had been reared in unfavorable environments. (One might suspect that their ancestors included Watusis or other towering Africans.)

Where the current growth changes are most marked is among the children. Studies made in Boston and Milwaukee, for instance, show that as compared with boys of the 1880's nine-year-olds today are almost 4 inches taller and fourteen-year-olds, almost 5 inches taller. In

* A "profile" on Robert J. Cousy, a professional basketball star, in the *New Yorker* Magazine in 1961, referred to the fact that ". . . Cousy is virtually a dwarf among those in the trade because he stands only 6 foot 1½ inches in his sweat socks." The article pointed out that in compensation for his "shortness," Cousy has huge hands, unusually large arms, and remarkable vitality. However, the much greater ease with which basketball's seven-footers can jump up and literally "dunk" or stuff the ball into the basket, is bolstering the argument that the present 10-foot-high basket rim—set long ago in a time of much shorter players—should be raised by 1 foot or more.

Eugene, Oregon, a recent crop of fifteen-year-old boys averaged 5 feet 8 inches in height—½ inch taller than the average for full-grown World War I American soldiers. But part of this big difference related to another phenomenon: the fact that children grow at a faster rate than they once did, and mature sooner physically, so that at any given age before puberty they are more advanced in relative growth than children of previous generations. By the same token, though, their growth slowdown and stoppage comes earlier. Thus, where at age fifteen a boy today might be 5 inches taller than his great-grandfather was at the same age, his adult height might be no more than 2 or 3 inches greater.

But when we compare mature heights, past and present, there is no doubt about a considerable stature increase. American inductees of World War II averaged about two-thirds of an inch taller than those of World War I, who, in turn, were about an inch taller than their fathers. At most American colleges now, women and men students are two or more inches taller, as well as considerably heavier (this especially), than their predecessors fifty years ago. (In the Yale University class of 1967, the average freshman on entering—at the age of 18½—was 5 feet 10⅓ inches tall and weighed 160 pounds.)

Yet marked differences in stature still persist among various groups—classified by race, occupation, social and economic level—and it is not always easy to determine how far these are caused by environmental influences and how far by genetic factors. For example, children of Italian-American stock are shorter than children of Scandinavian stock. But these two groups differ not only in heredity, but in living conditions, diet and habitat (with more Italians concentrated in Eastern city areas and more Scandinavians in Midwestern farming areas). However, when Italian and Finnish boys living in much the same circumstances in Hibbing, Minnesota (where most of their fathers were miners) were compared, the Finnish boys still averaged considerably taller.

On the social and economic basis, American boys whose fathers are in the upper-level groups average 1 inch taller than sons of laborers. But there also may be ethnic or racial differences between those in the two groups which may be acting together with environmental factors; for by and large more Americans in the favored groups come from the taller European stocks, while those in the less favored groups come more often from the shorter stocks of Eastern or Southern Europe, or latterly, from Puerto Rico. At the same time, racial differences in stature, as between Negroes and Whites, or Whites and Mongolians, must also be viewed in the light of environmental differences. A case in point is that of Japanese stature.

While the new generation in Japan is taller than previous ones, Japanese young people who have grown up in the United States are taller than their cousins in the Orient. However, the young Japanese-Americans in California are still not as tall as the White children in the same state. Also of interest is the fact that while the sons of the Emperor Hirohito of Japan may be considered as having had the best of diet and medical care, the Crown Prince is barely 5 feet 5, and his brother is under 5 feet 3. Remembering that their royal parents also are very short (by American standards), one might guess that heredity was a major influence here.*

Granted that external environments may direct human stature upward or downward by a certain number of inches, it would nonetheless appear that in a given period and place, large differences in stature between persons of different stocks, or between members of a given family, may be dictated by hereditary factors. These, however, as suggested before, refer not only to specific "stature" genes, but to the general makeup of the individual (as, for instance, one's sex), and to a variety of genes that may be involved in the growth process. For example, individual differences in the rhythm or rate of growth may be in considerable part genetic, even though such factors as diet and disease may also affect growth. Thus, some children who grow much faster than the average during one period may grow less than the average during another period. There also is great normal variability in the time at which puberty may arrive for either a boy or a girl. Some are "early bloomers," some "late bloomers," and one girl may menstruate for the first time at nine, another at sixteen, without there being need for alarm in any of these cases. So, too, one boy's time clock is geared to make him shoot up much earlier than another. While pathological conditions may sometimes cause deviations from the averages for puberty onset, heredity may also play a part in this.

But with regard to young people in general, as we have previously observed, the environmental effects on growth rates are clearly seen in the earlier maturation which has accompanied the increase in stature. Among American girls today the onset of puberty (or menstruation)

* The possibility that the Japanese may carry a majority of "short-stature" genes, but a considerable proportion of hidden "tall-stature" genes, is suggested by the fact that many towering Japanese wrestlers have resulted from generations of breeding for size among families whose men were in this profession.

has been arriving at an average age of about thirteen years—perhaps a year sooner than it occurred among their mothers or grandmothers, and two years earlier than among girls a hundred years ago. As further evidence that better environments bring earlier blooming, girls from the more favored groups (now as in the past) have been achieving puberty about a year before those from underprivileged groups. Here we may pause to quash a popular but persistent fallacy that Negro girls mature earlier than White girls, or that girls in India, or among primitive peoples in the tropics, mature earlier than those in Europe and the United States. The facts are quite the reverse. It is the White girls, and those in the more advanced countries, who have been achieving puberty earlier, for the same simple reason that crops which receive better nutrition and care come up earlier than do less well-cared for ones.

Also important with respect to growth and stature among different groups and peoples is the fact that heights may not be achieved in the same way. Some "growth" genes affect body development uniformly, whereas others specialize with respect to certain segments—limbs, torso, neck, etc. For example, in the very tall African Negroes (and in some of our strapping basketball stars), a good part of the height derives from storklike legs. Again, among some short strains, such as the Japanese, heights may be cut down sharply by stubby legs and short necks, although torso lengths may be average. In certain families there are special characteristics in the sizes and shapes of chests, necks, legs, hips, any or all of which may be gene-influenced; and where body segments of different kinds are combined, there can be considerable variation in height among members of a family even though their key "stature" genes may be the same.*

When we take into consideration, then, all of the many genetic and environmental factors we've discussed, it is evident why the question of stature inheritance is extremely complex. Nonetheless, on the basis of a number of authoritative studies, these general conclusions are justified:

> The "tallness" genes are most likely to be recessive, the "shortness" genes to be dominant. Thus, any two normally tall parents are probably both carrying "tallness" genes and can usually count on having *all* tall children. (As in the Franklin D. Roosevelt family and the royal family of Sweden.)

* When parents are broad-chested, their children are likely to develop faster and be taller and heavier—during the growing period, at least—than offspring of narrow-chested parents, according to a 1960 study made at the Fels Institute.

Short parents may be of several genetic types: (1) Where both are of consistently short ancestral stock and probably carry predominantly "short-stature" genes, in which case all their children will probably be short; (2) where both parents are of ancestry with mixed "stature" genes, in which case they may have children of various heights, grading up to quite tall; (3) where parents are short not because of their genes, but because their growth was suppressed by adverse environments, in which case all their children will tend to be taller than they.

Where one parent is tall and the other (particularly the father) is short, there is a greater probability that any given offspring will be short rather than tall. But from what has been brought out about the trend toward increased stature in recent years, there is more and more likelihood of short parents having children considerably taller than themselves, especially if the parents were foreign-born or had been reared under adverse conditions.

In estimating the future height of any given child, one should keep certain reservations in mind. First, remembering that there are individual patterns of growth, one should not be too hasty about concluding that tallness or shortness lies ahead merely on the basis of the child's stature at any age before puberty.* Another pertinent fact is that stature may increase in young men up to the twenty-fifth year and in women up to the twenty-third year (or, in occasional cases, even later). Also, regarding parent-child comparisons or forecasts when the parents are middle-aged or older, allowance should be made for the fact that after full height has been reached there may be a gradual decrease at the rate of from one-twentieth to one-thirtieth of an inch annually. (In part this is due to a shrinkage of the cartilage discs in the spinal column, in part to "hunching up," or stooping.) Thus, when you near sixty, you may have lost an inch or more of your maximum height, and when you near eighty, two or more inches.

Now one more question:

How far can human stature continue to increase? If you're visualizing future humans as story-book giants, scientists pooh-pooh the notion. Some authorities believe that, at best, in two hundred years the *average* for American males might reach 6 feet. But a restraining fact is that if human stature goes too far, it may be at the expense

* That children of tall parents either mature earlier or grow more during their growth spurt, or both, than those of short parents was suggested by growth studies at the Institute of Human Development (University of California). It is also considered probable that the amount of growth during the adolescent spurt is to some degree genetically determined.

of fitness. (There already is some evidence that American young people are not as strong as European children, or as Japanese, of their ages.) To quote anthropologist Harry Shapiro (himself well over 6 feet), "Man is a more efficient organism if he isn't excessively tall, so it's hardly likely he'll ever reach dinosauric proportions." If man *did*, it might lead to extinction, for there is evidence that too marked a speed-up of growth in animals tends to shorten their lives. We do know that the few human giants who attained heights of 8 feet or over have all been pretty weak and defective individuals and have died in their twenties or early thirties. (There's a dim suspicion now that Goliath in the Bible may have been one of these "rubber" giants, and a push-over for David.) In any case, no change in outward conditions could carry our heights much further than the potential limits fixed by our genes.

THE FORM DIVINE, OR OTHERWISE. Here's where we *really* have a problem in establishing heredity, for much more than in the case of stature, innumerable environmental influences are all entangled with gene-workings in shaping human physiques. How unpredictable people's shapes may be is apparent when we not infrequently see Miss Americas and movie beauties spring from parents with contours like well-filled potato sacks. Moreover, while stature is virtually set after puberty, you yourself know (alas!) how inconstant may be your own figure at any stage of life.

Despite this, attempts have been made to classify human shapes according to basic types (the emphasis being rather more on the underlying structure than on the upholstery). One of the first and most widely used classifications was that devised by Dr. E. Kretschmer:

The *pyknic*—short and fat, thick neck, protruding abdomen, barrel-shaped thorax.

The *leptosome*—elongated, tall and slender, with long legs, hands and feet, long face, narrow chest.

The *athletic*—the intermediate type, with broad, square shoulders, muscular limbs, large hands and feet.

The *dysplastic*—the "mixed-trait" type, having parts of one type, parts of another, which characterizes many young people.

(Another and more detailed system of classification, devised by Dr. William H. Sheldon, will be dealt with and illustrated in Chapter 37.)

An assumption going with somatotyping is that while environmental factors may modify or alter human shapes, persons as a rule tend by inheritance to be of one or another of the body types described, or of some gradation in between. Moreover, there has been the theory that the leptosome (or asthenic) type is characteristic of the Nordics, the pyknic type of the Alpine peoples, and the athletic type of the Dinaric peoples (southeastern Europeans). This becomes rather dubious when we see that all the types are liberally represented in all races and nationalities, and that where a particular body form does seem to characterize the members of one large group, similarities in environment may often explain these differences just as easily as might genetic factors.

However, when the classification is applied to persons living under the same conditions, there does seem to be evidence that body types, or degrees of slenderness or obesity, tend to run in given families through genetic influences. This is quite consistent with what one finds in different strains of domestic and experimental animals. As noted by Dr. Jean Mayer, Harvard nutritionist, "Farmers have recognized and utilized the genetic determinants of obesity for thousands of years. When animal fat is desired, certain strains predisposed to adiposity are selected in preference to other strains." Plumper pullets and porkers and fatter steers have been bred in this way. In laboratory animals, obese strains of mice have now been developed which grow enormously fat on the same diets that keep other mice slim. Genetic obesity is also found in the Shetland sheep dog; nor need one point to the wide varieties of shapes that have been achieved in breeding the numerous dog strains.

With respect to human obesity, psychological factors and overeating have been stressed as the principal causes. Nonetheless, it is apparent that, under the same conditions, some people put on weight much more readily than others, and studies have indicated that tendencies in this direction run in families. Among Massachusetts high school students, it was found that less than 10 per cent of the children of normal-weight parents were obese, but the proportion rose to 40 per cent if one parent was obese, and to 80 per cent if both parents were. Another study, of Iowa girls, by Drs. Lee G. Burchinal and Ercel S. Eppright (1959), showed that those who were "fatties" as a rule tended to have overweight parents, suggesting that constitutional factors, and/or family eating habits, might be more involved in causing juvenile obesity than emotional disturbances. (But even inordinate appetites may be traced in some cases to genetic conditions.)

However, the theory that overplump children eat too much and scrawny ones not enough has been challenged by Dr. Penelope S.

Peckos, Boston nutritionist. She found that when children (aged six to fourteen) of different physical types were compared with their food intake, the fatties as a group were actually eating less than the skinny ones. While in some cases overeating may be the cause of obesity, or undereating of leanness, Dr. Peckos concluded that, in general, the factors in fatness are (1) inherited differences in the way food is utilized and converted into fat, (2) differences in activity, and (3) differences in how body heat is lost or conserved. Further supporting the case for inherited tendencies have been other findings that identical twins tend to be much more alike with respect to obesity or slenderness than are fraternal twins.

Various theories have been offered as to how obesity is inherited. As one possibility, there may be some gene action which reduces the amount of food required per pound of gain and increases the proportion of gain that is fatty tissue. Normally, there is a balance between the intake of food and the amount of it that is utilized by the body. But in constitutionally fat people the body's regulating devices may be out of order, causing a good part of the food to be converted into fatty tissue. Glandular disorders (sometimes present at birth) may also be involved in many cases of obesity. Whatever the processes, it has been suggested that in some families slenderness may be caused mainly by recessive genes, and obesity mainly by dominant genes. Thus, as in tall and short stature (and allowing always for environmental influences), slender parents as a rule have slender children, whereas fat parents—presumably those with hidden "slenderness" genes—while tending to have children who are fat, also, as a rule, may have others who are slender.*

Again, as with stature, predictions for children's body form must take into consideration important changes that have been occurring in recent generations. Children today can be expected to be not only taller, but also heavier at each age, on the average, as compared with those of previous periods. At the same time, there have been changes in body proportions. For one thing, sons and daughters tend to have relatively broader shoulders, narrower hips and longer limbs than the parents of their sex. As judged by present standards, there also is no question that

* Just in case you think *you* are fat—or for others interested: The fattest human being on record was a Florida woman, Mrs. Ruth Pontico, who weighed 772 pounds, with a height of 5 feet 5½ inches. She was a circus fat lady like her mother who had weighed 720 pounds. Mrs. Pontico, who died in 1942, aged thirty-eight, had started her career by weighing 16 pounds at birth and 50 pounds at the age of one.

most sons and daughters have better figures than their parents had at their ages.

But with every individual of any generation, what was and is most important in determining shapes and physiques—as any bright child of three can discern—is the person's sex. For Nature is ever mindful of this in constructing and upholstering bodies. First, there are the general differences, with the male skeletal and muscular structures larger and heavier throughout. Then there are all the differences in the fleshy formations, such as the breasts, and those in the hip region, thighs, legs, etc., with the female tending to have more fatty tissue everywhere. Not only are there genes governing all the detailed sex characteristics, but the degrees in which they are accentuated may also be influenced by heredity. As is the case with lower animals, the "sex" genes in one human strain may be more potent than in another and may send the sexes further apart in their development.

For example, there are certain African tribes where the females have quite small breasts and narrow hips, bringing them closer to the male form, while in other tribes (as well as in most White strains) the females have much larger breasts and hips, differentiating them much more from their males. Shapes of breasts may also vary considerably as a result of genetic factors. But in both sizes and shapes of breasts, environmental influences, as well, may have important effects. (There is some indication that American young women are tending to develop larger breasts than those of their mothers, but this awaits confirmation.) However, while living habits, occupations, the extent of childbearing, etc., may help to exaggerate or reduce the basic bodily sex differences, degrees of civilization do not have too much to do with this, inasmuch as primitive peoples the world over have every range of sex differentiation. If anything, civilization seems to have reduced the differences, for anthropologists report that in all the prehistoric races the sex differences in body structure and size were much more marked than they are in modern men and women.

In establishing the sex of skeletons, the bone experts go principally by these clues: the hip structure of females is wider, shallower, smoother and more forward tilted than that of males; while in males the bones of the limbs are relatively longer, thicker and heavier, and the "knobs" are bigger. Also the chests of males are larger, the shoulders broader; their hands and feet are relatively longer, and the finger and toe bones are heavier and blunter. There are also various differences in the skulls (as we noted in the last chapter).

One interesting *average* sex difference (for it doesn't always apply)

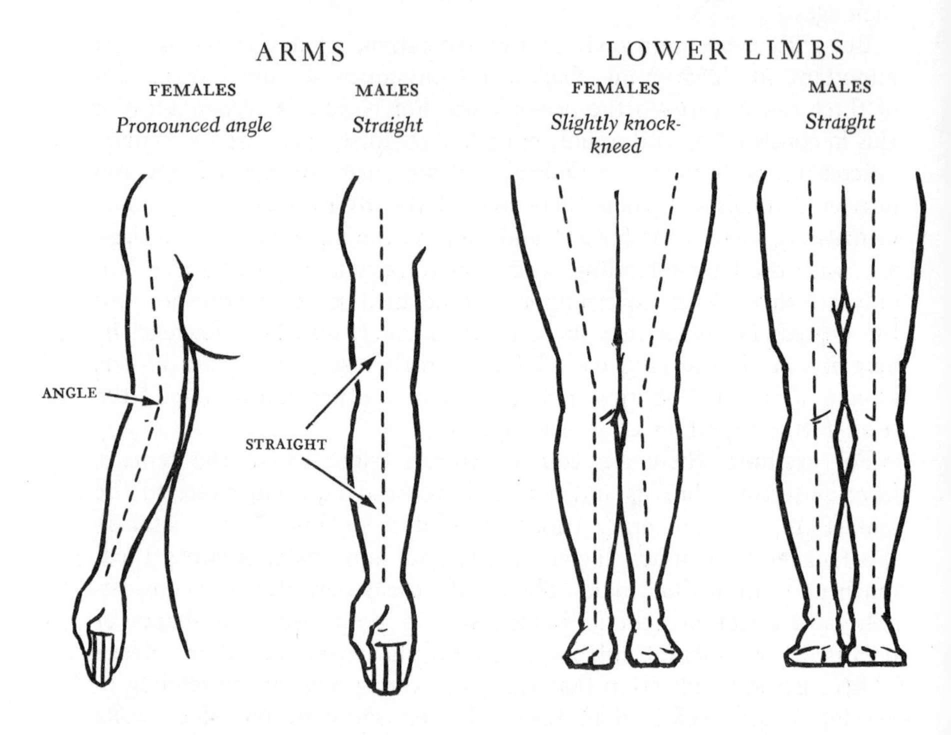

is in the *angle* at which the upper and lower arms are set together at the elbow, and the thigh and leg at the knee. You can test this for yourself by standing alongside a person of the opposite sex (with both of you preferably in bathing suits), in front of a full-length mirror. When you hold your arms straight down with palms facing front (as shown in the illustration), you will find as a rule that the female arm has a much more definite bend at the elbow whereas the male arm is much straighter. Similarly, when the thighs are placed together, you will see that the female tends to be much more knock-kneed.

Another difference in body contour is in what has been called "the more pronounced southern exposure" of the female, due both to the greater angle made by her pelvis and the incurve of the back, and to the proportionately greater amount of fat and muscle distributed there. The degree of this protuberance can be determined by heredity, and where it occurs in an extreme form is known as *steatopygia*. This condition is highly prevalent among women of Hottentot and Bushman tribes (where it is greatly admired) and is occasionally found among American Negro women.

In noting all the many special sex characteristics in body form, it must always be kept in mind that any genes involved in producing them can be carried equally by both sexes (except insofar as males receive only one set of X-chromosome genes to the females' two). This means that inheritance of any particular type of female breast, hips or other sex detail can come as easily through the father as through the mother, and any special male characteristic can be transmitted through the mother as well as the father. Thus, in many cases, especially where genes in the father are dominant, a girl may develop a figure much more like that of her father's female relatives than that of her own mother; and a young man may have body-form traits much more like those of his mother's father or brother than those of his own father. All of this may be evident to breeders of livestock or poultry, who know that desired traits in prize cows or bulls may be passed on through parents of either sex, or that super-"egg-laying" genes of a prize biddy may be passed on through her rooster son to his daughters.

It is the difference in the male and female *internal environments*—chiefly the genetically produced glandular differences—which causes the same genes for sex characteristics to produce contrasting results. So another point is that, in either sex, degrees of physical "masculinity" or "femininity"—or any particular sex characteristic—may be modified in their development by the workings of the individual's sex glands. Where something happens in an individual to upset radically the hormonal balance proper for his or her sex—particularly if this occurs before puberty—feminization of the male's figure, or masculinization of the female's figure, may result, with, in either case, an approach toward the "neuter" type. (The abnormal cases will be dealt with more fully in Chapter 27.) Lesser shifts in the hormonal balance may lead only to some underdevelopment of the breasts or hips in females, or some overdevelopment of these parts in males. During old age, too, there is often some degree of masculinization or feminization in the physical sex characteristics.

Head to Toe. The shapes of heads may have little importance for average readers, except when they go into hat stores. But to anthropologists this has long been a fascinating subject for research, one thought being that head shapes might prove useful in the classification of races and subgroups. This objective has been somewhat dimmed by findings in recent years that in every race and in almost every large group of people there are many forms of heads and that to some extent head shapes may be modified by environment. Nonetheless, the most important details are determined by genes.

Roughly speaking, people are identified as round-headed or long-headed (as viewed from the top). Your hat-store man might say "broad oval" or "long oval." The scientist would say *brachycephalic* or *dolichocephalic*, and with very detailed measurements would give you your exact cranial index down to several decimal points. On a racial basis, the broad oval ("round") head is predominant among the Mongoloids, the long head being rare among them; while among Negroes it is the long head which is common, the round head rare. Whites, taken as a whole, have among them every type of head shape in about equal proportions, although in different stocks there are strong tendencies toward one type of head shape or another.

Quite a number of genes are involved in determining head shapes, some genes working toward roundness, some toward length, some toward size, and some toward special shaping. The complexity of the factors makes difficult any accurate predictions of what children's head shapes will be. But, in general, the genes for the round head tend to dominate those for the long head.

The inherited tendency toward one type of head shape or another asserts itself quite strongly from birth onward (although, with the same genes, boys' heads start off by being somewhat rounder than those of girls and end up by being somewhat longer). Among certain primitive peoples, heads are artificially shaped by various appliances, but in less marked ways head-shaping may be somewhat influenced in our own world by prenatal factors and later, by diet and habits of sleeping, eating and talking. The principal head-shape change in the United States has been a certain degree of lengthening which has come largely with the increase in stature.

Differences in *head sizes*, grading from fairly large to fairly small within the normal range, also result from an interplay of heredity and environment. Where head size has special importance is in extreme deviations from the average in young children, which may sometimes indicate glandular or mental abnormalities. However, Drs. Henry K. Silver and William C. Deamer, who worked out scales of head-size norms for babies, reported that in most cases what appears to be an abnormally large or small head may be simply a hereditary characteristic, and if the head continues to grow at a normal rate, there need be no cause for alarm. They also reported—which should be no surprise to mothers—that boy babies have somewhat larger heads than girl babies, averaging about ¼ inch more in circumference at birth and about ½ inch more at the age of two.

From the head down to the toes, there are innumerable other details governed by heredity—the size and shape of hips, chest, thighs, legs and arms, fingers and toes, muscle formations, and so on. Each of these characteristics may have social importance, where they are judged in terms of attractiveness. They may also have much practical importance, where they affect the capacities for various kinds of physical performance. It is with the latter thought that geneticists have given most of their attention so far to inherited defects or abnormalities in these structural details (to be discussed in Chapter 23). Otherwise, questions regarding the inheritance of beautiful legs or other points of interest to beauty judges (professional or amateur) must for the time being go unanswered.

We must emphasize again, however, that occupations and living habits can account for at least some of the difference in bodily details seen among individuals. If the son of a blacksmith has brawny arms like his father, it doesn't necessarily follow that these arms were "inherited," for they may just as well have been developed by working in the smithy. So, too, when we take note of certain physical characteristics common to families of tailors, policemen, farmers and miners, we must not confuse the results of similar working and living habits with those of heredity.

In your own case, then, you may find that many of your bodily characteristics which may have seemed an integral and inevitable part of you can likewise be ascribed to conditioning. On the other hand, the possibility that heredity is involved becomes of interest when you try to guess whether these traits will also appear in your children. This is what we're about to help you find out.

CHAPTER FIFTEEN

What Will Your Child Look Like?

We have gone far enough in identifying genes linked with various characteristics so that, given certain facts about you and your mate, we can make some fairly accurate predictions as to what your children may look like.

Were we able to breed people as the geneticist breeds flies, we could make many more predictions, with greater accuracy. By constant breeding and inbreeding, geneticists have established strains of Drosophila, ranged in rows of bottles in their laboratories, whose genes they know almost as well as the chemist knows the makeup of his various compounds. In fact, with almost the same precision that the chemist mixes compounds, the geneticist can "mix," by mating, two flies of any strains and predict the types of offspring that will result.

We cannot, of course, ever expect to do anything like that with human beings. Pure strains of humans cannot be produced, like flies (or like horses, dogs or cats), by long inbreeding of parents with children, brothers with sisters, etc. And where flies can have three hundred offspring at a time and more than two generations to a month, human couples do not average more than three or four offspring to a marriage and only three or four generations to a century.

So, genetically, in most respects we humans are unknown quantities. With regard to your own genes, you can only make guesses, but in this you will be helped considerably not merely by the characteristics which you yourself reveal, but by those which appear in your parents, grandparents, brothers, sisters and other close relatives. As was noted in the chapter on eye color, if you are dark-eyed, the chances of your carrying a hidden "blue-eye" gene increase according to the number of your relatives who have blue eyes, and their closeness to you. Furthermore, if you marry a blue-eyed person and have a blue-eyed child, then you know definitely that you carry a "blue-eye" gene. On the other hand, if two, three or four children in a row are all dark-eyed, the presumption grows that you do not have a "blue-eye" gene.

Likewise, where both parents are dark-eyed, the appearance of a blue-eyed baby is quite good proof that both carry hidden "blue-eye" genes. But if all the children are dark-eyed, it still might mean only that *one* of the parents has no "blue-eye" gene.

These qualifications hold for every case where persons have some characteristic due to a *dominant* gene (dark hair, curly or kinky hair, a Hapsburg lip, etc.) and wish to know what chance they have of carrying a hidden gene which might produce a different trait in their child.

But before we try to make any predictions these facts should be clear:

All forecasts of the types of children people will have are based on *averages determined by the laws of chance* (or, more technically, the "laws of probability").

Wherever *dominant* and *recessive* genes are involved, it is like tossing up coins with heads and tails. Toss up coins long enough, and the number of heads and tails will come out even. So if you are carrying one dominant and one recessive gene for any characteristic, were it possible for you to have an unlimited number of children, you'd find that *exactly half* would get the dominant, half the recessive gene.

With *two* parents involved, the results will be like those obtained in matching coins. This, of course, conforms with Mendel's laws.

When we think in terms of the *characteristic* produced, the expected result in mixed matings will be that the dominant characteristic (dark eyes, dark hair, etc.) will show up *three out of four times*, the recessive only *one in four* because it requires a *matching* of the recessive genes.

Of course, where one parent carries *two* dominant genes, all the children will show the dominant trait. Where one parent carries a dominant and a recessive, and the other parent *two* recessives, *half* the children, on the average, will show the dominant trait, half the recessive.

But here is something else to bear in mind:

Wherever it is a question of a child's getting one gene or another, or having such and such a characteristic, *the odds for every child are exactly the same.*

THE LAWS OF CHANCE, OR PROBABILITY

If you toss a coin with two different sides, one "head," one "tail"—

The odds are exactly even that it will land

"Heads"
1 in 2 times

"Tails"
1 in 2 times

Similarly, if you carry a mixed pair of any gene, one "dominant," one "recessive"—

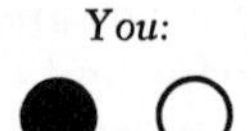

When you mate, the odds are exactly even that any child will receive

The Dominant Gene
1 in 2 times

The Recessive Gene
1 in 2 times

If you and another person each toss a coin

The odds are exactly

1 in 4 times Both Heads

1 in 4 times Both Tails

2 in 4 times One "Heads" One "Tails"

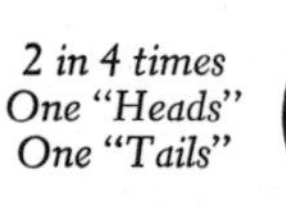

Similarly, if you and your mate each carry a mixed pair of genes for some trait

You:

Your Mate:

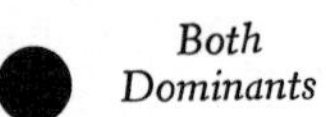

With every child the odds are exactly

1 in 4 times Child will receive — *Both Dominants*

1 in 4 times Child will receive — *Both Recessives*

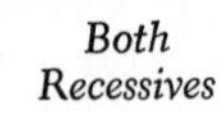

2 in 4 times Child will receive — *One Dominant One Recessive*

No matter how many times coins are tossed, the odds will always be exactly the same for the next toss.

No matter how many children you have or how many "in a row" are the same type, the odds will be exactly the same for the next.

Some gamblers might dispute this, but if you toss up a coin one time and it comes up heads, that does not mean that the next time there is any better chance of its coming up tails. *There is the same fifty-fifty chance on each tossup.* Even, if through an unusual "run," there would be ten heads in succession, on the eleventh toss there would still be an exactly even chance for either heads or tails. (This applies to dice, roulette or any other game of chance. Many a gambling addict has lost a fortune trying to disprove it.)

So, let us say, if the odds are even for your having a blue-eyed child, and your first one is brown-eyed, that doesn't mean that the odds are any better that the next will be blue-eyed. Even if four or five children in a row are born with brown eyes, there is still that same fifty-fifty chance, no more and no less, that the next child will have either brown or blue eyes.

But perhaps we need not have gone into all this. With regard to an expected baby's sex, we noted (Chapter 7) that the chances are almost equal for the arrival of a boy or a girl. Yet, as can be seen in many families, there may be a succession of four, five or more boys (or girls), with no reason why this cannot be due purely to chance. Nor does the fact that one has had four or five boys in a row in any way better the odds that the next baby will be a girl, or vice versa.

In the boy or girl question, it is a simple matter of one or the other. But in the case of features or form—in fact, of any detail of the body—we must contend with innumerable variations, often involving many complex gene combinations, plus, always, the possible influences of environment. Thus, the degree of precision in forecasting any given trait depends, first, on how directly it is produced by heredity; second, on how many genes are involved in producing it; and, third, on whether any of these genes are "eccentric" ones which do not always work in the same way.

By these principles, you will find that the forecasts for eye and hair color, and for hair form, will come closer to the mark than those for the shapes of nose, ears, eyes and mouth, or for stature or body build. On the whole, however, if you and your mate conform to the averages, the forecasts here presented will prove fairly dependable. Where there seem to be exceptions, these need in no way represent upsets to genetic theories, for the expert might easily be able to account for them.

Anyway, do make allowances for the exceptions and, whatever happens, don't blame us (or the geneticists on whose studies the following Tables are based) if the baby does not turn out the way the forecast indicates.

And now to Sir Oracle!

HOW TO USE THESE CHILD-FORECAST TABLES

1. If this is to be your first child, find out as much as possible about what genes you and your mate may be carrying by consulting the detailed treatments of each feature in preceding chapters and by studying other members of your families.* Make allowances for all characteristics influenced by environment.

2. If you have already had one or more children, study each child for additional clues about your genes.

3. Remember that no matter how many children you have had, or what they look like, the odds that your next child will receive a given characteristic are exactly the same as if it had been the first.

4. In consulting the tables, look for your own characteristic in either of the "parent's" columns. (They each apply equally to father or mother.) If you and your mate are of different types, look first for the type most pronounced—the darkest coloring, the most extreme hair form, etc.

5. Remember that these "forecasts" are based on averages in large numbers of matings. Any one child may be an exception.

6. Wherever age is a factor, make due allowances for its future effects or changes that may be expected to take place. (This applies particularly to hair color, hair form, eye color and nose shapes.)

7. Under no circumstances should doubts be raised about a child's parentage purely on the basis of this or that unexpected physical trait—eye or hair color, a type of feature, etc. Many other factors must be taken into consideration. (For parentage tests, see Chapter 45.)

* In the following pages "family" refers not only to parents, brothers and sisters, but also to grandparents and other close relatives.

HOW TWO HOMELY PARENTS MAY HAVE A BEAUTIFUL CHILD

FATHER

Bald

Murky-green eyes; lashes lost through disease

Misshapen mouth due to bad teeth

Bad nose due to accident

MOTHER

Dull, dark straight hair

Dull-brown eyes; drooping eyelids

Bad skin (local disorder)

Protruding underlip

BUT *they may carry and pass on to their child hidden genes for . . .*

Blond, curly hair

Blue eyes

Long lashes

Pretty nose

Cupid's-bow mouth

Lovely complexion

Result: A beauty contest winner

HOW TWO HANDSOME PARENTS MAY HAVE A HOMELY CHILD

FATHER

Curly, black hair

Large, dark eyes, long lashes

Well-shaped mouth and chin

MOTHER

Wavy, blond hair

Blue eyes, long lashes

Regular teeth, pretty mouth

BUT *they may carry and pass on to their child hidden genes for . . .*

Dull-brown, straight hair

Murky-green, small eyes with short lashes

Protruding jaw and teeth

Other irregularities, added to by environmental factors

Result: An ugly duckling

EYE-COLOR FORECAST

IF EYES OF ONE PARENT ARE:	IF EYES OF OTHER PARENT ARE:	CHILD'S EYES WILL BE:
BROWN (or *BLACK*)		
Type 1. If all this parent's family were dark eyed	× *No Matter What Color*	Almost certainly dark
Type 2. Where some in this parent's family have lighter-colored eyes (gray, green or blue)	× *Brown,* Type 2	Probably *brown,* but possibly some other color
	× *Gray, Green* or *Blue*	Even chance *brown* or lighter color (most likely like that of lighter-eyed parent)
GRAY or GREEN	× *Gray, Green* or *Blue*	Probably *gray* or *green,* but possibly *blue.* (Rarely brown)
BLUE	× *Blue*	Almost certainly *blue,* especially if both parents are light-blue-eyed. But note exceptions in Text, p. 77.
ALBINO (*Colorless*)	× *Normal*-eyed parent of any eye color	*Normal,* leaning to shade of normal parent's eyes, *unless* this parent carries hidden "albino" gene, when 1 in 2 chance of child's being albino
	× *Albino*	Definitely *albino*

EYE-SHAPE FORECAST

Width: If just one parent has wide eyes, child is quite likely to have them, too; if both parents, the chance is much greater.

Slant: If one parent has slant eyes (but not of Chinese type), child will not be likely to have them unless slant eyes also appear in the family of other parent. If, however, a parent's eyes are of the Chinese, or Mongolian (epicanthic-fold) type, there is great likelihood the child will have them.

Lashes: Where just one parent has very long lashes, child has a more than fifty-fifty chance of inheriting them.

HAIR-COLOR FORECAST*

IF ONE PARENT'S HAIR COLOR IS:	IF OTHER PARENT'S HAIR COLOR IS:	CHILD'S HAIR COLOR WILL BE:
DARK (*BROWN* or *BLACK*)		
Type 1. Where all in this parent's family had dark hair	× *No Matter What*	Almost certainly *dark*
Type 2. Where there are lighter shades among others in this parent's family	× *Dark*, Type 2	Probably *dark*, but possibly some lighter shade
	× *Red*	About equal chance (a) *dark* or (b) *red-brown* or *red*, with (c) some slight possibility of *blond*
	× *Blond*	Probably *dark*, but possibly *blond*—rarely *red*
RED	× *Red*	Most probably *red*, and occasionally light brown or blond
	× *Blond*	Even chances (a) *red* or (b) *light brown* or *blond*
BLOND		
Type 1. If medium shade	× *Blond*	Fairly certain *blond* or, rarely, *brown*. (*Red* possibly if this shade is present in either parent's family)
Type 2. If flaxen or white	× *Blond*—Flaxen or white	Certainly *blond*, but with shade of darker parent apt to prevail

* In forecasting hair color of a child, remember to be guided by what color hair you and your mate had originally, in your earlier years. Refer to Chapter 11.

HAIR-FORM FORECAST

IF ONE PARENT'S HAIR IS:	IF OTHER PARENT'S HAIR IS:	CHILD'S HAIR WILL BE:
CURLY		
Type 1. If all in this parent's family are curly-haired	× *Any Form,* except kinky or woolly	Almost certainly *curly* (rarely any other)
Type 2. If some wavy or straight hair in this parent's family	× *Curly,* Type 2	Probably *curly,* possibly *wavy* or *straight*
	× *Wavy*	Even chance (a) *curly* or (b) possibly *wavy* or occasionally *straight*
	× *Straight*	Probably *curly* or *wavy,* possibly *straight*
WAVY		
Type 1. If no straight-haired persons in this parent's family	× *Wavy* or *Straight*	Most probably *wavy,* seldom *straight*
Type 2. If there are some with straight hair in this parent's family	× *Straight*	Even chance *wavy* or *straight* (rarely anything else)
*STRAIGHT**	× *Straight*	Almost certainly *straight*
KINKY		
Type 1. Where all in this parent's family are kinky-haired	× *No Matter What Hair Form*	Almost certainly *kinky*
Type 2. Where other hair forms appear in this parent's family	× *Curly* or *Wavy*	Even chance (a) *kinky* or (b) *curly* or *wavy;* rarely *straight*
	× *Straight*	Almost same as above, but with greater possibility of *straight*

Woolly: While fairly frequent among Negroes, it is rare among Whites. Where, however, it appears in even one parent, there is a fifty-fifty chance that any child will have woolly hair.

* If the parent with straight hair is Mongolian (Chinese, Japanese, full-blooded Indian, etc.), *all* the children will probably be straight-haired, regardless of the hair form of other parent.

FORECAST OF FACIAL DETAILS

NOSE

(Nose shape is not inherited as a unit. Different characteristics of the nose may be inherited separately, from either or both parents. Environmental factors also have great influence.)

Generally: Where both parents have about the same type of nose, a child in maturity will have a similar type.

But: If just one parent has a *pronounced* type of nose—very *broad,* or very *long,* or very *high and narrow-bridged,* or *hooked,* or *pug-shaped,* or *bulbous-tipped,* etc.—while the other parent has a *moderate* nose, there is a greater than even chance that the child's nose will eventually resemble that of the parent with the more extreme type. This applies especially where a marked nose form has appeared in several generations of a parent's family.

EARS

Large: If just one parent has large ears (very long, or very wide), the child will be quite likely to have similar ears.

Cup-shaped: If in one parent, about a fifty-fifty chance for the child.

Affixed Lobes: If only one parent has *affixed* ear lobes, or *absence* of lobes, and the condition does not appear in the other parent's family, there is not too much likelihood that the child's ear lobes will be of these types.

MOUTH

Lips: If just one parent has *thick* lips, the child will probably have them, or at least thicker than average lips.

Hapsburg (protruding) lip and jaw: If one parent has this, child has an even chance of inheriting it.

Chin: If one parent has a receding chin, a child is not likely to have it unless a similar-type chin has appeared in the family of the other parent. This is also true if only one parent has a very narrow or pointed chin. But if one parent has a dimpled or cleft chin, there is quite a strong chance a child will have the same.

Dimples in cheek: If one parent is dimpled, the child is very likely to be, also.

Teeth: If either parent has unusually shaped teeth, there is a considerable chance they will appear in a child. (Inheritance of normal teeth shapes hasn't yet been worked out, but forecasts for dental abnormalities are given in the Appendix, in the Wayward Gene Tables.)

FORECAST OF BODY BUILD

STATURE

Both parents tall: The child in maturity will almost certainly be tall, or taller than average.

Both parents short: The child will probably be inclined to shortness, but may possibly be taller than the parents, and even very tall.

One parent tall, one short: The child will probably incline toward the *shorter* parent.

(Where parents are short, make every allowance for the possibility that their full growth potentials may have been impeded by environmental factors.)

BUILD

Both parents slender: The child will be more likely to be the same than if both parents are fleshy. But build is a highly variable characteristic, dependent on so many conditions and genes that it can hardly be predicted.

(For the inheritance of "abnormal" conditions and characteristics of all kinds, in features, form and appearance, see later chapters.)

CHAPTER SIXTEEN

What Makes Us Tick?

YOU HAVE SEEN what produces your external appearance. But you are much more than a hollow doll with such and such kind of eyes, hair, skin, etc. While your looks may be extremely important, your real importance as an individual lies in what is within your shell: your organs—brain, nerves, heart, lungs, glands and other functional parts. These are "what make you tick" and they are what account for the greatest differences between individuals.

In fashioning and constructing every one of our organs, our genes are constantly at work, and we know that individual differences in these organs are often inherited. But the task of identifying them is vastly more complicated than it was in the case of features, for we are here dealing not with easily recognizable characteristics but with *functions and effects*. In that regard, mere appearances are of very little help to us, for in very few cases have we yet been able to establish by mere surface inspection the nature of the important organs and their hereditary aspects.

The glands, for instance, form a group of organs which hold special interest because almost every peculiarity in humans is being ascribed to them these days. When people talk of "glands," they do not mean such old standbys as the liver and kidneys, or the gastric and salivary glands. They refer to the "ductless" (endocrine) glands—the pituitary, thyroid, parathyroids, pancreas (one part), adrenals, suprarenals, pineal, thymus, and the testes or ovaries. These introduce into the blood certain all-powerful substances called "hormones," the effects of which are often confused with the direct action of genes. (See Chart, p. 179.)

Comparing our glands with those of lower animals, we can see quite clearly that general glandular construction and functioning are determined by heredity; and also, that in given cases of individuals or of humans of different strains, a vast number of glandular characteristics show distinct hereditary influences, such as rates of growth and development, age and onset of menstruation, of puberty and of change of life, and of many continuing processes of the body. But glandular differences among individuals may also be conditioned by environment. It is only, as a rule, when there are marked abnormalities in the *effects* of the glands, repeatedly following the same pattern, that the hereditary differences may reveal themselves.

Then there is the brain, the most important single organ in the body. We can again assume that if the enormous differences between the human brain and those of lower animals are inherited, as we know, then smaller differences or variations in brain structure between one human being and another may also be inherited. Thus, Professor K. S. Lashley of Harvard has said, "Even the limited evidence at hand shows that individuals start life with brains differing enormously in structure . . . unlike in number, size and arrangement of neurons as well as in grosser features. Some of the more conspicuous structural differences can be shown to be hereditary, and there can be little question that some of the lesser variants in cerebral structure are likewise hereditary."

So, too, it may be possible to show that differences in form and structure of other organs—heart, liver, lungs, stomach, etc.—are inherited, as we can prove with lower animals. But little has been done as yet to identify and classify any of these inherited differences in *normal* humans—not even as between such distinct types as tall, White American intellectuals and primitive, pygmy Hottentots. Only in the case of easily recognized *abnormalities*, or defects in functioning, have we much knowledge regarding inherited individual differences in given organs, or clues as to the specific genes involved.

"Normal" and "abnormal," by the way, are vague words wholly inadequate to express what we mean. "Abnormal" means "not normal"; but "normal" cannot be defined except in relation to some standard that in itself is usually highly variable. For instance, if a man eats 3 pounds of meat at a sitting, we'd say he has an "abnormal" appetite; but suppose that man were 7 feet tall and weighed 300 pounds? An appetite abnormal for others would be normal for him. Four feet six inches would be an abnormal height for a man in northern Scotland and 5 feet 8 would be normal; but 5 feet 8 inches would be an abnormal height among Pygmies, whereas 4 feet 6 would be normal. In other words, an abnormality is a deviation from some arbitrary standard which may vary according to the point of view. It should not be confused with a defect, for an abnormality may be favorable or unfavorable. An idiot is abnormal; so also is a genius. Also, even what is considered a defect may sometimes be so only in one time and place, but not in another. (All of this will be dealt with in much greater detail presently.)

But the question which looms largest with regard to human differences in what makes us tick is the one which inevitably pops up, like a heckler, in any discussion of genetics:

"Which is more important, heredity or environment?"

We have tried to bring out that both forces go hand in hand in shaping anyone's life and that consideration of one without the other is impossible. You will understand, therefore, why the geneticist counters with, "Which is more important, the fish or the water in which it swims?"

However, when a person says, "I don't believe in heredity. I believe in environment!" he is being downright silly.

Perhaps you are familiar with this old chestnut: "What do elephants have that no other animals have?" The answer is "Baby elephants"—not trunks!

Obviously, where humans are also most unique is in their having human babies. Even more obviously, this is not because of anything special in their environments which is not accorded to other animals. No amount of similarity in environments, prenatal or otherwise, could cause a chimpanzee to develop into a human being (although a man may sometimes make a monkey out of himself).

First and foremost, what causes us to develop as human beings, different from all other animals, is heredity. Environment can only enable our hereditary potentialities to assert themselves, or can modify or prevent their development. But once our genes are allowed to go their course, there is not too much that environment can do to alter the general *physical* directions taken by members of our species. Under infinitely varied conditions, all the way from cave men down to modern men, the physical and organic makeup of human beings has remained remarkably constant, with their similarities far overshadowing their differences.

Thus, what we really have in mind when we argue about "heredity versus environment" is always the relatively small differences between one person and another within special areas. As biological mechanisms, which all human beings are primarily (regardless of what spiritual qualities they may have), it is apparent by now that they are constructed by their genes in an infinite variety of ways, with many minute shades of difference. Our problem is to find out how important are these differences with respect to specific functions—especially those which cause one individual to be much more efficient and successful in life than another. For it is only when we concentrate on specific points that we can begin to identify the relative influences of the two forces of heredity and environment.

For instance, during the World Wars of our time, millions of young men were killed in battle, and vast numbers of other young people died in bombings or were exterminated in Nazi concentration camps. Which factor was more important in bringing on early death, heredity or environment? The question is easily answered.

On the other hand, there are conditions due to simple recessive genes which will cause a baby to die before it is three years old, no matter what care is given it. Which is more important here, heredity or environment? That, too, can be answered easily.

Many physical defects in people—facial disfigurements, bodily distortions, crippling conditions, loss of eyes—can be attributed solely to accidents or other environmental mishaps. Many others, as you will presently be reading, can be blamed entirely on defective heredity. In such clear-cut cases we are quite justified in speaking of *either* heredity *or* environment.

But still other conditions, both advantageous and disadvantageous, can be seen as resulting from special hereditary factors in combination with special environments. And, finally, when we begin to deal with the composite of many traits and circumstances—with the *sum total* of any person's life—the constant interplay of heredity and environment becomes increasingly apparent. Here truly, as with the fish and the water, you cannot ever disassociate your own biological mechanism, or that of anyone else, from the surroundings in which it operates, and from all the forces which control or influence its workings throughout its existence.

So, in seeking to find out why people tick the way they do, we are required, first, to concentrate on one trait or function at a time (which is how old Mendel proceeded). And, second, to make sure that with respect to this trait or function individuals either have *exactly the same heredity,* in which case we can attribute any differences to environment, or that they have *exactly the same environment,* in which case we can attribute their differences to heredity.

But with some individuals we can go even further. Instead of applying these principles to only a few of their traits, we can apply them to their total makeup, and their lives as a whole. For, in these individuals, to whom we will now turn, Nature has provided teammates who can offer innumerable fascinating comparisons.

CHAPTER SEVENTEEN

Twins and Supertwins: I

The Types, and How Produced

EVERYWHERE IN THE WORLD, and as far back as records can show, twins have been objects of special interest. Mythology and folklore have surrounded them with auras of mystery, superstition and innumerable fanciful notions, beliefs and assumptions, valid or otherwise. But regardless of one's knowledge or views, there are few persons so blase that they will not turn for a second look—and a smile, chuckle or comment —when twin babies are wheeled by.

What concerns us here, however, is not the emotions aroused by twins, but their scientific importance. For to the geneticist, medical man and psychologist, twins offer bountiful material with which to study many of the most significant aspects of human heredity. In turn, the scientific findings have helped parents of twins to better understand these children, twins to better understand themselves, and people in general to acquire greater insight into the relative effects of heredity and environment on their own lives.

Haven't you sometimes thought, "How would I—or someone else with my same heredity—have turned out under different conditions? Or, under the same conditions, but with a somewhat different heredity, what sort of person would have resulted?"

That could be answered—or at least, partly answered—in this way:

1. If there were *two* of you to start with and each were exposed to different conditions; or

2. If you started life with somebody else at the same time *within the same mother*, and after you were both born, developed under approximately the same conditions.

Is either of these situations possible?

Yes, for Nature has most thoughtfully provided us with *twins*, who, willy-nilly, are ideal subjects for such experiments.

For the first experiment we have the so-called "identical" twins; for the second, "fraternal" twins. The two types differ in these respects:

Identical twins are the products of a single fertilized egg, which, after it begins to grow, gives rise to two individuals. Each has exactly the same hereditary factors as the other, so identical twins are always of the same sex, among other things, and usually—though not invariably—tend to look very much alike.

Fraternal twins, on the other hand, are the products of *two entirely different eggs* which happen to have been matured simultaneously by the same mother and then fertilized at approximately the same time by *two entirely different sperms*. They therefore may carry quite different assortments of genes and need be no more alike than any other two nontwin children of the same family—as often as not, in fact, being of opposite sex.

In other words, identical twins are, from the standpoint of heredity, virtually the *same individual in duplicate*.

Fraternal twins are genetically *two different individuals* who merely through chance were born together.

The important distinction between the two types of twins has not always been fully recognized. In former times, twins were considered identical if they were of the same sex and resembled each other fairly closely. Even when the one-egg and two-egg distinctions became known, mistakes in diagnosing them were frequently made (as they still are being made), because of erroneous assumptions about how the twins of each type developed prenatally. One persistent belief has been that identical twins always share a single fetal bag (*chorion*) and one placenta, whereas fraternal twins always have entirely separate fetal bags and placentas. But as is now certain, this is far from always true. Sometimes identical twins may grow apart with fully separate placentas; and fraternal twins, in almost half the cases, may have placentas and fetal bags so fused as to look like one.

To elaborate, when identical twins are formed by the splitting of a fertilized egg (or of the embryonic cell mass growing from it), the two resulting individuals usually develop within the same outer bag and share the same placenta, although they are encased in different inner sacs (*amnions*). But in about one in four cases when a fertilized egg divides to form identical twins, there is a separation of the two halves, so that the twin embryos move into the womb independently and become implanted at different (but usually adjoining) points. They can

HOW TWINS ARE PRODUCED

"IDENTICAL" TWINS

Are products of a single sperm *and a* single egg.

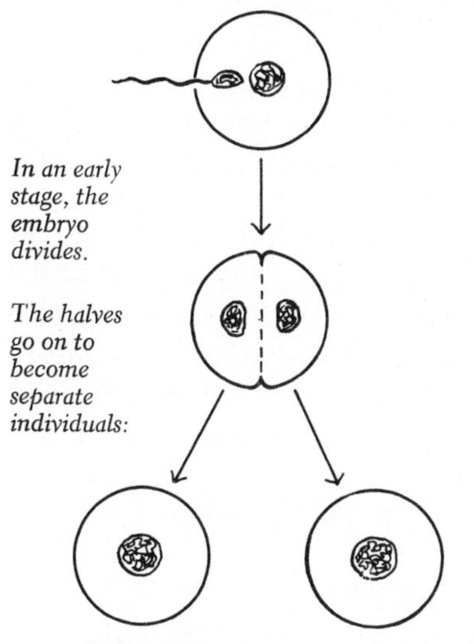

In an early stage, the embryo divides.

The halves go on to become separate individuals:

*Having come from the same sperm and the same egg, these two twins carry the same chromosomes and genes:**

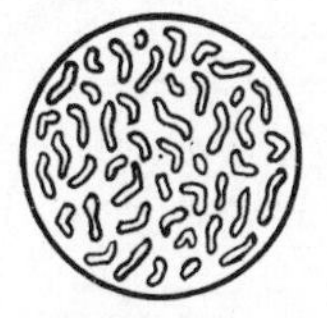
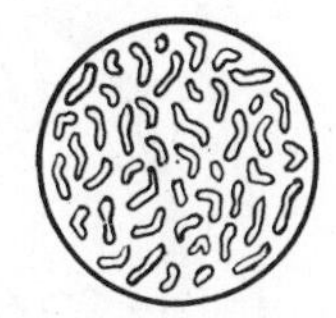

*Thus, identical (one-egg) twins are always of the same sex**

. . . Either two "look-alike" boys

. . . or two "look-alike" girls

* For exceptions, see Text.

"FRATERNAL" TWINS

Are products of two different *eggs, fertilized by* two different *sperms.*

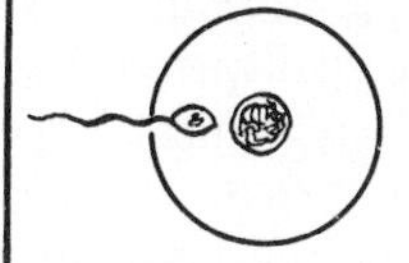
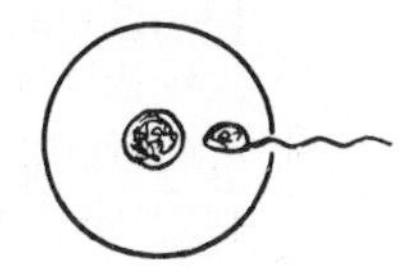

They go on to develop into two different individuals:

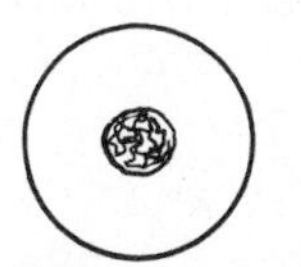
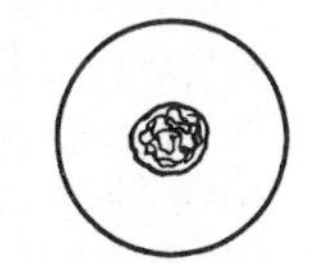

They carry different assortments of chromosomes and genes:

Thus, apart from having been born together, fraternal twins are no more alike than any two children born separately in a family. They may be:

Both of the SAME *sex*

Two boys

. . . or two girls

Or a MIXED *set*

. . . a boy and a girl

then develop with different chorions and placentas (much as might fraternal twins from two different eggs); and although the placentas may fuse, it may often be impossible from the surface evidence at birth to tell what types of twins they were. (Only tests made later can establish this, as we shall presently see.)

Nevertheless, regardless of whether identical twins arise from an early division and share the same placenta, or arise from a later division of the embryonic cell mass and have separate placentas, they are genetically just as identical, since all of their cells carry duplicates of the same genes that were in the single fertilized egg from which both developed. The evidence for this may become more and more apparent as they grow. For if anything else were needed to prove not only that they carry the same genes, but how closely these genes (or any genes) ordinarily follow the precise and intricate paths laid out by their blue-

HOW TWINS USUALLY DEVELOP

(See Text for other ways)

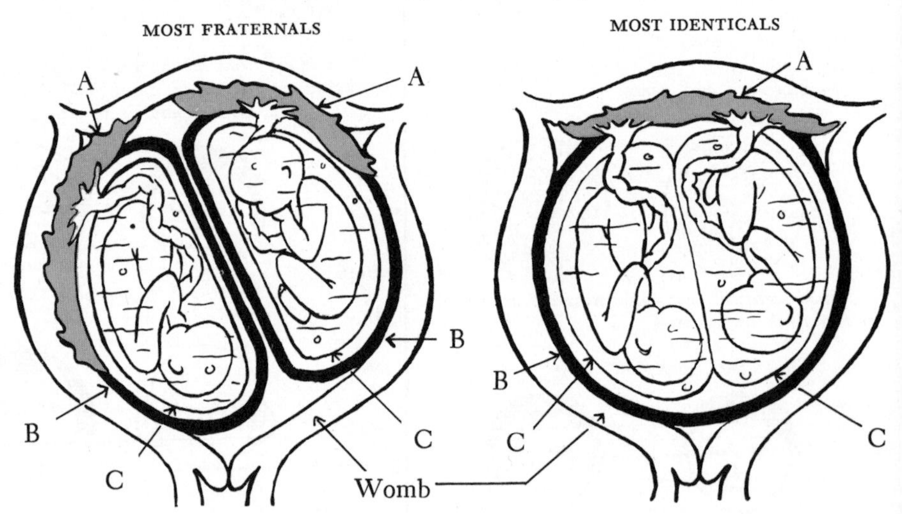

A. *Separate placentas*
B. *Separate outer bags (chorions)*
C. *Separate inner sacs (amnions)*

A. One *common placenta*
B. One *common outer bag*
C. *Separate inner sacs*

prints, it can be seen in the amazing way in which most identical twins resemble each other, down to many of the tiniest details, and often throughout their lives.

Yet it need not follow that identical twins must inevitably be identical in appearance, constitution, health or behavior. They often are not, because differential influences of environment, from the beginning of their development, are always important. Prenatally one twin might be at a slight disadvantage in receiving nutrition, or might suffer some accident or upset in development, so that the other twin might be born bigger or healthier. Following birth other environmental situations may result in some handicap for one of the twins, and this may be carried on throughout life and produce other differences, physical and psychological. To the extent, then, that the conditions and experiences of two identical twins may have been or continue to be dissimilar at any stage, differences of various kinds and degrees may appear. In extreme cases, as when one of an identical pair is stunted, wizened or warped by some deformity, disease or accident and the other is not, the two might look less alike than do many fraternal twins.* Thus the term "identical," while still popularly useful in distinguishing twin types, must be understood to mean only *identical in heredity*, and not necessarily in appearance. That is why geneticists prefer to speak of one-egg and two-egg twins, or, in technical terms, *monozygotic* (MZ for short) and *dizygotic* (DZ for short)—from *mono* for one, *di* for two, and *zygote*, the fertilized egg.

As for fraternal twins of the same sex, the degree in which they may or may not differ from each other, and be recognized as *not* identical, depends on the extent of the differences both in their genetic makeup and in their environments. Just as with any two nontwin children of the same parents, fraternal twins, in terms of averages, could be expected to have 50 per cent of their genes in common—or, conversely, to differ in half of their genes. But it could easily happen that in the dealing out of parental genes, two fraternal twins might receive much less than the average quotas of duplicates and be very dissimilar. Another pair of fraternals might receive much above the expected average quotas of matching genes and look almost like identicals. (This might be so, especially, if the parents themselves were closely related, or were otherwise genetically similar in many traits.) But the chance that any two fraternal twins would be as completely alike in genetic makeup as identicals are is fantastically remote (one in tril-

* For even more startling differences possible in one-egg twins, see page 149, and Chapter 20, page 194.

lions). Nor would the "togetherness" of fraternal twins in the womb necessarily heighten their chances of being more alike than two separately born siblings. Quite often development at the same time could increase the differentiation between twins (fraternals or identicals) because of prenatal competition for the food supply and the added risk of an accident to one or the other.

Thus, with some identical twins looking less alike and some fraternal twins of the same sex looking more alike than one would expect, it can be seen how confusion as to twin types may often exist. Sometimes the parents themselves can resolve their doubts by checking on all of the surface traits discussed in preceding chapters which are known to be hereditary: coloring of eyes, hair and skin; hair form; and basic details of features. Any significant differences in these respects would stamp the twins as fraternal, although similarities would not necessarily prove them identical. But in a good many cases there may be enough uncertainty about the type of a twin pair so that professional investigation or testing is needed.

Blood tests are usually the most effective means of establishing twin types. These will be dealt with fully in Chapter 28, but at this point it can be noted that many different blood elements are determined by heredity, and any differences between twins in any one of these will prove them fraternal. But, again, similarities need not prove them identical.* Some experts also find value in a study of the fingerprints, palm prints, and foot-sole prints of a twin pair. The patterns are often very much alike in identicals (and rarely so in fraternals). However, the fingerprints of identical twins are *never exactly the same,* as the FBI has proved in a number of cases. (See illustration for one example.) †

If all the combined criteria and twin-type tests so far mentioned are still inconclusive (which happens in rare cases), the *skin-graft* test offers a sure method of proving that twins are or are not identical. Be-

* A new "twin-typing" blood test being developed involves putting together in a test tube the lymphocytes—white blood cells—from two twins. If the twins are fraternal, the dissimilar lymphocytes will stimulate each other to begin enlarging, dividing and multiplying. If the twins are identical, there will be no such reactions.

† In a recent study reported by Dr. Irving I. Gottesman of Harvard, the fingerprints of sixty-eight pairs of twins of the same sex, not indentified as to type, were shown to an FBI expert (John Douthit). He was able to pick out correctly through the prints 88 per cent of the twin pairs who were identicals, and 68 per cent of those who were fraternals. This would offer corroboration that while most identical twins have quite similar fingerprints, some do not, and some may have as much difference in their fingerprints as do many fraternals.

IDENTICAL TWINS–DIFFERENT FINGERPRINTS

TWIN NO. 1

TWIN NO. 2

(Enlarged prints of left index fingers of each)

Contradicting reports that two identical twin ROTC cadets from Texas had identical fingerprints, the FBI provided an analysis proving that the prints were definitely not the same (as they never are in identical twins, nor in any other two persons). The prints showed these differences in classification:

Twin No. 1: $\frac{\text{11 O 25 W } \textit{IMO} \text{ 16}}{\text{S 27 W 00I}}$ Twin No. 2: $\frac{\text{16 O 25 W I00 18}}{\text{M 27 W I00}}$

The key of twin No. 1 is 11, that of Twin No. 2, 16. Twin No. 1 has a whorl-type pattern with an outer *tracing; in Twin No. 2 it has an* inner *tracing. Other differences appear in the ridge details and whorl shapes. The fact that one print is darker than the other is not significant, being due merely to different inking impressions (Data, FBI Law Enforcement Bulletin, February, 1953.)*

cause identical twins are completely alike in the hereditary makeup of all their tissues and body chemicals, it is possible for the skin or flesh from one twin to be grafted onto or into the other, with assurance that the graft will "take"—just as if it were from another part of the same person. But when attempted with fraternal twins, the skin graft will slough off before long, unless there is very special prior treatment. (Further details about transplants in twins, including those involving whole organs, such as kidneys, are given in Chapter 19.)

A dramatic example of the use of twin-type tests was the case of the "mixed-up" twins in a Swiss village some years ago. On the same night, in the little village hospital, twin boys were born to one mother, and a third boy was born to another mother. The parents who were rearing the "twins"—a pair quite different in appearance—found out when the two began attending school that there was another boy there who looked remarkably like one of the twins. The suspicion grew that there had been a mixup at the hospital. A team of experts was called in, and by thoroughly examining the three boys and making blood and skin-graft tests, they proved conclusively that the "outsider" boy was indeed the true twin, while the supposed other twin was the child of a different mother.*

Before leaving the "twin-typing" tests, we might mention two kinds of evidence, *mirror-imaging* and *opposite-handedness*, which were formerly considered of much more value in helping to establish identical twinship than they are now. What is most interesting about these traits is that they involve not similarities between identical twins—as are stressed in other tests—but *differences*. In mirror-imaging some mark or bodily detail may appear on the right side of one twin and on the left side of the other (as in a mirrored reflection of the same person). This quite frequently occurs in identical twins with respect to the hair whorl at the top of the head (in one, clockwise; in the other, counterclockwise); in some quirk in the teeth, an ear, or another feature; in the site of a birthmark; or in fingerprints, or palm or footprint

* How did the reunited twins and the third boy fare thereafter? The author has been kept in touch with their story by two of the three Swiss experts who handled this case, Drs. A. Franceschetti and D. Klein (the third was Dr. P. Namatter). For a while it was very difficult for the true twins to adjust to each other and for the two exchanged boys to adjust to their new homes and new parents. The "ex-twin" continued to see his former "twin," and the couple who had previously been his parents continued in their affection for him. But in time the paths of the true twins and that of the unrelated boy diverged. (One reason may well have been that the twins were raised in much more favored circumstances, their parents being quite prosperous and educated, while the other boy was reared in underprivileged conditions, off and on by his widowed mother and by relatives.) The twins went on to college (studying for a year in the United States) and to professional careers. The "ex-twin," whom they see occasionally, is a postman in their native Swiss village.

patterns. Internally, in rare cases, some organ in one of the twins, such as the heart or stomach, is in a reverse position from the normal and from that of the other twin. While mirror-imaging in one or more characteristics may appear in about 30 per cent of all identical pairs, the traits involved are often so minor as to be detectable only by specialists; and since the transposition may sometimes be seen equally in fraternal twins or in two nontwin children of a family, its importance in establishing identical twinship is very limited.

Opposite-handedness—one twin left-handed, the other right-handed—has been even more discounted now as a special characteristic of identical twins. A recent thorough study by Dr. Torsted Husen, of handedness among all twins recruited for Swedish army service during a five-year period, showed that the proportion of twins who were left-handed (6.2 per cent) was only slightly greater than the proportion among singleton draftees (4.8 per cent).* Further, the frequency of left-handedness among fraternal twins was almost the same as among identicals. In another study, among American twins, Dr. Arthur Falek found that the proportion of "lefties" in either identical or fraternal twin pairs was not sufficiently greater than in singletons to justify the belief that handedness is significantly related to twinship.

Whatever the frequency, how can either opposite-handedness or mirror-imaging occur in identical twins when they have the same genes? For one explanation we must go back to the initial point at which a single egg or embryo divides to form two individuals. If the separation between the two halves takes place in the very earliest embryonic stage, before any body differentiation has begun, twins may be as identical as possible. But if the division takes place later, when the potential two sides of the body already have begun to be laid out, the twin from one side might develop differently in a number of ways from the opposite twin. Or, with respect to mirror-imaging and handedness, just as the cut halves of an apple may show the same details on opposite sides, the halves of a human embryo may also show certain physical characteristics on opposite sides.†

* Estimates of the incidence of left-handedness vary considerably in different populations, having been reported as high as 10 per cent, or more, in some groups. Among army men in the United States the incidence of "lefties" has been given as 8.6 per cent.

† However, it is also characteristic of many inherited physical anomalies and defects that a given trait may sometimes appear only on one side (either right or left) of an individual, or sometimes on both sides, or it may not

"MIRROR-IMAGING" IN IDENTICAL TWINS

Various physical traits—inherited or otherwise—may sometimes appear on opposite sides of the twins.

1. Hair whorl *at top center of head: Counterclockwise in one, clockwise in the other.*
2. Larger eye *on right of one, on left of the other.*
3. Mouth *tilted up, and crooked* tooth, *on right side of one, on left side of the other.* (*Mirror-imaging may also occur in other features.*)
4. Birthmark *on right shoulder of one, left shoulder of other.*
5. Handedness: *One right-handed, one left-handed (but this occurs in only a minority of identical twins).*

show at all (even though the person is carrying the gene or genes for it). This could happen in the case of two identical twins—just as with two non-twin siblings or with other individuals carrying the same gene—so that the two might or might not show the trait in different ways.

One unfortunate oddity in the production of identical twins can occur when there is an incomplete separation of the halves of the original embryo. This may result in the development of *Siamese twins* (so-called after the two most famous ones, born in Siam, who were circus attractions a century ago). Depending on the stage and place at which they failed to separate, such twins may be joined at birth at any of various points and in many degrees—at their heads, chest, hips, sides, buttocks—and there have been cases of some with two heads and one body, or with one head and a double set of limbs and organs. Few of the most abnormal Siamese twins achieve birth or survive long thereafter; but among those whose attachment is not too deep-rooted or extensive, operations now can often produce successful separation.*

To the freak situations may be added the seemingly incredible ones where two babies born to a mother at the same time are really *not twins,* or two babies born at different times *are* twins. As one possibility, a mother may be carrying two babies who were conceived days or even weeks apart. In most actual cases of this kind the mother is found to have a two-compartment womb. Also, in a few instances, babies that were twins who developed at different rates (and perhaps one in a Fallopian tube, the other in the womb) have been born days, weeks or more than a month apart. Not least of the phenomena have been fraternal twins with different fathers—a possibility if a mother, after producing two eggs—has successive relationships with two different men in a matter of hours. In one such case twins born to an amiable Chicago widow were proudly claimed by each of two boarders as being "his." A court contest led to blood tests which proved that one man had sired one of the twins, his fellow boarder the other. (Readers should be cautioned that "two-father" twins are exceedingly rare and that the mere fact of twins looking very different, and one not resembling a husband, is no cause for suspicion. As stressed in previous chapters, unusual combinations of genes from both parents may result in a child who has little resemblance to either parent or to other siblings.)

Next, we turn to the multiple births which in general arouse the most attention, those of the *supertwins*—triplets, quadruplets and quintuplets. Having seen how two-at-a-time children are produced and what

* Another rare oddity in identical twinning, recently discovered, is that in which one-egg twins, while theoretically "identical" genetically, might yet not be carrying precisely matching chromosomes and genes and might even be of *opposite sex*—or one might be a mongoloid idiot, the other normal. (This is discussed in Chapter 20, page 194.) If the hypothesis were stretched to the furthest limits, it would not be impossible for this to occur in Siamese twins under some fantastic combination of circumstances.

the distinctions are between the identical and fraternal types, we can readily understand the basic facts about the three, four or five-at-a-time babies. In most respects, they are "twins—only more so" ("or much more so").

With supertwins, as with twins, one can have an identical set, all arising from a single fertilized egg, or a fraternal set, arising from different fertilized eggs. But there also may be combinations of the two types in one set. Here, briefly, is how the different types of supertwin sets may be produced:

With identical triplets or quadruplets, the first stage is the same as in the production of identical twins—the splitting of a fertilized egg (or the twinning of the embryonic cell mass). But now one of the halves may split, or twin again, to form two individuals, while the other half "stays put"—and identical triplet embryos will thus result. If both twin halves split or twin again, to form two individuals each, the result will be an identical quadruplet set. Two such sets, the Harris quads of Chicago and the Axe quads of Lima, Ohio, were born in the United States in 1963. (We'll deal with quintuplets later.)

With *all-fraternal* triplet or quadruplet sets, there is simply (or should one say, "simply"?) the requirement that the mother produce either three or four eggs at a time and that each egg be fertilized by a separate sperm and then go on to develop through to birth.

In the *mixed-triplets* sets, there are two separate fertilized eggs to start with; and one of these splits to form an identical pair, while the other egg produces a single individual who is a fraternal in relation to the two identicals. In the *mixed-quadruplet* sets, there are these possibilities: Starting with *two* fertilized eggs (1) each egg may split to form an identical twin pair, with the partners of each pair being fraternals in relation to the two members of the other pair; or (2) one of the two eggs may produce identical triplets (by the process previously mentioned), whereas the second egg may stay put, producing a fourth individual who is fraternal in relation to the other three, as they are to him or her. Another quadruplet possibility is that *three* eggs may be produced and fertilized with one of the eggs splitting again to form identical twins, the other two eggs staying put, so that the resulting foursome will consist of two identicals and two fraternals.

As to the sex of triplets or quadruplets, it is obvious that the all-identical sets must consist of all boys or all girls. But where there are all fraternal, or mixed-fraternal and identical sets, one may find various combinations of boys and girls.

And now to *quintuplets,* the stars of the twin and supertwin world.

HOW "SUPERTWINS" MAY BE PRODUCED

1. *Triplets*

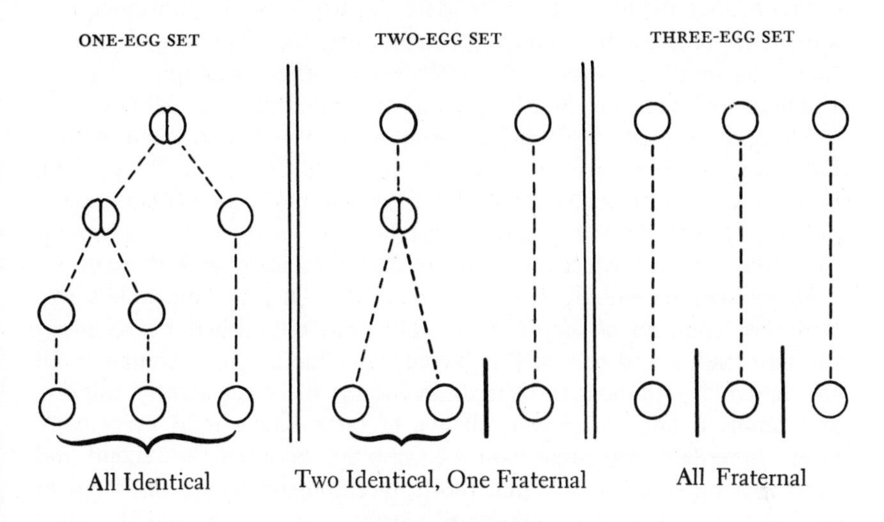

2. *Quadruplets*

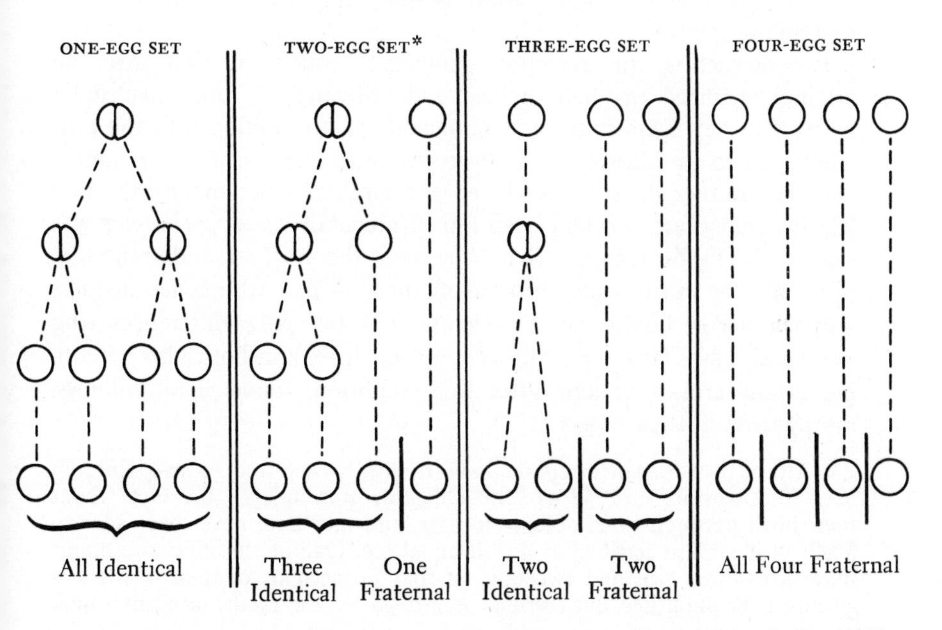

* A two-egg quadruplet set consisting of two identical pairs is also theoretically possible, but no such set has yet been found (see Text).

3. *Quintuplets*

Many possible types may be produced, ranging from all five identical (a one-egg set, such as the Dionnes) to—theoretically—all fraternal (five eggs), with various combinations of identicals and fraternals in two-egg, three-egg and four-egg sets, derived through steps much as in triplet and quadruplet formation.

We won't go into too much general detail about them because, in *authenticated* medical history to date, so few sets of quintuplets are known to have survived intact to birth: first, the *Dionne quintuplets,* then, almost thirty years later, in 1963, two other quintuplet sets, the Fischers, of Aberdeen, South Dakota, and the Prietos of Venezuela.*

Many readers will recall the sensation which was caused when the five Dionne girl babies arrived in a little town in Canada in May, 1934; how for days and months the world's attention focused on them as daily bulletins reported their progress; and how for years as they grew up their nursery was a Mecca for tourists streaming in to peek at them.

What was paticularly amazing about the Dionne quintuplets was that they were an *all-identical* set. This was established by scientists through studies and tests which showed that the five were similar in all the hereditary traits investigated, including blood elements, coloring and bodily details. Moreover, all five at birth had a mild (very common) inherited foot abnormality—"webbing" between the second and third toes on each foot. When tooth development began, the teeth of all five followed similar timing and patterns, and all showed the same mild kind of malocclusion (failure of the upper and lower teeth to fit together properly).

Reconstructing the evidence, geneticists concluded that after an original fertilized egg had divided and redivided to form quadruplet parts, three of these four parts remained set, to develop into the girls who were to be christened Yvonne, Annette and Cecile. The fourth part, in another division, produced the girls who became Emilie and Marie. That these two had been linked prenatally in a special way was suggested (1) by the fact that they were the smallest and frailest at birth; (2) by mirror-imaging between them in hair whorls, handedness and manner of holding an object; and (3) by ocular differences—the two being more far-sighted than the other three, and both also remaining mildly cross-eyed long after this condition, found in all infants, disappeared in their sisters.†

* Although newspaper reports have frequently referred to five children born in Argentina in 1943 as "the Diligenti quintuplets," the facts about their birth have remained obscure to date, and they have never been scientifically studied and certified as a quintuplet set. One of the mysteries about these five—three girls and two boys—is that no account of their births was given out or published until several months after the event, or events, took place.

† Going into more technical detail about the Dionnes, Dr. Norma Ford Walker has written to the author: "From evidence given by the French-Canadian midwives concerning the structure of the birth membrane (which

As for the Fischer quintuplets of South Dakota and the Prietos of Venezuela (all of whom reached their first birthdays intact as sets in September, 1964), the Fischers have been reported as a three-egg combination, of three identical girls and two fraternals—girl and boy. The Prieto quints, while all boys, are also a mixed set, but of what kind, genetically, has not yet been reported at this writing. In any case—and whatever other fivesomes may yet come along—it is highly doubtful that any multiple birth ever again could bring or carry with it the interest and excitement engendered by the Dionnes. It is not only because these were the *first* quintuplets to have survived—and may still be the only all-identical set to date—but that drama has continued to attend their lives into their mature years (as will be told in Chapter 37). Additional facts about the Dionnes, and about twins and supertwins in general, will be told in the next chapter, and in other parts of our book.

---

the midwives floated out in a tub of water), it is highly probable that there were only two amniotic sacs, each containing a single umbilical cord. One cord branched into two parts, the other into three. A single chorion surrounded the two amnions. This indicates that the early twinning occurred in the embryonic cell mass, after the chorion had formed. Each twinned mass was then surrounded by an amnion and within this protecting sac, one embryonic cell mass twinned again (producing Emilie and Marie), the other split into three (producing Yvonne, Annette and Cecile)."

CHAPTER EIGHTEEN

Twins and Supertwins: II

Incidences and Implications

HAVING LEARNED HOW TWINS and supertwins are produced, you may wonder now what practical meaning the facts about them may have for you.

If you are an expectant or prospective parent, you may think, "What are *my chances* of having twins?"

If you are a twin yourself, you may ask, "How can the scientific findings help me and my twin, specifically?"

Or if you have no personal connection with or interest in twins, you may still wonder, "Just what have the studies of twins contributed to the knowledge of human heredity, and perhaps to my own welfare?"

With regard to the first question—the chances of having twins—no doubt every mother, at the outset of pregnancy, has speculated on this possibility. For some mothers the possibility may seem much greater than for others, as indeed it is. But why and how? The answer lies in a variety and combination of influences.

Most important in a mother's twinning chances at any pregnancy are (1) her age and (2) the number of children she's previously borne. Taking age alone, the likelihood that twins will arrive is slimmest for mothers under twenty and greatest—rising with advancing age—for mothers in the thirty-five to forty-year-old group. (Thereafter the chances begin dipping again.) So, since the two influences—advancing age and more previous pregnancies—usually go together, the statistics show these extremes: For first-time mothers, aged fifteen to nineteen, the average chances of having twins are less than 5 in 1,000 births, or about 1 in 200. For fifth- or sixth-time mothers, aged thirty-five to thirty-nine, the twinning chances are at least four times as great—over 20 per

THE DIONNE QUINTUPLETS

First there were five—

The Dionne quintuplets on their first birthday, May 28, 1935. The birthday cakes show their respective names. By coincidence, Marie and Emilie (misspelled on her cake), who were paired in development—as told in the Text—are shown here side by side.

—Then there were four

The surviving Dionnes at the grave of Emilie, who died August 6, 1954, at age of twenty. Left to right, Marie, Yvonne, Cecile and Annette. (Details of the Dionnes are given in Chapters 17, 18 and 37.)

Photograph under direction of the author by M. Lasser.

A REMARKABLE "TWIN" FAMILY

This unusual and charming family group comprises the identical-twin Rubin brothers, married to identical-twin sisters, with one couple parents of identical-twin daughters, the other of a son born four days earlier. The twin couples have shared the same home on Long Island, New York, and raised their children as brother and sisters, *which genetically they are. For, while legally the girls and the boy are "double first cousins," the fact that their parents are genetic duplicates makes them biologically as much sisters and brother as any children of the same family. Actually, the children until they were eight years old did not differentiate between the two sets of parents and, while knowing which were their true parents, continued to refer impartially to either woman as "mother" and either man as "father."*

(Identification: Standing, rear, at left, Benjamin and Sylvia Rubin, parents of Carol and Linda, seated. Rear, right, Ruth and Hyman Rubin, parents of Edward, center.)

1,000 births, or about 1 in 50. The comparative twinning rates for mothers of all ages and pregnancy groups are given in our accompanying Chart. (Regarding the chances of having supertwins—not shown in the Chart—the influences of maternal age and previous childbearing are even more striking. Triplet births are six times as frequent, and quadruplet births twenty times as frequent, among mothers over thirty-five as among those under twenty.)

Another important fact is that it is chiefly the *fraternal*-twinning chances which are affected by the mother's age and previous childbearing, whereas the chances for having identical twins are only slightly affected. Thus, as our Chart shows further, when twin births are classified by type, the fraternal-twinning rate for first-time mothers under twenty is only 3 pairs per 1,000 births, but rises to 17 per 1,000 for fifth-time mothers aged thirty-five to thirty-nine. On the other hand, the incidence of identical-twin births moves up only slightly between the two maternal extremes, from 3.5 pairs per 1,000 births among mothers under twenty, to 4.5 identical pairs to mothers aged thirty-five to forty-five.

Not knowing the foregoing facts, if one were to hear it said, "The chances of women's bearing twins may be greatly affected by their education, religion, viewpoints, husband's income and social levels," the statement would seem preposterous. Yet if we stop to think that any or all of the factors mentioned have an influence on when a woman begins and stops having children, and on how many children she has, we can see—in the light of facts previously brought out—that these social factors can in turn affect her chances of bearing twins. So, too, in any particular group or country, social and cultural factors which lead to less childbearing by the older, more "twin-prone" mothers and increased childbearing among the younger, less "twin-prone" mothers, will reduce the incidence of twinning; and since fraternal twinning is chiefly affected, a decrease in the relative incidence of fraternals will mean a rise in the relative proportion of identicals among the twins born.

How this has actually worked out in the United States with the trend in recent decades toward earlier childbearing (going with earlier marriages) and fewer children at the later, most twin-prone maternal ages, is shown in the government's statistics on twins: In the quarter century from 1935 to 1960 the twinning incidences among Americans as a whole dropped by over 10 per cent—from 11.4 pairs per 1,000 births to 10.4 pairs (or from 1 twin pair in 88 births, to 1 in 98). Accompanying this drop—mainly in fraternal-twin births, as we said—the ratio of identicals among all twins born rose from 30.7 per cent to 37 per cent.

A MOTHER'S TWINNING CHANCES*

. . . And how they are influenced by her age and previous childbirths

An expectant or prospective mother should note where the vertical line for her age group is cut by the dotted line for previous childbirths, then look across to the number at the right. The chances the twins will be identical are shown in the shaded lower part of the graph.

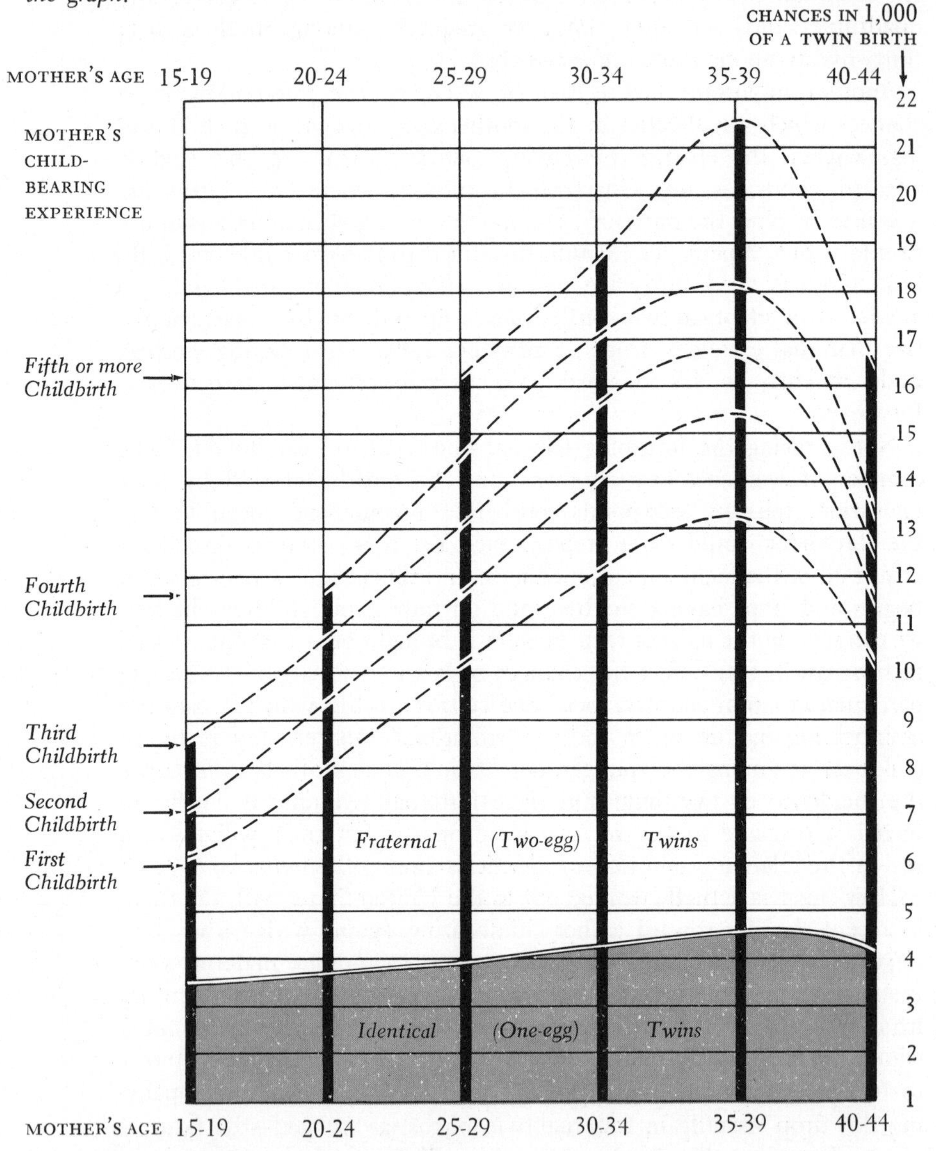

* Figures given are for White mothers. Among American Negro mothers twin-bearing chances are about one-third higher, and among Mongolian mothers (Japanese, Chinese) about one-third lower than those of Whites. The racial differences are mainly with respect to fraternal twins. (See Text.) Chart adapted from "Bio-social effects on twinning incidences," by Amram Scheinfeld and Joseph Schachter, *Proc. 2nd. Intl. Congr. Hum. Genetics,* Vol. 1, 1963.

(Still earlier in the century the identical-twin ratio was reported as only 25 per cent.) The triplet incidence in the United States also fell during the twenty-five-year period from 1 in 8,400 births, to 1 in 11,450 births.*

In Europe, likewise, social changes which have affected mothers' ages and childbearing rates have caused a drop in incidence of twinning in most countries. So, too, existing differences in reproductive patterns (relating to maternal ages and birth rates) might explain why twins appear so much more often in one country than another as shown in our Table on page 159. For example, in Ireland, which currently has the highest European twinning incidence—15 twins per 1,000 births—women marry at the latest ages (over twenty-six, on the average) and then have the highest birth rate in the most twin-prone categories. In Spain, at the other extreme, the low twinning rate (only 9 pairs per 1,000 births) has been ascribed largely to more childbearing by younger, less twin-prone mothers and by marked birth limitation in the older, twin-prone maternal brackets.

Apart from, or in addition to the other factors mentioned, a mother's *race* has a great influence on her twinning chances. On the average, Negroes have by far the highest incidence of twins, and Mongolians (Japanese, Chinese, American Indians), the lowest, with Whites in between. The racial twinning differences may be accentuated or diminished according to the makeup of given racial strains and the relative ages and childbearing rates of mothers. Thus, American Negroes, who are racially mixed to a considerable extent, and whose reproductive patterns and rates are not too greatly different from those of Whites, currently have a twinning rate about 35 to 40 per cent higher than the White rate, or, roughly, one twin pair in 72 to 75 births, compared with the White incidence of less than one in 100. But among African Negroes there may be up to one twin pair in 30 to 35 births and in one large tribal group, the Yorubas of Nigeria, the extraordinary rate of one twin pair in 22 or 23 births has been reported. Japanese have an incidence of little more than one twin pair in 160 births.†

* The facts here given about the changing incidence of twinning in the United States as related to social changes were brought out by the author, in collaboration with Joseph B. Schachter of the U.S. National Office of Vital Statistics, in a paper presented at the Second International Congress of Human Genetics, in Rome, 1961. A second paper, by the author, dealt with the chaning twinning incidences in Europe and with racial differences in twinning rates.

† The racial differences are even more marked with respect to supertwins. Negro mothers in the United States, in relation to their numbers, have

But what is most important about the race differences in twinning is that they relate almost entirely to the production of *fraternal* twins, for in all racial groups the incidence of identical twins is close to the same—varying (sometimes in any one group from time to time) from about 3.5 to 4.5 pairs per 1,000 births. However, the contrasts in fraternal-twinning rates are enormous. At the highest extreme, the Negro tribes in Africa referred to may have an incidence of up to 35 or even 40 fraternal pairs per 1,000 births; and at the lowest extreme, the Japanese are known to have scarcely more than two fraternal pairs in 1,000 births. But the theory that it is the smallness of Japanese women which accounts for their low twinning rate is challenged by the fact that they produce just as many identical twins as do Negro or White mothers.

Apparently, then, the marked racial differences in fraternal-twinning rates appear to be due largely to hereditary factors. This brings us to another possibility: Does heredity also influence the individual twin-bearing chances (of having identical as well as fraternal twins) for mothers within any racial or other group? Actually, there always has been a strong belief that the tendency to twinning "runs in families." Offering support for this are family trees fairly studded with twins, in comparison with other families where no twins have been known for generations. Cited further are the records of individual mothers who have produced as many as six or seven pairs of twins each, or several pairs of twins plus a few sets of triplets and quadruplets. It would seem hard to ascribe these cases purely to statistical coincidence. What is also known is that among various species of lower animals which normally produce only one offspring at a time—cattle, sheep and goats, for instance—twins occur regularly in some strains much more frequently than in others; and in one species of armadillo (the nine-banded), identical quadruplet sets are the rule.

But finding proof of hereditary factors in human twinning, and identifying them, has been difficult. For one thing, since identical and fraternal twins are produced in entirely different ways, the factors influencing their production, whether hereditary or environmental, or a combination of both, must be different for each type. In identical twinning, any hereditary influence would involve some faculty in a

---

borne triplets almost twice as often, proportionately, and quadruplets almost four times as often, as White mothers. In quintuplet deliveries with one or more live births among them, the Negro ratio has been even higher, although no full Negro quintuplet set has survived to date.

TABLE I

Incidences of Twinning

1. *United States: Decline in incidence of twinning, accompanied by increase in ratio of identical (one-egg) to fraternal (two-egg) twins, among Whites, from 1918 on.**

Year	Twin pairs per 1,000 births (both types)	Ratios, identicals per 1,000 fraternals	Per cent identicals among all twins
1918	11.0	442:1,000	30.7%
1938	10.4	534:1,000	34.8
1958	9.8	589:1,000	36.7
1961*	9.5	—	—
1962*	9.25	—	—

2. *Ethnic and Racial Differences in Twinning Rates: For all twins, and by types of twins, in various countries during comparable years.*†

	TWIN PAIRS PER 1,000 BIRTHS		
	Identicals	Fraternals	Total
WHITES			
United States (1956-58)	3.8	6.1	9.9
Spain (1951-55)	3.2	5.9	9.1
Norway (1946-54)	3.8	8.3	12.1
Italy (1949-55)	3.7	8.6	12.3
Denmark (1951-55)	3.8	10.2	14.0
Australia (1947-49)	3.8	7.7	11.5
Israel (1952-57)	3.8	7.3	11.1
MONGOLIANS			
Japan (1956)	4.0	2.3	6.3
NEGROES			
United States (1956-58)	3.9	10.1	14.0
African Yorubas	5.0	40.0	45.0
Johannesburg Negroes	4.9	22.3	27.2

* Data from U.S. Natl. Off. Vital Statistics. Information needed to estimate identical/fraternal ratios not gathered after 1958, but scheduled to be available again in 1965.

† Table taken from "Bio-social effects on twinning incidences," Scheinfeld, Amram, and Schachter, Joseph. *Proc. 2nd. Intl. Cong. Human Genetics*, 1961, Vol. I, pp. 300-305. (Data for African Negroes from Bulmer, M. G., *Ann. Hum. Genet.*, 24, 1960.)

fertilized egg which would cause it to split apart at an early stage. Such a faculty could come through genes from the father as well as from the mother, or perhaps, necessarily, from the two parents. Environmental influences, theoretically—some chemical produced by the mother, or some disturbance at or following conception—might also cause eggs to split and produce identicals. But on this basis it would be hard to explain how identical-twinning rates could be so much the same throughout the world, among all races and groups and in all environments, and could be so little affected by maternal aging, as compared with fraternal twinning.

For fraternal twins, a prerequisite is that the mother must produce at least two eggs simultaneously (one from each ovary, or two from the same ovary). But, as we've seen, this is influenced environmentally to a very great extent by the mother's age and previous childbearing. The father, too, might enter the picture, because there is some evidence that once two eggs are present, certain men may have more than the average probability of siring fraternal twins, presumably because their sperms are more than ordinarily active and virile. In any event, the strongest argument for hereditary factors in fraternal twinning is its markedly and consistently different incidence among racial groups, which leads to the assumption that among individual parents in any group such hereditary factors may also be at work.

Yet how explain the cases where *both* fraternal and identical twins have been born to the same couple—sometimes two or three pairs of each? Or where a mother bears a mixed supertwin set—triplets or quadruplets—with combinations of identical and fraternal members? One theory is that the same mother might carry both kinds of "twinning" tendencies. Another theory is that the predisposition to *conceive* either or both types of twins is widespread—occurring in perhaps one in every four or five mothers—but that the *capacity to carry twins through to birth* is present in some of these mothers much more than in others. If this were so, the same mother would have an increased chance of producing both fraternal and identical twins.

Altogether, the possible role of heredity in twinning remains speculative at this writing. What can only be said is that, on the basis of findings to date, once a mother has borne one set of twins (and particularly if they are fraternals), her chances of having another set have been variously estimated as from three to ten as great as the average twin expectancy for a mother who has not had twins before. The sister of a twin-bearing mother also has an above-average chance of herself

HEREDITY IN TWINNING

How the "twinning tendency" may work

In producing FRATERNAL twins:

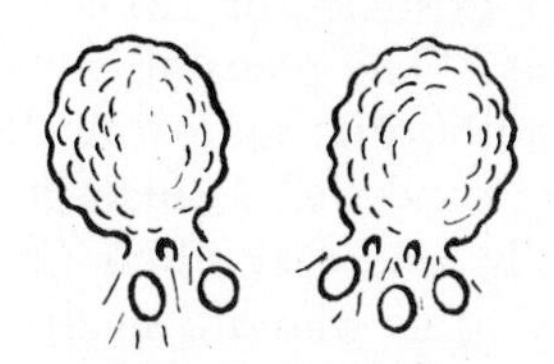

(1) Mother's ovaries *may be extra active and often produce two or more eggs a month instead of one.*

(2) Father's sperms *may also be extra active and fertile, increasing chances that if two eggs are waiting both will be fertilized.*

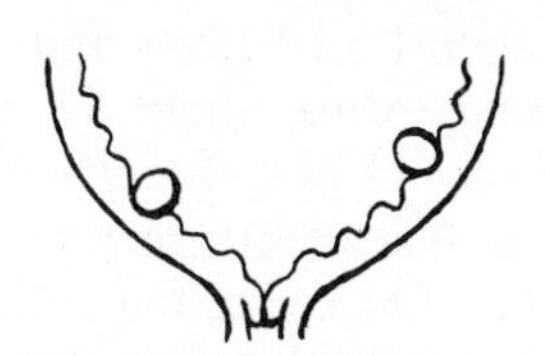

(3) Mother's womb *may provide an especially good environment in which twins can develop safely.*

In producing IDENTICAL twins:

A special inherited tendency for eggs to divide may result from . . .

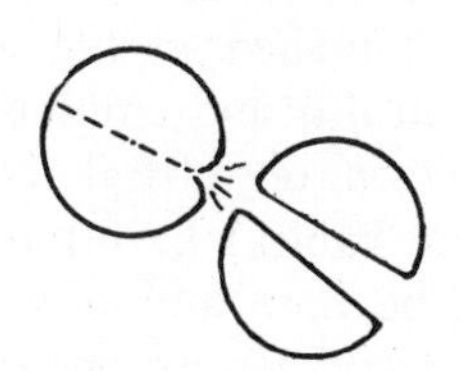

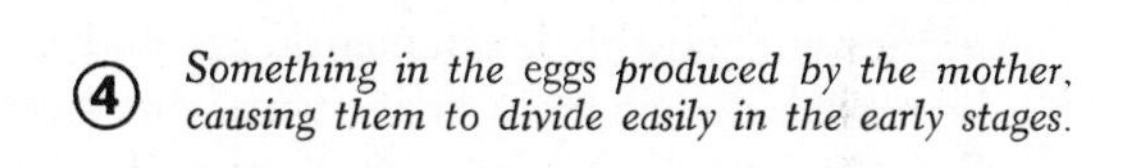

(4) *Something in the* eggs *produced by the mother, causing them to divide easily in the early stages.*

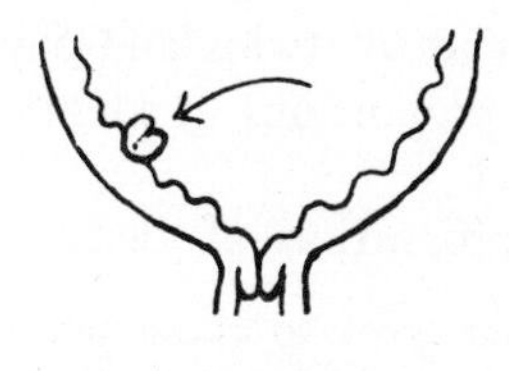

(5) *Something in the environment of the mother's* womb, *acting on the egg to cause division.*

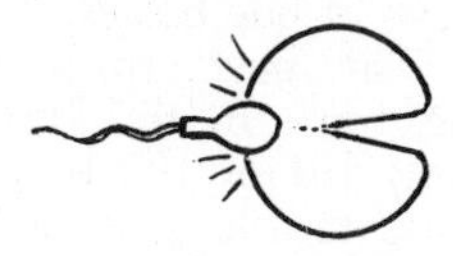

(6) "Twinning genes" *in the father's sperms, combining with "twinning genes" from the mother to impel division.*

producing twins. Also, mothers who themselves are twins tend to be more "twinning prone" than those who are singletons.

At any rate, the mother's ability to carry twins through to birth, whether or not influenced by heredity, can often be of decisive importance in view of the fact that the prenatal existence of twins is so much more precarious than that of singletons.* The prenatal casualty rate for twins is extraordinarily high, some authorities estimating that about three to four times as many twins are conceived as manage to achieve birth. (If both are not killed initially, one may often be an immediate casualty; and, undoubtedly, many of us who are singletons started off with a twin partner.) At birth the delivery risk for each twin is also increased much beyond that for a singleton. Further, since more than half of the twins are prematurely born—eight times as many as among singletons—and since "preemies" have below average survival chances, there is an extra toll of twins following birth. All of this is compounded in the case of the supertwins, almost all triplets and all quadruplets being immature; and where among twins, about 12 per cent of those approaching or achieving birth do not pull through, the fatalities among triplets from birth through the first month may reach 30 per cent, and among quadruplets, 50 per cent. Once past the critical period, however, twins and supertwins have very much the same chance for good health and longevity as do the singly born.

With twins and supertwins being so vulnerable to prenatal and birth hazards, it follows that the numbers of those launched in life are dependent to a considerable extent on maternal health and childbearing conditions. Thus, improvements in prebirth care, obstetrical techniques and postbirth safeguards (as with incubator babies) have much increased the chances that twins conceived will be born and survive. But how much this has offset, or will offset, the trend toward decreasing twin and supertwin birth rates, for reasons previously given, remains to be seen. The likelihood is that the incidences of twins and supertwins in the world will be less, although the proportions of identicals among those born will be greater than in the past.

Incidentally, with regard to the relative proportion of twins and

* Boy twins as a rule are more vulnerable than girls to these prenatal hazards (in accord with the generally greater weakness of boy babies under adverse circumstances, mentioned in Chapter 7). This may explain the lower ratio of boys among twins than among singletons: In the United States, for example, among twins born there are only 101 or 102 boys to 100 girls, compared with a ratio of almost 106 boys to 100 girls in the single births. Among triplets and quadruplets the ratio of boys is even less, with girls outnumbering boys.

supertwins born, Nature seems to be following an amazing formula. Known as "Hellin's law," after the German scientist who first noted the phenomenon in 1895, it apparently works this way: That whatever the incidence of twins may be, that of triplets would be the square of this, of quadruplets the cube, and of quintuplets, the twin figure carried to the fourth power. Thus, if the incidence of twins is one pair in 100 births, that of triplets should be 100 × 100, or once in 10,000 births; of quadruplets, 100 × 100 × 100, or once in 1 million births; and of quintuplets, 100 × 100 × 100 × 100, or once in 100 million births. (This formula applies only to multiple sets at birth, and not to the numbers surviving; for, as we've noted, the survival chances for the multiple-born decrease with the increasing sizes of sets. With quintuplets the formula becomes academic, since to date so few full sets have survived.) *

Why Hellin's law should work is still not clear. But that it does work in a general way is shown by the actual figures for supertwin births. In the United States, for instance, during the decade from 1948 to 1958, with 38,239,794 births, twins occurred on an average of once in about 95 births; triplets occurred once in 10,593 births, where the Hellin formula prescribed one set in 9,025 births (95 × 95); and quadruplets occurred once in 910,471 births, where the Hellin's-law expectancy was once in 857,375 births (95 × 95 × 95). At other periods, or in other countries as well as in different racial groups, the actual supertwin rates might be somewhat nearer to or further from the Hellin formula, but usually the figures are sufficiently close to give the law significance.

Having discussed the chances of having twins, we turn to our remaining question: What practical benefits have the twin studies had or can they have?

First let us see why it is so essential to know whether the members of a twin pair are identical (from one egg) or fraternal (from two eggs).

Inasmuch as identical twins have exactly the same heredity, whatever differences there are between them must be due to environment (whether in prenatal life or thereafter). Thus, if the two members of an identical pair differ significantly in any way—for example, if one is be-

* *Sextuplet* births also have been reported at rare intervals (usually in remote regions where the facts could not be checked), and a few cases have been listed in the medical literature of former periods. But for lack of clear proof, some authorities question whether a sextuplet birth has as yet occurred, although it is not unlikely that sextuplets may have been or may be conceived.

hind the other in growth, in health or in mental development—it is apparent that something in the environment is responsible; and if the cause is established, remedial steps may be taken. Again, if the symptoms of some disease known to be hereditary appear in one of an identical pair, there can be immediate alertness to provide early preventive or corrective treatment for the other.

In the case of fraternal twins, since they may be very different in heredity, there is no reason to expect them to be alike physically or mentally. If one of a fraternal pair is sturdier, bigger, healthier or brighter than the other, it may be for the same reasons that any two children in a family can differ in these respects. And if one has an hereditary affliction or deficiency (of any of the kinds to be discussed in the chapters which follow), it may offer little menace to the fraternal partner—or at least no more so than if a singleton sibling were affected.

Where pairs of identical twins are compared with pairs of fraternal twins, further light can be thrown on the relative influences of heredity and environment. We already know that *heredity isn't everything,* since identical twins are never exactly the same in all respects. We also know that *environment isn't everything,* since fraternal twins are not nearly as alike as identicals, even though they have similar environments before birth and are often reared under equally similar conditions after birth (particularly where parents mistake them for an identical pair). So what scientists have sought to learn is just *how much more alike* identical twins are, on the average, than fraternal twins. This has been one way of estimating the relative influences of heredity and environment in the development of a great many physical, mental and behavioral traits, as well as in the causes and workings of many diseases and defects. Indeed, we can thank twins—and the scientists who have studied them—for many of the facts already presented or to be presented later in this book.

Undoubtedly, much more could be learned through twins and supertwins if scientists could deal with them under fully experimental conditions. For example, consider the Dionne quintuplets again: Geneticists, doctors and psychologists were allowed to study them for only a few years—from birth to the age of three or four. Even in that time much was learned of scientific importance, for it was clearly shown that despite their having identical heredity, and being reared and nurtured in as similar a hothouse environment as any five little girls had ever had, they still developed recognizable differences in physical makeup, behavior, personality and achievement (though the differences may have been slight compared with the similarities). After the scientific studies ceased, the facts about the quintuplets which came out through newspaper reports showed how their lives and paths continued to diverge. One sad, climactic event was the breakup of the set of five by the accidental death of Emilie in 1954, from suffocation during an epileptic fit.* Since then, three of the remaining four have married and had children—Cecile in 1961 producing a pair of fraternal twins—while Yvonne has become a nun (which Marie, and then Emilie before she died, had for a time also set out to be).

Many a geneticist might have wished, secretly, that for the sake of science the five Dionnes could have been separated at birth, and adopted and reared under very different conditions, to show how far identical hereditary factors might work out in different environments: one quint to be reared with a millionaire socialite family, another in the worst slums, a third in the home of a college president, a fourth with the most backward tribe in the jungles, and perhaps the fifth in just an average middle-class American home. This could be only a scientist's wild dream. But situations not so remote—although nowhere near as decisive—have arisen naturally in many instances where identical twins have been separated and reared in different homes and environments from infancy onward. What has been learned from studying such twins will be told in later chapters.

But the Dionnes themselves have already proved this much: That starting with the same genetic plot, environment can write out five stories in quite different ways and that the substance and nature of an individual is never exactly repeated. So it could have been told at the beginning to each of these five—as to other members of identical twin or supertwin sets, or to any one else—what was said in the very first lines of our first chapter:

"Stop and think about yourself: In all the history of the world, there was never anyone else exactly like you, and in all the infinity of time to come, there will never be another."

* The fact that Emilie had been afflicted with epileptic seizures from childhood on was kept secret by her family until it was divulged by the four surviving quints in a magazine article in 1963. (See Chapter 37.) The other four showed no symptoms of the condition.

PART II

Diseases, Defects and Abnormalities

CHAPTER NINETEEN

Your Chemical Individuality

IT MAY BE HARD to think of yourself as a chemical factory, but that's what you are: a vast chemical plant, with untold numbers of interacting production units, far more complex, and with an infinitely more varied output, than any of the world's greatest chemical-industry giants.

This chemical plant which is yourself is something more than a producer of compounds and substances. It is also a *living being;* and it is unique and special among all other living beings. Turn again to our opening statement, "In all the history of the world there was never anyone else exactly like you. . . ."

What this actually means, biologically speaking, is that there never was, is not now, and never will be another person with your precise chemical makeup—or *biochemical individuality*. The way that your body utilizes and converts the chemicals in your food and drink into other chemical products, or chemically dictated functions, makes you different from any other human being. This was summed up in a verse, quoted by the late Dr. George W. Gray:

It's a very odd thing—
As odd as can be—
That whatever Miss T. eats
*Turns into Miss T.**

Dr. Gray went on to say, ". . . the biologist learns that not only is Miss T. unfailingly able to convert steak and potatoes into the unique

* From the poem "Miss T." by Walter de la Mare. Reprinted by permission of the Literary Trustees of Walter de la Mare and The Society of Authors as their representative.

pattern of the tall, angular blond woman that is Miss T., but every one of the billions of cells that make up her body carries the individual design that marks it as exclusively her own."

So, too, in your case, the steak and potatoes—or lamb chops, pie, ice cream, or whatever else you eat—ends up as something different in you than in the other people at your table who might have eaten exactly the same food. What accounts for this "biochemical individuality" of yours?

In Chapter 9 we noted that the basic chemical stuff of life has been identified as deoxyribonucleic acid (DNA)—that it is of this chemical compound that the heredity-determining elements of the chromosomes and genes are composed. It was also brought out that the differences in the way genes work to "tap out" instructions to the cells housing them result from different chemical codes embodied in their particular segment of DNA. To understand this more clearly, let us take a closer look at the DNA molecule.

In its atomic composition, as we said earlier, the DNA molecule is something like a spiral staircase, or a flexible, twisted rope ladder. Between and connecting the two very long outer sides, or strands, throughout their length, are a series of steps or rungs. Each of these consists of a pair of contrasting chemical bases, selected from two *purines*, *adenine* and *guanine*, and from two *pyrimidines*, *thymine* and *cytosine*. In any given step adenine is always paired with thymine, and guanine with cytosine; although in successive steps, sometimes one member of a pair comes first, sometimes the other. Thus, four types of DNA steps are possible (referring to each substance by its first letter): A-T, T-A, C-G, and G-C (see illustration).

At first thought, with just these four steps, it might seem that the variations in the DNA molecule would be greatly limited. But consider the possible arrangements of successive steps, for example: TA, CG, AT, GC, GC, AT, TA, CG, AT, GC, CG, TA, etc., etc. How infinite can be the variations becomes apparent when we find that a single DNA molecule is so enormously long that it might have millions of steps, following one another in endlessly varied ways.

Remember, then, that each section of the DNA molecule—and therefore, every gene—has a different arrangement with respect to the "TA" or "AT," and "CG" or "GC" pairs. So we can see how each gene, in turn, can be endowed with the capacity to send out particular chemical instructions by its own special code. If two given genes in the same individual were delegated to two different assignments, they would differ greatly in the arrangement of their steps. To illustrate, the following

DNA: THE MOLECULE OF LIFE

Shown diagrammatically is a section of DNA—Deoxyribo-Nucleic-Acid—the basis of a gene

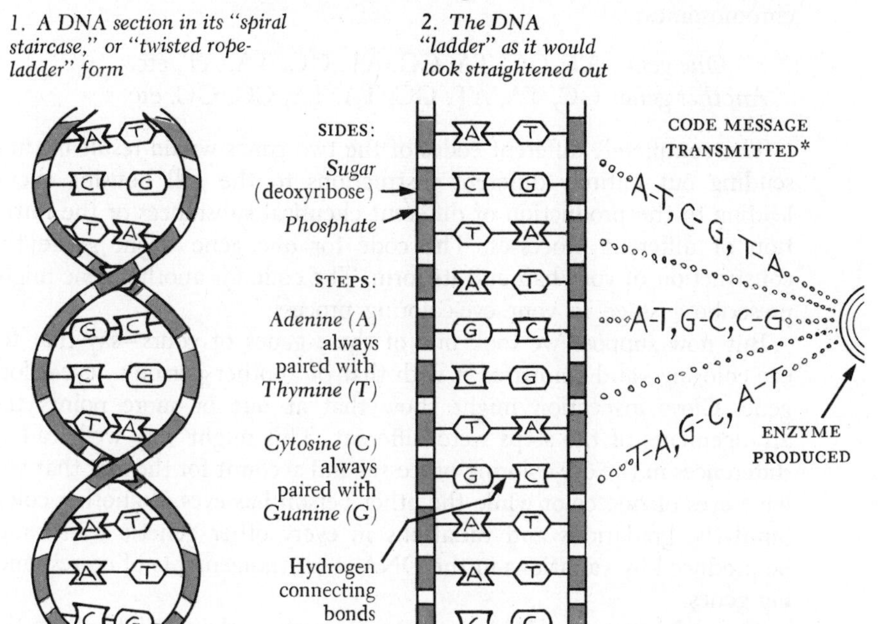

HOW A DNA MOLECULE REPRODUCES ITSELF

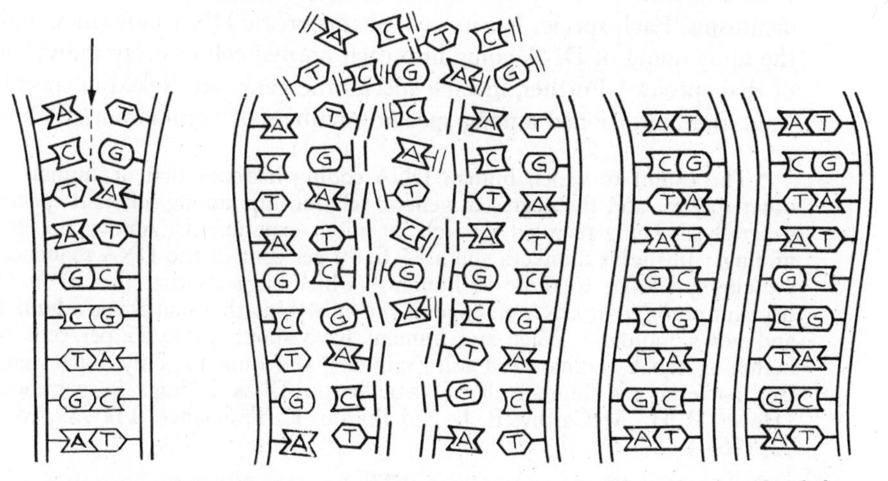

* Actually, the code message is in terms of half steps, transmitted by only one side of the DNA ladder at a time. The message is then carried out by an "RNA-messenger" strand which is formed alongside the DNA ladder. For more details, see Text.

(hypothetically) might be short sections of two of your genes, either from different parts of the same chromosome, or from two different chromosomes:

One gene: AT, GC, TA, CG, AT, GC, TA, AT, etc.
Another gene: GC, TA, AT, GC, TA, TA, CG, GC, etc.

The completely different codes of the two genes would result in their sending out entirely different instructions to the cell housing them, leading to the production of different chemical substances or the initiation of different processes. The code for one gene might guide the construction of your hair and its form. The code for another gene might prescribe a stage in your eye-coloring process.

But now suppose we took one of these genes of yours—say that for eye coloring—and compared it with that of another person's "eye-color" gene. Close inspection might show that at one or more points the arrangements of the steps were different. This might very well lead to differences in the eye-coloring processes and account for the fact that you have eyes of one color while the other person has eyes of another color. Similarly, gradations and variations in every other genetic trait could be produced by variations in the DNA step arrangements of corresponding genes.

The differences in DNA patterns not only make possible all of the individual genetic differences among human beings but also account for the differences between human beings and all other living things, both animal and vegetable, and, in turn, the variations within each species. For human beings have no monopoly on DNA. It also is the basic life stuff of every other animal, as well as of plants and microorganisms. Each species has its own characteristic DNA molecules, with the same quota of DNA going into each normal cell of every individual of that species.* Further, in each species the genes are linked in specific ways and apportioned among specific numbers of chromosomes. Thus,

* The extent to which human DNA coding matches that of animals of other species—and the degree of genetic relationship among different species generally—is being revealed through ingenious experiments. Among the first findings: In rhesus monkeys about 85 to 90 per cent of the DNA sequences are closely similar to those in humans (which suggests that the 10 to 15 per cent of *different* DNA messages is the only thing that makes you a human and not a monkey). Mice and humans have about 20 to 25 per cent of matching DNA codings, and fish (salmon) and humans, only 5 per cent. But bacteria and humans have virtually no DNA codings in common. (Hoyer, B. H., McCarthy, B. J., and Bolton, E. T. *Science*, 144:959, May 22, 1964.)

DNA DIFFERENCES IN DIFFERENT GENES

...And how their workings are affected

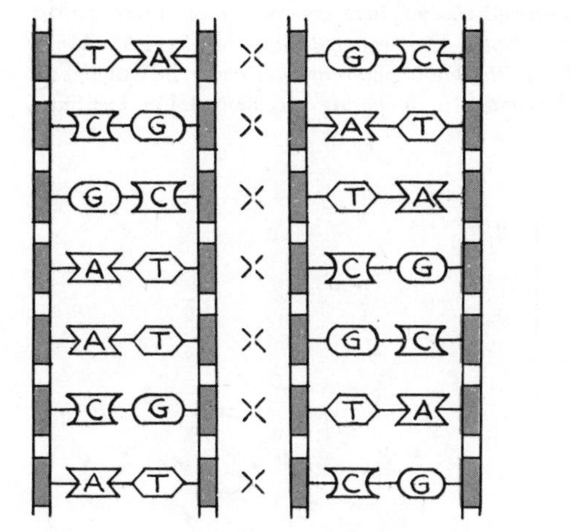

1. Entirely different genes

Hypothetical DNA sections of genes for two different functions. Code arrangements of T-A, C-G steps are entirely unlike.

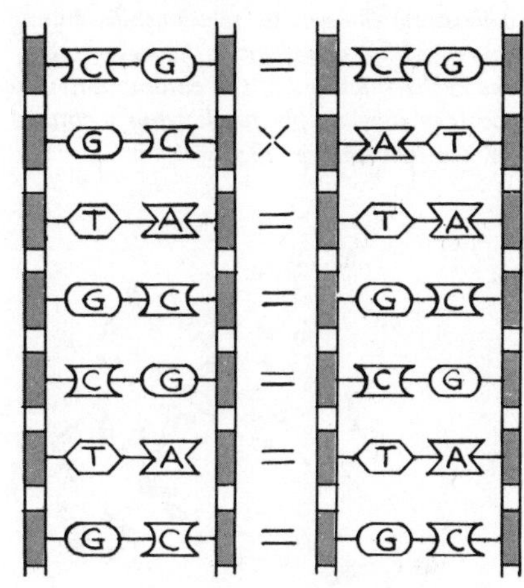

2. Matching genes, different effects

Hypothetical DNA sections of genes for the same function, but differing in one step *(a G-C substituted for one A-T step). The single difference may result in a marked change in effects.*

compared with the human's 23 pairs of chromosomes fruit flies may have 4 pairs; a cow, 30 pairs; a mouse, 20 pairs; a dog, 39 pairs; one species of monkey, 17 pairs, and another species of monkey, 33 pairs. (It will be noted that the *numbers* of chromosome pairs need not be related to the size of the animal, nor do more chromosomes imply more genes. In one species there may be many more genes per chromosome than in another.)

What is most striking about the chromosomes—and offers proof of how much alike all living and growing things are in the basic mechanisms of their heredity—is the remarkable similarity in appearance between the chromosomes of one species and another. In fact, as you can see in the accompanying illustration, only an expert could tell by looking at isolated chromosomes whether they are those of a gorilla, a rabbit, a hamster, or a human being.

The foregoing gives us only a general insight into the nature of the chromosomes and their chemical composition. Still waiting to be clarified are the precise means by which the genes in any species transmit

WHICH ARE THE HUMAN CHROMOSOMES?

To illustrate the genetic relationships among different species, here are six sets of chromosomes: human, gorilla, chimpanzee, hamster, zebra, and rabbit. Can you tell which are which? (Variations in the thickness of the chromosomes shown in the photographs are not necessarily significant since they may simply result from technical differences in preparing specimens.) For identification, see footnote, p. 175.

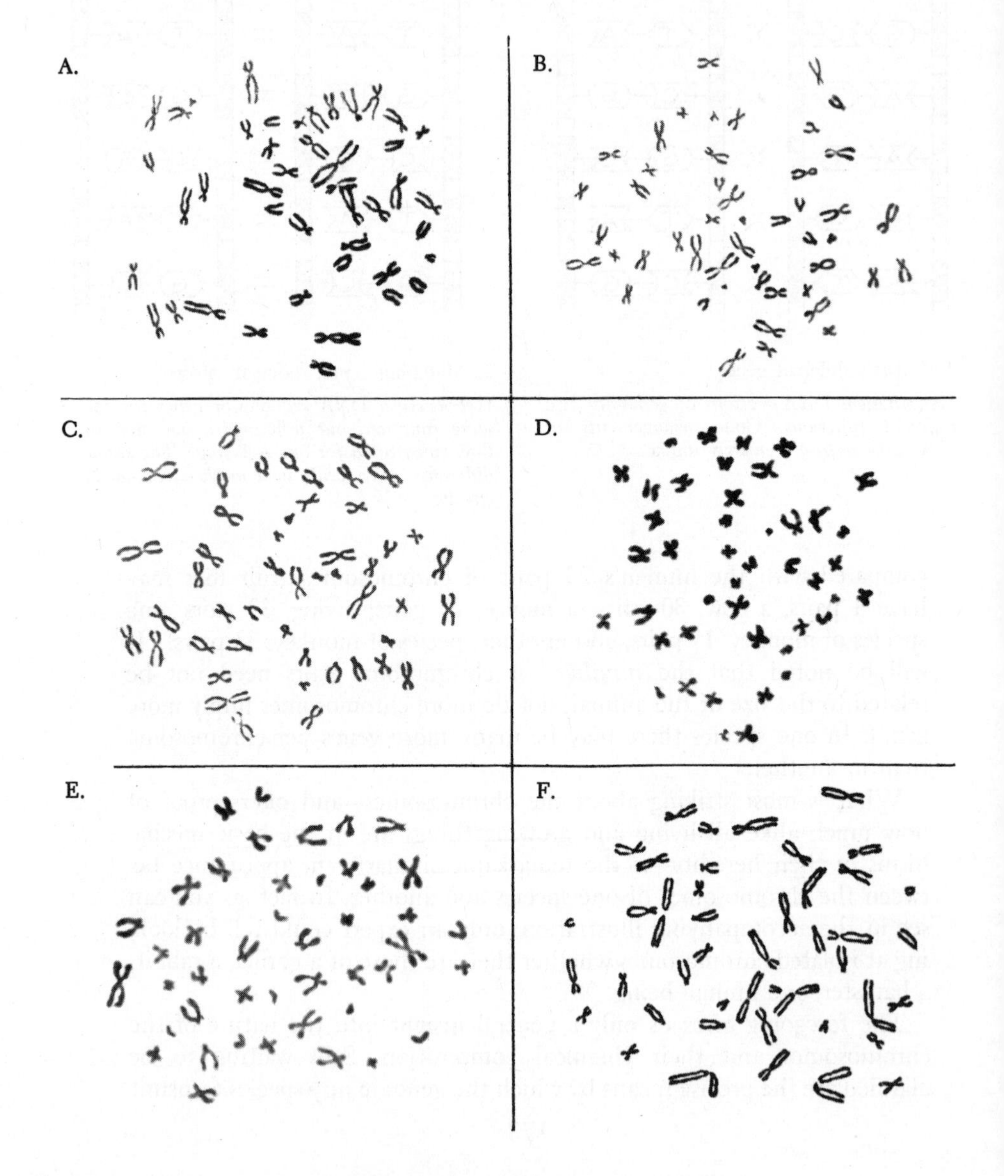

their coded instructions and cause them to be carried out. We can get some notion of how gene codes work by likening them to the coded punch cards, tapes or other code transmitters used in many automated industrial processes. By such codes the preparation of the most complex chemical compounds, or the production and assembling of highly complex mechanisms, can be directed from one step to another.

As an intermediary in carrying out the DNA orders of the gene, there is another related substance called *ribonucleic acid* (RNA), which is found both within the cell nucleus and outside of it in the general cell material, or cytoplasm. Formed along one or another side of the DNA ladder, the RNA molecule is a single strand, and thus, unlike the double-stranded DNA, consists of a chain of only half steps. Further, in the process of RNA formation, each thymine base is replaced by another base (not too different), *uracil.* Accordingly, a DNA message section which on one strand might read "A-C-T-T-G-A," etc., would be translated by the RNA into "A-C-U-U-G-A," etc.

At any rate, RNA may be considered the medium through which DNA gives its genetic orders to the cell. To state this another way, DNA is the chemical master which lays down the genetic law, while RNA is its creature which enforces it. Once a gene's DNA has given its instructions to the RNA "messenger," the latter moves out into the cell and directs the formation of particular substances, or enzymes, from the specific little chemical building blocks (amino acids) that are available there.

In its chemical behavior the gene might be likened to a catalyst (as noted in Chapter 9) which acts on an existing substance or compound so that, in combination with other materials, it forms a new compound. This behavior may go on in chain reaction to induce the production of

---

On page 174, the chromosomes of a human male are shown at the lower right (F). The others, with the normal number of chromosomes for each species given in parentheses, are: A. Golden Syrian hamster, male (44); B. Chimpanzee, male (*Pan troglodytes*, 48); C. Gorilla, male (lowland species, 48); D. Damara zebra, male (44); E. Domestic rabbit, female (*Oryctolagus cuniculus*, 44). Individual chromosomes of any species may be almost indistinguishable from some chromosomes of other species, including man, but when complete sets are compared, the human, gorilla, and chimpanzee sets have far more in common with each other than with those of any other species. (An expert, however, could easily distinguish the human set from that of any other primate.)

The sources of the photographs, page 174, are: Dr. H. L. Cooper (golden hamster); Dr. H. P. Klinger (chimpanzee, gorilla, human); Dr. Kurt Benirschke (zebra); Dr. Susumu Ohno (rabbit).

additional compounds and products; and so the process may continue until the end result prescribed by the gene code has been achieved. The actual sequence of chemical events has already been traced in the case of certain genes, as we shall see in later chapters.

Stated in more detail, the prevailing theory is that each gene controls or affects the production of a specific enzyme involved in a given process. Some genes are *structural*—concerned with the production of specific proteins; other genes are *regulatory*—each affecting the *rate* of production of a specific protein, which is to say that, while a given "structural" gene may determine the type of protein, the accompanying "regulatory" gene may decide how much of this protein is produced: (1) a "normal," or average, amount; (2) an above-normal amount, or too much (if the regulatory gene is hyperactive); or (3) a below-normal amount, or too little (if it is hypoactive). Any marked overproduction or underproduction of a given protein may result in an abnormality. Likewise, any initial failure or error in a "structural" gene may result in failure to produce a protein substance normally needed by the body, or may result in production of the wrong substance; in turn, the "regulatory" gene may have its workings disrupted and may further compound the error.

Still hazy are the theories as to just how the same aggregation of genes at one location in the body will direct the production of bones; at another point, that of internal organs; and at another, that of facial features—not to mention particular shapes and structures of bones, organs and features. The barest hint of how this happens might come if you think of a familiar substance, such as an open can of paint: Allowed to stand a while, the exposed part will become rubberlike; with more time the surface will shrivel up and a pattern of wrinkles and cracks will form. Likewise, at other points in the can, from top to bottom and from the sides to the center, there will be big differences in the density, composition and form of the material.

In a very rough way this should suggest how, given the same materials, different types of cells may be fashioned at different locations in the body or perhaps even in the egg; and how, in turn, as the natures of the cells are differentiated, and different types of chemical environments are created within them, certain genes may be stimulated to do particular jobs called for, while other genes may be kept inactive—although in some other cells these latter may be the busy ones. In the vast population of genes each one, in one way or another, or at some time or another, has its job to do: sometimes a job in which it stars, sometimes one in which it is a member of a large team. In any case, what your genes do, separately or collectively, gives you your chemical individuality.

Outwardly, your distinctive looks—face, coloring, body form—provide abundant evidence of the individuality of your chemical makeup, for your features are no less chemical end products than are your bones and skin. But this external chemical individuality of yours need not be due only to your genes As noted in previous chapters, your features, stature, body form, and many other aspects of your appearance have been considerably influenced in their development by environmental factors.

The best proof of your inherited biochemical individuality can be found in your tissues and blood. As one indication, tests of your various blood-group substances, taken together, will show that they differ from the blood-group combinations found in most other persons. However, this still leaves a considerable number of individuals in the world with blood groups matching your own. Only when we include all of the other chemical substances in the body tissues—known or not yet identified—can it be shown conclusively that, chemically, you are unlike anyone else.*

The best existing proof of this chemical uniqueness can come through making skin grafts or tissue transplants. If part of the skin on your body were badly burned or injured, it would be possible to take a patch of skin from some other part of your body and graft it onto the damaged spot. It would adhere and grow there and soon become a permanent part of the surrounding skin. This is because the tissues throughout your body, under the supervision of identical genes, are chemically the same. But if another person's skin—even that of your father, mother, brother or sister—were used in the grafting process, it would not "take" permanently (although it might act temporarily as a protection) and before long would slough off because the tissues were chemically different from your own. The one exception to this would be if you have an *identical twin*.

As noted in Chapter 17, one-egg twins are never exactly alike externally. Nor need they be alike in all aspects of their body chemistry

* Evidence of how human biochemical individuality asserts itself externally has long been given by the ability of keen-scented dogs to distinguish among the body odors of different persons. Testing this ability experimentally, Dr. H. Kalmus showed that hunting or police dogs could distinguish promptly among the scents of siblings or other members of a family but were at first confused by the similar body odors of identical twins. However, after training and more familiarity, the dogs were able to recognize and act on even the slight differences in the twins' scents.

and functioning, inasmuch as these can be affected by environmental differences in their diets, habits, diseases and experiences. Yet none of these environmental influences can alter their *genetic* chemical likeness. In every basic respect the substances and tissues of identical twins remain so much alike chemically that not only patches of skin, but whole organs, can be transplanted from one to another and can grow there as permanently as if the grafted part had come from the recipient's own body. This chemical likeness has been demonstrated most dramatically in recent years by the many operations in which a kidney has been transplanted successfully from one identical twin to his critically ailing partner. Such transplants also have been made in some cases between fraternal twins, and on occasions, between nontwin members of a family. In the latter cases, the transplants were attempted after first altering the chemistry of the recipient's tissues in some way (as by heavy radiation) so that they would "ignore" the fact that the grafted tissues were foreign to their own. However, even in these cases, so far, most of the operations have fallen short of success, and how permanently successful other organ grafts between nonidentical twin individuals will be is still uncertain.*

The most striking ways in which your chemical individuality asserts itself is through your *endocrine* (or ductless) glands and their hormones. If you think of all the processes in your body as a chemical symphony, these glands would be among the leading players, not only interpreting the notes composed by your genes, but giving them added twists of their own. Our accompanying Chart shows how each of the endocrine glands has its own designated functions, although also interoperating with the others. Following are more detailed descriptions of what the individual glands do and how they may be involved in var-

* The greatest immediate hope for better "takes" in kidney transplants between persons other than identical twins may rest in *tissue-typing* before the operations: that is, in identifying beforehand by skin-graft tests whether the prospective donor of an organ is among the one in six or seven persons at large whose tissues may be compatible (or almost identical-twin like) with that of the recipient's in terms of immunological response. As Dr. John P. Merrill of Boston (a pioneer and leader in this surgical area) has reported, it is likely that the persons who previously survived longest after kidney transplants from nonidentical twins had done so because, quite by chance, they got organs from persons with matching tissues. (Most amazing of the possibilities—currently in the experimental stage—is that organs or other body parts from *lower animals* may be transplanted to human beings after the inherent chemical conflicts have been neutralized. Among the transplants have been kidneys from chimpanzees and a baboon, which managed to take root and function in human beings for a number of weeks.)

THE ENDOCRINE (DUCTLESS) GLANDS

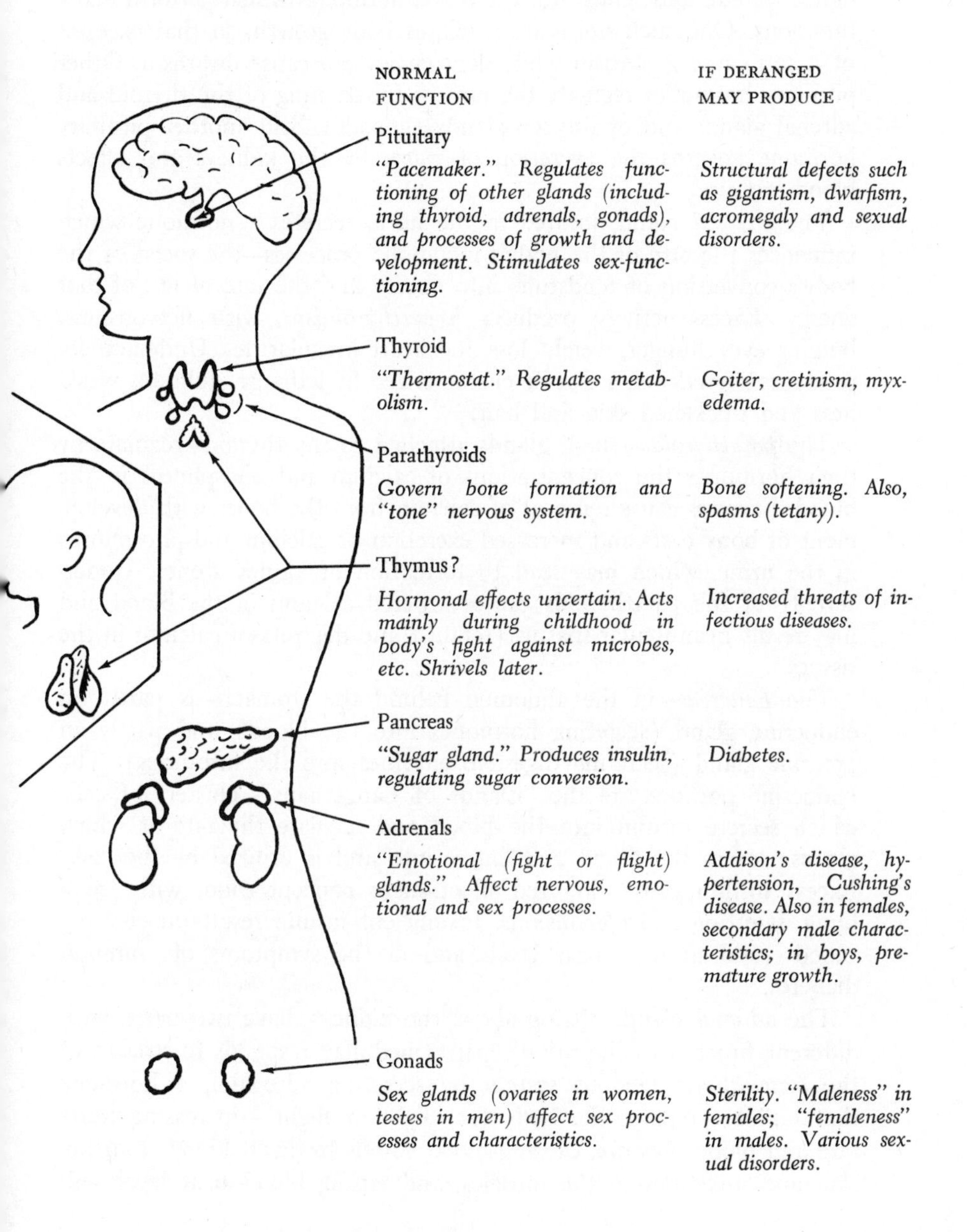

ious diseases and defects to be discussed in succeeding chapters.

The *pituitary* (or "master") gland, located on the underside of the brain, secretes and sends into the blood hormones which perform many functions. One such hormone regulates body growth, so that excesses of it can cause gigantism while deficiencies can cause dwarfism. Other pituitary hormones regulate the rate of functioning of the thyroid and adrenal glands and of the sex glands (gonads). Still another pituitary hormone governs the excretion of water by the kidneys and affects blood pressure.

The *thyroid* gland, located in the neck, secretes a hormone which influences the rate of the body's metabolic processes—the speed of the body's conversion of foodstuffs into energy and the rate of use of that energy. Excess activity produces *hyperthyroidism,* with nervousness, bulging eyes, hunger, weight loss and heart irregularities. Underactivity produces *myxedema,* a disease characterized by lethargy, dullness, weakness and thickened skin and hair.

The *parathyroids*—small glands attached to the thyroid—regulate by their hormones the concentrations of calcium and phosphorus in the blood. Excesses cause removal of calcium from the bones with development of bony cysts and increased excretion of calcium and phosphorus in the urine, which may lead to formation of kidney stones. Underactivity of the parathyroids causes lowered calcium in the blood and may result in muscular spasms (tetany) and deposits of calcium in the tissues.

The *pancreas*—in the abdomen behind the stomach—is partly an endocrine gland (secreting hormones into the blood) and partly an *exocrine* gland (secreting digestive enzymes into the intestines). The endocrine portions are the "islands of Langerhans"—clusters of cells which secrete insulin into the blood and regulate the rate at which glucose leaves the blood and enters into and is utilized by the cells. Excess insulin causes lowered blood-sugar concentration, with trembling, sweating and convulsions. Insufficient insulin results in elevated blood- (and urine-) sugar levels and in the symptoms of common diabetes.

The *adrenal glands,* sitting above the kidneys, have two parts, with different functions. The middle part (medulla) responds to stresses of the sympathetic nervous system by secreting adrenalin, a hormone that rapidly prepares the body for "fight or flight"—increasing heart rate and blood pressure, causing blood vessels to divert blood from the skin and intestines to the muscles, and raising blood-sugar levels—all in very short order. The outer part of the adrenals (the cortex) secretes several hormones. One, very similar to cortisone, regulates inflammatory and allergic reactions; aids in repair of injured tissues; and acts on salt, water and sugar metabolism in the body. Another, aldosterone, is concerned mainly with salt and water excretion. Still another is closely related in action to the male sex hormones. Derangements of the adrenal cortex can cause Addison's disease, Cushing's disease, and masculinization of females, among other conditions.

The *thymus,* while listed among the glands, may or may not produce any hormones (this has not yet been established) but is highly important in the body's defense mechanisms. Located at the base of the throat, behind the breast bone, the function of the thymus was entirely unknown until recently. One mystifying fact was that, after increasing in size from about a third of an ounce in newborn babies to an ounce or more in twelve-year-olds, it begins to shrink as maturity nears, and returns to baby size in older adults. Why this is so has become clear with the finding that during the early and childhood years the thymus is highly active in producing special white-blood cells (lymphocytes), which then migrate to the lymph nodes throughout the body and the spleen, where, serving as "master cells," they stimulate the production as needed of substances (antibodies) to fight off invading microbes or other hostile intruders. As childhood advances and the major work of the thymus is apparently done, the shrinkage of the organ follows. Whether it serves any function in adults is still in doubt, but there is some indication that it may continue to play a part in the body's immunizing mechanisms under certain conditions (including, perhaps, that presented by excessive radiation).

Finally, we come to the *sex glands,* most dramatic of all the glands in their effects. It is mainly through the sex glands that one of the most significant aspects of your chemical individuality—your sex, with all of its biological manifestations—asserts itself. If you are a female, you will be producing relatively more of the *estrogens* (the so-called female hormones) and less of the *androgens* (the male hormones). If you are a male, it will be the reverse. These chemical differences, together with others that are set in motion by whichever combination of "sex" genes the individual carries, have far-reaching effects. Not only do they induce in females such distinctive processes as menstruation, egg production, pregnancy and the menopause, but they also cause them to be chemically different from males in many other ways. How this leads to further significant differences between the sexes in degree of susceptibility to various diseases and defects, in longevity, and in the manifestation of other genetic traits, will be told in Chapter 23 and at other points in our book.

In both sexes, *age* also plays a part in a person's chemical makeup, for this may undergo successive changes throughout life. What you could eat as a young child (for instance, a diet largely of rich milk), your system could no longer tolerate in your adult years. That midnight snack of fried hamburger with onions and a chocolate malted, which was almost a sedative in your college days, could give you a sleepless night in your fifties. But regardless of age, some people go through life with "cast-iron" stomachs, able to eat and easily digest almost everything, while others must rigidly restrict their diets. Some are like Jack Sprat, who could eat no fat, and others like his wife who couldn't (or wouldn't) eat the lean. So, too, one person can take a prescribed drug with no side effects, while another will react violently to it. (Among the drugs to which marked inherited differences in reaction have been identified are *insulin,* used in treating diabetes; *isoniazid,* used in treating tuberculosis; *primaquine,* an antimalarial drug; and *suxamethonium,* a muscle relaxant.) All of this may provide evidence of individual differences in body chemistry, although to what extent genetic factors are responsible and to what extent the causes are environmental—through conditioning, habit, bodily upsets or mental states—is by no means always clear.

Where genetic differences in the chemical makeup and functioning of individuals can be most readily identified and have their most serious effects is in specific diseases, defects and functional eccentricities. It is to these that we now turn.

CHAPTER TWENTY

The Wayward Genes

No reputable manufacturer of any complex mechanism—an automobile, a space vehicle, a computing machine—would think of delivering his product until it had been thoroughly tested and was as free from flaws as possible.

Unfortunately, Mother Nature, the greatest producer of complex mechanisms, exercises no such care with regard to many of the human machines. We see, alas, that she creates individuals with every conceivable kind of defect or abnormality. In many cases this is the result of accident, or of being given bad materials to work with or bad conditions to work under. In other cases the abnormalities might seem to be almost deliberately produced, perhaps as by-products of a constant process of experimentation in some vast evolutionary scheme which we can as yet only dimly understand.

Actually, it is futile to compare the human body with any machine which we ourselves have constructed, for the finest automobile, airplane or scientific instrument is but the crudest of mechanisms compared with the human body and what it can do. Moreover, while some human beings do have serious hereditary flaws, what should never cease to impress us is how amazingly free of them are the great majority of us. Nor should we forget that many human ills which in former times were blamed on Nature—or on bad heredity—have now been found due wholly or in large measure to bad environment and are being steadily reduced by improvements in living conditions and medical care. Even where heredity is to blame in one degree or another, these environmental improvements have done much to lessen the threats of many conditions.

This having been said on the brighter side, we turn to the distressing fact that there are a great many inherent human defects which assert themselves regardless of what we can now do about the environment. In addition, there are many other defects and diseases which, under the same conditions, will afflict some individuals more readily and more seriously than others because of inherited weakness or susceptibility. No human being is completely without a few hereditary imperfections. Fortunately, these for the most part are so slight as to cause little damage or inconvenience. Some of them, however, are serious enough to interfere with important functions; to produce abnormal appearance which may make social adjustment difficult; or, sometimes, to cause premature death.

To all of those genes responsible for harmful or detrimental conditions, we have given the name *wayward genes*. And only when one or more of these wayward genes is involved in producing a defect, disease or abnormality can such a condition be considered, in the scientific sense, as *hereditary*.*

This latter fact cannot be stressed too strongly. One of the greatest mistakes in the past—one still commonly made, even by some doctors—is to confuse *congenital* or *familial* with *hereditary*. Sometimes these terms may apply to the same conditions. Often they do not. In other words, the mere fact that a condition is congenital (present at birth) or that it is familial (running in a family) is no proof that it is hereditary, for there are many environmental effects that show themselves at birth or that consistently run in families. Conversely, a condition may be hereditary and yet not show itself at birth or for many years thereafter and, also, may never before have appeared in the individual's family (as, for instance, when rare recessive genes are brought together or new wayward genes are created).

One of the commonest errors is that of speaking of congenital syphilis as "hereditary." *Syphilis is not, never was, and never can be inherited.*

A dramatic episode from my own experience bears on this point. Some years ago I was conducting a baby contest in a Midwestern state, the ostensible purpose of which was to select the most perfect babies. (A more important objective was to have the little ones brought in for careful medical checkups.) With a first prize of $1,000 and other

* Exceptions are certain conditions caused by chromosome abnormalities or derangements, to be discussed later in this chapter.

awards as lures, thousands of babies were entered, and after weeks of examination the field was narrowed down to a dozen, each of whom was now in line for a prize.

I was present as an observer at the final judging by a group of leading pediatricians. Suddenly one of the specialists, who was examining a two-year-old girl baby, whispered excitedly to his colleagues, "Congenital syphilis." No one else had previously detected the almost imperceptible symptoms. The child had already been publicly proclaimed as among the winning finalists. Nothing could be done now but to award her the last major prize—a cash certificate and a big silver cup proclaiming her as "one of the most perfect babies in the state." And then, after the mother had glowingly received it, came the awful moment when she was taken aside and told that her beautiful little girl had a venereal disease—and even more, that the child had *acquired it from her*.

For the only way that a child can be born with syphilis is through *infection by the mother*. The mother herself may have had the disease to begin with, or she may have been infected by her husband during conception or later. But the germs had to be present in her if her child was born with syphilis. All this is also true of gonorrhea or any other infectious disease. A father, no matter how diseased he himself may be, cannot possibly transmit the germs for his condition through his genes (nor can the mother through her genes). However, germs could easily be carried with the sperms in a man's seminal fluid. Thus, after the mother was infected, the disease could have been transmitted to the child at some prenatal stage, or during the birth process (this occurs more often in the case of gonorrhea). Happily, under modern methods of prenatal care, delivery and treatment, it is possible to prevent such infection of babies in almost every case.

We have dealt at some length with syphilis because, ever since Ibsen's drama *Ghosts* startled the world, the belief has been widespread that the disease can be inherited and that a "syphilitic taint" may persist in a family for generations. Attending this was the theory, called "blastophthoria," that syphilis, as well as drunkenness, drug addiction and depravity, might in some way permanently affect the germ cells in a family strain and produce progressively worsening effects from generation to generation. But as we have seen in previous chapters, no disease or evil habit can alter the *genes* in the germ cells, and therefore there can be no such thing as an "acquired" hereditary taint.

When a child does come into the world with some serious acquired defect or abnormality, the question as to whether it should be termed "congenital" or "inherited" may seem to be mere quibbling. But this is emphatically not so. For if a condition is congenital but *not hereditary,* the afflicted individual can grow up and marry with no fear that his children will inherit his defect. If it is a disease which has been cured (as syphilis can now be cured), he will certainly not pass it along. But where a condition is hereditary, *no matter whether the parent has been cured of it or not,* and no matter how healthy he or she may be, there is always the possibility that the wayward genes involved will be passed along and reproduce the condition in some of the offspring. Thus syphilis and many other acquired diseases, congenital or otherwise, may be wiped from the earth in a few generations merely by modern methods of treatment and prevention. Inherited conditions may persist forever, for they could be eliminated only by preventing all carriers of the wayward genes from breeding, an almost impossible prospect.

In the case of *familial* conditions, it is equally important to distinguish between those which are hereditary—running in families because the same genes are passed along in successive generations—and those which are purely *environmental*—recurring in a given strain merely because the same bad conditions or harmful influences continue to prevail. This latter applies to many infectious diseases, many physical defects, and certain types of mental abnormality and abnormal behavior which once were thought to be hereditary. We know also that dietary deficiences—the lack of needed foods and vitamins—can explain repeated occurrences, in some underprivileged strains, of a great many diseases, disorders and physical defects.*

But there is another side to the picture. For just as grave mistakes were made in the past in designating conditions as hereditary when they were wholly or chiefly environmental, so mistakes were made and still are being made in dismissing conditions as purely environmental when heredity may have much to do with their causation or development. The extreme environmentalists of today (including those who

* A new distinction is being made between the terms "hereditary" and "genetic," which formerly were used interchangeably. As the term is now used, a condition is *hereditary* if it is caused by genes, and if these genes, present in eggs or sperms, can be passed on from parents to offspring. However, in recent years various conditions (to be dealt with later in the chapter) have been traced to whole chromosome abnormalities which may arise only in body cells. While these conditions are *genetic,* in that they involve the person's genetic material, they cannot be called *hereditary* if the genetic abnormalities involved do not occur in the sperms or ova and are thus not transmissible to the person's offspring.

see almost every disease as psychosomatic) tend to ignore the fact that the line between heredity and environment in disease is seldom clearly drawn. Many conditions depend for their expression upon adverse heredity *plus* adverse environment; nor does failure to pin down a condition to a simple gene formula mean there is no evidence of hereditary influence. In many of the major conditions, as we shall see, the genetic mechanisms may be very involved, just as environmental influences may be.

What has become increasingly evident is that a person's whole genetic makeup—and not merely a specific gene combination—may be important in relation to almost every disease. Even when a condition is largely environmental, the biochemical differences among individuals, of which we spoke in the last chapter, may cause them to react to it in different ways. The most striking distinction in this respect, as we also noted previously and will bring out in more detail later, is that between *males* and *females*. Except for certain glandular conditions, and those which by their nature are peculiar to females, males are far more vulnerable to almost every serious disease and defect. In some instances the determining factors are largely or entirely genetic, in some largely or entirely environmental; but in general the greatest influence in making boys and girls, or men and women, react differently to many diseases is the fact that they are genetically different in their construction and chemical functioning. Throughout your own life, your sex may be a major factor in determining which of various diseases will strike at you, and in what way, and how severely.

What is true, then, of a male and a female in their different reactions to disease may be true, in varying degrees, of any two persons who are significantly different in genetic makeup, regardless of their sex. No matter what the condition, whether traceable directly to gene action, or whether caused primarily by an infection or some other outside influence, the risk of its manifesting itself may be much greater for one individual than for another, or for members of some families or stocks than for others. In other words, while the *external environment* always plays a part in determining how healthy or diseased individuals may be, or whether they will survive or perish at a certain time or age, a factor fully as great, if not greater in importance, is the state of each person's *internal biochemical environment*. (The term "host factor" is now used to describe the individual's degree of susceptibility or counteractivity to given diseases.) And this, in turn, is much influenced by the person's genes and the extent to which they do or do not function properly.

Now to the question, "How do the wayward genes produce their effects?"

For the answers, we must return to our previous discussions of the chemical composition of genes and the manner in which each gene carries out its prescribed functions according to the code embodied in its DNA "steps." Suppose, then, that the steps in a given gene have been shaken up, altered or damaged at some time, so that thereafter it sends out chemical instructions that are partly garbled, incomplete or otherwise incorrect. This is what characterizes any gene we call "wayward." Depending on the job to which it has been assigned and how much it strays from the normal prescribed course of action, a gene's mistakes may result in anything from very slight and almost imperceptible effects to consequences of the most serious nature. A gene may "overwork" (as when it produces six or seven fingers instead of five, or an excess of some chemical substance). Another gene may "underwork" (as in constructing a hand with missing fingers, or leaving some organ incomplete, or not producing enough of some important hormone). Or sometimes a gene delegated to a vital job may fall down so badly, or go on the rampage to such an extent, as to kill the person who is host to it.

In many ways, also, wayward genes may be as capricious as human wrongdoers. They may act abnormally and do bad things only under special circumstances, or only in the company of certain other genes; they may be just a little bit bad, or very, very bad; and they may go wrong in early life (as "juvenile delinquents") or not until later life; or sometimes, unexpectedly, they may not start acting badly at all. Even in members of one family, the same wayward gene or genes may sometimes produce a particular effect in one individual and a different effect—or none at all—in another, or may act at different ages. The fact that this occasionally applies to identical twins—who in so many way, genetic and otherwise, are more alike than other siblings—indicates the extent to which differences in environment can foster, modify or suppress the action of many genes.

In further assessing the behavior of wayward genes, one must always bear in mind that these genes, like those for normal traits described in preceding chapters, must be classified according to their degrees of dominance and recessiveness. To review and enlarge on what was brought out earlier, here are the basic facts about the wayward gene types:

A *dominant* gene is one which, singly, passed on by just one parent, can produce a given defect in a child. The presence of a dominant-gene condition in either parent implies a fifty-fifty risk of its reappearing in any offspring.

A *recessive* gene is one which must be coupled with a matching gene—that is, there must be a *pair* of such genes, one coming from each parent—in order for the effects to show in a child. If both parents are afflicted by a simple recessive condition (and each therefore carries two of the genes), every child of theirs will receive the two recessive wayward genes and can develop the disease or defect.*

A *sex-linked recessive* gene is one found in the X chromosome, and when wayward in its action, it strikes most often against males. The reason is that a male has only one X, so if any of its genes are bad, the effects will show in him. But in a female who has two Xs, any wayward gene in one X will in most cases have its effects blocked by a corresponding normal gene in her other X. (The sex-linked genes and conditions will be dealt with at length in Chapter 23.)

However, whether wayward genes are dominant, recessive or sex-linked, they may be qualified or governed in their action by environmental factors or by certain other genes. This will account for the fact, previously noted, that wayward genes sometimes seem to be capricious and unpredictable in their behavior. To describe the degree of regularity with which a gene works, geneticists use the term "penetrance." If a gene—a single dominant, or a recessive in a double dose—manifests its effects in all individuals who carry it, the gene is said to have "100 per cent penetrance"; if its effects show in 75 per cent of the cases, "75 per cent penetrance"; and so on. Thus, a single dominant gene with 100 per cent penetrance may cause the same defect (or quirk) to appear in successive descendants of a family, generation after generation. On the other hand, if a gene has only partial penetrance—75 per cent, 50 per cent, or whatever—its effects may be discontinuous and may not show through in all persons who inherit it. In other words, the *trait* (but not the gene) will sometimes skip a generation.

* The line of demarcation between recessive and dominant genes is no longer considered as clearly defined as it once was thought to be. Many recessive genes formerly believed to be entirely suppressed in the single state by their dominant partners can now be shown to produce some detectable effect. (We will see examples of this in a number of diseases and defects, where the ability of a recessive to reveal its presence may have vast importance in identifying "carriers.") The possibility that every recessive gene may in time be found to have some detectable effect has made some geneticists question whether the whole concept of dominance and recessiveness should not be abandoned or greatly modified.

HOW GENETIC PEDIGREES ARE ILLUSTRATED

Square: *Male*

Circle: *Female*

Plain outline square or circle indicates individual who does not have the trait in question.

or *Solid black indicates individual with the trait in question.*

or *Heavy dot indicates person who* carries *the gene for given trait but does not show trait.*

1. Simple Dominant Pedigree*

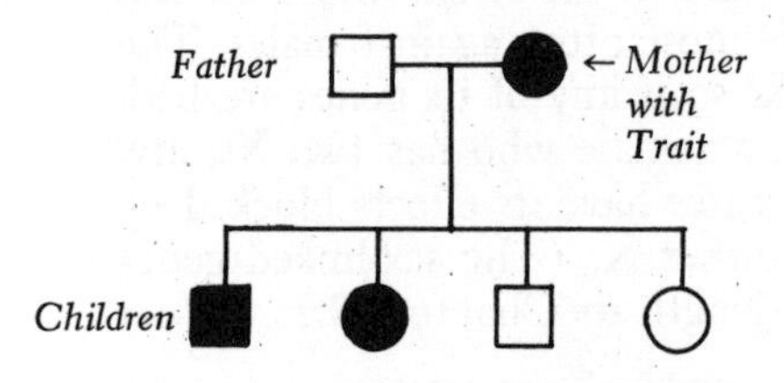

Any parent carrying dominant gene shows the trait, and average of one in two children will receive the gene and also show the trait. If parent has two of the dominant genes, every child will show trait.

2. Simple Recessive Pedigree*

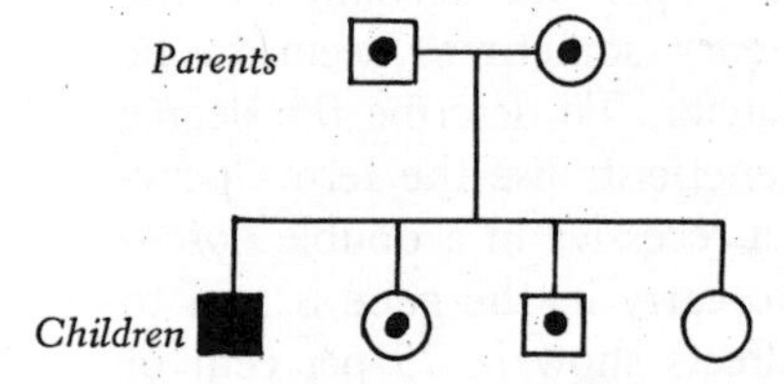

Both parents here carry same recessive gene. Average of one in four children will receive both *genes and have the trait. One in two children will receive only one gene and be carriers, but without the trait. One in four children will be entirely free of the gene.*

3. Simple "Sex-Linked" Pedigree*
(Gene Carried on "X" Chromosome)

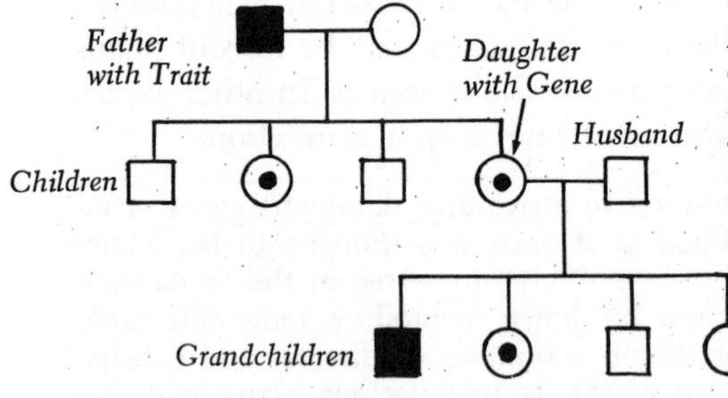

Father with trait passes on gene to all *daughters. Each is a* carrier, *and if she has children chances are one-in-two any son will receive gene and have trait, like his grandfather; and one-in-two any daughter of hers will be a* carrier, *like herself.*

* Details regarding other types of dominant, recessive and sex-linked genes will be found in the Wayward Gene Forecast Tables in the Appendices.

Expressivity describes another variable characteristic of genes: the capacity of the same gene to express itself in different ways or degrees in different individuals. For example, in some of the hand abnormalities we'll deal with, the same gene in the same family may produce much more marked or severe effects in one member than in another, or even in one hand than another of the same person. These variations in behavior of the same gene should not be confused with different gradations of a similar condition produced by quite different genes. Also to be noted is that where approximately the same condition crops up in different families or individuals at the same or different stages of life (as in certain eye defects, or muscular conditions), it may be because of different genes which are distinguished from one another not by their end effects but by the fact that they achieve the same end results by different means.

Tracing the chemical behavior of wayward genes, medical geneticists have been coming ever closer to finding out in what ways many of these genes make their mistakes through what are called *errors of metabolism.* In fact, in the case of certain genes it is now possible to identify the precise nature of their chemical errors and to pinpoint the stage at which these occur. For example, there is one gene whose function is to form an enzyme which directs the conversion of the chemical phenylalanine, a food ingredient, into an essential substance called tyrosine. When the gene fails to do this properly, an abnormal amount of phenylalanine is set loose in the blood and spinal fluid; and as one of a number of ensuing chemical effects, the brain of the individual is impaired. (This condition, called *phenylpyruvic idiocy,* or *phenylketonuria,* will be discussed more fully in Chapter 27. Other errors of metabolism will be dealt with in Chapter 25.)

Another way in which genes can work chemically is by affecting the nature or amount of the hormonal secretions produced by the endocrine, or ductless, glands. The most common results of specific malfunctions of these glands, whether due to heredity or not, were listed in the chart on page 179.

In some cases a single wayward gene produces only one significant effect, in other cases several effects. But often a whole cluster or series of defects (a *syndrome*) can result directly from the action of a single dominant gene or a pair of recessive genes. (Consider the plight of the unfortunates who get blindness, deafness, deformity and mental defects, all in one genetic package—as in *amaurotic idiocy* [Tay-Sachs disease], or in some forms of *retinitis pigmentosa.*) As a rule, syndromes result when the gene failure relates to a key chemical process, affecting many parts of the system, and begins producing its effects early during the body's development—sometimes in the egg stage. But whether genes are wayward in a big way, or a small way, or work singly or in teams, one may well ask: What caused them to depart from the normal or customary gene path?

A few pages back, we referred to the possibility that DNA steps in a given gene may be "shaken up, altered or damaged at some time." This actually describes a gene *mutation* which, as we briefly noted in Chapter 9, can occur in a number of ways. In recent years the public has been made aware of how gene mutations can be caused by X rays or by atomic radiation. But long before there were any X-ray devices or atomic bombs or atomic tests, gene mutations were being induced through natural radiation (cosmic rays, radioactive rocks, etc.), through various chemicals, or, most often, through *spontaneous* changes in genes without any known outside influences.

At any rate, it is through successive and repeated gene mutations that all the existing varieties of genes have come into being (as will be explained at length in Chapter 40, "Evolution"). Moreover, all of the processes by which genes have mutated are still going on; and, what is most important with respect to wayward genes, through an entirely new mutation a given hereditary disease or defect can occur in a person even though neither the condition nor the genes for it had ever before been present in his parents or ancestors. Such mutations, in fact, may account for a considerable proportion of the new cases of the most serious hereditary conditions (for example, *hemophilia*, or *genetic muscular dystrophy*), particularly those causing early death, the genes for which would be gradually eliminated were they not continually replenished by new mutants.

But mutations of individual genes are by no means the only way in which serious genetic afflictions can arise for the first time in members of a family. *Chromosome alterations* can also be the culprits. What these may be, how they occur, and the specific types of damage they can do, began to be revealed only in recent years. Earlier, studies of fruit flies had shown that various types of chromosome abnormalities in individuals—extra chromosomes, a missing chromosome, a "crippled" chromosome, two paired chromosomes sticking together, or a part of one chromosome hitched on to another—could cause defects and abnormalities of one kind or another. But evidence that this also was so in human beings did not come until 1959, when new techniques made it possible to identify and study human chromosomes with great precision. Suddenly, where before one could only *deduce* that certain

CHROMOSOME ABNORMALITIES

In the normal reproductive process, a sperm and egg each carry one *of every 23 pairs of the parent's chromosomes. But in abnormal situations:*

ABNORMALITY	PARENT A'S GERM CELL	PARENT B'S GERM CELL	RESULTING INDIVIDUAL
1. Extra chromosome* *(a) Germ cell of one parent carries* both *of a pair of chromosomes instead of just one.*			Triploidy *(or trisomy)—three instead of two of a given chromosome*
(b) The extra chromosome is attached to one of the other chromosomes			*Apparently normal (46) chromosome count, but actually one extra*
2. Missing chromosome* *One of parent's chromosomes had been lost when germ cell was formed*			Monosomic *individual, lacking one chromosome of a given pair*
3. Deletion *Part of a chromosome broken off and lost during egg or sperm formation*			*Individual lacks portion of genes in one chromosome*

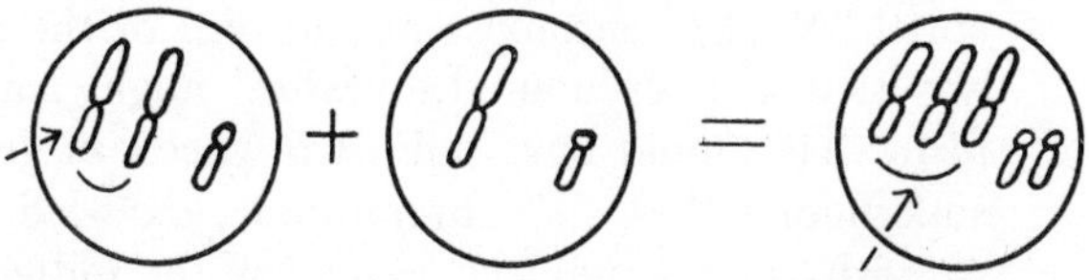

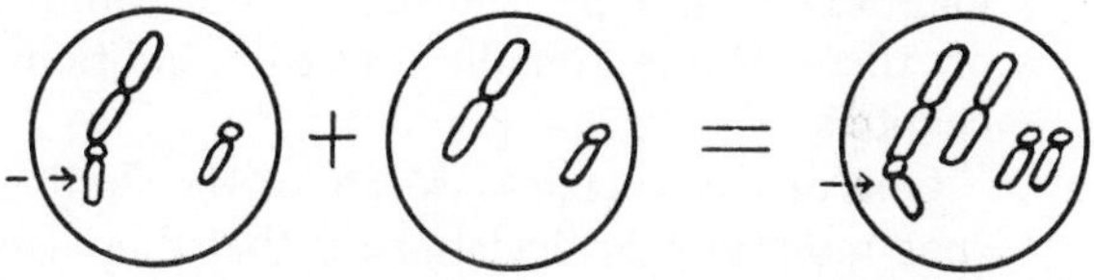

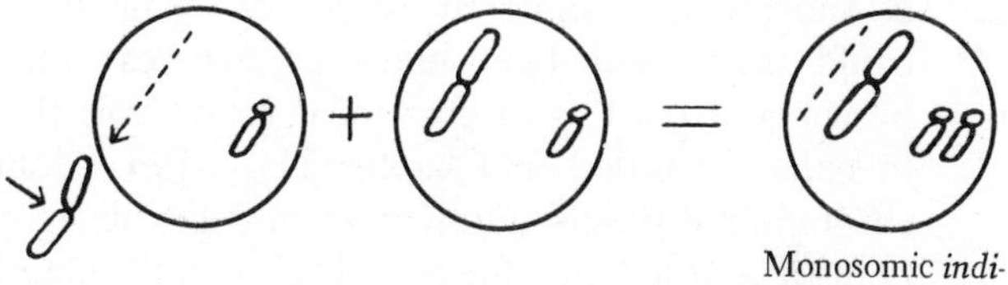

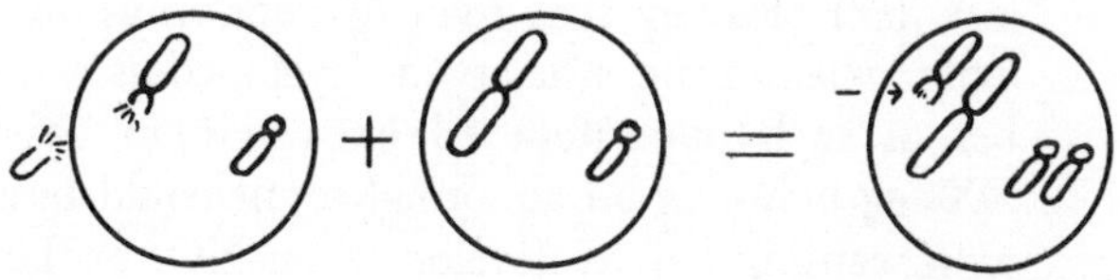

* If the extra or missing chromosomes are sex chromosomes, various abnormalities in sex development—Klinefelter or Turner syndromes, etc.—may result. (See Chapter 27.) If the extra chromosome is No. 21, Mongoloid idiocy may result (Chapter 26). Another sex-chromosome abnormality, mosaicism, is illustrated and discussed in Chapter 27.

human aberrations might be caused by genetic abnormalities, one could now, in some cases, *actually see* these genetic abnormalities under the microscope!

The first startling finding was that *mongoloid idiocy,* long one of the greatest medical mysteries, develops as the result of an individual's being started off at conception with an extra chromosome: three of the small "No. 21" chromosomes, instead of the normal two. Much in the way that a superfluous third wheel might completely disrupt a mechanism that should have only two wheels in gear at a certain point, the superfluous "No. 21" chromosome, crowded into a fertilized egg, apparently throws the development of the individual askew in many ways. Further studies showed that various other once perplexing conditions, including several serious types of faulty sexual development (the Klinefelter and Turner syndromes, etc.), were due to chromosome abnormalities of the types shown in the accompanying illustration. (Details of these chromosomally caused conditions will be given in later chapters.)

One of the amazing aspects of the findings about chromosome and gene mutations or deviations is that they have shown it to be theoretically possible sometimes for even identical (one-egg) twins to be *genetically different*. Such a situation could occur if, after an egg divided to form twins, a genetic change occurred in one of the halves but not in the other. A mutated "eye-color" gene in one twin, for example, could make that twin blue-eyed, whereas the other twin could be brown-eyed (just as in occasional individuals the two eyes are different in color, as noted in Chapter 11). Most astounding, if a whole sex chromosome should go awry when a fertilizing egg is dividing to form twins, the result may be as radical as that one of the "identical" twins would be a *male,* the other a *female*. This could happen with an XY egg if, after the chromosomes doubled and the egg divided, one half lost its Y, leaving that twin to develop as an XO, or Turner-type, female, whereas the other twin developed as a normal XY male. A number of such opposite-sexed one-egg twin pairs have been found. In XX-egg twinning an abnormal result could be the loss of an X by one twin, causing her to develop as an XO, or Turner-type, female, while her XX twin would be a normal female. Added to other afflictions in such a situation, the abnormal sister would be exposed to the effects of any wayward recessive genes in her single X chromosome (for instance, a "color-blindness" gene), just as would be a male with only one X; whereas her twin sister, with a safeguarding additional X, would have the same reduced chance of developing sex-linked conditions as would any other normal female.*

Our knowledge of the role of inheritance in diseases and abnormalities will undoubtedly continue to increase in scope and accuracy. Yet there will remain for us this question: "What good is it to know that a disease, defect or abnormality is inherited?"

As we proceed further to discuss specific conditions, it will become clear that the facts regarding hereditary influences can indeed be of great value in many ways: in helping doctors to diagnose various afflictions, alerting them to their appearance in families, enabling them to take important preventive measures; in giving persons who have serious inherited defects, or who have close relatives with such defects, an awareness of what hereditary dangers do or do not confront their children; and in aiding medical researchers to determine the causes of many baffling ailments, thus opening the way to treatments and cures.

Again, while the ensuing chronicle of diseases and defects may seem foreboding, you need not find it too upsetting. On the brighter side you will note that the great majority of the conditions which are most strongly hereditary—and particularly those which are most menacing—affect only a small number of individuals, whereas the most prevalent of mankind's major bodily ills are largely due to or greatly influenced by environmental factors. Moreover, the effects of increasing numbers of the hereditary conditions can now be overcome or controlled, as in the case of such ailments as diabetes and other chemical disorders, defects of vision and of hearing, some of the blood diseases, some of the skin afflictions, and so on.

In reading through the chapters ahead (and in checking the Wayward Gene Tables in the Appendix), it is not unlikely that you will find various conditions which have personal meaning for you either because you yourself are affected, or because they are present, or threaten to appear, in members of your immediate family. Wherever this is true, you might be cautioned (if you are not a physician) to draw no fixed conclusions nor make any decisions before or until you have consulted with some competent medical authority. In every case, it is important

* Further indication that one-egg twins—and particularly supertwins—may not always be identical in their genetic makeup or chemical reactions has come from studies of the nine-banded armadillo, an animal which regularly produces one-egg quadruplets. Dr. Kurt Benirschke has shown that the four presumably "identical" quads often do not accept skin grafts among themselves, possibly because of chemical differences engendered during the prenatal stages.

to establish, first, that the condition which concerns you personally is actually the one which is discussed or listed in this book, for in many instances an hereditary condition may be paralleled by one with quite similar symptoms which is *not* hereditary. And, second, it should not be assumed that some condition is not hereditary simply because it is not mentioned in this book. We have not sought to list every known wayward-gene condition, no matter how rare or minor; nor would the character of this book warrant such a listing. Moreover, it is to be expected that from time to time additional conditions will be identified as hereditary. However, every effort has been made to present in this book the essential facts about all diseases, defects and abnormalities of major importance, or of interest to a considerable number of readers, which at this writing are known or suspected to be influenced by heredity in any significant way.

CHAPTER TWENTY ONE

The Big Killers

ALMOST EVERY DAY you read that such-and-such a disease is the "No. 1 Killer" or the "No. 2 Killer"—or some other rank—among the disease enemies of human beings. What interests us here is the part that heredity may play in these most menacing conditions. Of first significance, then, is the fact that the ranking order of the killers has undergone a striking change within a few generations, for this would not have been so *if heredity played the major role* in the chief causes of death.

Let's look at the "box score" of deaths from the ten principal causes in the United States in 1960, as shown in our Table. *Heart diseases* stand clearly at the top, claiming more than twice as many victims as the No. 2 killer—*cancers*—and more than three times as many as the No. 3 cause of death, *cerebral hemorrhage*.

But at the beginning of the century, in 1900, the listing was radically different. Then the No. 1 killer was *pneumonia and influenza*, with a death rate of 202 per 100,000—over five times the 1960 rate. No. 2 in 1900 was *tuberculosis*, with 194 deaths per 100,000. By 1960 this same disease had dropped far out of the first ten, down to No. 15 among the causes of death, with a rate of about 6 per 100,000—3 per cent of the 1900 rate. But *heart diseases* in 1900 were only fourth on the list, and *cancers*, eighth on the list. Other rankings in 1900 were as follows: No. 3, diarrhea and enteritis; No. 5, cerebral hemorrhage; No. 6, nephritis; No. 7, accidents; No. 9, diphtheria; and No. 10, meningitis. (All of the foregoing estimates and comparisons are approximate, because diagnoses and census death records in 1900 were not always on the

Table II

The Biggest Killers

Deaths from twelve principal causes for males and females (Whites) in the United States (1960)

Cause of Death	Males *Rate per 100,000*	Females *Rate per 100,000*
Diseases of heart	454.6	306.5
Cancers and other malignancies	166.1	139.8
Cerebral hemorrhage ("stroke")	102.7	110.1
Accidents	70.0	31.5
Influenza and pneumonia	39.2	29.0
*Certain diseases of early infancy**	38.6	26.2
General arteriosclerosis	20.0	21.8
Suicide	17.6	5.3
Cirrhosis of liver	15.6	7.4
Other diseases of circulatory system†	14.0	8.4
Diabetes mellitus	13.7	19.1
Congenital malformations	13.3	10.8

* Refers to the aftermaths of birth injuries, asphyxia, infections of the newborn, extreme prematurity, etc.

† Includes hypotension (extreme low blood pressure), diseases of the capillaries, diseases of the lymph nodes, etc.

Source: *Vital Statistics of the U.S.*, 1960, Vol. II, Sec. 1, pp. 1-5, 1-26*ff*. U.S. Dept. of Health, Education, and Welfare, 1963.

same basis as in later decades.)*

The sharp change in the order of the "killers" has been due principally to two related factors. One is the vast improvement in living conditions and in medical care, especially with regard to the prevention and treatment of infectious diseases (the latter factors account for the big drop in the tuberculosis death rate, for the drop in meningitis deaths to 3 per cent of the 1900 rate, and for the almost total elimination today of deaths from diphtheria). These environmental improve-

* Additional uncertainties arise in comparing the mortality rates of Negroes with those of Whites for given causes. The diagnoses of causes of death reported for Negroes have often been unreliable, especially in the past, and cannot be equated precisely with the data for Whites. This, principally, is why our Table of "Big Killers" lists only the rates for Whites. However, the main points regarding comparative Negro and White mortality rates and life expectancies are given in Chapter 29, and these may be considered dependable.

ments, bringing their greatest benefits to infants and children, have permitted ever-greater numbers of individuals to live well into the mature years. And this, in turn, explains the second factor: With the number of middle-aged and older persons steadily increasing (from a proportion of 18 per cent of the population aged forty-five and over in 1900, to 30 per cent by 1960), there has been a corresponding increase in deaths through diseases which come mainly in the later ages, such as heart conditions, cancer, cerebral hemorrhage and diabetes.

What meaning does this change have with respect to heredity? Obviously, the hereditary makeup of our population has not changed in any important way. Only the environment has changed. But the more we eliminate the worst hazards of environment and the more we equalize conditions for all persons, the more the *inherent differences in individuals can assert themselves.* The more important, then, also, becomes the role of heredity in creating distinctions, large and small, between human beings with respect to what diseases and defects afflict them, and how serious these turn out to be.

In short, it is increasingly easy to see not only where heredity is almost directly responsible for certain conditions, but where and how it exerts varying degrees of influence on many others, including some of the major diseases in which the role of heredity was previously obscured. As an introduction to what will follow, we might make these general classifications:

Most directly inherited among the "killers" (with environment playing only a small part in causation) are perhaps the majority of cases of common diabetes, some of the very rare forms of cancer, and various conditions to be discussed in later chapters.

Conditionally inherited (where, with the given wayward genes, a person will develop the affliction under certain adverse circumstances) are some types of heart and arterial diseases (including arteriosclerosis, and possibly rheumatic heart disease), and a number of metabolic disorders.

Influenced by heredity, in some degree or another, may be virtually all of the other major diseases. But there are important qualifications. In the major heart and arterial conditions the evidence for some probable hereditary role is very strong, whereas in common cancer the *possible* role of heredity is still uncertain. In the case of the infectious diseases, while there certainly are not and could not be any genes for tuberculosis, pneumonia, syphilis or any other germ disease, it is quite likely that inherited constitutional weaknesses may make some individuals an easier prey than others to given infections.

All of this will be clarified when we take up specific diseases in detail. But note, first, that we are confining ourselves in this chapter to the major diseases which are killers and are rating them in terms of how many lives they claim each year. If we were to rank diseases as enemies in accordance with how much unhappiness they cause, or what their effects are on society, the order of importance might well be different. Topmost among mankind's principal enemies, with heredity playing a prominent role, one might have to place mental diseases and defects. Also, in evaluating diseases as major or otherwise, it would be necessary to consider not only how many people they afflict, but *when* they strike and what damage they do, even if they do not kill. By this standard, a condition which strikes at the very young, bringing a lifetime of disability, would be regarded differently from one which does not afflict or carry off a person until the twilight years; and a disease which brings death suddenly and painlessly might have to be weighed differently from one which is attended by years of extreme suffering.

So, as we discuss the various diseases from the standpoint of heredity, it is for you, the reader, to evaluate them in your own way, according to their meaning for you and for those close to you.

HEART AND BLOOD VESSELS. There is hardly a family in which some member or immediate relative isn't afflicted by, or hasn't been carried off by, one or another of the heart or blood-vessel diseases. Most common and serious of these are ***hypertension*** (high blood pressure), ***arteriosclerosis*** (hardening and thickening of the arteries), and ***rheumatic heart disease***. In all of these conditions heredity, directly or conditionally, may play some part. Even when a non-hereditary condition such as syphilis is involved, the general genetic makeup of the individual may be a factor in the way the heart or other organs are affected. But also ever present in the heart conditions are environmental influences. In some of the ailments to be discussed, defects present early in life may lead to serious heart disturbances later, but only under adverse environmental conditions of diet, stress, infections, etc. Thus, even when individuals—sometimes in the same family and sometimes identical twins—have the same defect, there may be great variability in the way it does or does not ultimately manifest itself in the heart ailments, depending on similarities or differences in given environmental influences.

Hypertension, when not an effect of other diseases, is more likely to occur in persons with a family history of the condition than in those without one. Although no specific genes for this disease have been identified, there is a strong possibility that a predisposition to it may be hereditary. (Aggravating influences may be nervous strain, intense living, and adverse personality factors.) Where hypertension is most serious is in the threat that it may lead to (1) heart disease (enlarged heart and heart failure) or coronary artery disease, with blood clots; (2) apoplectic stroke (brain hemorrhage); or (3) kidney diseases.

The problem of identifying the role or roles of heredity in hypertension is complicated by the fact that the condition is so widespread and variable, and that the different possible genetic and nongenetic causes are so often mixed together. However, one way in which inheritance is believed to induce hypertension is by affecting the manner in which the systems of different individuals utilize salt and water. The level of blood pressure is closely related to the retention and excretion of these substances, and if they are not adequately dealt with by the body because of some genetic abnormality, high blood pressure may result. Sex differences also are important in hypertension, for although the incidence of the disease itself is considerably higher among women, it is less often fatal for them than for men.

Arteriosclerosis, responsible for from 25 to 40 per cent of the heart deaths, involves the thickening and hardening of the walls of the arteries, with a resultant loss of elasticity and interference with the blood circulation. A common and serious type of this condition is *atherosclerosis,* in which fatty deposits encrust the artery walls. Blood clots may then form on these fatty "plaques," reducing or completely shutting off the blood supply to vital organs. When this occurs in the coronary arteries on which the heart muscle depends for its blood, there can be a sharp chest pain (*angina pectoris*), or, if more severe, a heart attack (*coronary thrombosis*). Arteriosclerosis may also affect the blood vessels in the brain, especially in the case of elderly persons, and produce strokes (*cerebral thrombosis*); or it may affect the arteries of the lower limbs, causing pain on exertion, and in severe cases resulting in gangrene.

Heredity may foster the development of atherosclerosis through at least two kinds of genetic "errors" in the metabolism of fat. One of these conditions, *hypercholesterolemia,* causes a marked rise in the individual's level of the fatty substance cholesterol, crystals of which then accumulate in the walls of the blood vessels. In a second condition, *idiopathic essential hyperlipemia* (sometimes present in children who have peculiar attacks of abdominal pain), there is a genetic failure to clear the blood properly of the dietary fats (or *lipids*), and again fatty particles are deposited in the blood-vessel walls. Also, in both of these conditions, one visible outward effect may be the appearance of fatty deposits under the skin at various points of the body (called *xanthelasma*) and sometimes on the eyelids. Dominant or semidominant genes are involved in both of these conditions. In the case of hyperlipemia, one gene for the condition will predispose the individual—but mainly if a male—to the danger of early heart attack (before age forty). This gene action may be involved in as many as 15 per cent of all heart attacks. Where an individual—male or female—receives two of these genes, atherosclerosis generally develops in childhood and may prove fatal before the age of twenty.*

If left untreated, the foregoing defects of fat metabolism can cause not only atherosclerosis, but also—particularly in men—coronary thrombosis at an early age. In women with these defects, atherosclerosis usually develops to a milder degree than in men, and if coronary thrombosis does occur, it usually comes later, generally after the menopause, when the hormonal state undergoes changes. Before then, apparently, a woman's high level of estrogen (the female hormone) in some way combats atherosclerosis. Thus, one of the treatments for this disease involves administration of estrogen, although in men it may have undesirable side effects. Among other possible genetic reasons why men are more vulnerable to atherosclerosis is the fact that, at birth, normal male infants already have thicker coronary artery walls, and consequently a narrower blood passage, than females. This, coupled with the apparent protective effect of the female sex hormones, may further help to explain the much lower frequency of coronary atherosclerosis in women than in men.

Congenital heart defects, which—if not fatal in early life—often give rise to serious heart conditions in later years, are perhaps in most cases due to prenatal accidents or influences of a purely environmental nature (for instance, German measles in the mother, or some of the other causes discussed in Chapter 6). However, some types of congenital heart defects may possibly result from gene failures, or (as in mongoloid idiocy and some of the abnormalities of sex development) from chromosome deviations. Whether the causes are environmental or genetic, the end results may be the same. The heart is formed during the third to fifth week of embryonic life, and among the things that can go wrong at this stage are failure of the walls separating the heart chambers to close properly; malformation of the valves governing the

* A genetic predisposition to atherosclerosis may come from still another direction—by way of *diabetes* (i.e., the common form), which, among its effects, also produces elevated levels of fat and cholesterol in the blood. (Diabetes will be discussed later.)

blood flow through the heart; and faulty construction or misplacement of the heart's blood vessels. Where such defects are traceable to failures of specific genes (or are among the effects of some genetic syndrome), one frequently finds the same congenital condition appearing in children of certain families over successive generations. However, if the causes arise through whole chromosome abnormalities (which may occur, as explained in Chapter 20, prior to or during egg fertilization), the resulting heart deformities and other effects usually prove so severe as to bring death well before the reproductive age is reached, and thus such defects are not likely to be passed on.

Rheumatic fever (or childhood rheumatism), which mostly affects children and younger adults, may possibly be influenced in its appearance by genetic tendencies, although this remains uncertain. One of the most puzzling of diseases, it is characterized by joint swellings and pains, chorea (St. Vitus dance—occurring more often in afflicted girls than in boys), fever, and inflammation of the heart. It claims as its victims one in every 100 school-age children and is the fifth leading cause of death among those aged five to fourteen. It is estimated that about 2 million Americans have had or will have an attack of rheumatic fever during their lives and that probably half of these will suffer some residual heart damage as a result of the disease.

As to its immediate cause, it is now widely believed that rheumatic fever may follow a streptococcal infection, with the fever and symptoms being a sort of allergic reaction to the invading bacteria. However, there is a possibility that this reaction, and the disease, is most likely to develop in individuals who have a special genetic susceptibility. Some evidence for this comes from studies showing above-average incidences of rheumatic fever in certain families, and among relatives of affected individuals. Connected with these findings is the theory that the susceptibility to the disease may be caused by a pair of recessive genes. In any case, the incidence of the disease can be much reduced by prompt treatment of streptococcal infections with penicillin and by other preventive measures.

Varicose veins, a condition which involves a weakening of the walls of the veins, leading to their enlargement, appears only in some cases to be directly linked with heredity. The ailment is quite common, especially in women after childbirth and in persons of either sex whose occupation requires continual standing. The predominant or precipitating causes as a rule are external, but in certain families varicose veins have been found to be transmitted through an irregular dominant gene.

CANCER. The greatest of all disease mysteries is presented by cancer. Thousands of medical and scientific detectives throughout the world have been working to throw light on this disease, or group of diseases, and to track down any hereditary factors that may be involved. As yet we know only that heredity is directly responsible for just a very few rare types of cancer—not necessarily related to the others—and plays a part in certain additional types that are more prevalent. But in the common cancers, although most authorities believe there is some degree of hereditary influence, just what the influence is and how it may operate is still a matter of conjecture.

Why it should be so difficult to clarify the role of heredity in common cancer, when in every community large numbers of people are afflicted or killed by it annually, becomes clear when we consider its many baffling aspects. Cancer is not a specific disease. It is a general term for malignant areas in body tissues or organs where cell growth is speeded up abnormally and the cell mechanisms are so disrupted that they cannot perform their proper functions. There are many different types of cancer, each with its special peculiarities and its particular sphere of action, but all cancerous growths have this in common: Their abnormal cells do not partake in the orderly processes of the body, and as they multiply they also become outlaws which invade, starve and destroy the neighboring cells and tissues. Eventually they break loose, and by way of the lymphatic fluids and blood stream, spread throughout the body, wreaking general havoc and ultimately bringing death to the host individual.

How do these cancer cells originate? Not so differently from the ways in which human outlaws and killers might be launched on criminal careers: through some inciting influence in the environment; through an inner instability; or through a combination of adverse internal and external forces.* Among the environmental factors which can cause cells to become cancerous are hundreds of irritants called *carcinogens*, including a great variety of chemicals and abrasive dusts, as well as radiation particles. Also strongly suspected now as a cause of human cancers are *viruses*, since certain viruses have been found capable of inducing

* "It now seems clear that cancer, regardless of its inciting 'causes,' is ultimately the result of a change in the genetic material of the cell. Any change in the cellular blueprint, it is now known, brings about a change in cellular function which means simply a change in one or more of the cell's enzymes." (From the Sloan-Kettering Institute for Cancer Research, Progress Report XVI, July, 1964.)

cancer formation in experimental animals (although only under prescribed conditions). In some cases it is believed that the virus remains dormant (in a "sleeping" state) and is aroused to cancer-inducing action only when the cells are disturbed by radiation or various chemicals. Whether there is such a connection between viruses and human cancers awaits proof.

But whatever the outside stimulus, the likelihood exists that some degree of *inherent instability* in the cells of some persons may make them more than ordinarily prone to the development of cancerous growths.

Heredity might then enter the human cancer picture in a number of ways:

(1) Through inheritance of specific abnormalities of certain tissues at given sites, which might render them unusually sensitive to cancer-inducing influences.

(2) Through the general biochemical makeup of certain individuals, as governed by combinations of genes, which might make their bodies at any location more than ordinarily susceptible to cancerous changes.

(3) Through hormonal action. It is known that the induction and growth rates of certain cancers are influenced by the body's hormonal secretions. In certain strains of mice, abnormal amounts of certain hormones can induce malignant changes. If this is true also in human beings, hereditary factors might cause the hormones of some individuals to be produced in excess and thus, be more "carcinogenic" than those of other persons. So far, the relationship between hormones and human cancers has been demonstrated only in the case of the sex hormones, primarily by their use in cancer treatment. Breast-cancer growths in women before the menopause can sometimes be slowed by administering male hormones and by removing the ovaries or other endocrine glands which produce female hormones. In men, cancer of the prostate can be slowed in growth by administering female hormones and by removing the testes, the main source of the male hormones.

Most recently linked with certain cancers are various abnormalities in the *chromosomes* (too many or too few, breakages, malformations, etc.). Such abnormalities are frequently found in human cancer cells, but it has not been established whether they are at any given time the causes, rather than the results, of cancerous growths. To date no specific chromosome abnormality has been found associated with any particular type of cancer, except in the case of one form of blood cancer, *chronic myelocytic leukemia.* In this disease (which differs from the more common *acute leukemia* in that it occurs mainly in older people and takes longer to produce serious effects), one of the chromosomes (probably a "No. 21") in the affected cells is often found to be slightly smaller than the normal size, suggesting the possibility that a part is missing.* If this chromosome abnormality is present only in the individual's white blood cells and not in the germ cells, it would not be hereditary. (See page 328.)

But applying any of the foregoing facts or theories about cancer inheritance to a specific person or type of cancer is far from simple. For example, take *breast cancer.* A grandmother, a mother and a daughter all have had or have breast cancer. Is this not proof of inheritance? No. Where any type of cancer is very common, as this cancer is, its appearance in a number of members of the same family might be no more than a coincidence; or, even though the cancers in the case cited were all eventually focused on the breast, the cancer cells might have originated at different points and through different means. True, in animal experiments strains of mice have been bred in which specific genes—usually only under prescribed conditions—cause spontaneous cancers of the breast to develop in a high proportion of the females. But neither the breeding involved nor the conditions required to produce such cancerous strains has been or can be paralleled in human beings.

Thus, proof of inheritance of human cancers must come through other means. One approach is to study families in which cancer has appeared. Such studies have shown that among close relatives of persons afflicted with cancer, the same kind of cancer—but not other types of cancer—occurs considerably more often than it does on the average in the general population. This has been found true of breast cancer, prostatic cancer, and cancer of the stomach and intestines. The suggestion is, then, that any inherited human predispositions toward cancer are specialized, each kind working to make only certain tissues or organs susceptible to malignant change but not increasing the chances of developing other types of cancer. It should be stressed, however, that, even where the familial susceptibility to a special type of cancer has been demonstrated, the risk of developing that cancer may not be much greater for members of the family than for persons in general. For instance, the risk of breast cancer is about 1 in 100 for the average woman, but is twice as great—1 in 50—if a woman's sister or mother has had breast cancer. Yet even for this woman with an afflicted close relative, the odds are still 50 to 1 *against* developing breast cancer.

* This abnormal chromosome is sometimes referred to in the genetic literature as the "Philadelphia chromosome," after the city where it was first identified by the investigators, Drs. P. C. Nowell and D. A. Hungerford.

In a second approach, through the facts about cancer in twins, some studies have indicated that when one twin has cancer the chances that the other will develop the same cancer (and at nearly the same time) are considerably greater when the two are an identical pair than when they are fraternal. However, recent findings in Denmark tend to minimize the importance of the twin relationship in cancer. Specifically, geneticists Bent Harvald and Mogens Hauge, after checking completed histories of many hundreds of twins in whom cancer had appeared, found that the proportion of cases among identicals in which both members of the pair were afflicted, and/or had cancers of the same site, was only slightly higher than among fraternals. The investigators therefore concluded that while it is logical to assume that genes determine at least some of the characteristics of living tissues which allow malignant growth, the evidence from their twin study suggests that either with respect to cancer in general, or the appearance of cancer at specific sites, the genetic influence is negligible compared to the effects of other factors.* (As to other studies which have shown a significantly higher concordance of cancer in identical twins than in fraternals, it has been suggested that the greater similarity in the environments of the identicals may have contributed to the difference.)

The origin of cancer is often most puzzling when it appears in very young people; for despite the fact that cancer is spoken of as a degenerative disease, it is by no means confined to the middle-aged or elderly. Indeed, among children in the United States aged from one to fifteen, cancer now kills more than does any other disease. The cancers afflicting these children include malignant tumors of the blood, bone, skin, eyes and kidneys. Sometimes such cancers are found to have begun developing long before birth and may or may not have been influenced by heredity.

Further suggestive evidence that genetic factors, in interoperation with environmental influences, are involved in the development of cancers—and of particular types of cancer—comes from comparison of the cancer rates for women and men. In every type of cancer the incidences for the two sexes differ considerably, with the male rate being

* Some of the figures reported by Drs. Harvald and Hauge were these: In the twin histories studied (1,528 identical sets, 2,609 fraternal, born in Denmark between 1870 and 1910), cancers had occurred among 164 of the identical sets but had developed in both members in only 21 cases, and in these cancer of the same site appeared in only 8. Among the 292 fraternal sets with one or both afflicted, cancer had appeared in the two members in 48 of the cases, and of these, 9 had partners with cancer of the same site and 39 with cancers of a different site. (Reported in *J. Am. Med. Assoc.*, Nov. 23, 1963.)

higher for all except the cancers specific to women, such as cancers of the breast and the womb. Although occupational hazards may account for part of the higher rate among men in cancers of the skin, lungs and throat, there seems to be little question that the greater *inherent* susceptibility of males to most diseases also applies to cancers. (The possible influence of the sex hormones on cancer was touched on previously. But there is a strong possibility that the whole genetic constitution of males and all aspects of the male chemical makeup and functioning, as differentiated from that of the female, have a bearing on relative cancer incidences and fatalities in the two sexes. This will be dealt with further in Chapters 22 and 29.)

Even less clear are the facts about variations in cancer incidences among people of different races or ethnic stocks. For instance, statistics may show that the general cancer rate for Negroes in the United States is much lower than that for Whites. But this doesn't mean that Negroes are any more resistant to cancer. Proportionately fewer Negroes may die of cancer because fewer among them live into the older ages when cancer is most likely to develop. As life spans of Negroes have lengthened, their cancer rate has gone up. So also, the much higher cancer mortality for persons of some ethnic stocks (Irish, German, English, etc.) as compared with others (Italians having the lowest rate) must be viewed in the light of a great many complex factors before any conclusions can be drawn.*

With respect to particular types of cancer, one again finds significant racial and ethnic differences. Some of these can be linked with known environmental causes. For instance, in Egypt there is an unusually high incidence of cancer of the bladder, and this has been traced to the presence of a particular parasite found only in this part of the world. In China, a high frequency of cancer of the liver has been traced to another parasite whose eggs irritate the liver cells. Similarly, differential exposure to various other environmental influences may account for greater or smaller incidences of specific cancers not only among peoples of different countries, but among various groups in any given country Thus, cancers of the skin, mouth, throat or lungs may have above average incidences among workers who are continually exposed to

* These reservations also apply to findings (reported in 1963) that Polish-American females had a risk of stomach cancer almost three times that of other foreign-born females; Italian-born males had a risk of cancer of the colon 2.7 times as high as that of other foreign-born persons; while among German-American males the risk for mouth cancer was only two-thirds that of other foreign-born persons. (These figures are based on a study of more than 35,000 patients at the Roswell Park Memorial Institute of Buffalo, New York.)

abrasive dusts, as in mining, quarrying, pottery making, metal working and glass blowing, or to chemical irritants, as in various chemical industries. Also, cancers of given types may vary in incidence as between persons living in large manufacturing centers or in rural farm communities. In short, where persons of different stocks are distributed disproportionately in particular environments in which cancer incidences are slanted in given directions, the apparent racial or ethnic differences in rates and types of cancer need not at all be related to heredity factors.

Another question involving environmental cancer causation which has been much discussed is the relationship between cigarette smoking and cancer of the lung. While the debate about this is currently still going on among doctors and scientists, the weight of opinion seems to be that cigarette smoking does contribute to the development of lung cancer, although in what way, specifically—through nicotine, coal tars or other irritants—has not been established. Theoretically, heredity may be involved (1) by making the lung tissues of some persons genetically susceptible in above average degrees to cancer, and/or (2) by way of genetically influenced personality factors which might make some individuals more than ordinarily likely to become heavy cigarette smokers.

For the public in general, *radiation* now poses or will pose one of the greatest environmental dangers of cancer induction. The first warnings of this were given many years ago, when it was found that X rays, if not properly controlled, could unbalance and injure normal cells and cause cancer. (As noted in Chapter 20, X rays can also cause mutations in the genes of germ cells, giving rise to various hereditary defects and diseases.) Thus, many doctors and radiologists of earlier years who did not protect themselves against X rays subsequently became cancer victims, and even today the incidence of leukemia (a blood cancer) and of skin cancers is well above average among those who have worked frequently with X-ray apparatus or who are unduly exposed to industrial radiation. (Some readers may remember the tragic case of a group of girls in a New Jersey watch factory who worked at painting luminous dials—pointing the brushes with their lips—and who all developed cancers and died within a space of twenty years.) Nor need it be added that atomic bombs (if they are ever widely used) or atomic plants for peacetime purposes (if proper safeguards are not set up) may set loose huge quantities of radiation which might loom up as a new colossus of cancer danger.

However, radiation also has a bright side in relation to cancer, for it offers a frequently effective means of treating the disease. Just as radiation can disrupt normal cells, it can also, if focused properly, destroy malignant cells. Even more effectively, surgery, alone or in combination with radiation, can eradicate localized cancers, and now also chemical and hormone treatments sometimes provide means of arresting cancerous growths. This is why it is so important to be alerted to the risks of developing cancer and to the knowledge of any possible hereditary factors. When cancer was considered a hopeless condition, there might have been little reason to alarm individuals about its threats to them. But as medical science continues to make steady progress in finding the causes of cancer and the means of preventing or combating it, the more that laymen as well as doctors know about cancer, the closer we may come to reducing its hazards.

DIABETES. Playing and living in a special camp in the Catskills each summer are several hundred children with diabetes. They are some of the thousands of children throughout the country who are similarly afflicted. Half of the children in the camp (maintained by the New York Diabetes Association) come from underprivileged homes, where the means are not available to give them the special care that they require. But the disease is just as likely to afflict children in wealthy homes. For there may be nothing unusual in the early diet or living conditions of a child to account for the onset of diabetes. Where there is no evidence to the contrary, one must conclude that these children are victims of *diabetic heredity*.

What applies to these children also applies in large measure to older diabetics. Although increasing age makes it more likely that unfavorable diet or other factors may initiate or worsen the disease, diabetes in most cases is primarily the result of an inherited predisposition. In large measure this "sugar sickness" (diabetes mellitus) is caused by the failure of the pancreas to secrete sufficient *insulin*, vitally necessary for the proper conversion and utilization of sugar in the body processes. In the diabetic condition the sugar accumulates uselessly in the blood and is excreted in the urine (hence, a urinalysis can reveal diabetic symptoms). Unable, then, to obtain needed energy from sugar, the body shifts to the utilization of fats at an abnormal rate, which may cause a series of degenerative changes, and ultimately—if not checked—will bring death.

In its manner of inheritance, the tendency to diabetes is believed due to a pair of recessive genes. Evidence for this comes from many studies of families where diabetes has appeared, and from twin studies. When one identical twin has diabetes, the other almost invariably has the disease or shows a tendency toward it. In fraternal twins, the simultaneous appearance of the disease is no more frequent than in two nontwin siblings of a family, as governed by the risks of recessive diabetic inheritance.

Incidentally, to return to the diabetic cases in children, diabetes may sometimes involve a unique reversal of the usual procedure of forecasting inheritance of a disease. In most instances, where diabetes (or any other inherited condition) has occurred among parents, doctors are on guard for its appearance in children. But in certain cases where there is a child with diabetes, careful examination of the parents may reveal, unexpectedly, that one or the other, or both, have the disease itself or symptoms of its oncoming. (Ordinarily, however, inasmuch as diabetes is recessive in inheritance, a child could have the disease without its being present in either parent, even though each one was a carrier of the diabetes gene.) Especially interesting is the fact that if a baby is unusually large at birth (ten pounds or over), it may often indicate that the mother is diabetic or prediabetic. As for the baby, the prenatal effect of the mother's condition on its size need be only temporary, nor would the child have diabetes unless it had received a gene for the disease from the father as well as from the mother.

So far we have dealt only with the hereditary aspects of diabetes. It is important to point out, however, that in all cases of diabetes the severity and the time of onset of the disease may be governed by many environmental factors, including unfavorable diet, stress, injuries, and perhaps adverse emotional traits.* In the absence of the required environmental push, many persons may never develop the disease at all, despite their carrying the genes for diabetes. Of special importance here is the fact that diabetes is one of the few major diseases in which women are more vulnerable than men, deaths from diabetes at any given age claiming about one-third more women. Heightening women's risks from diabetes are adversities of childbearing.

What is certain is that under equal environmental conditions heredity does single out some individuals and families far more than others for the development of diabetes. Of those afflicted with it in the United States, one in four has diabetic relatives, a far higher proportion than in the general population. (The official reports give the total of *known* diabetics in the United States as about 1,500,000, or about 1 in 110 persons. But a great many diabetes cases are unreported, and it is clear

* It has been claimed that environmental factors by themselves—that is, without any hereditary predisposition—can produce diabetes in some persons. But there is as yet no proof that this actually has occurred, except where the pancreas has been damaged by some other disease, or has been removed.

that vast numbers of persons are diabetic without ever becoming aware of the fact.)

As with cancer and heart disease, the incidence of diabetes is rising as more persons are projected into the later ages when the degenerative diseases most often develop. At the same time, insulin, special diets and other treatments have been keeping more diabetics alive, and for longer years. But even though medical science has been making life increasingly easier and safer for diabetics, the continuing dangers of diabetes should not be minimized, nor is undue optimism regarding its control warranted. For instance, even though the immediate dangers of the disease can be reduced, there may often be progressive damage to the blood vessels, especially those in the eyes. Unless means are found not merely to treat diabetes but to *cure* it, this disease will continue to be—and increasingly as the number of those afflicted grows larger—one of the worst of those which can be blamed on wayward genes.

It should be kept in mind that we have been speaking of the common *diabetes mellitus*, or true diabetes. Several unrelated hereditary conditions, some of whose symptoms approximate those of diabetes, are often confused with it and cause unnecessary fear. One, *diabetes insipidus*, is related to true diabetes only in name and in the fact that, as in diabetes mellitus, there may be abnormal thirst and frequent urination. This condition has several forms, some hereditary, some not. The hereditary types of diabetes insipidus, the effects of which are generally not severe (except for possible inconvenience), may be caused by a dominant gene or by either of two sex-linked genes. The more common environmental condition is usually caused by an injury to the pituitary region of the brain, through accident or through another disease.

Sugar urine (*renal glycosuria*) is the condition in which there is an excessive amount of sugar in the urine but with none of the harmful effects which attend a similar symptom in diabetes. It is inherited, probably, as a dominant.

TUBERCULOSIS. Here is a disease regarding which most of the facts are on the optimistic side, as was brought out earlier (page 197) by the figures showing how enormously the death rate for this former No. 2 killer has dropped in half a century.

What is most significant is that for a long time tuberculosis was thought to be in large measure hereditary. Once it became known that the disease was caused by a germ (the tubercle bacillus), that it was most prevalent in slum areas or wherever living conditions were the worst, and that it was spread by personal contact, the factor of heredity began to be minimized. Nevertheless, there is now considerable evidence that even where the exposure is the same, the hereditary factors may influence the chances that TB will develop, or, if it has taken root, the rate and nature of its progress and the possibilities of cure.

Pointing to inherited degrees of susceptibility to tuberculosis have been studies showing that in experimental animals (chiefly mice and rabbits), some strains are unusually susceptible to the disease; and that among human beings living under the same conditions, the disease appears much more frequently in members of some families than in others. Some years ago it was also found that, if one parent had tuberculosis, the chance of its appearing in a child was over 60 per cent greater than if neither parent had the disease; and, if both parents had the disease, over 300 per cent greater. These findings (which raised questions about the role of environment in the parent-offspring incidences) are considered much less significant than the facts regarding tuberculosis in *twins*. Studies of hundreds of pairs in various countries, including the United States, have shown not only that the dual incidence of tuberculosis in identical pairs is significantly much more frequent than in fraternal pairs, but that there also is much more similarity in how, where, and with what severity the disease takes hold.*

While it is more than likely that individual human beings do differ genetically in their reactions to tuberculosis (as to almost any other disease), it is another matter to prove that there are such inherent differences among human *races*. For example, a common belief is that Negroes are much more susceptible than Whites to tuberculosis because—allegedly—they have "weaker respiratory systems." Undoubtedly, the Negro death rate for tuberculosis is far higher, currently more than three times the White rate. But also, without question, the general living conditions of Negroes are much worse. Wherever their conditions improve, their tuberculosis death rate drops and may fall even far below what the White rate was in former times. To illustrate, a check which I made in Milwaukee two decades ago revealed that the tuberculosis mortality rate among Negroes inhabiting a slum area of the city was fifteen times as high as the general rate for Whites of the city. But a generation before, the rate among Whites (mostly foreign born) who lived in that very same depressed environment had been almost exactly the same as it was among these Negroes who had succeeded them there.

Again, in 1910 among Whites in the United States as a whole, the

* Added to the evidence about probable genetic differences in susceptibility to tuberculosis are recent findings that one of the drugs most effective in treating the disease—*isoniazid*—is rapidly inactivated by many persons, who may inherit this chemical response through a semidominant gene.

annual tuberculosis death rate was 146 per 100,000—*nine times* the Negro rate today, a half-century later. No one would suggest, on the basis of the comparative tuberculosis rates, past and present, that Whites of former years were genetically inferior to their descendants two generations later—or to Negroes today. Until the environments of Negroes are fully equalized with those of Whites, statistics provide no real basis for conclusions about inherent racial differences with respect to tuberculosis.

Nonetheless, within any racial, ethnic or other group, individual genetic differences in resistance to tuberculosis or its cure will continue to be important. It is estimated that today about one American in five harbors the tuberculosis germs in latent form and that tens of thousands may develop the disease under adverse circumstances. A further disquieting fact is that new strains of tuberculosis bacilli are developing which are resistant to the current drugs. Thus, despite the enormous strides that have been made in combating tuberculosis, its defeat is by no means in sight. And so long as the tuberculosis bacilli are with us and complete and speedy cures for the disease are not possible for everyone, the hereditary differences with respect to tuberculosis resistance factors will continue to be important.

Pneumonia and influenza, although today far outranking tuberculosis as causes of death in the United States, have been given very little study from the standpoint of heredity. But while no specific evidence has been found, the presumption must exist that, as in all infectious diseases, the general genetic makeup of the individual may have a bearing on susceptibility to pneumonia and influenza.

GOITER. In the common, simple form of this disease, which is attended by a swelling of the thyroid gland in the throat, the role of heredity is still open to doubt. This is because development of the disease is so largely dependent on a specific environmental factor: lack of sufficient iodine in the food or drinking water, with a resultant strain on the thyroid gland whose function it is to abstract and utilize this material.

Yet there is also the possibility that an inherent thyroid weakness of certain individuals may prevent them from making full use of the iodine supply that is available. Thus, in regions where there is an iodine insufficiency, as in inland places far from the sea, only some of the persons—often only certain members in a given family—develop goiter; and in regions where iodine is plentiful, there still are some individuals who develop the disease. It is on this premise that the theory of goiter inheritance rests. Some studies have suggested a qualified dominant gene, others, recessive genes, as responsible for common goiter susceptibility. But the evidence is inconclusive.

On the other hand, there is no question but that the major threats of goiter can be overcome easily by insuring sufficient iodine in diets. In one experiment some years ago among thousands of Ohio schoolgirls living in the so-called "goiter belt" of the Great Lakes region, goiter was averted for all but a handful of those who were given supplementary iodine for several years; whereas in another group not receiving this iodine supplement, many hundreds of girls developed the disease. However, there are several rare types of hereditary goiter in which treatment with iodine is not effective because of a block in the utilization of the chemical to form thyroid hormone. Individuals so afflicted must be treated with the hormone itself.

What is most striking about common goiter is its far higher incidence among females—at least four women being afflicted to every man. The reason for this is still not clear. But one of the most unfortunate consequences of thyroid deficiency in a mother may sometimes be the birth of a *cretin* (a type of mentally defective child which will be discussed at length in Chapter 26, "Slow Minds").

Exophthalmic goiter (or Graves' disease) differs from common goiter in a number of respects, chiefly in that it results from an overactivity of the thyroid for which the cause is not known and has more serious consequences, taking the lives of several thousand women (and a fifth as many men) in the United States annually. Protruding eyes are a characteristic symptom, in addition to the glandular swelling. The disease occurs most frequently among young adults of highly nervous temperament and of narrow, light build (but these physical factors may be either cause or effect). As in common, or simple, goiter, inheritance of the disease has been claimed by way of qualified dominant genes, or of recessive genes, but, again, adequate proof is lacking.

Digestive diseases. There are a great many diseases associated with the digestive organs and processes, some high on the list of "killers"; but with the exception of diabetes, none of the common ones can be conclusively linked with hereditary factors.

Ulcers of the stomach and intestines, growing in incidence and as causes of death, are often blamed largely on excessive nervous strain, but this may be going too far. Among the reasons for suspecting some constitutional influences (related to the internal environment) are the following: the fact that the incidence of the disease is so much higher among men—with deaths four times those among women; that the disease appears to go more often with certain types of body build; that it is found on occasion in young people; and that the proportionate incidence of duodenal ulcer is greater among people with blood group O. But while some investigators also report a higher than average incidence of ulcers in certain families, the existence of an hereditary factor remains unproved.

Among the kidney diseases, *nephritis* (in the common and serious forms which are a leading cause of death) is usually traceable to environmental factors, and little, if any role, can be ascribed to heredity. Only certain rare forms of kidney disorders, among them *polycystic disease of the kidneys*—especially serious for pregnant women—have been proved to be hereditary.

Certain diseases of early infancy and *congenital malformations,* Nos. 5 and 9 on the country's current "Big Killer" list, are catchall categories for a great variety of conditions. (See footnote to Table II.) Most of these diseases of early infancy are environmental, as has been shown by the marked reduction in infant deaths during the past half-century (from one in ten live-born babies dying within their first year in 1915 to a rate now of less than one in forty). However, a great many of the deaths resulting from congenital malformations or defects of certain kinds are due in part to heredity and will be discussed in later chapters.

This, then, completes our discussion of inheritance in the major fatal diseases. Genes are involved, however, in many other menacing conditions, fatal or otherwise, which remain to be discussed. What we have seen so far is that fairly direct and simple inheritance is the chief factor in only one of these big killers—diabetes. In the major heart and arterial ailments, heredity appears to be no more than a partly contributing or predisposing factor, with environment playing the principal role. In common cancer any possible hereditary influence is vague and shadowy. As for the infectious diseases, heredity can never involve more than degrees of susceptibility which need be reckoned with only so long as particular bacteria or viruses are allowed to attack and to survive.

But whatever the disease or the part played by wayward genes, one must always bear in mind that some individuals are inherently more vulnerable than others. In this regard nothing is more important than the person's *sex,* which can not only influence the general course of any disease, as has already been indicated, but can largely determine whether or not certain hereditary diseases and defects will manifest themselves at all. This fascinating aspect of heredity is our next topic.

CHAPTER TWENTY TWO

The Poor Males

"IT'S A MAN'S WORLD!" Women have been saying this, and flattered men have believed it, no doubt from the beginning of time.

But in one important respect it's all wrong: In health and physical well-being—as related to lesser threats not only from the major diseases we've discussed, but from the great majority of others—the human female from *before birth* and throughout life today is favored far above the male.

The reasons for much of this discrimination are now clear. First, there are the general sex differences in bodily makeup and chemical functioning, which are known to endow the female with advantages in resisting or overcoming most diseases. Second, the male is much more likely to be victimized by specific and directly hereditary diseases and defects, as we'll presently explain. Third, there are environmental factors, such as differences in occupation, habits and behavior, which expose the male to greater hazards. But these differences in the paths taken by the two sexes are in themselves, to a considerable extent, outgrowths of inherent sex differences which affect their ways of living. In short, the margin of advantage which women have over men with regard to physical defectiveness or mortality is much less due to environmental influences than is generally supposed.

Long before anyone can talk of males leading "rougher and faster" lives the discrimination is apparent. Even in prenatal life, as we noted in Chapter 7, the male is more vulnerable to almost every adversity, a much higher proportion of males than females being carried off before birth. Further, more males die a-borning, more come into the world with congenital abnormalities, and in the first year of life, the average death rate among boy infants is markedly higher than among girls —currently by about 34 per cent among White infants of the United States.* (The accompanying Table shows the excess of boys claimed by the leading causes of death in infancy.) It would be foolish to try to attribute the greater mortality of boy babies to any environmental discrimination—such as that mothers might give the boys less care than girl infants receive. Nor can we see anything but some inborn difference in the fact that when boy and girl infants are involved in exactly the same accidents, such as falling down stairs, or being in an auto collision, the chances of fatality, as statistics show, are considerably greater for the little boys.

As childhood proceeds, the margin of higher male mortality drops for a short period (between the ages of one to five), then begins growing ever greater until, at the ages of fifteen to twenty-five, the death rate among males is more than double that among females. Thereafter, all the way until the very old ages, the excess of male mortality ranges from about 65 per cent to 95 per cent above the female rate. Undoubtedly, environmental factors play a growing part in creating most of this enormous disparity between the sexes after early childhood, for it is obvious that as the years go on the sexes become more and more differentiated in their life experiences, largely to the disadvantage of males.

Nevertheless, there is every reason to assume that the initial pattern for greater male susceptibility to most diseases and defects, set in prenatal life and infancy, continues to be operative. What is most striking is the way in which this pattern has become increasingly pronounced in modern environments. As the environmental hazards and death rates for both sexes have continued to be reduced, the females have benefited progressively more than the males; and as the activities, habits and ways of living of the two sexes have become equalized (with special dangers to women, such as those from childbearing, largely eliminated), the greater has grown the disparity in death rates between them.

One can only conclude, then, that *under like conditions, females are*

* Among the few congenital abnormalities which discriminate against females are *congenital dislocation of the hip* which afflicts about eight times as many girls as boys. The explanation may lie in a difference between the sexes in the structure and prenatal development of the hip joints and hip area, which makes it easier for a dislocation to occur in the female before birth. However, there is evidence of an hereditary vulnerability for this malformation, inasmuch as among identical twins, when one is affected, so is the other in about 40 per cent of the cases. Another congenital malformation discriminating against females is *spina bifida cystica* which afflicts about twice as many girls as boys.

Table III

How Male Babies Get the Worst of It

Sex differences in principal causes of death during infancy (first year of life) among White babies in the United States, 1961

Cause of Death	Male Infants *Rate per 100,000*	Female Infants *Rate per 100,000*	Excess of Boy over Girl Deaths
Birth injuries, postnatal asphyxia, etc.	750.0	518.8	44.6%
Infections of newborn	97.0	67.5	43.7%
Other diseases peculiar to early infancy and immaturity	755.8	572.9	31.9%
Congenital malformations	403.2	342.8	17.6%
Pneumonia and influenza	179.8	141.5	27.0%
Gastritis, enteritis, etc.	33.9	27.3	24.2%
Hernia and intestinal obstruction	25.0	18.4	35.9%
Meningitis	17.7	13.1	35.1%
*Other infective and parasitic diseases**	19.0	12.8	48.5%
Major cardiovascular-renal diseases	11.0	9.4	17.0%
Accidents	79.5	62.6	27.0%
ALL CAUSES†	2,546.8	1,920.0	32.7%

* Figures include totals for scarlet fever, diphtheria, whooping cough, polio, measles and other infective diseases. (Data for Table from *Vital Statistics of the U.S.*, 1961, Table 5-12, rates for 59 selected causes. Publ., 1963.)

† Includes causes of death not listed separately.

better adapted to cope with most human afflictions because they are *genetically* better constructed and have a more efficient chemical system.

To account for this in some detail, we go back to the moment of conception and the initial genetic difference between the sexes: the fact that the female is started off with two Xs, the male with one X, plus a Y. We have seen how this throws the development of the sexes in different directions and leads to differences in sex glands, sex organs, muscle and bone development, body form and other structural characteristics. But we know also that, at the same time, different biochemical environments are created for the two sexes, the clearest evidence for which is their relative outputs of sex hormones—in the female, proportionately more of the estrogens, in the male, more of the androgens. The female has been shown to be biochemically more variable in many respects than the male, perhaps because of changes in body chemistry that take place in her during menstruation and childbearing. (In a sense the female may be said to have a "better stocked internal medicine cabinet" than the male.) The same chemical variability, then, may make the female better able to adjust to bodily stresses and accidents and to rally defenses against infectious diseases or other adverse outside influences.

This general advantage of females may largely account for whatever inborn sex differences contribute to their lower mortality rates from most of the major ailments which we listed under the "big killers."* By the same token, a few disadvantageous factors in the biochemical makeup of women may account for the higher incidence among them of such diseases as diabetes and goiter. However, one must never overlook the important, if not predominant, influence of environment on the causes of death in the two sexes. The effects of environment are clearly shown in the varying death rates for men and women from given causes in different countries. As the most notable example, in the United States heart diseases claim about twice as many men as women between the ages of forty and seventy-four. But in other countries the excess of male heart deaths has ranged from a high of 118 per cent more males in Australia down to only 13 per cent more males in Italy. Since there is little likelihood that the populations of these countries could differ very much in any hereditary factors involved in heart conditions, one must look for environmental causes to explain these great sex differences. What these causes are remains a mystery, but if they can be identified, knowledge of them will go far toward preventing and combating the heart ailments.

In addition to whatever general genetic disadvantages males may have, they are discriminated against by a special handicap: the fact that Nature has *shortchanged* them in some of their genes. We refer to the *sex-linked* conditions (briefly discussed in Chapter 20), which result from genes carried in the X chromosome. It will be recalled that at conception the female is started off with *two* X chromosomes (one from each parent), while the male receives only the single X from his mother, plus the very small Y from his father; and if any wayward genes are in the male's X chromosome, they are usually far more dangerous for him than they would be for his sister, even if she got

* The role that female hormones may play in heart disease is indicated by the report that when a large group of men who had had heart attacks were given small daily doses of natural female hormones (estrogens), the death rate from the disease among them after a five-year period was about half that within a control group of similarly afflicted men not so treated. (In some cases this treatment may result in some "feminizing" physical effects.)

the very same X. This is always so when the defect-producing gene is recessive, as it most often is. Here's why:

When a female gets a recessive wayward gene in one of her X chromosomes, the chances are that there will be a normal gene for the job in her other X. But if a male gets such a wayward gene in his single X, he's in a bad spot, because there is no corresponding gene in his very small Y chromosome to do the job. Inasmuch, then, as there are a great many recessive genes in the X chromosome which every so often are wayward, males, by and large, are directly exposed at conception and thereafter to many more special defects and afflictions than are females.* At the same time, in the fact that the male's X can come only from his mother, we have the answer to what was long a mystery—why certain diseases and defects are *transmitted to sons only by way of their mothers.*

Hemophilia. Most celebrated and dramatic of the sex-linked conditions is this bleeding disease, caused by the failure of a certain X-chromosome gene concerned with the blood-coagulating process to do its work properly. The result in the past for those with the classic or common form of hemophilia was usually death in early life. With modern methods of treatment and with unusual care, hemophiliacs can be kept alive into maturity—but often precariously, and with the danger of becoming crippled by bleeding into the joints. While hemophilia is comparatively rare (estimated incidence about 1 in 25,000 persons), it has been given much prominence by this remarkable fact: A single gene for hemophilia, passed on through Queen Victoria to one of her great-grandsons, the last little Czarevitch, may well have been a contributing factor in bringing on the Russian Revolution and in changing the course of the world's history.

As records show, it was because of their son's affliction that the credulous Czar and Czarina became victims of the designing monk Rasputin, who held out hopes of a cure through supernatural powers. From Rasputin, as from a spider, spread a web of intrigue, cruelty, demoralization and mass indignation which helped to precipitate the collapse of the empire and all the subsequent political developments. *If*

* That there is a greater prenatal hazard in having only a single X chromosome is further shown by this fact: In poultry, where the sex-determination mechanism is the reverse of that in human beings and other mammals (as reported in Chapter 7), the *females* having only one X whereas the males have two Xs, the embryonic deaths are much higher among the females.

THE ROYAL HEMOPHILIA PEDIGREE*

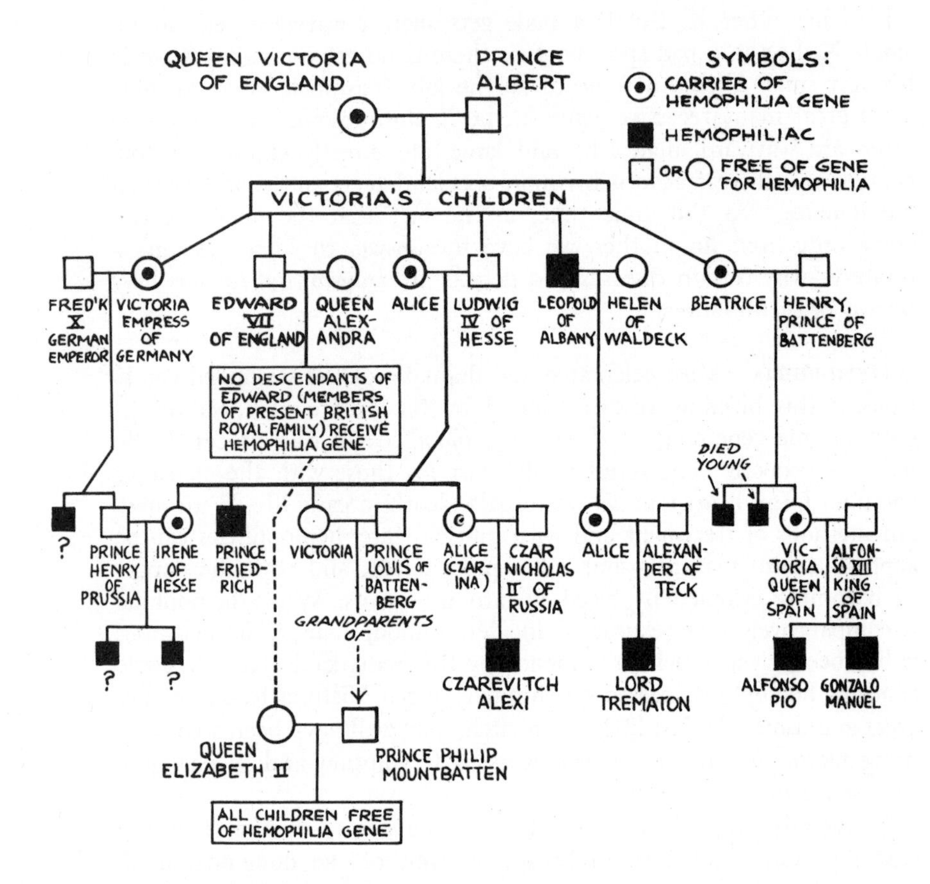

* Only some of the members of each of the families represented are shown here: For example, Princess Margaret, who, like her sister, Queen Elizabeth II, was free of the hemophilia gene and would not pass it on to her children, is omitted. (Chart based on diagram and data compiled by Dr. Hugo Iltis.)

the Czarevitch hadn't had hemophilia, *if* his parents hadn't become the prey of Rasputin, *if* Rasputin hadn't demoralized the court. . . . Thus a momentous structure of "ifs" can be built up, like an inverted pyramid, resting on that infinitesimal bit of substance, constituting a single gene, which found its way to one sad little boy.

In all, ten of Victoria's male descendants suffered from hemophilia, and at least seven of her female descendants were carriers of the gene. Of Victoria's five daughters, three—Victoria, Alice and Beatrice—were carriers, as were two granddaughters, the last Czarina, who bore one hemophiliac son and the last Queen of Spain, who bore two. (Of the latter, both died from bleeding, one as a child, the other in maturity following an automobile accident.)

What of the present British royal family? The fact that Queen Elizabeth and her husband, Philip, Duke of Edinburgh, are *both* descendants of Queen Victoria created worry in some minds at the time of their marriage that they might have a hemophiliac son. Happily, geneticists were able to dispel these fears, inasmuch as the dread gene was not received by the great-grandfather of Queen Elizabeth, King Edward VII, nor, obviously, by her husband, Philip. The arrival of the present healthy Prince of Wales and of his two healthy younger brothers supported the scientists' opinion. Similarly, no danger of hemophilia confronted any son of Elizabeth's sister, Margaret, who in 1961 also gave birth to a healthy boy.

Of special interest to geneticists is the fact that the particular hemophilia gene passed on by Queen Victoria appears to have arisen through a *mutation* either in her or in her mother or father. It is precisely in this way that many other cases of hemophilia are known to arise. Were it not for this constant production of new hemophilia genes, the disease might long ago have been eliminated through the deaths of afflicted males before they reached the reproductive ages. (The possibility of ordinary hemophilia appearing in a female, and facts about special types of hemophilia and other bleeding diseases, will be discussed in Chapter 28.)

COLOR-BLINDNESS. This is by far the most prevalent of the hereditary sex-linked conditions afflicting chiefly males. As almost everyone knows, color-blindness in its common form is the inability to distinguish between red and green as *colors*. This does not mean—to answer a frequent question—that color-blind motorists can't tell the difference between a red traffic light and a green one (they can, because the one appears to them different in intensity from the other). Nevertheless, the flat colors of red and green, as seen on objects or in nature, may not be so distinguished; and it is for this reason that so many color-blind men have been barred from the Air Force and from those branches of the navy, army and merchant marine in which good color vision is essential. In many types of civilian work where color matching and good color perception are essential, color-blindness may also be a barrier to employment.

Normal color vision is made possible by a grid of little cones, set close together inside the eye, at the rear of the retina, which act to receive and filter out color impressions, and transmit them to the brain (somewhat in the way that a color-television camera registers and transmits, and a color-television set receives, color images). But in color-blind individuals the cones in the eye are deficient in their color-receiving and transmitting mechanisms. This deficiency is caused by a wayward gene which, as in hemophilia, is carried in the X chromosome. If the one X that a male receives has the "color-blindness" gene, he will be color-blind. And again, as with hemophilia, since a male's X chromosome can come only from his mother, it is only through mothers that color-blindness is transmitted to sons.

Where a woman carries one X with a defective color-vision gene and her second X has a normal gene, there is a fifty-fifty chance that any son will be color-blind. But if the mother herself is color-blind, because she carries two defective genes—which happens to less than one woman in 200—every one of her sons is almost certain to be color-blind.

What about the daughters? Only if the father is color-blind and the mother is a carrier of the gene, or is herself color-blind, will a daughter be color-blind. But there is a possibility that even with one defective gene a woman may be slightly red-green color-blind, recent studies having shown that in some cases the normal gene does not quite overcome the influence of the defective one, which may then act to give the woman poorer than average color perception. This has significance because in these cases the women who are carriers of the "color-blindness" gene may be identified.

Can environmental factors, as well as heredity, produce color-blindness? Various theories to this effect have been advanced, among them that a vitamin-A deficiency can cause defective color vision, or that dietary treatment can overcome it. But this has not been proved. Moreover, while color-blind individuals may be trained to discriminate more sharply between red and green on the basis of shade differences, there is little evidence that their actual color vision can be permanently improved, not to say cured, by any known treatment.

The incidence of color-blindness varies considerably among races and ethnic groups. In different White populations, the proportion of color-blind men may range from 5 to 9 per cent. Among American White males, about 4 per cent are definitely red-green color-blind, and

SEX-LINKED INHERITANCE

If a defective gene is in the X chromosome (as in color-blindness)

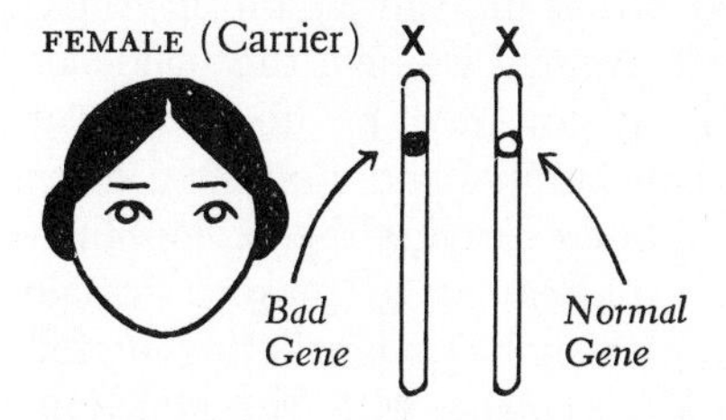

With two Xs, a female carrier usually has a normal X gene to block the bad one and is herself normal.

MALE (Afflicted)

X

Bad Gene

Y

With only one X, male has no normal gene to block the bad one and develops the defect.

WOMAN CARRIER'S SONS

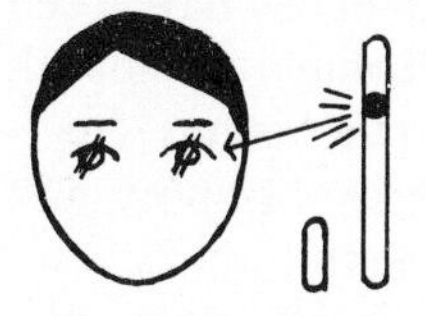

One in two gets mother's bad X gene, and has the defect.

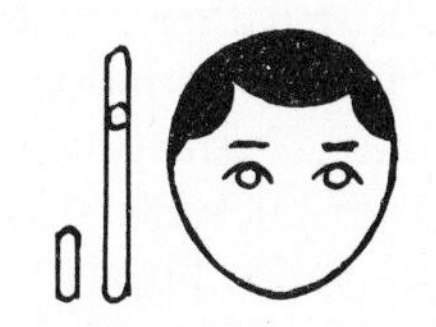

One in two gets mother's normal X and is not defective.

AFFLICTED MAN'S SONS

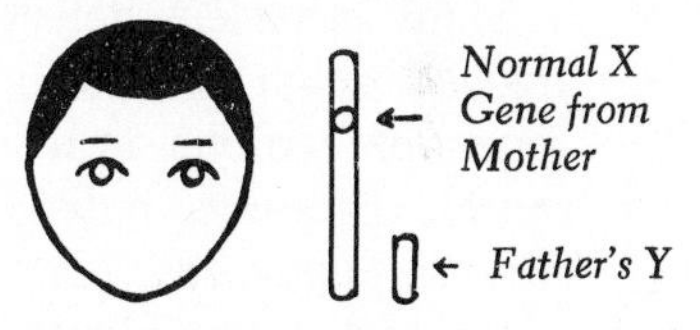

No son gets father's X, so every son is free of the defect and cannot pass on the bad gene.

WOMAN CARRIER'S DAUGHTERS

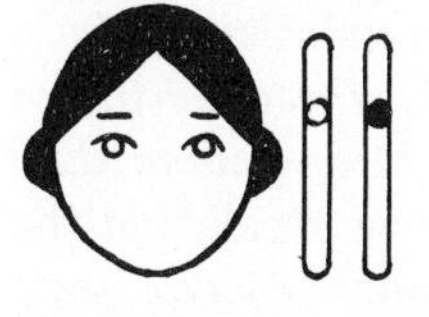

One in two gets mother's bad X, and is a carrier like mother.

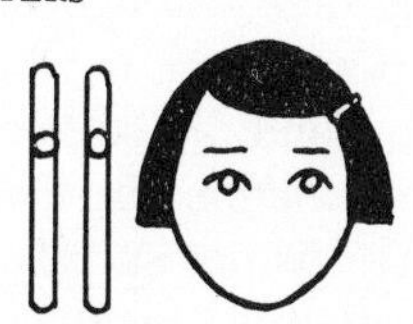

One in two gets the normal X, and cannot pass on defect.

AFFLICTED MAN'S DAUGHTERS

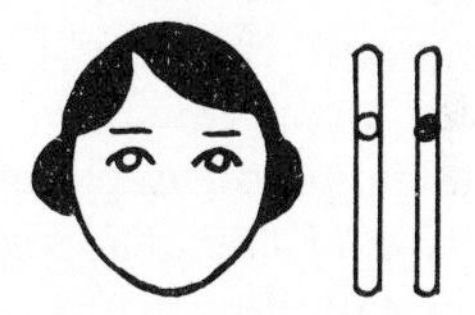

Every daughter gets the father's bad X, and is a carrier (like woman at top left of page).

AFFLICTED FEMALE

Only when a female gets an X with a bad gene from both parents will she develop the defect.

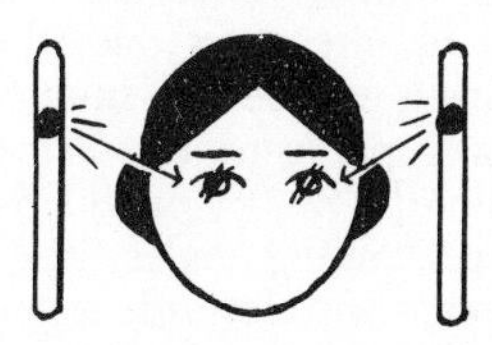

Every one of her sons will have this same defect. Every daughter will be a carrier of the gene.

another 4 per cent are partially deficient with regard to red-green or some other type of color vision. In American Negro males, color-blindness is reported only about half as frequent as in Whites, and in American Indians, only a fourth as frequent. Among women, the incidence of color-blindness is always small and is governed by how frequent the condition is among the males of their families and groups. In other words, the chances that two X chromosomes carrying the "color-blindness" genes will come together—the requirement to make a female color-blind—depend on the proportion of such Xs circulating in the population, as revealed by the number of males with the affliction. Thus, in a group where 5 per cent of the men are color-blind, only about one woman to twenty men would be afflicted; in a group where 9 per cent of the men are color-blind, the chances that two Xs with "color-blindness" genes will come together are greater, and about one woman to 11 men would be afflicted.*

Many other eye defects which strike particularly at males have been tracked down to "sex-linked" genes. Included are some but not all forms of extreme near-sightedness, oscillating eyes (*nystagmus*), eye-muscle paralysis, enlarged cornea, defective iris, optic atrophy, and *retinitis pigmentosa* (in which the retina fills with pigment).

Altogether, eye defectiveness, *including blindness*, is far more prevalent among males than females. While only a few of the known sex-linked conditions directly produce blindness, some hereditary influences certainly contribute considerably to the much higher incidence of blindness among males. Even if we make full allowance for the more frequent loss of eyesight among men through injuries and accidents, it is significant that the blind population in the United States starts off with about one-third more blind boys than girls, *three-fourths of these childhood cases originating before the fifth year, and more than half being of prenatal origin.* This is not surprising if we recall that the male in early life is inherently more vulnerable to almost every disease and defect. There is thus reason to believe that the situation may also apply to blindness, even where there is no inheritance (as in congenital blindness due to syphilis or to some other infectious disease).

* There is a simple formula for calculating the chances that a female will have any sex-linked defect, due to an X-chromosome gene, when its incidence among males is known. Since the female requires two of the same gene, whereas the male requires only one, the incidence among females of any given X-gene condition will be the square of its incidence among males of their group. With a 5 per cent incidence among males, the incidence among females would be $.05 \times .05$, or .0025—one-twentieth as much; with a 9 per cent incidence among males, the female incidence would be $.09 \times .09$, or .0081—about one-eleventh as much.

Other sex-linked conditions. *Speech disorders* provide another category in which males are overwhelmingly in the majority, but in this case there is as yet no clear proof that heredity is involved. We know only that stuttering is from five to ten times more common in little boys than in girls, the ratio increasing with age. *Reading disorders* (such as dyslexia) are about four times more common among boys. While psychologists are inclined to attribute many cases of both speech and reading defects to early emotional disturbances or personality disorders, it is open to question whether environmental factors alone can explain away all the cases and the whole big difference between boys and girls in the incidence of these defects. Some inherent male weakness may well be involved. (There will be more on this in Chapter 24, "Functional Defects.")

We have by no means presented all of the hereditary conditions which chiefly or exclusively afflict males—for instance, the peculiar sweat-gland defect which makes the victim pant as dogs do, and a form of muscular atrophy where the man can't stand properly and appears to have a drunken gait. Many other known sex-linked conditions are too rare to warrant listing here, and there is every certainty that further research will lengthen the list.

When we look for sex-linked conditions discriminating against females, we find very few indeed. This can be explained by the fact that such conditions have to be due to *dominant* genes on the X chromosome. In most of the other X-chromosome conditions (hemophilia, color-blindness, etc.) the wayward gene is recessive, provided there is a *normal* gene to block it (as is usually the case in the female with her extra X, but cannot be so in the male, with his single X). However, if a wayward gene in the X chromosome is *dominant* even when coupled with a normal gene, it will most often menace females, because, with their two Xs, they have twice as much chance as males of getting such a gene and can receive it from either parent. While a mother with a dominant X condition can pass it on to a son as well as a daughter (since she transmits one of her two Xs equally to each), a father's single X can go only to his daughters (his sons receiving his Y). Thus, if he is afflicted with a dominant X-chromosome condition, every daughter, but no son, will be affected by it. This explains the much higher incidence in women of several conditions, among them one type of defective teeth with brownish enamel and one form of chemical deficiency which can cause a girl to develop a peculiar type of rickets resistant to vitamin-D treatment.

A provocative recent theory (of Dr. Mary Lyon) is that a female's two Xs are not both operative in any one cell but that in each cell only one X functions while the other is repressed. Which X becomes the boss in a given cell and which one is inactivated is a matter of chance, the theory holds, so that in some cells the functioning X is the one derived from the father, in other cells, the X from the mother. This would mean that if one of the female's Xs carried a wayward sex-linked gene, and the other X a normal gene, the effects of the wayward gene would not be suppressed throughout the body but would be asserted in the cells where the other X and its normal gene were inactivated. For instance, if a female received one "color-blindness" gene and one normal one, the rear of her eye would be a *mosaic* of part color-blind cells, part color-seeing ones, and she might then have defective color vision to a greater or lesser degree. Again, in the case of a genetic chemical abnormality, such as hemophilia, where a vital clotting substance is missing from the blood, the cells where the wayward gene-carrying X was the active one would fail to produce the given substance, but the other cells, with the functioning normal X gene, could be sufficiently productive of the substance to meet the body's needs. If the theory is confirmed, these facts would apply to a variety of sex-linked conditions.*

Whether the little Y chromosome carries any genes for specific diseases or defects remains a question. The fact that the Y plays a potent role in the sex-determination process (as told in Chapter 8) would strongly suggest that some of its genes may not be confined in their functions solely to the sex traits and may be involved in other activities (just as are many genes in the X chromosome). If there are any Y-linked wayward genes, they would show themselves in conditions transmitted only by fathers to sons. It was long thought there were a number of such conditions, all extremely rare, among them one called "porcupine skin" which causes quill-like outgrowths on the body. But searching inquiry has so far pointed to only one abnormality—extensive *hairiness on the ear*—as being probably a Y-linked trait. Another former

* In female identical (one-egg) twins, the results of X inactivation (if proved) would be of particular interest. While the twins' two Xs would be the same (one from the father, one from the mother), the probability would be that the pattern of active and inactivated Xs throughout their bodies would not be the same. Thus, with respect to traits governed by "X" genes, many of the cells at given locations in one twin would differ from the corresponding cells of the other twin. Further, the biological differences between female identical twins would therefore usually be greater than between male identicals, who, carrying the same single X chromosome, would show no differences in their cells with respect to X-influenced traits.

belief, that certain wayward genes might be carried equally in both the Y and X chromosomes, is still to be verified.

Finally, there are genetic conditions which discriminate chiefly against males because, while females can get the same genes, it is usually only in the male chemical environment that the effects assert themselves.* Such genes, or conditions produced by them, are called "sex limited," or "sex controlled." In the case of various normal traits we have seen how the same genes, although transmitted by and present in both sexes, show their effects only in one sex or the other: "breast-shape" genes, manifesting themselves only in females; or "beard-pattern" genes, manifesting themselves in males. That there would also be some wayward genes with similar sex-slanted effects appears probable; but while some of these genes may be involved in a general way in major conditions which are sex discriminating, only a few with extreme and clearly evident sex-slanted effects have been identified in human beings. Several of these (including *gout*) will be discussed in later chapters. At this point we will turn to the best-known of the sex-limited (or sex-controlled) conditions: one which in its physical effects is not serious—no man gets sick or dies from it—but which, nonetheless, occasions untold psychological distress, leads to the wasted spending of millions of dollars annually and is otherwise sufficiently worrisome to many men to warrant considerable attention. We refer to—

BALDNESS. There comes a time when the hair on the head of the rugged male begins to loose its hold like the seeds of an autumn dandelion, presently to be gone with the wind.

Dejectedly, fearfully, he watches the teeth of his comb, as if they were those of some devouring monster, gobbling up more and more of his precious locks. What causes this?

In some instances, it is true, falling hair is the result of a disease or scalp disorder. In the great majority of cases, however, the hair of a man healthy in every way falls out for no apparent reason. This almost never happens to a woman; and what adds to the mystery is that even if a woman has the same scalp condition or disease that a man has, rarely does she lose her hair to the same extent. Is it (as far too many people still believe) that through the ages men have cut their

* One recently identified hereditary condition which afflicts mainly females is *kuru*, a peculiar disease prevalent among natives of the Fore tribe of New Guinea. Called the laughing disease (because of its symptoms, although it is fatal in its effects), kuru appears to be transmitted by a gene which acts as a dominant in females and a recessive in males.

hair short, have worn tight hats, have been more negligent than women in taking care of their hair and scalps—or have overworked their brains? These and similar explanations have as much evidence to support them as the theory that if you cut off the tails of puppy dogs for a number of generations, their descendants will be born with stub tails.

The explanation has been found in a particular sex-limited or sex-controlled gene, through which ordinary *pattern* baldness—the kind most common in men—is inherited. Since this gene (like other sex-limited genes) is carried not in the X chromosome, but in one of the twenty-two other chromosomes, it can be inherited equally by a man or a woman. But it doesn't *act* the same way in both. It behaves like a *dominant* in a man, only *one* gene being required to produce baldness. In a woman, however, the gene acts as a *limited recessive:* She must receive *two* "baldness" genes before she will be affected, and, even then, only partial baldness or merely a thinning of the hair may result. Why?

The best theory is that the glandular makeup of the two sexes governs the way in which the gene does or does not express itself. In a woman, apparently, the lack of an excess of male hormones (androgens) and the absence of their effects keeps the hair from falling out even where the inherited baldness tendency is present. In a man, the excess of male hormones makes the hair follicles particularly vulnerable to the action of the "baldness" gene. Evidence to support this has come through studies and experiments by Dr. James B. Hamilton. Ages ago Hippocrates observed, "Eunuchs do not grow bald." Checking on this, Dr. Hamilton selected a group of men in a Midwestern institution who had been demasculinized as youths through accident or injury or who for biological reasons had failed to mature sexually. Not one case of baldness had appeared among them. But when male hormones were administered to the group, the hair began to fall out in many of the men, particularly those from families in which baldness was common. On the other hand, Dr. Hamilton found evidence that no amount of male hormones will produce baldness in men who lack the hereditary tendency to baldness.*

* As further evidence of the role of sex hormones in male and female hair growth, women with a certain tumor of the adrenal gland which results in a high production of male sex hormones not only may become bald, but may also develop such other typical sex-limited traits as beards and mustaches. Upon removal of the tumor, these masculine traits disappear. Another pertinent fact is that in many women the hormonal changes that accompany childbearing bring an increase in hair loss—often quite heavy—

HOW COMMON BALDNESS IS INHERITED*

"BALDNESS" GENE

In Men: *Dominant. One produces baldness.*
In Women: *Recessive or completely suppressed. Two such genes needed to produce any degree of baldness.*

"NORMAL-HAIR" GENE 

In Men: *Two such genes, one from each parent, can insure normal hair retention.*
In Women: *Only one such gene needed to insure normal hair growth (barring certain diseases).*

MEN

Type A

TWO "BALDNESS" GENES
(All of this man's sons will be bald, and if wife is Type A, *daughters also.)*

Type B

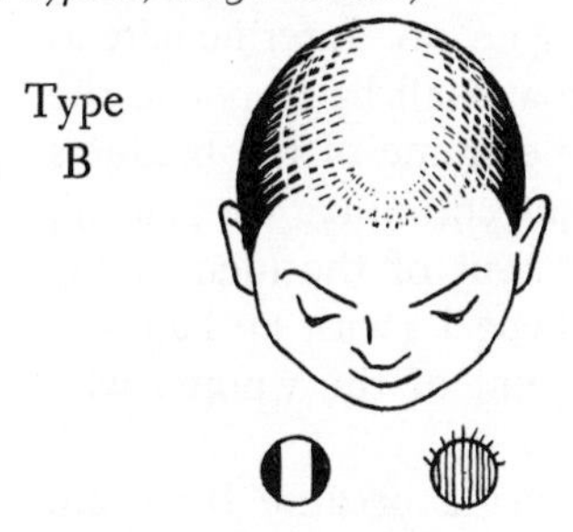

SINGLE "BALDNESS" GENE
Marked baldness, but possibly not as much as in Type A. *(Some of this man's sons may have normal hair.)*

Type C

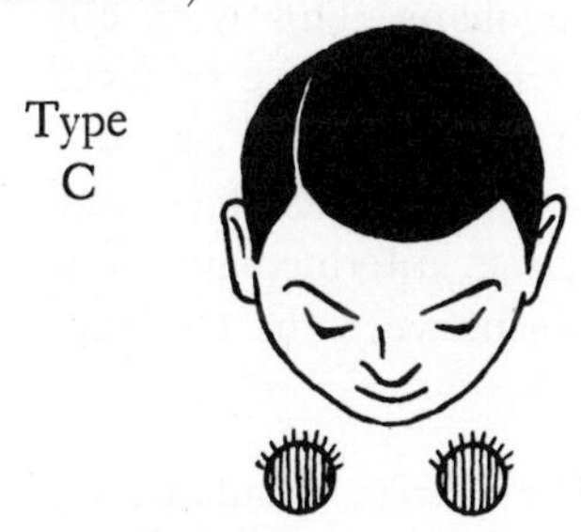

TWO "NORMAL-HAIR" GENES
No baldness in this man (nor in sons, unless wife is Type A *or Type* B*).*

WOMEN

Type A

TWO "BALDNESS" GENES
Produces (with aging) sparse hair or partial baldness in women. (All sons will be bald.)

Type B

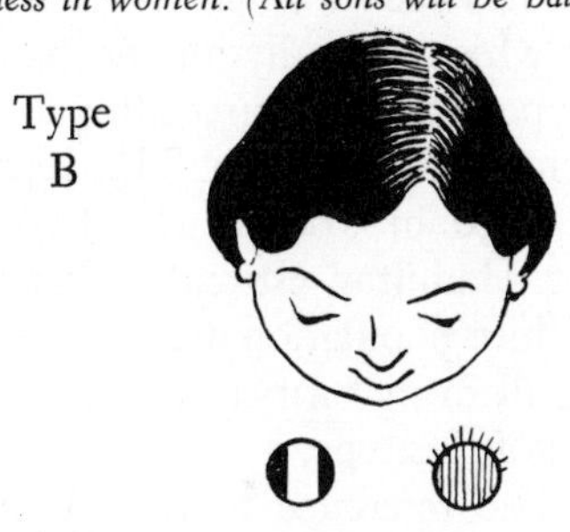

SINGLE "BALDNESS" GENE
No baldness, but possibly some hair-thinning. (At least one in two of sons will be bald.)

Type C

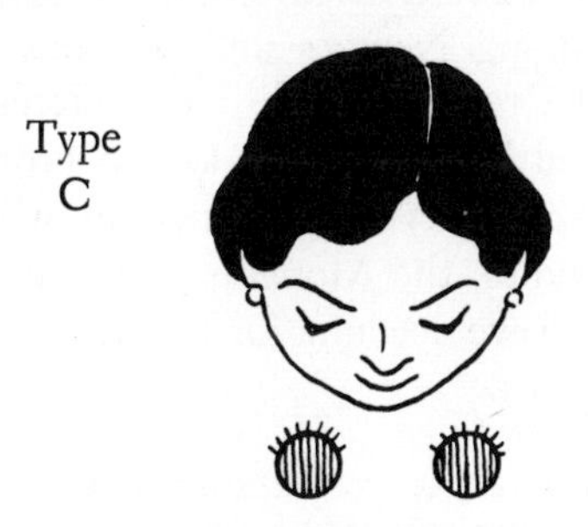

TWO "NORMAL-HAIR" GENES
Full hair growth in this woman (and no baldness in her sons, unless husband is bald).

* Reference in all cases is to inherited pattern baldness, the most common type, and exclusive of hair loss caused by diseases.

The foregoing makes hash of the popular notion, perhaps derived incorrectly from the Samson episode in the Bible, that a large amount of head hair is correlated with virility. Actually, we have the contrary. If the "baldness" gene is present, the most feminine man—with a deficiency of male hormones—is most likely to retain his hair; the most masculine man—with an excess of male hormones—is most likely to lose his hair. However, without knowledge of the hereditary tendencies, the presence or absence of hair on a mature man can have no meaning in this sense. Similarly, there is little basis for the belief that "grass doesn't grow on busy streets," although if recent theories hold up—that thin scalps grow bald sooner than plump ones and that nervous strain may hasten hair fall—intellectual men, who tend to be both leaner and more the worrying kind, may lose their hair more quickly. But always, only if the "baldness" gene is present.

Degrees and specific types of pattern baldness, with different ages of onset, are probably also inherited. (Precisely how hasn't yet been worked out.) However, Dr. Hamilton has made eight classifications, ranging from Type 1, where there is no hair loss at all, through to Type 8, in which virtually all the hair is lost as time goes on. Intermediate is Type 4, with hair receding at the temples, and a small bald spot in the center of the crown. This represents about the extreme of the baldness reached in women who have inherited the susceptibility. In general, Dr. Hamilton estimates that only about 4 per cent of the men in the White population are lucky enough to be of Type 1 (with no hair loss at all in maturity), as compared with 20 per cent of the women who are of this type.

The reference to the White population is made because there are important hereditary differences among racial groups with respect to balding tendencies. Whites are the most susceptible to baldness (those of eastern Mediterranean stock most of all). There is less baldness among Negroes and still less among the Mongolians. This is exactly the reverse of the race differences with respect to the degree of facial and body hair, for, as was brought out in Chapter 13, "The Features," it is the Whites, generally, who have the most hair on their faces and bodies, the Mongolians the least, with the Negroes intermediate. How can we account for this inverse *racial* correlation between the tendency

---

during the period from one to three months after delivery. Similarly, the waning activity of sex hormones in aging women may often result not only in balding, but in the growth of coarse hair on the face and in the loss of rounded contours.

to lose hair on the head and to grow it elsewhere? It doesn't apply to individuals within a race, because studies show that men who retain their head hair may be as likely as balding men to have heavy beards and hairy chests. (Eunuchs, previously discussed, are the exception.) Nor have hormonal tests of Whites in comparisons with Mongolians or Chinese shown any special differences in amounts of the male hormones, which bring out the balding tendency. The indications are, therefore, that the "baldness" genes are inherited independently from those governing facial or body-hair growth and that it is merely a coincidence that both types of genes are present in different proportions among the racial groups.

At any rate, can we predict whether a given man will become bald? Within reasonable limits, yes. If a man's father was bald, there is at least a fifty-fifty chance that he also will become bald eventually. This chance is increased if baldness has been common among men in his mother's family (her father, brothers, etc.). Specifically, chances of baldness inheritance depend on whether a man's parents are of the "*two* baldness-gene," "*one* baldness-gene," or "*no* baldness-gene" types, which may be roughly guessed at by observance of his close male relatives on both sides, including older brothers. Where either father or mother carry *two* "baldness" genes—and particularly, in infrequent cases, where both do—a son reaching maturity had better plan on having a shiny pate or wearing a toupee. (And a woman with baldness rife on both parental sides should not expect to have luxuriant tresses or to qualify for the role of Lady Godiva.)

In addition to the ordinary pattern baldness, there are a number of less common types of baldness, whole or partial, which have an hereditary basis. These, however, are not due to "sex-limited" genes and affect females as well as males. But always keep in mind that there are a number of non-hereditary causes of baldness, among them various infectious diseases, fevers, diabetes, certain glandular and nervous ailments, and local scalp disorders. These latter do *not* include ordinary dandruff, or even *seborrhea*, a disorder of the oil glands in the scalp, which—despite what baldness institutes or hair-drug advertisements may imply—have not been proved to cause baldness.

There seems now to be virtual agreement among recognized authorities on hair that ordinary pattern baldness is the result of a natural process in many individuals and that no *cure* for it is yet known. In fact, advertising of any common-baldness "cure" is currently forbidden by the United States government, heavy fines having been assessed in recent years against institutes and concerns violating this injunction.

The whole situation with respect to baldness might be summed up with a story from the author's own experience. During a period when I was studying in Paris, I grew concerned about my undue hair fall and went to see one of the outstanding hair specialists of France. After an assistant had examined me thoroughly and written out a report, I was ushered into the sanctum of *le Professeur* himself. The great man glanced at the report, looked at me, then smilingly bent over. In the center of his head was a very large bald spot!

"*Voila*," he said. There was an eloquent silence. I sadly asked, "*Il n'y a rien à faire?*" ("Is there nothing that can be done?")

"*Oui*," answered the great specialist. "*Il faut choisir vos parents!*"

Which means, simply, "The one thing to do is to pick your parents." (He might have added "*—and* your sex.")

While we make light of hereditary baldness—inasmuch as it has no relationship to a man's health—the collective evidence about all the other inherent greater disadvantages of males as compared with females has highly serious implications. Further, we have the situation mentioned earlier, that as environments have improved, the genetic disadvantages of males have been asserting themselves more and more. All of this (which will be amplified in Chapter 29 on longevity) should prompt us to give decent burial to the age-old fallacy that men are hardier than women, and, particularly, that little boys are "sturdier" and require less attention than their (presumably) "more delicate" little sisters. Bigger bones and heavier muscles do not necessarily mean greater resistance to disease and death. People might do well to recall the parable of the sturdy oak and the frail reed, and what happened to each in the thunderstorm.

ALBINOS

They are found in all racial groups. Here are some striking examples.

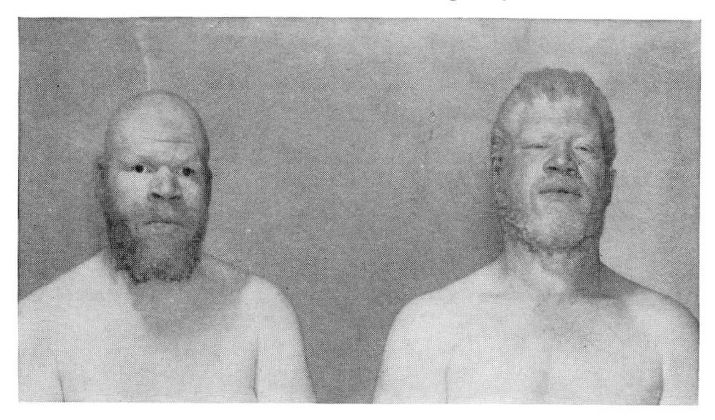

NEGRO ALBINOS (*above*). *Born of black parents in Virginia and otherwise clearly Negroid, these brothers, aged 56 and 57, have very white skins, pale bluish eyes, yellowish hair, and* nystagmus (*oscillating eyeballs, typical of many albinos*). *They have appeared in circus side shows under the names of Iko and Eko.* (*Photo courtesy Rex D. Billings, Montreal.*)

PANAMA INDIAN ALBINO (*above*). *This girl is one of the many Cuni Indian albinos of the San Blas, Panama, province, which has the world's highest incidence of albinism—close to 7 per 1,000.* (*Photo by Dr. Clyde E. Keeler.*)

JAPANESE ALBINO (*right*). *A unique photograph of a 9-month-old Japanese albino boy, made for the author by Dr. Koji Ohkura, Tokyo. It is rarely possible to photograph Japanese albinos because of extreme sensitivity about their condition and the common practice of dyeing their hair as a disguise.*

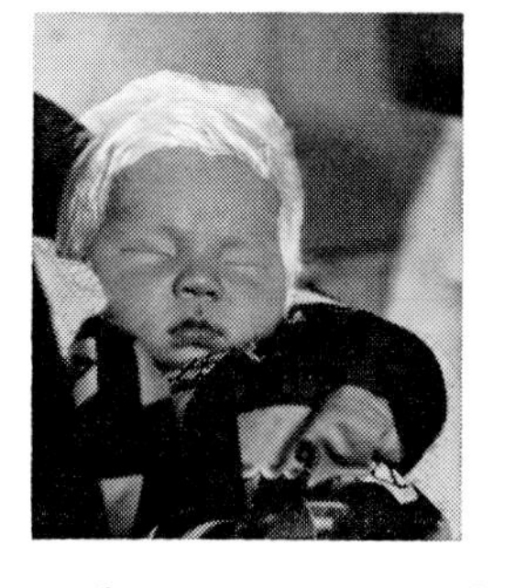

CAUCASIAN ALBINO *from India* (*extreme right*). *A 28-year-old mathematics professor, this handsome albino is from a Brahmin-caste family whose other members are preponderantly dark-skinned and intensely black haired* (*although of the White race*). *Among his noticeable albino traits are oscillating eyes* (*nystagmus*). (*Photo courtesy of Dr. K. R. Dronamraju.*)

HOPI INDIAN ALBINOS (*left*). *A rare photograph, made in 1885 in a Hopi village in Arizona, of two albino children, their albino father and their normal mother. From the same or a related Hopi strain have come many other albinos, including the twenty-two* (*in the Hopi population of 5,000*) *currently living in Arizona.* (*Photograph from Smithsonian Institution archives.*)

Photographed under direction of author by M. Lasser. Courtesy Ringling Brothers.

HUMAN SIZES

Little Ones—Big Ones—Average

Shown here are two types of little people, together with a giant, a fat lady, and, for comparison, an average man at upper right (the author, 5 feet 9 inches, 145 pounds). In front, first four, are the "Dolls" (née Schneider), a rare family of midgets or "Lilliputians"—three sisters and a brother, aged (from left) 38, 50, 34, 48, born in Germany of normal-sized parents whose four other children were also normal-sized. Next to the midgets are two unrelated achondroplastic dwarfs (Frankie Saluto, Jimmy Armstrong) born in the United States. The giant, Jacob Nacken (7 feet 4 inches) when photographed was 44 years old, vigorous (a veteran of the German World War II army) and apparently of the hereditary, non-pathological type, extreme tallness having run in his family. The fat lady (Irene Perry, born in Detroit), weighing 540 pounds, may be of the glandular but nonhereditary type, having come from a family all normal in size.

CHAPTER TWENTY THREE

Structural Defects

WE ARE SO ACCUSTOMED to seeing human beings who are almost uniform in their construction that we can hardly help staring (unless we're very polite and mature) at anyone who deviates noticeably from the expected pattern.

The surface abnormalities, being most easily detected and classified, have been most intensively studied from the standpoint of inheritance. In fact, the very first condition in humans for which Mendelian inheritance was established was *brachyphalangy* (stub fingers), which, back in 1904, a few years after the science of genetics was born, was proved to be a dominant condition. Not long after, *polydactylism* (extra fingers and toes) was also shown to be due (in some families) to a dominant gene. But almost two thousand years before, Pliny, the Roman author, had reported the case of a nobleman with six fingers and six toes, "and his daughters likewise, whereupon they were named *Sedigitae*." (C. Plinius Secundus, in his *Natural History*.)*

At circus or carnival sideshows you can see some of the more startling examples of wayward gene caprice—midgets, freaks with distorted bodies or features, Negro albinos, "India-rubber" men, and other types we shall deal with. A great many of the freak individuals result not from heredity but from abnormal conditions before birth or after. However, many hundreds of inherited structural abnormalities have now been identified. We shall confine ourselves here to those which are most common, conspicuous or otherwise important.

* An earlier reference to this condition, though not linked with heredity, will be found in the Bible (II Samuel 21:20), which speaks of a six-fingered, six-toed warrier.

STATURE. Midgets, or "Lilliputians," always have been of engrossing interest. Technically known as *ateleiotic* dwarfs, they stop growing early in childhood because of an inherited deficiency in the pituitary, or "growth" gland, and do not reach beyond 3 feet 6 inches in height.* (One of the smallest midgets, Martina de la Cruz, 21 inches tall, died in Georgia in 1949 at the age of seventy-four.) Where they differ markedly from other dwarfs is in their having normal proportions, which gives them their appealing doll-like appearance. In contrast are the *achondroplastic* dwarfs (the waddling little bow-legged men of the circuses) who have normal-sized torsos but are dwarfed because they have stunted arms and legs. (Also, usually, oversized heads and pug noses.) Moreover, where midgets generally are sexually underdeveloped or sterile, the achondroplastic dwarfs are normal in all respects except body form, their abnormality being caused, apparently, not by any basic glandular disorder but merely by peculiar workings of their "structural" genes. (While in prenatal life achondroplastics have a high mortality rate, once they survive infancy they are as sturdy and as long-lived as normal persons.) †

In manner of inheritance, also, there are marked differences between the two dwarf types. The genetic mechanism responsible for midgets probably involves two separate rare dominant genes, although in some families inheritance seems to have been simple recessive. However, in the infrequent instances where midgets are able to have children they are not likely to reproduce their type, their offspring tending to be normal; and so it is from normal parents that midgets almost always derive. In achondroplastic dwarfs, however, inheritance usually is more direct and may be due to a single dominant gene, with a possibility in some cases of recessive inheritance. The fact that in many cases there are no pre-

* In a milder form of pituitary dwarfism, individuals may grow to 4 feet 6 inches, putting them in the midget class.

† That the great French painter, Toulouse-Lautrec (1864–1901), may not have been dwarfed because (or merely because) of "a childhood accident which broke both of his legs," as his family said, is now considered likely. One theory, suggested by the apparent shortness of his arms as well as of his legs, together with other characteristics shown in photographs, is that he was an achondroplastic dwarf. Another theory is that if his legs were so easily fractured in childhood as to stunt his growth, he may have had the hereditary condition of *osteogenesis imperfecta* (brittle bones), described later in this chapter. In either case there would have been reason for the noble Lautrec family to attribute their scion's condition purely to an accident.

vious achondroplastics in a family may indicate that the gene for this condition often arises through a new mutation. Whatever the causes, achondroplastic dwarfs are much more common than midgets, whole families of them occurring here and there.

Pygmies are in a category different from either of the dwarf types mentioned. These undersized peoples, found largely in Negroid groups of Africa, and in New Guinea, the Philippines and Malaya average about 4 feet 6 inches in height. They are generally regarded as merely products of "small stature" genes, the theory that some pygmies are achondroplastic dwarfs having been questioned.

Before taking leave of the undersized people, it is worth noting that parallels to all these human types are found throughout the lower animal kingdom. In dogs, bulldogs and dachshunds are achondroplastics (with the characteristic stunted and bowed limbs); and Pekingese, Chihuahuas and other "toy" types are midgets. Achondroplastic sheep—the Ancon short-legged breed—were in favor many years ago among canny New England farmers because they couldn't jump fences. Dwarf species also have been bred or pop up spontaneously in almost all species of domestic animals, and occur regularly among birds and fish.

At the extreme opposite pole of stature abnormalities are *giants*. Although seven-footers may be direct results of "tall stature" genes, the cases where men reach heights of nearly 8 feet or more can be ascribed almost invariably to some early derangement of the glandular system, usually extreme overactivity of the pituitary. Of this type was Robert Wadlow (the Alton, Pennsylvania, giant) tallest of all humans known to medical science, who reached a height of 8 feet 10 inches and was still growing when he died in 1940, aged twenty-two. Gigantism in dogs, such as Great Danes and St. Bernards, is not an analogous condition, because while unusually active glands probably contribute to it, this gland action is hereditary (as it apparently was not in Wadlow's case) and is linked with other genetic factors for bigness.

Extreme obesity, of the circus "fat lady" kind (distinguished from the ordinary obesity discussed in Chapter 14), also is a glandular disorder, but whether heredity plays any role in this is uncertain. Since cases are on record of mother-and-daughter circus fat ladies, the prenatal glandular influence of the mother on the child may have been involved.* (This would be especially true if the mother were diabetic.)

* Among the mother-daughter circus fat ladies (as mentioned on page 118) were a Florida woman who, when she died in 1942, had weighed a record 772 pounds—with a height of 5 feet 5½ inches—, and her mother, who had weighed 720 pounds. (In all medical annals until then scarcely twenty

INHERITED HAND ABNORMALITIES

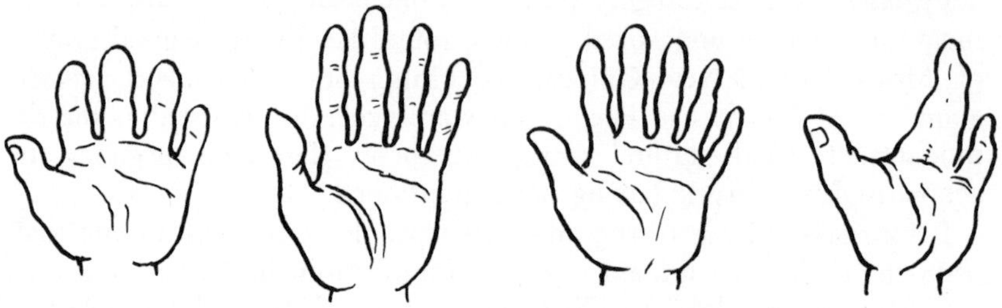

STUB FINGERS	"SPIDER" FINGERS	EXTRA FINGERS	SPLIT HAND ("Lobster claw")
Middle joints missing	*Abnormally long and thin*	*Usually one, but sometimes two*	*Fingers fused*

NOTE: Any one of the above hand conditions may be accompanied by a similar condition in the feet of the individual. All are dominant or partly dominant in inheritance. (See Text.)

Myxedema, a milder type of obesity, with thick dry skin and sluggishness, is the result of a thyroid deficiency and may be associated with goiter. Usually it comes on in adult life (most frequently among women), but it may also occur in children. Here, too, the possible role of heredity is in doubt.

HANDS AND FEET. Just about every conceivable kind of hand and foot abnormality may be hereditary. We have already spoken of stub fingers and extra fingers, but dozens of other hereditary hand abnormalities have been identified—claw hands, spider hands, fused finger joints, webbed fingers, crooked fingers, paddle-shaped thumbs, missing fingers or thumbs, double joints, and so on. (Some of these are depicted in our illustration.) In many instances, the abnormality is duplicated in the feet. What is most important is that in almost every one of these conditions the abnormality or anomaly can be due to a *dominant* gene, providing for direct transmission from parent to a child. However, genes causing these conditions may vary in their action in different individuals,

---

persons were listed as having exceeded 700 pounds in weight.) Claimed to have been the heaviest man in history was a carnival freak, Robert Huges, who was reported as weighing 1,065 pounds when he died in 1958 at the age of thirty-two. Previously, the heaviest man on record had been listed as Miles Darden of Tennessee, who had reached the weight of about 1,000 pounds, with a height of 7 feet 7 inches, when he died in 1857.

possibly influenced by other genes or external factors, so that in the same family varied forms of hand or foot abnormalities may occur. Also not to be overlooked is the fact that similar congenital abnormalities may be due to prenatal developmental accidents.

In rare instances, genes seriously defective in their chemical functions may cause complete absence of hands or feet, or of most of the arms and lower limbs. A few cases are known of families with several of their members so afflicted. But these cases have been wholly overshadowed by the vast numbers of congenital amputations and deformities produced environmentally through the use of the drug thalidomide (as told in Chapter 6). That the environmental condition was akin to the genetic one (a *phenocopy*, to use the technical term), has special significance. Since thalidomide is known to have produced its damaging effects during a short period of early prenatal life (from about the tenth to the forty-second day) when the arm and leg buds were developing, it might be assumed that much the same kind of abnormalities could be produced if, during the same period, wayward genes caused equivalent chemical upsets in the developmental processes. However, even in the thalidomide-caused abnormalities, genetic factors could conceivably play some part; for, inasmuch as four out of five of the mothers who took the drug during the crucial stage of pregnancy produced wholly normal babies, the possibility is seen that gene influences might be involved in making only a minority of mothers and/or their fetal infants especially vulnerable to the adverse effects of this drug (and perhaps of certain other drugs as well).

The extent to which *congenital clubfoot* is hereditary is still in doubt. The term applies to a number of conditions which, if and when inherited, probably are caused by recessive genes. In other cases various prenatal upsets may be involved.

SKELETAL. "Brittle bones" (*blue sclerotic,* so called because one of its attendant effects is bluish eye-whites) is a not uncommon condition in which the bones are so brittle that they may break at the slightest strain. (In one classic case reported in Norway a young man broke his ankle when he turned sharply to look at a pretty girl.) This condition, which may also be accompanied by deafness and defective teeth, is inherited as a qualified dominant. Another condition thought to be hereditary in some cases is *spina bifida*—"cleft spine," which permits part of the spinal canal to extrude—the result of incomplete construction of the spine and spinal cord. Some *hunchback* conditions are due to this defect, but many, especially those dating to the pre-antibiotic era, are the result of tuberculosis of the spine and are therefore nonhereditary.

Cleidocranial dysostosis, a bone and skull condition in which there is incomplete bone development, particularly of the roof of the skull and the collarbone (so that the shoulders can be brought together in front), is ordinarily inherited as a dominant. The rare individuals of this type may have fairly normal life spans. Another condition is *microcephaly,* in which there is an arrest of brain and skull development, leading to idiocy. (See Chapter 26.)

Cleft lip (harelip), with or without *cleft palate,* and cleft palate by itself, without harelip, have been mistakenly thought of as forms of the same condition. While in both conditions there is a similar failure of the parts involved to fuse prenatally, the causes, and the possible roles of heredity and environment in the accompanying defects, may be quite different. The two sexes also are affected in different ways. Cleft lip *and* cleft palate occur more often in boys (in about 1 in 1,000 at birth). Cleft palate by itself is more frequent in girls. While these conditions often show strong evidence of heredity, the specific methods of inheritance remain uncertain, except for a few very rare types. One assumption is that inheritance of the common forms may be *polyfactorial*—that is, due to a *group* of genes—but often depending on adversities in the prenatal environment. How important even minor environmental differences before birth may be in causing these defects to develop or be suppressed is proved by the fact that among identical twins, where one has harelip, with or without cleft palate, in about a quarter of the cases the other has not, or shows only slight symptoms of the condition.

TEETH. Most teeth defects are clearly due to environmental factors, as is shown by the increase or decrease in tooth decay that goes with variations in diet, eating habits, and dental hygiene. However, in contrast to most environmental ills, dental caries (the technical term for tooth decay) tends to be higher in groups with advanced living standards than in underprivileged groups because the diet of the more favored groups includes richer foods, more sweets and more refined and canned foods, whereas the poorer groups have sparser but more natural diets. This was strikingly illustrated during World War II, when restricted diets in war-torn and occupied countries brought a marked decrease in tooth decay among children. After the war ended and previous diets were restored, the dental caries rates rose sharply to former levels. But other types of tooth defect—including malformed, crooked and missing teeth—show a higher incidence in underprivileged groups where dental care is minimal.

Nonetheless, in any given group and under similar conditions, some individuals may be predestined by their genes to run up greater dental bills than others. For example, in a community where fluoride, a tooth-decay preventive, had been added to the drinking water, dental disease was still considerably greater among children of parents who had bad teeth than among those from families with good teeth, leading to the inference that some hereditary factor was involved. Again, among experimental animals under identical conditions, some strains are highly susceptible to tooth decay while others are highly resistant. In human beings it is thought that the hereditary tendency to tooth decay, or conversely to its prevention, may work through chemical substances in the mouth which either stimulate the growth of tooth-attacking bacteria or inhibit their development. A fortunate few persons—one in a hundred—may be completely immune to dental decay.*

Shapes and sizes of teeth are much influenced by hereditary factors (as noted in Chapter 13). By the same token, many specific defects or peculiarities in teeth are also traceable to heredity. Some of the most serious ones are by-products of other conditions we've discussed, such as "brittle bones," cleft palate, and albinism. But many defects in tooth structure or development are inherited independently.

Absence of certain teeth may be among the defects caused by gene failures. In some cases, none of the *incisors,* or only the upper incisors, appear, while in still other cases the *molars* may not appear, or the roots of some teeth may be absent. But *extra* teeth may be inherited too, and so may a hastening of tooth development which gives some babies teeth at birth. (For the manner of inheritance of these and other teeth

* The belief that tooth decay is genetically influenced in its suppression or appearance gains support from the finding that mongoloid idiots have unusually sound teeth and that this may be related to special biochemical factors accompanying their genetic abnormality. In one group of 136 mongoloids, aged three to thirty, Drs. M. Michael Cohen and Richard A. Winer (Tufts Dental School) found that 96 per cent had no dental decay, despite gross neglect of mouth hygiene. Analysis of the saliva of the mongoloids showed significantly above-average levels of sodium, calcium and bicarbonate. In another study, reported by Dr. Carl J. Witkop, Jr. (National Institute of Dental Research) it was found that among normal children, those with the inherited ability to taste a certain chemical substance ("PTC," to be discussed in Chapters 24 and 36) have a 28 per cent lower tooth-decay rate than the hereditary nontasters. The possibility is thus suggested that among individuals in general, genetic differences in biochemistry may have a bearing on relative risks of developing dental caries.

abnormalities, see the Wayward Gene Tables in the Appendix.)

Malocclusion—the improper fitting together of the upper and lower teeth, endangering healthy dental development—may be partly due to faulty chewing habits, but heredity may also be involved. (A mild form of malocclusion was found in all five of the Dionne quintuplets.) Dr. J. H. Sillman, who has followed the dental development in many children from birth to early maturity, has reported that the pattern for malocclusion may often be present at birth—definitely at 1½ years of age—and has suggested that it may be hereditary.

Among inherited *enamel* defects are some in which the enamel is soon lost or pitted and others in which it is discolored. One condition producing brown teeth—through a dominant sex-linked gene carried in the X chromosome—was discussed in the preceding chapter. Another rare condition, *porphyria,* produces reddish teeth (as well as other defects). *Opalescent dentin,* in which the teeth are almost as transparent as plastics, and break or wear away easily, is also hereditary. It should be borne in mind, however, that many cases of defective dentin or enamel, especially where they are prevalent in a community, are due to such extraneous causes as harmful chemicals in the water or faulty diet. Attention to these factors has greatly reduced the incidence of teeth defects in many places.

SKIN. Abnormalities in people's outer wrappings are so easily recognized that, like eye defects, they have been intensively studied and a great many of them have been established as hereditary. Most of these inherited skin disorders or abnormalities are not functionally serious but derive their importance from their effect on the social and economic life of the individual. Such conditions as scaly skin, shedding skin, elephant skin or blotched face (large birthmarks) may greatly interfere with the victim's chance of employment, social adjustment and marriage. Further, some hereditary skin conditions may be accompanied by serious bodily defects, and a few may even prove fatal.

Among the oddities is *rubber skin.* The sideshow "India-rubber" men may have skin so elastic that it can be pulled out five or six inches on the chest or forearms and snapped back. This is possible because of an increased amount of elastic tissue, a condition produced by a rare qualified-dominant gene. In still another odd condition, *Darier's disease,* there is perpetual "goose flesh," due to pinhead-sized gatherings of skin at the mouths of the hair follicles. Inheritance is apparently dominant.

Birthmarks are, of course, the most familiar of skin peculiarities, but only in the rarer instances do they deserve special attention. Everyone has some type of birthmark, and most people have many. Often these are hereditary, with the genes dominant or partly so in the common forms and with the same type of birthmark appearing in the same position in successive generations, at birth or in later years. Where there is danger is in a few kinds of birthmarks which may be forerunners of cancer. These are not the common ones which are hairy, raised or warty, or merely discolorations in the skin. The dangerous birthmarks are generally the rare ones which are a *combination of smooth, hairless, pigmented and not raised,* or which change color (turning black suddenly), or which increase in size, become ulcerated, or bleed. While you should not give way to unnecessary fears about birthmarks, wherever there is any doubt, consult your doctor.

Of the cancerous birthmarks, only a very few have been proved to be hereditary. Most serious are malignant freckles (*xeroderma pigmentosum*), dark, freckle-like inflammations (affected by exposure to sunlight) which appear in infancy or childhood and are produced by recessive genes. They usually bring early death. *Coffee-colored spots,* with nerve-end tumors (*neurofibromatosis*), which appear at birth or in childhood over various parts of the body and spread later, may also give rise to cancer, as well as to blindness, paralysis and various internal effects. The condition is dominant or partly so in inheritance.

Large hereditary birthmarks (*naevi*) may be of various types, some due to blood-vessel swellings, some raised, some merely pigmented patches of the skin. A familiar kind is the "Nevus of Unna," on the nape of the neck. Most remarkable of all large birthmarks on record is that of a woman in the Soviet Union half of whose body, completely down from head to foot on the left side, was one huge birthmark. Such a mosaic condition is not hereditary, however, being produced in the very earliest stage of development by an upset in one half of the fertilized egg just after it has divided. Similar upsets in cells during later stages of embryonic development can produce other nonhereditary birthmarks. Also to be noted is that mosaic conditions of various kinds can occur when genes or chromosomes which are active in some cells are inactive in others, as in the mixed eye-color condition—one dark, one light (Chapter 10); the alternate X condition (Chapter 22); or the "sex mosaics" (Chapter 27).

Small fatty growths or yellow patches on the skin and other areas (*xanthomatosis, lipoid proteinosis,* etc.) may be by-products of inherited metabolic disorders which will be discussed in the next chapter.

Among the serious skin oddities is *absence of sweat glands.* The men afflicted are sometimes called "dog-men" because, not being able to perspire, they pant like dogs when overheated (although it has recently been shown that dogs do have sweat glands). This condition, which is often accompanied by an incomplete set of teeth, sharp-pointed, and by hair and growth defects may be due to sex-linked genes (which explains why most of the victims are men).

Other hereditary skin conditions include the following:

"Scaly" skin, in which the skin is dry and scaly, easily inflames and continually sheds. This appears in infancy and is dominant or sex-linked in inheritance. In *"horny" skin* (recessive) the skin is hard at birth and becomes cracked at the joints. (Sometimes it is accompanied by small deformed ears.) The condition may subsequently correct itself but, if it worsens, is sometimes fatal. *"Elephant" skin,* an extreme form of "horny" skin, causes premature birth, with death following shortly. This, too, is a simple recessive.

Psoriasis, a quite common variant of "scaly" skin, in which there are mottled, scaly patches, decreasing with age, may be inherited only as a *susceptibility* (through an irregularly dominant gene), the condition developing under certain adverse environmental influences. This is in line with the belief that there are inherited degrees of skin sensitivity or predispositions to skin irritations of general or specific kinds.

(*Shedding skin, blistering skin, light sensitivity,* and a number of other rare skin conditions are briefly dealt with, under "Skin," in the Wayward Gene Tables in the Appendix.)

Albinism, mentioned in earlier pages, is primarily a skin-pigment defect which is now classed as an "inborn error of metabolism" leading to interference with the processes of pigmentation. The common type, due to recessive genes, occurs among all peoples, light-skinned or dark (as it also occurs in most animals). The highest incidence of albinism is among the San Blas Indians of central Panama (7 in 1,000), but the condition is most striking when found among Negroes (see photo, p. 234). *Albinoidism* is a mild form of albinism, in which the skin and hair are nearly unpigmented at birth, but some pigment develops later; the eyes are not affected as in albinos. The condition is inherited through a qualified-dominant gene. *Partial albinism* has several manifestations. In the piebald type (dominant) there are stripes on the individual's back, sometimes with white patches elsewhere. (In *vitiligo,* with somewhat similar patch effects, heredity is doubtful.) Mildest and most familiar of the partial albinism conditions is *white forelock* (dominant), in which there is a blaze, or patch of white hair, above the center of the forehead, usually present from birth on. (In recent years many women have regarded such an unpigmented lock of hair as attractive and have imitated it by bleaching where it was not naturally present.)

HAIR. The most common hereditary hair defect, *baldness,* has already been discussed. Some other hair conditions (for which essential details are given in the Wayward Gene Tables in the Appendix) are complete or partial hairlessness, beaded hair, and excessive facial hair. Also hereditary may be *premature* grayness (differing from ordinary graying), in which the head hair begins to turn gray in adolescence, but with no effect on the life span. (Note, too, that natural gray hair in maturity has no relationship to health or lack of it.)

NAILS. There are many inherited nail defects and anomalies. The more serious may be inherited independently. Among these are *absence of nails* (partial or complete); excessively *thick* nails, sometimes with thick skin on palms and soles; *hooked* or *incurved* nails; *small, thin, soft* nails; *flat* and *thin* nails; completely *milky-white* nails; and *bluish-white spots* on nails. All of the foregoing conditions, present at birth, are inherited through dominant or partly dominant genes. (But absent nails also can be inherited through recessives.)

Other nail peculiarities may be by-products of certain diseases. In cirrhosis of the liver, for instance, the nails often are very flat and may also be clouded. "Spoon nails" may be associated with a form of anemia. Separation of the outgrown nail from the nail bed (*onycholysis*) may result from hyperthyroidism, among other causes. Such nail changes as thinning, brittleness, and ridging may accompany arterial diseases. In fact, medical diagnosticians are reported as giving increasing attention to fingernails as indicators of various systemic diseases.

CHAPTER TWENTY FOUR

Functional Defects

THINK OF ALL the different working mechanisms there are in your body: your circulatory, respiratory, digestive, muscular, nervous, glandular and reproductive systems; your sensory equipment—for seeing, hearing, tasting and feeling; and finally, your brain. Then think of all the possible things that can go wrong or be wrong to begin with in any one of these intricate mechanisms. That will give you a clue to the endless number of *functional defects* that can be produced by bungling or shiftless genes. At the same time, as you read about these conditions in the pages that follow, never cease to be impressed with how lucky you are that so very, very few of all these possible defects are present in you!

In the preceding chapter, we were concerned with *structural* defects —abnormalities in the skeletal and also outer structures of the body. Now, in this chapter we shall turn our attention to other kinds of defects—those in either the construction or functioning of given organs of the body, which prevent them from performing properly or normally.

EYES. The human eye, because of its importance and accessibility, has been studied so extensively that hundreds of hereditary eye defects or anomalies (unusual conditions) already have been brought to light. Some of these may cause the individual only minor difficulty except in extreme cases (marked color-blindness, for example). But other conditions produce seriously defective vision ranging up to total blindness.*

* *Total blindness* refers to complete absence of vision. *Legal blindness* is defined in the United States as (1) inability to read ordinary newsprint even with the aid of glasses; or (2) no more than 20/200 vision in the better

Functional Defects

In the great majority of cases, environmental factors are directly or predominantly responsible for blindness, but the picture has been undergoing rapid changes. While many of the previous environmental causes of blindness—notably infectious diseases and accidents—have been eliminated or greatly reduced in seriousnesss, the incidence of blindness has been steadily growing. From this we may infer that the role of heredity is becoming increasingly important. The reasons may be found in the continually rising age level of the population, which has been increasing both the frequency and the proportion of cases of blindness resulting from general diseases such as diabetes and arteriosclerosis, as well as from such specific eye diseases as glaucoma and cataract.

With respect to blind children, the story has changed even more radically. In previous times, large numbers of children entered the world blinded prenatally or at birth by infectious diseases such as syphilis or gonorrhea; and a great many more were subsequently blinded by diphtheria, smallpox, typhoid or trachoma. These causes of blindness have almost been eradicated in the United States and many modern countries. So also has been a unique condition, *retrolental fibroplasia,* which, during a period of years prior to 1954, blinded thousands of premature incubator babies, including many twins. This affliction was regarded as a consequence of premature birth until suddenly revealed as due to excessive oxygen in the incubators. Although with today's precautionary measures new cases of this disease are rare, an estimated 10,000 of its previous victims remain in the permanently blind population (accounting for perhaps 25 per cent of the blind children of school age), and thousands more may have been left with some degree of visual impairment.

But again, since the proportion of childhood blindness due to diseases has greatly decreased, the proportion of *congenital* blindness has increased, from 9.5 per cent in 1940 to about 14 per cent in 1960. Of these cases of blindness at birth, roughly a third are believed due to hereditary defects which result from the failure of wayward genes to guide the proper construction of the iris, retina or some other vital part of the eye mechanism. Many other cases of blindness may develop following birth—in later childhood or in maturity—as an aftermath of any of a number of hereditary eye defects affecting the lens, eye nerves,

---

eye, even with glasses—that is, ability to see at no more than 20 feet what a person with normal sight can see at 200 feet. However, new types of lenses are improving the vision of many persons formerly classed as legally blind.

or eyeball. Furthermore, besides the conditions which are more or less directly identified with specific gene action, there are blinding eye afflictions which result from diseases influenced in some general way by hereditary factors. Hence it may be difficult to make any precise estimate of the role that heredity plays in the total cases of blindness or other visual defects.

Cataract, or opacity ("cloudiness") of the lens, is today the chief cause of loss of vision, responsible for close to 18 per cent of all blindness in the United States. Very mild forms of cataract, which never become serious, occur in a great many people. Some congenital cataracts may be due to prenatal infections or certain diseases in the mother (German measles, for example); and some cataracts appearing in later life may also be due to infection or disease coupled with the degenerative effects of old age. But authorities are convinced that most congenital cataracts and many of those which develop spontaneously before old age are *hereditary,* often through a dominant gene. (In one pedigree, among 100 descendants of a blind woman, one-third were blind through congenital cataracts.) The form and severity of cataracts, and the time of their onset—birth, childhood, puberty or middle age—are usually similar within a family. However, it should be stressed that the major effects of cataract can be averted in a great many cases by timely operation, which is why foreknowledge of the hereditary threats for any given person is important.

Glaucoma, now accounting for over 13 per cent of all cases of blindness, involves an abnormal pressure of the fluid inside the eyes which can lead to blindness unless treated early and effectively. If glaucoma occurs in children while the eyes are still growing, enlarged eyes may result; if in adults, the eye is already formed and does not enlarge. The infantile and childhood glaucoma conditions are comparatively rare and are due to recessive genes. Glaucoma occurring among adults is perhaps in most instances not hereditary, but authorities feel that heredity may be involved in more of the adult cases than has previously been suspected. (Relatives of glaucoma patients show a high incidence of the condition when subjected to special testing—a reason why, if glaucoma has been found in one adult of a family, others might well have their eyes checked for possible symptoms.) Where heredity is a factor in adult glaucoma, the genes may be dominant or sex-linked.

Among other hereditary causes of blindness are these:

Retinitis pigmentosa, in which there is a gradual degeneration of the rods and cones in the eye mechanism, accompanied by deposits of pigment in the retina, leading to marked visual defects and pos-

INHERITED EYE DEFECTS

Some of the Points at which "Eye" Genes May Go Wrong

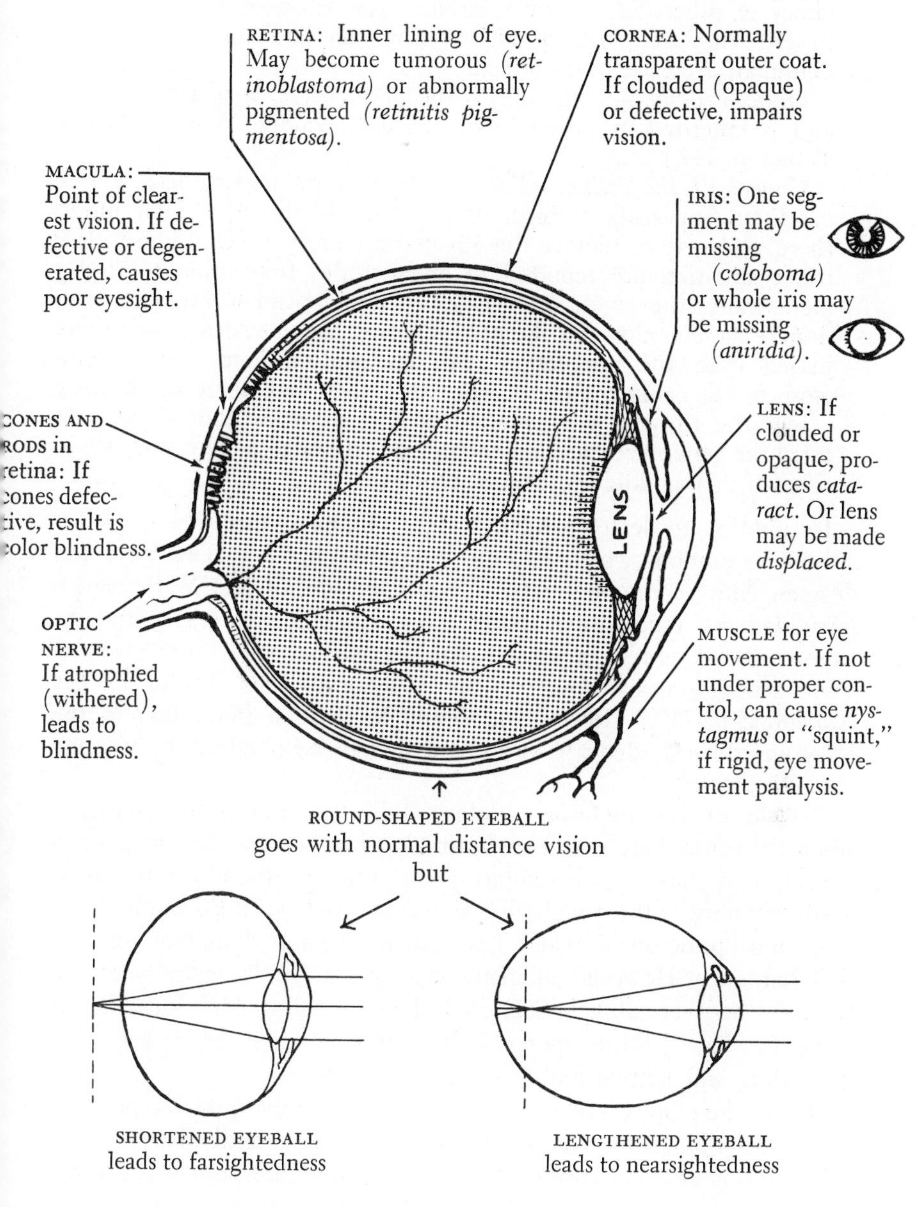

(Also, eyeball may be made abnormally undersized [microphthalmia])

sibly to blindness. In the early stages the disease is manifested by night blindness (inability to see in dim light). Retinitis pigmentosa occurs in a number of forms, sometimes accompanied by deafness, idiocy or other defects; with various ages of onset, from early life to late maturity; and through various methods of inheritance—dominant, recessive or sex-linked.

Optic atrophy, a withering of the optic nerve, is of several types and is inherited in various ways. (See Appendix, Wayward Gene Tables, p. 744.)

Cancer of the retina (glioma retinae or retinoblastoma), a rare condition appearing at birth or in early childhood, is fatal unless there is surgery to remove the affected eye or eyes. (Children in this tragic situation are reported in news stories from time to time.) Genetically, the condition usually is dominant, but the fact that, despite its fatal element, the gene involved has not been largely eliminated from the population and has continued to appear, with close to the same frequency, suggests either that many of the new cases arise through gene mutations or that many individuals who carry the gene never develop the disease, although it may crop out in some of their children who receive the gene.

In addition to defects which produce blindness, there are many hereditary conditions which impede or limit vision in various ways and degrees. Most common by far are *astigmatism* (defective focusing); *farsightedness*, especially if extreme and present from birth; and *extreme nearsightedness* (myopia), sometimes associated with eye tremor. (For genetic mechanisms see the Wayward Gene Tables in the Appendix, p. 745.) However, it should be kept in mind that in the foregoing defects adequate vision usually can be obtained by wearing corrective lenses.

Certain of the *dyslexias*—reading difficulties found in children—when the causes appear to be constitutional rather than merely psychological, may have an hereditary basis. One of these conditions is "mirror-reading," the peculiarity of seeing (and writing) words in reverse and upside down, which has been reported as dominant-qualified in inheritance. However, in many instances a child's inability to distinguish between alphabetical look-alikes (reading "b" for "d," or "was" for "saw") is a temporary failing quite common among beginners in reading and writing and is remedied in time.

Of the hereditary eye defects which affect only special aspects of vision, *color-blindness* (discussed in Chapter 22) is the best known. Also limiting vision are some *night-blinding* conditions, much less severe than retinitis pigmentosa, in which the defect is restricted chiefly to the inability to see in dim light. While some night-blindness cases have been ascribed to vitamin A or B deficiency, these are clinically different from the hereditary forms.

A number of common eye defects result from weaknesses or defects in the eye muscle or nerves. *Cross-eyes,* the most familiar defect of this kind, is very often (but certainly not always) hereditary. *Eye tremor* (*nystagmoid, or oscillating,* eyes), if present at birth, may also be hereditary. *Eye-muscle paralysis* (inability to move the eye at birth, or later) may sometimes be the result of injury or infection but can also be inherited. *Drooping eyelids* (ptosis), when present from birth, is an hereditary dominant.

Faulty construction of various parts of the eye itself characterizes many conditions. In the cornea one may find an opaque ring over the iris; or a cone shape, causing extreme astigmatism. Or there may be an abnormally large eyeball, which need not affect vision. There may also be abnormally small eyes (microphthalmos), which do affect vision (in extreme forms of this condition, the eyes may be absent altogether). Among other genetic causes of microphthalmos may be the presence of an extra chromosome, which produces this defect as part of a certain syndrome. *Dislocated lens* may be another hereditary condition, in some cases associated with abnormally long fingers, toes and limb bones (the *Marfan* syndrome).*

There are many other inherited eye defects and anomalies too rare to justify discussion here, but the interested reader will find the more important ones listed in the Appendix and adequately treated in some of the books cited in our Bibliography.

Ear and hearing defects. Estimates of the proportion of deafness cases that are primarily or largely hereditary range anywhere from 10 to 45 per cent. This uncertainty is due, first, to the varying interpretations of deafness in different studies and the failure, often, to distinguish among those born totally deaf, those who become deaf later,

* That Abraham Lincoln may have had the Marfan's syndrome has been suggested by several medical specialists. Among Marfan symptoms noted in Lincoln were disproportionately long extremities and hands; excessive leanness; a thin, elongated head; malpositioned, large ears; a sunken breast; and spiderlike legs. Also, in accord with the theory of inheritance (dominant-qualified), the Lincoln lineage reveals Marfanlike traits in scattered kin of the President, including several recent descendants of these kin who definitely have the Marfan condition. However, there is dispute as to whether the inheritance stemmed originally from Lincoln's paternal or maternal side. (See article by Dr. Harold Schwartz, *J. Amer. Med. Assn.,* Feb. 15, 1964.)

and those who suffer severe hearing loss and become partially deaf only with advancing years. Second is the fact that environmental influences are heavily involved in many of the hearing defects, and that in tracing back pedigrees of deafness one can never be too sure about the primary causes. Such conditions in the mother during pregnancy as syphilis, German measles and other infections can produce congenital deafness in her offspring. Also, in childhood, serious deafness can result from various infectious diseases—meningitis, scarlet fever, mumps—or from their treatment with certain drugs (such as streptomycin). Because these early hazards were much more prevalent in former years, we can assume that proportionately more cases of congenital and childhood deafness were once environmental, whereas today a larger proportion than before is due to hereditary factors or influences. This is much like the situation with respect to blindness. In addition—again as with blindness—the rising age level of the population has caused an increase in the types of deafness, developing in middle age and later, in which hereditary factors may play a considerable part.

Congenital deafness, when hereditary, may result directly from defective gene action in constructing the ear mechanism. (Sometimes this may be part of an inherited syndrome, as in retinitis pigmentosa and deafness, or hereditary goiter and deafness.) If a child is deaf at birth, the condition is especially serious because speech impairment is a natural consequence, inasmuch as inability to hear before the speaking stage develops makes it extremely difficult for the child to learn to imitate sounds. Formerly, congenital deafness usually meant *mutism* as well, to the point where "congenital deafness" and "deaf mutism" were used as synonymous terms. But in recent years new teaching techniques are enabling a large proportion of the congenitally deaf to learn to talk fairly well.

Today, there may be close to 150,000 persons in the United States who were born deaf or who lost their hearing before they acquired speech.* Of the cases of congenital deafness now arising, it is believed that anywhere from 40 to 90 per cent may be ascribed in some degree to heredity. There is still some doubt as to the genetic mechanisms involved, but it is known that many different types of genes (most recessives but some dominant or qualified-dominant) can produce

* The total number of deaf or partially deaf runs into many millions. In the forty-five to sixty-four age group, the number of persons with hearing impairments is estimated as 52 per 1,000, in the sixty-five to seventy-four group, at 130 per 1,000; and in the ages after seventy-five, about 250 per 1,000, or 1 in 4.

congenital deafness. This may explain how children with normal hearing are often born to parents both of whom are apparently genetically deaf; for in these cases it is probable that a child has received one type of recessive "deafness" gene from the father, another type from the mother; whereas to produce deafness it would have been necessary to get the *same* gene from both parents.

Some indication as to the relative influences of heredity and environment in congenital deafness is given in studies of afflicted twins. These have shown that among the identical pairs, where one twin was totally deaf in infancy, the other was also totally deaf in 60 per cent of the cases and partially deaf in another 28 per cent of the cases. Among fraternal twins, where one was totally deaf in infancy the other was so also in only 20 per cent of the cases and partially deaf in 15 per cent of the cases, but otherwise the second twin had normal hearing. While these facts regarding twins point strongly to hereditary factors, the environmental influence is also clearly indicated by the 40 per cent of the identical-twin cases in which the twins differed with respect to congenital deafness—in 28 per cent one twin being totally deaf and the other only partially so, and in 12 per cent the other twin having normal hearing.

However, one can conclude that certain types of deaf people have a very much greater than average chance of producing deaf children; and also, that where parents are closely related (as in cousin marriages) there is a considerably increased possibility of gene combinations for deafness coming together in a child.

In maturity, the most common of the serious ear defects is *middle-ear deafness* (with estimates of those affected in the United States running as high as several millions). Often it results from injury or infection, but the hereditary form, *otosclerosis*, is caused by the deposit of soft and spongy, and hard (sclerotic), bone in the capsule of the inner ear. This usually develops in the late teens or early twenties, becoming more serious in later maturity, especially among women after childbirth. In general, the incidence of otosclerosis in women is twice that in men, leading to the theory that the female hormone (estrogen) may be a stimulus in the development of the condition.

Where, in either sex, heredity is the predisposing factor, most cases of otosclerosis are due to recessive genes (one pair or more), while some cases appear to be dominant. *Inner-ear,* or *nerve, deafness,* another type, which as a rule begins at about middle age, is dominant in many families. Fortunately, new operational techniques have made it possible to improve hearing in many individuals with otosclerosis. Moreover, in most cases improved hearing devices now enable the partially deafened person to hear, often quite as well as the individual whose hearing is normal.

Word deafness is more a defect of brain-functioning than of the ear, as the hearing is normal but the individual is unable to understand the meaning of the sounds. It affects many more males than females, appears in infancy, and may be inherited through a qualified-dominant gene.

In the outer ear there are several hereditary abnormalities which have no or little effect on hearing, among them "cup-shaped" ear (ear turned in) and imperfect "double" ear (one or both ears being doubled). However, complete absence of ear, usually on one side, is frequently accompanied by hearing loss in the affected ear. *Ear pits*—minor ear anomalies discussed in Chapter 13—usually are independent of hearing loss. But there is at least one clear-cut pedigree where ear pits have occurred in conjunction with deafness.

SPEECH DEFECTS. Disorders and defects of speech, ranging from lisping and stuttering to serious functional disorders which make speech almost unintelligible, afflict vast numbers of people. (It is roughly estimated that of every thousand Americans, at least fifty have some definite speech disorder.) But despite the widespread incidence of these defects and their repeated occurrence in several members of the same family, the question of their inheritance is extremely uncertain—except, of course, for those speech disorders which are the direct consequence of other hereditary defects, such as cleft palate.

The most common speech disorder, *stuttering*, or stammering (the terms are used interchangeably), which afflicts about one person in a hundred, has some puzzling aspects. Many psychologists incline toward the theory that it is almost entirely the result of emotional and psychic disturbances, dating from early childhood (inner conflict, overanxiety, insecurity, overdomination by the mother, sexual factors, etc.). But to imply that every person who stutters is or was emotionally distrurbed or psychically maladjusted is manifestly unfair in view of various unexplained facts. For example, under war strains some American soldiers were found to have developed speech disorders or to have had existing ones worsened though seldom permanently. Yet, as always, one asks why, under the very same stresses, did only a certain few of the men develop speech defects when the great majority did not?

On the other hand, findings in a number of areas support the belief that organic factors, possibly genetic, are involved in stuttering. First, the disorder tends to run strongly in families, one study having revealed that among close relatives of persons who stuttered, one in three also stuttered, far higher than the general average. Second, among twins, where one of an identical pair stutters, the other does so almost invariably; whereas among fraternal pairs the double occurrence of stuttering is the exception. Third, and perhaps most striking, those who stutter or have other speech defects include far more boys than girls—in certain groups studied, up to ten times as many. While it might be argued, as a presumable cause, that boys usually are under greater emotional stress than girls, there is not enough evidence to explain away the sex differences in speech disorders on a purely psychogenic basis. What can't be overlooked is that the great preponderance of boys among children with speech defects is consistent with findings of greater male defectiveness in a wide variety of other categories, as noted in previous chapters.

But if there is some hereditary transmission (sex-influenced or otherwise) of the tendency to stutter, the gene pattern for it hasn't yet been identified. Likewise, the assumption of some specific organic causes of stuttering or of other speech defects remains theoretical except in the conditions noted before in which the speech disorder is a by-product of some other affliction. Some authorities believe the answer to stuttering may be found in an inherited defect, such as delayed development of the speech-controlling areas of the brain, which, under special stress, would lead to a speech impediment, particularly during the first years of learning to talk. Strengthening this theory is the fact that symptoms of stuttering often show themselves in infancy and in some cases may even appear to be of prenatal origin.

The long-held belief that left-handedness and stuttering have some relationship is now open to doubt. For one thing, the reported proportion of left-handed persons among stutterers (about 7 per cent) has now been shown to be hardly if at all greater than in the general population. Also deflated is the previous notion that stuttering could be induced to develop in left-handed children by forcing them to write with the right hand. Challenging this, a study in St. Louis showed that the vast majority of left-handed pupils who had been taught to write with the right hand had not developed any speech defect. Equally significant is the fact that many stutterers who are left-handed were never forced to write with the right hand and that the big majority of stutterers who are right-handed never had the left-handed tendency. At any rate, whatever the cause of stuttering, it is heartening to note that modern speech corrective methods are now able to cure or reduce the impediment in many of those afflicted, especially if they are treated during childhood.

Cluttering—unrelated to stuttering though often confused with it—is another disorder which involves rapid, confused and jumbled speech and which has been reported as running in families. It also is much more common in boys than in girls. A mild form of cluttering, to which many persons are addicted, consists of transposing words or syllables, as when a United States senator told a colleague he would have "opple amportunity" to reply and referred to the "Chief Joints of Staff." Such word mixups have been called "Spoonerisms," after the Reverend W. A. Spooner, Warden of New College, Oxford, who became famed for making them frequently in his sermons (which may or may not have been a device to keep his listeners awake). In reviewing the findings about the more serious type of cluttering, Drs. Ruth and Harry Bakwin have noted that those afflicted—unlike stutterers—improve their speech when they strive to be careful; and also, that psychogenic factors have not been credited with playing an important role in cluttering, adding weight to the theory of some organic, if not hereditary, involvement.

Allergic diseases. Almost everybody is allergic to something, and if you've ever had skin tests made, you've probably been surprised at some of the findings. There was the American soldier who had to be discharged from service because he was found allergic to the khaki uniform. There was a Pittsburgh girl who advertised the sale of her new Angora wool cape because her boy friend was allergic to it. And there was the man whose sneezing fits every Sunday morning were traced to his allergy to the colored inks in the comic section. These allergic oddities could be extended endlessly, to substances and influences of every conceivable kind, to a large variety of foods and drugs, to alcohol and tobacco, to cats and dogs, and to almost anything you could think of.

Allergies become serious when they are tied up with asthma, violent hay fever or eczema, or when they greatly interfere with a person's eating or living habits. (Such serious allergies occur in about one in every ten persons.) There is also the danger that severe allergies in children may interfere with food utilization sufficiently to retard growth, weaken bones and muscles, cause many intestinal and skin disorders, and also lead to behavior problems. In extreme instances very severe, sudden and overwhelming allergic reactions may produce collapse, shock and even death. Such cases have occurred when individuals highly allergic to bee stings have been bitten and when highly penicillin-sensitive persons have received injections of the drug.*

Most allergic manifestations involve a chemical interaction in a person's system between unwelcome substances, called *antigens,* and the counteracting substances, called *antibodies,* which the person produces to combat the invaders. There are many types of antibodies, behaving in various ways. When the clash between the invading antigens and an individual's antibodies has an unfavorable or harmful effect on the system, it is spoken of as an *allergic reaction,* or *hypersensitivity.* The disturbed tissue may be the skin, membranes of the nasal passages (in hay fever) or the lungs (in asthma), or other tissues, each linked with some characteristic allergy.

Since the body's chemical makeup and processes are intimately involved in any of the allergic reactions, it might be expected that heredity would play some part in them. Thus, some people may have an inherited predisposition to become sensitized (that is, to produce antibodies) more easily, or to a greater variety of substances, than other persons; or hereditary factors may influence the way in which a particular organ or tissue area is singled out and affected by a given antigen-antibody reaction. For example, the well-known familial tendency to asthma may reflect an increased genetic susceptibility of the lung's mucous membranes. Hay fever, various skin rashes, and perhaps migraine headaches, all of which tend to run in families, may similarly reflect special types of hereditary predisposition.

In addition to the afore-mentioned allergic illnesses, there is a group of relatively rare, but genetically interesting conditions called the collagen diseases, because they seem to affect the body's connective tissues, which are composed to a large extent of the substance collagen. Each of the diseases in this group (listed in the Wayward Gene Tables in the Appendix) may produce its effects on the collagen in a particular organ or part of the body—the kidneys, lungs, skin, intestines, joints, nerves, blood vessels, muscles, and so on. However, these diseases may also merge into one another, so that a person may have one collagen illness at one stage and another of the illnesses at another stage. In all of the collagen diseases there may be allergic symptoms, often serious. Another aspect of these diseases is that they may cause the body

* A possibility, now under investigation, is that allergy to cow's milk proteins may be the cause (or one principal cause) of the mysterious, sudden and unexpected deaths of thousands of infants every year. In support of this theory are the reported findings that such deaths very rarely occur among breast-fed infants.

to lose its ability to distinguish between its own substances and foreign substances, so that it fights itself chemically. That heredity may be involved is suggested by the fact that in families of persons afflicted with a collagen disease there is an above average incidence of the same condition or a related one; moreover, the unusual types of antibody proteins often found in the blood of persons with these diseases may also be seen frequently in healthy close relatives.

Rheumatoid arthritis, also classed with the collagen diseases, is an affliction of the joints which occurs more frequently among females than males. Going with the disease may be an abnormality in the gamma globulin blood element which appears to be hereditary in many if not all cases. *Still's disease* is a variant of this condition occurring in children.

Extreme sensitivity to cold characterizes the diseases called *cryopathies.* One of the best known is Raynaud's syndrome, in which fingers may turn white on brief exposure to cold air or water. This condition affects seven times as many women as men. While ascribed to a mild abnormality in the nerves controlling the blood circulation of the fingers and toes, it is also believed that psychological factors may be involved. Some of the other cryopathies are related to abnormalities in certain blood elements which cause blood clumping and thickening at low temperatures. These cryopathies, and Raynaud's, may be genetically influenced. But cold diseases may also be by-products of chronic infections, anemias, allergies and cancers. Among the serious aspects of extreme cold sensitivity, sudden exposure to cold water may cause severe histamine shock and sometimes death. Certain deaths attributed to drowning, heart attacks or muscle cramps may in reality be due to such shock reactions suffered by cold-sensitive persons on diving or swimming in very cold water.

Migraine ("sick headache," usually on one side) has sometimes been attributed to an allergic reaction, but this theory is in question. In any case, migraine is strongly influenced by heredity. Most theories point to a dominant gene as the culprit. The ailment involves periodic swellings of blood vessels in the brain, caused by some chemical disturbance, possibly influenced by hormonal action (and perhaps with a peculiarity in the blood vessels themselves as a predisposing factor). That the chemical functioning of individuals may be important in migraine is indicated by the fact that women (among whom it occurs twice as often as among men) tend to suffer their worst attacks during the menstrual periods when marked shifts in the body's chemistry take place.

One of the most severe of the hereditary allergic conditions, fortunately quite rare, is *angioneurotic edema*, a periodic sudden swelling of the skin or mucous membranes, which may cause death if it occurs in the larynx or vital organs. Its onset, when it is most serious, is usually in childhood or at puberty. This condition discriminates against males, with about 50 per cent more boys than girls affected. Inheritance is believed to be dominant.

Having taken note of the hereditary factors in allergies, we may ask how far *psychosomatic influences* may be involved in causing these conditions or worsening their effects. Many physicians do hold that there is a psychosomatic element in many cases of common allergy. But to call an illness psychosomatic is not to say that "It's all in your mind." What is implied is that as a result of emotional stress, possibly combined with some adverse physical influence, an organ of the body is sufficiently disturbed so that actual damage or disease results. Continued emotional stress might then compound and perpetuate the ailment. But this by no means rules out the possibility that a genetic predisposition may pave the way for the psychosomatic influences to exert their effects. In the case of allergies, then, the genetic factors may govern the antigen-antibody reactions, and the psychosomatic factors, by way of nerve connections or hormone secretions, may pick out a particular target.

If the incidence of allergy is indeed higher than average among neurotic persons, it is possible that the allergic predisposition preceded the neuroses—that continual affliction with a severe allergy helped to bring on an emotional disturbance. But why are so many persons allergic only to *specific* substances, often with no awareness of what causes their trouble—especially in the case of some infants who are allergic to cereals or to other ordinary foods? Why do we so frequently find the same type of allergy running in families—a reaction particularly to sauerkraut, mushrooms, eggs, or strawberries, or to one drug, such as aspirin? And why, without psychosomatic factors, do we find among lower animals some strains which are considerably more allergic than others?

So we come back to the strong likelihood that peculiarities in body chemistry, and hence, predispositions to develop allergies of various or specific kinds, may be inherited—some authorities believe in about three-fourths of the common allergy cases (with purely "acquired" allergies contributing the rest). The fact that there are so many different types of allergic reactions, and that different causes may produce the same symptoms, has complicated the problem of tracing the heredity of these ailments in given families. However, while there may be doubt about the precise methods of inheritance, the evidence is strong enough to suggest that in families where allergic conditions are prevalent, pregnant women or nursing mothers should take extra precautions with their diet, should be alert for allergic symptoms in their babies and children, and should see that apparel and surroundings are as free as possible from known irritants. It should also be kept in mind that, if and where tendencies to allergy are inherited, different members of a family may not be allergic to the same foods and that foods which may be good for most children (such as cream, orange juice, cereals, chocolate or raw fruits and vegetables) may be bad for some.

INBORN ERRORS OF METABOLISM. This term as was noted in Chapter 20, "Wayward Genes," is applied to conditions in which there is a specific shortcoming in the body's metabolic processes (that is, in the necessary conversion of certain chemical substances into other substances). In a number of conditions these metabolic errors are especially clear-cut and directly traceable to the failure of some given gene or genes. Here we will deal with a few of these conditions. (Others will be discussed in later chapters, or listed in the Wayward Gene Tables in the Appendix.)

Gout—that old affliction of Dr. Samuel Johnson and other historical personages—has long been suspected of being hereditary, but only recently has the manner of its inheritance been revealed. It appears that the predisposition to this ailment is caused by a metabolic defect, due to a dominant gene, which raises the level of uric acid in the blood (*hyperuricemia*). Yet all individuals with this inherited predisposition, and with high uric acid levels, do not develop gout or its worst effects (including gouty arthritis and deposits under the skin known as "tophi"). Still a puzzle, further, is why gout afflicts twenty times as many men as women, although the gene for the predisposition is received equally by both sexes. What might be suspected is that, as with various other conditions, the male chemical environment tends to heighten, and the female chemical environment to suppress, the effects of the genetic predisposition to this ailment. (Most cases of gout in women occur after the menopause.)

Urinary excretions of various abnormal substances have been traced to inherited metabolic errors. One of these abnormal substances is found in persons afflicted with *phenylketonuria*, a disease with serious mental retardation, to be discussed in Chapter 26. Other conditions in which there are abnormal urinary excretions are listed in the Appendix Tables under Metabolic Disorders: 2 and 4 (pp. 738, 739).

A number of inherited disorders of *fat metabolism* may also lead to serious or mild effects. The most common of these conditions, the abnormal metabolism of *cholesterol,* which is involved in arterial ailments, was discussed in Chapter 21.

The disorders of carbohydrate or sugar metabolism include *diabetes* (dealt with in Chapter 21), and *galactosemia.* The latter is a recessively inherited defect in which there is a gene failure to properly metabolize the sugar galactose, one of the ingredients of milk. In infants this defect may result in damage to the nervous system and various organs, leading in most cases to mental retardation, convulsions, failure to grow, and usually early death. But this need not happen, for early discovery of the illness and replacement of milk with foods not containing galactose can bring about normal development. As the child grows older and healthier, changes in the metabolic system may permit a more normal diet.

The *glycogen storage diseases* are a group characterized by enzyme deficiencies in the metabolism of carbohydrates. One of the results is the accumulation of the substance glycogen, in excessive amounts (and/or abnormal forms) in various organs, injuring them and interfering with their normal functions. Inheritance in some of these defects is recessive, but in others the genes responsible are as yet unidentified.

Cystic fibrosis is an inherited metabolic disease which strikes young children, killing more under age fifteen than does rheumatic fever, polio, or diabetes. Because of the inborn chemical failure involved, many of the individual's secretory glands—particularly in the pancreas and the lungs—do not function properly. The result usually is early fatality (up till now no real cure has been found, although some improvement in the condition can often be made through treatment). Cystic fibrosis is inherited as a recessive, occurring in about 1 in 3,000 infants at birth. However, a great many normal persons—perhaps 1 in every 30—carry a single gene for the condition. Such persons, although otherwise unaffected, may sometimes be identified by an abnormal concentration of salt in the sweat secretions. This fact can alert these persons (and their doctors) to the chance that they may have a child with the disease, and if this happens, can permit treatment as soon as possible.

TASTE AND SMELL. Hereditary deficiencies—or differences—in these senses probably exist in many forms and degrees; but so far the only clues have come from tests with certain chemical substances. Most familiar to many students who have taken such tests is "taste blindness" to PTC (short for *phenylthio-carbamide*), a substance which tastes bitter to about seven out of ten persons, but not to the others. The inability to taste this substance is usually inherited, through recessive genes. But there is a possibility that the trait is caused in other ways or modified by other factors. Supporting this assumption is the fact that in occasional pairs of identical twins, one can taste PTC while the other cannot; further, there is a greater relative incidence of hypothyroidism (under-production of the thyroid hormone) among people who are nontasters of PTC than among the tasters. Race differences with respect to PTC tasting have also been reported (as will be enlarged upon in Chapter 42), but here again there is some question as to whether the statistics may not reflect environmental influences on taste sensations as well as variations in the way they are described among different human groups.

Hereditary "smell" deficiency of a serious kind has been established only in *anosmia* (total absence of smell), a very rare condition reported as dominant in some families. Also, with possible inheritance, it has been found that some persons with otherwise normal noses can't detect the odor of a skunk. But who would call that a defect?

MUSCLES AND NERVES. *Inherited* muscular defects, while not too common, include some of the most serious of the hereditary afflictions. These conditions may result from defective gene action in not properly constructing given muscles or sets of muscles, or in not providing for their adequate nutrition, leading to *atrophy*, or shriveling. Since muscle activity is controlled by nerves, it is often some nerve defect which inhibits the functioning or causes the withering of specific muscles, so it is not always easy to tell in a given case whether the fault lies with "muscle" genes or "nerve" genes. Moreover, the complex nature of many of the muscular disorders makes it difficult to classify or describe them, or to trace pedigrees; and, besides all this, many conditions which appear to be similar to the hereditary ones may be of purely environmental origin, due to birth injuries, accidents, or diseases. Thus, our discussion will be limited to those muscle and nerve conditions whose inheritance has been most clearly established, and which are most significant. (Other conditions in this category, in which there is any element of heredity, are listed in the Appendix, pp. 742–3.)

The hereditary nerve conditions are of several types, of which *Friedreich's ataxia* is the best known. This comes on in childhood and is symptomized by wobbly gait and speech defects. It is inherited as a recessive. In some families a milder form occurs with onset at about age twenty-one. This form is dominant. One form of childhood paralysis (*spastic paraplegia*) is of various types in inheritance and is more common in males. *Periodic paralysis,* characterized by intermittent attacks during which the muscles remain soft (flaccid), comes on in youth and is inherited as a dominant. It is unique in that, whereas most other types of paralysis are due to some disorder of the muscles or the nerves controlling them, familial periodic paralysis arises through a peculiarity in the body's chemical balance—an attack being brought on by a sudden drop in the concentration of potassium in the blood. Periodic paralysis differs further from other neuromuscular disorders in that persons afflicted usually are quite well between attacks; and when an attack does come, it can often be ended quickly by administration of potassium.

Inherited *muscular atrophies,* with shriveling or degeneration of the muscles, are also of various types. The best known is *peroneal atrophy,* which comes on in childhood and affects the feet, spreading to the calf muscles, and perhaps later extending to the hands as well. Other muscular conditions include *Thomsen's disease,* with muscle stiffness and slow movements, which comes on in childhood; and *myotonia atrophica,* affecting adults, in which there is slow relaxation after muscular contraction. (In this latter condition the afflicted persons if they shake hands may find difficulty for a time in letting go.) Finally, there are the serious *progressive muscular dystrophies,* in which there is muscular wasting, leading to weakness and invalidism, nearly always lasting for years. Conditions of this type currently affect about two hundred thousand persons in the United States, half of them children between the ages of three and thirteen. The hereditary nature of most forms of the disease is all too tragically illustrated by their frequent occurrence in the same family, some families having three or more children afflicted. The condition occurs in various hereditary forms, with different ages of onset and different genetic mechanisms—dominant, recessive and sex-linked.

Missing muscles is a not uncommon phenomenon. To anatomy students who can't find one muscle or another in a human specimen it may be explained that complete, or partial, absence of particular muscles is often due to inheritance (dominant). On the other hand, added muscles may also be present in some individuals through inheritance.

Some of the other conditions that affect muscular action are primarily nerve defects or by-products of other defects. For those of special interest, see again the Appendix.

Epilepsy. In many respects the "falling sickness," as it has been popularly known, is still a mystery. This might seem strange in view of its having perhaps the longest recorded history of any of the mental or nervous afflictions, with countless millions numbered among its victims through the ages. In the United States and other countries it is estimated that at least 1 in 200 persons has epilepsy in some form or degree and that many times more persons have a predisposition which could develop into epilepsy under certain adverse circumstances.

Because it may affect behavior in a peculiar way, and sometimes (though not necessarily) impairs the mind, epilepsy in the past was all too often associated with insanity; and in most mental institutions epileptics were indiscriminately herded together with the insane. In some benighted localities this is still done, but fortunately the tendency more and more is to regard epileptics as primarily victims of nervous disorders who should be treated as such; and only where there is some accompanying serious impairment of the mind (in a minority of the cases) is the need felt to institutionalize them. (The possible relationship between epilepsy and mental defect will be discussed in Chapter 26.)

It should be understood at the outset that epilepsy is not the name of a single disease but is a general term applied to a symptom, or symptoms, of a variety of convulsive disorders which may otherwise be completely unrelated. In fact, one should more properly speak of the epilepsies, classifying them according to the manner in which they arise—through organic abnormalities, diseases, injuries, tumors or chemical disturbances affecting the brain and nervous system. What they have in common is only the epileptic symptoms—sudden loss of consciousness, or loss of control over the nervous and muscular system—manifested in dozens of abnormal behavior patterns. The lesser patterns, without convulsions, come under the heading of *petit mal,* the more serious under *grand mal.* In this latter class are the violent fits, with falling, foaming at the mouth, and so on—an unhappy spectacle which many readers, perhaps, may have witnessed. Often the epileptic does not fall but becomes pale and stares blankly, being wholly or partly unconscious meanwhile. During such states he may commit strange and irrational acts and not be aware later of what he has done. (In some instances, among unenlightened peoples in the past—and still in many primitive tribes—epileptics were regarded as divinely inspired individuals whose strange behavior was dictated by holy spirits.)

On the question of inheritance, we may say at once, *there is no conclusive evidence that epilepsy in its common forms is directly inherited, but there is much to suggest that some predisposition may be hereditary.* Only in a very few rare diseases in which epilepsy is a concomitant has direct heredity been clearly proved as the cause. One of these is *myoclonus epilepsy*, which comes on in childhood and differs from the ordinary type in that there are intermittent spasms and tremors between seizures, and when seizures occur there is no loss of consciousness. This condition appears due to a single pair of recessive genes.*

But if we keep in mind always that the epilepsies are merely symptoms of many unrelated conditions, proof of inheritance in rare forms has little meaning with regard to other and more common forms. Again, a clear distinction must be made between epilepsy which results from obvious injuries, or diseases, and that which seems to appear spontaneously with no known outside causes. Thus, even though epilepsy may occur in several members of the same family, or with frequency in a pedigree, one must be quite sure that a common factor is involved before drawing any conclusions regarding inheritance.

However, there are some significant findings. The incidence of epilepsy is about five times higher among close relatives of epileptics than in the general population, and where one member of a pair of identical twins is epileptic, so is the other in about 70 per cent of the cases (four times the correlation in fraternal twins). But especially striking are brain-wave studies which show that the abnormal wave patterns characteristic of most epileptics also occur in some degree in a very high percentage (as much as 60 per cent) of their close relatives, even though the majority of these persons are themselves not epileptic; that where a child is epileptic, one or the other parent is likely to have an "epileptoid" brain-wave pattern; and that in identical twins, if one is epileptic and the other not, the same brain-wave pattern is nevertheless present in the nonafflicted one as in the other. (See Illustration.)

Thus, most authorities believe that *epilepsy is not inherited as such, but as a tendency or susceptibility* by way of some weakness or defect in the brain or nervous system, and that the condition does not assert itself unless there is a push given in the wrong direction by some acci-

* Another special type of epilepsy reported as hereditary in some cases is focal or partial epilepsy, marked by spasms in only one area of the body—mainly on one side and limited to one group of muscles; usually the disease appears in mid-childhood, subsiding after a few years. While such a condition can result from environmentally caused brain damage, recent evidence suggests that in some cases it may be due to a dominant gene of low penetrance, the susceptibility being revealed by certain brain-wave abnormalities. Thus, among afflicted individuals a large percentage of the parents and siblings have these EEG abnormalities although seldom manifesting the epileptic disorder itself. (Reported by Drs. Patrick F. Bray and Wilmer C. Wiser, University of Utah, 1964.)

EPILEPTIC BRAIN WAVES*

(A) Identical epileptic patterns of identical-twin sisters, aged eleven, both with frequent petit-mal seizures. (In each case the initial short section shows the patient's normal wave, followed by the extreme "spike-wave" pattern coinciding with the first few seconds of a period of unconsciousness.)

CONSTANCE (epileptic) KATHRYN (epileptic)

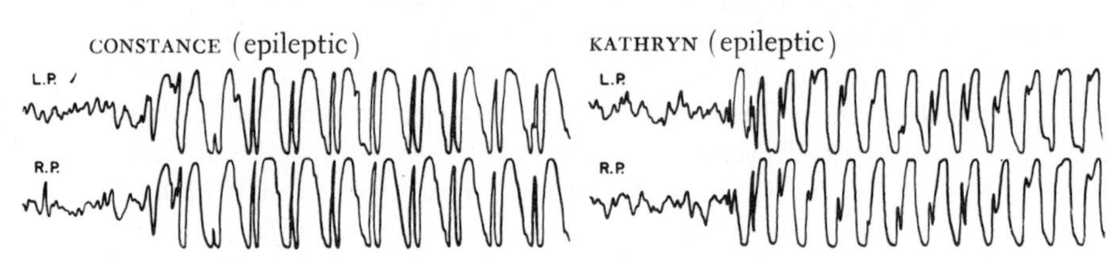

(B) Similar abnormal patterns of another pair of identical-twin sisters, aged ten, but only one of whom has epilepsy (5 to 10 petit-mal seizures daily), the other girl being as yet outwardly normal.

ALICE (epileptic) ARLENE (nonepileptic)

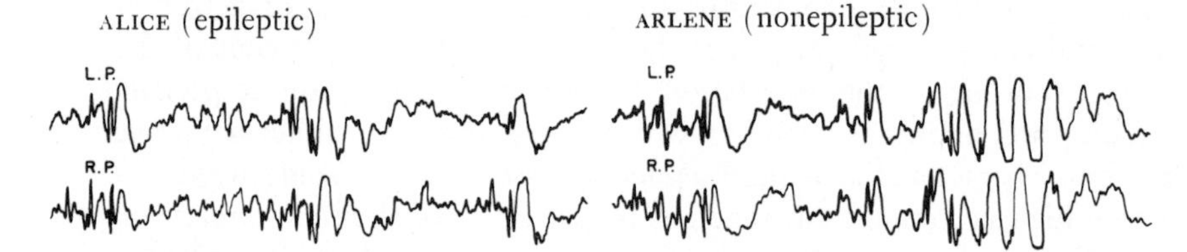

* Tracings and data prepared for author by Dr. William G. Lennox. Cases were discussed by Dr. Lennox in his book (with Margaret A. Lennox), *Epilepsy and Related Disorders*, Little, Brown, 1960: Vol. 1, Chap. 17. (For further evidence of inheritance of brain-wave patterns, as shown by remarkably similar EEGs of long-separated identical-twin women, see illustration, p. 486.)

dent or injury (prenatally, at birth, or later) or by some infectious disease, metabolic disorder, tumor or other adverse influence. A strong enough stimulus, such as passage of electric currents through the brain, can lead to convulsions in everyone; but in some individuals there may be genes which, by producing slight structural or chemical changes in the nervous systems, might cause them to have seizures under mild stimuli which would not affect normal persons.* But only guesses

* In mice there are certain hereditary strains so sensitive that the individuals go into convulsions and die when exposed to loud sounds from which they cannot escape, although other mice are left unaffected by the same stresses. (See Chapter 36.)

have so far been made as to the genes involved in producing the "epileptic tendency."

As to environmental factors, there is evidence that unfavorable prenatal conditions—particularly in older mothers—will greatly increase the chance of a child's being epileptic if the tendency is present. Also indicated in some studies is a relationship between *migraine* and epilepsy (the two conditions go together in many individuals or families), so that if parents have migraine, the risk of a child's being epileptic is increased. In sum, evidence from various sources suggests that in the case of an average epileptic parent there is about a 1 in 40 chance that any given child also will be epileptic, but this chance is lessened if acquired factors contributed to the parent's epilepsy, if his or her seizures began late, or if the family histories on both sides are apparently free of epilepsy cases.

As to the possibility of curing epilepsy, promising results have been achieved through treatments with drugs and diets, reducing or in some cases almost eliminating the seizures in many epileptics. But experts have warned that this does not constitute a cure, that in many cases present treatments do not help, and that in all cases where one might be dealing with hereditary epilepsy, the afflicted person, if married or of marriageable age, should give careful thought to the risks to prospective children. (See Chapter 45.)

CHAPTER TWENTY FIVE

Sick Minds

IN THE SEVENTEENTH CENTURY three brothers came to the United States from England each bringing with him one of the most terrible of all known wayward genes.

Through these three well over a thousand persons in this country have since come to as horrible an end as anyone could imagine—death from Huntington's chorea.* In addition, many thousands more who inherited the disease from other ancestors have met a similar fate.

A single dominant gene causes this condition.

A man (or woman) to all appearances normal, perhaps even brilliant, goes on into maturity with no sign of any impending doom. Then, quite suddenly, usually in his thirties, he begins to disintegrate. His speech becomes thick, his brain and nervous system go to pieces, his body collapses. In a few more years the individual is a helpless wreck, lingering on until he is carried off by a merciful death. *And no cure for this condition is now known.*

But that is not all. A person with a gene for Huntington's chorea may marry and raise a family before the disease strikes. Because the gene is a dominant, all the children of the victim are thus suddenly confronted with the one-in-two possibility that they likewise may be doomed, with as yet *no certain way of knowing beforehand* whether they are or aren't—or, if they themselves have married young, whether they have transmitted the gene to some of their own children.

* The disease is named after Dr. George Huntington who reported on it in 1872 after he, with his father and grandfather, also doctors, had practiced medicine at East Hampton, Long Island, for a combined period of 78 years. During that time they were able to identify and follow up the progress of the disease in an extensive pedigree.

The drama in this gruesome situation was recognized by the late Eugene O'Neill when he used it in his memorable play, *Strange Interlude.* As he wrote to me in 1938, the episode in his play was based on an actual case history of the disease (the name of which he hadn't known at the time) occurring in a family apparently of the New England pedigree. But it should be noted that additional pedigrees of Huntington's chorea, traceable to various other ancestors, have been found in Michigan and Minnesota and in many other countries. In fact, the disease is not confined to any human ethnic group or race, for it also occurs among Africans, Asians and South Sea Islanders. Current estimates are that Huntington's chorea, or the gene for it, is present in from 5 to 6 persons per 100,000. If so, in the United States alone there may now be close to 9,000 individuals who already have the disease and, in addition, perhaps two or three times that number who are carrying the gene and, if they survive into late maturity, will develop the disease itself.

Akin to Huntington's chorea is a rare condition, *Pick's disease* (lobar atrophy), which sometimes has appeared to be inherited through a dominant gene, although its exact method of inheritance is still to be established. In this disease there is a gradual deterioration of the brain in middle life, with symptoms of progressive dementia and severe speech disturbances. Eventually the victim cramps up in bed, almost in the position of prenatal life, and before long, dies.

But lest the reader be unduly alarmed, let us emphasize that the two conditions mentioned have little relationship to other forms of mental disease, and that no *common type* of "insanity" is known to be inherited so simply and directly through dominant genes.

However, in terms of persons affected, there is vastly greater threat and tragedy in the common forms of mental disease, such as *schizophrenia* (*dementia praecox*) and *manic-depressive insanity.* It is in connection with these conditions that the question of inheritance has the greatest meaning both for individuals and for society.

The statistics on the mental disorders are appalling. Rough estimates are that close to one in ten persons in the United States is or will be, at one time or another, afflicted by some mental disorder requiring hospitalization.* This means that there is hardly a family which does

* The uncertainty regarding the definition and the incidence of mental illness was pointed up with the publication in 1962 of the controversial "Midtown Manhattan" study (formally entitled "Mental Health in the Metropolis"). Under the initial direction of the late Dr. Thomas A. C. Rennie, a group of investigators presented evidence of the mental states (as judged by questionnaire responses) of 1,660 adults living in the heart

not have some member or some very close relative among the mental victims. Half of our available hospital beds at this writing are occupied by mental disease patients, and many millions more outside of hospitals are receiving or are in need of mental treatment. If we also consider all the manifold effects of the mental diseases—their victims' loss of normalcy during much or all of their lives, the tragedy to their families and the staggering costs for their care (in the United States, more than a billion dollars in tax funds annually)—we realize that these conditions are far more serious in their total impact than are any of the killer conditions, including heart disease and cancer. And as we are going to learn, the *present case for heredity* in these mental conditions is much stronger.

But let us first make certain general distinctions between the mental diseases ("insanity") and the mental deficiencies (feeblemindedness, idiocy, and so on):

Where a person is insane or partly so, it is because of some disorder or derangement of the brain which warps his thinking and behavior.* His condition need in no way be related to his intellectual capacity (for he may even be of the very highest intelligence), as it results from some disruptive change, sudden or gradual, in the orderly workings of his thinking mechanisms.

Where a person is mentally deficient, however, it is generally because he starts out with a *defective* or *weakened* brain—not a brain gone wrong, but one slowed down in its workings because of imperfect construction, abnormal chemical influences, or prenatal or postnatal mishaps, accidents or disease. (The mental defects will be discussed fully in the next chapter.)

Now we turn to detailed consideration of the various mental diseases and the relative roles that heredity and environment may play in them.

---

of New York City, with these startling conclusions: that close to 23 per cent of those in the sample were definitely impaired and in need of mental treatment; that another 58 per cent showed mild or moderate symptoms of mental or emotional illness; and that only about 20 per cent were completely symptom-free and fully well mentally. However, the methods of diagnosis and the conclusions drawn were strongly criticized by some authorities, and much doubt was expressed as to how far the facts from this limited metropolitan sample, whatever they might have been, could be applied to the general population.

* "Insanity," the old general term for mental diseases collectively, has fallen into disfavor with psychiatrists, but because of its continued widespread popular usage, will be employed here from time to time. Also to be noted is that "insane asylums" now are almost universally called "mental hospitals."

Are we going crazier? This first question might well be asked if we compare past and present figures for the incidence of mental disease and find, with full allowance for population growth, that there now are proportionately at least *thirty-five times* as many persons in our mental institutions as there were a century ago. But for most of this increase the reasons are obvious: The facilities for identifying and caring for the mentally diseased have been vastly expanded, and this fact alone would account for many more people being in such institutions. In addition there are some new factors: the growing proportions of older people who have automatically swelled the numbers of *senile psychotics;* the mental casualties of the two World Wars and the Korean War; and, on a broad scale, the speed-up in living, greater concentration of people in cities, and severer demands of modern society and employment, which have made it more difficult for those mentally unstable to adjust to the situations confronting them.

Just as there can be no accurate comparisons between the relative numbers or types of mentally sick people now and in the past, because no comparable statistics are available, it is also impossible to give any accurate answer to this next question:

Are there more or fewer "mental" cases in the United States than in other countries?

All we have to go by are the *recorded* numbers of cases in each country, which often proves little. For standards of psychiatric diagnosis vary, and there may be enormous differences between one country and another in the relative proportions of mentally diseased persons who are identified and classified as such and who are hospitalized. For example, if the United States has twenty times as many persons in mental institutions as Japan, it may mean only that our facilities for diagnosing and institutionalizing mentally sick persons are that much more extensive. Similarly, comparisons of psychiatric care between the Soviet Union and the United States have shown that while the Soviet mental hospital population is much smaller, it is so, in part, because more patients are cared for *outside* of institutions—and perhaps also because the United States can afford to care for more mentally ill persons both in and outside institutions. Further, there are marked differences in the proportions of mental diseases of different types represented in the Soviet hospitals (relatively fewer schizophrenics, and proportionately many more manic-depressives, neurotics and alcoholics) as compared with those in American mental hospitals.

As a matter of fact, wherever careful investigations have been made, using the same psychiatric yardstick, all the common mental diseases have been found in people of the most diverse cultures, down to the simplest primitives. There may be differences in the proportions in which given mental diseases appear; and in some cases certain peculiar mental disorders which are brought on by infectious diseases, dietary deficiencies, or native drugs may be confined to particular human groups. But again, standards for what constitutes abnormal mental behavior can vary greatly among different peoples. The type of man considered insane in one group may be regarded as merely eccentric, or even "inspired" in another group.

Whatever may be the relative numbers of the mentally ill in different places or periods, there is little question that environmental factors have played and do play an important part in bringing on or worsening mental diseases. At the same time, the evidence has been increasing that *heredity*, too, is a significant factor in most, if not all, common types of insanity, and the principal cause in many cases. In fact, many authorities now maintain that *no person is likely to become insane unless he has a certain constitutional vulnerability to a given mental disease.*

You frequently have heard it said, "She went crazy because . . ." of this or that (like Ophelia in *Hamlet*), or "He was driven to insanity by. . . ." But how much can psychological stress by itself do to produce mental disease? And why is it that in adverse situations to which a great many persons are exposed, only some become insane? These questions have been investigated in many studies.

Take the mental casualties among American soldiers in recent wars. Even in the worst of the battle situations, only a small proportion of the men broke down mentally, and some did so under the mildest stresses. Why? Army psychiatrists believe that most of these men who cracked up were *more vulnerable* than others, or were already mentally unstable before they entered service; and that even in ordinary civilian life, many of them eventually would have become mental cases.

Or take the German air attacks on England during World War II. The frightful rain of Nazi bombs might have been expected to cause many mental breakdowns, but surprisingly the number of hospitalized mental casualties during the period dropped even below prewar standards (some psychologists thought this was because of the feeling of togetherness and the sense of belonging engendered among the British people, which often helps to suppress psychotic tendencies). Or, most terrible of all, consider the Nazi concentration and extermination camps, where the horrors equalled or exceeded anything ever before imposed on human beings. Yet, while many of the victims may have lost their minds, we know that innumerable persons emerged from these excruciating ordeals broken in health, but with their sanity still intact.

So even under the worst conceivable stresses, there seem to be selective factors which single out for mental breakdown only certain individuals and not others. This is particularly evident in regard to the traumatic experiences, or serious emotional conflicts, in early childhood, which often are given as reasons for later insanity. But in any or all of the situations cited as predisposing to mental instability, only some children are affected (often only a single child in a family), while most others emerge with normal adjustment. Similarly, while stresses on a woman undergoing a difficult childbirth or the menopause are often popularly blamed for her "going insane," the fact is that only in exceptional cases do mental breakdowns accompany these experiences, which leave the vast majority of women mentally unscathed.

Another question is whether insanity can result purely from certain bodily ailments, such as arteriosclerosis, or from old-age changes, or from syphilis, alcoholism or drug addiction. This will be discussed more fully later, but here it can be said that all of the foregoing conditions are associated with mental disease in only a minority of the cases; and that, as with psychological stresses, individual differences, probably influenced by heredity, have much to do with whether a given physical condition will affect a person's sanity.

In short, even when we give full weight to the importance of environmental influences, the foregoing facts and many others which follow have strengthened the belief among psychiatrists that in almost every case of mental disease some hereditary factor is involved, either as a direct cause or indirectly through increased weakness under stress. It should be clear that such an hereditary predisposition need by no means doom a person to insanity (in fact, in a large percentage of cases it does not) or leave him without hope if a mental disease does develop. Increasing knowledge of the prevention and cure of mental diseases has made the situation considerably brighter. But by this very fact it has become more important to be on the alert for prepsychotic symptoms at the earliest opportunity, which is why the knowledge of hereditary predispositions may be so valuable.

Theoretically, it would be surprising, indeed, if there were not many types and degrees of inherited abnormalities affecting the human brain and its functions, as there are with respect to other organs of the body. However, it has been extremely difficult to identify the specific physical causes of given mental diseases (as you might trace a malfunctioning of your automobile to a faulty spark plug, or a clogged cylinder). Nevertheless, many promising clues have been emerging as to possible ways in which human brains can go wrong.

Chemical pathways to mental diseases are the ones now being most actively explored. Spurring this type of research has been the finding in recent years that a variety of drugs (reserpine, chlorpromazine, mescaline) are capable of inducing, reducing or aggravating the symptoms of the common mental diseases. An assumption might be that if chemicals administered from *outside* of the body can affect the manifestation of these diseases, then, perhaps, chemicals from inside the body, arising through natural processes in the individual, might likewise be involved in the manifestation of specific mental diseases. Some indications to this effect have come from experiments in which extracts of blood and/or body fluids from schizophrenics have been injected into both human and experimental animal subjects and are reported to have caused in them schizophrenic-like mental symptoms. It might then follow that the production of body chemicals predisposing to mental disease could result from the action of wayward genes.

Another indication of some difference in the physical functioning of the brains of mentally diseased persons comes through electroencephalograms (EEGs, or brain-wave recordings). Every person's brain continuously sends out rhythmic electrical impulses, but the brain waves of many persons with mental disease, even in preliminary stages, follow quite different patterns from those of normal persons. Added importance has been given to this by evidence that brain-wave patterns are to a great extent hereditary. (How this applies to epilepsy was told in Chapter 24.) Where parents are mentally abnormal, a far higher than average percentage of the children have abnormal EEGs, even though these children themselves have not yet shown (or may never show) symptoms of mental disease; and, conversely, parents of mentally diseased children are very likely to have abnormal EEGs, even though the parents themselves may not be psychotic.

But before one can link mental diseases with specific causes—hereditary and environmental—it is necessary to distinguish each type of these diseases from the others. Psychiatrists have made great strides in this direction.* In earlier years there was constant confusion and in-

* The growing knowledge of mental and behavioral disorders and defects has led to clearer classification of the specialists who deal with them: the *psychiatrist*, the *neurologist*, the *neuropsychiatrist*, the *psychoanalyst* and the *clinical psychologist*. The first three are *always* physicians—the psychiatrist one who specializes in *mental diseases*, the neurologist one who deals primarily with *nervous-system disorders*, and the neuropsychiatrist one who combines both these practices. On the other hand, the psychoanalyst, who confines himself largely to *personality disorders*, chiefly the neuroses, need not always be a medical man. (The great majority of psychoanalysts, however, are psychiatrists, although the majority of psychiatrists have not taken the special training to qualify them as analysts.) Finally there is the clinical

accuracy in trying to establish hereditary patterns for any mental disease because all of the diseases were often lumped together under the general heading of insanity, or, when distinctions were made, different investigators often did not make them in the same way. Today, although the classifications are still not clear cut, the great majority of mental patients can be put into quite definite and distinctive categories, these being principally the *schizophrenics,* the *manic depressives,* the *senile* (or "old age") *psychotics,* the *psychopaths,* and those with *organic brain disease* such as may be produced by *alcoholism* and *syphilis.*

From this point on we shall take up these different types of insanity in detail, and largely in the order of their seriousness and prevalence.

Schizophrenia, also called *dementia praecox,* has these names for two reasons. "Schizo-phrenia," Greek for "divided mind," is so applied because a victim of the disease has in many respects a "split personality," a cleavage in the mental functions between the normal and abnormal. In the second term, the "praecox" (Latin) refers to the precocious aspect of the disease, the fact that it often appears at puberty, or before—much earlier than any of the other common mental diseases. "Schizophrenia," however, is the term now most commonly used.

The first symptom of this disease is usually introversion, a withdrawal by the individual within himself. In its various phases, it may involve shyness, timidity, aloofness from others, lack of interest in outside things, and, as the disease progresses, states of ecstasy and hysteria, sometimes with suicidal attempts, or a growing emotional dullness and a withdrawal into the dream world of childhood. There are also subclassifications of schizophrenia: (1) the *paranoid,* quarrelsome and with delusions of persecution; (2) the *hebephrenic,* acting as if in a silly stupor, withdrawn into himself, and smiling quietly; and (3) the *catatonic,* going into tantrums or becoming rigid, in an "I won't" attitude. Then there are the milder and simpler schizoid types, in which there are only occasional periods of abnormal or freakish behavior, but with no mental deterioration, often characterizing extreme cranks and erratic individuals.

Schizophrenia is by far the most serious of the mental diseases, not merely because it is the most prevalent (accounting for more than half the cases in American mental institutions), but because it is the most difficult to cure and the most likely to induce dangerous or criminal

---

psychologist, whose function, if he works in the field of mental abnormality, is chiefly one of classifying the various cases, testing, and doing some corrective work with or training of defectives.

behavior. Moreover, because of its early onset, schizophrenia exacts the greatest toll in years per individual, sometimes almost the entire life span. While it usually first manifests itself between the ages of eighteen and thirty-six (later in females, as a rule, than in males), there is a growing indication that many cases of schizophrenia begin in very early childhood—sometimes, it is reported, in infancy—with the possibility that a child may be mentally abnormal even at birth. One school of opinion (notably among the psychoanalysts) ascribes *childhood schizophrenia* largely to early environmental influences (especially to something adverse in the mother-child relationship). But many psychiatrists hold that there is no convincing evidence that environmental factors can solely or primarily cause the disease in young children. On the contrary, extensive studies of schizophrenic children by Dr. Lauretta Bender and others have indicated that whatever the environmental influences may have been, these children begin life with some distinctive inborn weakness which predisposes them to the disease. Specifically, Dr. Bender has concluded that "no child can develop schizophrenia unless predisposed by heredity," although the disease itself may be precipitated by some psychological crisis and shaped in its pattern by various environmental factors. She and others also believe that just as the childhood schizophrenics grow up to be adult schizophrenics, many if not most adult schizophrenics might have been recognized by careful diagnoses as having shown preschizophrenic symptoms in childhood.*

The fact that in both children and young adults schizophrenia so often appears spontaneously, with no special environmental influence to explain it, adds strength to the belief that it has some inborn basis. (Whether the disease as it appears in children is the same as that in adults has not been fully established.) Among current theories—all *yet to be fully proved*—are these: (1) that there is a deficiency in the schizophrenic's adrenal glands or blood serum which in times of stress prevents his responding with increased output of needed stabilizing

* Vulnerability to schizophrenia may sometimes be detectable in children as early as the age of one month, according to psychiatrist Barbara Fish (New York University School of Medicine). After a ten-year-long follow-up study, she reported that the vulnerability showed itself in some of the children observed by such symptoms as ". . . unusual sequences and combinations of retardation and precocity in all fields—motor, perceptual, language and social." Three children, who had revealed the pattern of "abnormally uneven development" when one month old, proved on examination nine years later to have "pathological disorders of thinking, identification and personality disorganization." (The mothers of the two most severely affected of these children were themselves schizophrenic.)

chemicals as would normal persons; (2) that the pituitary gland is overactive in a large proportion of schizophrenics; (3) that the schizophrenic's liver is unable to carry out the normal process of transforming certain toxic ingredients of foods into harmless, essential substances, but instead allows these poisonous ingredients to get into the nervous system; and (4) that schizophrenics have a metabolic error in the production of a chemical called "serotonin." All or any of these biological factors, if indeed abnormally present, could affect the mental processes. As additional inferences that schizophrenics are slanted biologically away from the normal, there are tentative findings that their fingernail capillaries tend to be of unusual types and that their *fingerprints* show a high incidence of certain unusual patterns—almost complete circles, tented arches, and so on—which, since these patterns appear at birth, might help to identify individuals predisposed to the disease.

Whatever the outer or inner influences in schizophrenics may be, there is a growing belief among authorities that (as with the previous observation about the disease in children) *schizophrenia does not occur unless certain inherent predisposing weaknesses are present.* Evidence for this comes from many significant findings. Especially striking are the studies among twins which have shown that where one of a pair of identical twins has schizophrenia, the odds are very great that the other twin also will have the disease. Close agreement on this point is found in the twin studies in the United States by Dr. Franz J. Kallmann and his colleagues, and in England by Dr. Eliot Slater, both showing that schizophrenia occurred in two twins together in about 85 per cent of the cases where they were identical pairs, but in only 15 to 17 per cent of the cases where they were fraternal pairs—or little more frequently in two fraternal twins together than in two singletons of the same family. These figures are for twins past puberty and into the ages beyond. But among very young twins, also, the situation is much the same. Dr. Kallmann's studies showed that where schizophrenia occurred during childhood in one of an identical pair, it appeared in the other in 70 per cent of the cases, but among fraternals, in only 17 per cent.

Even if one were to assume that environmental influences are often more alike for identical twins than for fraternals, this by itself could hardly explain the vastly higher correlation in the appearance of the disease among them, or the many strange cases in which both identicals of a pair developed schizophrenia at exactly the same time and with the same outcome.* On the other hand, the exceptions—the instances

* The theory that one identical twin can "catch" schizophrenia from the other—a situation sometimes referred to as *folie à deux*—is discounted by

in which schizophrenia appears in one identical twin but not the other (about 15 per cent of the cases, as noted above)—prove that heredity by itself certainly is not always the sole or direct cause of the disease. Environment, too, must be involved, probably to some extent in most if not in all cases.

Other studies have shown that where schizophrenia has appeared in one member of a family, the chances are much above average expectancy (more than four times as great) that it has appeared or will appear in other close relatives. Where parents have the disease, it is estimated that the chances of a child's being affected are nineteen times greater than average, and for grandchildren, nephews and nieces, five times as great.

But finding the genes responsible for schizophrenia or for the predisposition to the disease is another matter because of the complexities of this disorder, its various forms and times of onset, and the ever-present factor of environment. An assumption at present is that *inheritance in schizophrenia may come in an irregular way through several interacting genes, possibly a pair of recessives plus one or more secondary genes, all governed in their expression by certain environmental factors as yet not identified.* More simply stated, for a person to inherit schizophrenia, or the tendency, the genes would have to come from *both parents,* and even then, the disease need not appear unless the individual is exposed to certain outside influences—probably some special psychological or physical strain. The severity of the condition and the time of its onset may depend upon the outside influences as well as on differences in gene action. (Treatments and possible cures for schizophrenia are discussed at the end of this chapter.)

Manic-depressive insanity is less prevalent than schizophrenia and differs from it in its symptoms, time of onset, method of inheritance, and susceptibility to treatment. The disease gets its name from the fact that its victims go from moods of mania to depression in alternating spells. At milder stages or in milder forms (referred to as *cyclothymia*), manic-depressives reveal themselves by their tendency to become overelated or overdepressed, to overact and to talk too much, and by their difficulty in settling down to ordinary routines and social relationships. When those afflicted display this behavior in an extreme form and are

---

Dr. Kallmann. Nor are cases reported from time to time of two singleton siblings, or two other persons in close contact who became insane successively, considered proof that mental disease is catching. Significant in this regard was a situation involving a mixed set of triplets—two identical sisters and a fraternal brother—where both identicals were found to have schizophrenia at age thirteen, whereas their fraternal brother showed no evidence of psychosis.

no longer able to keep contact with reality, they give way to serious maniacal outbursts and require institutionalization. Further, the manic-depressive insane may often recover after a comparatively short period of treatment (sometimes no more than a few months or a half year), but (as with schizophrenics) individuals apparently cured may, under undue stress, suddenly crack up again and commit dangerous acts.

Manic-depressive insanity usually comes on during maturity—in middle life and occasionally in the older ages or in the late teens—although never as early as schizophrenia. But, again as in schizophrenia, while adverse environmental influences are involved in bringing on manic-depressive insanity, *a predisposition toward it is hereditary.* Here again studies of twins are significant: Where one identical twin is a manic-depressive, the other has been reported so also in about 95 per cent of the cases; whereas in fraternal twins, the dual appearance of the disease is stated as occurring only slightly more often than in any two children of the same family. Also numerous pedigrees tend to establish the inheritance of the disease. However (and again as in schizophrenia), there still is uncertainty as to the genetic mechanism involved, although some outstanding authorities believe *that the predisposition to manic-depressive insanity is inherited through semidominant,* or irregularly dominant genes, the irregularity being due perhaps to the required presence of certain other genes and, in addition, to certain adverse factors in the environment.

In a comparison of the chances of inheriting tendencies to schizophrenia or manic-depressive insanity, there is another significant difference. We have pointed out that in schizophrenia *matching* genes for the predisposition probably come from both parents. There is doubt as to the situation in manic-depressive insanity, but on the basis of the gene mechanism believed involved, it is possible that two or more *different* genes may combine to cause the predisposition, so that the contribution from each parent need not be precisely the same. In either mental disease, the hidden genes in the population are prevalent enough so that where one parent has the disease there is always a chance that the other parent, though normal, may supply the other gene or genes required to produce the disease predisposition in a child. However, the chances differ for the two diseases. In schizophrenia, where only one parent has the disease, the chance of a child's being similarly affected is reported as roughly about *one in ten;* in manic-depressive insanity, where only one parent is afflicted, the chance of a child's being similarly afflicted has been reported as high as *one in three.* Another difference is that, where schizophrenia is more likely to crop out and has an earlier onset in males than in females, the reverse is true for manic-depressive insanity. In either case, estimates of the chances of having inherited the predisposition must be qualified by the sex and by the age of the individual. On the latter point, if a person in whose family schizophrenia has appeared is well on into maturity with no sign of mental abnormality, there is much less chance of his being afflicted with the disease; in manic-depressive insanity, however, one must wait longer to be sure if there is a threat.

But this is especially important: Whatever the genes involved, *the manic-depressive tendency is very probably inherited separately and independently from that for schizophrenia.* Hence, if insanity appears or has appeared on both sides of a family *in the two different forms,* the chances of its occurring in any given child are not nearly so great as if the mental disease on both sides were of the same type. On the other hand, it is possible that *both* types of mental disease may run in the same family (in view of the high general incidence of these conditions). So, in tracing pedigrees or drawing conclusions from them, it is hardly enough to say, "There is a history of insanity," since it is important to know precisely what form or forms of insanity one is dealing with. It has not been established whether or not a mixed heredity, with one type of insanity appearing on one parental side and another type on the other, tends to heighten the hazards of becoming psychotic.

Involutional melancholia, often thought of in the past as a form of manic-depressive insanity (and possibly also related to schizophrenia), is now classified as a separate type of mental disease. Although its symptoms include spells of depression, there are no contrasting "manic" or highly excitable periods as in the manic-depressive condition. The disease comes on usually in the older ages (on the average in the late fifties), and the profoundly depressed state of its victims is accompanied by extreme anxiety, restlessness, agitation, delusions and impairment of memory. That hereditary factors are involved in this condition is again indicated by the studies of twins among whom it has appeared. According to Dr. Kallmann, if one identical twin has involutional melancholia, the other will have it also in 60 per cent of the cases, whereas in fraternal pairs a repeat occurs in only 6 per cent of the cases. But precisely how heredity works in this condition is not yet known.

Senile psychoses—the mental diseases that develop in old age (as opposed to ordinary weakening of thinking powers)—are accounting for ever growing numbers of mentally ill persons as the age level of the population rises. Yet why do only some of the old people become mentally deranged, whereas so many others live into the late eighties or even well into the nineties with their sanity intact? For instance, one has only to think of such great oldsters of recent times as George Bernard Shaw, Frank Lloyd Wright, Bernard Baruch, Winston Churchill, Konrad Adenauer, Somerset Maugham, and Grandma Moses. It is not unlikely that the brains of some individuals are genetically geared to function efficiently and normally for a longer period than the brains of others.

Once more, significant evidence from Dr. Kallmann's studies of aging twins shows that in those among whom senile psychoses had developed the double incidence in identical pairs was 43 per cent, but in fraternal pairs only 8 per cent. From another direction, family studies have shown that in persons with senile dementia there was or is an abnormally high incidence of a similar condition among their close relatives. For instance, the most recent study in Sweden showed that siblings of afflicted individuals have more than four times the average chance of developing senile dementia. The same study indicated further that the hereditary predisposition to senile dementia is unrelated to the tendencies to other mental diseases (particularly schizophrenia); nor do families in which senile dementia is frequent have any higher incidence of the other mental ailments than is found in the general population.

As for the means by which genetic predispositions to old-age psychoses may be produced, theories have been offered that certain metabolic processes affecting the brain cells (involving lipid, nitrogen or carbohydrate metabolism, or the utilization of glucose) may be more than ordinarily likely to begin going awry in some aging individuals because of instability in certain genes. However, since so many contributing factors may be involved in senile psychoses, it is doubtful whether any simple genetic mechanism could be found for most cases.*

The *psychopathic* (or *sociopathic*) *personality*, has been identified as a distinct type of mental aberration, midway between sanity and insanity. (One medical student defined a psychopath as a person "who acts crazy but doesn't talk crazy.") This is what makes the condition so dangerous, because the psychopaths, although mentally sick, may never be recognized by those about them as anything more than cold, selfish, emotionally immature and peculiar—but often charming—individuals. They may thus go through life unrecognized, causing innumerable heartaches, if not, as in many cases, actually committing criminal acts. The

* *Syphilis* now accounts for many thousands of cases of mental derangement in the later years. But although syphilis is an environmental disease, even in these cases, there may be degrees of genetic predisposition that control the extent of its influence on the mind. Considerable evidence exists to indicate that syphilis is much more likely to affect the brains and nervous systems of some individuals than others—men, particularly, as compared with women.

exact numbers of psychopaths are not known, but they may reach into the hundreds of thousands. (In World War II, thousands of American draftees were screened out as "psychopathic personalities," and many thousands more who saw service, and later entered mental hospitals, were classified as such.)

As diagnosed by psychiatrists, psychopaths may manifest no delusions or hallucinations, but they never "grow up" to be responsible persons. They live in the present and for themselves alone, have no firm emotional attachments or loyalties, are resistant to discipline, are chronic liars, and follow the early childhood pattern of getting things by ingratiation and extortion, going into tantrums if they do not get what they want. They may be found in every level of intelligence and often are extremely shrewd or even brilliant, rising to high eminence through dynamic acts which may take violently antisocial forms. Thus, Adolf Hitler, in the opinion of most authorities, was a psychopath to an intense degree. Many of the most vicious gangsters and criminals, as well as lesser offenders (particularly swindlers), are psychopathic personalities. Yet it should also be noted that where the peculiar energies of the psychopath are accompanied by great talent and directed into creative activities, very high achievement may result.

The exact causes of the psychopathic (or sociopathic) personality are still in question. Possibly there are quirks in the brain and nervous mechanism which inhibit normal emotional reactions. (Abnormal brain-wave patterns have been found in a very large proportion of the psychopaths.) The fact that the condition dates back to earliest childhood, that it cannot be explained easily by any unusual circumstances, and that it may persist throughout life regardless of any changes in the environment leads to the inference by many psychiatrists that heredity is involved. But precisely how remains a guess.

Not to be confused with the conditions just described are various other forms of psychopathic or sociopathic behavior which may result from physical factors, such as overactivity or underactivity of glands (thyroid, pituitary or sex glands), brain tumors, drugs, or accidents, which may produce sudden changes in behavior in individuals previously normal. But it is even more necessary to make distinctions with regard to the so-called "psychoneuroses" and other forms of personality disorders which fall outside the range of insanity. True, the view has been held that the neuroses are merely mild degrees of psychoses ("everyone's a little crazy—some are just more crazy than others"). This view is no longer tenable, for the evidence we have cited shows not only that psychoses are distinct diseases, probably with some organic basis, but that one type of psychosis may be a disease quite different from another type.

The *neuroses*, or emotional disorders, are also of many types: from compulsive neuroses, anxiety neuroses, hysterias, and hypochondrias, down to the plain garden variety of neuroses which almost anyone can develop under properly adverse circumstances. Any good book on psychoanalysis will describe these conditions in detail and tell what environmental factors may be involved in their causation. That heredity may also play a part in any of these conditions—by making it more likely for some individuals to become neurotic, and more severely so than others, under the same circumstances—is quite probable. Studies of highly inbred mice, rats, dogs and other animals certainly do indicate that there are inherited degrees of emotionality and of susceptibility to neuroses under certain enforced conditions. But in genetically diffuse human beings, neurotic behavior can be an outgrowth of so many diverse influences that it would be extremely difficult to isolate and identify specific hereditary factors for anything so general as "neuroticism" (even though some investigators have attempted to do so). We can say only that any two neurotic parents stand a good chance of having neurotic children, but to what extent this will be because of the home environment or of inherited predisposition, remains in the realm of conjecture.

Alcoholism could be added to the mental diseases or personality disorders, for it is now so regarded by psychiatrists. In fact, alcoholics are frequently placed in mental hospitals (although not so often in the United States as in some other countries). This prompts the question, "Does alcohol by itself produce insanity?" Many authorities now will say, "No." Certainly there are countless chronic alcoholics who go to their graves without ever becoming insane, which might support the theory that where alcoholism and chronic mental derangement do go together, a predisposition to insanity already may have been there, and, in fact, may have contributed to the alcoholism. Further, there is now some indication that inborn chemical factors play a part in causing some individuals to become alcoholics and causing others to resist developing the habit when exposed to the same influences. The same conjecture may perhaps also apply to *drug addiction*. However, because alcoholism and drug addiction, as well as other types of self-destructive behavior, are still punishable under our legal codes or are regarded as acts of willful wrongdoing under religious and social doctrines, we shall deal with them more fully in a later chapter, "Crime and Its Roots"—keeping in mind, moreover, that most authorities would place virtually all crime itself in the category of mental ailments.

We have seen that the mental diseases and emotional disorders—from the ominous Huntington's chorea to the mildest neuroses—must be viewed as comprising one of the direst sections in the catalogue of human misery. But much can now be said on the brighter side. For whether or not the causes of the most common mental afflictions have been fully identified, and regardless of what part heredity may play in these ailments, great strides have been made in reducing the severity of their symptoms and consequences. Particularly striking have been the results that have come through chemical treatments, notably with the tranquilizing drugs. *Chlorpromazine*, *reserpine*, and *tetrabenazine* (all marketed under various names) have proved highly effective in alleviating manic, paranoic and certain schizophrenic symptoms. Another group of drugs, the *psychic energizers*, can bring people out of deep spells of depression and melancholia and can inhibit suicidal tendencies. Sometimes tranquilizers and energizers are used in combination. *Insulin shock* treatments and, in extreme cases, *brain operations* have also proved highly effective in some cases of mental disease. And in the treatment of the neuroses and emotional disorders, which are predominantly psychological and do not involve pathological conditions of the brain and nervous system, psychoanalysts are becoming ever more discerning.

While none of the treatments yet found for mental diseases can be considered real *cures* (any more than insulin treatment can be considered a cure for diabetes), they have been enormously effective in reducing the severity of the symptoms and checking the progress of given diseases, to such an extent that it is now possible for many afflicted persons to leave mental hospitals and return to normal lives and occupations. (Treatment, however, may often have to be continued after release from a hospital.) This development is dramatically shown by the fact that since 1955, when the new drugs for mental diseases began to be used on a large scale, the previous steady annual increase in the population of the country's mental hospitals suddenly stopped, and there was for the first time a marked downward trend. With each year thereafter the number of persons discharged from mental hospitals has exceeded the number of new ones admitted.

Thus, the much overworked phrase "*New hope for*—," can be applied with every justification to almost all the major conditions we have listed under "Sick Minds." The more the mental diseases are revealed—like all other diseases—to be due not to mysterious and unearthly causes but to defects in construction or malfunctions of chemical processes, the closer we will come to making the hopes for combating them a reality.

CHAPTER TWENTY SIX

Slow Minds

"YOU'RE AN IDIOT!" or "—an imbecile!" or "—a moron!" is an expression which you may apply at times (lovingly or angrily, but never too seriously) to your husband, wife, sweetheart or best friend. You might imagine that you'd have no difficulty recognizing a *real* idiot, imbecile or moron if you met one. Quite possibly you'd guess right in many cases, but then only in a general way. For you'd have to be an expert to identify accurately the feeble-minded individuals of the different grades and types. And even the experts can make mistakes.

As noted in the last chapter, "mental deficiency" is a catch-all term for subnormal intelligence. About 3 per cent of our population in the United States (and probably in most other countries) falls within this category. But what do we mean by mental deficiency? A general definition, to quote the American Association for Mental Deficiency, would be "a person with a mind so retarded as to make him incapable of competing on equal terms with his normal fellows or of managing himself or his affairs with ordinary prudence." There need be no hesitancy about applying this definition to those in the lowest mental levels, the *idiots* and *imbeciles*. Our real difficulty comes as we go up the scale to the *morons* and try to identify and grade them. Here we are forced to accept rather arbitrary standards based on existing intelligence tests. A person scoring between 90 and 110 on the Stanford-Binet Scale is considered as having a normal IQ (intelligence quotient). But a score below 90 is rated in this way according to the scale:

90–80: Slow learner, or "dull"
80–70: On the borderline between dullness and subnormalcy
Below 70: Feeble-minded*

69–62: High-grade moron 62–55: Mid-grade moron 55–50: Low-grade moron	*Moderately mentally deficient*
50–20: Imbecile 20– 0: Idiot	*Severely mentally deficient*

It can be seen from these gradings that even if the tests are accepted as valid measurements of degrees of mental deficiency, there is much room for faulty classification. If an individual varies in his responses under different conditions or if his examiner makes a slight error, a shift of a few points in IQ can classify him as an imbecile rather than a moron, or as a moron rather than a dull-normal person, or as dull instead of completely normal. Beyond that, there is always the question of how accurately the tests measure *basic intelligence* (to be dealt with more fully in Chapter 31) and what allowances should be made for environmental factors. Furthermore, two individuals with the same low IQ may be deficient for quite different reasons or in different phases of the tests, or may vary greatly in their capacity to have their IQs raised by special training or treatment. These and other difficulties in the way of diagnosis and classification have greatly complicated the problem of establishing the inheritance of most types of mental deficiency.

Nevertheless, there is much to suggest that *genetic factors are either directly responsible for or play an important part in producing the majority of the mental defectives.*

Each type of mental defective, however, must be considered separately, for the factors producing one type seldom have any relationship to those producing another. This was largely overlooked in the early studies of pedigrees of mental deficiency, when idiots, imbeciles and morons were lumped together indiscriminately on the wholly erroneous assumption that intelligence was a unit and that all the different degrees of mentality were merely the results of variations in a few key genes.

Our first clear distinction, then, must be between the two main groups

* The term "feeble-minded" was formerly used (and occasionally still is) to designate mental defectives only above the grade of idiots or imbeciles, but the usual practice now is to apply the term collectively to all those below an IQ of 70 and to identify the upper defectives by the term "moron," as will be done here.

of mental defectives: the idiots and imbeciles in the one group, and the morons in the other. The first group is in the minority, comprising about 25 per cent of the mental defectives, of whom idiots represent only 5 per cent and imbeciles about 20 per cent. Perhaps 1 in 200 children of school age is in the idiot or imbecile class. (The proportion is considerably higher among children at birth, but by school age many have died, and many more also die before reaching or proceeding far into maturity, so that in the population as a whole perhaps only 1 in 400 is an idiot or imbecile.)

Morons, the second group, may number about 2 per cent, or somewhat more, of the general population. The most important differences between the two groups, however, are in their physical natures and in the factors causing them. Almost all of the idiots and imbeciles are *physically as well as mentally defective,* their mental impairment being accompanied by, or resulting from, some usually rare chromosomal, glandular, metabolic, structural or neurological abnormality which produces a variety of other effects. In this class are the mongoloid idiots, the cretins, amaurotic family idiots, microcephalics, and a small proportion of the mentally retarded epileptics.

The *morons,* on the other hand, are very largely *aclinical* cases—without any physical abnormalities, constitutional defects or symptoms of disease to explain their condition—their mental deficiency, so far as is known, being due directly to some basic slowness in the workings of the brain. Mental levels ranging from low to high may be as normal and natural, biologically, as stature ranges from very short to very tall. Again—in contrast to the idiots and imbeciles—the factors causing morons are much more common and ordinary, whether hereditary or environmental. However, there also are a minority of the morons whose backwardness may be accounted for by some nutritional deficiency, by various diseases, by reading disabilities, hearing or other special sensory motor defects, by emotional disturbances, or by lack of proper training. The proportion in this latter group is probably decreasing with improvements in living conditions, medical care and education.

Thus, whatever may be the factors that produce idiots and imbeciles, they are *different from and independent* of those responsible for morons; which is to say that idiots and imbeciles may appear in a family irrespective of the intelligence of the parents, because, as was noted, they are not primarily products of defective "intelligence" genes but of some over-all abnormality. Morons, on the other hand, being in most cases direct products of such defective "mental" genes, are far more likely to occur in families where low-grade intelligence is common than in families of normal intelligence. To emphasize further the distinction in the way the two main groups of defectives are produced, there is the paradox that two moron parents, generally, are not much more likely to have an idiot child than are two brilliant parents. This, however, requires some qualification, for among low-level families in depressed communities, where mating between defectives is combined with adverse environmental factors, there is a greater likelihood that mental defectives of all types will appear together.

Supporting the foregoing conclusions are innumerable studies made in the United States, Great Britain and other countries. Here, first for morons, are the principal findings:

> More than 75 per cent of the morons come from about 10 per cent of the population, from families with a high incidence of mental defect.
>
> Where one member of a family is a moron, the chance is about five times the average that another member will be similarly retarded.
>
> Where both parents are morons, the chances are anywhere from 60 to 75 per cent that any given child also will be mentally subnormal.
>
> In identical twins, if one is a moron, so is the other in virtually every case, although among fraternal twins this holds true in less than 50 per cent of the cases (some studies putting it at 25 per cent).

But with all these, and many more facts, geneticists still are not certain as to precisely what genes are responsible for producing morons. What can be said is that earlier conclusions that morons were produced by a pair of simple recessive genes are hardly tenable, at least in most cases, for if this were the rule there would not be so many instances of two moron parents producing normal children. At the same time, however, the fact that in such matings, collectively, a majority of moron offspring results and that mental defectiveness runs heavily in certain families for successive generations might suggest that not too many genes are involved. A commonly accepted theory, then, is that feeblemindedness in the moron grades is usually produced by multiple genes—a pair of recessives *plus* some supplementary gene or genes.

An alternative theory, in line with the previous comparison between intelligence and stature, is that "short mentality" (but only down to the moron grade) and "tall mentality" may be extremes of graded mental capacity controlled by a series of paired little "minus" and "plus" genes, and several key modifying genes. A person's mental capacity may then be equal to the sum of all the "plus"-gene effects. In turn, the more the genes of a given couple are on the "plus" side, the more likely their children would be to have good or superior intelligence; but these parents might still have some hidden "minus" genes to pass on, which, depending on how many of them were received, could make one child less bright than the others and, occasionally, subnormal. By the same principle, two dull parents, with mostly "minus" genes, would be most likely to have dull or mentally retarded children; yet these dull parents could still be carrying between them enough "plus" genes so that, if all of them come together by chance in a particular child they might cause him to be of good or even superior intelligence. This, then, could in part explain differences in intelligence on either the plus or minus sides between parents and children, or among children of the same family.

While we have dealt largely with the role of heredity in producing morons, we should not lose sight of the fact, already stressed, that some individuals are mentally retarded, or have been erroneously classified as such, because of learning obstacles, emotional blocks, or other factors unrelated to their inherent mental capacities. Throughout the United States thousands of retarded young people who might formerly have been consigned to the mental and social scrap heaps now are being salvaged by special psychological treatment and education.

More and more, it is recognized that IQ scores in the moron range, as in the normal ranges of intelligence, may fail to measure or reveal important differences in special abilities. That is to say, one child with an IQ of 70 may be dull-minded in virtually all ways, whereas another child with the same low IQ, and equally dull in most ways academically, may show talent for music or a flair for artistry, handcrafts, housework, or even for mathematical computation (the "idiot savant" type). Thus, it is often important to look beneath the wet blanket dulling a child's mind and see if there are not special capacities which could be developed.

Yet for the great majority of the *truly* dull-minded (especially where the defect is due to heredity), while IQs may be raised somewhat and better adjustment made possible, hope for great improvement is still premature. Any startling claim in that direction must be approached with great caution. For instance, some years ago there was much excitement over reports that a teacher, by some new training method, had succeeded in raising the IQs of a large group of feeble-minded children by an average of *40 points*. But, alas, an official investigating committee of psychologists could find little proof of the claims. Similarly, the reported successes with the *glutamic acid* treatment in raising IQs of retarded children were subsequently deflated. However, of a quite different nature, and much more likely to be effective, are chemical treatments in cases of low-grade mental defect (idiocy or imbecility) caused by specific chemical deficiencies or disorders, which will be dealt with presently.

IDIOTS AND IMBECILES. These unfortunates, we have already noted, are in a class entirely different from the morons in every important respect, but among themselves they also differ radically in the nature and cause of their conditions. The relative numbers of the different types are hard to determine accurately, because many are not entered in the records or are improperly classified, and those of some types die off at earlier ages and at much higher rates than others. A rough estimate might be that of the very low-grade mental defectives one encounters, about 20 per cent are mongoloid idiots; another 10 per cent are microcephalics; another 10 per cent embrace the cretins, the amaurotics, the phenylketonurics, the seriously retarded epileptics, the hydrocephalics, and certain other types; about 2 per cent are contributed by those with sex-chromosome abnormalities; and the balance includes those whose conditions are the results of prenatal upsets (especially in the prematurely born), birth injuries, infectious diseases, accidents, and so on. (Very few cases of idiocy, however—perhaps no more than 1 per cent—result from falls on the head or other such accidents in infancy and childhood.)

The *mongoloid idiot*, or imbecile, has been a focus of excited scientific interest ever since the long-mysterious cause of this defect was finally revealed (mentioned briefly in Chapter 20).*

First, let us note that mongoloid idiots, or "mongols" for short, are so called only because of the small, slanting eyes with inner skin folds, characteristic of their condition, which, while not the same as the "Mongolian eye" may give them a somewhat Mongolian look (Chapter 13); in no other way are they like persons of the Mongolian race unless they happen to be born among them. In fact, these strangely defective individuals occur among persons of all racial groups, including Whites and Negroes, and with perhaps the same relative frequency of about 1 in 600 live births.

Seeing the mongoloid idiot smiling, lively and cheerful, as he usually

* In place of "mongoloid" or "mongolian" idiocy many geneticists and physicians now use the term "Down's syndrome" (after Dr. John Langdon-Down, a London doctor who first described the condition in 1866). This is because of the feeling that "mongoloid" idiocy is not merely an improper name for this complex condition but leads to confusion with the correct use of the word as a racial designation. However, "mongoloid" idiocy is still so strongly identified with the abnormality in the popular mind and in much of the medical literature that it will continue to be used in this book as the preferred term.

is, you might well wonder what he has to be so happy about. For the very same mishap in development which retards him mentally may bring with it many other abnormalities often apparent at birth: bridge of nose flat and sunken, forehead generally large and misshapen, ears small and malformed, tongue fissured and perhaps protruding, voice guttural, and hands frequently deformed (fingers stubbed or webbed) and with peculiar palm patterns. Most important, mongols are usually dwarfed; frequently have serious heart defects; are particularly susceptible to serious infections; and the males among them, and many females, remain sexually underdeveloped. Not all of these physical abnormalities may be present in every mongol; nor is the extent of these abnormalities necessarily related to the varying degrees of mental retardation found among mongols, from low-grade moron to high-grade moron.

How can we account for the strange and large assortment of defects that may accompany mongolism? For many decades medical scientists had weighed possible theoretical causes—prenatal glandular deficiencies, shocks, peculiar accidents of one kind or another. Environmental causation was strongly suggested by the fact that the chances of a mongoloid-idiot birth went up sharply with a mother's advancing age, from an incidence of less than 1 per 1,000 among mothers under age thirty-four, to 3 per 1,000 at maternal ages thirty-five to thirty-nine, 7 per 1,000 at ages forty to forty-four, and the staggering rate of almost 30 per 1,000—almost one in thirty births—among mothers aged forty-five and over! Yet why was it so exceedingly rare for *two* mongol babies to be born to the same mother—unless they were twins? And if there were twin mongols, why were they almost invariably an *identical* pair, whereas in the case of fraternals, if one was a mongol, in almost every instance the other was not, even though both had shared the same maternal environment? This seemed to point to some genetic influence, and particularly—in view of the variety of physical abnormalities going with mongolism—to something occurring at the very beginning of prenatal life, perhaps in the egg stage.

So the growing suspicion among geneticists that a chromosome abnormality could be the cause of mongolism was finally proved correct in 1959 when a French scientist, Dr. Jerome Lejeune, while studying the chromosomes of some mongoloid idiots by the then newly discovered techniques, found the precise answer: In each cell of these defectives there was not the normal quota of 46 chromosomes, but 47. The extra chromosome was then seen to be one of the very small ones, subsequently called the No. 21 (as shown in our accompanying illustration). This type of chromosome abnormality (touched on in Chapter 20) may be produced by the occasional "sticking together" (nondisjunction)

PHYSICAL EFFECTS OF MONGOLOID IDIOCY

Down's Syndrome

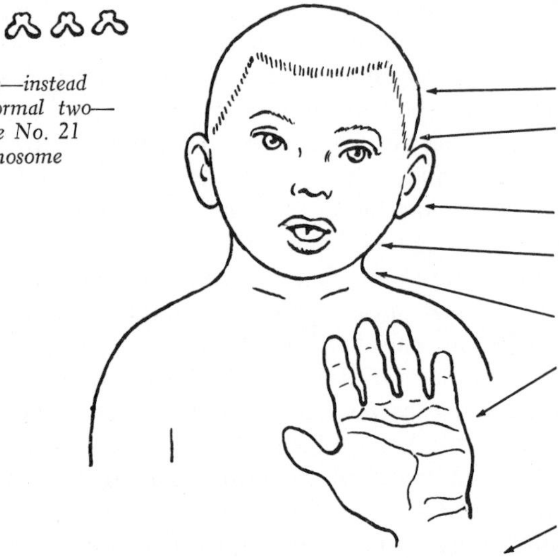

CAUSE

Three—instead of normal two—of the No. 21 chromosome

PHYSICAL PECULIARITIES

Stunted growth

Small, round head

Upward-slanted eyes (often crossed) set wide apart

Short, squat nose

Fissured tongue, Protruding underlip

Short neck

Hands broad, stubby, Abnormal palm lines, with creases almost straight across, In-curved, short little finger

Underdeveloped sex organs

of two of the No. 21 chromosomes during the process of egg or sperm formation, so that both of the chromosomes from one parent go into the same egg or sperm, instead of the customary single chromosome. When these two are united in the fertilized egg with the corresponding chromosome of the other parent, a baby would be started off with the trouble-making threesome (trisomy), which in some way not yet clear would throw the entire developmental process out of gear.

This, then, is how trisomy (the term for triplication) of the No. 21 chromosome produces mongolism. The chromosomal abnormality seems far more likely to originate in eggs than in sperms, because, for one thing, advancing age in the father has no such relationship as does the mother's age to the chances of there being a mongol baby. But just what it is in maternal aging that so strongly affects the conception and production of a mongol is not yet known. However, various other types of chromosome abnormalities, and the defects they cause, also increase in incidence as mothers grow older. This applies to many cases of the Klinefelter syndrome, as well as to syndromes associated with tripling of chromosomes in the Nos. 13–15 (D) and 17–18 (E) groups which cause multiple congenital defects.

Yet important as the maternal aging influence may be in mongolism, there must be other causative factors, since mongols can be born to mothers at any age. Among possible activating influences which have been suspected are faulty diet, glandular disturbances and abnormalities, infections, drugs and poisons. While specific proof for any of these as a cause of mongolism is lacking, it is not unlikely that, as with many other congenital accidents or mishaps, some environmental factors—whether those mentioned or others—may act at a vital time to derange the normal chromosomal processes in human egg or sperm production, with mongolism as an end result.

What of *hereditary* factors in mongoloid idiocy? There remains the possibility that whatever the environmental influences may be, certain wayward genes acting specifically on or in the No. 21 chromosome—and increasing the chances of two sticking together—could create in some women a special predisposition to produce mongol babies. (One theory is that the younger mothers who produce mongols would be most likely to have this predisposition, whereas purely environmental mongolism would be the rule among the older mothers.) If there are genes for mongolism, their effects are shadowy and not strongly manifested. This can be gathered from the fact that when a mother has had one mongol child, the chances of her having another are usually only slightly greater than the average risk of a mongol birth to any other woman of the same age. The only exceptions are in the occasional cases when a woman who is physically normal is found to be *chromosomally abnormal*: That is, one of her No. 21 chromosomes is seen to be hooked on to some other chromosome (often a No. 13). The hooked-on No. 21 will therefore go as a unit with the other one during the process of egg formation, and if the woman's normal free No. 21 also goes into the egg, on its fertilization by the father's sperm with another No. 21, there will be the triple No. 21 situation resulting in mongolism. The important fact about normal mothers who carry a hooked-up No. 21 is that as many as one in four of all the eggs they produce may be of the type that, on fertilization, can develop into mongoloid children.

In a few instances mothers themselves mongols have given birth to mongol babies, and at least one mongol mother has had two such babies. These events would be extremely rare, since mongoloid-idiot females, if they survive beyond puberty, are in most cases sexually underdeveloped and if not, are usually confined to institutions. Once proved fertile, however, a mongol mother could present a fifty-fifty risk of a mongol child with each pregnancy. (This is based on the fact that since in the normal process of egg production a mother contributes to each egg one of every pair of her chromosomes, if she has *three* No. 21's to divide, one egg might get two of these, another egg only one. On fertilization, each egg with two No. 21's, after receiving the additional No. 21 brought in by the father's sperm, would have three No. 21's and could develop into a mongol.)

In addition to mongolism, several other types of mental defect accompany conditions caused by chromosome abnormalities. Two of these are the Klinefelter and Turner syndromes, which result from abnormal numbers of sex chromosomes (to be discussed in Chapter 27). A few rarer conditions, in which mental retardation results from other types of chromosome abnormalities, also are known, and more in this category may eventually be revealed. In fact, a common feature of most of the conditions so far traced to chromosomal aberrations is that some degree of mental retardation usually goes with each of them. It would seem that in order for the human brain and nervous system to develop properly, the normal quota of 46 chromosomes is required and that any addition to or subtraction from this quota will interfere with full mental development.

Mental defects caused by chemical or metabolic abnormalities represent a different category, within which the condition called *cretinism* is the most prevalent and longest known. The cretin, whose mental age usually ranges from that of the imbecile up into the moron class, is a victim of prenatal thyroid deficiency, in many cases being born to mothers who are goitrous. This thyroid deficiency (sometimes characterized by total absence of the thyroid gland in the afflicted child) not only stunts mental growth, but dwarfs the body, so that at the age of twenty-five a cretin may look and act like a very dull small boy—one who also has a large head, pug nose, pot belly, thick lips and protruding tongue. Because, under like conditions, cretinism may run more in some families than in others, it has been suggested that it may have some hereditary basis—that in addition to the mother's thyroid deficiency there may be some inherited thyroid defect or special susceptibility in the child. Evidence for this, however, has been established only in a few rare types of cretinism where inheritance has been traced to recessive genes.

In the great majority of cases cretinism seems clearly environmental, due to an insufficiency or absence of thyroid hormone during early postnatal or even prenatal life. Thus it has been shown that very early thyroid treatment of the cretin—as soon as the condition shows itself (usually by the sixth month)—may help a great deal to restrain its effects. In some instances where doctors have been alerted by the previous birth of a cretin to a mother, the thyroid treatment of a succeeding cretinous baby immediately after birth may still further decrease the effects. However, follow-up studies of treated cases to date show that only a few children born as cretins attain a near-normal mentality, and all remain physically defective to some degree.

Amaurotic family idiocy is essentially an inherited metabolic disorder which causes the nerve cells of the brain and spinal cord to fill with fat, swell up, and produce not only idiocy, but blindness and paralysis, and in most cases, death within a few years. This form of idiocy is of several types, differing in time of onset, in seriousness, and in the genes responsible. Most common and earliest to develop is the infantile type (Tay-Sachs disease), which is due to a pair of simple recessive genes. In many cases where children are afflicted the parents' families are closely related. (The disease also has an above-average frequency among Jewish people of eastern Polish origin, although among other Jewish groups the incidence is about the same as for non-Jews.)*

Gargoylism is another unusual affliction resulting from a metabolic disorder (of complex carbohydrates—the mucopolysaccharides) which, as in amaurotic idiocy, affects the brain and nerve cells. It not only causes severe mental defect, but also produces extreme dwarfism, and other external and internal abnormalities, and is often fatal before puberty. The condition has been reported as of two types, one a simple recessive in inheritance, the other sex-linked.

Phenylketonuria is a rare type of mental deficiency, inherited through simple recessive genes and involving a failure in the proper metabolism of the amino acid *phenylalanine*. The condition is unique in that it can be detected in the first weeks of infancy (even before any mental defect appears or is recognized) by tests of the baby's urine. If a drop of the chemical *ferric chloride* is placed on the urine-wetted diaper of an affected infant, it will cause a green discoloration. There is some hope that phenylketonuria may be treated and its effects reduced by feeding afflicted infants with a diet free from the chemical substances

* A possibility held out by recent findings is that normal persons who carry a gene for amaurotic idiocy may be identified by certain peculiarities in their amino acids. If the evidence is confirmed, it will be possible to know in advance which individuals from families where the disease has appeared run a risk of having similarly afflicted children, particularly if they mate with carriers like themselves. (Problems faced by carriers of wayward genes for various conditions will be discussed in Chapter 45.)

which their systems cannot handle properly. Also important is that, while two recessive genes are required to produce the disease, a person carrying only one of the genes, even though normal in all other ways, may be revealed by a chemical test as having in a slight degree the metabolic deficiency previously mentioned. The test may be of great value in identifying beforehand prospective parents who are in danger of producing a child with phenylketonuria and in making possible preventive or precautionary measures.*

Galactosemia, which can be added to the list of metabolic disorders causing mental defect, was discussed in Chapter 24.

In still another category of mental defectives are those who can be identified by abnormalities in the sizes or shapes of their heads. Best known of these (because, alas, of having been so often exhibited among circus freaks) is the *microcephalic* idiot of the "pinhead" type, whose intelligence may range between that of an idiot and an imbecile. The condition results from an early stoppage of the growth of the brain and the skull enclosing it, the individual usually also being dwarfed. It is clear that some of the cases of microcephaly are inherited, through recessive genes, there being many instances of families with several microcephalic children. But the condition might also be caused by certain prenatal upsets, such as a pituitary deficiency, and in rarer cases, by careless radiation of the mother during pregnancy. Quite a number of such X-ray microcephalics have been reported in medical studies. While increased medical precautions have greatly lessened the chances of similar prenatal mishaps occurring through X-ray treatments, an entirely new threat has been posed by radiation from atomic-bomb explosions. As an ominous portent, a study following the Hiroshima bombing in 1945 showed that there were 7 microcephalics in a group of 205 children born to mothers who had been within a mile of where the bomb dropped, and in mid-pregnancy at the time. This incidence of microcephalics, vastly higher than average expectancy (less than 1

* *Histidinuria*, another recessively inherited condition, is sometimes misdiagnosed as phenylketonuria when the ferric-chloride test is made. Histidinuria is apparently due to absence of the enzyme histidase, needed to utilize the amino acid histidine in the diet. As a result, excess amounts of histidine accumulate in the blood, which, along with its breakdown products, are excreted in the urine. One of these products reacts in the ferric-chloride test to give a positive result like that in phenylketonuria, leading to the misdiagnosis noted. However, in histidinuria there is only mild mental impairment, if any at all, although there may be some impairment of speech development.

in 25,000 births) was in addition to other congenital abnormalities that resulted.

Hydrocephaly, or enlarged "water head," is another skull and brain abnormality often, but not always, accompanied by mental retardation. In this condition the brain fills with an abnormal amount of cerebrospinal fluid (due usually to defective drainage), swelling the head. While this may produce mental defect and convulsions, it frequently does not—especially if there is early treatment to drain off the fluid—and there are many cases of hydrocephalic children with a high degree of intelligence. Inheritance has been identified in only some of the cases, which have been ascribed either to a qualified-dominant gene (asserting itself only sometimes) or to a sex-linked gene. Among many domestic animals—dogs, rabbits, livestock, etc.—hereditary hydrocephaly has been clearly established, and where appearing usually proves fatal very quickly.

Epiloia (*tuberose sclerosis*) is a curious condition which produces malformations in the brain resembling potato roots, and also a "butterfly" rash on the face. Idiocy is among the most serious effects, another being epileptic symptoms (although this condition is unrelated to common epilepsy). Epiloia appears to be inherited as an irregular dominant, a fact which adds to its gravity, inasmuch as some persons carrying the gene do not develop the condition and may grow up to reproduce and pass it on to offspring who can develop it.

After what has been said about epilepsy, it might be added that while this affliction itself should not be listed among the mental defects—since epilepsy is primarily a nervous disorder (so discussed in Chapter 24)—one of its results in many cases, unfortunately, is mental retardation in some degree. Statistics on this point are not too certain, but by rough estimates perhaps only 10 per cent of the epileptics are mentally retarded enough to be institutionalized. Also indicated is that where persons with normal minds develop epilepsy as a result of a hereditary predisposition to the disease, it is a separate entity, and mental deterioration is much less likely to occur than in individuals in whom epilepsy is purely acquired as the result of some accident or injury, often at birth. Thus, the average IQ may be 96 for epileptics without brain injury, and 77 for those with brain injury. In other words, it is not epilepsy which causes mental retardation, but accidents of certain kinds which, in damaging the central nervous system, produce both epilepsy and brain impairment.

Cerebral palsy can be regarded as similar to epilepsy in that in most cases the condition is not accompanied by mental defect, but in some cases it is—if the same accident which caused the palsy affects the brain. Such an accident usually occurs before birth, or at birth, and is particularly likely to occur when infants are premature. This explains an unduly high incidence of the condition among twins, and is also one of the reasons why there may be a slightly greater number of mental defects among twins than among the singly born. However, cerebral palsy occurs little more often in both members of an identical-twin pair than in both of a fraternal pair, which reinforces the conclusion that heredity is not involved in its causation, except in occasional families where simple recessive genes are responsible.

Many other conditions, hereditary or otherwise, which produce mental defectiveness among their effects could be added to our list, but these are rarer ones which are of interest only to specialists and to the very few unhappy families concerned.

By and large, idiots and imbeciles are of far less concern either to their parents or to society than are the *morons*, because if the lowest-grade defectives survive much beyond puberty, they are generally kept in institutions and do not present the problem to their families or society of trying to adjust them to average schooling, jobs or social relationships. Even more important, most idiots and many imbeciles remain sexually undeveloped or sterile, or otherwise are in no position to reproduce themselves. Morons, on the other hand, contribute heavily to the problem groups since they are more numerous, live much longer—outside of institutions in most cases—and are under continuous handicaps in schools, jobs and social relationships. They are of great concern to society since they are likely to marry (or, if girls and women, to become pregnant outside of wedlock) and to produce a high percentage of offspring who, if not all genetically defective, will be improperly reared.

The "repeat risks" for mental defectives of various types—that is, the chance that a mother who has borne one mentally defective child may bear another like it—will be discussed in Chapter 45. The general problem of what should or can be done to reduce the incidences of mental defectives will be dealt with in our discussions of eugenics in Chapters 46 and 47.

CHAPTER TWENTY SEVEN

Sex Abnormalities

THERE IS THE FAMILIAR STORY of the husband and wife who could not decide whether they wanted a boy or a girl, and so got a child who was neither.

We are speaking not of a "male who behaves like a female," or vice versa (to be discussed in a later chapter), but of human beings who are *biologically* "in-betweens." Such individuals can be produced through some genetic or environmental upset which may also account for various other kinds of abnormalities or defects in sexual construction.

The Bible says, "Male and female created He them. . . ." There is no need to dispute that. But however distinctly the first man and woman might have been differentiated the one from the other, in the billions that are assumed to have sprung from them may be found every gradation of sexuality. In short, sex is a highly variable characteristic, and there is not always a clear-cut distinction between male and female.

Think back to the chapter, "Boy or Girl?" We learned that sex determination depends on which combination of sex chromosomes an individual receives at conception. An X and a Y produce a male. Two Xs produce a female.

This unquestionably holds for the vast majority of human beings achieving birth. But as with everything else in nature there can be exceptions, deviations and qualifications.

First, the X and Y chromosomes are not the sole arbiters of sex or sex characteristics. Second, occasional babies may not receive at conception a normal XX or XY chromosome combination (as noted in previous chapters). The consequences involved in either case may be considerable.

Proceeding with the first point, while the X and Y chromosomes carry the principal genes which start sex development off in one direction or another, there also are "sex-influencing" genes in the other chromosomes, received equally by both sexes. At any rate, all persons carry in themselves at the outset *the potentialities to develop characteristics of either sex.* This can be seen in the earliest stages of prenatal life when similar sex glands and rudimentary "maleness" and "femaleness" sex organs are found in both sexes. (See Chart, p. 47.)

In the normal XY fetus, the primary sex-gland cells develop into testes, and then, step by step, the male organs and sex characteristics develop to their fullest point, while the female organs remain in the rudimentary stage or disappear.

In the normal XX fetus, the primary sex-gland cells develop into ovaries, and then, step by step, the female organs and sex characteristics develop to their fullest point, while the male organs remain suppressed.

But at any point along the way, from conception onward (or possibly prior to this, when the sperm and egg are formed), something may go wrong with the process. Even when the sex of the individual appears to be set, the process may go into reverse gear to some extent or other. In fact, *the type of sexuality* of any individual can be changed in some degree or varied at every stage of life.

Among more elementary creatures, one can see rather amazing deviations from what we think of in humans as sex normalcy. Many lower organisms, such as the snail, earthworm, flatworm, or oyster (as well as most plants and flowers) are normally *double-sexed* in structure and function, either simultaneously or alternately. How can this happen? Strangely enough, the sex-chromosome mechanism in these lowly creatures is basically the same as in humans. They, too, start out with the potentialities of either sex. But in the snail or earthworm, the "sex" genes, instead of sending the individual off in either the "maleness" or "femaleness" direction, allow potentialities of both sexes to assert themselves and both types of sex organs to develop. In the oyster, however, the "sex" genes are so regulated that they allow first the male organs to develop, then go into reverse and make way for the female development, so that the oyster is capable of changing its sex from year to year. Or, as we might state it in verse:

The oyster leads a double life,
One year, it's husband, next year, wife.
It's both a father and a mother;

Sex Abnormalities

It's both a sister and a brother.
No wonder, if all this is true,
The oyster ends up in a stew!

Another double-sexed type, but strictly in the freak class, is the gynandromorph, or "half-and-half" creature: one side of its body may be of one sex, the other side of the opposite sex. Or, sometimes, the gynandromorph may be male in the upper half, female in the lower half, or vice versa. Such freaks are fairly common among insects which are normally single-sexed, including butterflies, moths, wasps, bees, flies, ants and spiders.

One way for a gynandromorph to be produced is by having one of the X chromosomes in a prospective female go awry or get lost during the very first stages of cell division after conception. In all the cells which subsequently develop, those with the original two Xs will produce female characteristics, while those with only the single X (in species in which the Y plays no part) will produce male characteristics. Another possibility, among species in which the Y is a maleness-determiner, is that, in an initially XY individual at conception, the Y may drop out of part of the cells during the first stages following fertilization. In this case the cells with the XY would produce normal maleness traits, whereas the cells with only an X would tend to produce femaleness traits.

Now in what way do such queer conditions apply to human beings? The true gynandromorphs—completely half male and half female—are believed to be found only among the more elementary creatures, such as insects, whose hormonal activity is limited and whose different organs are not so intimately linked together as our own. In human beings or other mammals, even if derangement of the sex chromosomes might take place, the sex hormones circulating through the body would tend to produce a blending effect throughout. Nevertheless, conditions akin to—if not actually—gynandromorphism, called *mosaicism,* can occur in human beings. The extremely rare persons who are mosaics may show a mixture of two types of cells, some with the XX chromosome combination and some of the XO type (where one of the Xs was lost); or a mixture of XY and XO cells (where the Y was lost in part of the cells).

Actual examples of some of the possible human mosaics described already have been found, with one or another kind of abnormal sex development. In at least one case, also, an individual has been reported with a female-type breast on one side and a small, male-type breast on the other (and also, with a partial beard on one side of the chin).

While circus freaks of the past who claimed to be half-man, half-woman, were probably spurious, it is not inconceivable that some, having female breast development on one side and lacking this on the other, might have been chromosomal mosaics.

The *true hermaphrodite*—the completely in-between individual, with *sex-glands* as well as organs and other characteristics of both sexes—is so rare that in all medical annals to date only about 150 cases have been reported. Such individuals may be produced if the rudimentary male and female sex glands and organs, present in embryos of both sexes, go on to develop equally (through some upset which results in a tie race), instead of following the normal procedure in which one of the sex-producing mechanisms takes precedence over the other. But no case is on record of any human being who had both normal ovaries and testes and *who was able to function reproductively as both male and female.*

Whether any of the double-sexed human individuals—those with sex organs and glands of both sexes—result from hereditary factors has not been established. However, there is evidence that in goats the frequent occurrence of hermaphroditism in certain breeds may be influenced by heredity, and recessive genes have been reported as involved. A strain of pure-bred mice with an unusual incidence of hermaphroditic individuals also has been reported. Among the fruit flies hereditary in-between individuals of various types can also be produced through breeding. But in human beings, individuals who might appear to be partly of each sex are most likely to be of the types we will discuss next.

If there is only a partial upset in early sex determination or sexual development, it can cause the much more common condition, in at least 1 in 1,000 persons, of *pseudo-hermaphroditism* or, more properly, *intersexuality.* In this type of abnormality, the individuals may have organs of both sexes, or sometimes male glands with female internal organs, or the opposite condition. Also resulting in time may be mixed secondary sex traits of both sexes (such as breast development, body development, and face and body hair). Thus, among intersexual individuals one may find males with undeveloped masculine genitalia and various degrees of female genital development, and females with undeveloped or incomplete female organs and with rudimentary—or sometimes fairly well-developed—male organs. Such intersexual conditions can arise in various ways; (1) through the action of specific *wayward genes,* (2) through *chromosome* abnormalities and accidents, and (3) through *hormonal influences,* usually in prenatal life.

Testicular feminization, a prenatal "male-into-female" transformation, is among the most striking of the intersexual conditions and one

HUMAN MOSAICS*

1. *Not Involving Sex Chromosomes (Autosomal)*

A.

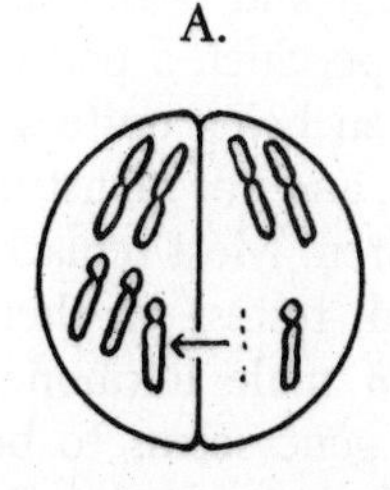

B.

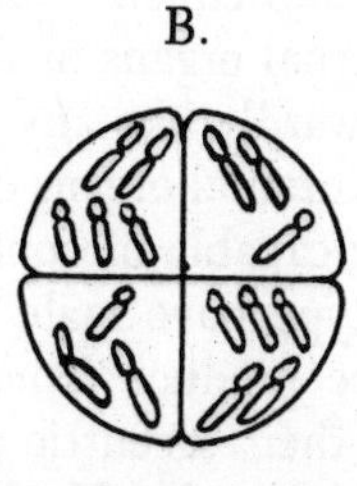

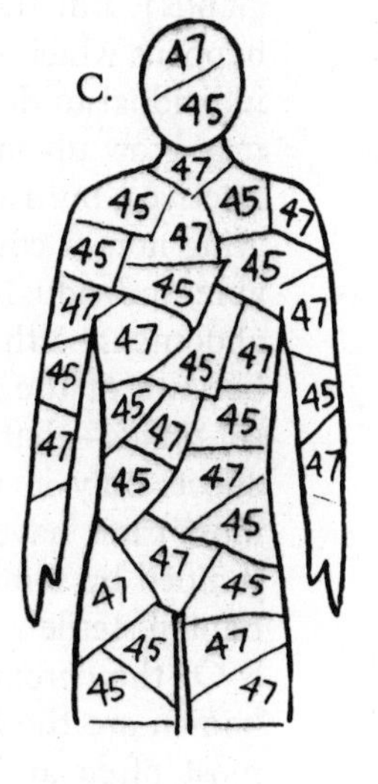

(A) When a fertilized egg with its 46 (doubled) chromosomes divides, an "error" may give one of the cells an extra chromosome lost by the other cell. (B) In ensuing cells, some will have 47 chromosomes, others only 45. (C) Resulting individual is a mosaic of "plus" and "minus" cells.

2. *Involving Sex Chromosomes*

XX-XO Mosaic

A.

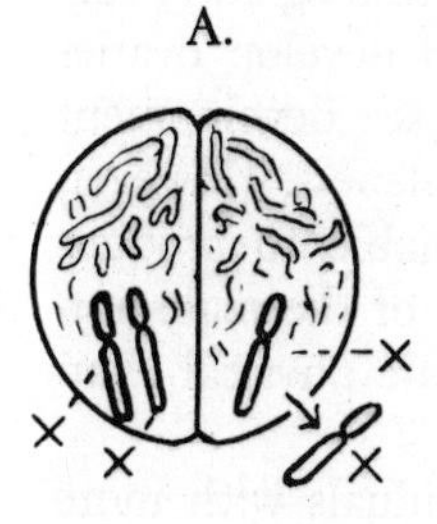

B.

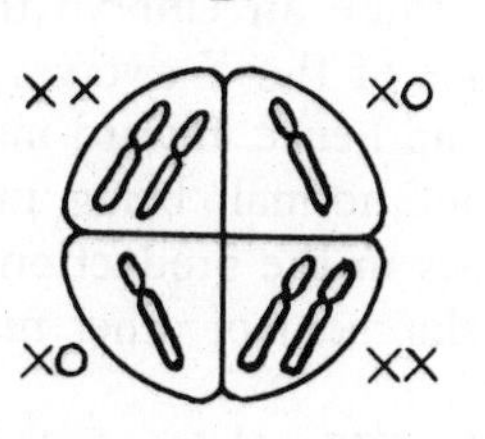

(A) In an XX egg at first division, one cell loses an X. (B) Resulting individual is a mosaic of XX (normal female) cells, and XO (abnormal female, Turner's syndrome) cells. Also possible: XXX-XO mosaic if an X lost by one initial cell goes into the other.

XY-XO Mosaic

A. B.

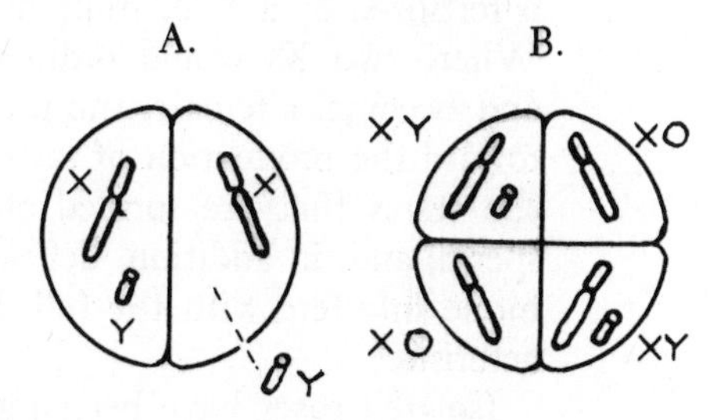

(A) In an XY fertilized egg at first division one cell loses its Y. (B) Resulting individual is a mosaic of XY (male) cells and XO (abnormal female, Turner's syndrome) cells. Also possible: An XXY-XO mosaic (part "Klinefelter male," part "Turner female") if fertilized egg starts off with abnormal XXY combination.

* Mosaicism of any type can occur not only after an egg's first division (when it would extend throughout the person's body), but at any stage or area of cell multiplication thereafter, so that in some individuals only fractional or scattered parts of the body are mosaics. The exact condition of the subject depends on the chance distribution of the different cells throughout the body, especially in the case of the sex-chromosome mosaics. For example, in an XY/XO mosaic, if the embryo's gonads (or germ cells) are primarily XY cells, a male will result; if they are primarily XO cells, a Turner-type female will result. The same basic mosaic patterns may therefore produce variable results.

of the few that can be directly caused by heredity. In this condition an individual conceived with an XY chromosome combination starts out, as might any XY male fetus, by beginning to develop testes (male sex glands). But then, under the influence of a misguided gene, something happens which suppresses subsequent male development and causes the individual to develop external organs of the female type. Such a person may grow up to look outwardly like any woman (often being quite attractive), but lacking the internal organs of the female, and not menstruating nor, of course, being capable of conceiving children. Most remarkable, individuals of this type have male glands (small testes) in their abdomens, although these produce female instead of male hormones because of the action of their eccentric gene. (This gene seems to be *sex-linked*—that is carried on the X chromosome but producing its effects only in males.) Despite their abnormality, these individuals in most cases have made satisfactory adjustments as women—particularly if aided by some surgery and hormonal treatment—although they must remain sterile.

Of the *chromosomally produced* types of intersexuality, the two best known are the *Klinefelter* and *Turner* syndromes. The Klinefelter cases most often are those of individuals with an XXY chromosome combination, resulting from the presence of two joined Xs in an egg which is fertilized by a Y, or from an abnormal XY sperm fertilizing an X egg. Where two Xs would ordinarily cause an embryo to develop ovaries and become a female, the presence of the Y swerves sex development toward the production of testes, and hence, toward maleness. However, the testes that are formed are not normal, being unable to produce sperm; and, in addition, deficiencies in the production of the male hormone interfere with the full development of other male physical characteristics.

Isolated cases have been found of Klinefelter individuals with more than two Xs, plus a Y—XXXY, XXXXY and XXYY combinations—due to different kinds of chromosomal upsets during egg and/or sperm production. There also are females with more than two Xs, having XXX or XXXX combinations. Such abnormal situations were first seen in fruit flies many years ago, and individuals having the extra sex-chromosome combinations were referred to as "supermales" or "superfemales." But these terms have proved to be misnomers for the human beings with extra sex chromosomes (as they may have been for the fruit flies), inasmuch as the individuals involved, far from being "super" for their sex, are usually defective as either males or females, and are often (but not always) sterile.

In the Turner syndrome, where individuals are of the XO type, the

CHROMOSOMALLY CAUSED SEX ABNORMALITIES

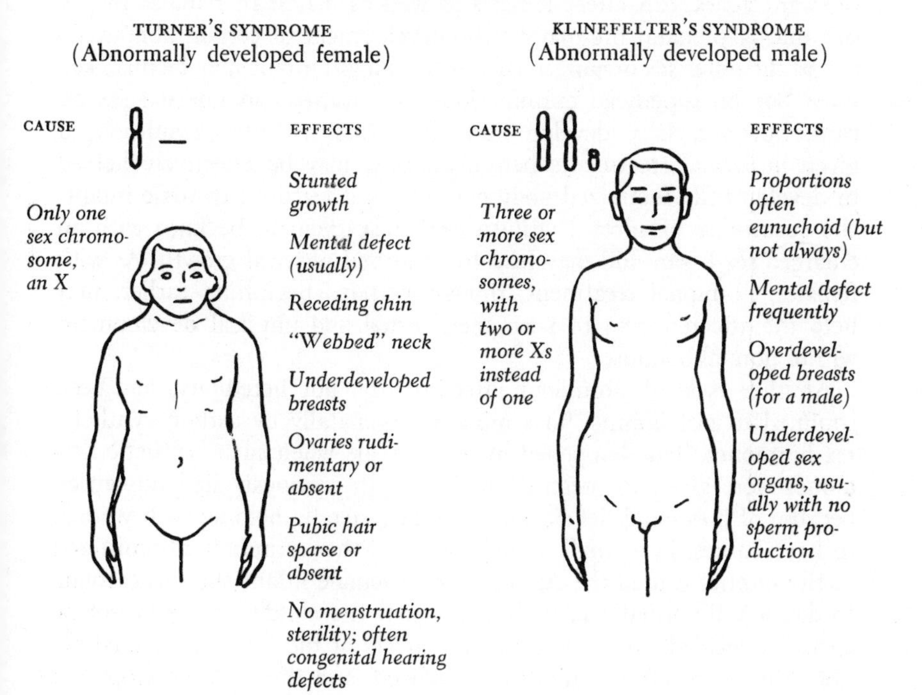

presence of only one X and the absence of a Y prevents the embryo from developing fully as a female, yet does not enable it to swerve toward maleness. Accordingly, the female sex glands remain undeveloped and the internal sex organs remain infantile.* The external organs, however, do take a female form. In addition to defects in sex development linked with sex chromosomes, there may be some due to abnormalities in other chromosomes. For example, the Triple No. 21 chromosome combination responsible for mongolism also causes sex underdevelopment in the affected males.

Among the hormonally induced intersexual conditions, a principal one is congenital *adrenal virilism* (*adrenal hyperplasia*), an abnormal-

* The rare instances in one-egg twinning in which one twin may be a Turner-type female and the other either a normal male (if the egg was XY) or a normal female (if the egg was XX) were discussed in Chapter 20, page 194.

ity of the adrenal glands which results in overproduction of one of the male-type hormones. The condition, sometimes caused by recessive wayward genes, can affect females as well as males. In females the result during prenatal life may be the development of certain abnormalities in the outer sex organs, so that at birth a girl infant may be mistaken for a boy on superficial examination. The ovaries and internal sex organs, however, may develop normally. Girls with the condition, if given hormonal treatments before puberty, may be effectively helped to develop fully feminized body contours and features. In male infants the occurrence of adrenal virilism can cause them to be born with an enlarged sex organ and may lead to abnormal physical growth. As with females, hormonal treatment, though perhaps beginning earlier, may help the affected boys to keep their sexual and physical development within normal bounds.

Another type of hormonal intersexuality, not hereditary, has been confined to girl infants "masculinized" prenatally by certain synthetic sex hormones. This happened in recent years when such synthetic hormones were given to pregnant women who previously had miscarried because of hormonal deficiencies. While generally helping such women to have babies, in a small number of cases the hormones administered to the mother caused the sex organs of a female infant she was carrying to develop abnormally in such a way that the child's sex could not be clearly identified. In many cases the effects of the initial hormonal abnormality wore off as childhood advanced, and the girl's development took a normal course by itself. In some cases, special hormone treatment or surgery was required. One odd fact about this condition is that the synthetic sex hormone administered to the mother which caused masculinization in a girl fetus was not a male-type but a female-type hormone. Also puzzling is that the same artificial hormone has not been shown to have any harmful effect on boy infants, who, almost always in the same situation, were born sexually normal.*

In any of the intersexual conditions discussed, the error of misclassifying a child's sex at birth and then rearing it as a member of the wrong sex can account for many of the reported cases of sex changing at

* Another situation in which a female, but not a male, is adversely affected by sex hormones in prenatal life is in the classic case of the *free-martin* among opposite-sexed twin cattle. In this circumstance, where the twins being carried are of opposite sex, the female can be so affected by alien sex hormones produced by the male twin as to be rendered sterile. Fortunately, *no* corresponding situation occurs in human girl-boy twins. A girl twin's sex development is not adversely affected by the presence of a boy twin, nor is she impaired in her later capacity to bear children.

puberty or even during adult life. This particularly applies to the surprising accounts of "women" athletes who later "became" men (although, in fact, they always were). One notable example was a Czech "girl" who won the women's 800 meter run in the 1936 Olympic games; another was a former holder of the English women's shot-put and javelin-throw championships.* Both of these persons subsequently underwent operations and treatments which brought them more in line with their true male sex. (The English athlete also married later. That heredity may have played a part in this individual's intersexual condition was suggested by the fact that another member of the family, a supposed sister, also turned out to be an intersexual male, who, after an unsuccessful operation to correct his condition, committed suicide.)

Many similar cases have been in the news—persons mistakenly reared and living into adult years as females who then dramatically began new lives as males, usually after operations and treatments. Within the past decade these included a former lieutenant-colonel in the British army woman's auxiliary corps who, at age forty-seven, underwent a legal change of sex and married a woman friend; a Scottish physician who had practiced as a woman doctor for many years until "she" had herself legally declared a male and then married his ("her") housekeeper; and another "woman" doctor—this one a member of the Irish nobility—who, in "her" thirties, after years of living as a female and being so listed in the peerage books, was reclassified as a man and then began serving as the bearded medical officer on a ship. In many of the cases where males have been mistaken for females at birth, the error could be traced to an extreme form of a male congenital abnormality called *hypospadias*. In this condition there is a misplaced opening of the urethra on the underside of the penis, and often a failure of the testes to enter the scrotum (the sack normally containing them) and of the two sides of the scrotum itself to fuse. As a result, the baby's sex organs as a whole might easily be mistaken for those of a girl. The abnormality in various degrees is sometimes hereditary, produced by a dominant-qualified gene.

In an opposite direction are the less common male-into-female changes, involving individuals mistakenly classified at birth as males

* These embarrassing situations led to the requirement in Olympic Games that all female entrants undergo a prior physical examination to certify their sex. Within the United States, also, since 1949, after suspicion had been aroused regarding several star members of women's athletic teams, similar medical sex certification has been required of all entrants into women's track and field events controlled by the International Amateur Athletic Union.

but subsequently found to be females. Most often such errors are traceable to conditions previously discussed—adrenal virilism or hormonally induced prenatal masculinization—which, among other things, result in the overdevelopment of a girl baby's clitoris so that it may be mistaken for a penis. (The mistake, we noted, usually becomes apparent at or before puberty, when the condition either corrects itself, or can be corrected by treatment.) But entirely different are the pseudo sex transformations sensationalized in the press some years ago, in which it was claimed that physically normal men (an American among them) had been "transformed into women" by operation and treatment. As conceded by the European doctors who performed the operations (not permitted in the United States), the procedure had involved merely castration and some alteration of the external sex organs, at the request of the men involved who had an "irresistible desire to think of themselves as women and to dress and be regarded as such." Akin to such men are women who are physically normal in their sex but have an irresistible urge to dress and pose as men, and sometimes to court and "marry" other women. But in both sexes this type of abnormality—that of wishing to be and act like one of the opposite sex—is primarily psychological and will be discussed in Chapter 40.

From what has been said about the truly intersexual conditions, it is evident that a person's *genetic sex*—as it should have been dictated by his sex-chromosome combination—may not always be the same as the sex apparently shown by outward physical development and traits. Nor can even a doctor always say with certainty when a baby is born, "It's a boy!" or "It's a girl!" So where doubt exists, the answer can now be given by *sex-identity tests.*

One, the *sex-chromatin test,* first developed by Drs. Murray L. Barr and E. G. Bertram, can establish quite simply in the vast majority of cases whether an individual is of the genetic XX or XY type. This test involves making a microscopic examination of some cells painlessly scraped from the inside of one's cheek. Where a person is a normal female, the two Xs in a cell will result in the presence of a clump of *chromatin* (chromosome material) at the rim of the cell. (See illustration.) Cells showing this mass and individuals having such cells are called *chromatin positive.* In individuals with only one X (true of the normal male), the sex-chromatin mass is not seen in the cells, which, accordingly, are designated as *chromatin negative,* as are their host individuals.

The "*drumstick*" *test* is another one which helps to establish sex identity. In this test, in which white blood cells are examined, the

SEX-IDENTITY TESTS

Where there is doubt, these tests can determine the genetic sex *of an individual, regardless of appearance or biological development.**

A.

Chromosome Test.
(With prepared white blood or bone-marrow cells.)
Genetic male always *has a* Y *chromosome plus an* X. *Genetic female has no* Y, *but if normal has two* XXs.*

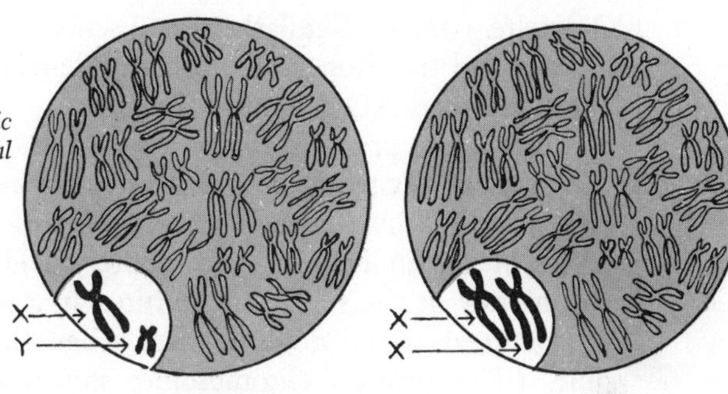

B.

Sex-Chromatin *("Barr-body")* Test.
(With cells scraped from inside cheek.)
Dark speck—the chromatin mass—shows up at rim of nucleus in typical female cells, but seldom appears in male cells. (Mass may be a "shriveled"* X*—the one inactivated by the other* X. *See Text.)*

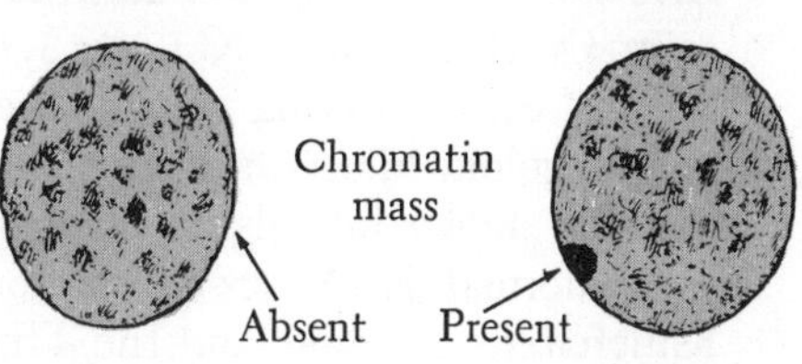

C.

"Drumstick" Test.
(With white blood cells at certain stages.)
Typical female cell shows small "drumstick" formation extending from the clumps of chromosomes. Normal male cells lack this "drumstick" (which may be an inactivated X, *like the chromatin mass above.)**

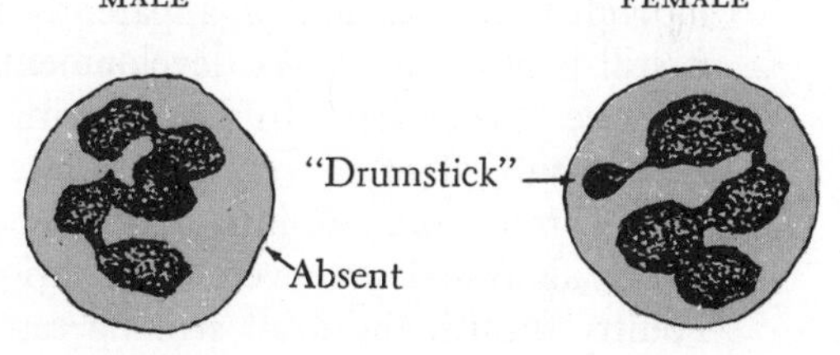

* In abnormal "Klinefelter males," with two or more Xs (plus a Y)—as revealed in the Chromosome Test—the cells will show chromatin masses and "drumsticks" like those of typical females (Tests B and C). By contrast, the cells of abnormal "Turner's syndrome females," with only a single X, will look like those of males in Tests B and C, lacking chromatin masses or "drumsticks." In hermaphroditic or intersexual individuals the tests will indicate their true genetic sex.

presence of two Xs is revealed by an odd little "drumstick" formation projecting from the cell nuclear mass. It is normally found only in cells of females, and not in the cells of males where there is only one X. However, neither the chromatin test nor the "drumstick" test by themselves can establish sex identity with absolute certainty, inasmuch as the chromatin-positive (XX) mass, as well as the drumstick, ordinarily characterizing females, are also found in the cells of abnormal XXY (Klinefelter) males. Similarly, the absence of both the chromatin-positive mass and the drumstick, normally characterizing males, also occurs in the abnormal XO (Turner) females.

A much more precise and conclusive sex-identity test, but more difficult to make than the preceding ones, is the direct chromosome test, which actually shows the sex chromosomes themselves. In this test, samples of the individual cells (white blood cells or those from the bone marrow or the skin), are cultured in a test tube and then treated and examined at the doubled stage when the chromosomes are most visible. (The author's chromosomes shown in the Frontispiece were cultured in this way.) This test can not only reveal whether individuals have normal sex-chromosome combinations (XX or XY) or abnormal ones (XXY, XXXY, XO, etc.), thus establishing their true genetic sex, but can also bring to light the chromosome abnormalities involved in mongoloid idiocy and other conditions. In a few striking cases the test has shown in cells of the same person the combined presence of an abnormal XXY sex-chromosome combination, characterizing the Klinefelter syndrome, and the Triple No. 21 chromosome combination of mongolism.

So far, we have dealt with sex abnormalities occurring prenatally—at conception, or at some later stage before birth. But even after an individual is born and is apparently normal for one sex or the other, it still is possible for sex development to go awry. Let's consider again the case of the oyster. In dealing with this eccentric bivalve, we touched on the subject of *sex-reversal*. While nothing like the oyster's *normal* change from one sex into another happens among animals of higher levels, abnormal sex reversals in various degrees may sometimes occur. Poultry furnish the most striking cases. In several instances, hens who had been peacefully pursuing their lives as egg-laying biddies suddenly found themselves developing into roosters, complete with hackles, spurs, tail feathers and lusty crows, and, sometimes, with male organs. In at least two cases such roosters (born hens) were credited with having actually *fathered* chicks after their transformation. All of these, or milder cases of sex reversal in chickens, pheasants, and other poultry are due to ovarian tumors or other internal upsets which suppress the action of the female hormones and shift glandular control to the male hormones normally present in the female in subordinate amounts. This makes possible the development of male organs in some degree. In laboratory experiments, various stages of sex reversal in hens have been induced by removal of the ovaries or by treatments with male hormones.

But although many a woman has said, "I wish I were a man," there is no chance that anything like the sex transformations of hens can occur among human females. The mechanism of sex determination for birds differs in certain ways from that for humans, and there is also, apparently, a difference in hormonal action which permits the balance from femaleness to maleness to be shifted more easily in birds.* What may and sometimes does occur in a woman, however, is a modified change in the secondary sex characteristics, such as development of hair on the face, deeper voice, and enlargement of the features, hands and feet. Most frequently responsible for these changes is a milder and late form of the hormonal condition called adrenal virilism, which we have previously discussed. But in mature women, the removal of the ovaries or the presence of ovarian tumors *rarely produces any noticeable masculinization.* Only in young girls, before puberty, does removal of the ovaries or impairment of their workings cause marked effects, such as failure of the sex organs and breasts to develop properly, or of the mature female bodily proportions to be achieved. Also in the prepuberty stage, tumors of the adrenal glands in a girl may produce various *masculine* physical characteristics, including a deep voice and excessive facial hair, as well as sex-organ abnormalities. (In boys, the same kind of adrenal tumors may result in other abnormalities, and if occurring in early childhood, may be one cause of the "infant-Hercules" type of premature physical development.)

The results of castration among human males (which also give evidence of what the sex glands normally do if present, and what happens if they're lacking) are very often misunderstood. It is only when the testes are removed *before puberty* that the more significant physical characteristics associated with eunuchs are produced. Castration after puberty, when the masculine physical structure already has been set, produces only gradual changes. (Nor is there any sudden turn in the

* For an explanation of how the sex chromosomes in bird species work in reverse from the way they do in human beings, with the XX chromosome combination producing males and the XY combination, females, see footnote, Chapter 7, p. 45.

quality of the voice.) But, even without removal of the sex glands, any glandular disorder or upset before puberty which inhibits the proper production and workings of the male hormones may cause a boy to grow up with *eunuchoid* characteristics in the female direction—large hips, narrow and sloping shoulders, absence of beard, sparse body hair, high-pitched voice, and so on. One quite common condition which occasionally can produce modified traits of this type is *cryptorchidism*, failure of the testes to descend into the scrotum, as they do normally, usually in prenatal life. While cryptorchidism has been found to be hereditary (recessive) in breeds of dogs, especially the dwarf types, its inheritance in humans has not been established.

In maturity, also, and especially in old age, changes in glandular workings may often cause men to develop more highly pitched voices or otherwise to take a mild turn toward feminization. But—contrary to popular fallacy—in neither sex, male nor female, does removal of the sex glands (testes or ovaries) *after puberty* necessarily interfere with sex functioning. Eunuchs, if so produced in maturity, may have sex relations like other men (a fact which challenges the assumption that castration of adult sex criminals—prescribed in some states—will prevent repetition of their crimes). The sex impulse in men and women —in its physical aspects—is engendered, and sex functioning is governed, not by the sex glands alone, but also by other hormonal action, and particularly by nervous and psychic stimuli.

Hardest to believe for most people, when they first hear it, is that *sterility* may be hereditary or inborn. This can happen in a number of ways:

(1) Through recessive genes, coming together—one each contributed by a normal parent—which produce defects in the sex organs or the functioning of the sex glands sufficient to prevent reproduction. (An example is congenital adrenal virilism, if not treated in time.)

(2) Through some of the chromosome abnormalities discussed earlier (the Klinefelter and Turner syndromes and mongolism).

(3) Through a dominant mutation in one of the parent's genes, which may carry sterility among its effects.

(4) Through certain hereditary defects and diseases, such as some forms of dwarfism, and one type of progressive muscular atrophy, in which partial or complete sterility may be a concomitant.

(5) Through mechanical difficulties in sex functioning and/or defects in the sex organs, which may cause total or partial sterility

(*agenitalism, infantilism, eunuchoidism*). These latter conditions have been suspected of being inherited in some cases, but the evidence is insufficient to warrant elaboration here.

Among the *secondary sex characteristics* (which are unrelated to sexual functioning), certain peculiarities, such as unusual hairiness in men, or unusual breast shapes in women, appear to be inherited, but they have as yet been given little accurate study. Extremely large or very small breasts, in women, often occur in successive generations of the same family and may well be hereditary, just as within the normal range different types of breasts characterize women of certain strains and tribes. Also believed hereditary, in both sexes, are *supernumerary* (extra) *nipples*, reported in about one woman in ninety and one man in sixty. Usually there are only one or two extra nipples, but some individuals have up to four and five pairs, arranged above and below the normal two. Most extra nipples are in retarded states of development, but occasionally some of these are functional. In some instances complete extra breasts are present. How these or extra nipples are inherited in human beings is not clearly known, but it should be kept in mind that the presence of several pairs of nipples is normal in all mammals that bear litters. (The occurrence of human triplets, quads and quints prompts the thought that extra nipples in women might be an evolutionary provision for human litters, too.)

That degrees of *fertility* in human beings may be influenced by hereditary factors is another possibility. In Chapter 18 we noted that heredity may account in large part for an extremely high production of twins and other multiples in some racial and ethnic groups, and a very low incidence of multiple births in other groups. It also is not unlikely that the prolific childbearing in certain human strains or families as compared with others (when no control over births is employed) may have some hereditary basis, just as the number of offspring in a litter or the frequency of reproduction among various species of other animals is apparently genetically controlled. No "fertility" genes however, have yet been identified in human beings.

The timing of sex development, quite clearly, is also influenced by genes. It is no accident that puberty and maturity come to most human beings at approximately the same time, as, among women, does the menopause. So, too, where in certain strains or families there is a frequent deviation from the average timing, with these events arriving markedly earlier or later, heredity may be responsible. Evidence is to be found in studies of female twins which have shown that the ages at onset of menstruation and the menopause are much closer for identical than for fraternal pairs.

But environmental factors must always be considered as affecting these biological phenomena. Favorable environments are clearly known to bring on earlier physical maturation in human beings and other animals (as they do in plants), while unfavorable living conditions are found to delay the onset of puberty.* Thus, with the great environmental improvements in recent times, puberty has been arriving earlier for girls in the United States and other advanced countries than for their grandmothers, and often their mothers—at an average age now of between 13 and 13½, or from 1 to 1½ years earlier than in former times. Likewise, within given countries, puberty comes earlier to girls in the more favored groups than for those among the underprivileged.

What, then of the longstanding and oft-repeated belief that "girls mature earlier in the tropics" (Africa, India, etc.) than in the temperate zones? Or that Negro girls mature earlier than Whites? These assumptions have been exploded by many studies. In the United States, for instance, puberty comes at just about the same time for Negro girls as for White girls living under the same conditions; but among underprivileged Negro girls, puberty comes later than it does for White girls in more favored circumstances. Again, negating the theory that hot climates make girls mature earlier, a study in Africa in 1950 showed that Nigerian girls averaged 14½ years of age at the onset of puberty, as compared with an average of 13½ years for girls in England.

In all groups, racial or otherwise, there are exceptional cases not only of girls but of boys who begin achieving puberty at abnormally early ages—sometimes as early as age two or three. (*Pubertas praecox* is the term applied when this occurs.) Often the unusual speed-up in development can be traced to glandular upsets such as might be caused by tumors of the pituitary or ovaries. In one such case, studied by Yale University experts, a New Haven girl reached puberty at age 3½, went through an apparent menopause at 13, and died at 18 following an operation for brain tumor. Another cause of premature sexual and physical development, particularly in boys, may be the *adrenal virilism* condition mentioned several times before. But in some cases abnormally early sex development or puberty does not appear to have any pathological basis. This may have been so in the celebrated case of the

* Conversely, it has been claimed that environmental improvements are bringing later menopause—that is, extending on the average the period during which a woman could bear children—but evidence on this point is not conclusive.

Peruvian girl who in 1939 gave birth to a child at the age of 5½. As for boys, a clear indication that heredity may be involved in extremely precocious puberty was given in the case of a Texas boy, son of an army Air Force officer, who at age two had well-developed male sex organs and pubic hair, and at four (when he looked to be seven or eight) had the muscular development of a young man and was shown to be producing sperms. Further investigation revealed that similar precocious development had occurred in the boy's father, grandfather, two grand-uncles and the great-grandfather from whom all were descended. This would strongly suggest dominant inheritance, limited to males (inasmuch as females in the family line developed normally). Also significant, as indicating that the effects of the gene or genes were confined to inducing sexual precocity, was evidence that all of the affected adult males were or had been healthy, strong and normal in appearance.

But in the foregoing cases, as in many other situations discussed in this chapter, an important question remains: How do deviations from average or normal sexual development affect the thinking, behavior and lives of the individuals concerned? This will be dealt with, among other things, in Chapter 40.

CHAPTER TWENTY EIGHT

It's in Your Blood

No FLUID IN THE WORLD is more important or dramatic than that which we call blood. There is nothing new in saying that it is the vital fluid of life. This has been known as long as there have been men to think about it. What is new is the increasing scientific awareness that human blood—your blood and every other person's—far from being just some red liquid lifestuff coursing through the body, is actually a composite of innumerable varied chemical substances, compounds and structures, each with special properties and often with the power to affect every aspect of one's existence.

Furthermore, the blood of one human being may not only be distinct from that of any other person in many ways, but in certain circumstances can prove highly dangerous to another person—even to someone referred to as of "one's own blood." That this was so began to be apparent long ago, when it was found that blood transfusions often ended disastrously. It wasn't so hard to understand this when the blood of some lower animal was transfused into a human being (as was sometimes done in the preceding century) or when the bloods of different species of animals were mixed. But why, when the blood of one person was transfused into that of another, were there also serious consequences in many cases—*even if it was the blood of a mother or father given to either one's own child*, or the blood of one brother given to another? The answer, we know now, is that it is as possible for two persons in a family to inherit different kinds of blood substances as it is for them to inherit differences in eye color and other physical traits.

This did not begin to become clear until 1900 (by coincidence also the year in which modern genetics was born), when Dr. Karl Landsteiner discovered the existence of the A, B, and O blood types (the AB type was found later). He showed that some persons' red blood cells carried one kind of biochemical substance (A), others carried a different substance (B), some carried both substances (hence being of the AB type), and many individuals carried neither (being designated as of the O type). Later it became apparent that these blood types were inherited, but it was a statistician, Dr. Felix Bernstein (neither a geneticist nor a medical man!) who first figured out, in 1925, exactly how the genetic mechanism works.

The main blood groups result from three variations of one principal gene. Two of these, the variant genes for A and B blood, are *dominant* over the gene for O blood, but are of almost equal strength in relation to each other. So here's what happens:

Gene *A* produces, principally, a substance known as *Antigen* A.

Gene *B* produces, principally, *Antigen B*.

Gene *O* produces neither antigen.

Every one of us inherits two of any of the above genes—one gene from each parent. We therefore may receive two genes of the same kind—*AA*, or *BB*, or *OO*—or mixed pairs, *AB*, *AO*, or *BO*. The *O* gene being recessive, an *AO* combination results in A blood, just as does the *AA* combination, while the *BO* combination produces B blood, just as does the *BB* combination. And, as the *A* and *B* genes are of almost equal strength, when these two are both present the blood will be of the AB type, carrying antigens of both kinds. At the same time, for whichever of the A or B antigens an individual does *not* carry, there is a corresponding inherited *antibody*, or clumping substance, which can combat the foreign antigen if it enters his blood stream.

Within the individual himself, it is of little importance which type of blood he has inherited. Trouble comes only if he is given a transfusion with blood that clashes with his own, producing clumping or disintegration of the blood cells. This is most likely to happen if a group O person is given bloods of any of the other types, or if A individuals are given group B blood or vice versa. On the other hand, O blood—against which there are no antibodies in A or B bloods—can be used with the least risk in emergency transfusions. In reverse, AB blood, while least useful in transfusions—having elements alien to the other three types—can nevertheless *receive* any of the other bloods most easily. (See our illustration.) But while these facts were considered sufficient to go by in former years, what has now become apparent is that clashing substances other than those produced by the

HOW THE A, B, AB, AND O BLOOD TYPES INTERACT IN TRANSFUSIONS

Each type of blood produces distinctive substances which are incompatible with those of other types. Here is what would happen in a transfusion if you were given (1) blood of the same type as yours or (2) blood of a different type.

1. COMPATIBLE TRANSFUSION

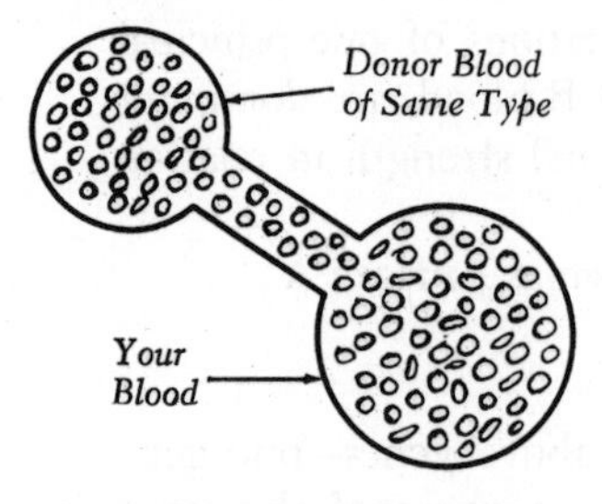

Blood of donor mixes freely and safely with your own to make transfusion successful.

2. INCOMPATIBLE TRANSFUSION

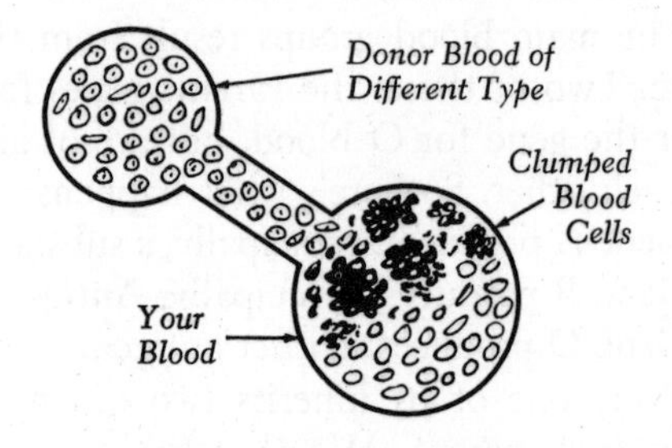

Donor's different blood cells are clumped (agglutinated) by your antibodies.

Worst reactions: A, B *or* AB *blood infused into* O *blood.* Next *in order of danger:* A *into* B, *or* B *into* A; *then,* AB *into* A *or* B. Less *danger,* A *or* B *into* AB. Least clumping: *when* O ("*weakest*") *blood is transfused into any of others.**

* Individual differences with respect to additional blood groups mentioned in the Text may be important in many transfusions, so that O is no longer considered the "universal-donor" blood; nor is merely matching A, B or AB bloods sufficient to insure protection. The special Rh problem in transfusions, with regard to threats to prospective children of Rh-negative women previously given Rh-positive blood, is discussed later in the chapter.

main blood types may exist in individuals and that the wisest procedure in transfusions is to give a person blood as much like his own as possible in every respect.

Where the "blood detectives" scored one of their greatest successes was in revealing that serious blood clashes can result not only from transfusions, but from conflicts in prenatal life between the hereditary blood substances of the baby and the mother. Here was the solution to a long-standing mystery, of why vast numbers of fetuses and newborn infants had been dying as the result of a certain blood disease—a serious type of anemia or jaundice called *erythroblastosis fetalis* (destruction of the fetal red blood cells) and now popularly known as the "Rh disease."

For the story of the Rh findings we turn again to Dr. Landsteiner, and also to two of his younger associates, Dr. Alexander Wiener and Dr. Philip Levine. It was the senior medical scientist and Dr. Wiener who in 1940 jointly discovered the Rh factor through their experiments with the *Rhesus* monkey (hence the Rh name) and then proved how this factor was inherited by human beings and the role it played in transfusions. It was Dr. Levine who proved that the Rh factor was responsible for the mother-child blood clash which produced the "Rh disease." Subsequent findings by Drs. Wiener and Levine, and by many other medical scientists (notably, in England, Drs. R. R. Race and R. A. Fisher) not only revealed the intricacies of Rh inheritance, but also showed how most of the Rh hazards could be prevented or overcome.

To begin with, in addition to the main blood-type substances, the Rh chemical substance is carried by the red blood cells of about 85 per cent of White persons but is absent from the blood of the remaining 15 per cent. (Among Negroes the Rh substance is carried in 92 per cent and is absent in only 8 per cent. Among Mongolian peoples—Chinese, Japanese and "pure" American Indians—virtually all have the Rh substance.) This Rh substance is produced through the action of a *dominant* gene. Whether a person inherits one such gene or two, his blood develops the substance, and he therefore is known as of the *Rh-positive* type. If, instead, a person inherits two *recessive* Rh genes and therefore does not develop this blood substance, he is known as of the *Rh-negative* type.

But it is almost entirely in prenatal life, as we said, that the Rh factor has its importance. For if a mother is Rh-negative and the child she carries is Rh-positive (through having received the gene for this blood substance from the father), this is what may *sometimes* happen:

1. As the developing fetus begins to produce the special Rh substance in its blood, some of it may filter through the placenta into the mother's blood stream.
2. The mother's blood begins to manufacture *anti-Rh* substances, or antibodies, to combat the invading foreign chemical.
3. Some of the antibodies from the mother then filter through the placenta and into the child, attacking and destroying its blood cells, or in other ways seriously impairing them.

But only in about one-sixteenth of the cases where the mother is Rh-negative and the child Rh-positive does this occur. In other words,

HOW AN RH-DISEASE BABY IS PRODUCED

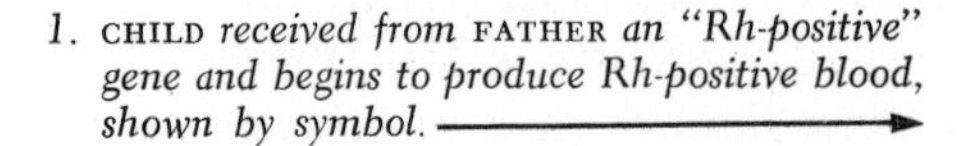

1. CHILD *received from* FATHER *an "Rh-positive" gene and begins to produce Rh-positive blood, shown by symbol.*

2. MOTHER'S *blood is Rh*-negative, *shown by symbol.*

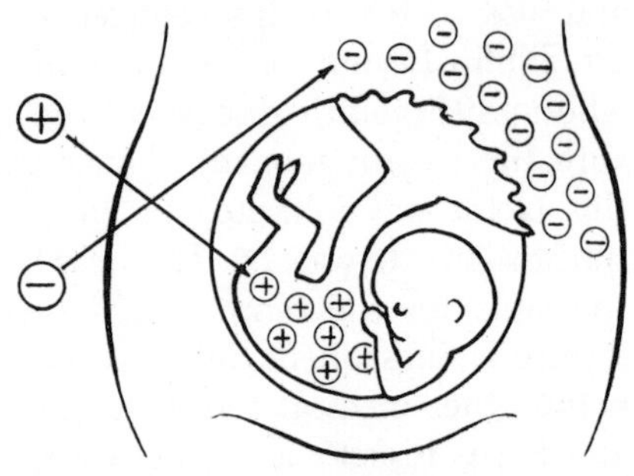

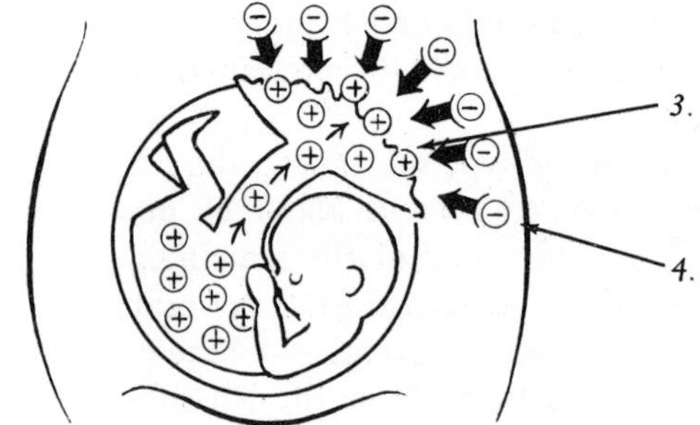

3. *Some of the* CHILD'S *Rh-positive blood substance travels through placenta into* MOTHER.

4. MOTHER'S *blood begins producing* ANTIBODIES *to attack hostile substances from baby.**

5. ANTIBODIES *from* MOTHER *enter* CHILD *and begin destroying its blood cells.*

* If Rh-negative mother has had previous Rh-positive pregnancies, or transfusions with Rh-positive blood, antibodies already are present.

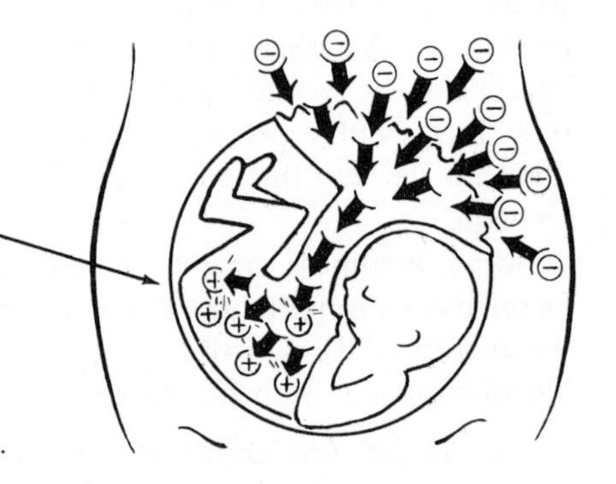

Chart prepared with aid of Dr. Alexander Wiener.

about one in every twelve pregnancies involves an Rh-negative mother and an Rh-positive baby, and yet babies with the "Rh disease" (erythroblastosis) appear in no more than about one in every 200 full-term deliveries. Why so few?

Here are some of the answers: The production and interchange of the clashing blood substances between mother and child does not always take place. There are differences in the types of Rh substances and antibodies, some being weaker than others (not inciting or not clashing with their opponents so violently).* *But most important, the amount of antibodies attacking the fetus and the stage in its development when the attack begins depend largely on what prior experience the mother has had with the foreign Rh substance, either through bearing previous Rh babies or having had transfusions at one time or another with Rh-positive blood.*

If an Rh-negative mother is bearing her *first* Rh-positive baby and an interchange of their blood substances does take place, the antibodies she produces—if any—are very rarely strong enough to harm the fetus. But suppose the mother is carrying her second or later Rh-positive baby. She starts out now with attacking antibodies already present in her blood through previous pregnancies, and these are poised to attack the blood cells of the new fetus at an earlier stage and in much heavier dosages than was the case with preceding babies. Such a heavy onslaught may be sufficient to kill the fetus, or if the baby achieves birth, to bring it into the world with its blood seriously damaged. By tests being made today doctors can tell at every stage of an "Rh" pregnancy how much antibody is being produced by the mother and whether danger exists.

But far more serious than previous Rh-positive pregnancies, as we

* For the technical-minded: There are various types of Rh-positive factors—the original Rh factor (the most common and the one responsible for most cases of erythroblastosis) and weaker ones. In Rh-negative bloods, similarly, there are several contrasting Hr factors, one or more of which may also occur in Rh-positive blood (for example, if the individual carries mixed genes, one positive and one negative). When blood tests are made, the various factors mentioned can be found in different types of Rh and Hr bloods. In designating the numerous Rh-Hr types, two different systems of nomenclature are in use. The system devised by Dr. Alexander Wiener uses the letters "R" (and what he regards as its genetic variants, R^0, R^1, R^2, R^z) representing the "dominants," and "r" (and its variants, r', r'', r^y) representing the "recessives." The other system, dvised by Drs. R. R. Race and R. A. Fisher of England, uses the letters "C-D-E" and "c-d-e" to represent different closely linked Rh-group genes. For example, persons of one of the Rh-negative types may be designated by the first system as "r-r," by the second system as "cde-cde." (Other negative types could carry combinations of "cde" with everything but a big "D.") Rh-positives, according to the first system would all have a dominant capital "R" component and by the second system would have any combination which included a big "D": cDe, CDe, CDE, etc. (Note: The existence of little "d" is still theoretical.)

said, would be the situation where an Rh-negative mother has had transfusions with Rh-positive blood prior to childbearing. (In the years before the facts about Rh factor were known, this must have happened quite frequently, as first shown by Dr. Levine.) In such an event, the woman would have developed antibodies in ten times greater abundance than might have resulted from carrying an Rh-positive baby. Thus, at any "Rh" pregnancy thereafter antibodies could already be there poised to attack the child.

Altogether, in former years about 40 per cent of all "Rh" babies born with erythroblastosis died within a short time, and many more deaths from the disease came—and still come—in prenatal life. But, happily, the dangers in the Rh situation have been radically reduced. For one thing, it is now routine for obstetricians to ascertain immediately whether there is an Rh danger in any woman's pregnancy. There can be only if the wife is Rh-negative and the husband Rh-positive. In that case the odds are strong that the baby will also be Rh-positive, and this calls for testing the mother to see if she has already produced anti-Rh substances, either through earlier Rh-positive pregnancies or through transfusions. Then, if danger does threaten, precautionary measures can be taken. Should the child be born with the "Rh disease"—that is, with gravely impaired blood cells—it can still be saved in most cases now by an immediate blood-exchange process, in which all of the infant's damaged Rh-positive blood is replaced by healthy new Rh-negative blood.* With the various precautions now possible, there need be little fear of serious consequences for an Rh-negative mother's first Rh-positive baby, and only slight danger for a second such baby.

The question is sometimes asked, "What happens if the *husband* is Rh-negative and the wife is Rh-positive?" In that case there can be no Rh threat to a baby, because even if it were Rh-negative it would produce no substances against which the mother's blood would react (except in extremely rare instances). Obviously, also, if both wife and husband were Rh-negative (a highly unusual situation) the baby would be Rh-negative, and there could be no prenatal blood clash.

We have by no means presented all the intricacies of the Rh factor and the "Rh disease." In rare instances complicated interactions between the different Rh subtypes in mother and fetus may produce the

* Why give an Rh-positive infant Rh-*negative* blood? Because when the damaged blood is drained off, the baby's tissue spaces still retain some of the mother's antibodies, which could continue to be damaging to Rh-positive blood, but not to Rh-negative. Later these antibodies cease to be a threat to the baby, and it can be safely left to make its own blood.

"Rh disease," even where the mother is primarily Rh-positive and the baby is also. But much more important, erythroblastosis or blood diseases very much like it may in some instances result in ways other than from the Rh incompatibility we have dealt with. In fact, there are indications that A-B-O blood group incompatibilities between mother and child may cause more prenatal casualties than does Rh incompatibility. At this writing some authorities attribute as much as a 20 to 25 per cent prenatal loss—though practically all of it in the earliest and undetected stages of conception and gestation—to A-B-O mother-and-child differences. In particular, where the mother is O and the child's blood is incompatible (especially A), serious disease may occur in up to 1 in 200 deliveries. On the balancing side, there is evidence that A-B-O mother-child incompatibility tends to block the circumstances leading to the ill effects of Rh incompatibility. On the whole, while the A-B-O mother-child blood clashes may be more frequent than those involving the Rh factor, the cases of babies who approach or achieve birth with serious blood impairment as a result of them are relatively fewer.

As a concluding note to the pathological aspects of blood groups, some degree of association has been reported between particular blood types and the susceptibility to such diseases as stomach cancer (claimed to be above average in persons with type A blood) and duodenal or gastric ulcers (above average in persons with type O blood). But this is not yet fully proved, nor are certain other reported blood group and disease relationships.

We come now to a variety of other blood groups or blood types which, like the A-B-O and Rh groups, occur in distinctive hereditary forms but are only rarely involved in any transfusion clashes, mother-child conflicts, or disease susceptibilities. Knowledge of these other blood groups can, however, be highly important in other ways, as will be noted presently. Among such blood groups are those which involve the substances called M, N, P, S, V, Diego, Lutheran, Lewis, Kidd, Duffy, and others. The best known of these, the M and N (discovered in 1927 by Drs. Landsteiner and Levine), are produced by two equally dominant genes, a child receiving a gene for one or the other from each parent.

It has been found that the A and B blood substances can be detected in the salivary secretions of some individuals, who are therefore known as *secretors*. This finding, together with the knowledge of the M, N, and other blood substances mentioned above, has so far proved of practical importance in extending the effectiveness of tests to establish nonparentage; to prove twins fraternal or identical; to identify criminals through blood stains; and so on. Not least, all of the blood group evidence has been proving increasingly helpful to anthropologists in tracing the possible origins and relationships of racial and ethnic stocks (as will be explained in Chapter 42).

So far we have dealt only with blood substances which have meaning as members of blood groups and in relationship to one another. But there is another aspect of the ingredients of human blood: their capacity, where not properly compounded or formed, or not produced in adequate amounts, to cause diseases or abnormalities directly. In all of these blood conditions wayward genes may be involved.

The *hereditary anemias* constitute one of the most serious categories of blood disease. We may note, first, that the more common anemias which people know about are not hereditary and involve deficiences in the quality or nature of the blood resulting from various diseases and infections or from malnutrition. However, there are some types of anemia which are directly due to inherited abnormalities in the blood cells; and what is especially significant about the two most prevalent types is that they are found concentrated largely among persons of certain races or ethnic stocks. Also, inheritance in both follows the same pattern: a single, semidominant gene causing mild, but recognizable effects; a double dose of the gene causing serious consequences.

Cooley's ("Mediterranean") *anemia* is found preponderantly among persons of Italian, Greek and Armenian descent. The single gene for the condition produces mild blood abnormalities; but two of the genes together (one from each parent) produce not only severe anemia, but skeletal abnormalities and mongoloid features, and usually result in death in early infancy or childhood. The genetic deficiency in this condition apparently involves an inadequate production of the essential substance hemoglobin.

The second race-slanted type of anemia, which afflicts chiefly Negroes, is *sickle-cell anemia,* so called because of the curious, twisted sickle-shaped red corpuscles resulting from it. The mild, one-gene, semidominant condition, in which sickle cells occur only occasionally, is produced by inheritance through just one parent.* But if each parent

* While the single sickle-cell condition ordinarily has little ill effect on the individual, it may become dangerous if he is exposed to high-altitude pressure. In that circumstance damage to the spleen (spleen infarcts) may result. Thus, service such as that in the army parachute corps presents special hazards to the 8 per cent of Negro males who carry the single sickle-cell gene.

RED BLOOD CELL ABNORMALITIES

As found in various inherited anemias

1. NORMAL RED CELLS
(for comparison)

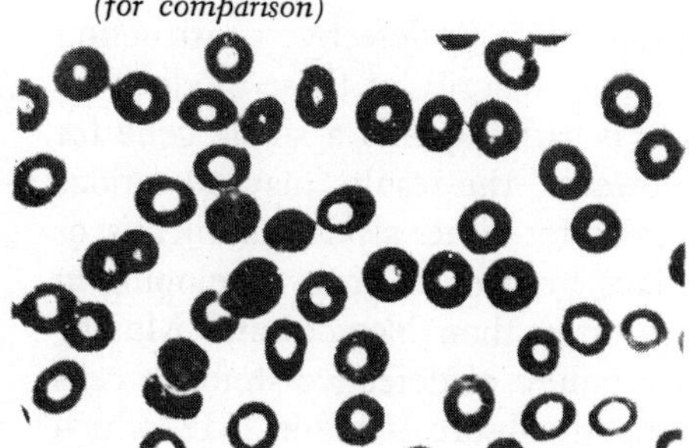

2. OVAL-SHAPED RED CELLS
(Ovalocytosis or elliptocytosis)

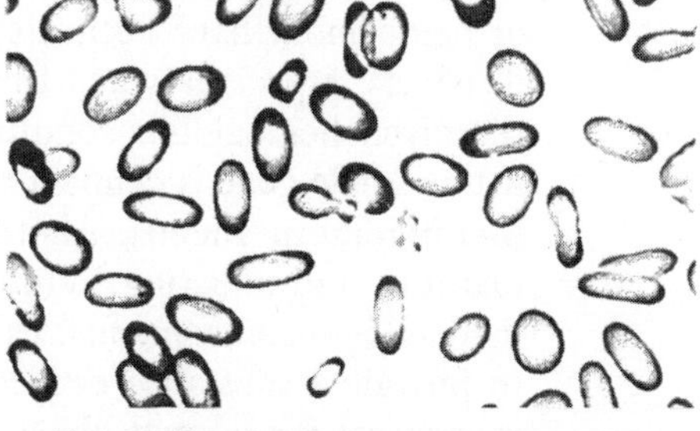

Above cells, while usually harmless, are sometimes associated with anemia, and may be present with other blood-cell abnormalities.

3. SICKLE CELLS IN SICKLE-CELL ANEMIA

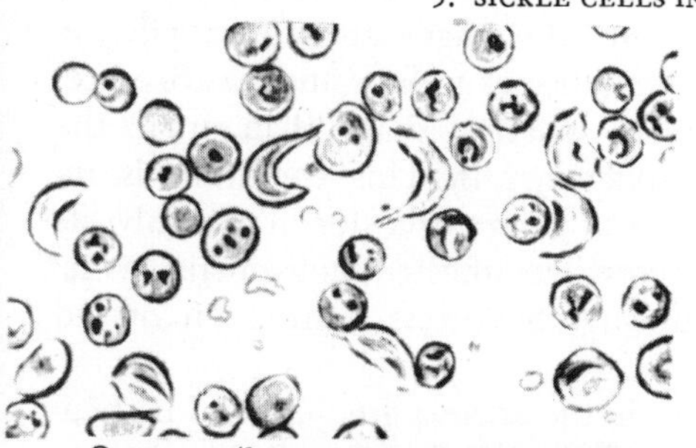

*a. One-gene (heterozygous): mild condition, with scattered sickle-shaped cells and other abnormal cells, but also normal cells.**

b. Two-gene (homozygous): an extreme sickling condition, causing severe anemia, usually fatal in childhood.

4. CELLS IN THALASSEMIA (COOLEY'S, OR MEDITERRANEAN, ANEMIA)

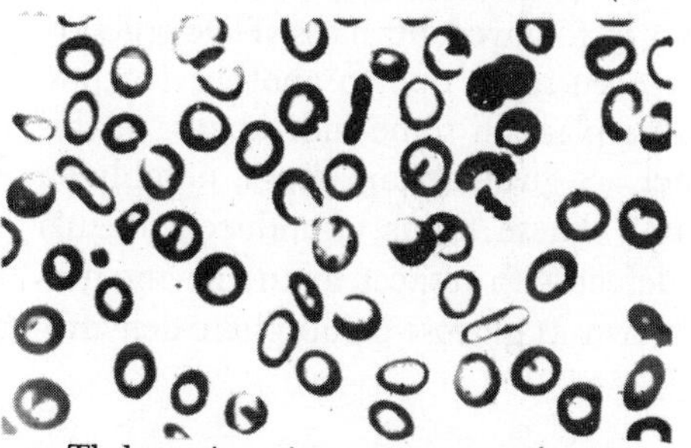

a. Thalassemia minor: *one-gene (heterozygous) condition, with scattered "target" cells among other abnormal cells, plus normal cells. Usually not harmful.*

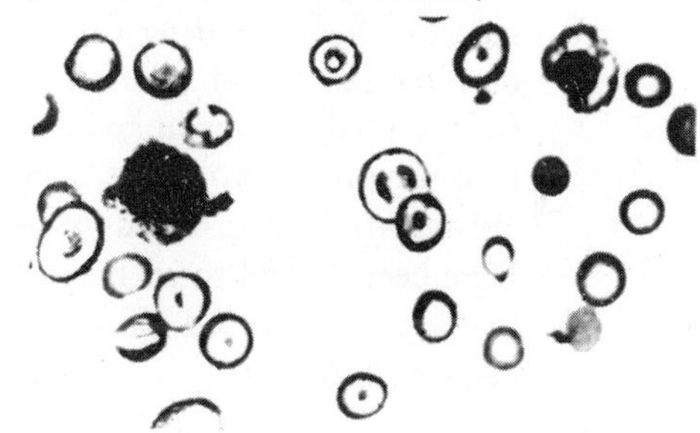

b. Thalassemia major: *two-gene (homogzygous) condition, showing red cell abnormalities, causing severe, usually fatal, anemia.*

* Original sickle-cell photograph made after special treatment of cells (by artificial deoxygenation) to bring out abnormalities.

Illustrations adapted from photographs by Dr. James V. Neel (Nos. 1, 3-b, 4-a, 4-b); Dr. C. Lockhard Conley (3-a); and National Institute of Health (2).

carries and passes on the same gene to a child, the double dose will cause extensive cell sickling and destruction, with death often following before long.

Many other inherited blood conditions due to defective construction of hemoglobin have been identified. The majority of these produce no ill effects, especially when the person is carrying only a single gene for the given hemoglobin condition. However, the results may be serious if this single gene is coupled with a gene for either sickle-cell anemia or Mediterranean anemia. Another serious form of anemia, developing in maturity and afflicting Whites more often than Negroes and Mongolians is *pernicious anemia,* caused by failure of defective stomach cells to provide a substance needed to properly absorb Vitamin B-12, which is essential for normal blood production. While known to run in families, its method of inheritance is not clear—possibly recessive or incompletely dominant.

Still another hereditary blood condition is *acholuric family jaundice,* in which the red blood cells are unusually fragile and are rapidly destroyed in the spleen, resulting in jaundice as well as anemia. Because, further, the red cells assume a more spherical shape than usual, the disease is also called *hereditary spherocytosis.* This condition is inherited as a dominant, and since it can be watched for in a family, its more serious effects may be prevented in affected individuals by removing the spleen and thus reducing the excess destruction of red blood cells.

One of the most unusual diseases in the anemia group, found particularly in the Mediterranean region, afflicts people for an odd reason—that its development is related to the eating of the fava bean, thus giving the disease its name, *favism.* In persons, most often White males, who are genetically susceptible (because of a certain sex-linked recessive gene), eating of the bean is followed by massive destruction of their red blood cells. A similar occurrence due to another (or possibly a related) sex-linked gene takes place in some individuals, in this case usually Negro males, when they are given certain drugs, notably an antimalarial drug called *primaquin.** These forms of induced anemia have been linked to an hereditary defect with respect to an enzyme normally present in red blood cells, known as glucose-6-phosphate-dehydrogenase (G-6-P-D).

* Among American Negro males approximately 14 per cent—one in seven —may have the inherited sensitivity to primaquin and related oxidant drugs, such as sulfonomides, sulfones, and nitrofurans. Through simple tests, doctors can screen out the susceptible males for whom these drugs should not be prescribed.

We proceed now to hereditary blood diseases which result from defects in the blood-clotting process. The body's mechanism for preventing excessive bleeding by this process is a highly complex one, involving many different factors in the blood (mostly proteins) and tiny cell-like bodies called platelets. Normally, the interaction of the platelets and the various protein substances in response to some injury to the circulatory system results in the formation of a firm blood clot at the site of damage. If any of the factors involved is missing, clotting may be delayed or prevented altogether, resulting in profuse bleeding. Thus, since many of these deficiencies may result from gene failure, heredity can be the cause of a considerable number of bleeding diseases.

Hemophilia—the most famous of the bleeding diseases—was discussed in Chapter 22. There we dealt with the common or classical form of the disease, now called hemophilia A. But other types have been identified, among them one called hemophilia B, or Christmas disease (after an English family of that name in which it was first discovered). Perhaps one out of six hemophiliacs has this type of the disease, but only by special tests can it be distinguished from classical hemophilia, although it is transmitted in the same manner, as a sex-linked recessive. The fact that the two types of hemophilia are distinct is shown in an interesting way: When transfusions are made, blood of a patient with Christmas disease will temporarily correct the clotting deficiency of a patient with classical hemophilia, and vice versa.*

One question long debated was whether a *girl* or *woman* could have hemophilia. Many authorities had considered this almost impossible, because, first, to produce such a female would require the mating between a hemophiliac man and a woman carrying a hidden "hemophilia" gene (the odds against this running into tens of millions); and, second, it was believed that if a female baby with two hemophilia genes were conceived, such a double dosage would prove fatal before birth or soon thereafter. But this view was suddenly upset by the finding in 1951 of several cases of women who *did* have hemophilia. One of these, a woman of Manchester, England, who at the age of twenty-four almost bled to death following childbirth, was found to have had a hemo-

* Another variant of hemophilia is *von Willebrand's disease*, also called vascular hemophilia or pseudohemophilia. It is sometimes confused with hemophilia A because of a similar although milder deficiency in the same blood-clotting protein. It differs genetically from the other form in that it is not sex-linked, but is inherited through a gene in a chromosome other than the X.

philiac father and—with evidence that her mother also carried the gene for hemophilia—a hemophiliac brother.

(Several less common and less serious hereditary defects in blood clotting are listed in our Wayward Gene Tables in the Appendix.)

In addition to the proteins required for blood clotting, the blood normally carries many other substances with specialized functions. Among these are the *gamma globulins,* assigned to the task of fighting off invaders such as bacteria and thus constituting an important part of the body's normal defense against infection. Certain individuals have an hereditary deficiency—almost a total absence—of gamma globulin (a condition called *agammaglobulinemia*) and therefore suffer from frequent severe infections, especially pneumonia. They can be helped by periodic injections of gamma globulins taken from normal individuals. Since the condition is due to a sex-linked recessive gene, it afflicts mostly males.

Another specialized protein in normal human blood is *ceruloplasmin,* the function of which is to carry essential copper molecules in the blood. When there is a genetic deficiency in the production of this protein (a recessively inherited condition called *Wilson's* disease), the copper absorbed from food lodges in the liver and in parts of the brain, causing defective liver functioning, tremors, and mental deterioration. Afflicted individuals can be helped by drugs which prevent the accumulation of copper in their systems.

Two other kinds of genetically determined protein substances in the blood are the *haptoglobins* and the *transferrins.* The haptoglobins have the task of "binding"—or holding in place—the hemoglobins from broken-down and aging red blood corpuscles, thus preventing the release into the system of certain substances, such as excess iron, which might cause damage to the kidneys or other organs. The transferrins are the normal iron-transporting proteins of blood serum, and these, too, are found in several genetically determined types. What effects different transferrins have on individuals is still being investigated.

The *white blood cells* may be defective in a number of ways. The most important and most serious type of defect in these cells is *leukemia,* mentioned in Chapter 22. Although some hereditary influence may be present in leukemia, there is no good evidence of any direct gene action producing the disease. When it results from disruptive changes in genes and/or chromosomes due to radiation, as noted on page 572, it may be considered an environmentally induced condition which while *genetic* need not be *hereditary*—that is, capable of being passed on—unless the change occurs in the reproductive cells of the person.

CHAPTER TWENTY NINE

How Long Will You Live?

"You'll go when your number's up!" is a popular saying with soldiers. Many other people share this fatalistic belief—some through their religions—that every person comes into the world with the time of his exit already stamped by fate.

We can doubt this, certainly, as applied to any exact hour, month or year. But in a very general way there is indeed a "time clock" which governs our stay on earth and which is set earlier for some individuals than for others. Thus, when you hear, "So-and-so is of long-lived stock," or "So-and-so is of short-lived stock," it may well have meaning with regard to inherited differences in potential longevity. Though not always!

Within certain limits the life span of human beings, as of all other species, seems to be set by inherited factors. The oldest living things are trees. Different varieties are characterized by different limits of longevity, the farthest limits, into thousands of years, being reached by certain cypresses and the American sequoias, with the granddaddy of all things now living believed to be an almost 7,000-year-old bald cypress in Tula, central Mexico. Soil, climate and other conditions are, of course, vitally important factors, but there is also something in the *nature* of trees that determines their potential age and that makes a tree of one variety live longer than another growing by its side.

Animals, whose biological mechanisms are much more complex and whose lives are far more hazardous, are narrowly limited in their life spans, but these limits likewise vary greatly with different species. Under the best of conditions, elephants can live to be no more than about 75 years; horses, up to 35; dogs and cats, about 20 to 25; cattle, about 30; parrots, about 45, and cockatoos, reputedly 70 or more. Wild animals and wild birds probably have much shorter life spans than domestic ones. Among fish, carp, pike and catfish may live to be upwards of 50 years, and some sturgeons are reported as having lived beyond 80 years. Of all lower animals, large tortoises or turtles appear to be the longest-lived, with recorded ages up to 150.

How long can a man live?

There is much reason to be dubious about ancient records of longevity. Figures in the Bible are now believed to be based on methods of calculating years which differed from those employed in post-Biblical times. For instance, the age of Adam is given as 930, of Methuselah as 969, of Cainan as 910, Jared, 962. But all of these were pre-Flood personages. After the Flood we find the ages cut to about one-fourth—Abraham, 175, Isaac, 180, Jacob, 147, Moses, 120; and in turn, even this second group of ages seems inconsistent with the later references in the Bible to "three score and ten" (or 70 years) as a full human life span—which indeed it was at a time when life was far more precarious than it is today.* At any rate, with respect to the phenomenal earlier Biblical ages, one theory is that they were derived through a miscalculation of a very ancient code, which, if properly interpreted, would have scaled Adam's age down to 96, and made Noah 48 when he built the Ark, not 600.

In comparatively recent times, the extraordinary ages have been credited invariably to obscure persons, born in remote localities or in periods when no reliable records were available. For example, England's champion oldster was Thomas Parr, a Shropshire farmer who died in 1635 at the reputed age of 152; but although this presumed achievement brought him interment in Westminster Abbey, later skeptics have suggested that his age may have been derived by crediting him with the birthdate of his father, or even grandfather. In 1956 an American side-show promoter produced a tiny, 4-foot-4 Colombian Indian who was claimed to be 167 years old. (He died the following year with his age unverified.) The Soviet Union has from time to time publicized cases of individuals reputedly 150 years, or older, and has reported phenomenal numbers of persons aged 100 and over—up to 144 per 100,000 population in one region, Nagorno-Kharabak, and (in 1959)

* The possibility of added years for those endowed with special vigor is referred to in the full quotation from PSALM 90:10: "The days of our years *are* threescore years and ten; and if by reason of strength *they be* fourscore years, yet *is* their strength labor and sorrow; for it is soon cut off, and we fly away."

almost 600 persons aged 120 or over in the country at large. But, oddly, the phenomenal oldsters and the high incidences of extreme longevity in the Soviet Union (as elsewhere) seem invariably to occur in the more remote and backward areas, while in the Moscow area, where data can be considered most reliable, only one centenarian per 100,000 population is reported.

Nor are the United States figures on extreme oldsters above suspicion. According to our official government reports, during the year 1961 there were 81 deaths at the reputed ages of 110 and over; of these, 23 were listed as at 115 and over (including seven at ages 120 and over). The government compilers could go only by data given on death certificates, but the dubiousness of the extreme-age figures becomes apparent when one finds that *more than half* of those who were listed as having reached or passed the age of 110, and *all seven* in the 120 and over bracket (two of these at age 125 and one at age 128), were *Negroes*—despite the fact that Negroes comprise less than 10 per cent of the American population and that their average life expectancy has long been very much below that of Whites.

In all, whenever whopping ages are reported, scientists and actuaries are inclined to attribute them to errors, hearsay testimony and the well-known tendency to exaggerate ages of centenarians—especially when no birth certificates are available.* The true situation about the current limits of human longevity may be indicated by the statistics from some advanced countries—such as the Netherlands, France and Great Britain—where there are reliable birth records going back a long time. In these countries recent figures show an incidence of only four or five centenarians per million population, and no individual attaining anything like the 115-plus ages previously mentioned. The actual facts in the United States are probably much the same. We know, moreover, that to date the all-time *authenticated* longest American life span was that of Mrs. Louise K. Thiers of Milwaukee, who died in 1926 at the age of 111 years, 138 days.† This was subsequently slightly exceeded

* Statistician Walter G. Bowerman showed that there is a direct correlation between the illiteracy rates and the numbers of centenarians reported for various countries and regions. Bolivia some years ago, with an 83 per cent illiteracy rate, claimed 75 centenarians per 100,000 population; New Zealand, with only 2 per cent illiterates (and with one of the highest life expectancies in the world), reported only one centenarian per 100,000 population.

† In 1960 there was much to do over the death of a Texan, Walter Williams, who was reputedly 117 years old and allegedly the "last surviving Confederate veteran of the Civil War." But although shortly before his

in Ireland by a spinster, Miss Katherine Plunkett, who died in 1932 at the age of 111 years, 329 days. One would have to go back a century and a half to find a person definitely proved to have lived longer: a Canadian, Pierre Joubert, who was 113 years, 124 days when he died in 1814.

Let us say, then, that the maximum human longevity is now considered to be no more than about 115 years. But as we know, very few persons have come anywhere near reaching this maximum, even though life expectancies have been increasing at a rapid pace for some time. Moreover, there is and always has been considerable variability among persons living in the same period and under the same conditions in attaining the prevailing average life expectancy. In some families most of the members for successive generations have lived or are living well beyond the average, while in other families most members have been consistently shorter-lived. This has led to the belief that the extent of longevity for human beings, first as a whole, and second, individually, is in large degree governed by hereditary factors, which, acting separately or collectively, can set potential limits to one's life span.

Perhaps, as with certain mechanical appliances, every human being starts life with a qualified guarantee as to how many years he can be kept going. We might call this "conditioned longevity." The word "conditioned" is extremely important, for nowhere does environment play so large a part as it does in relation to length of life.

Quite true, Mr. So-and-so may boast that he comes of long-lived stock and may produce figures to show that his parents and all his ancestors lived to the age of ninety. But let Mr. So-and-so, driving home from the club some winter night with one-too-many under his belt, try to round an icy horseshoe curve at sixty, and all the statistics as to his potential life expectancy may be of no avail.

In the matter of longevity we can think only in terms of broad general averages. You read everywhere that length of life has been steadily increasing, how in George Washington's time the average was about 35 years, how in 1850 it was 42, how in 1901 it was about 50 years, and how today it is 70. But this does not mean that because your parents lived to an average age of 75, you will live to be much older—90, say, or 100. What has been increased, through greatly improved hygiene, living conditions and medical treatment, is not the *potential longevity* of

---

death an investigation proved his reported age to have been much exaggerated and his service in the Confederate army to have been a tall tale, the story had been so long accepted that officials of the state and nation preferred to go along with the legend, and what amounted to a state funeral was held.

THE ADDED YEARS

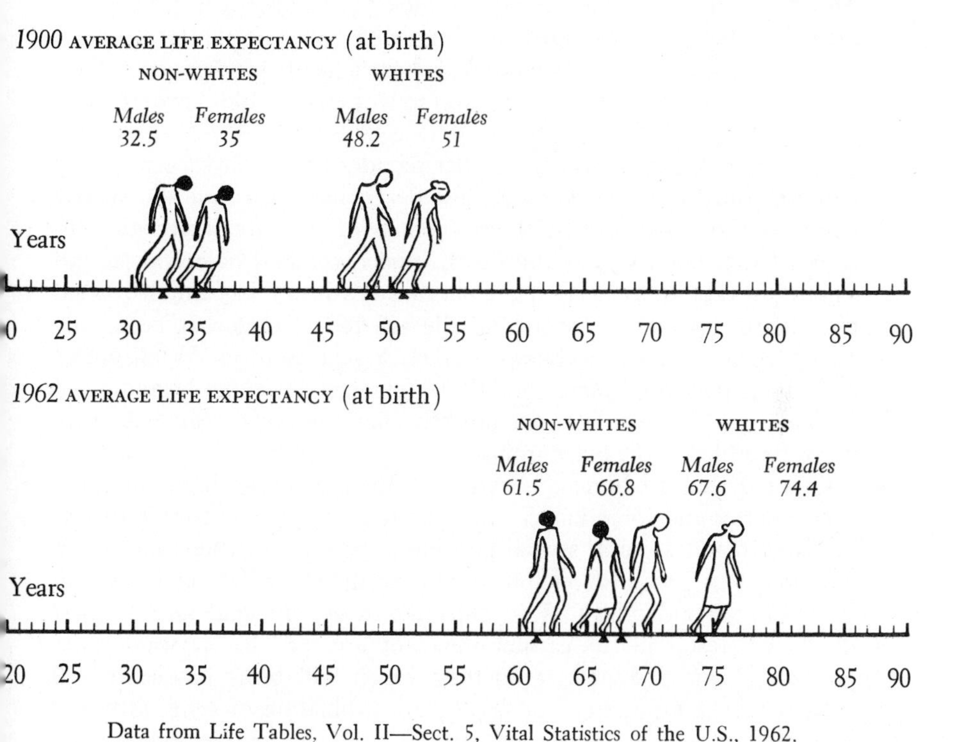

Data from Life Tables, Vol. II—Sect. 5, Vital Statistics of the U.S., 1962.

human beings, but their *chances* of achieving this potential.

The very young are the ones whose life expectancy has been increased most sharply through the tremendous reduction in death rates from diseases of infancy and early childhood. Where a century ago more than one in every six American babies died before reaching one year of age, and in 1900 one in eight died, today the infant mortality rate is only *one in forty;* and where in 1900 only 60 per cent of the American babies could be expected to survive up to or beyond age fifty, today 90 per cent can be expected to do so, according to current life expectancies.* The tuberculosis mortality rate has been cut to a

* That the United States has by no means gone far enough in reducing infant mortality is shown by the fact that in 1962 there were nine other

tiny fraction of what it once was, and many plagues and epidemics that formerly carried off vast numbers of persons have been stamped out. Antibiotics and other germ destroyers have saved the lives of untold numbers. Insulin treatment has prolonged the lives of many diabetics. And one could continue through the long list of the achievements of medicine and science—aided by improved living conditions—which have reduced human mortality. All of these benefits have projected into later life many individuals who formerly never would have reached maturity.*

But while the threats in the earlier decades of life have been greatly reduced, the further advance against death has been radically slowed down by the major and stubborn enemies of the later years—the diseases of the blood vessels and heart, cancer, cerebral hemorrhage, and so on. (In fact, from 1954 on the nation's death rate has almost leveled off.) Thus, for people beyond middle age today, there has been only a slight gain in life expectancy—scarcely a year more for White males and about two years more for White females—since the beginning of the century. However, it is not unlikely that additional years will eventually be added in the older groups.

Granted that all human beings have their *potential* limits of life—some shorter and some longer—by what means could heredity set these limits? It could do so in several possible ways, some of which have been mentioned in preceding chapters: (1) by direct "killer" genes, which may start an individual off so defective in certain vital respects that death will result in the earliest stages of life; (2) by wayward genes producing serious diseases or defects which will bring on death well before average expectancy; or (3) by a combination of genes setting a limit beyond which the various parts of the body will no longer work together efficiently, and a general breakdown will result. Supplementing this is the theory of physiologist A. J. Carlson that ". . . the hereditary

---

major countries which reported infant death rates lower than the American figure of 25 per 1,000 (or 1 in 40, as stated). The lowest rates were officially given in the Netherlands and Sweden as 15.3 per 1,000; but these rates, and those of several other countries, must be discounted by a few points in relation to the American figures because they do not include deaths of infants in the first day or days, as the American statistics do.

* In 1900, among American Whites, of every 100 males born, only 31 could have been expected to reach age seventy or over, compared with 53 in 1960; for females at birth, 35 out of 100 could expect to reach age seventy in 1900, compared with 72 out of 100 in 1960. Among Negroes, in 1900, the chances at birth of reaching age seventy were, for males, about 14 in 100, for females, 16; in 1960, the chances had risen to 39 in 100 for males, 50 in 100 for females.

TABLE IV

How Females Outlast Males

*Of every 100,000 males and females born in the United States, the following numbers survive at the successive ages given.**

At Age	Males	Females	Per cent More of Females
Birth	100,000	100,000	—
10	96,503	97,276	0.8% more
20	95,658	96,873	1.3% "
30	94,009	96,102	2.1% "
40	91,714	94,634	3.2% "
50	86,370	91,407	5.8% "
60	73,970	84,594	14.4% "
65	64,292	78,677	22.4% "
70	51,878	70,186	35.3% "
75	38,602	58,674	52.0% "
80	25,148	43,622	73.5% "
85	13,047	26,136	100.3% "

* Due to the higher ratio of boys at birth (see Chap. 7), the sexes start off unequally with about 105,000 males to every 100,000 females. The increasingly high mortality rate of males at each age wipes out their excess by the mid-thirties, and thereafter the proportion and the excess of surviving females grow steadily greater. (Data from *Vital Statistics of the U.S.*, 1962, Vol. II, Sec. 5, Table 5-3.)

time clock or power of living varies considerably in different organs of the individual, and since all organs are more or less necessary for living, the weakest organ becomes the weakest link and thus determines the life span of the individual."

Of the various "killer" genes, the most immediate and drastic in their effects are those known as the *lethals*. Usually they are recessives which, if present in a single state (that is, if an individual receives only one such gene) would have little effect, or, less often, serious but not fatal effects. They become lethal only if two come together, one gene contributed by each parent. A great many genes which work in this way have been identified in experimental animals and in domestic animals. Examples in human beings are believed to be double genes for *brachyphalangy* (stub fingers) and *telangiectasis* (a blood-vessel disease), there being no case known of an individual who had carried a pair of such genes and lived. Also lethal before birth is *anencephaly* (absence of brain), a not uncommon congenital malformation which in many instances may be caused by a pair of recessive genes.

Some other human genes established as causing death before birth or during childhood if inherited in pairs are those for "elephant skin" (*ichthyosis vulgaris*); Tay-Sachs disease, the infantile form of amaurotic idiocy; malignant freckles; one form of glycogen storage disease; progressive spinal muscular atrophy of infants; some forms of muscular dystrophy; and cystic fibrosis. Also generally lethal during childhood if two genes are present are *thalassemia* (Mediterranean anemia) and sickle-cell disease (sicklemia).*

But as Dr. Hermann J. Muller has pointed out, probably the great majority of persons possess at least one, and possibly three or four recessive genes, any of which, if inherited in a double dose (one from each parent), would cause early death. "That such genes are still not more abundant," said Dr. Muller, "is only due to the fact that, in the past, individuals getting them from both parents did die."

One might add that the term "conditioned" should precede the word "lethal" in many cases, for what may be lethal in one environment may not be in another, or may have different effects in different species. Among lower animals in a wild state, inherited defects or abnormalities in such physical aspects as skin covering (feathers or fur), teeth, skeletal structure, or digestion could greatly interfere with the capacity for survival and might bring death at an early age. Changing the environment or moving the individual from one place to another could make the same genes and their effect no longer lethal. Similarly, among human beings corresponding genes for such traits as defective teeth, skin, skeletal structure, or digestive peculiarities might also have been lethal under rigorous primitive conditions or might become so, but they would have only minor effects in modern civilized environments where protective and corrective measures are available. Further, some of the serious conditions previously mentioned, which are still classed as lethal, may in time, with medical advances, cease to be so.

In another category are the "delayed killers," which, while usually

* In technical terms, the "lethal" genes are those which kill an individual before birth or in infancy. Genes which bring death later, but at least before the reproductive period as a rule, are called "sublethals" (or "semilethals"). In addition to the lethal or sublethal conditions due to recessive genes, there may be some caused by dominant or semidominant genes. These "killer" genes can be inherited in two ways: (1) If the individual with the gene—who has the resulting condition in a milder than average form—manages to survive long enough to reproduce and pass the gene along (which sometimes happens with *epiloia*, for example); or (2) if the dominant lethal gene arises through mutation.

not fatal during childhood, or not necessarily fatal until well into maturity, nonetheless tend to hasten death by many years. Among them, under present circumstances, are a variety of inherited diseases and defects such as hemophilia, several metabolic conditions, Huntington's chorea, certain nervous disorders, and some rare tumors. To a more limited extent but for a much greater number of persons, the genes for diabetes and some of the heart and arterial conditions are also, obviously, life-shorteners. Indeed, any hereditary differences among families with respect to the major diseases discussed in our wayward gene chapters would have a direct bearing on the relative life expectancies of their respective members.

Finally, apart from the direct causes of death through specific diseases, there are the so-called natural factors which apparently set general time limits to how long anyone can live. Various theories have been advanced as to what these may be: wear and tear on vital organs, aging of the tissues, slowing up of the glands, lessened elasticity in the arteries, deterioration of the blood or brain cells, vitamin or oxygen starvation, enzyme deficiencies, waste-product accumulation, and chromosome mutations in the body cells or damage to the DNA. Wherever the responsibility may lie for the eventual body breakdown, it seems apparent that hereditary influences are involved. In fact, once the early environmental hazards are bypassed and persons reach the sixties and seventies where little has been done to lengthen life, individual differences in longevity are increasingly likely to be affected by heredity, although environment will still be important.

The first of the genetic factors that govern your own longevity are those which made you a male or a female. Throughout all your years your sex has been and will be the most important single influence on your life span, with the advantage greatly in your favor if you are a female. *Environment* has much to do with this—particularly under modern conditions; but, as we saw in Chapter 22, "The Poor Males," one ever-present advantage of females in longevity derives from the inherited sex differences which may make the male an easier prey for most of the major diseases or in other ways reduce his relative chances for survival.

At the very start of life, we may recall, many more males than females are killed off prenatally or at birth, and in the first year male infant deaths now exceed those of females by about 33 per cent. Thereafter, except for a period in childhood when the excess of male over female mortality dips to about 15 per cent, the male death rate jumps further and further beyond the female rate, and throughout most of their mature years men die off at a rate almost double or more that of women. For example, at age fifty the death rate per 1,000 White American men is now 9.4 per 1,000, but 4.6 among women; at age sixty the male death rate is 22.4 per 1,000, while for women it is 10.7; at age sixty-five it is 34.8 for males, 18.3 for females. Thus, where at birth the sexes start off with a ratio of close to 106 males to 100 females, the excess of males has shrunk away by early maturity, and the proportion of females steadily rises until, in the seventies, there are over 30 per cent more women than men surviving, and in the eighties about 60 per cent more. In the oldest ages, from about eighty-five on, there are twice as many women as men.*

In the accompanying Table you can see for yourself how the fact of your sex bears upon your life expectancy at every age (although keep in mind that we are always speaking of averages!). So one may ask, "Doesn't this prove the *natural* superiority of women"—at least in terms of inherent capacity for survival? Before jumping to conclusions we must give careful attention to environmental influences. An initial fact is that males are exposed to many special environmental hazards—in occupations, in war casualties, in a far higher accident rate, in harder and more careless living. Some of this may be an outgrowth of natural sex differences, some of it not.† On the other hand, women are by nature exposed to a special and exclusive hazard of their own—childbirth and its effects which, in a completely *natural* state, with unrestricted childbearing and no medical attention or precautions, would inevitably cause heavy casualties among them. The great reduction in the dangers of and from childbearing in the modern world, with a consequent enormous reduction in maternal mortality (to less than 1 per 2,000 births) and in the serious aftereffects of childbirth, must be considered an *environmental benefit* for women which has tended to

* In checking United States census figures for the relative numbers of men and women in the older-age brackets, one must allow for the fact that in former years a considerable surplus of men was continually brought in by immigration. This earlier male surplus as it has been carried along has reduced what would have been the excess of women in the existing age groups had the survivors been confined to the numbers of the two sexes born in the country. Conversely, in various European countries (Ireland, England, France, Germany, Russia) the excess of women among the aged has been increased by losses of men through emigration and wars.

† Among lower animals females also have been found to outlive males as a rule, when both are kept and studied under domestication, or in zoos and in laboratories. But it is still uncertain what the situation is among animals living in a natural state, where relative life expectancies would be affected by such factors as the greater average strength, fleetness and foraging ability of males and the risks that females run in maternity.

TABLE V

How Many More Years For You?

Here are the averages *of life expectancy at each age—which in your case may be more or less than the figure given, depending on your health, circumstances and other factors. (See Text.)**

Your Age Now	MALE More Years	FEMALE More Years	Your Age Now	MALE More Years	FEMALE More Years
15	54.9	61.3	50	23.2	28.3
16	54.0	60.3	51	22.4	27.4
17	53.0	59.4	52	21.7	26.5
18	52.1	58.4	53	20.9	25.7
19	51.2	57.4	54	20.2	24.8
20	50.2	56.4	55	19.4	24.0
21	49.3	55.5	56	18.7	23.2
22	48.4	54.5	57	18.0	22.3
23	47.5	53.5	58	17.3	21.5
24	46.6	52.6	59	16.7	20.7
25	45.6	51.6	60	16.0	19.9
26	44.7	50.6	61	15.4	19.1
27	43.8	49.7	62	14.7	18.3
28	42.8	48.7	63	14.1	17.5
29	41.9	47.8	64	13.5	16.8
30	41.0	46.8	65	12.9	16.0
31	40.0	45.8	66	12.4	15.3
32	39.1	44.9	67	11.8	14.6
33	38.2	43.9	68	11.3	13.9
34	37.2	43.0	69	10.8	13.2
35	36.3	42.0	70	10.3	12.5
36	35.4	41.1	71	9.8	11.9
37	34.5	40.1	72	9.3	11.3
38	33.5	39.2	73	8.9	10.6
39	32.6	38.2	74	8.4	10.0
40	31.7	37.3	75	8.0	9.4
41	30.8	36.4	76	7.5	8.9
42	29.9	35.5	77	7.1	8.3
43	29.1	34.5	78	6.7	7.8
44	28.2	33.6	79	6.3	7.3
45	27.3	32.7	80	5.9	6.8
46	26.5	31.8	81	5.5	6.3
47	25.7	30.9	82	5.2	5.9
48	24.8	30.0	83	4.9	5.5
49	24.0	29.1	84	4.5	5.1

SOURCE: *Vital Statistics of the U.S.*, 1962, Vol. II, Sec. 5, Life Tables, 5-4.

* Figures given are for *Whites* in the United States, based on life expectancies in 1962. For *Negroes* the expectancies are less (until about age 65) and can be calculated roughly in this way: At ages 15 to 29, for Negro males subtract 5 to 4 years from figures for Whites; for Negro females, subtract 6 years. At ages 30 to 39, for Negro males, subtract 4 to 3 years; for Negro females, 5½ to 5 years. At ages 40 to 49, Negro males have 3 to 2 years less, females, 5 to 4 years less; at ages 50 to 59, Negro males, 2 to 1 years less, females, 4 to 3 years less; at ages 60 to 65, Negro males, 1 year less, females, 2 to 1 years less. After 65 life expectancies for Negroes are about the same as for Whites, and, with advancing age, possibly a little greater. However, for younger Negroes today, the likelihood is that with increasing equalization of racial conditions Negro life expectancies will come steadily closer to those of Whites at all ages.

offset a *natural disadvantage*. Further, there is much to suggest that under modern conditions life has been made relatively easier for women as a group than for men, as compared with human existence in the past, where hardships and dangers were shared almost equally by the two sexes. In all, any inherent advantage possessed by human females in combating disease and stress would be much more limited in its effects under adverse natural conditions, and has been able to assert itself strongly only as environments have been made increasingly favorable.*

In proof of the foregoing, we can compare the sex differences in life expectancies in former and present periods, and in advanced and backward countries today. First, we find that women's markedly greater life expectancy is apparently a phenomenon of very recent times, for neither ancient records nor those up to almost a century ago offer any evidence that women outlived men. Second, in our own time, the female margin of greater life expectancy has widened increasingly in the United States and other advanced countries, from an excess of two or three years over male expectancy in 1900 to an excess of more than six years in 1960. But third, and most significant, in the more backward countries, the worse the living and health conditions (as reflected in higher death rates and lower average life expectancies for both sexes), the smaller is the excess of female over male life expectancy—to the point where it disappears altogether, or where male life expectancy is even higher. For example, according to the United Nations figures available in 1963, the life expectancies in Bolivia were exactly the same for both sexes—49.7 years; and in Guatemala, Ceylon, Cambodia and Upper Volta they were actually higher for males by a few months to a year. (In India, where previously the male life expectancy also had

* An unusual study of the life spans of Catholic nuns and Brothers, made by Father Francis C. Madigan and Dr. Rupert B. Vance in 1957, may throw light on what the comparative life expectancies of men and women might be if they lived under closely similar—but not *average* or natural—conditions. In this case the 30,000 males and females of the religious orders studied had almost the same diets, occupations—most being teachers—and habits (none smoking or drinking). The findings were that the nuns outlived the Brothers by a wide margin. At age forty-five, indicated by mortality rates in the two groups, the Sisters had an average remaining life expectancy of almost six years more than the Brothers (thirty-four years of additional life as compared to the men's twenty-eight). But the very fact that the nuns and Brothers were so closely matched in their environments, work and habits, which also meant not having the lives or experiences common to their respective sexes, would make doubtful the direct application of the findings to average women and men.

been higher than the female's, the figures have shifted to an expectancy of 46.6 for females, 45.2 for males.)

Further proof that the extra margin of female longevity is conditioned by how good the environments are is offered by the infant mortality rates. As data from many countries show, there is an interesting see-saw effect: Where the infant mortality rate is highest (reflecting the worst environmental conditions), the excess of male over female infant deaths is lowest—as in the mortality excess of only 4 per cent for male infants in Chile (1919), or 7 per cent in Peru (1957). Where the infant mortality rate is lowest, as in the United States, Norway and the Netherlands, the excess of male over female infant deaths is 30 per cent or more. Especially striking is how this see-saw has worked in the United States, with the excess of male-infant mortality (among Whites) having risen from 23.4 per cent in 1900, to 30 per cent in 1940, to 33.3 per cent in 1960, as infant survival conditions for both sexes have steadily improved. (The corresponding figures for Negroes are given in Table VI.)

It is clear, then, that the existing markedly higher life expectancy of females is not a fixed advantage dictated by Nature, but rather derives from a potential capacity which in modern environments is being given more and more chance to assert itself. Nor are the existing longer female life spans merely the result of women's reduced maternity hazards and easier lives, on the one hand, and the more hazardous, stressful or careless living of males on the other. For long before these differentiating influences are at work—in infancy, as we have seen—the greater capacity of the female to profit by general environmental improvements asserts itself.

While the differences in longevity between the two sexes can be seen as due both to hereditary and environmental factors, no hasty inferences should be drawn from this about the reasons for differences in longevity between other human groups—for instance, *racial* and *ethnic* groups. As between Negroes and Whites, it was easy, a few generations ago, to assume (as many did) that the Negroes' much higher death rate and shorter life expectancy were due largely to inferior heredity. The argument that Negroes had weaker respiratory systems or constitutions was offered to explain their higher death rate from tuberculosis and other infectious diseases. Now we can find a likelier explanation in the much inferior environments under which Negroes have lived; for with every step in their advancement, the margin of difference in longevity between them and Whites has been reduced.

In 1900 American Negroes at birth had an average life expectancy of

TABLE VI

The Infant Mortality "Sex Seesaw"

As death rates in infancy have gone down, the proportion of boy- to girl-baby deaths has gone up.

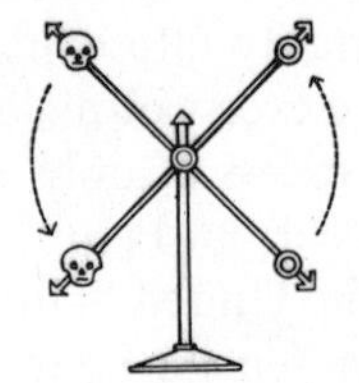

Here is how it has worked in the United States (among White babies):*

YEAR	DEATH RATE IN INFANCY PER 1,000 Boys	Girls	EXCESS OF BOY-INFANT DEATHS
1900	175.8	142.6	23.4% *more boys*
1920	103.7	80.3	29.1% " "
1940	56.7	43.6	30.0% " "
1960	26.9	20.1	33.3% " "

* Among Negro babies the death rates per 1,000 infants were, in 1900: males, 369, females, 299, an excess of 23.3% males; in 1960: males, 51.9, females, 40.7, an excess of 27.5% males. (Data from *Vital Statistics of the U.S.*, 1900-1940, and *Mortality Analysis*, 1960, Vol. II, Sec. 1.)

only 32.5 years for males and 35 years for females—*16 years less, for each sex, than the average for Whites* (which was then 48.2 years for males and 51 years for females). But in 1961 the Negro life expectancy at birth had risen to 61.9 years for males and 67 for females—only 5.9 and 7.5 years less, respectively, than the current rates for White males and females. One should also note that American Negro life expectancy is now not only much higher than the White life expectancy formerly was in the United States and Europe, but is actually higher than present White life expectancies in many countries. As American Negroes make continued economic and social advances, there is little reason to doubt that such differences as remain in life expectancy between them and their White neighbors will be much further reduced, if not entirely eliminated.

Among Whites themselves we also must be cautious about attributing to heredity the existing differences between persons of different ethnic stocks or nationalities. One used to speak of "hardy, long-lived old Americans" as compared with the "weaker immigrants." But as conditions improved for the newer arrivals and their descendants, the differences in longevity have been almost wiped out. Within the ranks of the foreign born, however, significant differences in death rates and longevity may persist. For instance, a study by Dr. Massimo Calabresi of the foreign born in New York State showed that those of Irish origin had very much the highest death rates, and those born in Italy the lowest; further, that the mortality differences between the two groups for specific diseases were in many respects duplicated among the residents in their homelands. While noting that subtle differences in diets, habits, occupations and activities undoubtedly affect the relative life spans of different ethnic groups, Dr. Calabresi also suggested ". . . there is no reason to doubt that there is a statistical difference in the frequency distribution of genes among different groups of the human species," which might apply to hereditary diseases and disease resistance generally and contribute toward the higher mortality rate among those of some nationalities as compared with others. (Racial and ethnic differences with respect to disease will be discussed further in Chapter 42.)

Nevertheless, there is much to show that throughout the United States, as in other countries, any marked differences in average life expectancies between or among persons of various groups, however classified, are governed far less by their ancestral origins and/or genetic makeup than by the relative conditions under which they live.

But what of the longevity differences among families and individuals in the same group, and in similar environments? How far can these differences be due to heredity? We have already discussed such specific hereditary influences on life expectancy as "lethal" genes and "killer" genes, which can bring death before maturity, and the genes involved in various major diseases which can shorten lives to one extent or another during later ages. These genes could quite easily be distributed among different families in varying proportions. But there may also be general genetic differences in the constitutional makeup and body functioning of individuals, resulting from the interaction of a great many (or even all) of their genes, which could shorten or lengthen their potential life expectancies. Some evidence of how genetic factors as a whole may influence heredity comes from our favorite subject of study—*twins*.

Over a period of fifteen years Dr. Franz J. Kallmann and his associates followed the life courses of 1,019 pairs of elderly identical and fraternal twins, all sixty years or over when the study began. At the end (in 1959), when in about half the cases both twins of a pair had died, it was found that the average interval between the deaths of identical twins was considerably shorter than the interval between the

INTERVALS BETWEEN TWINS' DEATHS
(At ages 60-69)

If one twin dies in his sixties, the other twin's death comes on an average *this much later:*

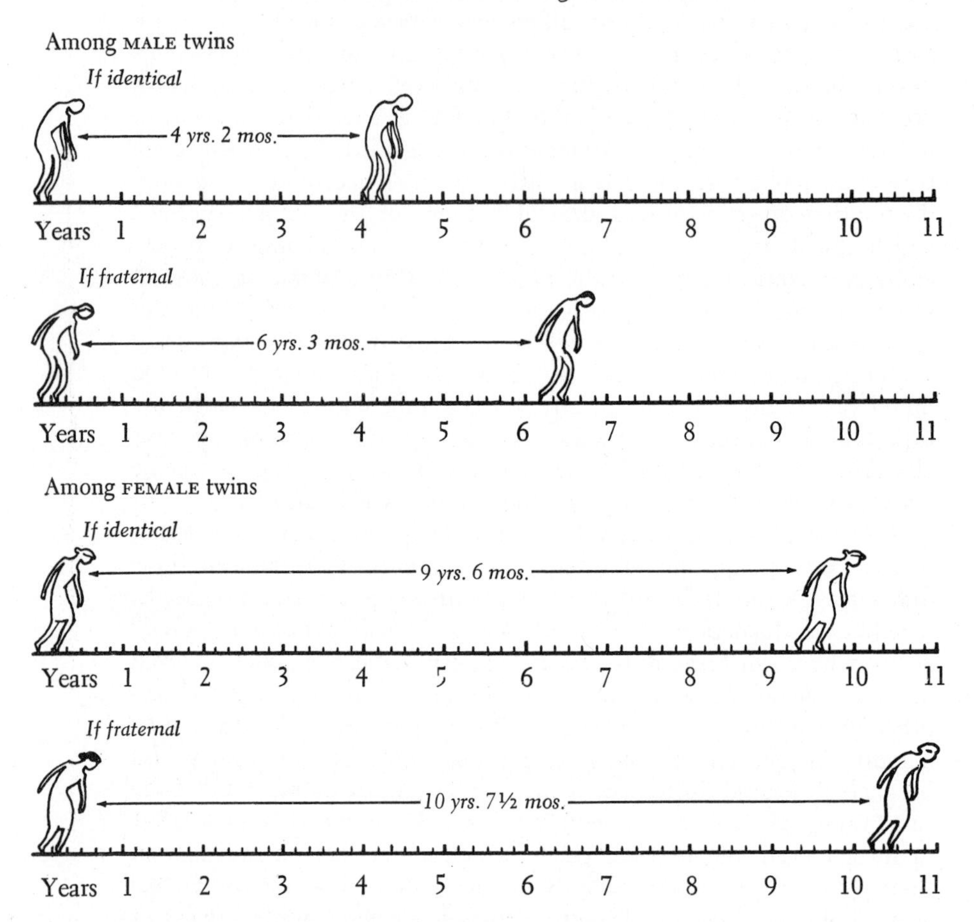

Data from study by Drs. Lissy F. Jarvik, Arthur Falek, Franz J. Kallmann and Irving Lorge. *Am. J. Hum. Genet.*, 12:2, June, 1960.

deaths of fraternal twins. The difference was most marked among male twins dying in their sixties, with an interval between deaths averaging about four years for the identicals versus six years for the fraternals.

As expected, the difference in intervals between deaths became smaller as twins grew older and their remaining years dwindled. On the other hand—again as expected—in the case of female twins, with much longer average life spans than the male twins (as with women in general), the intervals between deaths were greater, while the differences between identical and fraternal pairs in this respect were smaller. (The detailed figures are given in our accompanying Chart.) Dr. Kallmann also found that among the surviving elderly identical twins the symptoms of senility (enfeeblement and other defects of old age), or contrary healthy symptoms, were remarkably alike for both members of a pair, often even when they had been living apart, whereas little similarity in this respect was found in the fraternal pairs.

However, while these facts about twins may strongly suggest the importance of heredity in aging and longevity, two points should be noted: First, the greater closeness of identical twins in these respects may be ascribed in some degree to their having been more alike, as a rule, in their conditioning and habits (which also may or may not have been influenced by heredity), and having kept together more than fraternals. Second, and further qualifying the role of heredity, is the fact that, despite their having exactly the same genes, it is the rare exception for two members of even an elderly identical-twin pair to die at almost the same time; much more often, their deaths come years apart. (Nor should we forget that at the very beginning of life—before birth, at birth or in infancy—in many cases one identical twin may survive while the other does not.) Nevertheless, we know that the average differences in longevity not only between twins of either type but between two singleton brothers or sisters are smaller than between two unrelated persons. From this we can infer that heredity does lay out a pattern for longevity, although how far this pattern is carried out depends greatly on environment.*

Combining the influences of heredity and environment, collective studies by medical scientists and insurance company statisticians have shown that there is a significant correlation between the average life spans of parents and children. The findings have been such as these:

> Persons with the longest-lived parents have unmistakably lower mortality at every age and have a distinctly greater chance of reach-

* The case for familial and genetic influences on heredity gets further support from studies of lower animals. In breeds of mice, for instance, there are longer-lived and shorter-lived strains; and by crossbreeding, still shorter- and longer-lived strains can be easily selected out in the laboratory.

ing or passing the four-score mark than those with the shorter-lived parents. Among those who live to be ninety or over, about seven out of eight have had one or more long-lived parents, and a large proportion, two or more long-lived grandparents.

Longevity chances are greatest, and well above the average future expectancy for a given age (see Table V), if both parents were long-lived (eighty-five or over). But the chances are lowest, and below average expectancy, if both parents were short-lived (fifty-five to sixty-five, or less). Where parents' ages at death were in between the afore-mentioned extremes, or where one parent was long-lived and the other short-lived, a person's longevity chances are about the average for his age.

In all cases of short-lived parents, allowance must be made for deaths due to purely accidental circumstances unrelated to possible genetic influences or general health. In these cases, the longevity records of grandparents and of aunts and uncles might be consulted.

It is important that no conclusions should be drawn from family histories about anyone's longevity chances—yours or another person's—without first considering how such histories may have been influenced by environmental conditions. One hardly needs statistics to show that even with no differences in heredity, families which had lived for successive generations under depressed conditions would not have had the same average life spans as those enjoying more favored environments. We have seen proof of this in the comparative life expectancies of human beings of the past and those of the present, of American Negroes and Whites, and of inhabitants of backward countries in contrast with those of advanced countries. So, too, we can take families, or groups within our own or any other present-day population and apply another simple "see-saw" formula: *The higher the income and social level, the lower the mortality rate and the longer the average life span; the lower the income and social level, the higher the mortality rate, and the shorter the average life span.* Where this formula is most clearly applicable is in the infant mortality rates, which in almost any community are much higher among the socially and economically "have-not" families than among the favored families. In fact, the relative proportion of deaths at early ages largely accounts for the existing differences in average life spans between various human groups and families, just as it does for the longevity differences between former and present generations.

At any given time, however, the conditions under which individuals live may not be entirely unrelated to their hereditary makeup. Everything else being equal, people who are genetically handicapped, either physically or mentally, may be more than ordinarily likely to have inferior jobs and unstable marriages, and thus create inferior and life-shortening environments for themselves and their children. It is also possible that abnormal behavior tendencies, where and if influenced by heredity (as will be explained in later chapters), may have something to do with directing some individuals into the more hazardous activities and reckless patterns of living which lead to shortened life spans. On the other hand, persons who are genetically favored in capacity and who also govern their lives more intelligently may be given an extra push in the direction of longer lives and may in turn confer greater chances of longevity on their children.

How *occupations* may influence longevity is illustrated in the accompanying chart, "Jobs and Death Rates" (among White males, ages twenty to sixty-four). Scientists, teachers and social workers are in the most favored categories, with much below-average mortality rates (and hence, above-average longevity prospects). At the opposite end, among the less-favored occupations with the highest mortality rates, one finds miners (in the very worst position), musicians, tailors, taxi drivers, policemen and firemen. (The relative positions of other groups also are shown.) In most cases the reasons for the death-rate differences among given occupational groups may seem obvious, but certain qualifications and explanations are needed. For example, farmers may be in the more favored groups either because their work is healthier, or because men have to be healthier to go on working as farmers. In the marked contrast between artists (with moderate mortality rates) and musicians (with among the highest death rates), an important reason may be that the former tend to lead sedentary lives while the latter—particularly jazz musicians—usually lead highly unstable, nomadic and erratic lives. In fact, the differences in death rates among the occupational groups as a whole may be due only in part to the effects of the jobs themselves, but may be related even more to the personalities, habits, backgrounds, and living conditions of the men engaged in the respective occupations.* Further, such death-rate differences as there are among

* That a man's life expectancy may be affected more by the living conditions that go with his occupation than by the occupation itself is suggested by recent studies in England, where it was found that the wives of men in various job classifications had mortality rates proportionate to their husbands' rates. That is, wives of laborers had the highest death rates, wives of professional men the lowest, with other married women graded in between according to their husbands' occupations. Some occupations (such as mining) obviously have special hazards which may heighten the effects of the social and economic environment on the husbands, but not on the wives.

JOBS AND DEATH RATES

*How occupational groups of American men (ages 20 to 64, White) rank according to relative mortality rates, as shown by approximate positions of heads on the chart, grading down from most favored groups (top) to less favored (bottom).**

* While in general death-rate differences among occupations are correlated with life expectancies (lower death rates, better longevity prospects), these differences reflect not merely the effects of the jobs, but also of the varying backgrounds, living conditions, personalities and habits of those engaged in them. (All data from "Mortality by occupation and causes of death among men 20 to 64," Lillian Guralnick, *U.S. Vital Statistics*, Special Reports, Vol. 53, No. 2, Sept., 1963.)

the groups up to age sixty-four undoubtedly diminish in the later ages, when the major causes of death—heart conditions, cancer and other degenerative diseases—may become increasingly similar in all groups. However, even in the preceding ages, one can expect a continued reduction in the death-rate differences and life expectancies among occupational groups—as between professional persons and craftsmen or laborers, for instance—as new safety precautions, better and healthier working conditions, improved living standards and medical advances cause a further leveling off of the discrepancies between one occupational group and another.

Many other aspects of a person's makeup—whether genetic or environmental—and of his manner of living can have an influence on his longevity. Let us take up some specific questions.

Does your body build have any relation to your potential longevity? Yes, but mainly to the extent that it reflects good health or lack of it, or the efficiency or inefficiency of your body mechanism. Persons who are 25 per cent or more *overweight* have a death rate much higher than those of average weight, largely because marked obesity usually worsens or accompanies some disease which of itself would shorten life (as in heart and arterial diseases, diabetes, liver, gall-bladder and digestive ailments). Young adults who are much *underweight* also usually have a higher than average mortality, either because extreme underweight may betoken some disease or nutritional defect, or because it may make the person more vulnerable to infections and other life-shortening conditions. But at older ages to be somewhat underweight may be conducive to longer life.

As for stature, it, too, is usually correlated with longevity in relation to the accompanying health and physical condition of the individual. An extensive study by the Society of Actuaries (Chicago) has shown that short or very short men generally tend to have above-average mortality rates, but to what extent may depend mainly on how much overweight they also are for their size. Tall men or very tall men tend to have somewhat below-average mortality rates, but this may be counteracted if they are considerably overweight for their size, or if extreme tallness is traceable to some glandular disorder. However, in linking longevity with stature, it should be stressed that degrees of shortness and tallness—particularly as between groups—often reflect environmental influences, with shorter persons tending to come from poor early environments and taller persons from better environments. Thus, not the stature itself but all the influences contributing to its acceleration or retardation may have a bearing on life expectancy.

While we have also seen that there are hereditary differences in stature, it remains to be established whether, within normal ranges (that is, excluding dwarfism or gigantism due to genetic influences), "stature" genes and "longevity" genes in human beings are related in any way.

Does "faster living," strain or overwork shorten life? Some authorities hold that the rate of living does have a direct bearing on aging and point to the fact that in lower animals, those with the fastest pulse rates and metabolic activity (such as mice) have far shorter life spans than the slower and easier living ones (such as elephants). There is also evidence that undue stress may be an aggravating factor in heart conditions, ulcers and other life-shorteners. In any case, much may depend on individual differences, for when one considers the most arduous of occupations—that of President of the United States—the facts are far from clear. A number of presidents (not counting the four assassinated) did die much earlier than their life expectancy at the time of taking office—Harding, Coolidge and Franklin D. Roosevelt among the recent ones. But two other recent presidents, Hoover and Truman, can be listed among the very long-lived, and even Eisenhower (despite a heart condition and a major intestinal operation), has exceeded the average life expectancy. Nor should we forget that John Adams died at ninety, and Jefferson, Madison and John Quincy Adams in their eighties at a time when average life expectancies were very low. Perhaps, as with England's venerable Winston Churchill, some individuals thrive on stress—with heredity, quite possibly, aiding them to do so.

Is your blood pressure or pulse rate any clue to your length-of-life prospects? Yes. Mortality rises steadily and markedly with increasing high blood pressure, primarily because of heart-arterial-kidney diseases. But the effects of high blood pressure are much less severe in women than in men. *Low* blood pressure, however, unless it is extremely low, on the average may be actually favorable to longevity. In *pulse rates*, abnormalities are fairly common and in the majority of cases are not very prejudicial to life expectancies. But an extremely rapid rate—exceeding 100 per minute—may reflect some underlying disease, such as heart disease or hyperthyroidism—and may forecast some degree of life curtailment.

Does marriage increase life expectancy? Married men, it appears, do have average longer lives than single men (although some wag has said, "It only seems longer!"). In the United States the mortality rate is less for the married than for the single at all age periods, with the deaths of married men at ages twenty-five to forty-four averaging only half that of bachelors. But while the married men (despite their griping) generally receive better care and have a more favorable environment than the single ones, the statistics are loaded by the fact that the men who don't marry include proportionately many more who in the first place are sickly or defective, or mentally unstable or abnormal, or otherwise potentially shorter-lived. Among women this applies to a smaller proportion who remain single, for married women have an advantage in longevity over the unmarried only up to the age of forty, the mortality rates thereafter being practically identical for the two groups. In both sexes, divorced persons have higher mortality rates than those whose marriages continue, which may be attributed to worsening environmental influences following marriage breakups, plus the possibility that divorced persons have or develop less stable and healthful habits than the married ones.

Does drinking alcoholic beverages shorten life? Current medical thinking is that drinking in moderation does not affect longevity (except where there are certain diseases) and that even for persons with heart conditions an occasional drink may be beneficial in that it may help to relieve tension. *Heavy drinking* is quite another matter, being clearly responsible for many serious accidents, diseases, physical breakdowns and other life-shorteners. (For the possible role of heredity in alcoholism, see Chapter 38.)

How about smoking? While there is controversy on this point, current medical opinion leans strongly to the belief that *cigarette smoking* is one major cause of lung cancer and a definite aggravating factor in heart disease and chronic bronchitis. In general, heavy cigarette smokers are found to have a much higher mortality rate than nonsmokers. Neither pipe nor cigar smoking is implicated to any such extent, although pipe smoking is sometimes linked with lip and mouth cancer. But there undoubtedly are individual exceptions, many oldsters who are inveterate pipe smokers claiming that the habit has been relaxing and helpful. (The possibility that even the development of the smoking habit, and its degree, may be influenced by heredity, was suggested by British geneticist R. A. Fisher, after finding that identical twins—including some who had been separated a long time—were much more alike than fraternal twins with respect to smoking addiction or its absence.)

Do athletes die younger than nonathletes? A widely held belief is that the strain of athletics may cause enlarged and weakened hearts, but medical investigation has shown that "athlete's heart" is largely a myth and that even strenuous activity will not damage healthy normal hearts. There may be some heart enlargement in athletes, but there is no evidence that it causes disease or shortens life expectancy. A British study of Cambridge University graduates over a period of forty years, dating from 1860, revealed that those who'd gone in heavily for sports had lived just about as long as the nonathletic intellectuals. Other studies have indicated that men who do not exercise are more likely to die of heart disease than those who have engaged regularly in moderate physical activity.

Does where you live affect your longevity? The statistics suggest that in some regions of the United States, or in given areas of other countries, average life spans are greater than elsewhere. But the differences—for instance, as between one state or section and another—tend to be minimal and reflect primarily differences in living standards and in the makeup of the population with respect to older and younger age groups. There is little evidence that climate of itself has much affect on human longevity, except in individual cases where persons have diseases which are helped or worsened by specific climate conditions.

Can certain foods make you live longer? Although what is eaten provides the basic elements for good health and bodily functioning, there is no scientific proof that any specific foods will add to one's life span (despite the beliefs of enthusiasts for yogurt, black-strap molasses, pumpernickel bread, garlic, organically grown vegetables, and so on). In fact, among extreme oldsters one can find every conceivable variety of diet, governed not by selection but by what was available (and often including foods that would make a dietician shudder). Nor have vegetarians been shown to have longer lives than meat eaters, or vice versa. Before trying to link differences in longevity among various peoples and countries with what they eat, one must make sure that their diets are not concomitants of other factors—such as abundance or shortage of given staples, living conditions, and habits—which may be the real causes of the longevity differences.

Will your personality affect your life expectancy? Undoubtedly every aspect of your personality—your intelligence, temperament, disposition, social behavior, interests, sex attitudes—has some influence on your health and in turn on your life expectancy. Psychosomatic studies have shown that personality factors may be important to one extent or another in a great many, if not most diseases, including heart conditions, ulcers, diabetes and asthma, and are obviously involved in "accident-proneness." There seems little question that a well-balanced personality and good emotional adjustment may be an aid to longer living.

One could go on to discuss various other influences bearing upon human longevity, but enough has been said to suggest that there is probably no element in a person's life which is not to some extent involved. How important any single factor may be depends on individual circumstances. For remember that we have always been talking in terms of averages. You may know, alas, of relatives and close friends to whom the averages didn't apply. And you know of other persons who have gone on living in defiance of all the facts about life-shorteners.

In sum, how long you, personally, may expect to live depends on these principal influences: First, environment—the way in which you were started off in life and the conditions under which you lived thereafter and now live. Second, your inherited vigor or weakness (as applied both to specific diseases and defects and to general resistance factors), with *particular attention to your sex*. And third, *luck*.

Environment is placed first because, while many hereditary conditions may transcend it in cutting lives short at early ages, they are the great exceptions. Otherwise, the major differences in length of life between and among human beings—those of the past and present, and those of backward groups contrasted with advanced groups today—can be attributed far more to environment than to heredity.

Nevertheless, the more the big environmental disparities are reduced by giving ever larger numbers of persons the benefits of improved living conditions and medical care, the more the *hereditary* influences on longevity can assert themselves. (We already have seen how this has worked in women as compared with men.) For human beings in general, length of life may thus be governed in the future more and more by the potential limits set for us by our genes, both as individuals and as members of a species, in the same way that hereditary limits are set for the life spans of other animals.

Does this imply that nothing can be done to extend the existing limits of human longevity?

No, for in the opinion of most authorities, the human life span, on the average or with respect to individual extremes, can and very likely will be increased. But not too much! Prospects are bright that means will be found to add several more years to the average life expectancies of older adults by retarding or curbing degenerative diseases of the arteries, heart, kidneys and other organs, as well as lessening the menace of cancer. More radical means of extending life spans quickly and greatly, by some magic elixir or by some remarkable operation, are not in sight. True, innumerable, "rejuvenation" and "life prolongation" methods have been heralded over the centuries, and in our own time great claims have been advanced for monkey-gland operations, cutting and tying sex-gland ducts, injecting mashed-up sex glands, administering male hormones, the "Bogomoletz serum" (concocted by a Soviet scientist—who died in his sixties), and so on. But each method in turn has been tried and found wanting, and the Ponce de Leons remain as far from their Fountain of Youth as ever.

There is one other interesting possibility. Scientists have been making headway in keeping human tissues, and even sperms, preserved by refrigeration in a state of "suspended animation," capable of being defrosted and revived long after. So, possibly, one of these days some of us could be put in a deep freeze, and defrosted a century hence, to begin living and moving in another world, as characters in drama and fiction have already done.

For the present, it appears that only so much and no more can be accomplished to prolong the time that the whole human mechanism can be kept going efficiently. What could be gained by greatly increasing the proportions of centenarians and by extending the lives of a few individuals into the so-far legendary ranges, is another matter. It is one thing to enable more and more human beings to live out full lives into old age; it is another thing to prolong human existence beyond the limits where persons can no longer be productive or happy, and can continue to exist only as vegetables.

Perhaps, then, Nature may have been wiser than we think in setting the potential longevity of humans at certain limits. Whatever the future may bring, perhaps our immediate goal in a troubled world should be not so much to add more years to our lives, but to add more and better living to the years we already have.

PART III

Psychological and Social Traits

CHAPTER THIRTY

You as a Social Animal

WITH THIS CHAPTER, we enter a new phase of our book.

So far we have been considering human beings chiefly as biological mechanisms. Now we begin to think of you, and all the rest of us, as *social animals*—in the way that each lives and works in relation to others of his species. That is where we differ most among ourselves and are most distinctive among all other animals.

In purely physical construction and functioning there is nothing so special about us. From mice to elephants, all mammals have skeletal frameworks, sensory and internal organs, glands, and so on, which are essentially much like ours (except, perhaps, for our brains). Most important, heredity works almost as directly and conclusively with us as with lower animals in constructing our bodies according to the blueprints for our species and determining much of the physical variation among us. Environment, whether natural or man-made, is greatly limited in what it can do to or with our physical makeup, its strongest influence being in a negative direction, through its powers to thwart our development with disease, defect or death. But in the positive direction, despite all the changes in human environment over thousands of years, men today are organically almost the same as they were (even in their mental equipment); and in all the different existing environments, from the heart of Africa to New York's Park Avenue, men continue to be biologically amazingly alike in most major respects.

But the story changes completely when we consider man as a social animal. While other animals, too, have their social relationships, as one individual to another or in group organizations of herds, flocks, or colonies, it need hardly be said that in no other species is there anything remotely like the range of social differences found among human individuals, in behavior, intelligence, learning and achievement. True, among lower animals, one individual may be smarter than another in getting food and mates, or in evading enemies, and one may boss another. But taking the most advanced of them below the level of man—our distant cousins, the apes—there is no ape who amasses a billion coconuts while another has nary a one; there is no ape with prodigious learning while another is illiterate; there is no lady ape who sports a fancy wardrobe while another dresses in rags; and there is no ape who can give his offspring vastly more in worldly goods, education and opportunities than can another ape. What any lower animal is or can do in relation to his fellows is influenced to some extent by his environment, but in the main, both as an individual and as a member of his species, his behavior and achievements are rigidly limited by his inheritance.

With human beings, not only is the range of behavior development permitted by their genes infinitely greater, but the opportunities offered by their environments for creating differences among individuals is also infinitely greater. Making this possible above all else are the distinctive capacities humans have: for the use of speech and symbols to communicate with one another and to learn from one another; to alter their environments through inventions; and to pass along from one to another all the accumulated fruits of their experience. With the limitless degrees of these advantages being offered in human environments, it becomes apparent that no human being can be judged *individually* as a social animal unless we know how he has profited by or been influenced by his contacts with others. So the difficult job we approach now, in tracking down to their sources the many variations of human behavior or achievement, is that of disentangling all the complex hereditary factors from all the complex environmental ones.

We have seen that in our surface details—coloring, features and body form—and in our internal construction and many aspects of our physical functioning, heredity is either the direct cause or a dominant influence in creating many individual differences among us. To what extent the surface differences may affect our social relationships depends largely on how they are regarded by others. More important with respect to behavior, obviously, are any of the serious hereditary defects or diseases, and particularly those affecting the mind. But in these cases behavior differences result only indirectly, through inherited defects in the body mechanisms. What concerns us most is whether, in normal individuals and under normal conditions, there may not also be hereditary factors that can *directly* produce many gradations of behavior and mental performance and help to account, in one way or another, for the enormous social differences existing among humans everywhere.

Why do some individuals in any group scale the heights of success, while others sink into the mire of failure? Why, in the very same family, is one person brilliant, the other a mediocrity; one a musician or artist, the other an insensitive dolt; one member kind and happily adjusted, the other hateful and unable to get along with others? What has made *you yourself* so different in your job, thinking, personality and behavior from so many others with whom you grew up? Is it all a matter of pure chance, or environment? Or are there indeed different gene combinations for social qualities, as there are for physical traits, which *predestine* one person to be this kind of a social animal, another person that kind?

It really boils down to something you may have thought of many times: *Suppose that with the very same genes, you had been reared in an entirely different home, city or country, with different associates, education, and opportunities. Socially, how much would you still be like the person you are now?*

In the pages ahead light may be thrown on this question and many related ones. But as you will soon see, answers regarding the inheritance of social traits in human beings can be given with no such surety as were those regarding the inheritance of surface characteristics and organic makeup. The reason is not only that the social traits are so much more complicated. It is also and perhaps chiefly that the *measurements* of these traits are as yet in a rudimentary stage, and until the measurements are more accurate geneticists must continue to be greatly restricted in their study of human behavior and performance.

Thus, it can be said at once that in many respects we are now venturing into uncertain territory. Previously we have confined ourselves to the presentation, almost exclusively, of scientific facts. From this point on, as we begin to analyze the role of heredity in such variable and intangible human characteristics as mentality, personality, temperament, criminality, sexual behavior, and race differences, we will find our facts becoming more and more diluted with theory. Our genetic evidence, like a stream of clear, fresh water flowing into an arm of the sea, now begins to intermingle with theories of psychology, anthropology, sociology and even politics. It will be increasingly hard to filter out the facts from the theories—theories which, even in the case of leading authorities, are often tempered by unconscious prejudices or emotional reactions.

The social phases of human heredity are viewed through varied lenses. Some are rose-tinted—those of the optimists who see all the favorable possibilities in human genes; and some lenses are dark—those of persons who see human makeup shadowed by innumerable hereditary defects and inadequacies. There are the lenses which offer a long view of heredity, in relation to many other influences, and with insights into its future potentialities. There are the short-view lenses, which reveal only what is now known or seen and make persons say, "It's human nature—you can't change it." And, particularly with respect to man's social and psychological behavior, there are the extreme hereditarian lenses, which show all the differences in human minds, actions and achievements as due to heredity, while in sharp contrast there are the behaviorist lenses, which portray virtually everything human beings do or can do as conditioned by their training. The latter position was most strongly enunciated by the late Dr. John B. Watson, who, in his book *Behaviorism* (from which this school of thinking took its name), made the now classic statement,

> Give me a dozen healthy infants, well formed, and my own specified world to bring them up in, and I'll guarantee to take anyone at random and train him to become any type of specialist I might select—doctor, lawyer, artist, merchant, chief, and yes, even beggar and thief—regardless of his talents, penchants, tendencies, abilities, vocations, and race of his ancestors.

Finding where truth lies with respect to these conflicting views on the genetics of human behavior has been further complicated by another visual difficulty—the personal bias which in this context we may call "astigmatism." For not only laymen's opinions, but the very directions which researchers into human genetics may take, and the findings arrived at, may differ in the degree that the investigators themselves differ in personality, background and emotional makeup. The descendants of *Mayflower* stock and the offspring of recent immigrants, the hidebound reactionaries and the unrestrained progressives, the Russian, English, German and American social scientists, as groups often react differently toward the same evidence. Even in American colleges—in one faculty as compared with another—there are different schools of thought with regard to many of the vital points we are going to discuss.

Thus, you also will find yourself taking sides in the ensuing chapters. If you are socially and financially secure and come from a notable family, your reactions and conclusions will be different from those of the man at the bottom or of undistinguished stock. You may say of this, "That sounds unreasonable," or of that, "I don't believe it." And there may be no denying you the right to say it, for often the evidence may be open to various interpretations.

Nevertheless, among all the "ifs" and "buts" regarding the inheritance of human behavior traits there are a great many highly important facts that have been clearly established. If nothing else, genetics, in combination with its sister sciences, has in a few decades swept away a vast amount of rubbish regarding the nature and causes of social differences among human beings. Freed from the accumulated litter of past prejudices and misconceptions, the facts as we can see them now about ourselves and other people should do much to make every one of us a better social animal.

And now to our first major social question.

CHAPTER THIRTY ONE

IQs and Intelligence

HERE ARE TWO ORPHANED INFANTS, available for adoption at a placement bureau: They look the same and are certified to be equally sound and fit. You are eager to adopt one, and as there is apparently no choice, you are about to toss a coin to decide.

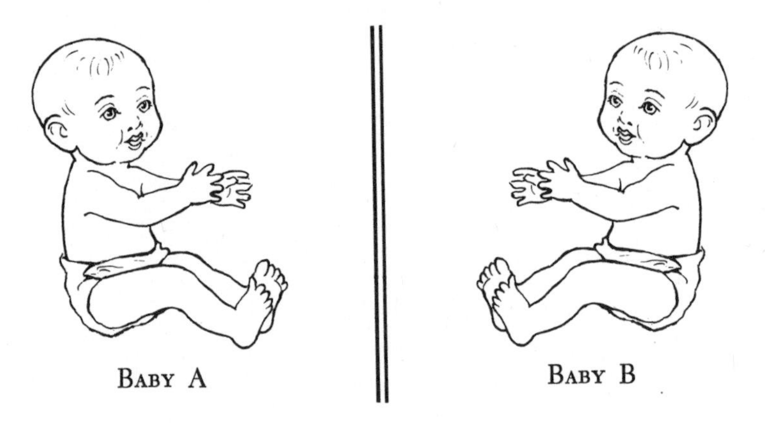

BABY A BABY B

Then an official tells you that baby A is the offspring of a charwoman and an illiterate day laborer.

Baby B is the offspring of a young woman writer and a young physician.

Would you still feel there was no choice?

Or would you pick Baby B on the chance that *it had inherited a higher degree of intelligence?**

* The question as posed, it should be stressed, is a theoretical one and may be largely academic to most Americans now seeking babies for adoption. The demand for these babies so greatly exceeds the supply that few childless couples are likely to make an issue of a baby's parentage so long as the child is adjudged healthy and mentally normal.

Around this question, or others closely related, centers one of the greater controversies in the study of human heredity. There are some authorities who say that no evidence exists which would justify a choice in favor of Baby B. There are many others who disagree with them.

What is the basis for the arguments pro and con? Let us start at the beginning.

We have seen that some types and degrees of *abnormal* intelligence—idiocy, insanity, etc.—can be inherited, but this tells us as little about the normal mental process as the throwing of a monkey wrench into any complex machine would tell us about its normal workings.

In lower animals, true enough, we know that a normal mouse inherits a certain type of brain, a cat another type, a dog still another, and so on. We know also that within each species itself there seem to be different degrees or qualities of brain activity. In dogs, breeds vary in alertness and capacity for certain kinds of performance. As one illustration, during World War II dogs of five specific breeds (and only these or their crosses) were accepted for military use: German shepherd, Belgian sheep, Doberman pinscher, collie and schnauzer. Also, training experiments with dogs at the Jackson Laboratories (Bar Harbor, Maine) have shown marked differences in the capacity of different breeds for given types of performance: fox terriers (best on tests calling for confidence in strange situations); beagles (good on leash control); African basenjis (best on tests requiring independent action); and cocker spaniels (excellent in obedience training.) Breeders claim that some dogs—for instance, Scotties and dachshunds—are usually much more intelligent than others, a particularly low rating being given to Russian wolfhounds. Geneticists have bred dull strains of rats and bright ones. Similarly, among many other animals—experimental, domestic and wild—what appear to be species and breed variations in capacity for mental functioning have been noted.*

If, then, degrees of normal intelligence are inherited among other animals, why not among human beings? All logic and all scientific generalizations lead to the assumption that there must or should be inherited differences in the construction and functioning of human brains just as there are inherited differences in all other aspects of our

* How are lower animals ranked in intelligence? Our distant relatives, the apes, are far in the lead, with the chimpanzees smartest of all. Below the primate level among mammals, most authorities have ranked, in approximate order, dogs, cats, elephants, pigs and horses. But recently, some scientists have placed the dolphin (also a mammal) at the top of the list next to primates. Birds—even the brightest—fall much below mammals in intelligence, and in descending order are reptiles, frogs and toads, and fishes.

construction and functioning. But the big question is: Is there a direct relationship between the kind of brain mechanism a person has and *what he is doing with it?* Or, to state the question differently, is a person's mental *performance* at any given time fair evidence of the kind of *basic intelligence* he has?

There's the rub! For should it be proved that the varieties and gradations of mentality revealed in the workaday, scholastic and social worlds are dictated to any great extent by heredity, then we'd be faced with the possibility that dullness, mediocrity and superiority are *inherent* in individuals and that failure and success may be largely *predestined.*

No wonder that beneath the placid surface of scientific research on this subject a struggle has been going on for years between the hereditarians and the environmentalists. The hereditarians are out to prove that certain individuals come to the top, like cream in a milk bottle, because of inherent superiority and that, graded according to their mental capacity, they tend to form classes at various intellectual levels. The environmentalists challenge this analogy. They are set on proving that while there may be some individual differences, the mental grades are in quite equal proportions within all groups, and that if we *shake the bottle and equalize the environments,* the levels of mentality in which classes of people now seem to be stratified would disappear.

How are we to settle this?

Our great difficulty is that intelligence can be measured today only by arbitrary procedures such as those set up by the current intelligence tests. We saw in Chapter 26, "Slow Minds," how the tests are used to determine degrees of mental deficiency: the IQs of idiots being under 25; of imbeciles, 25 to 50; of morons, 50 to 70; and of the borderline cases, 70 to 80. Above that we come to nondefective mentality, those with IQs of 80 to 90 being graded as merely dull and those scoring 90 to 110 as normal. Beyond this, there is a classification of scores from 110 to 120 as superior, 120 to 140 as very superior and 140 or over as genius (the latter term being used only in a technical sense, not implying that individuals with scores of 140 or more—of whom there are many—are actual geniuses).

It should be noted also that the intelligence tests used for children up to and including the age of sixteen are not the same as those for adults. These tests for children seek to establish how well a child is doing mentally in relation to the "norm" for his age.* But after the age

* The conventional intelligence quotient (IQ) is obtained by dividing mental age by chronological age: if a child aged eight has a mental performance up to the average for children of ten years, the 10 is divided by 8,

of sixteen it is assumed that intelligence doesn't increase measurably, so tests for the subsequent years are aimed not at giving a score for mental performance in terms of age, but in relation to the intelligence of other adults as a class. Thus, the "adult" intelligence tests are the same for late teen-agers and for those of all ages beyond, the theory being that any advantages of the younger people, in speed of response or more closeness to schooling, for instance, are balanced by the advantages of older persons with respect to acquired knowledge and experience. While the tests for adults (Wechsler, Otis, Army, for example) differ in many respects from those for children, the scores made on them by individuals generally do correlate rather closely with the scores made on earlier intelligence tests (with qualifications to be noted presently).

But how accurate are any of the tests in measuring a person's basic intelligence? As between someone with an IQ of 90 and someone with an IQ of 130, there can be little doubt that the tests do reveal a marked difference in mental capacity or performance. But with regard to smaller differences, IQ scores have much less significance. The same individual may score 108 one day when he isn't feeling well and 113 the next day; or he may be given one score by one examiner and a somewhat lower or higher score by a different examiner. It has also been claimed that the tests are not equally fair to all (especially as applied to those of different social groups or races or to rural as compared with city children); that they are not entirely consistent at the different ages; that they measure "classroom" intelligence much more than general intelligence. (We'll deal with these objections further along.)

Despite the foregoing and other criticisms, the standard intelligence tests are considered sufficiently reliable to be widely used in many important ways. If you have a child, and live, for example, in New York, that child may be placed in a class with average pupils, or backward pupils, or superior pupils, on the basis of its IQ. If you are seeking admission to universities, particularly to professional schools, your performance on intelligence tests can often decide your acceptance or rejection. If you were or are in the army, you know, of course, the part the tests play in assignments for special jobs or training. And in various corporations and institutions your test scores may be a factor in obtaining a position.

For all practical purposes, then, the intelligence tests must be seriously considered. But how significant they are in studying and establish-

---

giving 1.25, or an IQ of 125; if a child aged eight has the mental norm for age seven, the 7 is divided by 8, giving .875 or an IQ of 87.5.

ing the *inheritance* of intelligence is another matter. For only if the IQs are proved to reveal inborn capacities, little affected by environment, can they have meaning for the geneticist. As it is, so long as all the studies made to date bearing on the inheritance of intelligence are based on IQ scores, the interpretation of these studies or any conclusions regarding them must depend on the importance attached to the tests and to their actual or implied weaknesses.

Here is where the authorities fall out. You will not be surprised to learn, therefore, that the findings of various studies made on the same questions may differ markedly. All we can do is to summarize what are considered the valid studies, theories and opinions, on all sides. With respect to the two babies with which we started off, this will constitute the evidence.

Let us now follow the procedure of the investigators. First, they ask themselves questions. Then they set out to find the answers. We'll start with this:

1. *Do people of different classes or occupational groups show different degrees of intelligence?*

As measured by intelligence tests, yes. Extensive studies reveal that the higher up the social and economic scale one goes, the higher is the average IQ. That is to say, unskilled laborers and farmhands as a group have lower IQs than skilled laborers and farmers; above these rank skilled factory workers, white-collar workers and small businessmen; above these, semiprofessional people, bigger businessmen and managers; and at the top of the IQ structure are the professional men. Obviously, however, there is much overlapping among the groups, some unskilled workers having higher IQs than workers in levels above, and some professional men having lower IQs than persons in lower occupations.

What we may call the "rising to the top" theory holds that over a period of time, where opportunities for occupational movement exist, brighter persons tend to collect more in the professional and upper-occupational levels, the less bright in the lower working level groups. For example, studying drifts among high school graduates in Missouri rural areas and small towns, sociologists C. T. Philblad and C. L. Gregory concluded, "There seems to be a distinct tendency for the brighter boys and girls . . . to find their occupational levels in the higher prestige occupations . . . while those who perform more poorly on tests are more likely to become manual workers and farmers."

If we bear in mind that we are speaking always in *general terms* and allow for individual exceptions, it would seem that lower intelligence goes with so-called lower work. But does this mean that the unskilled laborer has a very low IQ because conditions have thwarted his mental development, or does it mean that he is an unskilled laborer because he has a low IQ? Which came first, the low condition or the low IQ?

Another question:

2. *Are the intelligence levels of children related to those of their parents?*

Yes. Children of a group of parents of high intelligence—on the average—are reported as having greater mental capacity than children of parents of low intelligence. Rarely, at best, where both parents are of inferior mentality, does a superior child result. Conversely, parents of superior mentality seldom have a child of very low intelligence, and if they do, it is usually because of some special defect—idiocy or some organic condition or prenatal accident (as was brought out in our "Slow Minds" chapter).

Comparing offspring of parents at various social and economic levels, one finds that children of unskilled laborers have the lowest IQs, children of professional men the highest, with those of the groups in between similarly correlated. Specifically, several studies some years ago showed these IQ averages of children grouped by their fathers' occupations: Professional, 116; semi-professional and managerial, 112; clerical, skilled trades, retail business, 107.5; semi-skilled, minor clerical, minor business, 105; slightly skilled, 98; day laborers, urban and rural farmers, 96. Surprisingly, even in the Soviet Union a study showed much the same general relationship between children's IQs and parental status.

We may note that the average differences in IQ between adjacent groups of children, as with their fathers, are not radically great, although the total difference, from the top level to the lowest, is about 20 points. However, when children are classified into the high IQ group, close to one-third of the brightest have been found to have fathers in the professional class, and barely 12 per cent, fathers in the skilled labor class.

One criticism regarding the above data is that the comparisons made are not between the mental *capacities* but between the *IQs* of children and the *assumed* mental levels of their fathers as reflected by occupational rank. If, as is claimed, the intelligence tests are weighted in favor of acquired knowledge, children from homes of poorly educated workers would be retarded in comparison with those from better educated and intellectually more favored circles. An additional criticism is that in the United States the intelligence tests may be slanted toward those of native stock as compared with those of foreign-born stock, and toward those of upper-middle class and more privileged groups, as compared with those of the underprivileged groups. For example, it has been found that if the phraseology of certain questions in the intelligence tests (involving familiarity with given objects or experiences) is changed and words with which underprivileged children are more at home are used, their scores go up. It is also significant that in many practical problems not represented in the intelligence tests, such as telling the right shoe from the left, putting on overshoes, or identifying and finding tools, young children from lower-level homes may do better than those from the more favored ones.

But allowing for the shortcomings in the tests, most psychologists would still say that they reveal a significant average difference in mental performance between children of upper and lower social levels. (As some evidence for this, on tests designed to be neutral—that is, not using language but involving little tasks with which youngsters of all groups should be equally familiar—middle-class children have still proved themselves superior to those from lower-level groups.) Assuming this to be true, we might again draw two contradictory conclusions: (1) that the children of parents in the lower intellectual levels have *inherited* their lower mentality; or (2) that the lower mentality has been thrust upon them by inferior environment.

Which is right? We go on:

3. *When children are taken away from their parents in infancy and reared elsewhere, do their IQs still show the influence of their heredity?*

According to some studies, yes; to others, no.

On the yes side, in various places tests have been made on large numbers of illegitimate children placed in institutions soon after birth. Most had never seen their fathers, who were of all mental or occupational levels. Yet, despite a similar environment for all the children in the institutions, their IQs differed markedly and were reported as conforming almost as closely, on the average, to the expected mental levels of their fathers as did the IQs of children on the outside. So, too, some studies of children adopted and reared in private homes have reported that the IQs of the children bear a much closer relationship to the levels of their true parents than to those of their foster parents. Other studies have contradicted this with findings that children, mostly of inferior parentage, who were placed in adoptive homes at early ages had IQs much less like the mental levels of their true parents and more like those of their foster parents and other children in the homes and groups in which they were reared.

Which sets of findings are correct? Psychologists have been ranged on opposite sides about this, depending on how valid they consider the particular studies. Perhaps the discrepancies might be explained by differences in the samples of the children studied—their age, parentage, how they were selected for adoption, by whom they were adopted and how they were reared; and by the possibility that the true parentage—or, at least, paternity—of a foster child (where knowing this is essential in a study) may often be erroneously given, or be very much in doubt. Also, where studies have indicated that the IQs and later achievements of foster children do lag behind those of other children in the same homes or types of homes, some psychologists have attributed this to inhibitions and maladjustments stemming from a child's knowledge of having been adopted, or the thought or awareness of not being fully loved and accepted by the foster parents.* On the other hand, it may be argued that under today's conditions the adopted child is likely to receive more than ordinary care, affection and attention from the foster parents, who tend to be above average in intelligence and circumstances.

We'll leave the preceding question open and move to this one:

4. *How far can training mold intelligence?*

Many studies show that education can increase the *scores* on intelligence tests, but whether or not this means that intelligence—manifested in all its various forms or as a whole—has been improved is another matter.

So once more, there are conflicting findings. Some investigators have reported that IQs of very young children can be increased materially if they are subjected in early life to stimulating training, as in nursery schools. But this has been found to apply chiefly to children from underprivileged homes. Among children from better backgrounds, nursery school attendance has proved to have little or no additive effect on IQ. Again, where stimulating early training does elevate IQs of young children, it is not certain that the effects are ingrained or lasting. Similarly, when children have been given special coaching in taking IQ tests, it has been reported as resulting in only temporary gains, which are all but wiped out by the time the child is given new tests at the next age level.

Above all, one must be cautioned that "before and after" intelligence tests made of very young children—say, under age five or six—are apt to be unreliable. The score of a young child on a given test at any time may be much influenced by his psychological and emotional state, response to the person giving the test, or home situation. Moreover, it

* An unusual twist was given to child-adoption cases when state welfare agencies in New Jersey, in 1960, brought action to have a four-year-old girl taken from her foster parents and placed elsewhere because her IQ—about 140—was much superior to theirs, and it was felt that they and their home environment would retard her development. The court decided, however, that the affection of the foster parents and the child's happiness with them outweighed the other considerations and justified their keeping her.

has been shown that very young children do not develop mentally (any more than physically) at uniform rates, even though they may have similar intellectual powers. (To quote Dr. Nancy Bayley, "Each child has a sort of inborn timetable of physical and mental growth. This timetable sets a basic pattern upon which an infinite number of outer forces act.") In some cases, then, the same child may be below average in IQ at one age but above average at another.

With increasing age, IQs tend to become more stable. Further, scores on intelligence tests may rise somewhat in relation to the amount of additional schooling. Thus, when the late Professor Irving Lorge tested a large group of New York men who as boys had had the same IQs in the eighth grade, he found that, twenty years later, those who'd gone on through high school now had higher IQs than those who'd stopped with grammar school; those who'd gone through college scored still higher; and the ones who had become Ph.D.'s scored an average of 20 points more in IQ than those who hadn't gone through eighth grade.* Wasn't this proof that intelligence can be progressively increased by training? Dr. Lorge didn't think so. What was shown, he felt, was only that *performance* on intelligence tests could be improved.

There were also these questions: Were the changes in test scores really or solely due to differences in education? Or was it possible that, if one of two boys with the same IQ at age thirteen stopped with the eighth grade while the other went on to become a Ph.D., they could have differed in their capacities for further mental development—just as, with the same height at age thirteen, one boy might stop growing and another go on to become much taller? Contributing further to rises or restraints in IQ scoring may be such influences as home and social attitudes toward learning, individual drives and personality factors. Not least may be the matter of *practice* in taking tests—coupled often with advancing education—which familiarizes a student with the technique of answering questions. This has been suggested as a major reason why American soldiers in World War II scored considerably higher on intelligence tests than did their fathers in World War I; or why in a number of colleges and high schools the IQs of present-day students seem to be higher than those of previous generations.

Conversely, it has been shown that lack of training can depress IQ scores. For instance, studies in Holland after World War II showed

* Similar findings came in a later study of Swedish army draftees by Dr. Torsten Husen, which showed that, in comparison with IQ scores made while in grade school ten years earlier, there had been an increase averaging 11 points for the men who'd had the most education, and even a slight decrease for those who'd had the least education.

HOW CLOSE ARE TWINS' IQs?

*Here are the IQ correlations (degrees of closeness) between pairs of twins of different kinds, as compared with those of other paired children.**

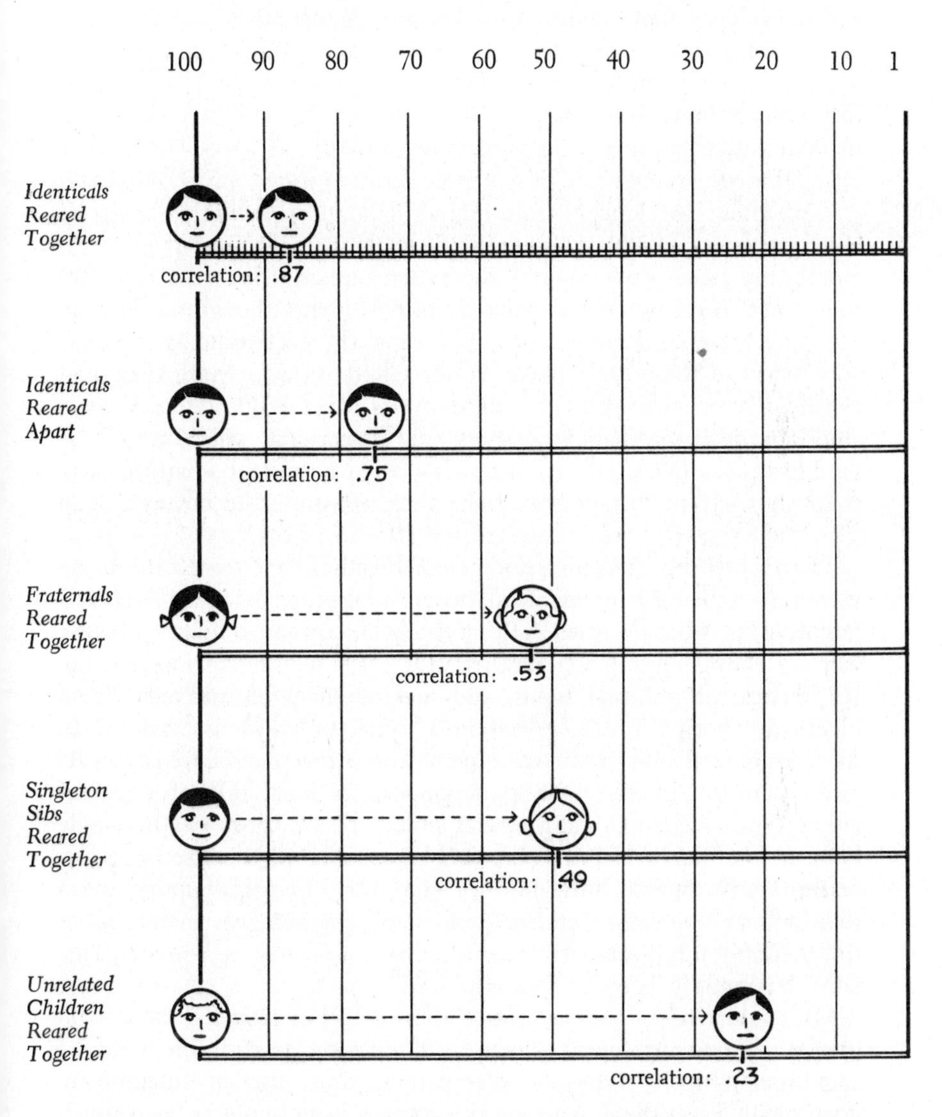

* Figures given are averages derived from an analysis of intelligence-test scores in 52 studies, covering thousands of twins and nontwin children, as reported by Drs. L. Erlenmeyer-Kimling and Lissy F. Jarvik. (*Science*, 142, Dec. 12, 1963.) "Unrelated children" refers to foster children of different parents reared in the same home. "Correlation" is further explained in the Text.

that in regions where for long periods schooling was seriously interfered with during the Nazi occupation, the IQs of children dropped about 4 points below the prewar average.

But evidence that training or other environmental factors can raise or lower IQ *scores* still does not answer our question about *inherent* mental capacity; for we know that in the same environments, with the same training, and often in the same homes, children may have marked differences in IQ. So, continuing, we ask:

5. *How far can heredity fix the boundaries of a person's intelligence?*

Once more we look to our human laboratory subjects—twins. If heredity alone determined one's intelligence, then identical twins, with exactly the same genes, should always have exactly the same IQs. But they don't. IQs of identical twins do often differ, although on an average they are considerably more alike than those of ordinary brothers and sisters in the same family (as our Chart shows). More than half of the identical twin pairs may differ by less than 5 IQ points, whereas among ordinary brothers and sisters only 23 per cent will so closely resemble each other. Further, more than 80 per cent of identical twin pairs score within 10 points of each other, as a rule, and no more than 5 per cent show IQ differences greater than 15 points.

That the effects of being reared close together have something to do with the similar IQs of identical twins is suggested by this: Even fraternal twins, who are genetically no more alike than ordinary siblings, still are considerably closer in their IQs. Also tending to increase the IQ likeness of fraternal twins, and making them less different from identical twins, is the fact that both types of twins are exposed to special prenatal and postnatal experiences which may have some retarding mental effects. During childhood, at least, IQs of twins of either type tend to average several points below those of the singly born, a condition which may in part be ascribed to the effects of premature birth and/or birth injuries (far more frequent among twins than among singletons) and in part—some psychologists maintain—to the mentally inhibiting effects of the way twins may be reared. (This latter assumption is by no means proved.)

Most interesting, then, are the studies of IQs of identical twins who were *not* reared together, but who were separated in early life, adopted and brought up by different foster parents, and often in different environments. Even these twins, on the average, were found to be as much alike in IQ, or slightly more alike, than *fraternal twins reared together*. Yet it also was found that if the homes and educational opportunities of separated identical twins had differed greatly, their IQ differences often were great. In the case of two identical-twin Canadian women, separated in infancy, one, reared in an isolated mountain district with only a few years of backwoods education, had an IQ of 92; the other, reared more favorably and a college graduate, had an IQ of 116—or 22 points higher than her twin. But in another case, of two identical British men likewise separated as infants and reared in very different environments, one rose to a brilliant position on a college faculty and the other remained a farmer; when both were tested the farmer was found to have almost exactly the same high IQ as the professor: 137.

On the whole, it would appear that under ordinary conditions—and sometimes under quite different conditions—the minds of identical twins are geared by their genes to perform in much the same way. While the instances in which identical twins have been quite wide apart in IQ show that heredity does not unalterably fix one's intelligence, the twin studies nonetheless indicate that heredity does play a great part in determining the potentialities for one's mental development.

Now let us try to find out what influences can hamper mental development.

6. *Does a poor body produce a poor mind?*

Or, conversely, does a healthy body make a good mind? To the surprise of most of us, there is little proof on this point. A first look at the evidence might prompt a "Yes" answer, for many studies have shown that, in broad averages, children with high IQs tend to be healthier, and bigger for their age, than children with low IQs. But analyzing further, we find that children who are in better physical shape generally come from homes which offer them not merely healthier living conditions, but also more educational and cultural advantages. What we must answer is this: Does the better physical environment of itself produce higher intelligence? Or, on the other hand, may not the better cultural environment that goes with it also be in part the result of higher intelligence of the parents transmitted to their children?

If we consider any large population as a whole (not including groups against whom there is special discrimination), there is now reason to question how great is the direct relationship between physical condition and intelligence. For one thing, the health and bodily development of children in the United States today (and of those in England and many other countries) is by all counts far superior to what it was in previous generations. *Has there been any over-all increase in intelligence?* Few educators would say yes. As mentioned previously, any slight increase in average IQs may be attributed to improved education and increased practice in test-taking. From another angle, we have experienced in this century several devastating wars and their aftermaths, which resulted in the undernourishment and health impairment of millions of children in many countries. And again there is no evidence of any decrease in average mental performance because of poor health, or, at least, of any permanent effects—certainly not if judged by the current achievements of the new generations in many war-torn countries. Among the worst sufferers of all were persons now in Israel who, as children, had survived the extreme physical drains of internment in Nazi concentration camps or other rigors. Regardless of the emotional impacts of their terrible early experiences, their mental acuity as adults was not shown to have been lessened.

Malnutrition, which differs from starvation in that it refers not to lack in quantity of food but in the food elements required by given individuals, has received special attention with respect to intelligence. The general findings seem to be that even where there is a dietary lack, whatever its physical effects on a child, intelligence does not seem to be much hampered. Some studies have indicated that when underprivileged mothers in certain communities (one in Norfolk, Virginia) were given vitamin supplements during pregnancy and later while nursing their babies, the IQs of their children were elevated above the average in their groups, at least up to the age of four. But among mothers elsewhere (including Kentucky mountaineer women), vitamin supplements given in the same way produced no IQ improvements in their children. This would suggest that it is only when mothers' diets are markedly deficient in specific vitamins or elements that the intelligence of their children may be affected. Even then, it has not yet been proved that the effects will be permanent, unless the deficiencies lead prenatally to some serious brain impairment. What also seems clear is that despite the wide differences in diets among peoples of the world—and in the sparsity and limitations in some diets as compared with the great abundance and variety in others—there is no evidence that food by itself, or eating habits, causes one people to be superior in intelligence to another.

To throw further light on the poor body–poor mind theory, we may ask:

7. *Does disease retard mental development?*

If we eliminate, obviously, mental diseases and deficiencies, the answer appears to be mostly "No," but sometimes, "Yes."

Consider *adenoids* or *bad tonsils.* Popular belief used to be that when dull-witted Johnny Jones, who had trouble with his adenoids, got them removed, he would brighten up immediately. But follow-up studies of children after adenoid or tonsil operations have shown no mental advancement.

Hookworm, formerly widespread in depressed areas of the South, also was thought to be responsible for lowered intelligence, but indications are that the real causes were other factors in the bad environments which produced both the disease and the low IQs. Similarly, in the case of *syphilis*, it has not been shown that the disease itself lowers intelligence (unless or until the brain is directly impaired), for the fact that children and adults with the disease tend to have lower than average IQs may mean only that syphilis is most likely to be acquired in depressed environments or in families where IQs are generally low.

However, certain diseases which attack the brain and central nervous system are known to diminish intelligence (sometimes but not always), among them spinal meningitis, cerebral palsy, hardening of the arteries, brain fever (epidemic encephalitis) and some types of epilepsy. Brain injury in prenatal life, at birth, or in early childhood may of course also impair mental functioning (as was noted in the "Slow Minds" chapter). But there is evidence that in the later years, after the brain has become fairly well "set," injuries to it, or even removal of part of it through operation, seldom have much effect on the individual's general intelligence, although there may be interference with certain of the thinking functions.

Physical defects may, however, adversely affect mental development by interfering with learning tasks and processes, as in cases where children have hearing, sight or speech disabilities. In recent years remedial programs for children with such handicaps have helped greatly to insure for them IQs in keeping with their real capacities.

It is also worth noting that history is starred with individuals who rose to the greatest heights despite almost every sort of disease or physical handicap, often present at birth (as we'll show in a later discussion of genius). Moreover, if we compare any individual with himself at different stages of his life, it can hardly be said that a man in middle life, afflicted with various ills and with a body on the downgrade, is less intelligent than he was in his mid-twenties, at the peak of his physical condition. Certain types of mental performance, particularly those relying heavily on memory, do tend to deteriorate somewhat with the passing years, perhaps partly because the mind as it matures becomes more selective about what it retains from the accumulating store of experiences and learning. But verbal facility, reasoning ability, and capacity for abstract thinking may increase well into middle age. Much depends on a person's earlier IQ. A number of recent studies have shown, in fact, that although persons of low IQ tend to become duller as they grow older, persons of higher IQ may become smarter. Even from age sixty on, individuals with IQs of 116 or over may suffer little or no drop in mental ability, whereas persons with IQs below 96 may show marked drops.

On the whole, then, there is little to prove that brawn and brain go together, or that mental capacity is dependent on physical well-being.

Leaving the question of health, we also find it difficult to establish any relationship in normal individuals between outward, or structural, physical characteristics and basic mentality. For instance:

8. *Does head size have any bearing on intelligence?*

Not that scientists can discover. Disregarding idiots or imbeciles with deformed skulls, exhaustive studies fail to prove that bigger heads or loftier brows mean higher IQs. Eskimos have larger heads than Whites, and the skulls of some prehistoric men had a bigger cranial capacity than the record head of modern times, that of Ivan Turgenev, the Russian novelist (2,030 cubic centimeters). As to the brain itself, the Wistar Institute of the University of Pennsylvania has a collection of two hundred brains of scholars, average persons and low-grade morons; but no scientist who has studied them has yet been able to determine how the brain of a wizard differs from that of a moron, in size or structure.

What has been said can also be used to refute the old theory that because women usually have smaller heads than men, they must also have "smaller" minds. From this observation we revert to our initial problem of choosing between two babies. That is, if your decision rested solely on which would be the most intelligent, would you be influenced by the fact that one was a girl, the other a boy? In other words:

9. *Is the intelligence of human females inferior, superior or equal to that of males?*

A lot of people may believe this is now a dated question, on the assumption that the intelligence tests have "proved" there is no difference in mental capacity between the sexes. For don't boys and girls have almost the same IQs? Yes, but what few people—even many teachers—do not realize is that *the standard intelligence tests have been deliberately designed to make the scores for the two sexes come out equal.*

The fact is that from the time the tests began to be devised, early in the century, important differences have been found between the sexes in the *way* that they think and in their mental performance on different problems. Girls are consistently better in those phases of the tests involving language usage, rote memory, esthetic responses (such as matching colors and shapes) and social questions (such as guessing ages or noting other personal details). Boys as a group are consistently superior in tests involving abstract reasoning, mathematics, mechanical ability or visualization of spatial relationships. So in an effort to make tests fit both sexes impartially, "girl-plus" items have been arranged to balance "boy-plus" items, *with certain questions on which those of one or the other sex were too far superior being eliminated altogether.* Thus, while the current tests are useful for measuring academic performance in various categories, they tell us little about the relative mental capacities of the two sexes.

Another important sex difference is in relative school achievement, if not also mental development, at successive stages. All the way through the elementary grades, girls, on the average, tend to be ahead of boys both in group IQ scores and in school grades. Going into detail on this point, educator Frank R. Pauly analyzed the IQs and class records of 30,000 elementary school children in Tulsa, Oklahoma. Girls, he found, were mentally more mature than boys at every age level. In the second grade, they averaged four months ahead of boys on the IQ tests; in the eighth grade, girls had an IQ lead of eight months; and in all the grades their total average marks were higher. However, as other studies have shown, by the end of high school boys on the average have caught up with or bypassed girls in both marks and IQs, and in college the men excel the girls. Yet in specific categories the patterns of thinking of the two sexes in college remain much as they were in childhood: The coeds generally excel the men in verbal abilities, modern languages (but not classical languages), literature, the arts, and some phases of the social sciences. Males are much ahead in the physical and mechanical sciences, in mathematics, and in subjects involving abstract thinking.

How account for this sex difference in kind of thinking and mental performance? Is it all environmental—due to upbringing, social conditioning and training for the different roles which males and females are expected to fill? Before you can say "Yes," unequivocally, you may have to explain why little boy and girl babies, with apparent spontaneity, may show certain differences in interests and reaction foreshadowing later sex characteristics in mentality and behavior (as will be enlarged upon in Chapter 39); or why males and females among lower animals —for instance, chimpanzees—from early life on, differ in various traits and kinds of performance.

Suppose we seek a *constitutional* basis for sex differences in human intelligence. It cannot be through direct heredity, for there are no differences in the "mental" genes which males and females receive. Indirectly, however, it is possible that the same genetic mechanism (XX versus XY) which produces the general sex differences in physical makeup and biochemical functioning may also produce a thinking apparatus in females somewhat different from that in males—in actual construction or in its way of responding to given stimuli. That is, even the same genetic mental equipment and thinking apparatus might work differently in a female than in a male.

One theory suggests that the boy-girl differences in IQ scores and school achievement may be correlated with the sex differences in rates of physical development. Biologically, as we have learned, girls are markedly ahead of boys in physical development from prenatal life onward, with a lead of about 1½ months at birth (in relative degree of physical development), the lead widening successively until girls reach puberty from one to two years earlier than boys. Dr. Pauly and others have postulated, then, that during the prepuberty years boys may not do as well as girls academically because in males at the same age level the rate of maturation is slower mentally, as it is physically; but that, just as boys on nearing and attaining puberty outdistance girls in stature and weight, they are "naturally" inclined also to outdistance them somewhat thereafter in mental performance, as a rule. To counter this, of course, are the arguments that any mental lag in girls sets in only because, with approaching womanhood, the awareness of their future roles as wives and mothers—and the feeling that boys don't want "brains"—dampens their zeal for academic achievement and makes them concentrate more on developing the feminine qualities which the world expects of them.

The foregoing explanations, remember, are still only theories. All we know is that, whether through conditioning or inherent direction (or a combination of both), the more that boys and girls become physically differentiated at puberty, the more pronounced become their differences in aptitude and intelligence tests, as well as in their ways of thinking with respect to problems in their everyday lives. Grown up, a man generally can outsmart a woman in some ways, but a woman can outsmart him in other respects (as you need hardly be told). Any attempt to reach a decision as to which has more *total* intelligence will get nowhere.*

* One of our leading intelligence-test devisers, Dr. David Wechsler, has stated the case this way: "The findings of the Wechsler Adult Intelligence Scale suggest that women seemingly call upon different resources or different degrees of like abilities in exercising whatever it is we call intelligence. For the moment one need not be concerned as to which approach is better or 'superior.' But our findings do confirm what poets and novelist have often asserted, and the average layman long believed, namely, that men not only behave but 'think' differently from women. This difference could probably be more clearly demonstrated if our intelligence scales included a greater variety of tests than now employed." (From Dr. Wechsler's *The Measurement and Appraisal of Adult Intelligence*, 4th ed., Williams & Wilkins, 1958.)

In short, so long as boys and girls and men and women are destined to be different physically and psychologically in innumerable ways, to be trained differently, to undergo different experiences, to live different lives, to have different incentives and objectives, we can reach only this conclusion:

No intelligence test exists which can precisely and accurately measure the relative mental capacities of the two sexes. And that being so, no one can say now whether the sexes are equal or unequal in intelligence, or whether either is superior to the other.

The difficulty in trying to measure the comparative mentalities of dissimilar masses of people is further emphasized in our next question: Suppose the *only* thing you were concerned with in adopting a child was his or her ultimate intelligence—and one child was *White*, the other *Negro*—or *Chinese?* To state the question more precisely:

10. *Do different races and nationalities have different degrees of intelligence?*

In *ways* of thinking, yes. In *amounts* of intelligence, no one can ascertain with scientific accuracy. (Note that these preliminary answers are remarkably similar to those concerning sex differences.) So long as there are no tests which can fairly measure the relative IQs of people living, learning and thinking under radically different conditions, any conclusions as to the relative *inherent* mental superiority or inferiority of one race or nationality as compared with another must remain highly questionable. As it is, we must bear in mind that the standard intelligence tests, which were devised by White Europeans and Americans to measure their own mental capacities, are apt to go badly askew and to be unfairly biased if applied to persons of other cultures without modifications.

Even within the same country, such as the United States, we must be cautious about interpreting the intelligence test scores made by individuals of different racial and national stocks, or in different regions, *if their home environments, cultural backgrounds and ways of thinking are not the same*—which is apt to be the case. This applies most emphatically to the much-discussed question of Negro versus White intelligence. True, in any given section of the country, the IQ averages of Negroes are considerably lower than those of Whites. But also, in any given section, the cultural environments of Negroes are relatively inferior. However, where the conditions for Negroes are better, their IQs are higher. And where conditions for Negroes are good, *their IQs may average even higher than those of Whites living where conditions are bad* (as in certain benighted mountain regions of the South, where IQs of White children may be 10 to 20 points below average). In fact, during World War II, the army rejection rate for substandard intelligence was much higher for southern Whites—about 10 per cent in North Carolina, Texas and Arkansas—than for northern Negroes, for whom it was only about 2.5 per cent in Massachusetts, Illinois and New York. (But within any given state the rejection rate for Negroes greatly exceeded that for Whites.)

What is also clear is that as the educational opportunities for Negroes have increased, ever larger numbers of them have been recorded in the upper IQ brackets, a great many now doing better than average Whites on intelligence tests. Especially significant has been the rise in the percentage of Negro children identified among those with IQs of 140 to 150 and over—technically in the genius class. While relatively far fewer than White children in this class, these gifted Negro children are found to come from much the same types of homes, in terms of economic and cultural status. Nevertheless, many brilliant Negroes have sprung also from the most benighted backgrounds. Most strikingly, in Africa, as educational gates and opportunities have opened for Negroes, outstanding political leaders and professional men have been emerging from hitherto primitive tribes.

Some investigators have stressed the fact that even when Negro children are in the same economic levels as Whites, and receive the same schooling, they still fall behind in average IQs and academic achievement. Others have pointed out that the social and psychological environments are nonetheless still different for Negroes as compared with White children and that subtle disadvantages, both in their homes and in the outside world, continue to curtail the Negro child's opportunities for equal mental development and to depress his incentives and drives toward achievement.

In short, although there are some who still believe that the continued lower average IQs of Negroes may be at least partly the result of heredity, the large majority of psychologists and anthropologists feel that this is without proof, and that not unless and until environments for Negroes—and the real life situations they are called on to face and must train themselves to face—are completely in line with those for Whites, can any convincing comparisons as to their relative mental capacities be made. This applies equally to attempts to weigh the intelligence of any other racial groups—Chinese, Tibetans, South Sea Islanders, Eskimos or whatever—against that of Whites.*

* Formal statements challenging and repudiating claims that Negroes (or members of other racial groups) have been proved inferior to Whites in innate mental ability, were issued in 1961 by the American Anthropological Association and the Society for the Psychological Study of Social Issues.

IQs and Intelligence

By now you have seen that while the intelligence studies have thrown doubt on many points previously taken for granted, few decisive answers have been given to the main question. So you may ask:

11. *Why cannot science tell us something more definite about the inheritance of intelligence?*

Chiefly because we haven't quite determined what intelligence is—particularly as a fixed, unitary trait—and haven't as yet (or may never have) any accurate means of measuring it.

The intelligence tests, as has been pointed out, were devised by educators principally for their own domain, the classroom. But academic intelligence and the practical intelligence required in the larger world outside are not always the same. Often the tasks and problems of everyday life demand mental attributes that the standard intelligence tests do not reveal—personality, character, will power, courage, enterprise, drive, curiosity, intuition, poise and the ability to understand people and to get along with them. Not least is the element of *creativity*. As educators have been recognizing more and more, the special qualities of mind and temperament which make for originality, enterprise and inventiveness may not necessarily be correlated with performance on intelligence tests or in classrooms.* Thus, two persons with the same IQ may differ widely in their capacities for achievement and success in life. In fact, a person with a high IQ may sometimes (though not ordinarily) prove less successful—actually because he is less intelligent in certain ways, or less socially adaptable—than a person with a considerably lower IQ.†

Some years previously a similar pronouncement was made by an international committee of social scientists on behalf of UNESCO (the United Nations Educational, Scientific and Cultural Organization).

* Awareness of this fact led to an important change in the rules and standards for selection of the National Merit Scholarship Awards. Where the awards—highly prized scholarships in the nation's leading universities and colleges—were formerly given mainly on the basis of high IQs and exceptional high school records, the rules were modified in 1960 so that about 20 per cent of the awards went to students with lesser academic achievement but who showed unusual creativity and other qualities in their thinking. (Special talents and aptitudes will be dealt with in the next chapter.)

† Mendel, father of genetics, twice failed in an important examination in botany at the University of Vienna, largely because he had been self-taught and his academic knowledge was deficient. He also failed in an oral examination in physics and the natural sciences. An 1850 report on this (the original of which has been preserved) says, ". . . though he has studied diligently, he lacks insight, and his knowledge is without the requisite clarity, so that his examiners find it necessary for the time being to declare him unqualified to teach physics in the lower classical school."

Not until we can define, weigh and measure intelligence accurately and fairly with respect to individuals in all environments can we have the basic material needed by geneticists to determine its degree and manner of inheritance. The first steps could well be to go back to Mendel's procedure of studying one measurable trait—or in this case one measurable aspect of intelligence—at a time. For a serious error previously was to try to study the inheritance of intelligence as a *unit*, when the evidence now shows that intelligence is compounded of many different elements or capacities, not necessarily related to one another. And if this is indeed so, each of these elements or capacities may be inherited independently, and we may have to track each down separately.

Preparing the way for the objectives indicated, psychologists have devised the factor-analysis technique of breaking up intelligence into its various functions or components. One of the first to do this, the late Prof. L. L. Thurstone, included among the primary factors verbal comprehension, word fluency, number facility, memory, visualizing or space thinking, perceptual speed, induction and speed of judgment. Subsequently, Dr. J. P. Guilford and others have extended the list with additions and subclassifications (to include creative thinking, originality, evaluation of judgment, sensitivity to problems, and symbol manipulation) so that to date close to sixty different factors involved in the human mental processes have been identified.* Whether heredity controls the mental factors, as individually designated, or given clusters or groups of factors remains to be seen. So far tests used in twin studies have shown that in such factors as verbal fluency, verbal comprehension, reasoning, space-sensing and mathematical ability, identical twins usually are much more alike than are fraternals.

While in general one type of mental ability is quite closely correlated with most other types, there are sufficient exceptions to support the theory that so far as heredity governs the results different elements in mental capacity may be inherited independently. For example, proficiency in language often is unrelated to proficiency in mathematics. Also, as even the brightest student knows, one study can be much more

* Dr. Guilford has suggested that the factors of intellect could be grouped into six categories, involving (1) memory for words, numbers, designs, music; (2) symbols in reading comprehension and arithmetic; (3) discovery—awareness of various kinds of things and sensitivity to problems; (4) production factors—conclusions, solutions or other outcomes of thinking; (5) divergent-thinking factors, concerned with a person's flexibility and originality; and (6) evaluation factors, involving logical evaluation and speed of judgment.

difficult for him than another, and even though he gets the same high grades in all, he may do so not because of equal capacity for each study but because of extra application to the weak spots. (The possible role of heredity in equipping individuals for some special types of mental performance and not for others will be discussed in the next two chapters.)

Assuming, then, that heredity lays the groundwork for intelligence as a whole, as well as for the different components of intelligence, how do genes work in this regard? According to the polygenic ("many-gene") theory, one would guess, first, that there are key genes which, by affecting the general development of the individual's brain and its functioning, set the approximate limits for his mental *potentialities*—but always subject to environmental influences (including incentives and opportunities). Second (again much influenced by opportunities and training), there may be genes which can accelerate or depress an individual's capacity for dealing with specific mental problems and meeting the requirements for achievement in one field or another. A person having the special genes for one type of thinking and performance might not have those needed for another aptitude or talent (as will be brought out in our next chapter). And always, in addition to the mental genes, there may be genes influencing various aspects of behavior and personality which play a part in the development of abilities of any kind.

The foregoing observations, remember, are still in the theoretical stage. So, too, are the concepts of precisely how the brain functions in general and in the performance of mental tasks. Professor Edward L. Thorndike believed that different mental abilities were governed not by different parts of the brain (as was once thought) but by different operations or "hookups" between the brain-nerve cells, sensory organs and other parts of the nervous system.* Going further, Swedish neurobiologist Holger Hydén has offered the theory that the brain carries out its functions under the guidance of codes embodied and imprinted in the RNA molecules of the neurons—the giant brain cells which, together with smaller "glial" cells sticking to them, transmit electrochemical impulses through an elaborate system of filaments. (For an explanation of RNA in relation to DNA, see Chapter 19, p. 175.)

* In another sense the brain may be regarded as a "thinking machine," or, as Dr. David B. Wechsler has stated, "The current definition of the brain that has met with greatest acceptance among the neurophysiologists is that it is essentially a screening and sorting organ, a complex electrochemical servomechanism that in some respects functions like a multiple television system with built-in screening generators and, in others, like a high-speed automaton computer."

The latter theory brings us back to the point that the human brain, despite its baffling intricacies and the intangible nature of what it does or produces, nonetheless—like all other organs of the body—must be *chemical* in its functioning as well as in its construction. Ultimately then, and fantastic (or prosaic) as it might seem, all of the human thinking processes—memory, reasoning, verbal expression, creativeness, foresight—may prove to be end products of the workings of enzymes, proteins, and other chemical substances activated and organized in large part through the DNA codes under the principal direction of "mental" genes. As for individual differences in mental performance, these, too, in considerable degree, may turn out to be aspects of the biochemical individuality we dealt with in Chapter 19.

For the time being, however, while we can assume that there are important hereditary differences in the mental endowments of individuals, we still are far from being able to identify, classify or measure them by any truly accurate, scientific standards. Nor can any geneticist yet venture to predict, as he might with many physical and pathological traits, the precise ratios and degrees of intelligence that can be expected in the children of parents with such and such IQ scores or levels (or kinds) of mental performance.

Which takes us back to the beginning of this chapter, and the question which you have been deliberating:

12. *What, then, is one to decide regarding a choice on the basis of intelligence between Baby A, the offspring of a charwoman and a day-laborer, and Baby B, the offspring of a young woman writer and a physician?*

Since we cannot find the complete answer in established fact, we might look for guidance to the opinions of leading authorities (geneticists, anthropologists, psychiatrists and psychologists) who have given detailed attention to all the evidence we have summarized. To the best of our knowledge, these opinions would take two main and divergent directions:

On the one hand, there are those who would say that no choice is justified, for these reasons:

We have no *proof* of the extent to which intellectual attainments are due to heredity or to environment. Even with the use of the present intelligence tests, faulty as they are, it is clear that what we call intelligence is greatly influenced by education and conditioning. We have no right, therefore, to compare by the same tests people whose environments are radically different and to consider that their relative scores have any bearing on their relative inherited mental *capacities.*

If we assume that there are genes which produce degrees of intelligence, in view of the complexity of the mental processes there would obviously have to be a great many such genes, which makes it difficult to conceive how—with the constant intermingling that has taken place among people of all levels—superior and inferior intelligence genes could have become noticeably segregated in different proportions within our different occupational groups, especially in so short a time.

Therefore, we have no basis for assuming that parents in the unskilled laboring group carry, or will transmit to their offspring, genes for intelligence inferior to those of parents in the professional class.

Accordingly, we are justified in concluding with regard to the intelligence of the two hypothetical infants that there should be no choice between Baby A, born of a charwoman and a day laborer, and Baby B, born of an authoress and a physician.

On the other hand, we know of authorities who would answer:

The view, that lacking clear scientific evidence of how intelligence is inherited, we are not justified in making deductions regarding it, is short-sighted and unwarranted. All the general findings of genetics point to the inheritance of degrees of intellectual capacity in the same way that other characteristics and capabilities are inherited.

Therefore, without knowing what the "intelligence" genes are, we may still rightfully assume that there are some which make for greater intellectual capacity and others for lesser.

Allowing for all possible powers of environment to depress or to raise intelligence, we know that many bright individuals born into lower social levels rise to higher levels; and many dull individuals born into upper levels sink to lower levels. Since this process has been going on throughout civilization, we may reason that in two large social groups differing radically in intellectual attainment, there would be more of the superior genes in the superior group.

The average IQ difference of 20 points between offspring of the unskilled laboring classes and those of the professional classes cannot be dismissed as without significance unless it is ascribed *entirely* to differences in environment. There is no proof that this is so. Knowing that in the same environment, even in members of the same family, great differences in intelligence exist, the burden of proving that environment alone is responsible for all these differences rests on those who make the assertion.

At the very best, in the situation cited, one can only say that Baby A might possibly be inherently as intelligent as Baby B. No one would venture to say, and not a single study has indicated, that children of unskilled laborers as a group would be expected to have a better intellectual heritage than those of professional people. On the other hand, many studies do indicate that there is a possibility, if not a probability, that the average child of professional people will turn out to be more intelligent than the average child of those in the lowest occupational groups.

The question thus becomes one of odds. Baby A might indeed turn out to be more intelligent than Baby B. But the odds are surely greater—although we cannot say how much greater—that Baby B, offspring of the authoress and the physician, would have inherited the better mental equipment.

Therefore, everything else being equal, on the basis of intelligence there should be a choice in favor of Baby B.

So here are two clearly conflicting interpretations of the same set of facts. How are you, the layman, to decide? For remember, the two theoretical babies are before you, and you can take only one of them.

You might beg the question by saying, "We are dealing, after all, not with objects but with human beings, helpless infants. So long as there is uncertainty, the humane thing to do, and the democratic thing to do, would be not to condemn Baby A as mentally inferior purely on theoretical grounds, but to give it the benefit of the doubt and consider it as equal." There is merit in this viewpoint, but our problem here is not a humanitarian one but essentially a scientific one. You are called on to decide, solely on the basis of the evidence, whether the offspring of the one set of parents would be likely to turn out more intelligent than the offspring of the other set of parents.

If you conclude that the facts presented are not conclusive enough to warrant a choice, and that you should leave the selection entirely to a toss of the coin, we can assure you that there are high-ranking authorities who will approve your stand.

But if you prefer to be guided by the weight of opinion, at least in a numerical sense, we may say this:

It is our belief that the *majority* of qualified experts of all kinds would subscribe to the second viewpoint previously stated and would unequivocally advise you to keep your coin in your pocket and to choose Baby B, child of the authoress and the physician.

CHAPTER THIRTY TWO

The Gifted Ones: I

Musical Talent

A LITTLE BOY, hugging a violin, walks out onto the stage at New York's Carnegie Hall. There is a flutter of applause from the thousands of persons filling the auditorium. The little boy tucks his violin under his chin and begins to play. The audience, skeptical, watches, listens. A tiny hand sweeps the bow back and forth, tiny fingers fly over the strings, streams of melody, now shrill, now full-throated, cascade forth. Already, in those first minutes, many mature musicians out front know that in all their years of study and work they have not been able to achieve such mastery. Soon they, and the others, quite forget that this is a little boy who is playing. As if drawn by invisible bonds, they are carried out of the hall, into the night, higher and higher. Then suddenly there is a burst of notes like a rocket's shower of golden stars . . . the music stops . . . and they are all back again in Carnegie Hall, incredulously storming with their bravos a little boy—a little boy who in a few hours may be crying because he isn't allowed to stay up and play with his toys.

The scene has been enacted a number of times in each generation, but not too many times, for little boys like this do not appear often. It may have been elsewhere than Carnegie Hall—in Paris, London, Vienna—possibly in your own town. And sometimes it was not a violin that the child played, but a piano. So Chopin, Mozart, Mendelssohn, Liszt, Schumann, César Franck, and of contemporary musicians, Heifetz, Rubinstein, Menuhin, Schnabel, Oistrakh, Richter, Stern, Kogan and Arrau, among many others, revealed their genius to the world as children.

In no other field of human achievement do the young so strikingly scale the heights and so easily outdistance great numbers of competing adults. One might concede that musical *performance*—the rendering of a musical composition on some instrument—is a special type of achievement which does not demand either full physical or full intellectual development (though both are essential for true virtuosity) and hence need not be beyond the reach of a child just because he is a child. But in musical *composition,* one of the highest forms of human creative expression, we also find amazing precocity: Mozart, at fourteen, with a score of symphonies and short operas already to his credit, offering his newest opera in Milan and personally directing the orchestra, the largest in Europe; Handel, by the time he was eleven, composing a church service every week; Mendelssohn, at seventeen—with years of composing behind him—producing one of his greatest works, the overture to *A Midsummer Night's Dream;* Franz Schubert, another veteran composer in his teens, writing, at eighteen, in a single day, one of his most notable songs, "The Erl-King."

To devotees of music, precocity among the great musical figures, not only of the past but of the present, is so familiar as to be taken almost for granted. But how explain this? Is it the flowering of some remarkable hereditary endowment? Or is it merely the prosaic result of some unusual type of early environment and intensive training? Each of these theories has had its advocates.

The subject isn't just one for the musician or the scientist. Music is now important in all our lives. Apart from performance for pleasure or profit, the question of whether you are or aren't responsive to music, and in what way or degree, may greatly affect your leisure activities, social relationships and personality, from childhood on. You may know all too well the difference it makes between thrilling to music or being tone deaf, between coming alive with rhythm on a dance floor or stumbling mechanically, between harmonizing joyfully with others around a piano or hanging uncomfortably on the fringes, trying to croak out a few off-key words. And if you're a parent, you can hardly escape the question of whether your child is or isn't musical, and why or why not, and what to do about it in either case. Thus, many of the points to be taken up in this chapter, on the nature of musical talent and the extent to which it is or isn't hereditary, will have a personal meaning for you.

To begin with, we may note that musical talent, as revealed through performance, is unique among other human talents in the way that the achievements of different individuals can be measured against one another. Exactly the same piece can be rendered by any number of musicians of all ages, backgrounds and nationalities (and, through recordings, even by the dead), and one does not have to be an expert to distinguish the superlative performances from the mediocre or inferior ones. No such easy method of comparison is possible in any field where creativity is the determinant—as in writing, painting, or musical composition—for creativity cannot be measured by any known yardstick, and relative judgments of it by contemporary critics are all too often wrong. Accordingly, so long as we confine ourselves to measurable talent for musical *performance* and the degree to which it manifests itself in different individuals and different families, we are in a much better position than in any of the other arts to distinguish the part played by heredity.

In former times, many believed that the inheritance of musical talent could be proved merely by pointing to such pedigrees as those of the Bachs, in whom great musical talent flowed in an unbroken line through five generations, or of the Mozarts, Webers and others. But these days, with a heightened awareness of the influence of environment, the skeptical might ask whether music-making did not run in certain families for the same reasons that watch making, pottery making or wine making ran in others, because individuals of successive generations were trained to carry on in the same way. The argument would have validity if it could be shown that any or every child could be trained to be a musician, and that only special environmental influences, and no special *inherent* qualities, were required for great musical achievement.

In the belief that many uncertainties could be cleared up by analyzing the musical histories and backgrounds of outstanding living musicians—not just a few, but a great many, fully representative of the whole number—a detailed study was made for my first book, *You and Heredity.* Among those queried were (1) thirty-five of the then outstanding pianists, violinists and conductors of the world (comprising, in the view of contemporary critics, a large majority of the greatest living virtuosi); (2) thirty-six of the outstanding singers, representing principals of the Metropolitan Opera Company; and (3) fifty students of the Juilliard Graduate School of Music, comprising a highly selected group of younger musicians and singers, still to make their marks.*

* In the original table of major instrumentalists there were included two then girl prodigies, mainly to give females more representation. These girls were Guila Bustabo (violinist) and Ruth Slenczynski (pianist). Miss Slenczynski stopped appearing in concerts at the age of fifteen, as a result, she wrote later, of revolt against parental compulsion, and did not resume professional performance until she was thirty.

In all three groups, information came directly from each individual (except in a few instances where supplied by some close relative), regarding the age at which his or her talent had appeared; how it had shown itself; the age when the professional debut was made (for members of the first two groups); and the incidence of musical talent in parents, brothers, sisters, and other near relatives.

One of the first points brought out was that, in almost every case, *musical talent expressed itself at an extremely early age.* Among the instrumentalists it appeared at an average age of 4¾ years for the virtuosi, and in the Juilliard group at about 5½. (The singers will be dealt with later.) This expression of talent was often in unusual ways. Artur Rubinstein, born into a very poor home in Poland where no musical instrument was to be heard, at the age of 1½ spontaneously began to sing little songs of his own making to express what he wanted or to designate various members of the family.* Eugene Ormandy at the same age knew all of the records of his father's hurdy-gurdy. Toscha Seidel, at three, would have tantrums when his uncle, a not-too-talented fiddler, hit a sour note. In other cases, talent was revealed by evidence of an acute musical ear (or the possession of *absolute pitch*), but in most cases it was simply through actual performance on the piano or violin. Any suspicion that the virtuosi had exaggerated the early onsets of their talents could be dissipated by published records showing that their professional debuts (not merely first public performances but the actual launching of their professional careers) had come at an average age of thirteen.

The histories of other virtuosi of the time not represented in our study showed similar precocity in almost every instance. Continuing to the present, the story has been the same. Among the newer virtuosi, almost all have made formal debuts as child prodigies, including such Soviet musical luminaries as violinists David Oistrakh and Leonid Kogan, pianist Sviatoslav Richter and cellist Daniel Shafran; and among Americans, pianists Lorin Hollander, Byron Janis, Shura Cherkassky,

* Violinist Josef Szigeti recently wrote the author that, while he had come from a musical family, "My musical ear and sense of pitch were not discovered—as one would presume—through my playing around and experimenting with the fiddles which were strewn all over. It was my singing along in my childish treble in instinctively 'right' harmony in thirds and in sixths, when one of our family sang the traditional Chanukah (or was it Seder?) songs, that made them prick up their ears and decide that here was still another youngster destined for the family profession!" (A full account of his life and career are given by Mr. Szigeti in his book, *With Strings Attached*, Knopf, 1947.)

Gary Graffman and Leon Fleisher; conductor Lorin Maazel; and violinist Ruggiero Ricci. Others, including Van Cliburn, would have shown the requisite precocious talent for debuts as prodigies had their parents and teachers been inclined to train and launch them as such.

Altogether, evidence from the past and present shows clearly that *great achievement in music is almost invariably correlated with an extremely early start.* Whether or not such an early start is a prerequisite for musical virtuosity, the question still is this: What accounts for the early start? Is it an unusual musical environment? Is it an unusual musical heredity? Or is it a special combination of both?

We may look for some answers in what the virtuosi instrumentalists (as well as those of the other two groups) reported regarding the incidence of talent in their families. As shown in our accompanying Summary Table, the majority had talented parents—one or both. Yet quite a number reported *no talent in either parent,* among them Toscanini, Schnabel, Smeterlin, Rubinstein and Rosenthal. Nor did the difference in the family backgrounds, or in the presence of talent in both parents, in one parent or in neither parent, seem to have anything to do with the caliber or quality of musicianship displayed by the individual. Some of the greatest virtuosi came from the humblest and least musical homes, where neither parent had talent; some of the lesser ones came from highly musical backgrounds, with both parents professional musicians. Corresponding situations were found in our vocalist and Juilliard groups and were consistent with the stories of many other musicians outside of our study.

Such a lack of direct and consistent correlation between musical achievement and background would suggest strongly that musical talent does not arise from any unusual *home environment,* per se. Further, the fact that a highly musical environment also (or alone) cannot produce talent was shown by data for the children of the virtuosi in our study, about one-fourth of whom were sadly reported by their fathers as being "without talent," and most as having no unusual talent, despite all the opportunity given them to develop it.

But if environment is not the determining factor in musical talent, can we show that *heredity* is? To do this, we must first offer evidence that talent runs in some unusual way in the families or near-relatives of those musically gifted. This is what our data did reveal. As our Summary Table indicates, talent or musicality in some degree was reported by the virtuosi instrumentalists in almost half the mothers, three-fourths of the fathers, and half the brothers and sisters; and in the other two groups, also, talent was reported in a preponderance of

Table VII

Music-Talent Study* Summaries

1. *Combined Data for Three Groups of Musicians and Singers*

	Virtuosi Artists (36)	Metropolitan Singers (36)	Juilliard Students (50)	All Groups (122)
Average Age Talent Expressed	4¾ yrs.	9¾ yrs.	5¾ yrs.	6⅔ yrs.
Mothers Talented or Musical	17 (47%)	24 (67%)	37 (74%)	78 (64%)
Fathers Talented or Musical	29 (81%)	25 (69%)	29 (58%)	83 (68%)
Brothers and Sisters, Total	110	103	72	285
Talented or Musical	55 (50%)	43 (42%)	51 (71%)	149 (52%)
Number with Talent in Other Kin	13 (36%)	16 (44%)	37 (74%)	66 (54%)

2. *Instrumental Virtuosi Who Participated in Study*

Abbreviations: comp.—composer; cond.—conductor; pian.—pianist; vio.—violinist.

John Barbirolli (cond.-cellist)
Harold Bauer (pian.)
Artur Bodansky (cond.)
Alexander Brailowsky (pian.)
Adolf Busch (vio.)
Walter Damrosch (cond.-pian.)
Mischa Elman (vio.)
Georges Enesco (vio.-comp.)
Walter Gieseking (pian.)
Eugene Goossens (cond.-vio.-comp.)
Percy Grainger (pian.-comp.)
Jascha Heifetz (vio.)
Myra Hess (pian.)
Ernest Hutcheson (pian.)
José Iturbi (pian.-cond.)
Fritz Kreisler (vio.-comp.)
Josef Lhévinne (pian.)
Yehudi Menuhin (vio.)
Nathan Milstein (vio.)
Erica Morini (vio.)
Guiomar Novaes (pian.)
Eugene Ormandy (cond.)
Gregor Piatigorsky (cellist)
Serge Prokofieff (pian.-comp.)
Sergei Rachmaninoff (pian.-comp.)
Artur Rodzinski (cond.)
Moritz Rosenthal (pian.)
Artur Rubinstein (pian.)
Artur Schnabel (pian.)
Toscha Seidel (vio.)
Rudolf Serkin (pian.)
Jan Smeterlin (pian.)
Joseph Szigeti (vio.)
Arturo Toscanini (cond.)
Alfred Wallenstein (cellist-cond.)
Efrem Zimbalist (vio.)

3. *Metropolitan Opera Singers Who Participated in Study*

Women: *Rose Bampton, Lucrezia Bori, Karin Branzell, Hilda Burke, Gina Cigna, Susanne Fisher, Kirsten Flagstad, Dusolina Giannini, Helen Jepson, Marjorie Lawrence, Lotte Lehmann, Queena Mario, Grace Moore, Eide Norena, Rose Pauly, Lily Pons, Rosa Ponselle, Elisabeth Rethberg, Bidu Sayao, Gladys Swarthout.*

Men: *Paul Althouse, Richard Bonelli, Mario Chamlee, Richard Crooks, Charles Hackett, Frederick Jagel, Jan Kiepura, Charles Kullman, Emanuel List, G. Martinelli, Lauritz Melchior, Ezio Pinza, Friedrich Schorr, John Charles Thomas, Lawrence Tibbett.*

* The study was made by the author for his book, *You and Heredity*, in 1938-39. Complete data and details of the study, including names of Juilliard students who participated, will be found in the afore-mentioned book.

the parents and siblings. Thus, in the families of all three unrelated groups of performers, differing in many ways but having the common denominator of marked musical ability, there was very much the same extraordinarily high incidence of talent.

But while the foregoing and other facts, brought out previously, are significant, much more is needed to prove that musical talent is indeed inherited. There must be a demonstrable *biological basis* for it; that is, there must be certain constitutional traits or capacities which would make possible such talent, and which could conceivably arise through the action or influence of specific genes. A prime requirement, then, would be to identify these biological ingredients of musical talent.

That is what a number of musicologists have sought to do. Pioneering in this work was the late Professor Carl E. Seashore who broke down musical aptitude (which is not quite the same as talent, as we shall see) into what he considered its components—the senses of pitch, time, intensity (loudness and softness), harmony, rhythm and tonal memory. For each of these senses he devised tests to discover to what extent they not only were present in individuals, but could be cultivated. Subsequently, additional tests were devised (Drake, Whistler-Thorpe, Lundin, Aliferis) in which musical aptitudes were grouped under such classifications as interval discrimination, mode (or chord) discrimination, melodic sequences, and musical imagery. Through the use of the various tests on many thousands of individuals in the United States and other countries, it has been possible to reach these conclusions:

The primary senses required for musical aptitude do appear to have a constitutional basis.

Each sense may be present independently of the others. (One person may have a keen sense of pitch and little sense of harmony; another may be blessed with a combination of all the senses.)

Training can develop any of these senses only to the degree that the capacity is *inherent* in the individual.

By the time a child is ten, his or her future musical performance can be quite clearly determined, and at sixteen an individual is musically set.

The musical aptitudes may be unrelated to intelligence: A highly intelligent person may be almost devoid of musicality, whereas a nitwit may be highly musical (there being many cases of morons who are fairly good musicians).* However, for the maximum development of musical capacity and success in studying and pursuing music as a career, more than ordinary intelligence may be required.

* One such remarkable case is discussed in a footnote in the next chapter, page 419.

One of the senses given most study is that of *pitch*, especially in its highest form, absolute pitch, which might be described as a "mental tuning fork." It enables the fortunate musician or singer so endowed to hit any note accurately, or to judge the accuracy of any note, without the aid of any instrumental cue. Often, among prodigies, it is one of of the first musical aptitudes revealed, but it is not essential to great musical achievement. While many of the virtuosi in our study reported having this gift, the large majority did not. That absolute pitch is inherited has been suggested in a number of studies, among them of families where the rare trait appeared for several generations (in Kirsten Flagstad's family, for instance). Also, among identical twins, both usually tend to be much alike in pitch discrimination or the lack of it, whereas there is little such correlation in fraternal twins.

Musical memory is another aptitude which gives strong evidence of being inborn in individuals in different degrees. Without a remarkable facility for remembering all of the nuances and complexities of musical compositions, it would be virtually impossible to achieve eminence in musical performance. Thus, musicologist Raleigh M. Drake places musical memory first among the requirements for musicianship and notes that from Beethoven through Toscanini and down to the great virtuosi of today, all without exception have had phenomenal musical memories. Toscanini, for instance, could memorize the score of a complete opera in one night and could play music he had not seen or heard for decades. Another titan among conductors, Sir Thomas Beecham, was known to have rushed to the podium at the last minute, picked up his baton and whispered to the concertmaster, "What opera are we doing tonight?" Not only musicians, but a great many other persons have the gift of quickly memorizing and accurately repeating complex melodies and songs by singing or whistling, this capacity usually being revealed in childhood. At the opposite extreme are the "Johnny-one-notes" who are unable to sing "Yankee Doodle" or any other simple melody correctly even after repeated hearings.

Studying musical memory, Dr. Richard Ashman found two contrasting pedigrees of Virginia mountaineers. In one, almost all the members for four generations showed great facility for recalling and singing songs (several having absolute pitch). Among members of a neighboring family, on the other hand, most for several generations could not learn to carry a tune, even with training. The facts, Dr. Ashman felt, seemed to point to a strong hereditary element in musical memory, its presence or absence in various degrees being conditioned by certain key genes. (Memory, incidentally, plays a large part in a number of other special aptitudes, including verbal ability, mathematical ability, and chess playing; but whether there is a common genetic basis for different types of memory is not yet known. The subject will be discussed further in the next two chapters.)

Tune deafness quite clearly gives evidence of being hereditary in many cases and in various degrees. According to Dr. H. Kalmus some types of the condition appear to be caused by a dominant gene.* The defect may be unrelated to ordinary hearing (tune deaf people often having acutely functioning ear mechanisms), nor has it anything to do with esthetic sensitivity in other ways. One theory is that tune deafness may be primarily a defect in pitch discrimination and in tonal memory. At any rate, Dr. Kalmus has noted that a gene for tune deafness, if brought into a musical family through marriage, could very well break the line of music-talent inheritance and could help to explain in some cases the unexpected lack of musicality in children of famous musicians and composers.

Another genetic peculiarity that might disrupt the musical-aptitude mechanism is the condition called "cluttering" (or *tachyphemia*), described in Chapter 25. According to speech expert Dr. Godfrey E. Arnold, this condition, revealing itself in a tendency toward jumbled and slurred diction, usually goes with lack of musical ability. Environmental factors do not seem to be the cause of this failing. Various other speech and hearing defects in which heredity may play a part can result in impairment of the musical senses.

Most elusive of the musical aptitudes and talents is that of composing. Composition on the grand scale (such as symphonies) will be left for discussion in the next chapter. Here we may note that even the composing of popular tunes may require some highly specialized and apparently inherited capacity. In an article about composer Richard Rodgers ("Oklahoma!," "South Pacific," "Carousel," etc.), music critic Winthrop Sargeant wrote: ". . . the melodic gift, as Igor Stravinsky recently pointed out . . . is a really mysterious and unanalyzable phenomenon. Some people have it; others don't. You can't teach it. . . . The ability to create successful tunes is a special talent, often lacking

* That there may be a sex-influenced factor in tune-deafness inheritance (or its manifestation) is suggested by a study in England in which it was found that among a sample of more than 16,000 children aged twelve to thirteen, teachers classified 7 per cent of the boys as tune deaf, in comparison with only 1 to 2 per cent of the girls.

among the most learned and technically accomplished musicians and sometimes found paradoxically among men of the most rudimentary knowledge." (The critic added that Rodgers' musical knowledge is "far from rudimentary," and that he appears to be able to turn his tune-making talent "on and off with an ease that few of his contemporaries possess.")*

From all the findings about the various ingredients of musical aptitude, we may gather, then, that there must be a special something in each person, from his earliest years onward, which governs the limits of his musical aptitude. But *aptitude* does not necessarily imply true talent in the way that musicians understand it. A person may have well-developed senses of pitch, rhythm, time, and tonal memory and yet be as mechanical as an old nickel-in-the-slot player piano. Aptitude is primary and essential equipment for the musician, but while, with it, he can develop technique, something more is needed for real talent and virtuosity. The lack of it may explain why many prodigies fail to pan out in their later years or why many musicians who can put on a dazzling display of technique leave their audiences unmoved.

The extra something required for talent is a lot harder to define or measure than is musical aptitude. It may lie, as Professor Seashore believed, "in the ability of the person to create and express subtle variations permitted within the limits of any musical composition." His nephew, psychologist Dr. Harold G. Seashore, described this special musical requirement as "the artistic deviation from the pure, the true, the exact, the rigid, the even and the precise"—in short, in esthetic departures from mechanical reproduction of the musical score, and in the coloring given the composition by the musician's individual subtle "errors" in pitch and nuances of timbre and tone. To illustrate, the senior Dr. Seashore showed with unusual graph-recordings that each virtuoso "deviates" in a characteristic way, and that such deviation must be dictated by extreme sensitivity, great emotion, and high intelligence (or, at least, a particular kind of "musical intelligence").

The afore-mentioned qualities, we may assume, are probably some of the components of great musical talent. But in addition there must be special traits or equipment for each kind of performer: For the pianist and violinist, unusual powers of muscular coordination and nervous stamina; for the singer, the right vocal apparatus, *plus* appearance and personality; for the conductor, prodigious musical memory, imagination, executive ability, and so on. Quite clearly, there can be no simple

* From a profile in *The New Yorker*, November 18, 1961.

formula for all types and degrees of musical talent. Nonetheless, just as there are certain basic requirements for aptitude, there are certain additional basic components needed to give a musician in any area that extra boost toward talent. As with the aptitude components, these talent components may very well have a *constitutional* basis and be in considerable measure products of heredity. If so, let us see how—theoretically—the gene mechanism for musical virtuosity might work.

First, before any talent can develop, there must be a foundation of musical aptitude. So an initial assumption would be that there are specific genes for each of the senses that comprise musical aptitude. To inherit the required combination of such genes (of the average type) would not be unusual, for they are apparently in wide circulation. Almost anyone who can carry a tune or learn to play a musical instrument must have them in some degree. Conversely, the less frequently encountered individuals who are tone deaf (or tune deaf), or lacking a sense of rhythm, harmony, or musical memory may be considered as having defective genes for the senses involved, just as color-blind persons have defective genes for color perception. If even a single one of an individual's musical aptitude genes were deficient, it could throw a monkey wrench into his whole musical machinery.

We can assume, therefore, that in a person who is predisposed to musical talent all of the required genes are present in good working order. But, apparently, to achieve virtuosity there must be in addition certain rarer genes which act either to intensify the effects of the more common ones or to produce some unusual supplementary effects. What clues have we as to the presence of such "extra talent" genes?

If we turn again to our study of the three groups of musicians and singers, an analysis of the incidence of talent in their families shows this:

Where both parents had musical talent, more than 70 per cent of the brothers and sisters of the individual reporting also had talent.

Where only one parent was talented, there was talent in 60 per cent of the brothers and sisters.

Where neither parent was talented, only 15 per cent of the brothers and sisters had talent.

In each of the groups the talent incidence followed this similar pattern: There were about 12 to 15 per cent more talented offspring resulting from "double-talent" matings than from the "one-parent-talented" matings, and a strikingly small proportion of talented children (other than the one studied) were produced where neither parent was talented.

The foregoing talent ratios roughly conform to what might be expected in the *polygenic,* or multiple-gene, mechanism of inheritance (explained in Chapters 26 and 31 with respect to intelligence). If we assume that the general and lesser musical aptitude genes were abundantly present in these families, the existence also of a few dominant "key" genes would make possible transmission of talent to children in the manners shown. If there were no more than two or three different dominant "key" genes required, a single parent with talent could pass them on in a cluster to several children. If both parents were talented and had the genes, more children would be likely to receive the required cluster. But if neither parent had the full set of "key" genes, but each had part of the set, the required talent cluster could still be assembled and passed on to a child, although the chance of this happening with two children in the family would be much reduced.

This, then, could largely explain how, on the one hand, musical talent has appeared profusely and successively in some families, for several generations (as in the case of the Fishbergs, shown facing page 408); and, on the other hand, how remarkable talent has cropped out in an individual neither of whose parents was talented and has then not reappeared in that individual's children—as with the Toscanini family. One of the greatest musicians who ever lived, Toscanini has a family history (as it was given to us) that showed no musical talent in either of his parents, or in his two sisters and a brother, or in any other close relative. Yet unquestionably his parents had carried between them the required "virtuoso-talent" genes. Possibly part of the special gene cluster had been carried along or gradually assembled on each side of the maestro's family for generations, the full cluster coming together for the first time in him. When Toscanini married, however, it was to a woman who was described as "mildly musical but not talented." Again it is not surprising—in view of how "talent" genes can be dispersed—that none of the maestro's three children (a son and two daughters) showed musical virtuosity, although one, Wanda, was regarded by her father as having an acute musical ear and marked critical ability. When this daughter married pianist Vladimir Horowitz, a new supply of great "talent" genes came into the Toscanini family, and among the maestro's grandchildren, only the daughter of the Horowitzes showed early talent for piano playing (which, however, she did not cultivate to any extent).

The facts about the offspring of other virtuosi instrumentalists who were married and had children at the time they were queried also were in line with what might have been expected. There were eighteen of these virtuosi parents—all men—and all but four had musically talented wives. With this heritage and background, it was not surprising that of their 37 children produced up to that time, 27 were reported as talented. However, in only one instance had the talent then carried to the point of virtuosity, in the case of Artur Schnabel's son, Karl Ulrich. One of Rudolph Serkin's sons, Peter, born later, made his debut as a piano prodigy a few years ago. Another virtuoso whom we've mentioned, but who was not in our study, David Oistrakh, has also produced a virtuoso son, Igor. But on the whole, the rare combination of "talent" genes, plus the special circumstances and personality factors required for the achievement of virtuosity, would make its occurrence twice in the same family exceptional.

Of equal interest, fully one in four of the children of virtuoso fathers, and, usually, talented mothers, did not show any musical talent, *despite having an intensely musical environment*. One can therefore say that while a favorable environment is unquestionably essential for the development of musical talent, even the very best musical environment—the most musical home and parents, the best instruction, the most avid interest and most arduous application—cannot *create* talent. At least, no one has ever shown that a child without evidence of inherent talent can be turned into a prodigy by training. A good many parents have tried it with their young hopefuls and have been disillusioned.

On the other hand, while an unfavorable environment can suppress musical talent, there are many instances of great musicians who evolved despite strong negative influences. Not infrequently neither parents nor teachers had recognized the future virtuoso's talents when he was a child, and often it was only on his own insistence, through his own efforts, that he was able to develop these talents. As one example, it would seem clear today that Leonard Bernstein—a virtuoso conductor and pianist, as well as composer—had all of the essential musical equipment to have been a child prodigy. But, as he has reported it, there was no awareness of this for many years. Not until he was ten was there a musical instrument in his home (an old upright an aunt wanted to get rid of), and only then was he allowed to take piano lessons; and, when the opinion was then expressed that he was a born musician, there was strong parental opposition to his training for a musical career (his father wanting him to go into something more "practical"). Contrary to the story with most virtuosi, therefore, Bernstein evolved gradually, his complete concentration on music not coming until after he was graduated from Harvard College (where, by the way, he had been rejected for the job of second Glee Club accompanist!). Here is a case where dynamic personality factors—and what could be called an unusually favorable "internal" environment—were able to push great talent through an unfavorable external environment.*

Where the "veto" power of environment—comprising perhaps internal as well as external environment factors—may best be seen as hampering full expression of musical talent is in *women*. Consider the very small percentage of women among the great virtuosi (with none, actually, rated in the topmost levels). Even more surprising is the almost total absence of women from the ranks of great composers. This certainly cannot be because women inherit any less talent potentiality than do men. We know that whatever genes for talent there are must be carried equally by women—and even in a greater amount if there were any such genes on their additional X chromosome. One explanation undoubtedly is that women never had, and still do not have, the same opportunity to develop their talents as men are accorded. Many repressive social and psychological factors could be cited. But there also may be something in the *biological makeup* of women, as a group—or, as said before, in their internal biochemical environment, as related to glandular and sensory functioning, energy and drive—which may act in an inhibitory way.

Asked why she thought there were so few other women virtuosi, Erica Morini, the violinist, wrote to the author, "In my opinion it is because only a few women have the great power of concentration, the strength and the energy required for such achievement. A complete absorption is necessary, and a readiness to give up the pleasures which most women seek." But even where all the qualifications are there, Miss Morini believed that prejudice against women musicians may be one of the principal reasons why, in the past two and a half centuries, records show only a few notable female violinists as compared with thousands of males. With regard to the similar paucity of women among piano virtuosi—despite the fact that many more girls than boys are started off on this instrument—an added explanation may lie in the extra stamina demanded for all-out and sustained piano performances.†

* Bernstein's unorthodox start may further explain why he is one of the few classical virtuosi who have also been successful in the popular-music field, as in his scores for such musicals as "On the Town" and "West Side Story." While we have given only scanty attention here to popular music—principally because no genetic studies have been made in this field—one cannot doubt that many great popular musicians have had the endowment for classical-music virtuosity, and would have achieved this had they not been swerved away by environmental factors and/or personal inclinations.

† Expanding on why relatively so few women musicians reach the top in the concert world, music critic Harold C. Shonberg (*New York Times*) has held that a major reason is the coolness of audiences—particularly of

However, Myra Hess, Guiomar Novaes and Jeanne-Marie Darré are among those who met these demands; and several other women virtuosi pianists, after brilliant concert careers, have been outstanding as teachers, notably Olga Samaroff and Rosina Lhevinne (the latter having had Van Cliburn and John Browning among her pupils). As another example, Nadia Boulanger, in addition to her achievements as organist, conductor and composer, has been rated the foremost contemporary teacher of composition, her pupils including such American composers as Aaron Copland, Virgil Thomson, Roy Harris, Marc Blitzstein, and, for a time, George Gershwin.

In the vocal field, where women are represented equally with men, there is no such sex discrimination as among instrumentalists, for the obvious reason that women and men singers do not compete on the same grounds, being cast in different roles and voice parts. For both sexes the veto effect of environment on singing talent is greater than it is on instrumental music talent, and it may work differently on women than on men. First is the question of voice, the *instrument* on which the vocal musician plays. At the outset, regardless of what talent may be displayed in childhood, singers are dependent on what happens to their voices during puberty. This explains, for one thing, the report by the Metropolitan Opera stars that their formal training did not begin until the average age of 15½ for women, and 16½ for men (although in both cases musical talent was reported as showing itself about seven years earlier). Regardless of talent, the singer is continually at the mercy of his or her vocal cords. His situation is unlike that of the violinist who, if a string breaks, puts in a new one, or if his instrument cracks, can get another.* Also, *looks*, as well as personality, and not least, just plain

---

women in the audiences—to female performers. This may relate to the feeling (shared by persons of both sexes) that the strenuous playing of musical instruments is "unfeminine." To quote critic Schonberg: "It is a historical fact that virtuosos are popular in direct ratio to the physical strength they employ. . . . Playing any instrument is a conflict in which the instrument [and the orchestra] must be dominated, and—generally speaking—men are better dominators than women, if only by virtue of their size and strength. . . . The more triumphantly an artist survives the ordeal, the greater the tension released in his audience. And that is why, by and large, men have an easier time of it. Many women play much more beautifully than men, but men can provide more the physical excitement that any audience, no matter how sophisticated, comes to share."

* The much greater role played by environment in singing than in instrumental careers is revealed in the relative toll taken by time and age. Among the instrumentalists in our study, virtually all have remained (or if deceased, had remained) professionally active throughout their years. But

luck, are far more important to the vocalists than to the instrumentalists. These and other outside factors may account for the public success of many mediocre singers, and the failure of many truly talented ones, to a much greater degree among women than among men.

The fact that the voice is used as the instrument on which the vocal musician performs suggests the possibility that, again unlike other musicians, the great singer may inherit both his instrument and the talent to perform on it. However, while there is good reason to believe that heredity is involved in the production of vocal mechanisms and capacities of all types, genetic studies to substantiate this have as yet been limited.* It is clear, of course, that the principal differences between male and female voices are genetically determined characteristics, resulting from the contrasting ways in which the vocal mechanisms develop from childhood on, but mainly at puberty under the influence of differential sex hormones. The larynx in the male throat then becomes considerably larger, and the vocal cords relatively thicker and longer, than those in the female throat. Castration of a male *before puberty* will inhibit further development of the male "voice box" and vocal cords, and hence of the normal deep voice of the mature male (a fact which accounts for the practice of castrating young male choir singers in Italy up to a century ago). After puberty, however, castration produces only mild, gradual changes in the male voice, although in the later years there is a natural tendency for men to develop more highly pitched voices.

With respect to vocal timbre, one theory is that bass voices in men and soprano in women are determined by a pair of one kind of "voice" genes; that a gene pair of another kind produces tenor voices in men and

---

among the vocalists none of the surviving women stars of twenty-five years ago remain in the operatic ranks, and only a few continue to make concert appearances. Time dealt more kindly with the male singers, most remaining (or having remained) active, if not in opera, then in other areas of singing well into middle age or beyond.

* The facts about singing capacities and qualities in birds may or may not have application to human vocal inheritance. The great differences among birds of different species and strains in voices, singing ability, musicality, types of songs and the urge to sing unquestionably are genetically determined for the most part, although training and imitation have been found to influence the singing expression of birds. Studies of canaries have shown that those of different species inherit tendencies to sing in particular ways. Even when individual canaries are hatched and reared in isolation, with no chance to hear other birds, they nevertheless develop the songs characteristic of their species.

VOICE INDIVIDUALITY

Shown here is how voices can be translated into distinctive "voice prints," such as these of the author and his wife each saying "Male and female voices differ." *Locations of patterns made by specific sounds and words, which guide voice-print experts in identifications, are noted below each print.**

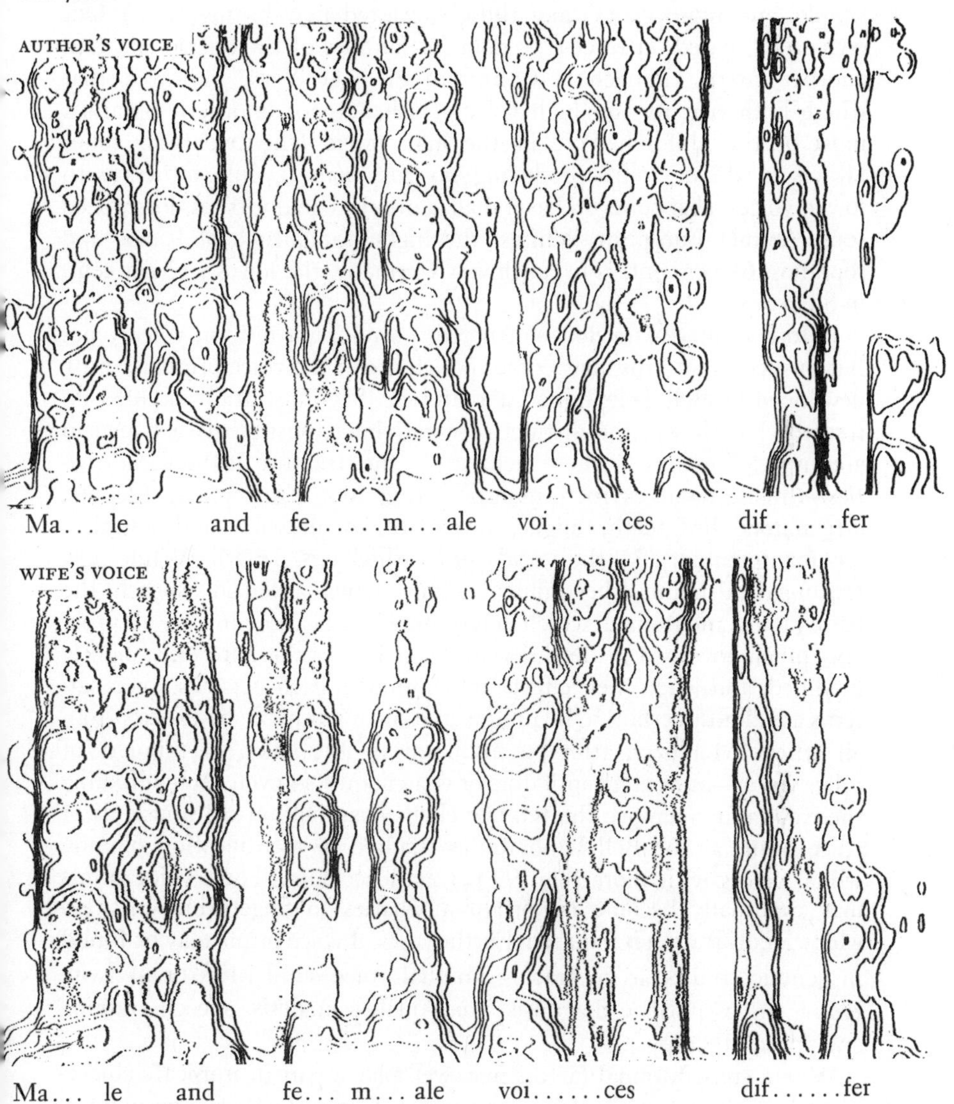

Note similarities in patterns on both prints made by the initial "M" sound, and, at right, by the word "dif...fer." The male-female clues include the lower pitch-frequency patterns on the male print (although differences here are narrowed by the author's voice tending toward the tenor, his wife's toward the contralto, producing much less contrast than between a bass and a soprano). However, the principal differences are unrelated to sex but stem from the individuality of the voices. *(Further details in Text, p. 404).*

* Voice contour prints prepared and data provided by Lawrence G. Kersta, Bell Telephone Laboratories, Murray Hill, New Jersey.

alto in women; and that a mixed pair of these genes—one of the first kind coupled with one of the second—produce baritone in men and mezzo-soprano in women. (The genetic mechanism is probably not so simple, and more genes than those suggested may be involved.) Our own study, based on voice-type data for families of the opera stars, and another group, the Schola Cantorum of New York, indicated that where both parents had high voices (father, tenor; mother, soprano), most of the children were in the high range (the twenty-six tenor fathers in these groups producing no bass sons); while fathers with lower voices (baritone or bass) had a marked majority of low-voiced sons. Among European Whites, the highest incidence of bassos and sopranos, 61 per cent, is in northwest Germany, the lowest, 12 per cent, in Sicily.

(More accurate knowledge of heredity's role, not only in voice types but in speech patterns, may come though use of the voice prints recently developed by Bell Telephone Laboratories. [See illustration, p. 403.] As research has shown, the distinctive vocal characteristics of each person are imparted mainly by these factors: (1) the structure of the *vocal cavities*—throat, nasal cavities, mouth—acting like organ pipes; (2) the *articulators*—lips, teeth, tongue, soft palate, jaw muscles—and how they are manipulated; (3) the *vocal cords*—thickness, length. While age, training, health and other influences may change a person's speech, the basic factors mentioned, which determine identification and are all probably much governed in their development by heredity, are only slightly modified; and when their effects are combined, each person's voice patterns are almost certain to be unique and distinguishable by experts from all others. Moreover, attempts at disguising the voice, or mimicry of other voices—as by stage imitators or ventriloquists—even though fooling the ear, cannot remove the tell-tale clues. Examination of thousands of voice prints at the Bell Laboratories resulted in identification of individual voices with more than 99 per cent accuracy. Thus, voice prints may eventually become important auxiliaries to fingerprints in areas where identification is required. Further uses of voice prints may be found in genetic studies, as in tracing familial voice similarities or in twin-typing, in psychiatric diagnosis, voice-quality analysis, speech training and speech therapy.)

Where constitutional factors may well play a part in musical achievement is in the selection of the instrument. Granted that the basic musical-talent requirements are much the same for virtuosity in any field, it still is apparent that each type of instrument may demand, for maximum success, certain special physical capacities (as well as temperamental traits) in varying degrees. (Think of yourself, for instance, in relation to playing a piccolo, a trombone, a bass horn, a harp, a violin, a piano, or a snare drum.) Sometimes the talented child may be started on the wrong instrument and may be impeded by it unless he's changed in time to what he's best suited for. A number of virtuosi who made their debuts as prodigies on one instrument did turn early in their careers to another, among these being the pianists Rubinstein and Harold Bauer, who began as violinists. Toscanini started out as a cellist, mainly because such a player was most needed at the time by his town orchestra; and had he not become a conductor, one might well speculate whether he could have equally fulfilled his virtuosity by continuing to play only the cello, or even another instrument. Also worth noting is that cellist Pablo Casals, when asked whether he would specialize on the cello if he could start over again said, "No, I would become a pianist. I became a cellist by accident."

There is, however, little foundation for the various popular beliefs that certain physical traits go with and are needed by instrumental musicians. One notion, for example, is that inheritance of a particular shape of hand (a "piano" hand, or a "violin" hand) predisposes one toward talent for the instrument in question. Actually, the basic shape of the hand or fingers has little to do with virtuosity, for every type of hand will be found among musicians, any manual characteristics they share being developed through training. While one hand may be structurally better than another for perfecting technique, an outstanding pianist, Josef Hoffmann, said, "My hand is a bad one for the piano—too small, my fingers not long enough for everything—and my technique is limited. I have pupils who have better technique than I." And need it be said that there's nothing to the notion that a certain type of face, or a certain kind of hair, goes with musical capacity? If musicians have more sensitive faces, credit their way of living. For the appearance of their hair (if they have any), credit their barbers.

One of the hardest questions to answer is, "*Don't people of certain races or nationalities inherit more musical talent than others?*"

The surface evidence might have seemed most convincing in former years: the Italians, with their unusually large number of great composers, singers and musicians, and their domination of grand opera; the Jewish people, with their remarkably high representation among the virtuosi musicians (most of the first-ranking violinists and a large percentage of the contemporary pianists, conductors and composers, are Jewish); the Negroes, with their many wonderful spiritual and blues singers, and jazz musicians and composers.

But all this may prove little with respect to inheritance, for, as most musicologists would point out, various environmental factors could account for marked differences among groups in musical expression. The Italians, with several centuries of unusual popular enthusiasm for opera, and concentration on this medium, could have been expected to produce opera composers and singers prolifically, just as Americans, with different environmental stimuli, became pre-eminent in jazz and popular music. But in recent decades a growing interest among Americans in opera, as well as in classical music generally, has led to their increasing representation among grand opera stars, virtuoso musicians, and classical-music composers.* Of special interest, Negro opera stars have begun to emerge for the first time, among them Leontyne Price and Grace Bumbry—one indication of how great talent among these peoples was previously suppressed.

Among Jews, it is argued, various cultural influences inspire an urge toward musical expression and a drive toward achievement in this field much greater than one finds in most other peoples. This might help to explain their exceptional production of prodigies and virtuosi among musicians.† Likewise, special aspects of the Jewish culture—sentimentality, humor, interest in the stage and entertainment—might account for the remarkable fact that perhaps a majority of the recent and current leading American composers of popular songs and musical comedies have been or are Jewish, among them Rudolph Friml, Sigmund Romberg, Irving Berlin, George Gershwin, Jerome Kern, Richard Rodgers, Kurt Weill and Frederick Loewe. So, too, special factors in the cultural environments of any other large groups of people, at given times, might explain to a large extent not only the amount of their musical expression, but its characteristic forms, types and moods—as, for instance, the music of different European peoples and of Asians and Africans.

* An amusing fact is that Americans had formerly been credited with so little talent for musical virtuosity that some of those aspiring for concert careers changed their names (if they sounded too American) and took foreign names—for example, Russian or Polish names, if they were violinists or pianists, and Italian names, if they were singers.

† Most surprisingly, even in the Soviet Union, despite the fact that Jews comprise scarcely 1 per cent of the population, and that for decades stringent official efforts have been made to suppress or restrict their special culture, there has nonetheless been an extraordinary outcropping of musical virtuosity among them. In fact, of the top Soviet instrumentalists who have appeared in the United States to date, most of them—including the Oistrakhs (father and son), Gilels, Kogan, Shafran, Ashkenazy, Oborin, Fliere, Gutnikov, Barshai and Zak—are Jewish (as confirmed for the author by impresario Sol Hurok).

Yet with regard to *degrees* of musicality, possible hereditary influences must also be considered. Theoretically, if genes for superior musical aptitude and talent are unusually prevalent in some families, it would be entirely possible, on a broader scale, for these genes to be more heavily concentrated in some races, ethnic stocks or nationalities than others (just as genes for certain blood groups, or for certain diseases and abnormalities, are disproportionately present in some racial and ethnic stocks). However, this is still only theory, for we can never be sure to what extent "musical-talent" genes are or are not present in different populations so long as they vary radically in their opportunities for musical expression. So far, tests for musical aptitude—not necessarily synonymous with talent, as was pointed out—among members of different racial groups, such as Negroes and Whites in the United States, have not revealed any significant differences in the basic sensory capacities (pitch, loudness, or time) or even in *rhythm.** (This finding comes in the face of widespread notions about the "natural rhythm" of Negroes.) Differences that were found could be explained on the basis mainly of training and conditioning in various areas of music. But whether the rare special capacities (or genes) for the highest types of musical virtuosity also are equally distributed among all peoples is a question which cannot yet be answered.

For the present, there is general agreement that the relative amount of musical *achievement* in any given human group—racial, national, social or however classified—is nowhere near what it could be if all of those with talent were properly identified and musically educated. To quote what Moishe Menuhin, the father of Yehudi Menuhin, wrote to me, "I am sure that there are many other young men now nobodies who might have become as great artists as my son if their talents had been immediately recognized by their parents and they had been given equal opportunities for training and development." But also, there always has been the awareness in the inner sanctums of music that only those individuals can be trained to virtuosity who start out with unusual equipment. Pianist Jan Smeterlin expressed the consensus of the musicians and singers in our study when he said, "Heredity without train-

* In one study of musical-rhythm aptitude among Negro, Anglo-American and Spanish-American high school girls in Arizona, sociologists Henry L. Manheim and Alice Cummins found that, if anything, the White girls had (or expressed) more rhythm than the Negro girls or those of Latin stock. It was strongly indicated that any group differences in musicality were much influenced by degrees of exposure to music and rhythm in or outside the homes (music lessons, records, television, dancing, church singing).

ing would not go far, but training without heredity would not go anywhere."

In sum, while all the finer shades of the question can be argued to infinity, it seems conclusive that musical talent, as a human trait, follows much the same pattern of development as do many other traits with an hereditary basis, namely:

What is inherited is a "talent susceptibility," dependent for its expression on the interoperation of many factors—some from within the individual (sensory, mental, emotional, physical) and some from without (home, social and cultural environments, opportunities for training and recognition). In some individuals, the susceptibility to musical achievement is so strong (talent plus *drive*) that even in a minimal or adverse environment it will assert itself. And also, just as there are some few individuals with inherited musical capacities, which can be developed to an astounding degree, there are others so limited or imperfect in genetic capacities that no amount of training can make them musical.

How does this apply to you? If you don't happen to be a musical virtuoso (assuming you would have wanted to be one), you may now be clearer as to why you aren't and can more definitely place the blame on your genes, your own efforts, or your environment. It may be too late to do anything about yourself; but if you're a parent, it may be important to consider the musical prospects of your child. From the evidence, you can see that even if you haven't any musical talent and your mate hasn't either, you may still have a child who is talented, so keep an eye open for that possibility and allow for its development. On the other hand, if you and your mate are highly musical, don't consider it unnatural for a child of yours to show no talent, and don't try to force musical training upon him or show disapproval. (Where there is doubt, a musical aptitude test is advised; and if no talent is revealed, try developing some other interests in the child instead.)

With awareness of the basis for musical talent there should be a wiser understanding of which children to train intensively and which ones to teach music to only for pleasure or cultural development. While many a musical genius may have been lost to the world through failure to receive training, there also has been many a tragedy because parents have tried to turn a Jimmy into a Jascha.

Photograph under direction of author by M. Lasser, New York, 1950.

A MUSICAL PATRIARCH

The 102-year-old flute player, Isaac Fishberg, shown here with five of his musician sons shortly before his death in 1951, was the progenitor of the largest musical pedigree in the United States and, probably, in the world. All of his twelve children were highly musical, seven of them being professional musicians (five, concert artists or with major symphony orchestras), the others amateur musicians; of 37 grandchildren, 14 were professional and 4 amateur musicians; and among the scores of great-grandchildren and later descendants to date, there are so many more professional and amateur musicians—with still others blossoming—that the family has stopped keeping count. (Numerous marriages with other musicians throughout the family line have intensified both the hereditary and environmental influences.) The elder Fishberg, himself the son of a violinist, came to the United States with his children from the Ukraine in 1920 and played professionally until he was 90. (His wife was also reported to have been musical.) The sons shown in the photograph are, left to right, Yascha, Mischa Mischakoff, Tobias, William and Fishel.

CHESS PRODIGIES: KINGS, OLD AND NEW

1. *Sammy Reshevsky, age 8, preparing to play Chess Master Edward Lasker at home of Chicago philanthropist Julius Rosenwald (center, looking on) in 1920. Others shown are Sammy's parents (behind him) and guests who withdrew when game began. (A few days later the author was present when Sammy engaged twenty players simultaneously in an exhibition match at Milwaukee.) After a period of retirement from public chess playing to complete his schooling (financed by Rosenwald), Sammy began formal competition and won the United States chess championship in 1936.*

2. *Bobby Fischer, at age 11, shown when first competing in a United States chess championship tournament, in 1954. (The absorbed onlooker is Chess Master Hans Kmoch.) Three years later, at age 14, he won the championship (which he still held in 1964) and at 16 became an International Chess Master.*

3. *Fischer (at age 18), the reigning United States chess champion, versus Reshevsky (at age 49), the long-reigning former champion (1936-1944 and 1946), in a non-title match in 1961. While he quite frequently won individual games from Fischer, Reshevsky did not defeat him in any championship tournament. (For further details of these and other chess prodigies, see Text, Chapter 33.)*

Mr. Lasker provided the early photo showing him with Reshevsky, and took the photo of Fischer at age 11. Photo of Fischer versus Reshevsky courtesy of *Chess Life*.

CHAPTER THIRTY THREE

The Gifted Ones: II

Special Talents

"You, too, can paint!" "You, too, can compose songs!" "You, too, can write stories!"

So you are told by the "Anyone-can-do" advertisements and the "How-to" books. They are correct—up to a certain point. Anyone can indeed learn to draw and paint, compose, write. But can anyone become a *great* painter? A *great* composer? A *great* writer? And if not, why not?

"You're born with it, or you aren't born with it." If it can be said of great musical talent, it can presumably also be said of capacities for great achievement in other arts and specialized fields. Proving this is another matter.

In our music-talent studies we confined ourselves, remember, to *performing* and *interpretive* ability, whereby innumerable individuals of all kinds and backgrounds could be compared by exactly the same standards—having them play the same compositions or sing the same songs, and then judging their relative capacities accordingly. But of all the arts, only in music—and only in musical performance and interpretation—is this possible. Had we ventured into musical *composition*, we would have been beset with many difficulties. For composing, as we pointed out, is a *creative* art, and as in any other creative art—painting, sculpture, writing—the most important elements are individuality, imagination, inventiveness, emotion, esthetic judgment and other such intangibles.

What, then, can we say about the inheritance of creative talents? As genetics can deal only with *measurable* traits, there must first be positive criteria for creative achievement and definite tests for measuring its potentials before any clear conclusions can be drawn as to the inherited factors. These tests are still to be perfected.* Nevertheless, many popular misconceptions can already be refuted, and much can be said which should help you better to understand the presence or absence of various types of talent in yourself or your children.

By such standards as we have, evidence of heredity in creative achievement rests principally on case histories of noted composers, painters, sculptors, writers and other creative artists which show in almost all of them patterns very much like those in the musical performers: the early appearance of talent; its cropping out in all sorts of backgrounds, often spontaneously, with nothing special in the environment to explain it; and its high incidence in certain families and strains. On the first point, it should be noted that while creative talents are usually revealed in childhood (though often unrecognized), it may take many more years for them to ripen than for musical talent. Quite definitely, biological and mental maturity is required to a greater degree for creative achievements, and more training and experience also are needed. *A significant fact is that in all history there is no record of a single important piece of creative work done by a pre-adolescent child.*

However, there is enough precocity to impress us in the fact that many composers have turned out notable works when they were still adolescents (as was brought out in the last chapter); that Michelangelo at eighteen was already close to being the greatest sculptor of his time; that Shelley, Byron and Keats were publishing poetry of great promise in their late teens; that Picasso and Dali (and earlier, Landseer and Ingres) were child-prodigy artists; and that many great books and paintings have come from individuals in their early twenties. With these in mind one might indeed suspect that the seeds for exceptional creative achievement are initially implanted by heredity.

Of the creative talents, that for musical composition gives the most direct evidence of hereditary influence, because of its earlier and more spontaneous onset and the specific aptitudes involved in it that seem quite definitely to have a constitutional and familial basis. In painting the evidence is less precise. There are innumerable instances, to be sure, of father and son, or brother painters, such as the Bellinis, the

* Some tests of creativity have been considered adequate enough for indicating strongly that creativity is often an elusive quality not measured by the standard IQ or aptitude tests. In fact, studies of adolescent students by Drs. Jacob W. Getzels and Philip W. Jackson have shown that many of the most creative individuals (as judged by originality, imaginativeness, cleverness, humor, and inventive thinking) failed to score in the upper IQ ranges, whereas many high IQ students were low on creativity by the standards applied.

Bassanos, Fra Filippo Lippi and his son Filippino, the Van Eycks, the Holbeins, and the Brueghels, or among contemporaries, the Wyeths—father Newell, son Andrew, and grandson Jamie; the Soyer twins, Raphael and Moses; and the Albright twins, Ivan Le Lorraine and Malvin. But such histories tell us little, for close contacts and similar training might account in large measure for clusters of painters within given families (though instances are exceedingly rare of two artists in the same family having attained notable distinction).

Nevertheless, there is some evidence that certain capacities making for the "predisposition" toward artistic achievement run in given families in more than ordinary degree. Professor Norman C. Meier of the University of Iowa, studying the backgrounds of hundreds of artists and art students, found in their families, going back for several generations, a much higher than average incidence of individuals with artistic or craftsman ability of one kind or another (wood carving, toymaking, instrument making, textile design, jewelry design, and so on). What is implied here is that a certain general type of hereditary endowment (composed of potentialities for various skills, traits and capacities) may run in families, thus providing a *predisposition* toward artistic achievement; that the development of this endowment would depend in part on environmental influences and in part on personal drives and inclinations; and that with the same endowment, while some individuals might go into the arts, others might enter entirely different fields (science, business, teaching, etc.) where the creative capacities might yet be put to good use.*

But much more research, with dependable tests for specific elements in artistic aptitude, is needed before we can reach positive conclusions regarding the relative roles played by heredity and environment in producing artists. For the present we might make these theoretical observations regarding the requirements for important achievement in any of the creative arts:

First, a person must have sensitivity, keen perception, imagination, emotion. (All great creative works, whether in music, literature, drama, painting, sculpture, architecture or ballet, embody the same basic prin-

* As examples, Leopold Godowsky, Jr., and Leopold Mannes, sons of two notable musicians (Leopold Godowsky, Sr., and David Mannes), and themselves of virtuosi caliber, turned from musical careers to research in color photography and became co-inventors of the Kodachrome process. In another case, that of the three sons of the American artist Lyonel Feininger, one of the sons (Lux) is a painter, a second (Andreas), an outstanding photographer, and a third (Laurence, a priest), a musician and choral composer.

ciples of composition, form, rhythm, and nuances of expression.) These capacities might be considered as resulting from some general kind of hereditary endowment.

Second, one must have the special sensory equipment required for pursuit of a particular art. We already have dealt with the primary musical abilities and have touched on some of the "art-craftsman" capacities. A painter, for example, must have a keen sense of color, but a color-blind person, gifted with the general art capacities, could still become an etcher, a black-and-white artist or a sculptor (although each of these fields might demand certain additional specialized capacities). So, too, we could identify the most important elements in the sensory equipment required for creative writing or any of the other arts. These elements might then in turn be related to special kinds or combinations of genes.

Third, a person's achievement in any art will be governed not only by the amount and kind of his artistic equipment, but by his drive, intelligence, personality, temperament and many other aspects of his social makeup and behavior. It is essential to keep in mind that the purely technical equipments, or aptitudes (inherent and acquired), are not the determining factors for the creative arts. Many mediocre composers, painters, and writers have much better technique than some of the great figures in their fields. (What constitutes superior technique, however, may be governed by arbitrary and changing standards. On the basis of their technique such painters as Van Gogh and Gauguin were classed as hopelessly inferior by their contemporaries, and many modernists and abstract painters of today are equally so scorned by the classicists.) But trying to relate to specific genes such intangibles as do make the great creative artists—"personality" or "temperament" and other characteristics—poses our greatest genetic problem (as we shall see in later chapters which deal with these traits).

And, *fourth,* there is our old friend, *environment,* with all that it can do to develop, suppress or divert creative capacities. A talent may be all dressed up with no place to go if it crops out in an environment where no one wants it. Again, under modern conditions, with insistence on specialization, a person with various talents is forced to decide early which talent to cultivate. The wrong choice may lead him to failure where the cultivation of an alternative talent might have brought great success. Few question that environment has stifled many more talents than have ever come to fruition.

If all the foregoing points are taken together, it is evident that certain combinations of genes, *plus* the required environment running in given families for a time, may produce many talented members; and that if special "artistry" genes of different kinds are present or absent in given individuals of these families, their talents might be expressed in different forms and careers. One member might show special talent for painting, another for writing, another for music, architecture, or acting, and so on (as can be illustrated by many artistic families, some of whom you yourself may know). But also, and always to be considered, are the personality factors and environmental influences which may cause a talented individual to take to one form of art or another or to turn away completely from any of the arts.

The close relationship among the various arts, and the existence of some general basis for aptitude in them, is further indicated by the fact that so many persons—perhaps most—who are talented in one field are usually also talented in other fields. We could fill pages with notable examples, grading down from the most versatile genius of all, Leonardo da Vinci (of whom we shall say more in the next chapter) to the countless multiple-gifted individuals of our own time. To mention a few, there are Deems Taylor, composer, novelist, critic, skilled cabinetmaker; Christopher La Farge, playwright, poet, architect, watercolorist; Noel Coward, playwright, song composer, actor; Zero Mostel, actor, comedian, painter; Al Capp, "Dr. Seuss," S. J. Perelman, Jules Feiffer, all cartoonists and writers (as was the late James Thurber); and Xavier Cugat, dance-orchestra leader and cartoonist (recalling, also, the great singer Caruso, who was also a talented cartoonist). Nor can one overlook the many musicians with talent for painting, among them Nathan Milstein, Efrem Kurtz, Harold Rome, Morton Gould and the late George Gershwin.

We could extend the list of the multiple-talented endlessly and carry it on to a relationship between the arts and other achievements: Benjamin Franklin, writer, statesman, scientist, inventor; Thomas Jefferson, who was not only all of these, but an architect and a talented violinist as well; or, in our own time, Winston Churchill, statesman and military genius, and also a great writer and a competent painter.*

When we turn to such arts as *writing, acting,* and *dancing,* genetic evidence definitely becomes fuzzy. No doubt some special inborn equipment is required for great achievement in each of these fields, but whatever hereditary factors there are appear to be much less specific than

* Even Abraham Lincoln showed another special talent—that of invention—when he took out a patent (in 1849) for a method for lifting vessels over canal shoals, which never came to fruition because canals shortly thereafter began being supplanted by railroads.

those for music or painting. Writers, for instance, might be said to be born to a lesser extent than they are made by their environments or themselves. One can, of course, point to many brilliant writing families—such American literary dynasties, for example, as the Jameses, Adamses, and Lowells, or among later generations, the La Farges, Terhunes and Benéts. But even if we assume that pools of hypothetical "writing" genes may have collected in these families, the strong influences of similar training, education and precedent can hardly be overlooked.

So, too, when in the theater reference is made to a "born actor," something might be made of the fact that there have been whole families of brilliant performers, such as the Booths, the Barrymores, (Lionel, Ethel, John, their parents and their uncle, John Drew), and many others. If there are indeed any "acting" genes, the intensive inbreeding in the theatrical profession, more than in any other, would lead to their concentration in given families. But there is also an equally great tendency in this profession for children to be trained or induced to follow in the parents' footsteps. Many chance elements in the theater also confuse the issue with regard to native endowments. The initial requirement of passing a test for "looks" must rule out a great many of those with the greatest ability for acting, while permitting many with mediocre talents to attain success (most notably in the movies). Here is one of the best examples of how environment may play havoc with hereditary capacities. What would happen to our painting and literature if those who wanted to be artists and writers first had to pass beauty tests?

But all other aspects of achievement in the arts, particularly with regard to the interplay of inherent factors and environment, are dwarfed by comparisons between the sexes. This is a most complicated matter to discuss and a ticklish one.*

If we consider now the sexes as two separate groups, the surface facts tell us that one group has been consistently far superior to the other in creative achievement. In the literary arts (novels, drama, poetry) the critics generally maintain that while many women have reached near-greatness, none can as yet be ranked with the greatest of the men. In painting, with large numbers of women now represented, the critics will again say, "There are many fine women artists. There has never yet been a single truly great one." Most puzzling of all, in music, one field in which countless women have been trained for generations, their *creative* efforts have been so negligible (as any musician will testify)

* The author dealt with this subject in great detail in his book, *Women and Men*, Harcourt Brace, 1944.

that there has never as yet been a single notable symphony, opera, concerto or other truly great piece of music produced by a woman. (Nadia Boulanger, referred to in the preceding chapter as the greatest modern teacher of composition, spoke of her own attempts at composing as "useless music.") Not even in the Soviet Union, which boasts of its equal opportunities for women, has there yet been word of a major musical composition by any female.*

Equally puzzling is the almost total absence of women among *inventors*, so far as records show. Some anthropologists claim that back in the dim past, the loom, elementary gardening devices and pottery making may have been invented by women in the course of their domestic pursuits. Yet in modern times one hears only rarely of an invention by a woman, and never more than a small item (the sapphire phonograph needle, for example, and a beer-can opener). And this despite the increasing contact by women with mechanical equipment in homes or on jobs.

How account for this? As was pointed out when we discussed sex differences in intelligence (Chapter 31), we must make every allowance for the environmental repression of women's capacities, through conditioning, diversion of energies into motherhood and homemaking, lessened opportunities, and prejudices. There is also the psychoanalytical theory that since women can fulfill their creative urges through the greatest of all experiences, childbearing, they are less impelled toward other forms of creative expression, whereas men, frustrated in biological creativity, are driven to express themselves in other ways. Yet, as with intelligence, we cannot ignore the possibility that inherent biological factors may also be involved in the sex differences in achievement. It has been shown that with respect to a great many physical traits the same

* In the United States, as well, the dearth of important compositions by women cannot be ascribed to mere lack of training or opportunity. More girls than boys begin taking music lessons; and graduate music schools (such as Juilliard and Eastman) generally have a larger enrollment of young women than young men. But while equal numbers of both sexes may start taking composition, the further the study advances, the greater is the proportion of girls or young women who drop out. A good indication of the situation has been the annual Student-Composers' contest conducted by the Broadcast Music Corporation. (Limited to classical-music compositions, the works are so submitted that neither the sex nor the identity of contestants is known to the judges.) In the nine years from 1952 to 1960, among 1,033 entrants, there were 180 young women—a little over 17 per cent—and among the 69 winners there were only 6 young women. Again, in the Juilliard School of Music, the proportion of males among students receiving high composition ratings and scholarship awards has been consistently higher than that of females.

genes work differently in the two sexes: in bodily growth and development, in biochemical functioning, in diseases and defects—with the female organism in most physical respects being superior to that of the male. Might it not also be possible that whatever genes are responsible for human inventiveness work differently in the two sexes, and even—by way of natural compensation—that males are given the advantage in the expression of these genes?

But only the barest suggestion is so far possible as to how internal biochemical influences may stimulate or repress mental performance between the sexes or among individuals generally. In the treatment of mental diseases and defects we have seen that certain chemicals can act as "psychic repressants" and others as "psychic energizers." (Psychiatrist Nathan S. Kline has told about "a fairly well-known young artist who had been unable to produce any canvases for over a year, but when treated with one of these chemicals, which seemed to 'break the dam,' he produced a profusion of oils, water colors and sketches.") That other chemicals and drugs, such as alcohol and narcotics, can slow up minds and inhibit creative performance, or conversely, at times, stimulate minds and performance, is all too well known. Thus, the possibility exists that men's brains may have more of the "psychic energizers," women's more of the repressants or inhibitors. The only available evidence we have on this point is that hormonal changes in women at various times (menstruation, pregnancy, menopause) do or can affect their mental states and performance. A further possibility, suggested by some authorities, is that the *cyclic* chemical changes in women may make them less able than men to carry on long sustained mental activity, and hence, to carry through long-range and extensively detailed projects.

At any rate, whether the causes are biological, psychological, social, or a combination of all three, it is apparent that talent and early promise in girls have a tendency to peter out sooner and oftener than in boys. For example, in the famous Terman studies of gifted children (to be discussed again presently), it was shown that among girls and boys who started out with the same genius mental scores, by the time puberty was reached the girls had begun to fall behind in relative achievement (often also showing drops in IQ), while almost three times as many boys continued on to high levels.

Bearing further on later achievement are the sex differences shown at the earliest ages in the *directions* taken by mental development. Girls (as you may recall from Chapter 31) show superiority in verbal and language tests, in perception of detail, in social awareness and in *understanding of people*. Could this not explain why, of all the arts, it is in literature that women achieve their highest excellence? Boys, on the other hand, show marked superiority—more and more as they mature—in mechanical, mathematical and spatial aptitudes, in *abstract* reasoning, and in inventiveness. All of these latter capacities play an important part in the planning and execution of complex musical compositions, dramas and paintings, and sculptural and architectural works.

Might not all of the afore-mentioned factors, then, coupled with the greater inherent physical drive of men, do much to explain male pre-eminence in the arts and most other fields? This and many related theories are, of course, necessarily tentative. Whatever the biological evidence, however, we can never ignore the fact that women's environments always and everywhere have been different from men's and more likely to repress achievement. Nor, lest men become too smug, should it be overlooked that in every field of the arts and many other areas of achievement, there always have been some women who surpass the majority of their male co-workers or competitors.

Thus, our conclusions here must be much as they were with respect to intelligence: No standards or tests exist by which we can accurately and fairly measure the *inherent capacities* for creative achievement of women in comparison with men; and there being little possibility that we can ever fully equalize their environments to permit of such standards or tests, we may never be able to make any conclusive judgments as to whether women as a group are inherently equal to, inferior to, or superior to men in their capacity for high achievement in any form of art or other area of endeavor.

When we move to such fields as *science, medicine, law, business,* and *trades,* it becomes increasingly difficult to identify the inborn ingredients of success but easier to recognize the influence of environment. For while there undoubtedly are great differences in inherited potentialities for high achievement in any of the fields mentioned, the reason why a given person goes into a particular one of these fields is more likely to be dictated by family background, social level and training, as well as existing conditions and inducements. For example, the fact that there are families where many of the men have been doctors, lawyers, clergymen or business leaders for several generations may have no more genetic significance than that there are families of fishermen, miners, railroadmen, plumbers or tailors—or where all the women have been weavers or cooks.

One must be especially careful, when looking at the upper and lower occupational levels in which a population is stratified, not to conclude that they must represent contrasting groups of "superior" and "inferior" capacity. It used to seem so, indeed. For did not studies once show that a large majority of the distinguished men in Great Britain, the United States and other countries came from the small, elite, upper-class and professional groups, whereas only a minority came from the far more numerous laboring, farming and small-business classes? But this has been rapidly changing. With a closer approach to equalization of opportunity and education, increasing proportions of persons from the once supposedly backward or inferior groups have been moving upward into the top achievement levels; and in the Soviet Union, and in many Asiatic and African countries, the overwhelming majority of present-day leaders in various fields have come from the former lower levels.

The possibility might still remain that genes for achievement, or for particular types of achievement, may not be distributed in precisely the same amounts or ways in all groups, but only if these groups have been shown to differ genetically in other respects. Leaving this for later discussion (Chapters 43, 46), we will for the present be on surer grounds by confining ourselves to the *individual differences* found within any group.

Are there, then, as in the arts, also inherent aptitudes for specific sciences and professions? Some indications that such aptitudes exist have been provided through a great variety of tests widely used for screening by colleges, industries and the armed forces. But these tests are still heavily slanted toward acquired knowledge, experience and other conditioned factors and often fail to reveal individuals with undeveloped capacities. Further, these tests are as yet seldom adapted for or used with children at ages young enough where *inherited* predispositions for future success in given sciences or professions can be clearly identified.

Mathematical ability, however, is one type of aptitude which does give evidence of being inherited fairly directly, as well as precociously. For one thing, college records show a high correlation in the mathematics scores of fathers and sons, and of brothers. More significantly, phenomenal mathematical ability often appears quite spontaneously in young children, in the same way that musical talent appears. (In fact, certain basic components of mathematical talent and musical talent are closely related.)* Many of the great mathematicians were prodigies,

* An example is physicist Donald A. Glaser, Nobel-prize winner in 1960, who had started out as a prodigy violinist, and later was a budding composer. Although he also had shown early talent for mathematics and science as well, it was not until his college years that he swerved to physics. Among many other scientist-musicians have been Albert Einstein and the three famed Comptons—Karl, Arthur and Wilson—all college presidents and rated "the three brainiest brothers" in the United States.

one of the greatest of all, Evariste Galois, having made many important contributions before he was twenty-one, when he was killed in a duel (in 1832). Gottfried Liebnitz, William Rowan Hamilton, Niels Abel, Karl Feuerbach and John von Neumann were also prodigy mathematicians. Among contemporaries in the field, one of our greatest, Dr. Norbert Wiener, was graduated from Tufts at fourteen, while the ranks of physics (closely allied to higher mathematics) abound in notables who showed precocity, Dr. J. Robert Oppenheimer and Professor Julian Schwinger being examples.

Giving further evidence of an hereditary basis, mathematical ability (like musical ability) may be present independently of other mental powers. Thus, many phenomenal lightning calculators are of mediocre intelligence, their powers arising chiefly from remarkable memories (as was the case with the celebrated Salo Finkelstein). At the extreme are the *idiots savants**—persons who are feeble-minded and yet can perform astounding feats of mental calculation. One such individual studied by psychologists was an eleven-year-old boy with an IQ of 50, who, in addition to his calculating feats, had unusual capacities for memorizing and playing musical compositions, and memorizing and spelling words without knowing their meaning.† All of this would suggest that mathematical or arithmetical ability (in a mechanical sense), or some unusual power of memory which makes it possible, is a unique trait which may be inherited apart from other capacities. Some authorities believe this

* A French term meaning "learned idiots."

† Three other remarkable *idiots savants* have recently been studied. Two of these were male identical twins in their twenties, with IQs of 60 to 70, who, when examined at the New York State Psychiatric Institute, revealed the same amazing capacity to instantly tell the day of the week of any future or past date in any year (going back centuries to long before there was any known perpetual calendar). Further, each showed a striking memory for important dates and for the weather on certain past dates, and—though having no arithmetical ability—each was able to calculate, for instance, how old George Washington would be if he had lived until the current year. Another amazing *idiot savant*, studied by Drs. Anne Anastasi and Raymond F. Levee—a thirty-eight-year-old, 6-foot 2-inch sturdy man with an IQ of 67 (the level of a ten-year-old)—combined great musical aptitude (including absolute pitch and marked musical memory) with an unusual memory for dates, biographical facts, places and past events. Psychologists Anastasi and Levee noted that in descriptions of other *idiots savants* in the literature, prominence is generally given to their penchant for the arts—painting, sculpture, drama and dance, as well as music—suggesting that artistic aptitudes may have a low correlation with the abilities that can be measured in intelligence tests and academic achievement (although for high achievement in the creative arts more than ordinary intelligence would be required).

may be through certain dominant genes, inasmuch as pedigrees are on record of unusual mathematical talent running through as many as five generations.

Chess prodigies offer additional examples of a specialized form of mathematical ability and memory which appears to have an hereditary basis. Many of the greatest chess players revealed phenomenal capacities at early ages, among them José Capablanca, Paul Morphy, Rudolph Spiegelman, Samuel Reshevsky, Larry Evans, and, most recently, Bobby Fischer—currently the top-ranking chess player of the United States. Fischer, one of the most phenomenal of all chess players (and rated by some as "the greatest who ever lived") began to play chess at six, became obsessed with the game at nine, and by thirteen was entering tournament play. At fourteen he won the United States championship and at fifteen became the youngest chess player ever to be designated as an International Grand Master. Apart from his chess virtuosity, Fischer has shown little other distinction and was an indifferent student, quitting high school as soon as he could, at age sixteen, to devote himself thereafter entirely to chess playing. His predecessor among the ex-prodigy chess champions, Reshevsky, who began his career at the age of eight, also showed no great promise in other directions, giving up his attempts to train for the rabbinate, and later becoming an accountant.*

These stories are not untypical, for very few of the great chess masters have been reported as otherwise distinguished. Apart from the fact that chess virtuosity requires specialization and constant application, it appears to involve capacities unrelated to other forms of achievement. Thus, Edward Lasker, himself a chess virtuoso, reported that studies of a dozen leading chess masters revealed unusual memory only for chess positions and no ability to think faster than average persons in other respects. The only distinctive capacities identified were those for objective and abstract thinking, highly disciplined wills and powers of concentration, good nerves, self-control, and confidence.

As in chess, unusual and special powers of memory may also be the

* Reshevsky's story has held a personal interest for the writer, who as a cub reporter many years ago had been assigned to spend a day with the prodigy shortly after his arrival from Poland at the age of eight, to go on an exhibition-match tour. The writer found him to be a rather average but undersized and high-strung little boy. Yet, as if by magic, little Sammy that night suavely and masterfully took on thirty adult chess players simultaneously and, moving from board to board with bewildering swiftness, in a few hours defeated all but two, with whom a tie was declared because he was getting sleepy.

principal explanations for the phenomenal scholarship of some children. Thus, one New York prodigy of my acquaintance, a young miss whose IQ at twelve was so high "it practically burst the IQ thermometer," had not only a remarkable memory but the ability to "see in chunks" (as she put it)—to read not by word or phase but by whole paragraphs at a time.* The two gifts in combination enabled her to take her studies and her passage through grade school and college in kangaroo leaps, but that's about all. Once through with her education she became a cashier, and at last report was a bright but not outstanding wife, mother and homemaker. (Similarly, many—but by no means all—of the "Quiz Kid" and adult "quiz show" notables during the period when such programs were popular were distinguished chiefly or solely by having acutely developed memories.) Among the erstwhile "Quiz Kids" who did have more to offer and who went on to high achievements were Dr. James D. Watson, one of the three Nobel-prize winners for work on DNA; Supreme Court Justice Byron ("Whizzer") White; and physicist Joel Kupperman.

The foregoing cases indicate that in many prodigies only *part* of the mentality may be prematurely developed. It is a serious mistake, therefore, to assume that a child of this kind has the adult intelligence which can come only with full biological and social maturity. By overlooking this, and trying to force a prodigy into levels to which he is not adjusted, great harm can be done.

One of the saddest examples was William James Sidis, son of a famed Harvard psychologist. At the age of 3½ young William began hurdling through public school courses with lightning rapidity; at six he produced a treatise on anatomy; at eight, a new table of logarithms; at fourteen, he lectured at Harvard on the "fourth dimension." And then, at twenty, he took the road to obscurity, and at the age of forty-six, an unemployed clerk, he died of a brain hemorrhage. Almost his last recorded words, in explaining his failure were, "You know, I was born on April Fool's day."

The case of Sidis may have been pathological and far from the rule, for, as we've seen, many prodigies have gone on to great success. One of these, as it happens, was another Harvard prodigy—Dr. Norbert Wiener

* The knack of "reading in chunks" is now being developed in many persons through special training techniques, with successes claimed in reading rates of as high as 5,000 to 10,000 words a minute. But whether these results are of the same order and kind or are as effective, continuous and lasting as the speed-reading capacities developed naturally by some individuals is still to be determined.

(previously mentioned), who, like Sidis, had a professor-father to push him along. But Dr. Wiener had expressed many misgivings about his own experiences and had extended the warning, "Let those who choose to carve out a human soul to their own measure be sure they have a worthy image after which to carve it, and let them know that the power of moulding an emerging intellect is a power of death as well as a power of life."*

The fact that forced development of an exceptional child may lead to tragedy is increasingly being impressed on parents and educators. Recognizing that, while prodigy children may have certain unusual capacities, they may yet be only average in other respects, educators tend now not to train them as superhuman freaks but to keep them at grade levels with children of their own age—although in special "high IQ" classes, wherever possible, and with extra outlets for their mental energies. How well this policy has worked out is shown by the famous Terman studies of a large group of gifted children and their subsequent careers.

Starting in 1921 with about 1,500 California boys and girls whose IQs averaged 150 (ranging from 140 to 200 or over), Professor Lewis M. Terman and his associates kept constant check on them for thirty-five years and were able to report fully on how the group turned out as adults. An unusually high percentage went through college and received Ph.D.s, and among the men a large proportion have achieved national—and sometimes, international—distinction in science, medicine, law, education, literature and industry. Of the women, while the majority became and remained housewives, a considerable number achieved marked success in the professions—a number in science—and in literature and other fields. Yet emphasizing that something more than a high IQ is needed for success, about 5 to 6 per cent of the gifted men failed of even moderate achievement, and some flunked out in college. (Ill health, emotional instability, lack of ambition or drive, or bad luck were among the reasons given.) But on the whole, the study indicated that very high IQ scores in childhood generally do betoken more than ordinary achievement and success in later life.†

Now to this point: *Are prodigies usually sickly, scrawny, squint-eyed,*

* From Dr. Wiener's book, *Ex-Prodigy*, published by Simon & Schuster, 1953. (Dr. Wiener died in 1964.)

† An interesting fact was that not a single painter, sculptor or musical composer of any note (not, that is, on a par with the scientists) emerged from this large high IQ group. As pointed out by psychologist Sir Cyril Burt, this would bear out other findings that music and art depend more on specialized abilities and less on general intelligence.

neurotic little kids with bulging foreheads but not bulging muscles? That was another long-popular notion blasted by the Terman studies. When originally examined, those in the gifted group proved to be actually superior in health and physique to average children, and taller and heavier for their age. (This has since been shown to be generally true of higher IQ children and may largely reflect the fact, brought out in our Chapter 31, that they come, as a rule, from more favored homes.) Thirty-five years later, as adults, the members of the Terman gifted group were found to be taller than average college men and women; to have a lower than average incidence of ill health, mortality, insanity, alcoholism, delinquency or crime; and to be otherwise fully as well adjusted emotionally, psychologically and maritally (having an even lower divorce rate) as persons in the general population.

Still another important point related to the 2,452 children born to the married members of the group by 1955: The IQs of these children averaged close to 133—extremely high, although about 17 points lower than the average of their parents at the same ages. However, this was in accordance with the established principle (Galton's "law of filial regression") that children of superior parents tend to move backwards toward the norm, in part because of mixed matings, and in part because if a combination of genes were involved, there would be a lessened likelihood of their coming together again in a child.

We have gone lightly on the matter of heredity, but it is well to note that psychologists who have worked with gifted children and prodigies are convinced that superior mental traits must be grounded, primarily, in superior hereditary endowment. Prodigies, it is held, do not have any uncanny powers that average persons lack, but merely have a bigger and better share of the powers, just as morons have a smaller share. To quote the late Professor Irving Lorge of Columbia University, "Superior intellectual ability is not a miracle. It is as natural as superiority in height or weight. Basically it is genetically constituted, but what the superior individual will do with his intellect will certainly be conditioned to a large degree by his environment and education."

From the principal points that have been made about gifted children and prodigies, it can be seen that much confusion has resulted from thinking of them as of one type, whereas, actually, they are of a variety of distinct types:

1. The *false prodigy*, who has no extraordinary mental equipment, but because of intensive early training by overzealous parents becomes a "mental athlete" who peters out later.
2. The *premature prodigy*, whose over-all mental equipment, while not basically superior, is merely speeded up in development—perhaps

biologically, somewhat in the way that some children achieve early puberty, so that he is superior mentally to other children only for a time, until his development stops and he levels off as merely an average individual.

3. The *limited prodigy*, who is endowed with only one or two special capacities, such as a remarkable memory or mathematical ability, which may enable him to become a lightning calculator, an extraordinary chess player or a phenomenal student, but not much else.

4. The *true prodigy*, born with a combination of unusual mental potentialities, which, from childhood on, gives him superiority over other individuals and which, given the right environment, training, and personality, would predispose him to outstanding success. In this latter class are all the many prodigies who made good brilliantly in innumerable ways. Besides those already mentioned, pages could be filled with other names: Michelangelo, Van Dyck, Spinoza, Pascal, Voltaire, Darwin, Disraeli, Osler, Goethe, Macaulay, Tennyson, Charlotte Brontë, and the hundreds of others who bespangle the world's honor rolls.

So if you have a prodigy in your home, or think you have, it is well to be clear as to the true nature of his gifts. It is also important to know that a child does not have to be a prodigy to be destined for greatness. This will be seen as we go on to more detailed consideration of those individuals—some prodigies, some not—who reached the very topmost rungs of human achievement.

CHAPTER THIRTY FOUR

Genius

THE MOST REMARKABLE letter of application for a job ever written came to an Italian duke late in the fifteenth century. It was from a young man who claimed that he not only was a wizard at devising and making every conceivable instrument of war—mortars, fieldpieces, siege apparatus, portable bridges and numerous contraptions hitherto unknown—but also:

"In time of peace, I could equal any other as regards works in architecture . . . sculpture . . . and painting."

Conceit? No. The job applicant was grossly understating his talents. He also was *a musician, poet, mathematician, and scientist* who in the years ahead would prove himself to be far in advance of his times as an *anatomist, botanist, geometrician, geologist, sanitary engineer and one of the most prodigious inventors the world has ever known;* and to cap it all, he would paint "The Last Supper" and the "Mona Lisa," among many other masterpieces.

This was Leonardo da Vinci, who has been called ". . . the most resplendent figure in the human race . . . the genius of geniuses . . . a legend in his own lifetime . . . a living miracle."

In Leonardo's life and career we could sum up almost everything brought out in preceding discussions on the role of heredity in human achievement. Whatever might be said of the influence of other factors, here was a man who beyond all question was *born* with the seeds of greatness in him.

The illegitimate son of a peasant girl and an unimportant Florentine notary, Leonardo began as a child to display dazzling genius in art, music and mathematics. At thirteen he was apprenticed to an artist, but most of his special talents he developed by himself, for in many of the fields in which he worked there was no one to teach him: He was the *first*. Look in your encyclopedia, or go to your library and read Leonardo's fantastic history. You will be hard put to explain him merely in terms of environment.

Or dip into the story of his contemporary, Michelangelo, who flashed out of a petty and decadent family which, baffled by his prodigy talents, tried to turn him into *anything but* an artist and sculptor—then regarded as low occupations. Here, too, was a man with the seeds of greatness so potent within him that by the time he had barely reached manhood he was already a towering figure in the world of art.

Then there was Shakespeare, who came into the world the very year that Michelangelo was leaving it. What parental influences, what early conditioning, what schooling can claim credit for his incredible achievements? And so with many geniuses in other fields—Plato, Sophocles, Spinoza, Newton, Lincoln, Edison and the few score more of similar caliber—who rose seemingly out of nowhere to the highest points in the human firmament.

But we no longer need to look upon these individuals as supernatural or inexplicable phenomena. We can find a reasonable explanation in facts previously brought out: that geniuses in all probability are no more than mortals endowed with rare and unusual combinations of "superior" genes. If certain genes can produce various degrees and types of mental capacity, talent and aptitude, then the highest forms of these genes, in given combinations, should produce increasingly greater and more unusual capacities. In other words, genius could come through the same *polygenic* mechanism of inheritance which was suggested as accounting for basic differences in intelligence; but in addition to the assortment of average "mental" genes, there would be a number of unusual "booster" genes which could propel the minds and capacities of those receiving them into the outer spaces of achievement.*

What geniuses have, to use a popular American phrase, is "the mostest of the bestest." The card player, thinking of genes as cards, might call the "genius hand" a grand slam in bridge, or a royal flush in poker. Or in another popular sense, one might say that the genius

* We might emphasize again here the important distinction between the term "genius" as technically applied in intelligence-test parlance to an individual with a very high IQ, and "genius" as referring to the extraordinarily rare individuals we have been discussing. One in a hundred persons may have an IQ of 140 or over, and one in a thousand an IQ of 180. But the true genius, in whom high intelligence is only one of the components, appears only once among millions.

has "hit the genetic jackpot." But neither in cards nor in any other game of chance is there such an infinite variety of combinations possible as with human genes—nor is there anything to correspond to all the environmental influences that can bear upon the "playing" of a "genius hand" once it is drawn by an individual.

However, the analogy with cards can prove useful in further clearing up the mystery of why genius appears so suddenly and then fails to reappear again in the same family. It may be a matter of shuffling and reshuffling of genes, which follows the same principle as in other inherited traits. In Leonardo's case, we might assume that in each of the packs of genes carried by his two undistinguished parents there were parts of the genius combination, ineffectual by themselves; the shuffle in the flash of Leonardo's conception brought together all the required genes; and in subsequent reshufflings never did this happen again. Leonardo did not marry (he had little interest in women and spurned the very thought of fatherhood), but even if he had had children, he could have given to any one only half of his genes, and his own combination could not have been duplicated. As it was, the other da Vincis, who had come from obscurity, went back to obscurity. Early in this century a genealogist discovered in the Florentine hills a direct descendant of one of Leonardo's half-brothers. He was a peasant, undistinguished by aught save his name: Leonardo da Vinci.*

So with Shakespeare or any other great genius, we can understand now why history gives us no record of any two similarly exalted individuals following each other in the same family. Yes, we have had successions of highly gifted relatives, in many fields, and particularly in music, as we've seen. But these seem clearly to involve capacities which require simpler gene combinations. The one-in-many-million (or billion) combination of genes—*and other special circumstances*—needed to

* Pierino da Vinci, a son of Leonardo's youngest half-brother, did achieve some fame as a sculptor (circa 1554). As told in the book by Antonina Vallentin, long after Leonardo's death the half-brother, Bartolommeo, "determined to make another test of the family's capacity to bring creative genius into the world." Recalling that Leonardo had been born of his father's mating with a servant girl, the half-brother, hoping to reduplicate the process, selected and had an affair with a young woman of similar type and the same peasant stock. When she obligingly bore a son, Bartolommeo proceeded to imbue him "with a love for art," and, as the author tells it, "chance or some obscure workings of heredity reinforced the influence of the Leonardo legend, and Pierino da Vinci became a sculptor of no mean ability." But he died young, and his modest efforts were "like a last momentary flicker of glory on the dried up Vinci stock." (From Chapter I. *Leonardo da Vinci*, by Antonina Vallentin, Grosset, 1956.)

make a Shakespeare or a Leonardo are far too complex to be so easily repeated.

In including "special circumstances" we always have in mind the many influences of environment which must go together with heredity in producing genius. Indeed, the more complex the genetic mechanism, the more likely it is to be governed in its expression by environmental factors. Thus, we need hardly point out that the presence of the requisite combination of genes would not insure the flowering of a genius. We have no way of knowing how many *potentially* great geniuses were suppressed in the world's history, or how many are being suppressed today, by adverse circumstances. But we can easily guess that they've been vastly more numerous than those who have managed to fulfill themselves.

One type of genius which is certainly dependent on special circumstances, perhaps more than any other, is that called "genius for leadership." In fact, from what was brought out in our discussion of leadership in the last chapter and because of the difficulty of associating leadership with any direct genetic influences, there is doubt whether men whose claim to fame lay only in their being powerful leaders can be placed in the same category with creative geniuses. The talents of a Leonardo, a Beethoven, a Shakespeare could have asserted themselves in any time or place. But what made it possible for most great military or political leaders—including Caesar, Alexander, Napoleon or Churchill—to rise to tremendous heights could have been only very special historical environments.*

Even when we confine ourselves to true geniuses, it should be emphasized that what is commonly meant by a favorable or unfavorable environment may not be applicable. At least, the popular theory has long been that a genius *has* to be a little odd and therefore flowers best under influences that are not entirely normal or healthy. Plato, for example, believed there were two kinds of delirium: "ordinary insanity," and "the God-given spiritual exaltation which produces poets, inventors and prophets." About A.D. 50 Seneca said, "There is no great genius without a tincture of madness." And in the seventeenth century, John Dryden wrote the famous lines,

> *Great wits are sure to madness near allied,*
> *And thin partitions do their bounds divide.*

* Lord Randolph Churchill had decided that Winston was not clever enough to go to the bar and that the army was the only career for a boy of limited intelligence. So in 1893 he was sent to Sandhurst, the military college.

Such generalizations by philosophers and poets of prescientific days might be passed over lightly were it not that during the past half-century many psychologists have claimed that genius and abnormality really do go together as a rule, or at any rate, that a perfectly normal environment is not the best one for a genius. In an extensive review of many studies in support of this theory (which they themselves were inclined to question), Drs. Anne Anastasi and John P. Foley, Jr., presented long lists compiled by various authorities of geniuses or near-geniuses reported as abnormal in one way or another. Here are just a few from the clinical roll call:

> Reputedly suffering from some form of *insanity* or *emotional instability* were Socrates, Sappho, Marlowe, Ben Jonson, Bunyan, Swift, Kant, Molière, Baudelaire, Pope, Nietzsche, Schopenhauer, Goethe, Goldsmith, Cowper, Byron, Scott, Coleridge, De Quincey, Southey, Shelley, Emerson, Poe, Victor Hugo, Tolstoy and Bismarck; Wagner and Smetana among many other musicians; and among artists, Leonardo, Tintoretto, Boticelli, Cellini, Blake, Landseer, Turner, Paul Veronese, Raphael, Dürer, Van Dyck, Watteau, Van Gogh, Rousseau, Modigliani—on to numerous contemporary notables (whose names we'd better not mention). Reported as *epileptics* were Mohammed, St. Paul, Julius Caesar, St. Francis of Assisi, Alfred the Great, Peter the Great, Napoleon I, Dostoevsky. And so on into other classifications, with all those who were physically deformed, diseased, degenerate, alcoholic, sexually abnormal and what not.

But impressive as such lists might seem at first glance, one must not forget that long lists also could be made of great men and geniuses who were *not* abnormal mentally or physically: Washington, Franklin, Jefferson and Lincoln, Disraeli, Gladstone and Churchill; Justices Marshall, Oliver Wendell Holmes, Brandeis and Cordozo; Darwin and Einstein; George Bernard Shaw, Sigmund Freud, Thomas Mann, Picasso, Toscanini—a normal list which could be extended to many pages. Nor need one omit from this list Franklin D. Roosevelt, for despite the emphasis placed on his physical handicap by some of his bitter opponents, he was already well on the way to political eminence—and he was unusually healthy and athletic as well—when infantile paralysis struck him at the age of thirty-nine.

Because the lives of outstanding men have always been subjected to much closer scrutiny than those of ordinary persons, there naturally would be a tendency to exaggerate their peculiarities. Also, there undoubtedly has always been the sour-grapes attitude of less gifted humans—the wish to explain achievement far beyond them as resulting from something morbid and undesirable. Thus, many early scientists were looked upon as sorcerers, and many geniuses were hounded or even executed because their supposedly supernatural deeds were attributed to evil and sinister sources. Today the tendency is to regard anyone who does something startlingly new—for instance, the extreme modernists in art, literature or music—as merely "crazy." But as Dr. B. Freedman stated it: "The insanity of one age has often proved to be the genius of its successor."

For many other reasons the estimates or opinions regarding abnormality among geniuses—most of whom lived long before there was any accurate knowledge in this field—are scientifically open to question. Suppose, for instance, that such and such numbers of geniuses *were* epileptic, crippled, tuberculous, or syphilitic: What meaning could this have without knowledge of what percentages of average persons at their time were similarly afflicted? Particularly in the case of the mental and nervous conditions, for which accurate diagnoses were hardly made before half a century ago, psychiatric judgments with respect to geniuses of the past must be heavily discounted. So even if a certain proportion of them were unbalanced, it would not be surprising in view of the high incidence of insanity and nervous disorders in the population at large.

Nevertheless, various authorities have clung to the belief that abnormality among geniuses was and is out of all proportion to the general averages, and have offered theories as to why this should be. From the standpoint of genetics, this merits attention because the possibility is raised that "genius" genes in many individuals may be dependent for their full expression either on certain additional genes for abnormality or, as said before, on certain abnormal influences in the environment (internal or external).

The German authority, Lange-Eichbaum (who estimated that only 10 per cent of the world's geniuses could have been considered normal and healthy) wrote, "Almost everywhere, and especially in the subjective fields of imaginative writing, religion and music, gifted 'insanity' gains the victory over simple, healthy talent. . . . The psycho-pathological is an excellent pacemaker for talent. . . . This does not signify that genius is itself 'insane,' but that the mentally disordered person is more likely than the sane person to become famous and to be elevated to the ranks of genius."

To quote from other treatises: "Men who are considered 'balanced' cannot produce great works"; "Insanity tends to stimulate 'creative talent' "; "The schizoid disposition is a necessary condition for artistic genius"; "Psychosis releases latent creative powers, frees one of inhibition." Dr. Abraham Myerson, one of the greatest psychiatrists, inclined toward these beliefs. Following a study of twenty brilliant families of New England which produced United States presidents, Supreme Court justices, governors, philosophers, writers, and many other notables, he said they "teemed with manic-depressive psychoses," which he believed were largely inherited and intensified by a high degree of intermarriage. His theory was that the disease, in which the individual swings from elation to despair, may have helped to stimulate energy and drive.

Another noted psychiatrist, Dr. Nolan D. C. Lewis, held that while insanity need not be a concomitant of genius, some kind of neuroses may be. "All great works in the world are the doings of neurotics," he wrote. "If a psychiatrist wants to do his bit for civilization, he should help men of talent to stay neurotic." It was on this ground that Dr. Lewis refused to treat a well-known woman novelist, bearing in mind some previous experiences. In one case, he reported, "a famous pianist came to me and asked to be treated for his trouble. I warned him, but he begged me to go ahead. Well, I have cured him, but he no longer is an artist of the piano. He is a fine mathematician." Similarly, Dr. Lewis said he cured a painter who subsequently became a photographer.

On the other side of the argument, many psychoanalysts, among them the late Dr. A. A. Brill, have maintained that disturbed creative persons who by analysis have been cured of extreme neuroses or conflicts, have gone on to even greater achievement. Support for this belief is to be found in a study in New York of a dozen artists and writers who were given psychoanalytical treatment. As reported by Dr. Edrita Fried, ten of them "overcame blocks in their work and improved their creative output." Moreover, many modern writers and dramatists have produced notable works after analysis—an outstanding example being the late Moss Hart, one of whose greatest productions, "Lady in the Dark," dealt, in fact, with psychoanalysis and was inspired by his own experience with it. In line with this, it has been held that for many creative persons, their work provides not only an outlet for inner frustrations and aggressions, but a means for self-analysis, and so helps to cure them of their emotional ills. Thus, the works of some geniuses of the past, before psychoanalysis existed, have been explained as actually forms of "do-it-yourself" analytical therapy.

That geniuses tend to be short-lived is another long-standing theory. True, many geniuses have died young. But others lived to ripe old ages—Sophocles, ninety; Titian, ninety-seven; Michelangelo, eighty-nine; Verdi, eighty-eight; Sibelius, ninety-two; Shaw, ninety-four; Goethe, eighty-three; Newton, eighty-five; Edison, eighty-four—and many lived well beyond their seventies. Considering the shorter life expectancies of former times, it seems clear that geniuses on the average lived fully as long as did their ordinary contemporaries.

Yet we cannot wholly dismiss the possibility that a certain measure of abnormality—or, less harshly put, a "marked deviation from the average" —is a frequent if not invariable concomitant of genius. Nor is it inconsistent with accepted psychological theory that deviations of various kinds may act as special stimuli to achievement. The individual who cannot easily "belong"—who, during the formative years is sickly or physically handicapped and unable to compete on equal terms with others, or who is socially rejected, or who in any way is peculiar or is so considered—may, *if gifted,* understandably develop a more intense drive to make a place for himself through distinctive work. This is what psychologists call "compensation."*

Lack of usual or normal outlets for the emotions or sexual impulses (or deviations in sexual feelings) may further cause an individual to act, think and work differently from others, especially in creative directions. In turn, the greater and more unusual his achievements, the more the individual would be set apart from others; and with increased social and psychological pressures upon him, the more likely he would be to develop eccentricities or even to crack up. Thus, it would hardly be surprising if a high proportion of geniuses did indeed turn out to be abnormal by conventional standards, even if they did not start out that way, and despite there being no genetic evidence connecting abnormality with genius.

Intriguing, too, is the theory that specific kinds of psychic or physical aberrations might explain the distinctive qualities of the works of various great artists. For example, the unusual muscularity of Michelangelo's figures (female as well as male) has been ascribed to his homosexuality. In Van Gogh's case, his peculiar swirling, broken, brush strokes, and extraordinary colors, are linked with his mental aberrations. (He wrote in one of his letters, "The more I become decomposed, and the more sick and fragile I am, the more I become an artist.") Utrillo, it has been pointed out, might never have become a painter—and perhaps the kind

* André Gide (himself among the deviates) wrote: "I believe that there are certain doors that only illness can open. There is a certain state of health that does not allow us to understand everything, and perhaps illness shuts us off from certain truths; but health shuts us off just as effectively from others, or turns us away from them. . . ." From *The Journals of André Gide,* Vol. III, 1928–39, translated by Justin O'Brien, Knopf, 1951.

of painter he was—if he had not been an alcoholic from the age of thirteen on. Illegitimate son of an unstable artist-mother, Susanne Valedon, he was induced to take up painting as a form of "occupational therapy" during a period when he was in and out of asylums for alcoholics.

Again, there are the distortions characterizing the works of various great masters, such as the elongated figures of El Greco, Cranach and Modigliani, and the broad figures and faces of Holbein. The explanation offered by some noted eye specialists is *astigmatism*, which causes persons to see upright shapes as thinner and longer than they are, and horizontal shapes as shorter and thicker. Testing the theory, British ophthalmic surgeon Patrick Trevor-Roper showed that when the paintings of the artists mentioned were viewed through astigmatism-correcting lenses, the figures appeared normal in shape. Moreover, he suggested that Constable's brownish trees were made so because he was colorblind; that Cezanne's "blurs" resulted from myopia; and that Monet's yellowish greens and purplish blues may have been caused by his cataracts. (It may be noted that heredity is or can be involved in all such eye conditions.) Commenting on these facts, an American eye specialist, Dr. W. G. Ridgeway, has observed that if many of the Old Masters had lived in the present time, they would have been fitted with corrective glasses in childhood, "leaving them perhaps with normal eyesight, but never then producing the strange and highly individual masterpieces that grace art galleries today."

Whether or not, like ocular correctives, "psychological correctives" for various personality disorders and deviations might have interfered with the particular achievements of many geniuses can continue to be debated. But the fact should not be overlooked that most geniuses had enormous general powers of thought, creativity, concentration and "the capacity for taking infinite pains," and that many showed versatility in a number of directions. Thus, while some physical or psychological quirk might have channelled the energies of a genius in a given direction, it is highly probable that, without this quirk, or in different circumstances, his general endowment might still have propelled him to high achievement in another direction.

If abnormality—or environment—cannot account for genius, another moot question is what type of unusual brain construction or functioning can explain it. British neuropsychiatrist Sir Russell Brain (what an appropriate name!) has speculated that genius may arise through a "specific nerve-cell pattern" which endows the individual with an unusual memory and with exceptional capacities for assembling and analyzing data, producing new combinations of thoughts, evoking images and arousing feelings. (Shakespeare, according to Sir Russell, must have had an "extremely rich arrangement of his nerve centers.") But—touching on previous discussions—the British expert has held that the same unusual neurological pattern could cause the genius to veer more easily into mental abnormality, instability or even insanity (applied especially to the manic-depressive state which is regarded as most closely associated with genius because of its cyclic swings between frenzied elation and sudden depression).

Another theory about the constitutional basis of genius relates it to *biochemical* factors. If it is granted that all human mental functioning and performance is basically chemical in nature, it follows that various internal chemical influences could contribute to genius by fostering and stimulating brain activity in unusual ways and degrees. Only scanty support for this view has come so far through studies of the effects of various drugs on mental and creative performance. A wide variety of these drugs are in use (some long known among primitive peoples) and are classified under such headings as *psychoactivators, psychostimulants* and *psychic energizers* (all acting to speed up mental processes); *hallucinogens* (producing hallucinations and fantasies); *euphoriants* (elevating moods abnormally); *disinhibitors* (removing ordinary and customary inhibitions); and others which, while their effects persist, may intensify drives, create feelings of omnipotence, or distort realities. Used mainly so far by psychiatrists in the treatment of mental diseases, some of the drugs also have been tried out on normal persons. In one experiment, New York scientists administered two of the drugs (*mescaline* and LSD—lysergic acid diethylamide) to a number of prominent painters, with the reported result that the artists had "delusions of color and music," "saw wonderful pictures in their minds," and for a time thereafter, "painted with greater freedom of expression of forms and color."* The effects of mescaline as a stimulus to imagi-

* A controversial theory is that the toxins generated by the germs of syphilis, tuberculosis and other diseases may (for a while or periodically) tend to speed up the workings of the brain cells. In support, psychiatrist Edward Podolsky lists among many geniuses with tuberculosis, Robert Louis Stevenson, John Keats, Anton Chekov, Goethe, Heine, Schiller, Rousseau, Molière and Voltaire; and among many with syphilis, Schopenhauer, Donizetti, Smetana, Baudelaire and De Maupassant. (But at the time these geniuses lived, both of the diseases—then almost incurable—were widespread in their groups, and their incidences among geniuses could have been only coincidental.)

nation were also dealt with extensively by Aldous Huxley, after his own experiences with the drug, in his book *The Doors of Perception.*

However, warning subsequently came from medical authorities that LSD and similar drugs which can alter sensory perception may also have the power to cripple the mind permanently in many persons. Further, without careful medical observation, there may be no way to distinguish between those who can take the drugs with relative safety and those who may be led by their use into prolonged and perhaps lasting psychotic effects. Especially decried has been the practice among college students of experimenting with the drugs; for it has been reported that even small amounts taken by some students resulted in hospitalization for long periods.

What the evidence about the "psychic energizers" and other chemical influences on thinking might suggest is the possibility that the genius is a person who has had a "very special prescription filled at Nature's genetic drug store"—one which could activate the brain without necessarily damaging it. But always confusing the issue, and making it difficult to find any common denominator for geniuses, is the fact that geniuses have been and are of so many types as regards their physical makeup, personalities and achievement.

After all that's been said, you might ask, "Would *I* have wanted to be a genius? Would I want my child to be a genius?" A "No" answer could have been given more readily in the past than today. For that geniuses do not *have* to be abnormal and need in no sense be regarded as freaks is proved more and more by many great personages of our own time who stand the best chances of being considered geniuses by posterity—such men in the normal category as we've previously mentioned, or others that you yourself can easily think of. Further, as the world becomes more understanding of men of genius caliber, and as it becomes easier for these men to lead comparatively normal lives, there undoubtedly will be less association between genius and abnormality.

Perhaps we may not see again such highly versatile or spectacular geniuses as those of the Leonardo type, for this is an age of specialization, and it is much harder for a man to achieve mastery in many fields at once than it was in earlier centuries. But what we can expect is that in every field, with increasing education and opportunities, there will be more and more men of genius who will have a chance to fulfill themselves and to make their special contributions. One thing of which there is no doubt is that genius need not be as rare as it has been, for all authorities are convinced that the seeds or genes for genius are present in abundance among all classes and all peoples, and that by proper cultivation of the soil for them—not an unhealthy soil, but a healthy social environment—a bumper crop of geniuses will result.

Summarizing the evidence regarding intelligence, aptitudes and talents, we will find that with all the complexities there is nonetheless a pattern for achievement essentially like that for other human traits we've dealt with: Degrees of susceptibility or resistance to achievement of various kinds undoubtedly are inherited by different individuals, everything else being equal. At the lowest extreme are those persons who, because of inferior mental equipment, are completely resistant to outstanding achievement in any environment. At the highest extreme are those started off with combinations of special and superior endowments which make them susceptible to very high and varied forms of achievement, if they are given any reasonable break by environment. Between these extremes, in gradations up and down, are the great mass of people with moderate degrees of capacity for or susceptibility to many forms of achievement—dependent on their environments—yet limited by various genetic lacks with respect to high achievement in special fields. Finally, while in all three groups the interplay with environment is of vital importance, its relative influence is not always the same, and, as we've seen, what constitutes a favorable or unfavorable environment may vary greatly for different individuals.

Most of us can never hope to be geniuses in any environment, and many of us may not want to be. But there is plenty of work of all kinds to be done in the world, and no one can say, in the long run, which individuals are *genetically* the most important. You may be a relatively obscure woman or man, and yet you may be carrying genes which, in combination with those of your mate, may result in a child destined for greatness. So it was with the parents of Shakespeare, Lincoln, Franklin and most other geniuses. What the final score may be on your genetic contribution to the world's advancement can be told only through your children, and theirs, and theirs.

CHAPTER THIRTY FIVE

Instincts and Human Nature

YOU SAY, "I did that *instinctively*," or "I have an *instinctive* dislike for such things," or "My *instincts* told me—." Or you may also say about many of your acts "It's *human nature*," or "I couldn't help it—I'm *only human*."

What is implied by these statements is that much of what we human beings do and feel is motivated by the workings of natural, inborn factors; and that, even though as a species we may be endowed with unique attributes of mind, soul and conscience, we nonetheless are constantly swayed—if not as fully governed as are lower animals—by the upsurging within us of biological forces.

Are these assumptions correct? Are there really inborn "behavior wheels" in you and other persons, geared by heredity to produce *instinctively prescribed* actions and reactions?

The question has been hotly debated by psychologists. Some—partisans of the behaviorist school mentioned in Chapter 30—have condemned the very word "instinct," refusing to concede that even lower animals have instincts. "Conditioned reflexes," "stimuli," "drives," "impulses"—yes, these are approved terms to account for lower animal behavior. As for instinctive *human* behavior—well! Students of social science who so much as mention human instincts to certain of the all-out behaviorist professors would be almost kissing good-bye to their chances of a Ph.D.

But there's been a change in the psychological atmosphere. Increasing evidence from experiments and studies has shown that there undoubtedly are specific patterns (or causes) of behavior in each species which arise spontaneously, independent of training, and which can be explained in no other way except as instinctive. Thus, a group of the nation's top experimental psychologists, after weighing the pros and cons some years ago, not only officially endorsed the use of the word "instinct," but agreed that if there were instincts in other animals, it was inconceivable that human beings would not also have them.* By the same token, scientists have felt that just as knowledge about the inheritance of physical traits in lower animals has done much to clarify human physical inheritance, so the study of inherited behavior patterns in lower animals could throw much light on the genetic sources of human behavior. This, then, will explain why the ensuing discussions deal so extensively at first with the facts about lower animals.

Yet it must always be kept in mind that whatever instincts or other inborn behavior tendencies may be at work in human beings, their acts and impulses are greatly influenced by countless environmental factors and the accumulated results of thousands of years of learning and experience. Each human generation derives something from the behavior of the preceding one, each individual from the behavior of those who train and surround him. So what you often think of as an instinctive act or feeling may be merely the subconscious clicking together of a series of *acquired* and remembered responses.

The facts are quite different for lower animals. Birds, mammals, fish and insects of each species display the same instinctive behavior patterns —the same ways of building nests or shelters, feeding, fighting, mating, taking care of their young, communicating with each other—as did their ancestors ages and ages ago. Experience or training isn't essential (although in some of the higher mammals, such as apes, various instinctive patterns may improve with the animal's practice). But for the most part all the important equipment for the lower animal's existence is inborn. Without ever seeing other examples or blueprints, beavers construct their dams, bees their combs, spiders their webs and birds their particular kinds of nests. So, too, each species spontaneously displays many other complex patterns of behavior. Nor does the lower animal know what it's doing, or why. A tame squirrel reared in a house by humans will put nuts behind the legs of a sofa and go through all the motions of burying them. A dog, before lying down on a carpet, may turn around several times as if feeling the ground for rough spots, or

* The authorities who issued the statement included Drs. Leonard Carmichael, Karl S. Lashley, W. S. Hunter, Calvin P. Stone, Frank A. Beach and Clifford T. Morgan, the first four of whom have all been presidents of the American Psychological Association. (Incidentally, an official journal of the association, *Psychological Abstracts,* now carries "Instinctive Behavior" among its regular subject headings.)

flattening grass to make a good bed. A young starling, laboratory-bred and fed, will go through the motions of catching and killing flies although he has never seen a fly before.

How are such patterns of behavior produced? In general, one might say by a series of specific stimuli and reflexes, giving rise to a sequence of acts, with the end result in similar situations much the same for members of a given species. What might be considered *inborn* is not a whole pattern of behavior as such, but the *tendencies* to respond in specific ways to specific stimuli.

Were the foregoing facts true of human beings, we would find a good example in a fireman in a firehouse, who, awakened in the middle of a night by a gong, jumps out of bed, pulls on his clothing and boots, and slides down the brass pole. The gong is the initial stimulus which arouses him to action. The presence and sight of his clothing and boots are further stimuli, leading to the reflexes of putting them on. The sight of the pole is still another stimulus, causing him to run to it and slide down. Were the tendencies to such a pattern of behavior *inherited* rather than trained, one would expect that the son of a fireman, who had never known his father nor anything about a fireman's routine, would nonetheless react to the whole situation in the same way: If transported in his sleep to a firehouse and awakened by a gong, he would automatically go through exactly the same sequence of acts; or, even if he were not in a firehouse, he might go through motions representing these acts.

In a not too dissimilar way, then, this may describe the workings of instincts in lower animals, with the gong and the other firehouse influences replaced by a variety of stimuli, external and internal, for different reflexes and reactions. Outside stimuli may be conveyed through the senses—a kitten seeing a dog for the first time, a rooster on first meeting another rooster, a wolf smelling sheep, or a bear feeling the approach of cold weather. Internal stimuli may come through hormonal action, various chemical states of the body—hunger, thirst, or fatigue. In the case of behavior patterns which follow cycles, such as mating, reproduction, nest building, caring for the young, the stimuli and their consequent reflexes may be governed by the production of certain hormones at given times and may also be influenced by climate and various other external influences.

Technically, it is more accurate to speak of instincts as *potentialities* for the development and expression of given forms of behavior, rather than as inevitable determinants of such behavior. This conclusion follows from innumerable laboratory experiments showing how behavior tendencies of lower animals can be much modified, repressed or distorted by training or individual experiences, and how a variety of reflexes can be conditioned almost any which way. And, of course, in our domestic animals—cats, dogs, horses—and in circus animals, we have constant proof of what training can do to shape behavior away from natural tendencies. (See what happens to "natural instincts" when a cat learns to snuggle up against a dog, or a lion refrains from snapping off the trainer's head put between its jaws.)

Most fascinating of the experiments in animal conditioning have been those involving *imprinting*. This term is applied to the training process by which a very young animal's behavior may be diverted from a normal pattern (such as the bond with its mother) to a quite different relationship. In a classical experiment of this type, Dr. Konrad Lorenz (who coined the word "imprinting") capitalized on the fact that a duckling or gosling will follow the first moving object it encounters after hatching and will thereafter regard it as its mother. Thus, Lorenz was able to condition a duckling to regard and follow him around as if *he* were the Mama Duck. Similar results have been achieved with geese and chickens, while guinea pigs and sheep also have been imprinted in various ways.

Yet for imprinting to be effective or enduring, it is essential to do it at the *critical period* in an animal's life—in chicks and ducklings no later than thirty-two hours after hatching, and preferably between thirteen and sixteen hours. For each species, and for each kind of conditioning, a given critical period exists, usually at a very early stage and relatively brief. How extensive and lasting the effects of the conditioning will be depends on the hereditary nature of the animal. For instance, Dr. Lorenz found marked differences in the effects of imprinting among various species of birds: while some species could be imprinted to follow humans or mechanical devices, others could only be imprinted to follow members of their own species. Further, within the same species, hereditary differences in these responses are found, some strains of ducks having been bred which are easily imprintable while others are resistant to this conditioning.

One of the more ingenious of the imprinting experiments was that in which psychologist Harry F. Harlow took infant rhesus monkeys, isolated immediately after birth, and conditioned them to regard as their "mothers" contraptions made of wire and foam-rubber forms covered with terry cloth and topped by wooden heads with crude faces. The "bosom" of each form was equipped with a feeding bottle and a nipple. (See Illustration.) Each little baby monkey proceeded to treat this

SUBSTITUTE "MOTHERS"

Experiments in changing natural tendencies of animals

MONKEY INFANT *conditioned from birth to regard as mother a contrivance made of plastics, terry cloth and wire. (Experiment by Prof. Harry H. Harlow.)*

A DUCKLING, *which tends to follow and regard as mother the first moving thing it sees after birth (normally its duck mother), is conditioned to follow and regard instead a man as its mother.*

contraption precisely as it would a real mother, feeding at its "breast"; snuggling up to it for comfort; rushing to it for protection when frightened; and if left alone without the dummy mother, becoming panicky and having tantrums.*

* How the effects of this extremely artificial conditioning of infant monkeys could carry into later life was shown when many of those who had had the dummy mothers failed to develop normal patterns of sexual behavior. In the case of the females among them, mating could be achieved only by exposing them to very experienced males; but when they bore offspring they evinced no normal maternal behavior. However, the effects of being raised with "cloth" mothers were found to be much reduced or overcome if the infants were subsequently given an opportunity for sex play with other little monkeys.

But significant as the imprinting and other conditioning experiments with lower animals may be, this fact should not be overlooked: The experimental conditions under which the results are achieved go quite counter to those which the animal would normally find and for which its instincts are designed. In a sense each animal's instincts are part of a *whole* which ordinarily will also include the anticipated instincts of its parents and others of its species, as well as various aspects of the physical environment into which it will be born. A duckling in the natural state can expect to be born into a "duck" environment and immediately after hatching to see a mother duck. The instinct to follow the first large moving object it encounters has been imbedded in its heredity *because* the first large moving object is overwhelmingly likely to be its mother. (If not—under natural circumstances—good-bye, duckling!) So, too, the baby monkey at birth has an instinct to cling to the closest large, soft object because that is almost certain to be its mother.

The duckling's mother and the monkey's mother will in turn be prepared by their instincts to immediately recognize and take care of their offspring in prescribed ways, so that the little one's instincts will successively and continually interact with theirs. Further, there will be others of the species, young and old, with whose instinctive habits the instincts of the new individual can interact. But the "package" contains still more. For the long-existing instincts of the species as a whole will have caused the preceding members to establish themselves in a particular type of environment, replete with both advantages and hazards, in which all the instincts of the newcomer can be most efficiently employed in obtaining food, finding or building shelter, avoiding enemies, mating, having a family and otherwise leading the happiest possible existence. If learning plays a part in developing and directing the animal's instincts, the "school" and the "lessons" are normally also largely prescribed and provided by the heredity of the individual's species, as well as by the confines of its ordinary or natural environment.

As a result of experiments like those described above, it has become apparent that conditioning, whether in nature or in a laboratory, does not erase an animal's instincts or create any new pattern of behavior; it merely modifies the way in which the instinct expresses itself, or diverts it in some other direction, as one might divert a stream into a new channel. Nor can conditioning go beyond a certain point. Helping to establish this (and thereby to give the *coup de grace* to the die-hard behaviorist theories) have been the Brelands, Keller and Marian, a husband-wife psychologist team widely known for their success in conditioning animals of many kinds to do tricks for educational and commercial exhibits. Chickens have been trained to dance and to give desired responses or manipulate apparatus by pecking on disks; and pigs and raccoons have been taught to put money into piggy banks. Thousands of individual animals have been conditioned in various ways. *But—!* The Brelands have been disconcerted to find that the conditioning process repeatedly broke down when their trainees began doing unexpected things or wouldn't stay conditioned.

For instance, pigs who'd learned to pick up and put coins in a bank to get food shortly began dropping the coins, rooting them along, tossing them into the air and *not* dropping them in the bank—even though this meant their going hungry. Why? Because this was a reversion to the pig's instinctive practices when handling food. The raccoon who began washing the coins—rubbing them together and holding on—was engaging in instinctive behavior going with removing the shell of a crayfish or breaking up some other article of food. The dancing chicken was exhibiting the "scratch" pattern of her species, and pecking at disks was a diversion of her instinct of breaking open seed pods. So when chickens suddenly began dancing or pecking when they shouldn't have done so, or other trained animals began acting in contrary and unexpected ways, one could indeed say, "Hey! Your instincts are showing." More precisely, the Brelands interpreted the results as "clearly a demonstration of the prepotency of the instinctive behavior patterns over those which have been conditioned." And turning their backs on their own once strong behaviorist convictions, they said, "It is our reluctant conclusion that the behavior of any species cannot be adequately understood, predicted or controlled without knowledge of its instinctive patterns."

Does the foregoing also apply to human beings? And if so, what is the genetic mechanism by which their instincts, equally with those of lower animals, could be produced? As a noted physicist, Jerome B. Wiesner, describes the brain and nervous system, they may be compared in certain respects with one of the huge electronic computing machines which have built into them prescribed programs for computation. In each species of animal there would be a different type of "inbuilt programming" (a synonym for *instincts*) to respond to specific needs and stimuli in given ways; but the hereditary programming with which each individual is born could also be reinforced or modified by information acquired through experience and learning. "While no one," he adds, "would suggest that people are 'precoded' to the same extent that lower animals are, it is very clear that human beings also are

DO BABIES INSTINCTIVELY "SENSE" HUMAN FACES?

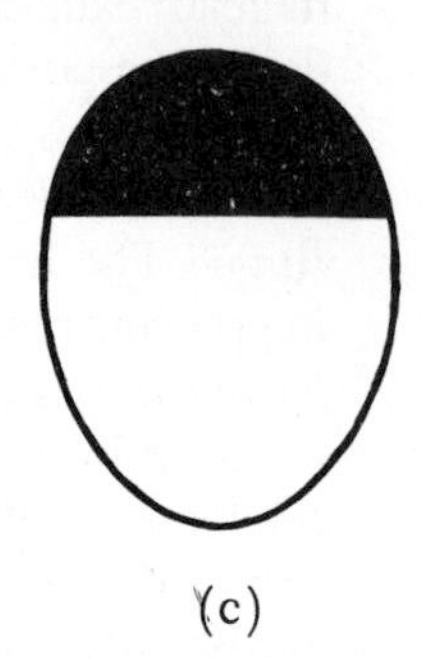

(a) (b) (c)

Above are depicted the three face-shaped cutouts with which psychologist Robert L. Fantz and his colleagues tested young infants from ages of one to fifteen weeks. The "real face" (a) held the attention of the infants—even the youngest—longer than the "scrambled face" (b), and they largely ignored the faceless plaque (c). The experiment suggested that there is an "unlearned, primitive meaning in the form perception of infants." (Adapted from article, "The Origin of Form Perception," by Dr. Fantz, *Scientific American*, May, 1961.)

born with a certain amount of programming built into their nervous systems."

The next question is how the "inbuilt programming" manifests itself in human behavior. The clearest evidence is in the actions of infants at birth and thereafter, with each response timed to appear when it usually would be most necessary. Newborn infants instinctively grasp and hold tight to the first thing they encounter—which normally would be a parent or another human being they could depend on. Next, the infant knows how to root around and find its mother's nipple; and it knows how and when to cry for attention and aid. Various additional aspects of the baby's behavior evolve in patterns almost as definite as those seen in the behavioral development of fledglings, kittens, puppies or the young of other species. Just as specific phases of physical development—from the growth of hair, the focusing of eyes, the hardening of different bones, the cutting of the first tooth (and of each tooth in turn)—proceed in given stages, so specific forms of behavior appear to be set off at appropriate intervals by an hereditary mechanism: reaching for things, paying attention to people, recognizing individuals, smiling, crowing, reacting to color, banging things together, taking part in play, and so on.*

* Experiments by Dr. Robert L. Fantz—one of which, infants' "sensing" of faces, we have illustrated—would tend to upset such notions as (to quote

Each of these behavior patterns emerges—or can emerge—without training, and in the baby's own good time. What is more, the norms for the prescribed patterns—the approximate age in weeks or months at which they appear, and the sequence in which they occur—have been found to hold for babies through successive generations, in all races and in the most diverse groups, civilized or primitive. Obviously, there can be and are variations among babies anywhere in the time and manner in which given types of behavior appear, just as there are with respect to physical traits. *Genetic* individuality, which leads to *biochemical* individuality, must inevitably also result in *behavioral* individuality as well —but always with environmental influences taking an active part throughout the route. That is why child-development experts advise against narrowing down too closely the estimates as to when a baby should begin doing this or that. It is only when a child's behavioral development departs markedly from the norm (as in the time for taking note of people, smiling, or playing with objects), that parents should be concerned. In particular, mental retardation and certain neurological diseases and defects can often be recognized when the behavioral time clock of the child runs badly behind, or erratically.

While hereditary differences contribute heavily to departures from the behavior norms, it is equally certain that environmental factors often account for much of the variation. But how far can the kinds of imprinting or conditioning we have seen employed with lower animals be effective with human beings? For instance, could a human baby be conditioned to accept a lower animal as its parent and to adopt the habits of that animal? Many believed (and still believe) that this could be so. As proof, accounts have been cited of *feral* (i.e., wild, animal-like) children—those allegedly lost or abandoned as babies and "reared by wild animals," with the result (usually, but not invariably) that they behaved like their lower-animal foster parents. Over the centuries, ever since Romulus and Remus—the legendary wolf-reared twins who founded Rome—we have had these stories. The following are some of the best known ones of recent years, which you may have read about, and can still encounter in certain textbooks and articles.

---

him) ". . . that the world of the neonatal infant is a big blooming confusion, that his visual field is a formless blur, that his mind is a blank state . . . and that his behavior is limited to reflexes and undirected mass movements. The young infant sees a patterned and organized world which he explores discriminatingly with the limited means at his command." From article by Dr. Fantz, "Pattern discrimination, etc.," in *Perceptual Development in Children*, Kidd, Aline H., & Rivoire, Jeanne L. (eds.), Int'l Universities Press, 1965.

There was, for example, the "baboon boy," allegedly found in 1903 in South Africa, "living with a troup of baboons and exhibiting baboon mannerisms." Later there were the two little "wolf girls" of India, Kamala and Amala, reputedly reared by a she-wolf and found "running on all fours, seeing in the dark, howling." Then there was the "gazelle boy," a wild Arab lad discovered "grazing and living with a herd of gazelles in the Syrian desert, and running at a speed of fifty miles an hour." (An American newspaperman subsequently ran a foot race with this boy—and beat him.) And there was Ramu, the "wolf boy" of Lucknow, India, who, when caught in the wilds in 1954, aged about nine, was reported to have "walked like a quadruped, lapped his food and water like an animal, and made incoherent noises like a wild beast." After a long period in a hospital, it was reported in 1960 that he had been transformed into a "semi-normal child, although lacking speech and unable to stand on his crippled legs."

This last observation provides a clue to the probable truth in all of these stories: That the child *was defective to begin with,* and, for this reason, was abandoned or allowed to go astray in the wilds by ignorant and benighted parents. (It may be no coincidence that most of the "wolf-child" stories have come from India, where child abandonment is common.)* In fact, in every instance where qualified authorities could examine a supposed feral child, or had accurate information, the child's symptoms were found to be those usually associated with certain mental or emotional disorders, if not also with congenital physical defectiveness. Of special interest, psychiatrist Bruno Bettelheim noted that the supposed wild-animal traits of many of the wolf children closely paralleled the behavior of children in a Chicago institution who were suffering from *infantile autism,* a severe emotional disorder believed precipitated in many cases by serious parental neglect or mistreatment. Children with this condition, well-known to psychiatrists, may show various animal-like traits long past babyhood, such as crawling on all fours, lapping up liquids from dishes or shoving food into their mouths with pawlike motions, tearing off their clothes, making grunting noises and unintelligible sounds, sniffing with uncanny sensitivity to smells, never laughing and being indifferent to human beings. Noting that such symptoms in a

* One of the most recent feral child cases, that of Parasram, the wolf boy of Agra, India, was personally investigated in 1958 by the late noted sociologist, William Fielding Ogburn, who happened to be in the region at the time the story was reported. Dr. Ogburn not only saw the boy and classed him as mentally abnormal, but also located the man who had found him, and who verified the fact that the whole subsequent story about the boy having been seen living with wolves was a fabrication.

child may be the result of people's inhumanity rather than of some lower animal's *kindness* toward it, Dr. Bettelheim concluded that "feral children seem to be produced not when wolves behave like human mothers but when human mothers behave like nonhumans."

The remaining possibility, that a young child *could be* reared by wolves or some other lower animals, appears incredible on many grounds. No baby could long survive under the care of wild animals, which would require, among other things, being suckled far beyond the animal mother's capacity, and then—still in the toddling stage—fitting into the life and habits of all the others in the pack. As to how solicitous wild, carniverous animals would be for a little human in their midst, a likely guess (remember Little Red Ridinghood's story?) is that the only love a wolf would have for a tender infant would be in the form of steaks and chops.

The serious and most important point of these observations is that immature human beings require for their survival a highly specialized type of care which can be given them only by other, older human beings; and that this care must be given them for a very much longer period than is needed by the young of any other animals. A further inference is that, even more than in the case of lower animals, the instincts and behavioral potentialities of the human offspring normally would come and must come "packaged" together with reciprocal instincts of the parents, and with various environmental essentials. No more than Nature planned or intended birds to be hatched in electrically operated incubators and reared by human beings, have human babies been designed to be born and reared in other than a human environment.

But all the special requirements of immature human beings and, most important, their prolonged period of dependency, provide, in turn, limitless opportunities for the shaping of their lives by their elders and their environments. Whatever, then, may be the inherited behavior tendencies with which any child starts off, as he grows older it becomes increasingly difficult to disentangle the natural or instinctive elements of his behavior from those which have been acquired. Yet if the behavior of human beings is much more malleable than that of other animals, this, too, is not unrelated to their hereditary equipment. For another important point is that heredity in large measure governs the capacity for *modification* of behavior traits in each species. Why are dogs so much easier to train than cats, or most other animals? Because dogs possess a greater variety and flexibility of inborn tendencies (or more responsiveness to social rewards), which can be directed through training into various specific patterns of response. This modifiability, or plasticity, of behavior development increases as we go up the evolutionary scale from the lowest to the highest animals—to apes, and finally to humans. (For an analogy in musical terms, where insects might have a behavior-potential range of one octave, human beings would have that of a symphony orchestra.) Thus, if the behavior of human beings can be so greatly shaped by environment, it is primarily because their genetic makeup endows them with so wide a range of capacities for development, and no less, for changing their environment to give still more play to their capacities.

How this interplay between individual inborn tendencies and environmental influences acts to shape specific behavioral traits of many kinds will be told in succeeding chapters.

CHAPTER THIRTY SIX

Social and Sensory Behavior

BOSS AND UNDERLING—Bully and Milquetoast—Belle and Wallflower—Lothario and Loveless—Fusspot and Stoic.

You know the types. But they aren't confined to human beings. Their counterparts exist among many other species of animals, from birds through dogs to apes. They reflect variations among individuals in the patterns of *social behavior*—primarily the manner of adjusting to fellow members of one's group—which characterize their species as a whole.

But in no species does social behavior count for so much and demand so much as it does among human beings. By the same token, the social behavior of people is the hardest to analyze in terms of basic factors and of the extent to which these interact with conditioning to determine the individual's success or failure as a social being.

In looking for the possible hereditary influences on this behavior, we must begin again with lower animals and the large variety of social drives or instincts found among them. These dictate, among other things, sexual behavior, parental behavior, collective behavior (formation into flocks, herds, colonies, or tribes) and the relationships of individuals within their respective groups. Moreover, each species shows variations in social patterns among its subspecies. Love birds, for instance, are such charming pets because of their remarkable male-female behavior—not found in most other birds—an instinctive tendency to pair off and, once a twosome is formed, to be unusually devoted and attentive to their mates. But even among love birds there are breed differences in types and gradations of the pairing pattern.

The endless varieties of social behavior among lower animals, no less than many other aspects of their behavior, can clearly be seen to arise through biological impulses. So, too, these impulses can be changed by alterations in biological states, either naturally or experimentally; for example, the sexual behavior of lower animals can be much modified by a mere shift in the balance and amounts of their male and female hormones. A virgin mouse, indifferent to young mice, can be made to develop a strong maternal feeling through injections of a certain pituitary hormone ordinarily produced during pregnancy. Or the relationship of one animal toward another can be completely changed through hormonal or other treatments.

But how far can parallels be drawn between the biologically determined social impulses of lower animals and those of human beings?

Two roosters, stags or stallions fight furiously over a female during their "romantic" period. Two gangsters will shoot it out over a dyed-blond moll; two college boys will come to blows over a coed.

All males among the lower animals are normally chivalrous toward their females, never fighting with them except under rare and unusual conditions. (If this weren't so, it would be perilous for the weaker females and for their species.) Men as a rule are equally reluctant to fight with women—the more primitive the men, the more rigid their restraint. The exceptions are found mostly among *civilized* men.

Male birds strut and display their plumage in the presence of females. Adolescent boys and men likewise go to extreme lengths to show off in front of their girl friends.

A mother bird will court death to protect her young. A human mother will do the same, without a moment's thought.

Sheep form into flocks, wolves into packs, elephants into herds. Men, too, everywhere, have the joining impulse, forming tribes, clubs, lodges, fraternities.

In the lower animals, each of the behavior patterns noted is unquestionably instinctive. Is the corresponding pattern of behavior among human beings also in any degree instinctive?

We can only make guesses. Away back in the earliest stages of human evolution, we might almost take it for granted that social instincts were at work in human beings as strongly as in lower animals. With no precepts, books or lectures to guide them, and minds only dimly aware of responsibility or of what was right or sensible, how else except through instincts could human beings have begun to live together socially, to pair off as male and female, mate, rear their young, form families? If there ever were such human social instincts, the genes for them can hardly have been bred out of us. We must still have them, and in forms not too different from what they originally were, even if we allow for occasional rare mutations. But otherwise, as has been stressed with regard to our heredity in general, no *acquired* changes in human social behavior—or the results of learning and experience through the ages—could of themselves have produced any changes in our genes.

We may assume, then, that our original "social-behavior" genes (or the genes for chemical activity which affects social behavior) are for the most part still present and active. However, it is far harder to identify them or to show what they are doing than it is to demonstrate that human social behavior is influenced by training, psychological factors and the society in which individuals live. Think only of the behavior differences between groups of educated people and uneducated, civilized and primitive, or those in one country and another, despite their all having biological mechanisms—glands, organs, nerves and senses—which unquestionably are very much the same.

Yet in individual cases anywhere, it is apparent that the physical or chemical states of the body have important effects on behavior. (You know this in your own case, by the way you act when you're in good health and when you're sick, when you're feeling fresh and when you're exhausted.) The biological changes at puberty are accompanied by a host of behavior changes, and another set of behavior changes develops with aging. Among women, fluctuations and/or changes in the hormonal balances during menstruation, pregnancy and the menopause may influence behavior and moods in various ways. Different diseases also have characteristic effects on behavior. In fact, every aspect of a person's chemical functioning, normal or pathological, is involved in his behavior; and insofar as heredity influences one's chemical state (or biochemical individuality), it also will influence one's behavior patterns.

Can we then trace the chemical pathway between a given gene and a specific type of behavior? In the case of various hereditary metabolic disorders (as noted in the chapters on the wayward genes) we saw that there is a fairly direct link between the action (or failure) of the gene or genes responsible and the resulting effects on behavior. *Phenylketonuria, cretinism* and *amaurotic idiocy* are examples. There is also a strong possibility that if genetically determined chemical bases can be established for such disorders as schizophrenia, manic-depressive insanity and certain other mental diseases, direct pathways may be traced from the responsible genes to the chain of chemical effects leading to the specific types and degrees of abnormal behavior that characterize these conditions. But in many other hereditary conditions (such as diabetes, gout, asthma, allergies, serious eye and ear defects, and certain muscular disorders), which in one way or another affect an individual's behavior, innumerable additional aspects of a person's physical makeup and thinking processes may play a part along the route and affect the end results. Similar complicating factors may confront us when we attempt to trace behavior directly from specific genes to their effects on given glands and hormonal outputs, on specific areas of the brain, or on mechanisms of the nervous system.*

Whatever may be the precise steps from the genes to their effects, we can see the interplay between biological factors and normal behavior patterns most clearly in various sex differences, first, in the comparative behavior of males and females as individuals, and second, in the sexual relationships as regards the differential responses of the two sexes. So much has been learned about these behavior patterns and the influence of heredity on them both in lower animals and in human beings, that we will leave the full discussion for a later chapter. Here we will deal only with some aspects of the sex differences that are important in social behavior.

Parental behavior is one area in which males and females may exhibit distinctive tendencies. In lower animals this behavior has been found to be so strongly governed by instincts, in detailed and specific ways for the two sexes of each species, as to make us suspect the existence of similar instincts in some degree among human beings. Of course, we think of mother love, paternal protection, and parental devotion as spiritual qualities in people, and to a considerable extent they may be. But the qualities also exist in lower animals, where the biologist regards them unsentimentally as merely the end results of a series of physical stimuli and reflexes. In fact, the parental impulses of lower animals need not at all be identified with their own offspring. You've seen pictures of mother cats nursing puppies, baby squirrels, baby monkeys or other creatures. Much though we'd like to read something beautiful and noble into these situations, the mother cats are being completely selfish. Once a cat has given birth to her young, a whole biological pattern of urges is set going within her. Her full breasts *require* draining, and she needs and likes the feel of warm little bodies snuggling against her.† Possibly the first human mothers (to go back again to

* One theory, suggested by Dr. Benson E. Ginsberg, is that genes may in part influence behavior through the enzyme control of metabolic and energy-yielding processes in the nervous system, with variable effects on the nerve impulses and motor reactions.

† When the cat's nursing process stops, there is no evidence of her further attachment to the young ones; and if they are of an alien species, as much hostility might develop between her and them as if she had never been their wet nurse. In one unromantic case, when the little ones were mice, the cat

the earliest dim stages of our beginnings) were also moved by the same purely physicial impulses and reflexes. But today so many psychological and social factors condition women toward or away from motherhood that their degrees of maternal behavior can hardly be attributed to simple instincts.

Nevertheless, we say of this woman, "She's the *motherly type,*" or "She's a *natural-born mother,*" and of some other woman that she isn't. Has science offered any support for this? Dr. David M. Levy, among those who studied human maternal behavior, answered, "Yes." He reported finding that degrees of "maternal impulse" could be correlated with various physical characteristics, such as breast type, hip form, and menstrual functioning. Further, he held that certain behavior characteristics in young girls could often reveal those who as adults would tend to be highly maternal, and those who most likely would not be. Studies of lower animals also have shown that the maternal drive is much stronger in some females than others. In some species the maternal instinct is almost entirely lacking—as in cuckoos and cowbirds, which leave their eggs to be hatched and their young to be brought up by other birds.* And in still other species—certain fish, for example—Mama goes off blithefully as soon as her eggs are laid and leaves the hatching and/or care of the young entirely to Papa. (Does this sound reminiscent of some "poor fish" human husbands?)

Dominance and *submissiveness* are among other individual differences in social behavior that appear in lower animals. This was first recognized in the peck order in groups of poultry or birds. When any two members of the same bird species are put together—two hens for instance—it doesn't take long to determine who's boss. One bird does the pecking, the other submits and backs away. In a flock of hens, each bird soon finds its place in the peck order dominating those below it in the scale and being dominated by those above. Thus, a "peck hierarchy" results, grading from the one hapless bird at the very bottom, pecked by all the others, up to the one bird at the top who rules the roost. A corresponding situation has been found in many other species, including apes, mice, and to some degree, dogs.

But it has also been established that the peck order within a group can be changed by adding certain hormones to given individuals (through injection) or subtracting them (through castration). In one

---

foster mother didn't wait until they were weaned, but did with them what comes naturally with a cat and mice.

* In domesticated fowl, strains have been selected for nonbroodiness because ordinary maternal responses would cut down on egg production.

DOMINANCE IN MONKEYS*

—And how ranking order can be changed by brain operations

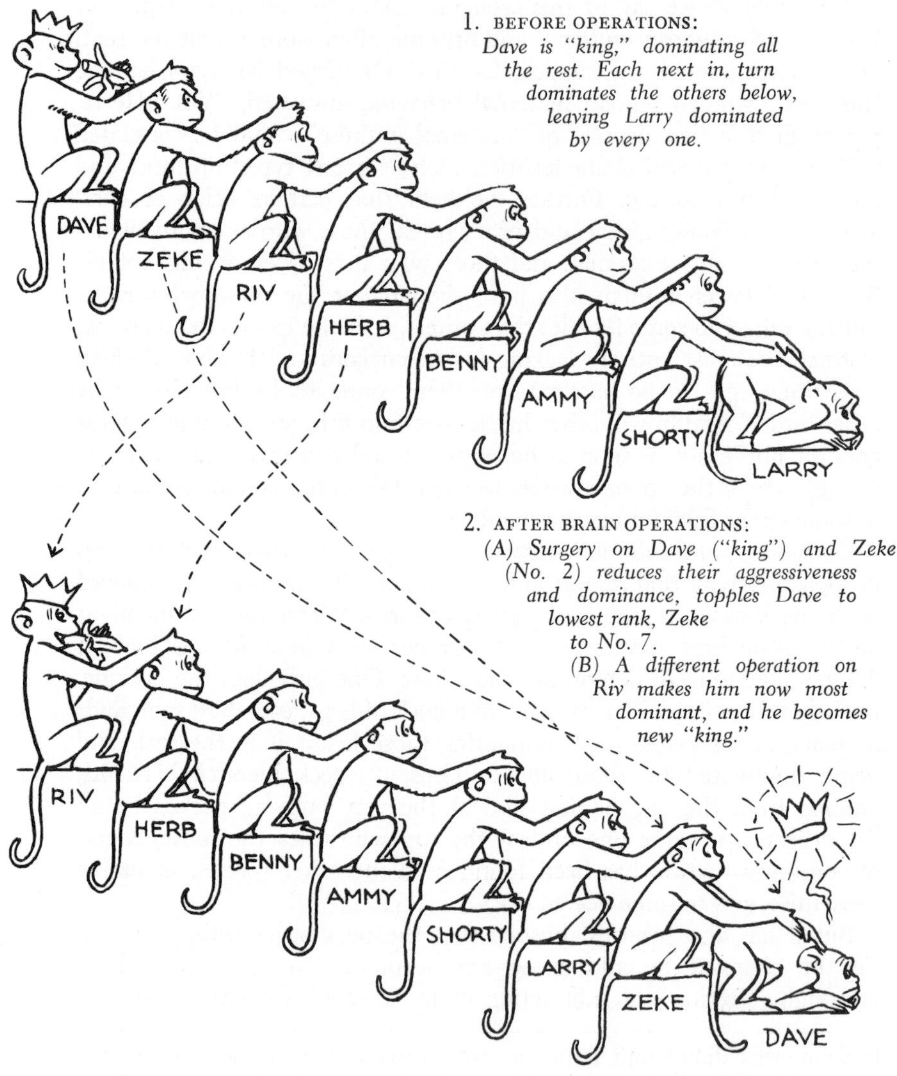

* Suggested by experiments and illustrations of Drs. H. Enger Rosvold, Allan F. Mirsky and Karl H. Pribram.

experiment, reported by Dr. Alphaeus M. Guhl, the lowliest hen in a group of eight, when injected with male hormones, rose rapidly in the peck order, squelching one rival biddy after another until she finally subdued the former top hen and became the new queen. In another experiment (by Yale University psychologists) the dominance order among eight young rhesus monkeys was radically changed after brain operations were performed on the three bossiest ones, including the "king," who thereupon fell to almost the lowliest position. (See illustration.)

All of this is far from saying that in a natural state the degrees of dominance or submissiveness are determined by inherent hormonal or neurological differences. Chance events might frequently cause one individual to gain an initial advantage and then keep it. Even less among human beings can we equate dominance-submissive relationships with biological influences. Some authorities have claimed that persons with greater hormonal activity—or at least, with more intense sex drives—are more likely to be of the dominant type. But there is every evidence that in human groups superiority complexes and inferiority complexes—determining who dominates whom—are much more likely to be conditioned by social, intellectual and psychological factors. (Whether or not the psychological states of individuals are in turn conditioned by genetic factors is another matter.) No less important is the fact that in the human world, with the scope of contacts and situations enormously greater than among lower animals, an individual could be dominant in one situation or relationship, and submissive in another—even with respect to the same persons. (Consider the case of a big, tough top sergeant who after the war finds himself in a lowly civilian job bossed by a fragile, mousey ex-private from his former company.)

Are degrees of *sociability*, *aggressiveness*, *passivity*, and *fear* inherited? Studies of lower animals do show hereditary differences in such traits and may or may not offer clues as to similar differences among human beings. The following summaries are based on findings at many institutions, including the Jackson Memorial Laboratory where behavior reactions of dogs and other animals are being extensively studied.

SOCIABILITY. Different species of animals, and strains within each species, reveal varying degrees of "sociability"—the capacity to fraternize or get along with those of their own kind or with members of other species. The extent to which sociability can be bred in animals is best illustrated among dogs, first in relation to each other, and then in relation to human beings and other animals. As Dr. J. P. Scott has reported, terriers, trained to attack game, are also more apt than other dogs to attack one another; but hunting dogs, such as hounds, bred and trained to work in packs, are most likely to avoid fights among themselves. In response to human beings, it goes without saying that dogs as a species tend to form closer attachments than do any other animals and that they can be bred and trained both to be devoted and protective to persons they know and vicious to strangers. Among feathered species, the best known and most striking contrasts are in poultry, where selective breeding has produced gamecocks which display far more aggressiveness and courage than do common domestic fowl.

Varying degrees of sociability or aggressiveness likewise characterize other species, and any zoo provides evidence that some animals can get along peacefully in one enclosure with their fellows and with assorted creatures of other species, while animals of some other kinds must be penned up individually. Even in supposedly peaceful species, genes can be selected and assembled through breeding so as to produce unsocial, fighting families or individuals. This has been done with mice, among which wild strains differ markedly from tame strains in fight reactions. And rabbits have been bred so vicious that they will bare their teeth and snap at human beings.

That heredity may work through the glands and hormones in producing variations in aggressiveness is one possibility. The adrenal glands, for example, are known to affect emotional states strongly; and it has been found that wire-haired fox terriers, among the most aggressive of dogs, have much larger adrenal glands than the less aggressive beagles or cocker spaniels. Also, among rats of the Norway breeds, the emotional ones have larger adrenal and thyroid glands than the nonemotional ones. (The relative influences of the male and female sex hormones on aggressive behavior will be dealt with in Chapter 38.) Another possibility is that degrees of aggressiveness may be governed by the way certain areas of the brain respond to given stimuli (the reactions, in turn, being influenced both by individual differences in brains and the nature of the stimuli, whether internal or external). As experimental evidence for this, electric charges conveyed to prescribed brain areas have caused ordinarily peaceful cats to turn into hissing, snarling, clawing animals—and, when the electric stimuli stopped, to become again friendly, purring pussy cats. Conversely, by stimulation of other brain areas ordinarily unfriendly and aggressive cats and monkeys have been made docile.

ACTIVITY. Dogs, again, offer the best examples of hereditary tendencies to degrees of activity under given types of stimulation. Terriers, originally bred to kill rats and other vermin, are among the most nervous and active of breeds, with wire-haired terriers being somewhat more so than Scotch terriers. Cocker spaniels are considerably quieter, while English pug dogs, bred originally as boudoir pets, are among the most passive and phlegmatic of canines. (In mice, also, active and inactive strains have been bred.) The varying capacities of dogs of different breeds for specific activities are well known: "retrievers," "pointers," "sheep-herders," "police" dogs, "seeing-eye" dogs and "carriage" dogs (these latter are the spotted Dalmatians, whose special love for movement adapted them for training to run beneath or alongside horses and vehicles, a quality which in former years made them popular as firemen's pets). Thus, while German shepherd dogs were selected by the United States Air Force for sentry work, setters and spaniels were rejected because of their tendency to dash off after game instead of attending to business. In all of these cases, training has merely served to bring out *inherited* capacities, which are believed related to special characteristics in the glandular, sensory and muscular mechanisms of each breed.*

FEAR. Instinctive fears among lower animals characterize each species, and as a rule are necessary for the animal's survival. In evidence are the countless instances in which animals of each species instinctively identify and recoil from specific natural enemies or specific menacing situations. Sometimes the reasons for the fear reactions may not be apparent. If elephants fear very small animals, perhaps it is because the little creatures might get into their trunks or their large ears. But why apes have a natural fear of inert, mutilated bodies—even papier-mâché models of these—is not known. A genetic basis for fear is shown by the breed differences in this trait found among dogs and other species. It has also been shown that fear reactions in lower animals can be greatly increased or diminished through training. In one case a sheep, bottle-raised in a sheltered environment, was fearless of dogs—and, so, alas, let itself be attacked and killed by one when other sheep ran away. Likewise, puppies, raised in isolation and protected against all

* Dr. J. P. Scott has reported to the writer, "What dogs of each specialized breed seem to possess is a *combination* of several simple traits which makes them easily trainable for complex behavior. For example, to be a good retriever a dog must show these tendencies: To follow a moving object, then to pick it up and carry it, then to come to the trainer when called, and then docilely to give up the object. All dogs show these traits to some extent, but some are more easily trainable than others. . . . We may guess that 'natural' retrievers are probably dogs with an exaggerated tendency to pick up things and carry them around, although they usually have had lots of practice before formal training in retrieving begins."

hurts, on release failed to show the usual fear reactions of other dogs.

Again, how far can the foregoing facts be applied to human beings? We know that among the people we meet there also are marked differences in degrees of sociability, aggressiveness, activity, and fear. As in lower animals, these traits may possibly be linked to some extent with genetic differences. But if behavior traits can be conditioned in lower animals, they certainly can be, and to a far greater degree, in human beings. Take *fear*, which persons may show in innumerable forms. Some of the human fears are physical, possibly arising through protective instincts similar to those in lower animals. But many fears are psychological and unique among human beings: fears of financial loss, failure, insecurity, sex, loss of status, disgrace, and so on. Such fears may be rational. Other fears—of high places, of enclosures, of cats, and a host of odd phobias—seem unrelated to any real threats. To discover how these fears evolve in individuals may be much less the task of the geneticist than of the psychologist or psychoanalyst.

Even women's fear of mice, often regarded as instinctive, may be largely conditioned. That it can be quickly "deconditioned" is shown by the ease with which high school and college girls, working in laboratories, learn to handle these rodents. (In New York, one day, the author saw two little girls walking along a street, each happily leading a white mouse by a string.) Even the fear of death—considered strongest of the self-preservation instincts—may be equally conditioned one way or another, as has been shown all too often by men in warfare.

Various other social instincts have been studied in lower animals: *communication, exploration, hoarding, cleansing,* and so on. All of these can be recognized as having parallels among human beings, and in our species as a whole there may be some underlying instinctive basis for such behavior patterns. But it seems far more likely that individual differences observed among people with respect to these patterns (the "talker" and the "close-mouth," the "explorer" and the "stay-at-home," the "hoarder" and the "spendthrift," the "compulsive washer" and the "unwashed") can be explained much more readily by the psychoanalyst than by the geneticist.

Leaving our discussion of instincts, we turn to the *sensory* differences among individuals—in capacities for tasting, smelling, feeling, hearing. What these have to do with social behavior becomes apparent if you stop and think how an acute sense of taste may stamp one as a "fusspot," or—flatteringly—as a "gourmet"; how an acute sense of smell might cause a person to shun certain places, jobs or people; or how differences in sensitively to sound, touch, or colors may affect everyday behavior in innumerable ways. But where heredity begins and training ends in determining the sensory reactions of human individuals is not easy to establish. This is best illustrated in the *taste for food.*

Among lower animals, each species is marked by instinctive preferences for or aversions to specific foods, and this is essential for the species' existence. Rats, for instance, when given access to a variety of foods, will usually select those which are best for them. Chickens, dairy cattle, dogs, cats—and pigs, too—are discriminating with food. As for human beings, while their menus cover a vastly greater range than those of any other species, there is much which lower animals eat that people cannot or will not. Instinct may sometimes be a factor in this. for even the youngest babies will often show good sense about rejecting harmful foods.* At the same time, people, more than lower animals, are apt to disregard their food instincts by eating what isn't good for them or rejecting food that is good for them or, at worst, not harmful. In this, training plays a major part. Do you personally abhor the thought of eating fried caterpillars, grasshoppers, monkeys, dog meat, snake steak or raw whale blubber? Your disgust is almost entirely conditioned, for to many people in the world these foods are delicacies. But if, within your own group or family, you can't stand the taste or smell of specific foods which others relish, your distaste may sometimes be due to a psychological quirk and sometimes to an inherent biological reaction. For genes have much to do with each person's sense of taste, and in some cases with taste eccentricities.

In general, tasting sensitivity involves reactions to the four basic taste qualities—sour, salty, bitter and sweet. The reactions vary in different species of animals, and also among individuals within a species—including human beings. The best evidence that heredity may govern such taste sensitivity has come so far from findings about the way persons taste certain chemicals, particularly *phenylthiocarbamide* (PTC). Perhaps in a high school or college biology class you've been asked to

* This was tragically illustrated in 1962 in the case of a group of infants in a hospital in Binghamton, New York, who were fed formulas prepared by mistake with salt instead of sugar. It was noted that the infants, only a few days old, seemed unusually finicky when fed, and were gagging, and trying not to swallow; and only after six of the fourteen infants had died did a chance event reveal what the sensory instincts of the infants had been trying to convey—that there was salt in the formula in sufficient amounts to be lethal to newborn infants. (The error was traced to the fact that the large sugar and salt containers in the hospital had stood side by side, and the labeled lids had been transposed.)

chew a snip of paper impregnated with this chemical. To most persons it tastes bitter, but to a minority it is tasteless. Explaining this, geneticists have shown that the ability to taste PTC is usually inherited through a dominant gene. A person with one such gene would be a *taster* of the chemical. A person receiving two recessive "minus" genes for the trait would be a *nontaster.** However, gradations of the PTC-tasting ability can be found, suggesting that certain qualifying genes may be involved. This could explain why even among persons with the dominant gene, some can taste PTC much more acutely than others, the reaction to the chemical occasionally being so strong as to cause extreme nausea. That the PTC-tasting sense may also be influenced by the chemical states of the body is indicated by the fact that identical twins are not invariably alike in PTC tasting; that the incidence of tasters is somewhat higher among women than men; and that persons with certain types of thyroid diseases (nodular and diffuse goiter) have a considerably higher than average chance of being nontasters. (On the other hand, persons afflicted with cystic fibrosis have a much more developed sense of taste and smell for many substances than have normal persons. Patients with adrenal insufficiency also have an increased taste sensitivity.)

Racial and ethnic differences in the proportions of PTC tasters and nontasters are also of interest. The incidence of tasters ranges from 96 to 98 per cent among American Indians and African Negroes, from 90 to 94 per cent among Chinese, and from 60 to 70 per cent among Europeans and White Americans. However, since there may be considerable variation within each race or racial subdivision, the extent to which these tasting differences are genetic or environmentally conditioned may be in doubt.

Various other chemical-tasting reactions which seem to be inherent—but if so, not unrelated to PTC tasting—have been reported. Vinegar (or acetic acid) is described by some persons as tasteless, by others as sour; mannose (a type of sugar) is described by some as sweet, by others as bitter or salty; sodium benzoate and sodium nitrate, usually considered salty, have been variously described by individuals as bitter, sour, sweet, or, sometimes, tasteless. (Even water—distilled or ordinary

* Much as with other recessive and dominant traits, where both parents are nontasters, it usually means that each carries only the "nontasting" gene for PTC, so that all their children would probably be nontasters also. If the parents between them carry both "tasting" and "nontasting" genes, their children could be mixed with respect to the trait. But as explained in the text, the genetic factors and their workings in PTC-tasting inheritance are not clear-cut, there being, for instance, cases of parents who are both nontasters producing a child who is a taster.

tap—which is tasteless to most people, is bitter to some individuals.) These reactions to given chemicals would suggest that inherited taste differences could contribute to likes or dislikes for such foods as sauerkraut, spinach, green or red peppers, and strong cheese, as well as for certain soft drinks or liquors. (One finding is that the more sensitive one is to bitter substances, the more foods one is apt to dislike.) In fact, while psychological factors may also be involved, many of the aversions shown by children or adults to "my goodness! but it's *good* for you!" foods may be due to something more than mere crankiness or fancied aversions and may in fact be a rejection of something in conflict with the individual's biochemical makeup.

The *sense of smell* is in many ways paired with the sense of taste (as you've learned when you've had a bad cold). Human beings normally have the capacity for detecting and identifying an enormous range of smells—at least 10,000 distinct odors, authorities say—in varying degrees of intensity.* Among lower animals, species differ considerably in smell acuity, but each is equipped by instinct with the particular favorable and adverse smell reactions essential to its survival. In human beings, also, such instinctive discrimination must underlie many generalized reactions to "wholesome" smells, in contrast to the "menacing" smells of spoiled food, noxious gases, smoke, and so on. While again, as with tasting, training unquestionably plays a large part in developing the human capacity to identify and react to specific smells, we can assume that there are hereditary gradations among people in "smelling quotients"—or what scientists call "olfactory thresholds"—ranging down to the total absence of smell (*anosmia*), which is due to a dominant gene.

Looking for some physical basis for differences in smell acuity, Dr. John L. Fuller has reported that among breeds of dogs, which differ greatly in smell-detecting ability, there are marked variations in the sizes and construction of noses. The nasal capacities of hunting and tracking dogs are more than ordinarily large, by contrast with those of such reputedly poor sniffers as Pekingese and flat-nosed English bulldogs. (Whether the relative sizes of noses among human beings have any bearing on smelling capacity is yet to be learned.) However, Dr. Fuller has also shown in the case of dogs—regardless of breed—that temperamental qualities (including a strong motivation to seek people or game) and capacity for training may be as important as structural equipment

* The primary human organs of smell are the two lobes called *glomeruli*, one in each nostril, and each consisting of about 2,000 bundles of smell-detecting neurons hooked up to the brain in various intricate ways.

in developing the sense of smell. Thus, it is especially likely that among human beings such psychological traits as sensitiveness and fastidiousness, as well as neuroticism, may combine with training to accentuate the sense of smell of individuals, while other psychological traits might dampen it. (Schizophrenic children have been reported as having a more than ordinarily acute sense of smell.)

Sensitivity to sound is another trait which in various degrees is governed by hereditary factors and also influenced by psychological factors. In mice it is well established that heredity by itself makes some strains so sensitive to sound that they will be thrown into convulsions (called *audiogenic seizures*) by certain loud or high-pitched noises. In fact, some mice can be killed by sound. One of the most fascinating experiments the author has seen was at the Jackson Memorial Laboratory when these supersensitive mice were first exhibited. There was a metal washtub with an electric doorbell attached. Mice of ordinary strains, placed in the tub, would merely run to the center and huddle there when the bell was set off. Mice of the special strain, placed in the same situation, would dash madly around, go into convulsions, pop up into the air and drop dead—all in half a minute. Other mouse strains have since been bred with high and low incidences of convulsions or survival after attacks induced by exposure to loud sounds. As yet no genetic counterparts have been found in human beings. But who knows? If you're one of those whose teeth are set on edge when someone scratches on glass with a diamond, or who are driven frantic by incessantly ringing bells and clackety-clack voices, you may be genetic kin to these sound-stricken mice.

Pain sensitivity, or "pain-proneness," also exists in different degrees among people. In occasional instances where a mother gives a child a hard spanking and says, "This hurts me more than it does you," it may be literally true. For the mother may feel her hand getting sorer and sorer, while the child may not feel a thing if he or she is among the rare individuals born with insensitivity to pain. Persons of this type may pick up searing hot cooking utensils without realizing they are scorching their hands, or—without anesthesia—can smilingly undergo tooth-nerve canal drilling, or various operations, which others would find excruciatingly painful. And there was the case of a comely coed, reported by Duke University psychologists, who since childhood had never batted an eye when she was bruised, burned, pinched or stuck with pins, and who, after an auto accident, blithely went off to a college dance unaware that she had a badly fractured ankle. What role heredity might play in such congenital insensitivity to pain, or in producing varying degrees of pain sensitivity (or *pain thresholds*), is still to be determined. In at least one case, that of a Boston woman who never experienced pain—she had borne seven children without feeling any labor pangs—the insensitivity was a family trait, present in various relatives and in two of her children. Two other children were exceptionally tolerant of pain, whereas the remaining three were normal in their responses. But as in other sensory reactions, temperamental traits and conditioning bear importantly on how individuals react to pains they do feel—whether like a spoiled brat or a Heap Big Indian Chief, or with enjoyment (as in the case of masochists).*

Sensitivity to cold may stem from various causes, some of them hereditary. Those with pathological symptoms (*cryopathies*), were discussed in Chapter 24, page 258.

Various other sensitivities and senses, inherent or conditioned in different degrees, are found both in lower animals and in human beings. The *sense of direction* (which perhaps should be placed under "instincts") is revealed most acutely among birds, who, with changing seasons, can fly to and from the same nesting place, thousands of miles each way, nonstop, over vast stretches of ocean and land. Theories have been offered that birds are guided by landmarks, stars, the sun, or weather currents, with experienced adults leading the novices. But at least one experiment, made with teal ducklings by ornithologist William J. Hamilton, III, has indicated that instinct alone may guide birds' migration, since these ducklings, kept behind after the others had left and set loose later, found their way by themselves to the proper destination over unfamiliar terrain and open water hundreds of miles distant. Other migratory creatures which periodically go on long journeys to pin-pointed destinations—salmon and eels, for instance—may also be guided in considerable measure by instincts, interacting with the stimuli of currents or other influences. And the "cat who comes back" might be guided not so differently from the way the homing pigeon is. Are there then also among human beings variant types of instinctive "steering mechanisms" which explain why one person has a remarkable sense of direction, while another seems to have almost none? Geneticists may tell us some day.

* How dogs can be raised without awareness or understanding of pain has been shown in experiments by Dr. Ronald Melzack. Scotch terrier puppies which had been isolated from birth on and protected against all hurts showed little reaction when pricked with pins and would poke their snouts into match flames and sniff them repeatedly, whereas normally reared littermates, after once touching a flame, would quickly shy away from it thereafter.

The *sense of time* similarly awaits further study. This sense is so strong in certain persons that, like living alarm clocks, they can awaken themselves at any desired point on the dial. Other individuals seem to be totally lacking in this sense. But so far the only assumptions about genetic influences are based on the proved existence in lower animals of "biological time clocks" which regulate the onset and termination of such acts and habits as migration, hibernation, mating and nesting. These "clocks" have been shown to be operated by various hormonal and sensory stimuli, under genetic control, which interact with given external stimuli (such as temperature, weather, changing day lengths, light and darkness). Possibly, then, human "time clocks" may operate in somewhat the same way, although powers of concentration, training and mental discipline may contribute greatly to individual degrees of time sense in people.

The *sense of spatial visualization*—the capacity to judge space and dimensions—is basic equipment among lower animals and vitally important in many ways, as in leaping across gaps, springing upon prey or deciding when an enemy is too close for safety, and also in estimating the relative sizes of objects and articles of food. Experiments have shown that human individuals, also, have this sense in varying degrees. Some indication that heredity contributes to these variations is given by findings that identical twins tend to be much more alike than fraternal twins in space-visualizing capacities.*

What of the so-called "sixth sense"—the presumed ability to "read" other people's thoughts, or the supposed uncanny faculty of sensing when some distant loved one is sick or has died, or when danger is impending? We've all heard about, or believe we know or have met, persons so endowed. Is there truly such a "sense"? Some psychologists believe there is, and have given it the name of *extrasensory perception*

* "Seeing with the fingers" may prove one of the most remarkable of human senses—that is, if its reported existence in some individuals is substantiated, and if it is proved to exist in varying degrees among people generally. The presumed phenomenon first was publicized in the Soviet Union in 1963, with the claim that a twenty-two-year-old girl had revealed the ability to "read," while blindfolded, an ordinary printed page by running her fingers over the letters, and also to be able to identify colors merely by fingering pieces of colored paper or cloth. Subsequently, in the United States, it was reported that a Michigan woman also could identify colors by touch, but when she was given successive tests, the results were inconclusive. If the seeing-with-fingers sense does exist (which is still to be confirmed), a theory offered to explain it is that certain fingertip cells—with variations in genetic degrees of acuteness—may react to color emanations through touch in somewhat the way that cells in the eye react to color through sight.

(ESP)—a mental capacity going beyond the ordinary physiological senses. While a great many experimenters, referred to as *parapsychologists,* claim to have proved the existence of this faculty and of persons possessing it in remarkable degrees, most scientists at this writing are still dubious about the reported ESP findings. This is not to deny the theoretical possibility that brains of individuals might conceivably act as transmitters and receivers of electric messages through brain waves, and that two given persons might be especially well attuned to each other in this respect. Immediately, then, one thinks of *identical twins,* who long have been credited with mystical powers of mental communication. If twins really have these powers, it not only would strengthen the case for extrasensory perception but would be evidence that heredity is involved in it. But, alas, not even Dr. Joseph B. Rhine—founder of the ESP school—has verified the popular notions. As he informed the author (in 1958):

> Nothing outstanding has occurred in any single case of identical twins tested so far. The averages on the extrasensory tests were approximately the same whether the sender and receiver were identical twins, fraternal twins, singleton siblings, or simply friends.*

Whatever else may yet be learned about our sensory capacities, there is no reason to expect that human beings will be found to be endowed with any sense that is supernatural or unique to our species. The facts already brought out indicate that every sense present in human beings also is present in lower animals, and vice versa; even more, just as all details of our physical makeup—our skeletons, features, organs, chemical functioning—can be recognized as merely variants of those in other animals, so any inherited behavior tendencies we have may prove to be only variants of those in lower animals.

But where the human senses and behavior patterns are indeed unique is in the extent to which they can be developed, modified, diverted, utilized and interpreted according to the conditioning and dictates of individuals and groups. How much more difficult it is, then, to attempt to trace a direct line between "behavior" genes and behavior patterns in human beings than in lower animals will be seen now as we take up such complex traits as personality, criminality and sexual behavior.

* Extrasensory experiments usually are conducted with a deck of special cards, devised by Dr. Rhine, carrying an assortment of five symbols. The person tested tries to "read" the cards, sight unseen, as they are turned up by the experimenter. When the right guesses go much beyond chance expectancy over a protracted series of tests, the person is credited with having extrasensory perception.

CHAPTER THIRTY SEVEN

Personality

IF HUMAN BEINGS in real life were like the characters in comic strips, fiction, movies or television plays, we could unhesitatingly make these assertions:

All fat people are jolly, frank and easily moved to laughter or tears.

All blond women are either dumb and frivolous or cold, calculating and morally loose.

All black-haired men with swarthy skins and sharp noses are villains.

All redheads are hot tempered and passionate; all men with high foreheads are intellectual; those with receding chins are timid wishy-washies; those with big flopping ears are fools.

The list could be extended endlessly to include the various features and looks that are supposed to indicate jealousy, meanness, trickiness, criminality, aristocracy, amorousness—in fact, the whole gamut of human personality traits.

But are there actually such correlations? If there were, then knowing that heredity has much to do with shaping one's appearance, we might go further and ask whether heredity does not also make a person act the way he looks.

The answers given today are not quite as clear-cut as those formerly offered. It still can be said that physical features and personality are for the most part not related and that people can't be glibly classified into personality types by their looks. Yet there are many ways in which facial and bodily details can offer clues to personality. In earlier chapters we noted how various inherited abnormalities can have specific effects on behavior and temperament; also how, even in normal individuals, glandular workings and other body processes may further influence behavior. To the extent, then, that specific genes or gene combinations may simultaneously affect both outward construction and mental and emotional functions, surface traits may sometimes be linked with personality traits. This isn't to say, however, that the popular notions regarding looks and personality are correct. Let's see where the fallacies lie.

A girl is blond, blue-eyed, beautiful. Does a certain kind of temperament, character, personality, go directly with that combination? Or rather, does not beauty in a girl evoke responses from others which tend to mold her personality in special ways? "Oh, isn't she darling!" "Oh, isn't she lovely!" she hears others say of her from childhood on (regardless of what a brat she may be!). Maturing, she is ever conscious of the effect she has on admiring males, employers, even other women. She has continually to act the role of being beautiful and may be elevated into positions which go beyond her capacities. A whole cluster of special personality traits may result. So to the extent that their experiences and effects are alike, all beautiful girls of a given type would tend to develop many of the same traits. The mistake would lie in assuming that the genes which produce their looks also produce their personalities.

Consider the very tall girl. If her growth comes early, she must begin dressing as a woman before her classmates do. Looking like a grown-up when she's psychologically still a young and inexperienced girl can produce a sense of maladjustment. Then the man problem comes along. The very tall girl encounters difficulty in finding dancing partners, boy friends or mates. She tends to become more retiring and reserved. And being tall, she's expected to act with dignity and restraint, even if she doesn't feel like it.

Very small girls, on the other hand, tend to develop quite different traits. In buying clothes they must patronize the junior-misses' or even children's departments. Men good-naturedly toss the little girls about, bigger girls baby them, people constantly jest about their "pint size." They're expected to "act cute." These and other factors may make the small girl hypersensitive, high-spirited, high strung.

The very fat girl, always a target for pleasantries, may build up a defense by being the first to laugh at herself. You may have observed how at costume parties the big, fat girls so often dress in "kiddy" costumes, exposing fully the plumpness of their arms and legs, just as do the "fat ladies" in side shows. Since they never feel quite at ease, it isn't surprising that fat girls as a class develop a number of special personality traits. At the same time, there may be traits originally present which are in part responsible for their obesity. Thus, says Dr. Hilde

Bruch, "The reason so many fat girls and women stay fat is that they started as timid persons and came to depend upon their bulk as a defense against men, sex, and assuming the responsibilities of grown-up womenhood." Overeating, Dr. Bruch believes, may be a substitute for love, security, and other satisfactions.

The effects of extreme homeliness (or what is so regarded) in a girl should be obvious. The indifference of men, the condescending attitude of other women, and the greater difficulty in finding various jobs and making a place for herself in society may well account for the development of certain personality traits. Yet most people confuse the resentfulness, the touchiness, the antisocial attitude often displayed by homely girls with something basic in their makeup.

Psychologists can easily extend this type of "character diagnosis" to women of many other physical types, and, of course, to men. Very short men, like very small girls, tend also to be oversensitive and eager to dominate (the "Napoleon complex"), and their social lives and careers—and hence their personalities—may be affected by their short stature in innumerable ways. Very tall men, on the other hand, are expected to be in a commanding position, and if they don't measure up may develop inferiority feelings more quickly than men of average size. Very handsome men, to whom many things come too easily—especially the attention of women—may often turn into ne'er-do-wells. Again, great physical prowess in a man, or, on the other hand, sickliness or physical inadequacy, may each be correlated with a type of personality.

All of these cases (which may conjure up pictures of some of your friends, or perhaps of yourself) show how looks can influence personality. Accordingly, if outward physical details are due to certain genes, we might think that a child who inherited genes which cause him to look like his father would automatically grow up to act like his father. In other words, that people who look the same would act the same. But this is hardly true, for two reasons: one (harking back to old Mendel and his peas) is that genes of all kinds may be inherited independently. In the very same family, two individuals who received many of the same "feature" genes may yet have entirely different gene combinations for intelligence and behavior. The second reason, touched on before, is that the effects of looks on personality are largely governed by how the looks are regarded by the individual and those around him.

Consider our tall persons again. Were they to find themselves in a society of uniformly tall people, the very tall girl no longer would be maladjusted, nor would the very tall man develop any special personality characteristics merely because of his height. Already, in our own and many other countries, with the marked increase in stature during the past several generations, tall girls or men who formerly would have been gaped at are now accepted as commonplace. On the other hand, many of those now viewed as "shorties" and suffering personality hurts because of this—men 5 feet 5, girls 5 feet tall—would have been considered quite average a half-century ago in the United States, as they are still so regarded in Latin-American and Asian groups.

So, too, how fat girls are appraised depends on where they live. In many parts of the world—in some countries of the Orient and in certain African tribes (where adolescent girls are put in "fattening pens" and milk-fed like prize porkers)—the plumpest women are considered the most glamorous and move with the same sense of queenly importance as do the belles of our own world. In these other places it would be the slim American fashion model who would develop an inferiority complex and who might tend to have many of the same personality traits which we here associate with very fat women. Similarly, various details of a woman's figure have been or may be viewed in radically different lights—the ample bosom and the small bosom, the broad hips and the narrow hips, the "piano" legs and the long, tapered legs—with sharply contrasting effects on the girl or woman involved.

Why specific types of looks are favored, and others are scorned in given groups, stems from complex social and psychological factors which we can touch on only lightly. For instance, the divergent attitudes about fatness in women are related by anthropologist Hortense Powdermaker to the availability and significance of food and eating in different cultures. In many tribal societies where there is frequent or constant scarcity of food and fear of famine, she observes, "it is not surprising that some degree of obesity is often regarded with favor." Conversely, in our own economy of plenty, the svelte female figure is the current standard of desirability and elegance. Further, the stress on slimness (and on dieting to reduce) is most marked among the upper-level and wealthier women who could eat all they wish, and least marked among underprivileged women with meager allowances for food. In the male ranks, too, where once the multimillionaire tycoon was symbolized by a bulging waistline (called at the time a "corporation"), he is now depicted as athletic and streamlined.

Many of the interpretations given to looks have no more than an allegorical basis. Take *redheads.* Some years ago the late Eleanor Roosevelt wrote in her newspaper column that she'd just seen an old portrait of George Washington showing him with red hair. She observed, "I understand better now why he held out at Valley Forge." Possibly Mrs. Roosevelt was jesting; but most people would accept this as further proof that red hair is linked with fighting qualities. Why? Most likely because red suggests fire; hence, red hair, fiery nature. Similarly, why have so many of our traditional stage and story-book heroines been depicted as fair-haired and fair-complexioned, and the villains and witches as dark-haired and swarthy? Mainly because the lighter hair and skin are closer to white, which has symbolized purity ("fair-haired" is also applied to a favored person) whereas darkness has connoted mystery, death, evil (the Devil) and sinister influences. Further, in the United States, the darker, swarthier type was originally more often that of the foreigner who differed radically from the prevailing native American stock—and the human tendency is to view the stranger with suspicion. One would suspect, then, that where people are preponderantly black-haired and darker-skinned, the villains might be those of opposite type. An amusing example is provided by the fact that in the current wrestling matches in New York City, where about three-fourths of the regular fans have been Puerto Ricans and other Latins, the "heroes" generally are the short, dark and swarthy wrestlers; and the "villains" are fair-skinned, taller, and light-haired. (If a man cast as the wrestling "villain" is not a natural blond, he may use peroxide, it has been reported by *The New York Times* wrestling editor.)

It would not be hard to find the allegorical roots for other feature connotations: close-set eyes supposedly indicating closeness, or cupidity; wide-apart eyes, openness, frankness; small eyes, rattiness, slyness; a sharp, thin nose, sharpness, meanness; a protruding chin, forwardness, bravery; a receding chin, timidity, cowardice. So, also, with the meanings attached to the shapes of limbs, ankles, hands or feet: thick ankles or wrists meaning a person is "thick," or common, and slim ones indicating inherent gentility. An even worse fallacy is to attribute such differences to either peasant or aristocratic ancestry. It is true that persons working at heavy tasks develop heavier bone structures. But the idea that these acquired characteristics can be passed along by heredity from generation to generation has, as we know, been exploded. Highly bred race horses and greyhounds may indeed have slim graceful legs (though how about equally aristocratic bulldogs and dachshunds?). No human beings, however, have ever been bred like lower animals to combine both physical features and behavior traits in the same strains.

Unfortunately, what people *think* certain features are supposed to indicate can have far-reaching effects—so much so that one may question whether it is always desirable *"to see oursels as others see us!"* Many a homely girl has gotten her man because she did not *see* herself as unattractive, and many a man has won success—with a woman or in a career—because he didn't view his physical traits as adversely as others did. But most people are not so immune to popular notions about their looks. One of the nation's leading businessmen, who has a chin which recedes so far it might almost be mistaken for his Adam's apple, has told how he had to battle constantly not only to overcome the prejudices of other persons about this feature, but to sell himself on the idea that he wasn't a weakling.*

Among young persons, psychologists report serious personality disorders in sensitive boys and girls resulting from physical deviations, or from facial peculiarities. (Fairy tales and children's stories, picturing the menacing, hateful characters as having warped bodies and ugly faces, continue to foster the prejudices of children regarding looks.) Being nicknamed "baboon face," "eagle beak," "fish mouth," or "donkey ears" may leave a sharp imprint on a child's character. In some cases where facial deformities have been deemed a factor in criminal behavior, plastic operations have been performed on features of delinquents or convicts prior to their release from a reformatory or prison, with the result, it is claimed, that a much higher than expected number of these individuals have subsequently gone straight. Among the more unusual plastic operations have been those to reduce breast sizes on self-concious young *men* who have had abnormal breast development. Also in the unique category was the case reported from Sweden, in 1961, of a comely miss who, suffering social difficulties and personality disturbances because at age sixteen she towered to 6 feet 1 inch, had herself shortened by 2 inches in an operation which involved cutting and rejoining her upper thigh bones.

To emphasize again that it is not a given physical trait itself but how it is regarded which affects personality, we have the situation involving the *Mongolian eye fold* (illustrated in Chapter 13). For there have been and are operations sometimes to *remove* it, and sometimes to *create* it, depending on cultural attitudes. Thus, in Japan, where, in recent years there has been a tendency to emulate things Western, many young women have had operations to reduce the Mongolian fold and make their eyes look rounder and more open. But in China and Hawaii, one hears that other young women, of mixed White and Chinese or

* So seriously was the presumed association between features and character taken in some business circles, *Fortune* Magazine reported in 1953, that many employers were guided in hiring men by the "Merton system of physiognomy"—classifying personalities according to numerous details of nose, mouth, jaw, ears, head formation, and so on. The "system" (devised by a medical artist, Holmes W. Merton) may still be in use in some firms, although the magazine stated it was never scientifically checked and validated.

Japanese parentage, have felt themselves discriminated against by their Oriental relatives because of *not* having the Mongolian fold, and so have undergone operations to create it.

Plastic surgery on features and figures may have a thoroughly practical meaning for entertainers, models, cosmeticians and others whose work is dependent on their appearance. But it is another matter when plastic surgery is sought to change features which the individual merely *thinks* are responsible for his or her lack of social and professional success. This is especially applicable to men seeking cosmetic surgery, who are more likely than women to be motivated by inner conflicts, often related to sex problems. To quote psychiatrist Joost A. M. Meerloo, "Plastic surgery often intervenes in a complicated psychological battle, and so makes it more difficult to accept the fate of one's face." With persons of either sex, psychiatrists warn that where there are serious underlying personality disorders, external physical changes will not correct them, and that letdowns often follow the high hopes for what plastic operations can accomplish.

We have been stressing mainly how attitudes toward looks can affect personality. But since heredity has much to do with one's looks, can it not also affect one's personality at the same time? Specifically, might not the same genes which act to mold features and contours also help to shape personality and temperament? This, in fact, is the theory in back of a number of systems which have been devised to establish correlations between human physiques and given clusters of personality traits.

The belief that "one's carcass is a clue to one's character" goes far back into history. Hippocrates, father of medicine, and then Galen, another Greek physician, speculated on this point. Centuries later, Shakespeare had Caesar say,

> *Let me have men about me that are fat;*
> *Sleek-headed men and such as sleep o'nights;*
> *Yond' Cassius has a lean and hungry look;*
> *He thinks too much: such men are dangerous.*

Oddly enough, when a few decades ago attempts began to be made to correlate body build and temperament on a scientific basis, old Shakespeare was given support. The fat men were roughly identified as of the "placid, sociable" type; the lean men as of the "mental, hesitant" type. Best known of the earlier classifications were those of Kretschmer (referred to in Chapter 14): the "pyknic" (rounded, thickset), the "leptosome" (tall, "stringbean"), the "athletic," and the "dysplastic" (mixed characteristics), each regarded as linked with different personality tendencies. In recent years the system of body-temperament classification has been much amplified by Dr. W. H. Sheldon. He, too, has roughly classified people into three main groups but has also allowed for various gradations and combinations of the elements of these groups. Here are the main Sheldon classifications and traits which he and others have reported as going with them:

The *Endomorph:* soft and round physique, with tendency to obesity and general muscular relaxation. Inclined to be home-loving (and food-loving), placid, sociable.

The *Mesomorph:* athletic, muscular, heavy-boned. Associated with fighters, heroes, leaders, lovers of power and action, thrills, exploits.

The *Ectomorph:* the "string-bean" type, flat-chested, linear physiques with stringy muscles. Persons of this type are said to tend toward the "mental," to be overly sensitive to their environment, to be introverted, less aggressive, hesitant, not at ease with people, blushing readily.

Granting that persons don't fall neatly into one body-type category or another, Dr. Sheldon devised a system of scoring individuals on each of the three primary or basic components of physique rated on a seven-point scale, with the results being determined by the balances and ratios among these components. One could then have such contrasting combinations as "7–1–1" (predominant endomorphy), "1–7–1" (predominant mesomorphy), "1–1–7" (predominant ectomorphy), and "3–4–4" (balanced physique—most frequent among college men). Correlating varied classifications with personality, Sheldon reported a marked contrast between the "long-legged, round-shouldered, completely soft and effeminate boy," and "the ruddy, powerful, barrel-bodied boy who is unusually full of bounding energy."

How close the Kretschmer or Sheldon systems come to establishing direct relationships between body build and temperament has remained debatable. The validity of such relationships might depend, first, on the extent to which skeletal and muscular structures are shaped by inherent glandular and other chemical influences, and, second, the extent to which these same influences have a bearing on personality development and formation. That hormones and other aspects of the body's chemical functioning are important in shaping features and physique has been brought out previously, particularly with respect to sex differences. But one also knows that two women with the same physical contours can still be very different in personality, or that a woman with the most feminine contours might still be very masculine in personality, just as a man with the most masculine and athletic build may sometimes be highly feminine in behavior. Similarly, in persons of either sex,

THE SOMATOTYPES

(As classified by Dr. William H. Sheldon)

Which type are you?

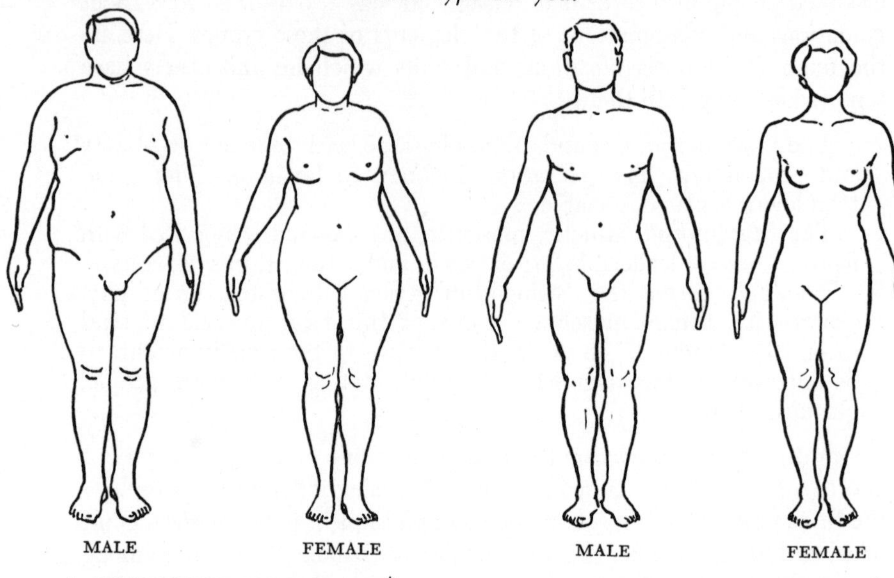

A. PREDOMINANT ENDOMORPHY*

B. PREDOMINANT MESOMORPHY*

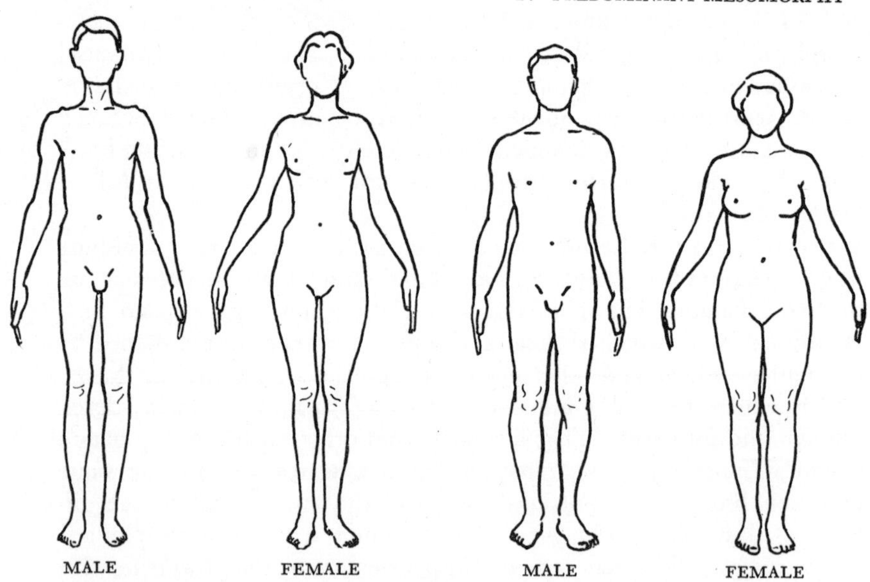

C. PREDOMINANT ECTOMORPHY*

D. BALANCED PHYSIQUE*

* Actual measurements on the Sheldon scale of the figures shown are as follows: (A) Male: 7-4-1; female: 7-2½-2½. (B) Male: 1½-7-1½; female: 3-5½-3. (C) Male: 2½-1-7; female: 4-1-7. (D) Male: 4-4-4; female: 5-3-3½. For personality traits ascribed to each type, see Text.

almost any given aspect of physical makeup can be found associated with different traits of personality.

Thus, many authorities regard the Kretschmer, Sheldon and other classifications as offering only very loose and undependable criteria of personality and temperament. However, there are others—especially among psychiatrists—who believe these classifications may be of considerable value, when combined with other evidence, in diagnosing certain mental diseases and personality disorders. For instance, schizophrenics have been reported as most often of the leptosome type, manic-depressives of the pyknic type. Further, we have seen in preceding chapters that various metabolic disorders affect both mental workings and physical traits at the same time. But in the case of normal persons, we must stress again that any facts about *average* correlations between body builds and temperament may have no bearing on the traits of given individuals; and even if certain physiques and certain personality traits went together in members of given families or groups, it would be no proof of the dual action of the hereditary factors involved, inasmuch as similar habits and conditions might just as easily be responsible.

Another way in which constitutional factors—hereditary or otherwise —may affect personality is in the rate of a young person's development before and during adolescence, as compared with others in his or her group. In a two-part study, psychologist Mary Cover Jones first observed a large group of adolescent boys and noted that those who had shot up first and achieved puberty earliest tended to be the most admired, poised, relaxed, and good-natured. By contrast, the late maturing boys tended to be more talkative, self-conscious and eager to get attention. Years later, when the boys had become men in their early thirties, Dr. Jones checked on them again and found that while the two groups now seemed more or less alike physically, the adolescent stamps put on their personalities by their being early or late maturing had largely remained. Similarly, other studies have brought out that girls who reach puberty earliest show certain personality differences from those who mature later, resulting from their being regarded as more grown up and trying (or being impelled) to act like women sooner. Not only are these earlier blossoming girls most admired and looked to more as leaders by the other girls, but they are apt to win more attention from boys and usually begin dating sooner and more often. (We might note that differences in time of achieving puberty within a group may be linked with heredity only when other factors—diet, health, living conditions—are about the same for all the members.)

There are still other ways in which genetically influenced biological processes or traits may be related to personality. *Coloring* is one possibility. Previously we emphasized that the popular linkage between given kinds of hair color, eye color and skin color is largely allegorical. But scientists do not rule out the thought that if pigmentation of one type or another results from gene-controlled chemical action, the same chemical action may also in some degree affect the sensory mechanisms and, hence, personality traits. Actual evidence for this is as yet limited and applies mainly to lower animals. Horse breeders have long claimed that the palomino—the lightly pigmented "golden horse"—is "gentle and tractable," and that this is also true, more or less, of mountain-bred, dappled gray horses. In contrast, wild bays, duns and sorrels are said to be "spirited and difficult to break in." With regard to mules, geneticist Clyde E. Keeler quotes a Georgia preacher as saying, "A gray mule is de lastin'es mule dey is . . . he jes' takes it easy. But a *brown* Texas mule, he jes weahs hisself out rushin' aroun'!" (Whether this is anecdote or science remains to be proved.)

Various correlations between coloring and temperament are also reported in dogs, cats, rats, cattle and other animals. But little specific evidence for any such correlation has yet been found in human beings, other than where deficiencies or abnormalities are involved, as in albinos, or in sufferers from certain diseases, such as *phenylketonuria* (where there is a tendency toward lighter pigmentation), or *Addison's disease* (where the skin may become bronzed). It also has been claimed that redheads are more susceptible to anesthetics than persons with other hair coloring. But no one has proved that blond women are like blond horses in being "gentler" and "easier to tame" than brunets.

A major reason why the color findings regarding lower animals cannot be applied to our own species is that in all the cases noted the animals have been bred intensively to the point where most of their genes are the same, and where a difference in the coloring genes may also significantly affect the workings of various genes concerned with behavior. Human beings, on the other hand, are all mongrels as compared with most domestic or experimental animals (a point we must always keep in mind) and differ among themselves in a vastly larger number of genes. The effects among people of one or two pigment genes, therefore, may be like a drop in the bucket compared with the effects of other genetic differences.

More important, human beings are vastly influenced by training and social factors and are unique among all other animals in the way that differences in coloring affect their attitudes toward one another and their feelings about themselves. This is most strongly shown among

Negroes wherever they are in a subordinate position to Whites. Everything else equal, Negroes with lighter skins have more favorable opportunities for jobs and social advancement than their darker-skinned relatives. A similiar situation now exists among Puerto Ricans in the United States—many of mixed Negro and Spanish stock—who, in their native island found little discrimination because of color. But on moving to the mainland, as noted by sociologist Clarence Senior, "the dark-complexioned Puerto Rican is puzzled and frustrated by color barriers," and, as one result, "the darker the skin of a Puerto Rican in New York City, the longer he is apt to cling to his mother tongue and culture, and the more slowly to become 'Americanized.'" Thus, as with hair or eye color, any significant relationship between human skin color and personality is likely to be overwhelmingly psychological.

So much for certain kinds of *looks* that affect behavior. But there is a reverse process. Many kinds of behavior, including habits of occupation and living, have specific effects on looks, ranging from the shaping and posture of the body as a whole, down to characteristic movements, facial expressions and mannerisms. Many of these effects come through what scientists call "muscle toning," the manner in which different sets of muscles become conditioned in their workings through repeated usage in similar ways. Every face, for instance, has beneath its surface an intricate network of muscles and fibers, acting variously to produce the expressions of smiling, laughing, weeping, frowning, affection, hate, chagrin, horror, and so on. Through use and habits starting from infancy, the facial muscles thus become "toned" in specific ways, giving the face as a whole a characteristic expression, even in repose, which may well provide clues to the individual's personality. So you're often very right in judging from a face that a person is kind, cold, mean, honest, alert, or otherwise. But don't trust these surface clues too far! From time to time magazines have published sets of photographs of criminals mixed up with those of highly respectable people, and readers have found it hard to judge which were which.

Likewise, mannerisms of walking, moving the hands, and posture develop largely through the toning of the muscles involved, and often provide insights into personality. (Perhaps your own manner of walking and moving may be so individual that friends can identify you, face unseen, from a long distance away.) But we must not confuse individual characteristics in bodily or facial movements with those developed in whole groups of people as a result of similarities in their working and living habits. Farmers, policemen, sailors, actresses, dancers, and prizefighters are examples. Even more emphatically, persons in any given country or locality tend to develop many of the same distinctive traits—of using their hands, walking, registering various emotions—which could be mistakenly regarded as hereditary. So, too, where members of the same family show similarities in such mannerisms, conditioning may be confused with heredity. But hereditary influences cannot be ruled out. We know that in lower animals of any species those of different breeds have characteristic patterns of behavior; and also, that in human beings, various hereditary disorders greatly affect bodily movements. This suggests that, in the normal range, milder-acting genes may affect human physical mannerisms in various degrees.

However, both in lower animals and in human beings, *unusual* physical mannerisms frequently are the result of psychological disturbances. Most often observed are the *tics*—involuntary movements of the hand or head, twitching of the eyes or mouth. Sometimes they are merely a means of relieving tension or a reaction to physical restraint. The prayer movements—head-shaking, rocking, weaving—of monks, Moslems and Orthodox Jews arise in this way. These movements may carry over into the everyday mannerisms of whole groups of people, and some may be catching—by children from their parents—illustrating again how the results of conditioning may be confused with inheritance. However, according to a British study, the fact that tics occur in 30 to 40 per cent of the close relatives of a person with such a trait does suggest some hereditary tendency. The study further showed that while children's tics usually are outgrown in time, in perhaps 6 per cent of the cases they may continue into adult life.

Hand movements, whether voluntary or involuntary, deserve special attention, for, unlike hand *shapes*, they may be importantly related to basic personality. Obviously, there are "expressive" hands, "sensitive" hands, "nervous" hands and "placid" hands; and various specific hand movements can be considered as revealing aspects of an individual's personality. So, too, particular hand movements may give doctors clues to many diseases (particularly, neurological and mental disturbances). Further, since fine hand movements are involved in *handwriting*, this has led to the new medical specialty of *grapho-diagnosis*, or the diagnosing of certain ailments by handwriting analysis.

But what of *graphology* itself—the "reading" of personality through an individual's handwriting? Even if we disregard the exaggerated claims of many charlatans in this field, there still is the question of how accurately and to what extent one's personality is revealed by the shapes, sizes, slants and fineness or thickness of written letters. Understandably, these details would be influenced by, and would throw light on, a person's education and training. Yet expert graphologists often are unable to tell whether the handwriting is that of a man or a woman or, sometimes, of an older or younger individual. As to the extent to which handwriting may be correlated with familial or inherited personality tendencies, it may be noted that even identical twins frequently differ in their handwriting, enough so that graphologists may not be able to recognize such twinship through handwriting samples. All this would prove that handwriting is indeed highly individual—which explains why banks pay so much attention to one's signature and why handwriting experts so readily identify forgeries. Perhaps with further development of graphology along scientific lines, it may become possible to correlate handwriting more closely with personality and with hereditary tendencies as well.

Palmistry is another matter. While palm-line patterns are in large measure determined by inheritance (identical twins usually being much alike in palm lines, as well as in foot-sole patterns), these patterns have not been found to have any relationship to inherent personality traits.* It is true that some clues to personality and to an individual's occupation and habits may be offered by the way that certain palm lines have become etched, or their patterns extended, through the use of the hands. But as to the "reading" in the palm of "love" lines, "success" lines, "life" lines, or "how-many-children-you-will-have" lines—science has just one word for that: "Bosh!" And in case you're interested, the same applies to *phrenology*—the "reading" of bumps on your head as clues to your character; and to *numerology*—linking the numbers in your birth date or the letters in your name to your personality and destiny.

No less pooh-poohed by scientists is *astrology*—the age-old system which deals with the supposed day-to-day influence of the stars and planets on the personalities and destinies of individuals, as well as on world events. While untold numbers of people take their astrological horoscopes very seriously, we can say only that astrology—not to be confused with *astronomy*—is regarded in authoritative scientific circles as completely in the realm of myth and fancy and its zodiacal symbols and interpretations as so much mumbo jumbo. Scientists have asked these questions: (1) Granted that identical twins, born at the same time and under the same stars, usually have similar personalities—but often far from the same destinies—why, then, do *fraternal* twins, equally born together, tend to differ greatly in personality—quite as much, ordinarily, as siblings with different birth dates? (2) If persons born under

* An exception is mongoloid idiocy, where among the effects of the chromosome abnormality causing the mental deficiency is a stubbing of the hands and certain peculiarities in the palm patterns. But any two mongoloid idiots might still differ greatly in various personality traits.

Libra (September 24 to October 23) should have musical ability—according to the astrologers—how explain findings by Dr. Paul R. Farnsworth that out of 1,500 musicians taken at random, fewer were born in this period or the period next to it ("Scorpius") than in any other period of the year? (3) What happens with the horoscope of a baby born prematurely through a sudden accident or delivered by a Caesarean several months before its scheduled time? Does its personality forecast change instantly? And by what physical means do the stars and planets accomplish this? Noting that no astrologers could properly answer these questions, a committee of leading astronomers, headed by Harvard Professor Bart J. Bok, which investigated the claims of the pseudo-science some years ago, issued the conclusion that "astrology is a magical practice which has no shred of justification in fact."

Handedness—primarily, being left-handed in a right-handed world—may conceivably have effects on personality *if* the individual is made to feel as a child that he is in some ways abnormal or is forced to use his right hand against his impulses. Even when freely allowed to write with and favor the left hand, the person is continually made conscious of his deviation by the fact that innumerable gadgets and contrivances are designed for right-handed people—including the broad writing arms of the seats in many high-school and college classrooms. But "lefties" aren't always so ignored, for a surprise finding by the writer on a recent trip to Israel was that each of the lecture halls in several educational institutions has a row of specially designed seats with the broad writing arm on the left side.

Other points about hands and arms given study are the manner of intertwining the fingers—right-hand ones over left-hand ones (the right thumb on top), or vice versa—and the manner of folding arms, right over left, or vice versa. In both cases the right-over-left habit has been found more common, and hereditary influences have been cited as probable. (The reader might like to check this with members of his family.)

Among voluntary movements that have been linked with heredity are several oddities in controlling the tongue muscle. One, the ability to turn up the side edges of the tongue, is present in about two-thirds of the population and usually is dominant in inheritance (but not completely so, since some identical twins differ in this trait). Another oddity is the capacity to fold the tongue into a clover-leaf pattern, which is possibly also a dominant trait. Much rarer (perhaps in one or two per thousand persons) is the ability to fold the tongue up and back when extended outside the mouth, this trait being recessive. The ability to stick one's tongue so far out as to touch one's nose may also be

THE GLIB TONGUE

Inherited oddities in ability to control tongue movements

(All usually dominant)

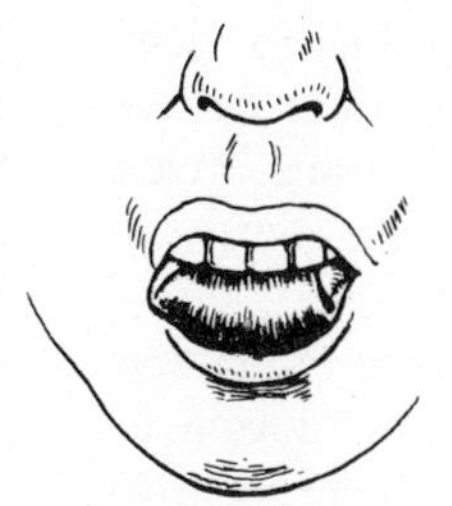

Folding the extended tongue tip up and in

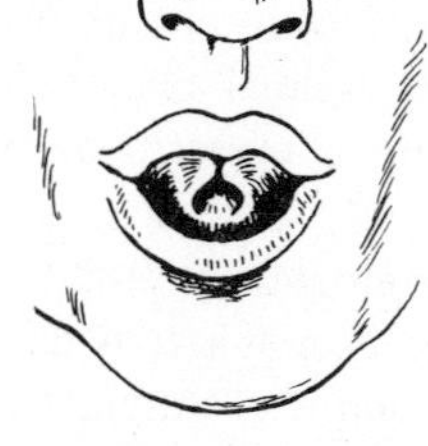

Rolling up the tongue in mouth

Folding the extended tongue tip up, also curling in sides

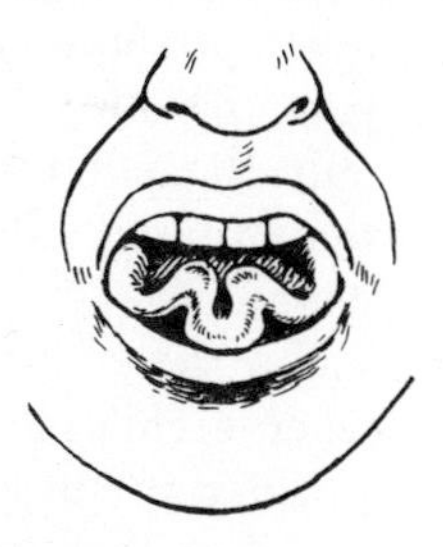

Curling up tongue in clover-leaf form

hereditary. The ability to move the ears, with or without the aid of the scalp, is another muscular oddity believed to have a genetic basis. But none of these traits can be said to have any special connection with personality—unless their possessors make a point of displaying them continually to others.

Come to think of it, we haven't yet given a *definition* of personality; and this is essential for the next phase of our discussion. Do you believe it is easy to define personality? Well, you ought to hear the top psychologists argue about it! First, there are important distinctions to be made between behavior and personality. As was brought out in the preceding chapter, behavior patterns—or the tendencies to evince them under given stimuli and conditions—are part of the hereditary equipment of all animals, with specific forms characterizing each species. (In human beings, it was noted, the behavioral traits that can be considered most directly genetic are many which appear in infants at birth or for a time thereafter, spontaneously and almost irrespective of their social environments.) But personality is something else—that manifestation of behavior, or the aggregate of various behavioral traits, which is a unique attribute of *persons*. While we may ascribe to lower animals certain humanlike traits, they cannot have "personalities." (Your dog may have a "dogality," your cat a "catality" if you wish to think of it that way.)

The big question is whether a baby comes into the world with the bud of his or her personality already formed, or whether the newborn infant in this respect has a blank area, which can develop a personality only through interrelationships with other *human beings*. There are many who hold to the latter conclusion. They maintain that while a newborn infant is a human being, and was one from the moment of conception, it is not a *person*—not yet having had social contact with other persons—and is therefore lacking in personality. From this viewpoint no one could inherit a personality trait but at best could possess only the potentialities for developing various personality traits, given the required stimuli and experiences in a prescribed human social environment.

But there is another school of psychological thought. This holds that when a child comes into the world it is not a neutral mass devoid of personality but already has many personality tendencies which begin to manifest themselves without awaiting social interaction. Thus, many observers have noted—as mothers could confirm—that one child from birth onward shows such and such traits, another child quite different traits. Moreover, there is evidence that individual personality patterns which emerge in infancy may continue to assert themselves later—traits such as aggressiveness, alertness, perseverance, tendencies toward smiling, laughing, crying, and other social responses. In one of the best known studies, Dr. M. M. Shirley recorded and analyzed the individual personality traits of a large group of babies. Years later, when the children had reached an average age of eighteen, another investigator, Dr. Patricia Neilon, studied them again and found that the basic personality patterns, established in infancy, had persisted in a marked degree in most cases, although in some individuals more of the original cluster of traits was retained than in others.

Further supporting the belief in the persistence of congenital personality traits were findings at the Fels Research Institute (Antioch College), where, over an extended period, records were kept of the prenatal reactions of many children to jolts, loud sounds, and other stimuli. After the children's birth and on through their adolescence, their personality traits were studied. Unexpectedly, those who had been the most active as fetuses usually turned out to be the most timid, submissive and anxiety-ridden individuals in their group. From this Dr. Lester W. Sontag concluded that fetal reactions in general may be related to the inherent makeup of the individual, as well as being influenced by the prenatal environment, and may foreshadow degrees of social anxiety persisting many years after birth.

In any event, even if a child does enter the world with preshaped personality tendencies, no one would doubt that innumerable outside influences thereafter can act to modify or warp such tendencies in significant ways. What is in dispute is the potency of various environmental factors and how far parents, teachers and others can mold the child's personality by the methods they pursue. Here we hark back to our discussion of imprinting and conditioning in the preceding chapter. With regard to lower animals, we noted that the effects of early conditioning influences—usually involving extreme and unnatural experimental measures—were limited by the range of instincts and behavioral tendencies present in the animal; further, that the effects often did not persist after the conditioning ceased and sometimes were erased when the animal was allowed to "be itself." What can be said about the early conditioning of young human beings?

The question is more than academic. For, during recent decades (and especially in the United States) parents and educators have been whirled around neurotically in a vortex of theories about child rearing, often jerked from one extreme to another within a short space of time. There were the restrictive theories (tied up with behaviorism), prescribing rigid schedules of feeding, toilet training, disciplining; the permissive theories (letting a child follow its own impulses); the controversies over the effects on the child of breast feeding versus bottle feeding, coddling and swaddling (or the opposite); the theories about birth traumas, emotional deprivation, parental rejection, maternal overprotection, paternal domination (or the reverse), birth order (being a first-born, youngest, or only child), twinship, sibling rivalry, and all the rest.

While all of the situations noted undoubtedly have effects on personality, there is much uncertainty as to whether these effects are as strong, sweeping and lasting as has been claimed. In a classic analysis of studies in this area, Dr. Harold Orlansky found many contradictions and little scientifically established fact. Drs. William H. Sewell and Paul H. Mussen, recording the experiences during infancy of a large group of Wisconsin children, saw no proof that their type of early training (breast feeding, demand schedules, gradual weaning, or the reverse) had much to do with their subsequent personality formation. Beliefs that lack of adequate "mothering" and "mother love" during early infancy (as with institutionalized children) would leave permanent scars on personality have been challenged by several investigators (Drs. Samuel R. Pinneau, Leon J. Yarrow, Lawrence Casler) who independently reviewed the studies from which the beliefs derived. Again, Drs. Harold Renaud and Floyd Estess checked on the childhood backgrounds of 100 young American army officers (averaging thirty-three years in age), who had been rated as "fully normal": i.e., well-adjusted socially and sexually, nine out of ten married and very few divorced, preponderantly conscientious husbands and fathers, and remarkably free of physical complaints and neurotic symptoms. Yet with few exceptions their childhood experiences had been little different in training and experiences during infancy or in parental discord, family tension, and sibling rivalry from those of a control group of psychotic patients.

We might conclude, then, that while current theories and recommendations regarding child rearing may provide valuable *general* guides and insights, they should not be accepted as scientific prescriptions which, if not followed in every particular, will cause irreparable damage to the child's personality development. Moreover, if one assumes that the shaping of a child's character depends almost entirely on little details of conditioning in infancy, and on its upbringing and the family environment, an enormous burden of responsibility may be placed unfairly on the parents. For one should not ignore the fact that much of what any individual child is, or can be made to be, may be governed by forces beyond the parents' powers to dictate or control. First, as previously indicated, most children may have much more "bounce" than they are credited with and can adjust to a variety of adverse early experiences (psychological as well as physical) without lasting damage. (To quote pediatrician Harry Bakwin, "There is every reason to believe that the run-of-the-mill errors in child rearing that parents make will be without serious consequences in the large majority of cases.") Also, as the child grows, many factors outside of the home sphere—in schooling, friends, social experiences—take part in personality development; and, unlike a conditioned lower animal, the growing human can understand and evaluate what has happened and is happening to him and need not conform meekly to every shaping influence.

But most important in our discussion, one must not forget that each child is an *individual*, whose response to any given environmental situation—so far as it can affect personality—will be governed in some degree by his own biological makeup and inherent tendencies. Whether or not there are specific "personality" genes, almost every gene concerned with producing a visible or functional hereditary trait may have some direct or indirect effect on the individual's psychological reactions. We have cited many examples with respect to behavior and the relationship between looks and personality. Obviously, also, any hereditary disease, defect or abnormality which makes the individual very different from the average in mental or physical performance must also lead to personality deviations. Likewise, various health factors—migraine headache, food allergies, ulcers, glandular diseases, vitamin deficiencies, digestive troubles—have been correlated with specific personality effects. Not only the hereditary traits of the child but those of the parents and the siblings are important in this. For whatever will affect the psychological atmosphere of the home can affect the personality development of each member.

But with such a mixture of other influences involved, where and how can we find specific proof of what heredity does or does not do in molding personality? Once more we turn to our old reliables, the *twins*. Many sets of identical and fraternal twins and supertwins (and mixed sets of the latter) have been studied with respect to personality traits.* Generally, identical twins have been shown to be much more alike in these traits than fraternal twins or than two singleton siblings. For example, testing eleven to fifteen-year-old twins of both types, Drs. Raymond B. Cattell, Duncan B. Blewett and John R. Beloff found that such traits as being easygoing, bold, sociable or withdrawn were present much more often in both members of an identical pair than in both members of a fraternal pair. Also, Dr. Steven G. Vandenberg reported much more similarity between identicals than fraternals in degrees of self-confidence and stubbornness and in response to music and color. In another study, reported by Drs. D. G. Freedman and Barbara Keller, eleven pairs of fraternal twins and nine identical pairs were observed in infancy; the differences in mental and motor abilities and personality development were found to be significantly greater within the fraternal pairs than within the identical pairs.

Before hurrying to ascribe all the similarities of identical twins to heredity, we must pause to consider that their conditioning also is usually much the same. This may be true as well for many fraternal twins in their early years, if parents do not or cannot distinguish between the two types of twins. However, as the twins grow and the resemblances of identicals become more obvious, there is a much greater chance that they will be kept or will keep together in their training and experiences. Complicating this has been the recent practice of trying to *de-condition twins away from similarity*, in the belief that their emotional, intellectual and personality development will suffer if their twinship is em-

* What had promised to be the most ambitious study of twins ever attempted was initiated in the Soviet Union in 1933, at Moscow's Maxim Gorky Institute, with many geneticists and psychologists from the United States and other countries joining Soviet colleagues under the direction of Dr. S. G. Levit. About 500 pairs of twins were assembled; and examinations, tests and experiments were begun. Then, suddenly, the whole project was quashed when Soviet political higher-ups became concerned that the findings about twins would give too much support to the importance of heredity and would challenge the Communist emphasis on environment. (The ensuing Lysenko period, with its dark days for genetics, is discussed in Chapter 41.)

LIVES DIFFERENT—BRAIN WAVES ALIKE

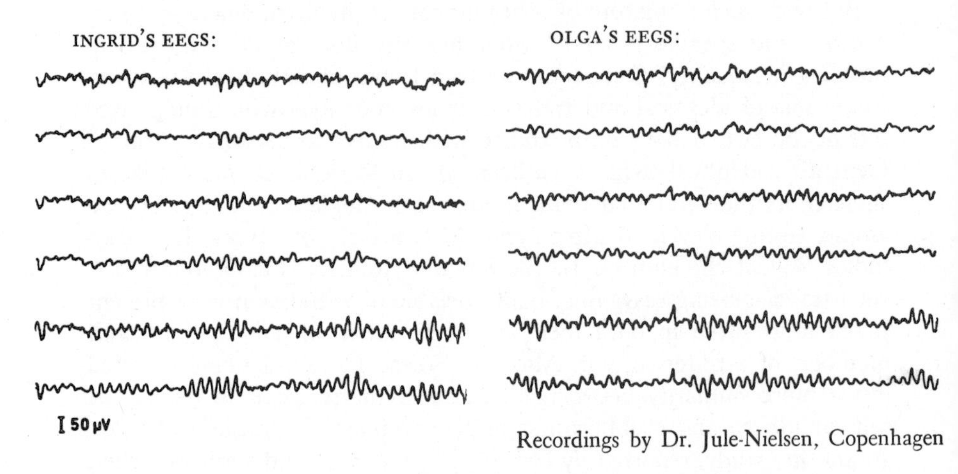

Recordings by Dr. Jule-Nielsen, Copenhagen

Above are shown the brain-wave recordings of two identical-twin Danish women who had been separated in infancy and not brought together until age thirty-five. Despite their very different environments and different personalities (see Text) the EEGs of the twins at age thirty-seven proved to be remarkably similar—evidence of the influence of heredity on brain-workings.

phasized and their individuality not cultivated to the fullest extent. While the general soundness of this policy for intensive application is yet to be scientifically established, it is nonetheless being followed by sufficient numbers of parents and educators to create doubt in many instances as to how far any similar personality tendencies of identical twins are or are not fostered and accentuated when they are raised together.

Thus, our most significant evidence about twins and personality has come from studies of twins who were *not* reared together—particularly those who, through parental deaths or adversities, were separated as infants, adopted by different couples, and brought up in different environments. Several scores of such separated pairs of identicals have been studied. Most interesting of the recent cases, reported on by Dr. N. Jule-Nielsen, were twelve pairs of adult Danish identical twins who had been separated in early childhood and not reunited until many years later, when they ranged in age from twenty-two to seventy-seven. (Dr. Jule-Nielsen kindly made available to the writer advance summaries of the twin cases. Of special interest were the brain-wave recordings of the separated identicals, which were found to be almost completely alike in all significant aspects. See in the accompanying illustration the EEGs of one of the pairs mentioned below.) As with many other long-separated identical twins, a number of those in the Danish group learned of each other's existence and were brought together only by the fact that their remarkable similarity in looks caused one to be mistaken for another. "Peter" and "Palle," one such pair, reunited at age twenty-two, had been reared apart from infancy in two very different types of homes in Copenhagen. One, much less favored, had received only an elementary schooling while the other had entered medical college. Yet when tested they were found to differ little in mental development, and to resemble each other closely in personality. As Dr. Jule-Nielsen reported, "Both lacked self-confidence, found it difficult to achieve emotional contact with other persons, and presented similar neurotic, hypochondriacal symptoms." Despite this, they took to each other so well that they left their respective homes and thereafter kept close together.

In another case—of the women identical twins "Ingrid" and "Olga," separated in infancy and brought together at age thirty-five—the personality patterns also were found to be similar, but their backgrounds had been so different as to provide no basis for mutual interests or continuing closeness. Still another Danish identical-twin pair, "Robert" and "Kaj," also separated soon after birth, had reached the age of forty before mistaken-identity coincidence reunited them. They, too, had had very different environments and experiences, but had turned out almost equally maladjusted. Kaj, from an underprivileged background, had served a jail sentence at age nineteen; Robert, much more favored in upbringing, had had no direct conflict with the law, but nonetheless showed a persistent antisocial streak. Tests revealed further similarities in basic personality patterns and the same concern about problems related to work, money and women (each having been married three times). Among the other Danish identical twins who had been reared apart for various long periods, Dr. Jule-Nielsen found that while there were usually many similarities in personality traits, degrees of difference in these or other traits as a rule could be related to differences in home conditioning and education.

The preceding facts and observations follow closely those brought out in earlier studies, made in the United States, of many other pairs of separated identical twins. One of the most remarkable cases, reported by the late Professor Horatio H. Newman, was that of the twins Edwin Iske and Fred Nestor. Separated in infancy and reared apart by different foster parents, each first learned he had a twin through a truly remarkable coincidence. Both had become interested in electricity and in their twenties were working as expert telephone repair men for the same company, but in communities a thousand miles apart. Eventually, traveling representatives of the company could not help noting that one repair man was the "spit'n' image" of another, and a reunion of the twins followed. Later, Professor Newman proved them to be amazingly alike in mental ability and personality. Moreover, he found that they'd been attracted to and had married girls of the same type, in the same year.

Another striking case was that of the twins "Millan" and "George," who were studied by Drs. F. E. Stephens and R. B. Thompson. These twins, too, had been separated as infants and brought up far apart under quite different conditions. When reunited briefly at age nineteen, they turned out to be remarkably alike not only in looks, but in temperament, intelligence, achievement, artistic leanings and athletic interests (both having won amateur boxing championships). Although they soon parted and resumed their separate paths, their twinship continued to assert itself in an unfortunate physical way: Four years later, while in military service, both almost simultaneously developed the same rare, hereditary crippling disease, *spondylitis.**

The cases of separated identical twins and the opportunities for studying them seem destined to grow fewer as the likelihood decreases that any such twins offered for adoption will be allowed to be parted—particularly in view of the readiness of many prospective foster parents to snap up both members of any twin pair available. (The author knows personally of a number of identical-twin pairs who have been adopted in the New York area within recent years.) However, even with identical twins who remain together, another kind of separation can occur—through a radical difference in health, physical functioning, or emotional experience—and the effects of this on the personalities of the twins in-

* Most recently appeared the story of two other male identicals, separated in infancy and reared by two different families—one, Italian, in New York State, the other Jewish, in Florida. Given the respective names of Tony Milasi and Roger Brooks, the two were unaware of each other's existence until, as in previous such cases, repeated instances of being mistaken for one another and a series of coincidences led to their reunion in 1963, when they were twenty-three years old. Not only were they amazingly similar in looks and physique (both 6 feet 3 inches tall, and each weighing just over 200 pounds), but when tested by psychologist Syvil Marquit of Miami, they proved to have nearly the same IQs, and the same marked aptitude for clerical work. However, their personalities were quite different—Tony "more extroverted and self-assured," Roger "more sensitive and impressionable"—not surprising in view of their marked differences in rearing and psychological environments. In any case, the new attachment proved so close that Roger went to live in the same city with Tony. (From an article in the *Saturday Evening Post,* March 21, 1964.)

volved can also be studied to advantage. For example, Dr. Franz J. Kallmann explored the case of 2 twenty-four-year-old New York male college graduates, one of whom (and he only) had long been afflicted with a spastic form of cerebral palsy. Nevertheless, although impeded in speech and movements, he had reached almost the same mental level as his normal twin, had almost as good a job, and was as well-balanced emotionally, as sociable, and as congenial with other persons. In another case, to check the effects of different emotional experiences, Dr. Kallmann studied two middle-aged male identical twins. One, after successive business failures and years of misery with a mentally diseased wife, had taken to drink. The other had enjoyed a happy home life and moderate business success. Yet, at the age of sixty, the twins revealed almost the same mental and emotional traits, and, said Dr. Kallmann, "their contrasting life situations have not obscured the remarkable similarities in their basic personality characteristics."

Supertwins also have provided special opportunities for personality studies. Five American quadruplet sets, in a number of combinations of identicals and fraternals, were tested by Professor Newman and Dr. Iva G. Gardener. In an all-identical girl set, the Morloks, who were studied at age ten, differences in personality were found for each, the biggest of the girls being consistently the leader, while the smallest—from babyhood on—was the quietest and least self-assertive. But in an all-fraternal mixed set, the Schenses (two fraternal boys, two fraternal girls), who were tested at age nine, the disparities proved far greater—in fact, with as much difference in mental ability and personality as would be shown by four singleton siblings in the same family. The Perricone quads, another four-egg set (though all boys), differed temperamentally even more than physically. In the Badgett girl quads (three identicals and one fraternal), the fraternal, genetically different sister proved to be distinctly much brighter than the other three and quite different in personality. But most interesting were the famous Keys girls of Oklahoma—the first quadruplet set to go through college. A three-egg set (two identicals from one egg, two fraternals from two other eggs), they had all been reared and educated closely together. But when given IQ and personality tests at the age of twenty-two, only the two identicals were found to be very much alike, whereas the two fraternal sisters not only differed greatly from the identical pair, but from each other.

Finally, we come to the all-identical Dionne quintuplets. Studied in their first few years by Drs. W. E. Blatz and D. A. Millchamp, the quints revealed differences in personality which were enough to set them apart as individuals. But the differences were perhaps minor as compared with those usually encountered among five singleton siblings and were probably overshadowed by their similarities. This supposition, in the absence of further psychological studies of the Dionnes (which were not permitted), has found some support in the periodic news reports about them through the years. Three of the sisters tried to become nuns, but only Yvonne has gone through with this. Emilie, after leaving and returning to a convent, died there in 1954. Marie, on giving up her try at convent life, failed in a brief business venture (a florist shop) and married in 1958. The previous year, Annette and Cecile were married, and they and Marie all now have children.

An insight into what had really gone on in the lives and minds of the Dionne quints during their cloistered childhood and adolescence, when they were in the charge of their family, was given in a magazine series published in 1963. The story they told was a sad one, of being reared in a harsh and restrictive atmosphere, with only themselves to cling to for true understanding and affection. Where equal heredity or equal environment could be credited with similarities in their personalities would now, therefore, be hard to determine. However, the surviving four quints reported that since being on their own, and with the marriages of three of them, their differences have become accentuated: that they have voted differently in elections and have thought differently on most subjects and occasions (which conceivably could reflect the differentiating influences of their husbands' thinking and their varying experiences). Nonetheless, the Dionne quints believe they still have a close psychic relationship, sometimes—they claim—one being able to tell what is in another's mind.

Reviewing all of the facts about twins and supertwins, what conclusions can we draw? First, it is apparent that heredity does play a potent role in shaping personality patterns, but not nearly as great as with respect to physical traits. For in almost all instances the resemblance in personality between any two identical twins—whether reared together or apart—has been found to be much less than their similarities in looks, anatomical structure and bodily functioning. Thus, as Professor Newman summed it up, physical characteristics are affected least by environment; intelligence more; education and achievement still more; and personality and temperament the most.

More specifically, with respect to the various factors involved in personality, the general belief among psychologists is that hereditary influences may be graded in this way:

Most likely to be influenced by heredity: Basic abilities, such as intelligence, speed of reaction, motor skills, and sensory discrimination.

Less likely to be influenced by heredity: Temperamental traits, such as emotionality, alteration or evenness of mood, activity or lethargy.

Least likely to be influenced by heredity (if at all): Attitudes, beliefs, values and other such characteristics in which training or conditioning are clearly major factors.

What we will need before we can report with more certainty on the inheritance of personality are better tests for measuring its ingredients than are now available.* This is not to minimize the value of such current tests as the Rorschach (seeing things in ink blots), the Minnesota Multiphasic Personality Inventory (MMPI), the Thematic Apperception Test (TAT), and many other projective tests in which persons are required to do things—answer questions, interpret drawings or stories, and so on. All have been found highly useful for revealing various normal and abnormal character traits, picking individuals for jobs, making rough predictions of marital adjustment, and assaying special aspects of personality. But they do not yet reach deep enough below the surface to identify basic ingredients of personality which can be directly correlated with genetic factors.

So for the present we can only *assume* that there are genes for normal personality traits just as there are genes for other aspects of human makeup and function, including certain abnormal personality characteristics. Where, in members of the same family, in a similar environment, there are great differences in personality, these might be ascribed at least in part to differences in gene combinations. We can also guess that some of the family similarities in personality are genetically influenced. But we're still a long way from identifying specific "personality" genes, gauging their effects or hazarding predictions as to what the personality of a given child will be on the basis of what we know about its parents. In short, heredity can never be considered as charting a fixed and definite course for anyone's personality. At the best, what anyone inherits is the potentiality for a wide range of personalities, the precise form into which a personality will "jell" being determined by many outside influences and chance events.

* To quote Harvard anthropologist W. W. Howells: "It is to be hoped and expected that traits of native 'personality' along with specific traits of intelligence eventually will permit identification and measurement, and genetic study of their variation, in precise terms. . . . It would appear inevitable that students of personality and culture . . . will not have all their solutions until they know more about the biological basis of personality, meaning its precise modes and ranges of variations, which will in the long run call for understanding of the genetics involved. . . ." (In *Yearbook of Anthropology*, William L. Thomas, Jr., Ed., Wenner-Gren Foundation, 1955.)

CHAPTER THIRTY EIGHT

Crime and Its Roots

IF YOU'RE A REASONABLY law-abiding citizen and you read about a particularly vicious crime, you may say, "I could never do a thing like that. I'm just not built that way!" Seeing a picture of the criminal, you may imagine that he even looks like a different sort of creature from you.

Whether you are or aren't right can be decided only by answering the question, "What makes the criminal?"

Is it heredity, with a special set of wayward genes for criminality? Is it environment, a result of bad early conditioning and warped moral judgments? Or does heredity load the gun, leaving it to environment to pull the trigger?

Here is a case where, perhaps more than in any other part of this book, you must be the judge and the jury.

The charge that criminal acts can be blamed on "bad heredity" in individuals or on "taints" running in certain families is as old as the hills. But the evidence is almost entirely circumstantial. Much of what has been offered in the past is, as lawyers would say, "incompetent, irrelevant, and immaterial." What can be offered today is still far short of conclusive.

To prove any inheritance of criminality there must be the assumption, first, that it is a distinctive and clearly defined type of abnormal behavior (which, as we shall see, is open to some question); and second, that this behavior has a biological basis—that it results from some specific physical or mental peculiarity in certain individuals which predisposes them more than other persons to become criminals.

Among lower animals we refer to tigers and wolves as "killers," jackals and buzzards as "ghouls," magpies and cuckoos as "thieves," leeches as "parasites," and so on. If these and hosts of other animals were judged by human standards, one would say they were all of them congenitally "criminal" or "antisocial."

But we do not hold the lower creatures strictly to account for what they do (although we may condemn them to death, nonetheless, when they menace us) because we do not credit them with the same sort of intelligence that we have, or with any will or conscience. Their acts are ascribed to "instincts," to "uncontrollable impulses"; we say they were "born that way." On the other hand, what the human animal does we like to think of as dictated by intelligence and reasoning powers. On this assumption, that what we do is done willfully, are based all our existing codes of law and morality with their punishments for bad behavior and rewards for good behavior.

Are we right? Is it possible that many human criminals, though entirely sane, are yet no more responsible for their acts than are lower animals? That some humans, too, are impelled by uncontrollable impulses and antisocial "instincts"? That they are born that way?

The theory that the criminal is a throwback to lower forms of humans—an individual in whom the "primitive urges" or "base animal passions" break through the civilized restraints—has popped up at various times. Toward the end of the nineteenth century the Italian criminologist, Cesare Lombroso, startled the world with his purported "scientific proof" that criminals, by and large, differed from normal human beings in various physical characteristics indicative of "degeneracy." Lombroso's theory was soon deflated by an onslaught of contradictory evidence. Years later, Professor Earnest Hooton again attempted to prove, on the basis of his studies, that criminals as a group tended to have in more than ordinary degree such characteristics as low foreheads, compressed faces, narrow jaws, very small ears, and extremes of broad or thin noses—all of which he regarded as evidence of "organic inferiority" and "primitivism." He also sought to show that specific types of crimes tend to go with specific physiques, and even more, with specific races or ethnic groups because of their biological differences. Dr. Hooton's contentions, too, shrivelled under professional attack, including evidence produced by a fellow anthropologist, Dr. Alex Hrdlicka, that measurements of 1,000 juvenile delinquents failed to show any significant physical differences between potential criminals and noncriminal types, or that any specific abnormality or defect went with any given kind of criminal tendency.

But the attempts to prove a relationship between criminality and physical makeup went on. Dr. William H. Sheldon claimed to have found that delinquent youths *did* differ on the average in their somatotypes from the law-abiding lads. Most delinquents, he asserted, tended to be short, stocky and heavy-buttocked, while the "good" youths, as a rule, were more often muscular, lean and long, and superior as physical specimens. Dr. Sheldon maintained further that the correlation between physical build and criminal predisposition was *hereditary,* and that it extended to differences among races and between constitutionally superior and inferior ethnic groups. Yet Dr. Carl C. Seltzer showed by other studies that delinquent youths not only were *not* constitutionally inferior to nondelinquents, in similar environments, but, on the contrary, tended to be superior in physique by ordinary standards—more masculine and better developed in their limbs and torsos. However, Dr. Seltzer did conclude that "personality traits which under certain circumstances predispose to criminality are correlated with certain somatic characteristics found in greater proportion in the delinquent than in the nondelinquent population."

It remained for the Harvard criminologist couple, Drs. Sheldon and Eleanor Glueck, to bring "constitution and crime" into clearer focus and to view it also in the light of social influences. Over a period of many years, extending into the mid-fifties, they studied the physical measurements, psychological traits and home environments of hundreds of delinquent youths, who were compared, boy by boy, with equal numbers of nondelinquents of similar age, background, IQ and ethnic stock. Although concluding that some *average* relationship did exist between physique and delinquency, the Gluecks stressed the qualification that many other influences, social and psychological, contributed to criminal behavior. As to physical factors, the Gluecks found the solid, muscular, energetic type (mesomorph) much more often among the delinquents (60 per cent) than among the nondelinquents (30 per cent). The leaner, longish, fragile and less vigorous type (ectomorph) was much more prevalent among the nondelinquents (40 per cent) than among the delinquents (14 per cent). The percentages of the other two types—endomorphs and balanced—differed little in the two groups.

How can we account for these possible correlations between physical makeup and a tendency to delinquency? The Gluecks reasoned that where environmental factors were unfavorable—disorganized and upsetting homes, an inciting social atmosphere—the boy who was physically vigorous and assertive (the mesomorph) would be apt to take out his tensions in antisocial acts unless his energies were channeled toward socially acceptable goals. Under the same conditions, the more delicately built, less vigorous and more sensitive youth (the ectomorph), even if he *felt* antisocial, would be less likely to *act* overtly and would be more apt to bottle up his emotions and become neurotic. (If he did do anything criminal, it would probably be in the more subtle categories, such as cheating, petty theft and forgery.) The youth of the soft, round, passive type (the endomorph) also would not ordinarily indulge in violent action but might be inclined to react toward adverse influences more phlegmatically than either of the other two types. But regardless of the physical type, the criminologists noted that delinquency could be caused in certain cases by many factors having little connection with body build—assertiveness, defiance, lack of self-control, feelings of insecurity, isolation, slum conditions, or parental neglect, harshness, alcoholism, and so on. Analyzing the various factors and traits, the Gluecks devised a "prediction scale" which is being used by many social agencies as an aid in identifying potential or prospective delinquents.*

In other words, while no physical factors or inborn tendencies *alone* would push a boy into delinquency, his being of one physical type might cause him to respond more strongly to certain unwholesome and stressful conditions and to have a greater "criminality potential" than a boy of another physical type. Such a hypothesis—since many of the physiological factors mentioned are influenced by heredity—could be consistent with the general theory of hereditary predisposition which we have seen applied to many diseases, mental disorders and other traits. So this leads to the question: If the criminally disposed individual can't be identified from his "outsides," can he be from anything *inside* of himself?

The first place to look for biological symptoms of criminal impulses would be in the brain and in evidence of its functioning. If criminality is regarded as abnormal or diseased behavior, it could arise in two general ways: either through a defective mind or retarded intellect, which would make the individual incapable of understanding when acts were wrong, or more likely to drift or be inducted into wrongdoing; or though any type of mental illness which might incite the individual to antisocial acts. To what extent, then, do these abnormal mental factors (with heredity often involved) contribute to the total crime picture?

With respect to mental levels, the over-all IQ in prisons and reformatories is undoubtedly below the average. Contributing to this is the fact that IQs are more likely to be repressed in the underprivileged environments which most strongly foster antisocial behavior; and that law violators of low intelligence are more likely to be caught and imprisoned than are cleverer persons who commit the same crimes. So to the extent

* The effectiveness of the Glueck scale appeared confirmed in an extended study by the New York City Youth Board (1964). This showed that, among 300 boys evaluated on the scale ten years before, forecasts of those who would turn out to be delinquent, non-delinquent or with an even chance of delinquency had been made with about 90 per cent accuracy.

that heredity accounts for mental retardation, it is important in this aspect of crime.

Mental diseases, however, are much more directly involved in criminality than are mental defects, and so, in turn, is heredity. Murder and other acts of violence not infrequently are traceable to the disturbed states of schizophrenia or manic-depressive insanity. But these conditions may account for less criminal behavior than do those classified as "psychopathic and/or sociopathic personality." As described in Chapter 25, the psychopaths are cold, calculating, untruthful, emotionally immature individuals, midway between sane and insane. Underlying their condition is a failure to grow up socially and psychologically and, despite their knowing right from wrong, an aversion to discipline, resentment of authority, and an inability to adjust properly to other people. But, whether or not heredity plays any part, it must be stressed that similar traits of emotional immaturity are found in many if not most habitual criminals who would be diagnosed as mentally normal and who appear to be largely products of unfavorable conditioning.

The attempts to distinguish between inborn and environmentally conditioned mental tendencies toward criminality have led to the studies of *brain waves* (electroencephalograms, or EEGs), of overt criminals and delinquents. That undue proportions of these individuals do reveal abnormal EEG patterns has been reported by many investigators. Further, psychiatrists Edward D. Schwade and Sara G. Geiger have noted that a characteristic erratic type of EEG (similar to but not caused by epilepsy) is repeatedly seen in children who are given to inexplicable outbursts of rage, menacing behavior, unruliness, irresponsibility, and mutilation of animals. Other investigators have reported finding abnormal EEGs in many youths and adults who actually have committed violent crimes; and psychiatrist Sherwyn M. Woods claimed to have observed the peculiar "6 and 14 brain-wave syndrome" (peaks at every sixth and fourteenth second in the EEG graph) in a significant number of adolescents whose behavior included fire setting and murder. However, the very expert who discovered the "6 and 14" pattern, Dr. Frederic A. Gibbs, said he himself had found no evidence of its being unduly present among murderers. Inasmuch, also, as most delinquents and criminals do not show any brain-wave abnormalities, their direct correlation with criminal behavior remains in dispute.

Nonetheless, the discussion of brain waves and criminality brings us again to *twins*. For brain waves of identical twins tend to be remarkably alike, even after they have been reared apart from infancy onward in different environments (see illustration on p. 486). If, then, certain types of abnormal brain waves did presage criminal tendencies, this would be one reason for assuming that identical twins would be inclined by heredity to be either both law-abiding or both criminal. What are the facts? Many studies of twins and crime have been made in the United States and other countries. Collectively, these studies have indicated that if one twin was criminal, so was the other in three out of four cases where the twins were identicals, but in only one-fourth the cases if they were fraternals. Further, the types of crimes committed by identical twins were usually the same, or closely related, as was not true among the fraternals. These findings are open to several interpretations:

> (1) The much greater likelihood of two identical twins' taking a criminal path together is evidence of their having the same inherited tendencies to do so. (2) The exceptions prove that even with the same heredity, but with slight differences in environment, one person can become a criminal and the other not. Or (3) comparing the identical and fraternal twins, while the identical pairs are much more apt to proceed together on a criminal path, it may be not only or because they have the same heredity, but because also, as a rule, they are more likely than fraternal twins to have had similar conditioning.

The principal difficulty in trying to track down hereditary factors in crime is that we are not dealing with a clearly measurable trait, and certainly not with a single kind of act. In fact, most authorities would agree that crime is not in the act itself but in the *nature* of the act, which can be judged only by the intent behind it and the circumstances under which it is committed. The act of killing another human being, for instance, is far from always being murder, as in the case of killings in the line of duty, by soldiers, police officials or public executioners. And when a killing does have the appearance of murder, there is always the question of "responsibility."* Today this often leads to a legal morass in which lawyers, judges, psychiatrists and jurors flounder about in confusion. What constitutes criminal responsibility—or legal sanity in the commission of crime—has been debated for centuries. In British and American courts a general criterion relates to the defendant's ability to

* The legal principle of criminal responsibility was first formulated in English courts of the thirteenth century, which ascribed *irresponsibility* to a person who committed a crime while "in a state of frenzy resembling that of a wild beast." Later, idiots also were included among the "irresponsibles," but the insane continued to be punished for criminal acts until 1843, when a jury acquitted a murderer named M'Naghten on the grounds of insanity. This precedent, known as the "M'Naghten rule," defined legal insanity as inability "to know right from wrong" at the time of the act and has continued to be widely followed in British and American courts. An additional decision, in an American case in 1954 (the "Durham" ruling), more clearly relieved an accused of criminal responsibility if the wrongful act was a consequence of psychosis or mental deficiency.

"understand the nature of his act" and to "distinguish between right and wrong." Just what this means is so uncertain—as can be seen in many a murder case in which opposing psychiatrists give radically different opinions—that committees of American jurists and scientists have been at work for years trying to formulate some solid and unequivocal standard.

Until "responsibility" is more clearly defined, there will be difficulty in deciding whether or not specific acts are truly criminal, and if, when and how they may be ascribed to hereditary abnormalities or weaknesses. A further complication is that even where crimes in any category are quite clearly rational and responsible acts, they may be so varied in nature and motivations (for example, murders) as to make it impossible to find any common psychological or biological basis for them. Not least, there is the question of establishing fixed boundaries between the criminal area in human behavior and the noncriminal area. For instance, it has been claimed that almost every one of us at one time or another has been guilty of some offense which, judged solely by itself, might have been technically punishable by a reformatory or prison sentence. Criminologist Austin L. Porterfield found this was true of nearly all the students in a Texas *divinity* college. (And have *you* ever snitched a towel, spoon or ash tray? Or gambled, or struck someone in anger? That—technically—might have gotten you six months or more in jail.) Not even the "urge to kill" can be regarded as peculiar to murderers, for, according to psychoanalysts, the impulse to kill somebody at some time is latent in all human beings.

Yet one can go too far in arguing that "everyone is 'criminal' in some degree." For it is mere quibbling not to distinguish between the large majority of persons who are only occasional transgressors in minor respects and the limited number who go to extremes in vicious and flagrant wrongdoing, who *carry out* acts for which others may have only impulses, and who consistently act in a way which by any civilized standards must be considered criminal. There is no reason why we can't draw a line here, as well as we can between persons who are occasionally and mildly sick and those who suffer from serious chronic diseases and disabilities, or between those who crack up temporarily and only under extraordinary pressure, and those who with little or no stress become insane. As with extremes in other traits, there can be some general understanding of what constitutes criminality; and whether it is regarded as a "disease" or a type of abnormal behavior, we cannot dismiss the possibility that heredity plays some part in it.

We have already noted that inheritable mental deficiencies and diseases may often be factors in antisocial and criminal activity. But also, among individuals classed as mentally normal, we must recognize (as was brought out in previous chapters) that there can be marked genetically influenced variations in emotionality, excitability, quickness to anger, aggressiveness, insensitivity, and other traits, all of which may be involved in given types of criminal behavior. For example, the "killing point" could be reached in some individuals much more quickly than in others; and so may the "assault" point, "rape" point, "burglary" point, etc. In other words, everything we've learned about individuality in human makeup and functioning would indicate that, among all persons, *degrees of susceptibility or resistance to criminal behavior may be inherited.* But how and to what extent is still largely speculative.*

Whatever may be the possibility that "heredity loads the gun" in crime, there is little question of the way in which "environment pulls the trigger."

Throughout the United States enormous differences in the rates for murders and other crimes are found in various sections, states and cities, which have little relationship to the ancestral origins or biological make-up of their populations.† Almost uniformly the murder rates for *Whites* in Southern states are far greater than in most northern states; and within the North itself, the murder rate in Chicago (among Whites, again) has been averaging annually about ten times that in Milwaukee, ninety miles away. Yet no one would claim that Whites of the South have a much more murderous heredity than those of the North, or that the people of Chicago are inherently ten times as murderous as those of Milwaukee. So with any communities or regions, in this country or abroad, which present radical contrasts in crime rates, the explanation lies very largely in environmental differences. Wherever living conditions are worse, poverty and ignorance proportionately more widespread, police

* One old theory which definitely can be squashed is that a "murder streak" can "skip" from an ancestor to a descendant. This was the theme of *The Bad Seed,* a highly successful novel, stage play and movie of the fifties, in which a nine-year-old multiple murderess was held to have inherited her sinister tendencies from a homicidal grandmother. The theory has no basis whatsoever in genetic fact.

† While recorded crime rates or official statistics do not necessarily represent the actual incidence of crime (as will be brought out later), there is little doubt that the very marked contrasts reported in crime rates among various sections, communities or groups of the country do generally reflect real differences to a considerable degree. This applies most strongly to homicide rates ("a dead man is an indisputable fact"), but decreasingly so to thefts, assaults and financial or other white-collar crimes which may be viewed, reported and dealt with very differently in one place than in another.

more inefficient or corrupt, courts more lax, *and people more tolerant of wrongdoing,* there the murder and crime rates will be higher.

Perhaps you remember this story:

In a "hell-roaring" mining town of Wild West days, there was a commotion one evening. An old Forty-niner stuck his head out of a barroom and saw his son being led to jail by the sheriff.

"Hey!" he called out. "What's my Willy done?"

The sheriff yelled back, "He got mad and killed one of them Eastern dudes!"

"Shucks," said the old-timer, going back to his tippling, "I thought mebbe Willy'd stole a horse."

This is not so far-fetched. Horse-stealing was a hanging offense in the frontier days, when a murder was often looked upon as an indiscretion. A not-too-dissimilar easy attitude toward murder on the part of juries today, coupled with lack of restriction against "totin' a gun," has a lot to do with high murder rates, not only in individual sections of the country but in the United States as a whole compared with European countries. Or, at least, the latter used to be true.

In the years preceding World War II the murder rate in the United States was far above that of almost any European country. For homicides (including manslaughter), the rate in large American cities was almost twenty times that in England, and about three to four times that in Germany or Italy, with the rates of other countries somewhere in between. But since the hereditary factors of the people in the United States could be no different, collectively, from those of their European progenitors, the conclusion was obvious that there was something in the environment of this country which tended to produce more murders. But think how quickly the situation could change!

In the space of a few years, Germany (some of whose authorities had previously pointed to the much higher murder rate in the United States as evidence of "barbarism" and "biological inferiority") shocked humanity with an avalanche of crimes more horrible in their cruelty and viciousness than the world had ever known. Psychologists may some day offer a full explanation of the Nazi reign of terror, which drew in among its worst participants persons of the highest cultural levels—doctors, scientists and educators included. Whatever the findings, this will stand forever as proof of how human beings—*almost any group of humans*—can react in an hysterically degraded environment. For, just as we said before that there can be nothing inherently more criminal in Americans than in Europeans, scientists would still say that there was and is nothing *genetically* more criminal in the Germans as a whole people than in any other national group.

The foregoing can be just as well applied to people of other stocks—racial, ethnic or national. Unfortunately, there has been a persistent belief in the United States that differences in crime rates among persons of different foreign origins were in large measure hereditary. This belief was one of the important reasons for limiting immigration and fixing quotas a few decades ago. For instance, it was claimed that immigrants from southern and Eastern Europe were by nature more hot-blooded and lawless and would have many more among them *inherently* disposed to criminality than those from England and the Scandinavian countries (or—ironically—the Germans). Statistics in former years did show that in certain large cities there was a considerably higher percentage of murders by Italians than by *native* Whites of some other stocks. But it also could be proved that the mass of Italians then lived under conditions more than ordinarily unfavorable, and that most of the murders among them were committed by transplanted members or affiliates of the old Sicilian Mafia, or Black Hand (later to be dubbed Cosa Nostra), who had no more in common with other Italians than some of our native gangsters had with other Americans. Even then, the general crime rate of Italian communities as a whole was lower than that among certain old-line, native-stock American groups in various benighted sections of the country. Further, with the successive upgrading of living conditions among descendants not only of Italian immigrants, but of those from other immigrant groups which previously had been suspect, the crime rates have dropped sharply to levels no different from the average among Americans in equivalent environments.

What of the belief that Negroes are *inherently* much more irresponsible and criminal than other peoples? By now it should be clear that if bad conditions could explain high crime rates among some groups of Whites, the immeasurably worse conditions under which most Negroes currently live could account quite easily for their still higher crime rates, without blaming their heredity or any inborn tendency toward irresponsibility. Also to be borne in mind is that the *recorded* crime rates of American Negroes may be greatly magnified by the fact that for any given offense individuals among them usually are much more likely to be arrested, convicted and heavily sentenced than are Whites, particularly in the Southern states.*

* The statements about Negroes and crime apply in large measure to the present high crime and delinquency rates of Puerto Rican groups living under extremely adverse conditions in New York and other large cities. However, not all the blame can be placed on the disruptive effects of transplantation from their native island to a hostile new environment. In many respects the transplanted groups had lived under almost equally bad condi-

In making comparisons previously between crime rates in the South and North, we referred (as you may have noted) to the rates among *Whites only*. For another wrong assumption is that the very high crime rates in Southern cities must be due mainly to their large Negro populations. While the higher percentage of Negroes—and accordingly, of the most underprivileged groups—does help to swell Southern crime rates, their presence is far from the sole determining factor. As Dr. Austin L. Porterfield showed some years ago with a crime index of the relative numbers of serious crimes for each state, there was no direct relationship between the size of the Negro population and the overall rate of crime. Mississippi, with the largest proportion of Negroes (about half the population) had a crime index almost 75 per cent less than Kentucky, with the smallest proportion of Negroes—only about one-seventh the Mississippi pecentage; and North Carolina, with a one-fourth Negro population (half the Mississippi percentage) had the highest serious crime rate in the country.* (The lowest crime rates were in New Hampshire, South Dakota, Wisconsin and Vermont.) But to whatever extent the relative numbers or changing proportions of Negroes among states contribute to their crime indexes, the results must be viewed as determined far less (if at all) by race than by environmental discrimination on the basis of race. As conditions for Negroes or for any other underprivileged and socially repressed group improve, their relative crime rates can almost certainly be expected to drop.

(One must note here that slum conditions, social repression and ignorance are by no means the only causes of crime; nor are parental rejection, broken homes, and similar explanations. All such easy answers fail to account for a multiplicity of other causes. Among the disconcerting facts is that delinquency and criminal behavior have been on the increase in the United States and other countries even though economic and social conditions have steadily improved, and that many delinquents —including thieves and killers—have been coming from economically secure, educated and religious homes and generally favorable environments. While failings of parents may sometimes or often be blamed for the delinquency of their children, it is as wrong and unfair to say sweep-

---

tions in their homeland slum communities and had there developed patterns of family disruption and high delinquency, as the author learned on a recent visit to Puerto Rico.

* The theory offered by Drs. Austin L. Porterfield and Robert H. Talbert was that Southern communities are much more tightly organized along class, caste, racial and kinship lines than those in the North, providing more occasion for conflicts and fostering more aggression.

A MURDER MAP OF THE UNITED STATES

Showing differences among the states in annual rates for murder and willful manslaughter. Pistol symbols, *in graded sizes, reveal at a glance the marked sectional differences.* Accompanying *figures for the respective states are the numbers of killings annually per* 100,000 *population, based on an average for* 1961-1962.*

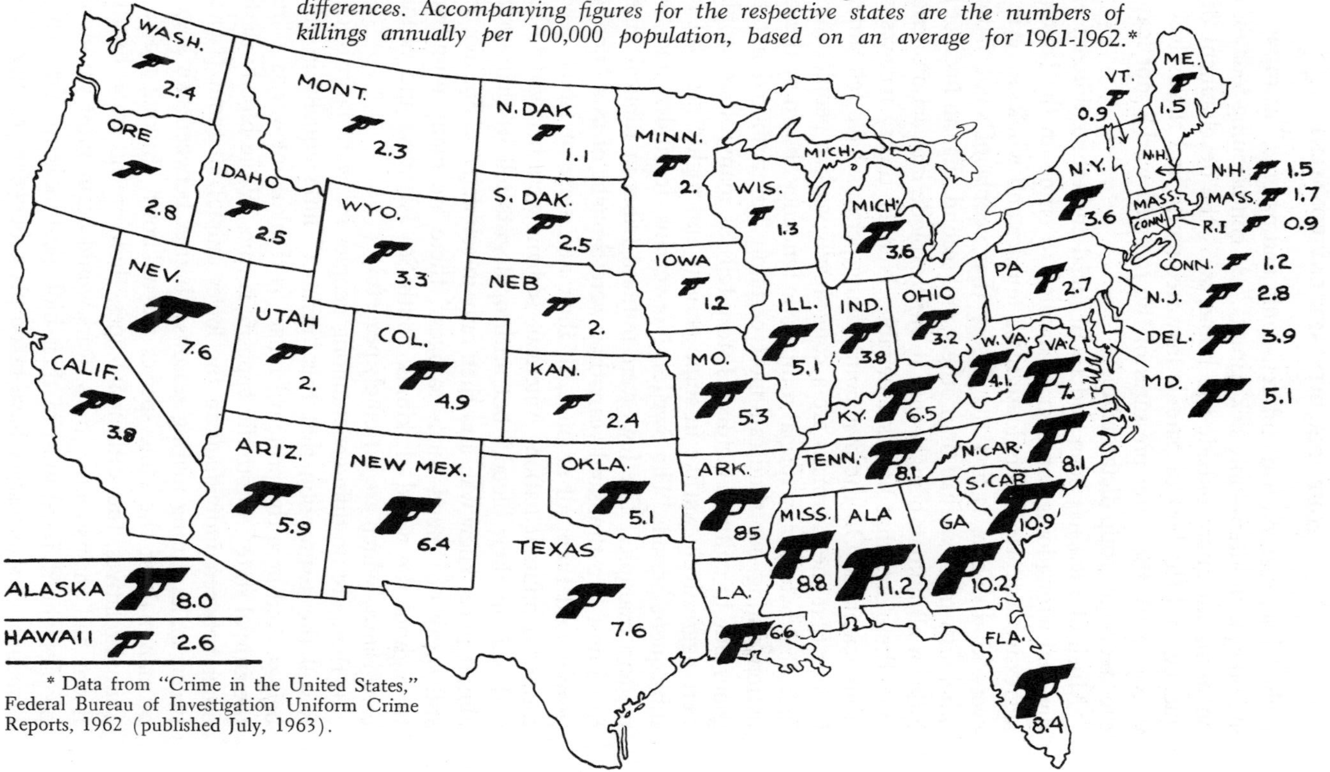

* Data from "Crime in the United States," Federal Bureau of Investigation Uniform Crime Reports, 1962 (published July, 1963).

ingly, "There are no bad children—only bad parents," as to say, "There are no sick children—only sick parents." As conditioning today is carried on more and more outside the home, so is the development of delinquency; and the best of parents may frequently find themselves unable to cope with the outer pressures of the frenzied world—not least being the "teen-age-group dictatorship.")

Much of what has been said about racial and ethnic differences in individual wrongdoing can also be applied to crime on the largest scale: aggressive war which lets loose mass murder and wholesale evil acts of every kind. There is first the theory that people of some nations are *inherently* more warlike than others; and second, that an *instinct for war* lurks in all human beings—that it must explode into action periodically and, therefore, that war is inevitable. Neither theory receives support from scientists. On the first point, many persons have felt impelled to believe that the Germans are a *congenitally* warlike people. But, as we indicated before, any special militaristic tendency shown by the German nation during our time can most certainly be attributed not to their genes but to cultural factors. History tells of many other peoples who were aggressively warlike in some periods and highly peaceful for long subsequent periods. The Swedes, Norwegians, Icelanders and Swiss are modern examples. The Jews were militant tribes in Biblical times; later for almost two thousand years they were thought of as a "nonfighting" people, and then, all at once, in their new state of Israel, Jews again formed a militant nation. American Indians were believed to be "born with war in their blood," but anthropologists tell us that some tribes almost never knew war, and then only in self-defense. So, too, while we still speak of "blood-thirsty savages," there are many primitive tribes throughout the world—Eskimos, South Sea Islanders and various African groups—which are extremely peaceful.

If there are no differences among peoples in "warring instinct," there is still the assumption that this "instinct" lurks equally in all human beings. And here the principal fallacy is thinking of war as a "return to the animal state." Quite the opposite is true. While bloody clashes may occur between *individuals* in the lower animal kingdom (typically between males during mating seasons), battles between two large groups of mammals of the same species are unknown. Among man's nearest animal relatives, the anthropoid apes, there are no wars. Furthermore, if war had an instinctive basis, wars would occur most readily between peoples who differed most in their biological makeup. Yet many of the bloodiest wars in history have taken place between the most closely related stocks (the wars among Indian, African, Arab and Mongolian tribes and peoples, the civil wars in the United States and Spain, and various European wars). In World Wars I and II the lineup of sides had nothing whatsoever to do with their ethnic or racial affiliations: In fact, the hated foes of one period became the brothers-in-arms of another.

When wars were dominated by hand-to-hand combat, it could be argued that they satisfied some uncontrollable "blood lust," activated by glands, instincts or other biological influences. But that theory now seems almost foolish in a time when personal and physical encounters have given way to push-button warfare, by which one mechanically kills masses of persons—unknown, unfelt, and *unhated*—vast distances away; and when any "instincts" or other biological urges toward warring have been replaced by arguments, debates, exhortations and conscription laws which cause reluctant men to give up peacetime pursuits and don uniforms. Indeed, rather than there being any "warring" genes, geneticists might be more inclined to believe that there are "cooperative" genes in men; for, ironically, the outstanding characteristic of modern war is the readiness to join together for some common purpose and presumed altruistic goal, at the sacrifice of individual interests and lives.

If there were any biological element in war—not in motivation, but in active participation—it would lie in the differences between males and females. Indisputably, the physical waging of war has always been almost entirely confined to males. (Stories of Amazons of the past have proved to be largely myths; nor do any new minor trends with respect to female participation in war alter the main implications.) Obvious reasons for this can be found in various biological differences which make men far better equipped for warring than women, and which would preclude any organized groups of women from doing battle with equivalent forces of men. But there also is the question again of whether men, *by nature,* are not more impelled and attuned to waging war and what goes with it than are women. For in all behavior associated with killing, violence or other crimes, males are and always have been far in the lead.

Consider the statistics: About twenty-five times as many males as females are in prisons and reformatories in the United States. In *convictions* for specific offenses the yearly average runs to about ten times as many men as women for murder and manslaughter, about seventy-five times as many for robbery, 120 times as many for burglary, and twenty-five times as many for forgery. However, women are by no means *that much* better, because the figures are thrown far out of line by the fact that women who commit equivalent crimes are much less likely to be arrested, convicted and sentenced than are men. For instance, in murder and manslaughter, the actual female rate is about one-fourth that of the male rate; yet, as noted, the females convicted for these offenses are only one-tenth the number of males convicted, and where they are sentenced it is usually for much shorter terms. Further, as pointed out by criminologist Otto Pollak, female criminality is grossly under-reported with respect to such acts as shoplifting, thefts by domestics, infanticide, poisoning of husbands and relatives, blackmailing and other offenses which can be concealed or which the victims cannot or will not report.

But even when full allowance is made for the greater leniency to women and for the probability that their wrongdoing may be more subtle and undetectable, there is no question but that anywhere in the world women commit vastly fewer criminal acts than do men. The evidence also is clear that wherever restraints against women are relaxed, and/or social conditions become disorganized, their relative crime rates shoot up. This has been happening in the United States and other countries. Moreover, where once murder was thought of as the typically masculine crime, the figures previously given show that the ratio of killings by women, as compared with those by men, is now relatively much higher than for other major crimes. (The knowledge that chivalrous juries usually acquit fair murderesses after they have dutifully recited, "He struck me . . . then everything went black. . . ." must be considered as having been a spur to many women with an urge to dispose of some irksome male.)

Another ominous development has been the springing up in New York and other large cities of vicious *girl gangs*, as adjuncts to boy gangs. Often these girls commit violence on their own, including assaults and thefts, usually with other girls, or women, as their targets. But for the most part they serve as weapon carriers, narcotic procurers, prostitutes and spies for the boys, and also act to abet and stir up violence between boy gangs—in these respects playing certain female secondary roles traditional in warfare. Whatever the increase in lawlessness among females, then, the basic sex differentiation seems to persist.

That conditioning may not be the whole answer for sex differences in crime is most strongly indicated by the general situation in the animal kingdom. Among lower animals, too, the males are more likely to indulge in acts of violence than are females. All the way from male mice, dogs and cats, through roosters and rams to bulls, bull moose and bull elephants, the male is more rambunctious and harder to restrain than the female. Nor do human mothers have to depend on psychological studies to learn that from childhood on boys show a greater tendency toward unruliness, aggressiveness and destructiveness than do girls. Thus, while as with other complex traits there can be no genes for criminal behavior carried by males and not by females, the possibility remains that over-all sex differences in biological makeup and functioning may in some measure contribute to a greater expression and degree of criminality by the males in any given environment.

Prostitution, however, is one type of wrongdoing which, for biological reasons, is almost exclusively female. (The only exception is prostitution among homosexual males.) Here is one of the clearest examples of wrongdoing that lies not in the act itself but in the psychological circumstances under which it is carried out and the way in which it is regarded by others. Not everywhere is prostitution considered wrongdoing. Nor, strictly speaking, can it be called abnormal. As the late Professor Robert M. Yerkes revealed in his notable study of chimpanzees, many ape females take advantage of the fact that their sexual capacities are much greater than those of males to wheedle food in exchange for sexual accommodation. He concluded, " . . . prostitution is a natural development among the primates."

No less among human beings, prostitution also is or has been regarded in many areas of the world not only as natural, but as an essential, acceptable and even honorable pursuit. In many European countries it had official sanction until recent times, and although now legally outlawed (in France and Italy only during the past decade) it has in effect merely been "swept under the rug." In the United States, too, while open red-light districts have long since been abolished, prostitution still has community approval in a few places, and in one degree or another exists almost everywhere. Doubtless, commercialized vice has diminished greatly in volume, mainly because of economic and social advancement, earlier marriages and other environmental changes. As to whether improved moral standards among women are a factor in this change, many feel that, on the contrary, it is a loosening of female sex codes which has diminished the demand for prostitution by making sex on a nonpaying basis more generally available.

Nevertheless, the changing social conditions and attitudes also have brought changes in the nature of prostitution and the character of the prostitute: that is, if we define her as one who deliberately makes a career of promiscuously exchanging sex favors for money or other gain, in defiance of laws and moral codes. Less and less is it becoming possible to look upon prostitutes as merely victims of poverty, ignorance or compulsion, or the products of broken homes and chaotic conditions. Today, in the United States and most other advanced countries, prostitution is becoming more clearly a debasing and venal practice, heavily allied with other areas of crime. Thus, the professional prostitute—whether low-level or elegant call girl—can usually be more clearly distinguished now from other women as a special type of social and psychological deviate, a coldly calculating, brazen, unprincipled sister-in-crime to the male whose lust for easy money must lead him in other sinister directions. Whatever the role of environment may be in prostitution, as in other areas of wrongdoing, the fact remains that in the worst environments only a limited number of women take to this calling; and in the best of environments, where good jobs or opportunities for marriage are plentiful, many women take to the profession out of preference.

The question, then, is whether there are in some women inherent quirks of one kind or another which impel them toward this behavior. Studies so far have proved only that popular assumptions as to what the quirks might be are in error. For one thing, there is the belief that prostitutes start off as women carried away by abnormal sexual desires. But many investigators have shown that relatively few prostitutes are oversexed—that, in fact, prostitutes tend more often to be *undersexed*, and that their comparative lack of emotion and their indifference to sex relations facilitate their taking up and carrying on their pursuit. Moreover, while psychological studies of prostitutes indicate that a higher than average proportion of them are mentally defective or psychopathic, many are of superior intelligence; many have good education—some among present call girls being college graduates; many have had good homes and conventional upbringing; and perhaps the great majority would be ranked as mentally normal. The answer to what makes the prostitute—if there is any single answer—remains elusive; and if there is any inherent constitutional basis for the "oldest profession," it has yet to be established.

If prostitution is peculiarly a female crime, *rape*—forcing sex relations on a member of the opposite sex—is peculiarly a male crime. Since this act is usually accompanied by violence, as prostitution is not, it is more likely to be linked with some emotional or mental disorder. On the other hand, sex crimes by men against children, where not involving rape, are much more apt to be due to low sex drives rather than to strong, violent sex drives—the offending men usually being emotionally immature and afraid of seeking relations with adult women. This also applies to such other types of male sex offenses as "peeping" and exhibitionism. As evidence, Dr. Paul W. Tappan reported that among male sex offenders diagnosed at a New Jersey clinic, more than half proved to be definitely sexually inhibited, 68 per cent to display personal inadequacy, and 91 per cent to be in some way emotionally immature. Dr. Tappan concluded that "the drive to sex crime is most often psychic rather than physiological, and in such cases must be treated if at all on the psychological level."

As in prostitution, then, any genetic factors in male sex crimes would be hard to identify and at worst may play only a vague and indirect part by way of producing abnormalities of sexual or mental functioning. (Homosexuality will be discussed in the next chapter.)

In a special category are those acts which do not necessarily involve offenses against others—though they can often lead to such—but which are primarily offenses against the individual himself. Included among these are *alcoholism, drug addiction* and *suicide.* While these acts are regarded as violations of laws, morals or religious codes (but not universally), the modern scientific view is that they are mainly expressions of personality disorders or mental illnesses. Whether they should be treated at all under the heading of crime is another question.

The theory that alcoholism has an hereditary basis starts off with the clear evidence that it occurs in certain families far more than in others. One could easily argue that any such family tendency is passed along, like other bad habits, through precept and association. But there are authorities who doubt that this is the complete answer. (From practical experience we know that some individuals enjoy the taste of whisky almost at once, while others can't abide it; that some people require a great deal of alcohol to make them drunk, and others get "high"—or sick—on one or two drinks; that some persons can't seem to stop drinking, and others can "take it or leave it" at will.) Thus, while there is general agreement that environmental influences may largely determine whether and how alcoholism develops, there is also the belief that differences in predisposition to alcoholism, as to many other conditions, may be governed by heredity. One possibility, long held forth, is that this predisposition may work through mental or emotional factors (as in the case of severe neuroses or psychoses). Another theory is that inherited peculiarities in the metabolic or chemical workings of individuals may be involved.

Dr. Roger J. Williams and his associates at the University of Texas have led in maintaining that alcoholism has some definite hereditary basis. In experiments with rodents they reported finding marked hereditary differences among strains in "alcohol appetites," which appeared to be related to differences in body chemistry and dietary "blockages" or needs. Corroborating this, Drs. Gerald E. McClearn and David A. Rodgers reported that when mice of five different genetic strains were given an opportunity to drink an alcoholic solution in preference to water, only one group proved to be consistent tipplers, suggesting that "alcohol preference may be determined by genetic factors." Still another theory is that alcoholism may be tied up with some form of allergy running in families—that there may be an innate susceptibility to alcohol, akin to an allergy to food or an idiosyncrasy to a drug, making alcohol and its taste and effects more attractive to them than to the normal drinkers. Likewise, with respect to psychological factors, degrees of predisposition to alcoholism have been attributed to emotional instability, insecurity, fear of sex, suppressed or latent homosexuality, and similar causes.

Whether or not any of the foregoing theories, if substantiated, will establish links with hereditary predispositions to alcoholism in human beings, there is no question that degrees of the drinking habit can be heavily conditioned by training and environment. We see this in the enormous variations in the incidence of alcoholism among persons of different nationalities, religions, geographic areas, and occupations, without regard to their hereditary makeup; in the radical increase in heavy drinking in certain groups where it was once unknown; in the increase of drinking among young people; and, particularly, in the ominous increase in alcoholism among women.* Essentially, then, whatever hereditary factors there may be in individual predispositions to alcoholism, its development and manifestation can be blamed largely on social and psychological influences.

Drug addiction seems to involve much the same basic factors as alcoholism. (In one group of drug addicts a third had started off as alcoholics.) However, there are certain differences in the psychological and emotional factors associated with the two habits. Principally, the much stronger legal and social pressures against drug addiction (in most civilized countries) and the far greater difficulty, expense and

* While the officially reported and currently accepted estimate is one female to five male alcoholics in the United States, this probably greatly understates the actual incidence of heavy drinking among women, according to Dr. Marvin A. Block, chairman of the American Medical Association Committee on Alcoholism. He maintains that a very large proportion of the women alcoholics never come to official attention, are less apt than men to seek help for problem drinking, are far more often shielded by their families from public disclosure, and, in the case of housewives (and unlike employed male alcoholics), can "get to the bottle" unobserved during daytime hours. Shifts in drinking from saloons to the home and more sanction of drinking by women have been major factors in increasing the incidence of female alcoholism.

furtiveness in the pursuit of the habit make it likely, as a rule, that drug addicts will start off as initially more neurotic and more indifferent to convention than alcoholics. It should be stressed here that mere exposure to narcotics (taking them a few times) does not of itself lead to addiction, as is shown by the many cases where persons receive opiates in hospitals or in other medical treatment without developing the habit. The important difference between these and the individuals who do become addicts is that the latter usually act on their own to take drugs—for curiosity, for "kicks," or for some assumed need—and in defiance of the law and of threatened consequences.* When drug-taking begins like this, the way is already prepared for the development of a habit. This is one big reason why drug addiction is so difficult to cure—much more so than is alcoholism. But there is also the possibility, as with alcoholism, that the systems of some individuals are more receptive or attuned to the effects of narcotics than are the systems of others.

Thus, there may be some degree of *predisposition* to drug addiction, through both psychological and biological factors, which conceivably might be influenced by heredity. But, unquestionably, environment is by far the most dominant influence in the development of the habit. In the United States, for instance, the incidence of the habit is in direct ratio to the availability of drugs and the opportunities or incentives to begin using them.† The number of addicts had reached a reported high of close to 200,000 in 1920. Then, under strong government repression

* Warning has been given by drug expert Dr. Frank C. Ferguson, Jr., (Albany [New York] Medical Center) that the danger of drug addiction is held out by the use of almost any drugs which affect emotions, including—in addition to opiates and alcohol—barbiturates, stimulants, bromides and tranquilizers. The choice of drugs generally is determined by the emotional needs of the individual: Stimulants by those desiring excitement; depressants by those seeking escape from worries and anxiety; narcotics by those who want to lose themselves. But, Dr. Ferguson believes, since addicts take drugs for their psychic effects, there must be an underlying psychiatric abnormality—"an addictable personality"—that makes the individual require such effects.

† The reported enormous incidence of drug addiction among medical men—estimated as up to 1 in 100, or twenty-five times that in the general population of the United States—gives evidence of how much the habit is correlated with the availability of the drug. Writing on the subject, Dr. J. DeWitt Fox of Detroit has ascribed development of the habit among physicians to a combination of professional stresses and accessibility to drugs. He classified doctor addicts into these broad groups: (1) alcoholic physicians; (2) tired doctors who habitually blot out fatigue with a narcotic; and (3) doctors suffering pain from disease, who overdose themselves with opiates.

and curbing of drug sales, it had dropped to 20,000 by 1945. In recent years the number had risen again to about 50,000, but—again as proof of the environmental influences in special areas—about 45 per cent of the addicts are concentrated in New York State, another 14 per cent in Illinois, 13 per cent in California and only 28 per cent in all the rest of the country. Four out of five of the American drug addicts are males, and about 12 per cent are under twenty-one years of age.

Suicide—the ultimate of "self-offenses"—is included in this chapter because under various religions (Catholic and Orthodox Jewish among them) and under common law in many countries, it is regarded as "willful destruction of life" and therefore as a crime, with attempt at suicide being punishable as a misdemeanor.* Psychoanalytical theory holds that in many cases suicide is indeed motivated not by the desire to die but by the desire to kill, which, for lack of some definable other victim, may cause the individual to act against himself. (It may be the active expression of a childish impulse to say to an offending grownup, "You'll be sorry when I'm dead.") Tending to back up this theory that suicide is often a substitute for murder is the actual see-saw fact that suicide rates generally are up when murder rates are down, and vice versa. Vermont, for example, with the lowest murder rate in the country, has the second highest suicide rate; Mississippi, with the highest murder rate, has one of the lowest suicide rates; and, on a racial basis, American Negroes, with a murder rate many times that of Whites, have a suicide rate only about one-third as high.

With regard to possible hereditary influences, the not infrequent instances of several suicides among close relatives would seem to support the belief that a tendency toward self-destruction runs in certain families. But one must consider the factor of coincidence. In view of the high incidence of suicide (about one American male in fifty eventually ends his life this way), Dr. Franz J. Kallmann has noted that mere chance

* The centuries-old British laws which made suicide a crime for violation of the Commandment, "Thou shalt not kill," and which prescribed various indignities and post-death penalties for suicides, had been largely repealed by 1900. However, a 1961 act of Parliament still provided that those abetting a suicide (survivors of suicide pacts, for example) could be punished by up to fourteen years in jail. In a number of states in the United States, attempted suicide is still legally a criminal act, and in Canada it is punishable by two years imprisonment. Orthodox Jewish tradition stresses that a suicide may have "no share in the world to come" (in the Talmud, the principle is applied even to Samson, despite the self-sacrificing motive of his act). However, the existing Codes of Jewish law provide that each suicide be judged individually according to the circumstances and the interpretation of the Rabbi.

could account for as many as 500 cases a year in which two brothers in a family committed suicide. Further, he and others have shown that double suicides among identical twins are extremely rare, which would discount the theory that any special hereditary trait or particular type of inherited personality deviation leads to a suicidal tendency. Nevertheless, other authorities have claimed that a high proportion of suicides may be linked with mental illnesses (chiefly manic-depressive insanity but also schizophrenia), and to the extent that it is a factor in these conditions, heredity may be an indirect influence in individual suicides and may in some cases account for repeated suicides in the same family. Moreover, in many cases where suicides are attributed to such causes as unrequited love, loss of honor, overwork, or failures of various kinds, it is possible that hidden elements of mental disease, with some hereditary basis, may have played a part. But the causes of and motives for suicide are so varied, and the range is so great, from rational and logical to irrational and unwarranted suicides, that the only common denominator for them may be some inner instability which propels the individual more easily over the brink when further living becomes seemingly unbearable. That there is any direct hereditary tendency toward self-destruction is highly doubtful.

The environmental elements in suicide can much more easily be identified. For one thing, the suicide rates vary enormously in different groups and countries, and at different periods, as determined on the one hand by religious and cultural attitudes, and on the other by degrees of stress. Among the Japanese, who regard suicide as an honorable deed (most strikingly shown by the kamikaze pilots of World War II), the rate is high—about 26 per 100,000 persons annually. But many readers may be surprised to learn that the suicide incidence is higher in several European areas, with the current rates (1959 World Health Organization data) being 34 per 100,000 in West Germany, 28 in East Germany, and 27 in Hungary. In Austria, Switzerland and Denmark the rate is between 24 and 22 per 100,000; in France, 17; in England, 13. Religion may have a marked influence on suicide rates, these generally being highest among Protestants, considerably lower among Jews, lowest among Catholics and almost negligible among Moslems. But other cultural factors may do much to modify religious influences, as is shown by the differences among predominantly Catholic countries, with Ireland having a suicide rate of only 2.5 per 100,000 (the lowest in Europe); Italy, 6; Portugal, 8; while in France (as said before), it is 17. Among the Jews in Israel the rate is about 7 per 100,000, which may be considerably less than that of their coreligionists elsewhere; and one need hardly add that during the Nazi terror reign the suicide rates among Jews in affected areas were understandably astronomic.

In the United States the national suicide rates reflect great differences among various groups, with religion, ethnic origins, social levels and other factors playing a part. But the most striking difference is between the Whites, collectively, with an average suicide rate almost three times the Negro rate—about 12 versus a little over 4 per 100,000 persons annually. This is strong evidence that suicide is not primarily the result of deprivation, harsh conditions, social rejection, or ill health, for in all these respects Negroes suffer more than do Whites. What apparently does count is the psychological capacity to withstand stress; and one may assume that long exposure to hardship gives Negroes the ability to "take it" better than Whites when adversity strikes. (But likewise, as previously mentioned, aggression, resentment and frustration in Negro individuals may more often be directed outward, through homicide, whereas in Whites it may more often be turned inward, through suicide.)

But within any group, an outstanding difference in suicide rates is that between males and females. Almost everywhere in the world the proportion of suicides among men is far greater than among women. In the United States the average ratio, for all ages, is about four males to one female. In the older-age groups, from seventy onward, the rate is six or seven times as high for men.* The reasons may include the more severe strains on men when they are failures, or are incapacitated, or "lose face"; their greater experience and familiarity with killing devices; the more frequent exposure of males to violent situations; and other social factors. But there also may be biological influences. For example, we have seen that men genetically are more likely than women to be afflicted by degenerative diseases or other ills which may make life seem unbearable; and that, by nature, in equivalent situations males are more given to aggressive and violent acts. That the sexes are distinguished more by the *carrying out* of the suicidal impulse than by the impulse itself is indicated by the fact that attempts at suicide which fail are far more frequent among women than men.

Thus, the general situation in suicide, as with alcoholism and drug addiction—all offenses against the self—is much as it is with offenses

* The only exception to the higher male suicide rate has been among fifteen to nineteen-year-olds in a few countries and in occasional recent years, where the female rate in this age group has exceeded the male rate. This has been true in Israel, Italy, Portugal, West Berlin, Venezuela and Ceylon.

against other persons: Both *inner* pressures and *outer* forces work together toward the end results; but the inner pressures, which might make some individuals more than ordinarily prone to wrongdoing under adverse influences, usually are subordinate to the outer forces.

To sum up the case of "Heredity and Environment" in crime:

Heredity might be a major factor in acts stemming from mental disorders or deficiencies with genetic elements. But if such acts were not considered rational and responsible, they would not be included in the crime category.

Rational acts which can be regarded as criminal are so enormously varied in nature and degree that it would be almost impossible to find any common basis for them with respect to presumed genetic predispositions. If there were any such common denominator, it might lie in the fact that willfully to do unjustified violence or harm to other persons, and to persist in flaunting *both* the dictates of one's group and one's own sense of what is right, may require a special type of psychological and emotional deviation. Conceivably this could involve genetic predispositions, somewhat as with other forms of human behavior. But, again, any genes for criminality would have to be so complex in their workings and so heavily enmeshed with environmental factors that pin-pointing them would be a formidable if not impossible task.

There can be little question about the influence of environment on crime. We have seen this in the enormous variations in the incidences of crime as a whole and of given types of crime within similar genetic groups; in the rapid shifts in crime rates from one generation to another, or within brief periods; in the sharp rise in delinquency in recent decades, among young persons of many different groups and backgrounds, throughout the world; and, not least, in the fact that "right" and "wrongdoing" among human beings rests largely on the shifting sands of social viewpoints, which can move persons readily in one direction or another.

The verdict?

Heredity, on circumstantial evidence, may be found to be at most an *accomplice* in some crimes and in some degree.

Environment is proved to be almost always the major villain in crime. If heredity does "load the gun," it is environment which largely determines the type of "gun" and its effects—from zip gun to atom bomb—and which "pulls the trigger."

To reduce serious human wrongdoing greatly in any category, there will have to be not changes in our genes but changes in the conditions, incentives and attitudes which aid and abet crime.

CHAPTER THIRTY NINE

Sex Roles and Sex Life: I

Normal

THERE IS A CLASSIC STORY about a little girl at an art gallery who studies a painting of the First Couple and asks,

"But which is Adam and which is Eve?"

"My goodness, can't you tell?" her governess exclaims.

"No," replies the little girl, "they haven't any clothes on."

The story has more than humorous implications. To the very young child, unaware of sex functions (and often of differences in sex organs), the sexes are distinguished mainly by how they dress and act. Later comes the perception of sex differences in body form, strength, hairiness, and so on. With puberty the knowledge of male versus female sex organs and reproductive functions assumes primary importance. But this sequence is actually the reverse of the order in which outward sex differences manifest themselves and are classified. Designated as the *primary* sex differences are those in the sex organs and mechanisms. The *secondary* sex differences are those in body form, breast, facial and body hair, and voice. A third class of differences might include some of the masculine-feminine behavioral characteristics first noticed by a child, assuming for the moment that these, too, have any biological basis.

Earlier parts of this book have dealt extensively with most of the afore-mentioned sex differences and have stressed their importance in growth, disease and defect, life expectancy, mental performance, achievement, and crime. In this and the following chapter we will concentrate on the two areas where human males and females may see themselves as most strongly differentiated—in *sex* behavior and *sexual* behavior. There is this distinction: The former refers to the behavior of individuals as members of their respective sex—the behavior of a girl as compared with a boy, of a woman as compared with a man. The latter refers specifically to behavior in which the *sexual impulses*—involving the sex organs—are expressed in one type of outlet or another. Both forms of behavior are, of course, closely related and are in turn interwoven with the whole fabric of the individual—physical, mental and emotional—and with almost every aspect of his or her existence and place in human society.

Thus, sex behavior and sexual behavior are perhaps the most complex of all traits and the hardest to analyze in terms of cause and effect. With the biological primary and secondary sex characteristics, heredity can be discerned as usually playing a major role, although environmental factors can be important as modifiers. But in sex and sexual behavior biology is so mixed up with psychology and sociology that the problem of sifting out hereditary factors may be the most difficult faced in this book.

To go back to Adam and Eve for the moment, it would be helpful, indeed, if we could observe some hypothetical male and female pair starting from scratch without any training whatsoever in sexual behavior—no precedents, customs, taboos or notions of how members of each sex were or weren't supposed to act. Would such a couple really be so alike in behavior that our little girl, seeing them alive (and *sans* clothes) would have difficulty in distinguishing one from the other? Or would their biological differences also have led to behavioral differences? No Adam and Eve being available, one alternative is to see what happens with an average boy and girl starting life together in our own world—say, "Johnny" and his little next door neighbor, "Mary."

Without for the moment going into the reasons, many studies have indicated (and many parents could affirm) that Johnny as an average boy baby will be more active, restless and aggressive than Mary and, as he continues to grow, will utilize his bigger and stronger muscles in more physically exacting and rugged play. Mary, on the other hand, will show greater deftness in her hand movements, will begin to dress herself earlier and more expertly than Johnny, will be more interested in what she wears and in the colors, and will engage in more sedentary play (*preferring* dolls to his mechanical toys). She will be more keenly aware of people and of details concerning them; she will be more jealous of attention, although she will be shyer in front of strangers; and she will be more modest in behavior. She will also be more nervous, *outwardly*, given to more nail-biting and thumb-sucking than Johnny. But Johnny may reveal more inward tension by his greater proneness to various nervous and emotional symptoms, such as tics, speech difficulties, bed-wetting, destructiveness, and temper tantrums.

These behavior differences are especially significant because they reflect much the same average contrasts in adult human males and females. We know, of course, that at a very early stage parents and others begin conditioning the boy toward "masculinity," the girl toward "femininity," through differences in dress, toys, games, examples, and discipline. At the same time, the "you must act this way" for girls and "you must act that way" for boys causes differentiating stresses which can affect the individual's whole psychological and emotional development. *But*—do children fall in with their respective sex-role and sex-trait conditioning only because they are taught to imitate the adults of their sex? Or might not the training conform, in some measure, to inherent early tendencies in each sex from which the adult patterns evolve?

Throwing light on these questions is the evidence regarding lower animals. For in almost every species one finds significant behavior differences between the sexes, starting at the outset of their lives. Most striking are the facts about our closest relatives among lower animals, monkeys and apes. From detailed observations at the University of Wisconsin laboratories, here is what psychologist Harry F. Harlow has reported:

> Male and female infant monkeys show differences in sex behavior from the second month of life onward. . . . Males threaten other males and females, but the females are innately blessed with better manners; in particular, little girl monkeys do not threaten little boy monkeys. The little females tend to be more passive. Grooming is much more frequent in them than in males. Caressing is both a property and prerogative of the females. Play behavior is typically initiated by males, seldom by females; and real rough-and-tumble play is far more frequent among the males and is almost invariably initiated by them. (From "The Heterosexual Affectional System in Monkeys," *American Psychologist*, Jan., 1962.)

Noting further how much a group of little boys and girls he saw at a picnic were like the monkey tots, Dr. Harlow concluded, "I am convinced that these data have almost total generality to man . . . and that these secondary sex-behavior differences are innately determined biological differences, regardless of any cultural overlap."*

* Observations about little "boy" and "girl" chimpanzees, conforming to the findings of Dr. Harlow but based on personal experiences of animal trainers, were reported in 1944 in the author's book, *Women and Men*. One of the trainers cited was Mrs. Gertrude D. Lintz, who long had made a career of training apes for exhibition purposes (her proteges having included the famed circus gorilla, "Gargantua"). Among the sex contrasts she reported were these: "Little ape females are more docile, gentle, and ever so much more sociable; the boy chimps are apt to show aggressiveness very early (they bite more, for one thing) and are more active, restless and

One possible biological basis for early sex differences in behavior is the fact that from birth onward the sex hormones are in production (with relatively more androgens for males, more estrogens for females) and are exerting their differentiating effects. In lower animals proof of this has come through repeated experiments in which changes in the hormonal balance from the male to the female direction, or vice versa, have produced changes in sex behavior. Nothing is more amusing than to see how young female chicks, injected with male hormones, become more aggressive, flap their wings and crow with canarylike squeaks. In reverse, young roosters if castrated lose their "cockiness" (literally). If full-grown hens are given heavy dosages of male hormones, or full-grown roosters of female hormones, a complete reversal of the normal sex pattern of behavior may develop. So, too, experiments with guinea pigs, rats and mice have proved that sex-behavior patterns can be much modified or altered through hormonal treatment. (One striking finding is that in dogs, early castration of male puppies may prevent development of the usual male reflex of elevating a hind leg when urinating and may cause the animal, instead, to take the female squatting position.) Added to the influence of the sex hormones, sex differences in the other hormones (pituitary, adrenal, thyroid) and in the entire genetic makeup of males and females must also contribute to their divergent behavior patterns.

Nevertheless, among human males and females it is far more difficult than in lower animals to draw a direct line from biological differences to behavior differences, for always in between, in people, are innumerable environmental influences. Most important is the fact that in every human group there is a prescribed behavior pattern considered normal for each sex, and that any great departure from this pattern will bring with it social disapproval and, often, punishment. The standards for sex behavior, it is true, may vary considerably from one group to another. But the main concepts of what is masculine or feminine, and what those of each sex should seek to be, have been surprisingly uniform throughout the world, among both civilized and primitive peoples. With masculinity are supposed to go such qualities as strength, virility, bravery, aggressiveness, enterprise, interest in mechanical things, outdoor

---

destructive. . . . The little girl chimp sits still and occupies herself much more than the male with activities not requiring a lot of moving around or strenuous exertion. She is more adept with her hands, it being far easier to teach little girl than boy chimps to thread a darning needle. The strangest fact is how the little girl chimps take an interest in clothes, loving to be dressed up and responding eagerly to colors. The male is utterly indifferent to what you put on him, and as like as not will rip his clothes off." (Does any or all of this sound familiar to mothers of human boys and girls?)

activity, and adventurousness. With femininity are linked domesticity, sedentary activity, softness, tenderness, emotionality, affection, and sentimentality. The patterns prescribed are certainly not always in accord with the facts—or with biology. But one must grant that they reflect, on the whole, traits which most women almost everywhere would like to see in their men, and most men would like to see in their women.*

At any rate, it is on the basis of the standards cited that various psychological tests of "masculinity" and "femininity" have been devised, such as those prepared by the late Professor Lewis M. Terman and his associates.† These tests call for responses which can be rated as "masculine" or "feminine." But as investigators have found (and most of us know), there are great ranges of masculinity and femininity within each sex, some males being more "feminine" in their responses than a good many women, and some women being more "masculine" than a good many men. In general, the highest "masculinity" scores on the Terman tests were made by college athletes and engineers; the lowest by clergy-

* Frequently cited as exceptions to prove that traits considered masculine and feminine have little relationship to natural factors are observations made by Dr. Margaret Mead many years ago regarding three small primitive groups in New Guinea, whom she visited in 1931. Dr. Mead had reported that among these three tribes, living not far apart, she found marked contrasts in the male-female behavior patterns: in the Arapesh, that both sexes were gentle, passive and "feminine" by our standards; in the Mundugomor, that both sexes were violent, aggressive and what we think of as "manlike"; and in the Tchambuli—a turnabout situation—that the men acted as we'd expect women to act, the women as we'd expect men to act. However, these personal and unique observations with regard to three small isolated groups, while interesting and provocative, were viewed by many authorities as hardly constituting a scientific refutation of the general theories about male-female behavior; and Dr. Mead herself, in a *New Yorker* article, by Winthrop Sargent, was said to have been "highly dismayed that feminists began vociferously citing her report [which she nowadays refers to as 'my most misunderstood book'] as proof that all the differences between the sexes are merely the result of artificial social conventions imposed by men."

† In addition to the Terman tests, various others have been in use to gauge the proportionate "masculinity" and "femininity" of children and adults. Many are verbal tests; others are drawing tests ("Draw a Person," "Drawing Completion," etc.) in which the sex of the figure drawn, and how it is drawn, are criteria; and still other tests make use of sets of abstract figures or special ink blots (such as those illustrated), which can be described in various masculine- or feminine-slanted ways. Masculinity-Femininity subscales, involving responses to female-slanted interests and preferences, also are incorporated in the Minnesota Inventory (MMPI) and the Kuder Vocational Preference Scale tests. Other M-F testings have been devised by Dr. David B. Wechsler in a work-interest test designed primarily for adolescents.

PICTURE TESTS FOR MASCULINITY AND FEMININITY

Reactions to the drawings below are regarded as throwing light on a person's "male-slanted" or "female-slanted" psychological qualities.

I. WHAT DO THESE PICTURES SUGGEST TO YOU?

1.
(a) baby
(b) bell
(c) idol
(d) incense

2.
(a) flame
(b) flower
(c) snake
(d) worm

3.
(a) boat
(b) door
(c) hat
(d) stump

4.
(a) couch
(b) cow
(c) deer
(d) horse

5.
(a) ax
(b) boat
(c) chopper
(d) moon

6.
(a) funnel
(b) horn
(c) jack
(d) vase

7.
(a) bow
(b) chain
(c) footprints
(d) tie

8.
(a) chimney
(b) coil
(c) smoke
(d) thread

II. WHICH PICTURES APPEAL TO YOU MOST? (PICK 4 OF 8)

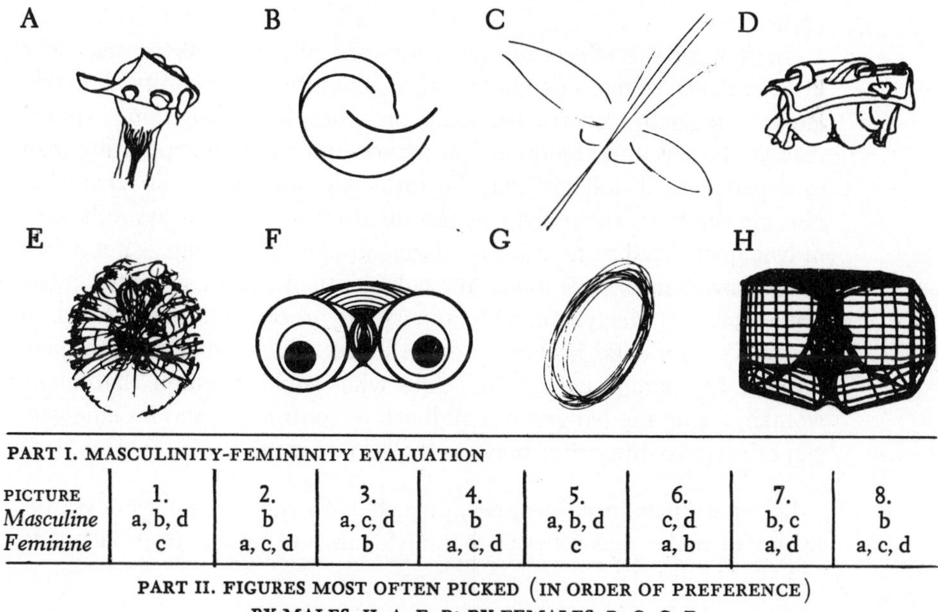

PART I. MASCULINITY-FEMININITY EVALUATION

PICTURE	1.	2.	3.	4.	5.	6.	7.	8.
Masculine	a, b, d	b	a, c, d	b	a, b, d	c, d	b, c	b
Feminine	c	a, c, d	b	a, c, d	c	a, b	a, d	a, c, d

PART II. FIGURES MOST OFTEN PICKED (IN ORDER OF PREFERENCE)
BY MALES, H, A, F, D; BY FEMALES, B, G, C, E.

Credits: Part I, Terman & Miles, "Sex and Personality." Part II, Tests by Drs. George Cerbus & Robert C. Nichols, using figures from "V-C Figure Preference Test," by Dr. Harold Webster, Copyright 1957.

men, artists and journalists, with male *inverts* (extreme types of males who act like females) being way over on the "feminine" side. A real surprise is that policemen and firemen were reported rather low in "masculinity" scores—presumably because they would tend to lack marked mechanical interests and financial objectives and would be especially eager for security and amenable to discipline—all rated as "feminine" traits. (Next time a cop acts tough to you, just smile knowingly.) Among women, the most "feminine" are reported to be domestics, the least "feminine" outstanding female athletes.

But the scores on any of the M-F tests, while possibly indicative of a person's behavior at a given time—as judged by prevailing standards—can hardly be accepted as an index of basic or inherent "masculinity" or "femininity." The relationship between M-F scores and occupation could mean either that persons with given traits are drawn into occupations which accord most with their masculine or feminine leanings, or that given occupations and activities tend to condition persons in one or another direction—or, perhaps, that both explanations apply. Whatever anyone's M-F rating and however it may be caused at a given time, it is by no means fixed and constant. Age alone can bring great changes in M-F scores, men becoming more "feminine" in their behavior as they grow older, while older women, too, become somewhat more "feminized." In this we have one of the few clear indications of biological influence on M-F scores, since with aging the sex-hormone balance is known to shift in both sexes, as a result mainly of a reduction in the output of the male hormone (androgen) relative to the female hormone (estrogen).

Apart from the effects of aging, outward physical makeup may give little evidence of one's psychological masculinity or femininity. A rough-looking, ruggedly constructed male or a heavy-set, masculine woman may yet be soft and "feminine" in nature; and a delicate-appearing man or a petite, baby-doll girl may be ruthlessly "masculine" in character. Nor are the traits themselves, as measured in the tests, always indicative of true qualities. The man who is daring in physical exploits is not necessarily braver, and hence more "masculine," than the artist, writer, musician, teacher or clergyman, who, in pursuing his career, is often called on to cope with mental hazards and challenges which call forth the highest type of courageous conduct. And who would question that many women, facing the dangers of childbirth or confronting serious emergencies of everyday life, are as brave as the bravest men?

So far we have been dealing primarily with the psychological sex behavior of males and females as individuals and not in relationship to

TABLE VIII

"Masculinity" and "Femininity" Ratings

*For groups of individuals in different classifications, according to average scores on the Terman-Miles "M-F" Tests**

"MASCULINITY" (LOWER ←----→ HIGHER)	"FEMININITY" (LOWER ←----→ HIGHER)
College athletes (highest "masculinity")	*Domestics (highest "femininity")*
Engineers, architects	*Stenographers, dress-makers, hairdressers*
High school boys (avg.)	*Women in arts*
Male college students (avg.)	*Housewives*
Lawyers, salesmen	*Business women, clerks*
Bankers, executives	*Professional women*
Dentists, teachers, doctors	*Teachers*
Mechanical occupations	*Nurses*
Clerks, merchants	*Women college students (avg.)*
Building-trades workers	*"Who's Who" women*
Farmers	*Women physicians and Ph.Ds.*
"Who's Who" men	*Male inverts*
Policemen and firemen	*Superior women athletes*
Journalists, artists	
Clergymen	

* The "M-F" tests show also that in males *age* is a factor, with the "masculinity" scores declining progressively as men grow older.

(SOURCE: Terman, Lewis M., & Miles, Catherine C., *Sex and Personality*.)

members of the opposite sex, particularly in a physical sense. Now we move into the area of *sexual* behavior, in which we are dealing with such relationships. It is in this area that the sex differences are most pronounced, for not only are the sexual mechanisms and functions of the male and female markedly different, but so also are the effects and meanings of their relationships to and with each other. At the same time, we find enormous variation in sexual functioning and expression among individuals of both sexes. Again, then, we have the question of what training contributes and what natural and inherent tendencies dictate in human sexual behavior, both in general and in individual terms.

It is an inescapable fact that, from the standpoint of Nature, sex is primarily a device for insuring the reproduction and continuation of the species. Toward this end males are equipped to produce sperms and females to produce eggs; and stimuli are provided for the sexes to join in fertilizing the eggs, and, after the eggs develop and the young are born, for the parents (in most species) to cooperate in helping their offspring to achieve maturity. The specific ways of attaining these objectives vary enormously among species, in accord with their biological makeup and the demands of their environments. Among most animals direct physical contact between opposite-sexed individuals is required, and this ordinarily involves active overtures by the male and acquiescence by the female. The female (in the natural state) usually acts in turn to excite the male's interest in having sexual relationships. Sometimes she does this automatically, when she is sexually receptive, by conveying stimulating odors, and/or by the appearance of her body, or by involuntary movements and other means. She may also signal her readiness for sex by gestures, postures and vocal sounds. The male of each species may correspondingly have his own way of reacting to and stimulating the female, through his body odors, actions, and vocal utterances.

Most—perhaps essentially all—of what has been said can be recognized as applying to human beings as well as to lower animals. (If the lower mammal female is equipped by nature to produce her own male-stimulating "Passionelle No. 5," the human female can get its equivalent at the drugstore.) Further, since the characteristic sexual-behavior patterns of all other species give evidence of being largely governed by heredity (i.e., genetically influenced stimuli and reflexes), one might guess that sexual behavior in human males and females also has a strong genetic basis. But, as with other behavior patterns discussed in preceding chapters, it is obvious that the line from biological impulses to sexual behavior in people can be nowhere near so direct as it is in lower animals. Let us try and see, then, how sexual behavior actually does develop among human beings and how social and psychological factors may modify natural tendencies.

First, are there *sexual* feelings in small children? It had long been thought that between birth and the approach to puberty there was normally just a neutral stretch with respect to sexual impulses—that young boys and girls were without sexual feeling. This thinking was changed with the formulation of the Freudian theory that sex life begins at birth, both subconsciously and physically. Closer observation showed, indeed, that even in infancy little girls and boys often manifest physical responses which are unmistakably sexual (resulting from self-manipulation or from friction with garments and objects). Any such early sexual activity, of course, is purely self-centered and of a reflex nature. Not until puberty nears do we ordinarily see evidence in young humans of any overt sexual feeling for, or move toward, sexual activity with members of the opposite sex.

By stages, the average "boy meets girl" story proceeds much as shown in the chart on the next page. First, in infancy, preoccupation with oneself. Then an indifferent stage, during the nursery or preschool period, involving play with children of both sexes. Then preference for play and companionship with one's own sex (due not merely to conditioning but to sex differences in strength, activity and other traits previously mentioned which might induce the sexes to move apart). There follows a stage of seeming antagonism between boys and girls, which might be only a cover-up for the attraction developing between them. Before long the sexes openly show interest in each other, at first collectively, then individually, with the girls taking the lead (consistent with their earlier puberty). Thereafter, starting with the teens, pairing up, and then (in our present society) "going steady" becomes the pattern. Concurrently, sexual intimacy in one degree or another is likely to develop.

Buy *why,* if the capacity for sexual feeling and response is present in early childhood, do boys and girls not evince it toward each other much earlier than they do? Is it because a period of biological indifference sets in? Or is it because of cultural repression?

If we observe the young of lower animals—kittens, puppies, lambs, not to mention little apes—we find no such prolonged period of latent sexuality. Active male-female sex play begins with them long before the equivalent of prepuberty or puberty in humans. Keeping pace with the development of the sexual stimuli, sex organs and sex mechanisms, the sex play proceeds by stages from mere imitation—or independent discovery—of adult male-female sexual behavior to fully mature sexual relationships. Most significant is the early sexual behavior of young chimpanzees, who are closest on the animal scale to human children. (The "childhood" period in chimpanzees, and apes generally, approaches the human developmental scale, in that chimpanzees do not achieve full reproductive maturity until the eighth year or later. Also highly important is that the sex organs and functions of apes—including menstruation in the females, which is confined to primates—closely resemble those in human beings.)

Summarizing studies of sexuality in young apes and monkeys, Drs. Frank A. Beach and Clellan S. Ford tell how very little males and females of these species indulge in sex play which reproduces all the sexual activities of their elders, so far as is physically possible. In little male and female baboons this is seen by the *ninth month* of babyhood, when their milk teeth have just appeared. One might assume, then, that young human males and females, too, if allowed freely to do so, would go through the same sequences of sex play, leading from imitative to mature sexual relationships. As a matter of fact, this does happen

STAGES IN HUMAN MALE-FEMALE RELATIONSHIPS

Infancy-Babyhood

Boy and girl interested only in themselves.

Early childhood (Nursery and pre-school)

Seek companionship of other children regardless of sex.

Ages 6 to 8

Boys prefer to play with boys, girls with girls.

Ages 9 to 11

Antagonism or indifference of boys to girls, but girls beginning to show interest in boys.

Ages 12 to 14

Boys and girls associate in groups. Some individuals begin to pair off.

Ages 15 to 16-17

Serious dating and going steady become common.

Ages 18-plus for females 20-plus for males

Mature one-and-one opposite-sex relationship, with intention of permanency.

in some very primitive groups where no barriers are imposed between little boys and girls, although in many other primitive groups the sexes are much more rigidly kept apart during childhood than they are in our own world.

We can assume, then, that biological—and also genetic—impulses toward sexual activity are present in very young human beings, but that whether or not such activity takes place may depend chiefly on social and psychological conditioning. We also can see that in the civilized world the conditioning is and has long been much less toward than *away from* any sex play or sexual activity before maturity. This is not without certain important effects. To go back to young apes and monkeys, Drs. Beach and Ford note that the early sexual practice is found to be essential in preparing them for successful adult sex life—particularly in the male's case. For when males of these species have been denied early sex experiences, they approach maturity as clumsy and frustrated lovers and mates. The females, surprisingly, have not been found to require preconditioning in sexual experience as have the males. As Beach and Ford describe it, in *Patterns of Sexual Behavior*:

> When fully receptive, naive [ape] females invite, accept and respond appropriately to the sexual advances of an *experienced* male. And if equally inexperienced males and females are put together after puberty, the female is obviously better prepared to carry out her part of the coital relationships. As a matter of fact, she not only performs all the necessary female responses, but may attempt to assist the male in the execution of his part of the pattern.

Whatever the extent to which the male-female sexual differences among apes are paralleled in human sexual behavior, it is evident that girls and women neither approach mature sexuality nor embark on it in the same way that boys and men do. For this, a complex interrelationship between biological factors and conditioned influences is responsible.

The sharpest initial distinction between the sex lives of a girl and boy comes with the first active personal experience each usually has with sexual functioning: In a girl, the onset of *menstruation;* in a boy, a *nocturnal emission* (wet dream). The girl's experience is a prolonged one, extending over a number of days; it causes discomfort, often tension and pain, sometimes embarrassment, and involves additional aspects of her physical and mental functioning; it heralds cyclical repetitions of the event, at fairly regular intervals for many years thereafter, with significant meanings if the event does or does not occur; and with this all, the momentous importance of the phenomenon in her whole social and sexual life, as symbolizing the acquisition of serious responsibilities to herself and others, is impressed on the girl (in many groups with special rites and taboos).*

Nothing approximating this initial experience of the girl, in nature or effects, happens to the boy. The nocturnal emission—ordinarily the first active functioning of his sexual mechanism—is a fleeting, minor episode of a few moments, involving an automatic sexual climax while asleep, with or without a dream of having sexual relationships. (Girls only rarely have such climactic dreams.) Not only, then, is this sexual experience of the boy wholly unlike what the girl undergoes, but it receives little or no outside attention (few parents even being aware when it has occurred) and, particularly, does not bring to the boy—as the first menstruation does to a girl—the momentous sense and feeling of being formally inducted into the world of adult sex life and responsibilities. Thus, in all respects, both through Nature's design and the social conditioning resulting from it, the approach to mature sexuality is a much more serious and important matter for the girl than for the boy.

When sexual relationships do take place, another marked distinction between the sexes is immediately apparent: the fact that before the human male can perform sexually, he normally must have the desire for the sex act and must be in the required physical state to make it possible. Among animals other than the primates (monkeys, apes and humans), the female, in turn, must be in a sexually receptive state before sexual relationships can occur. But this is not necessary for the primate female; hence, she can adjust much more easily to the male's sexual desires than he can to hers. Likewise, the completion of the sexual act by the male normally involves discharge of seminal fluid, whereas in the female no fluid is discharged (other than some mucous secretion). In all, then, the sex act demands much more from the male than from the female and limits much more his capacity for sexual performance. This explains, among other things (as noted in Chapter 38) why prostitution as a profession, in terms of opposite-sex relationships, is virtually impossible for a male.† But, most important of all, if the

* The causes of the pains and stresses of menstruation, which vary greatly among individuals, may be organic, psychological, or a combination of the two. As to behavioral effects, it has been reported that among schoolgirls there is a significant increase in misbehavior during or just before menstruation, and among women, a sharp increase in acts or crimes of violence, as well as in the accident rate.

† Noting that the primate female is physiologically equipped for far more sexual performance than the male, Dr. J. P. Scott has reported "In

sex act of itself as a physical function means less to the female, its possible consequences most certainly do not. For the male, the act ends with the completion of intercourse. For the female, it may be the beginning of a chain of biological and social events initiated by pregnancy.

Everything considered, it would seem doubtful that one could compare the sexual activities, responses and feelings of human males and females by the same standards. Yet this has been attempted repeatedly, in popular lore over the ages, and in professional studies during recent decades. Many readers will recall the Kinsey studies—by far the most extensive of all sex studies to date—which were made by the late Dr. Alfred C. Kinsey and his associates (Drs. Wardell B. Pomeroy, Clyde E. Martin and Paul H. Gebhard). Dealing with the sexual histories of over fifteen thousand American males and females, the Kinsey findings and conclusions have been widely discussed, endorsed, and criticized.* A basic question was how far the males and females who *volunteered* to give their sex histories were representative of the general population. Leaving this for discussion in our next chapter, we will concentrate here on what the Kinsey reports had to say about male-female differences in sexual activity and responses.

Many of the Kinsey findings conformed to those in previous studies, and often to what had long been popularly assumed, although the facts were pin-pointed in greater detail: that the sex life of the average male is much more extensive than that of the average female; that sexual activity tends to be more continuous in males, whereas females can and do maintain prolonged periods without sexual activity much more easily; and that males are more readily aroused sexually, especially by pictures, printed words or talk. The most important new aspect of the Kinsey studies was in setting up a system of "scoring" the sexual responses of

---

normal groups of howling monkeys and baboons, the receptive females go from male to male, leaving each one as he becomes satiated." (*Howling* refers to a species of monkeys.)

* The first published report of the Kinsey group, *Sexual Behavior in the Human Male,* appeared in 1948 and dealt with the sex histories of 5,300 American males. The second report, *Sexual Behavior in the Human Female,* published in 1953, presented statistical data regarding the sex lives of 5,940 females. These two reports had been regarded by Dr. Kinsey as only tentative and preliminary to the gathering and analysis of 100,000 case histories. But since his death in 1956, no additions to the previous data on male-female sexual behavior have been published, although in 1958 his associates at the Institute of Sex Research, which he founded, issued another report, *Pregnancy, Birth and Abortion* (see Bibliography).

individuals of both sexes by a simple standard under which each sexual outlet (or sexually climactic event), no matter how achieved—through masturbation, nocturnal emission, heavy petting, heterosexual (opposite-sex) intercourse, homosexual relations, or whatever—was given the same unit value: one *reported* outlet, one unit. And it was through this scoring that the Kinsey studies resulted in another set of conclusions:

That sexual responsiveness in males develops years earlier than in females (despite the female lead in biological maturation); that the peak of sexual activity (and of capacity), is reached much sooner in males—at about age seventeen or eighteen, versus almost thirty in females; that whereas average males have more sexual activity than average females, no male "sexual athlete" could compare in frequency of sexual performance and response with some females, who could have "twelve or more orgasms within minutes," and fifty or more per week (through intercourse), or who might achieve orgasms "up to 10, 20 and even 100 times within a single hour" (through masturbation); and that whereas impotency is extremely rare in males before old age, perhaps 30 per cent of the females are more or less sexually unresponsive throughout life.

This brings us back to the basic question of whether all human sexual responses *can be* reduced to the same statistical terms of "units" measured by one simple standard. For instance, can each sexual outlet of a teen-age boy—through masturbation, nocturnal emission, or some furtive petting episode—be judged equivalent to the full and emotionally satisfying sexual response of a mature and happily married man with his wife, or of a devoted wife with her husband? Or can one accept at face value the statistical conclusion that some women may have "a dozen or more orgasms" within minutes—each equivalent in unit value to the single sexual outlet the male mate achieves during the same act of intercourse? The mere fact that the male sexual response involves a specific physiological sequence, culminating with an ejaculation, whereas the nature of the female response—as to precisely how and where it occurs, and what it is—remains in dispute, must make us wonder whether a woman herself, or her husband, can ever or always be sure of how the sex life of one balances with that of the other.

The notion, developed in recent times, that the male and female sexual responses can indeed be fully equated has had various unfortunate results, as psychiatrists and marriage counselors have found out. For one thing, Dr. John L. Schimel has noted, the system of "double-entry sexual bookkeeping" has led many couples to feel that their joint sex life is a failure if the wife's response isn't fully (or almost always) synchronized with the husband's in nature, quantity, quality, precise timing, sensations and satisfactions. Whether or not this vague goal can be approximately reached in some cases, it probably cannot in most cases; and in any event, what constitutes the appropriate sexual response, or sexual satisfaction, for a given woman is a highly individual matter. This has particular meaning with respect to the use of the term *frigidity*. For another consequence of expecting women to respond sexually as men do is the tendency to label them frigid if they do not.

What frigidity in a woman really means is far from certain. As Dr. Lena Levine (a psychiatrist, gynecologist and marriage counselor) has pointed out, true frigidity designates someone who has absolutely no sexual desire or capacity for response, and this is rare. Far commoner is limited desire, or inadequate or infrequent sexual response. The reasons for and possible incidences of frigidity in various degrees will be discussed further along. In terms of sex differences, the point to be made is that, as with other aspects of sexuality, frigidity in women is not the same as *impotency* in men. While both involve inability to experience or respond actively to sexual stimulation, they may differ greatly to the extent that female and male sex mechanisms, organs and functions are different. Primarily, a woman can be wholly frigid and can still function sexually; she can *simulate* sexual responsiveness and provide full sexual satisfaction to the male; and, not least, she can fully carry through her reproductive role and bear children, no less efficiently than a nonfrigid woman. Very different is impotency in a male, which often can be present despite his having strong sexual feelings; moreover, it is immediately detectable, and, whenever and so long as it exists, is an effective bar to having sexual relationships and fathering children.

Another fallacy is in applying the term "change of life" equally to the *menopause*, in women, and to sexual modifications that take place in men after middle life. These phenomena, too, are unrelated physiologically, functionally and in their effects. One involves, chiefly, *reproduction*; the other, *sexual relations*. The menopause marks the formal end of a woman's reproductive capacity (and of menstruation), due to a series of hormonal changes which halt further production of eggs and the preparation of the womb for their reception. This event, usually in the mid-forties but sometimes as early as the mid-thirties (or earlier, if resulting from operative removal of the ovaries), may come at a time when the woman is yet in full vigor. It need not be correlated with capacity for or interest in sexual relations. In fact, for some women, previously inhibited by fear of pregnancy, the menopause may initiate heightened sexual responses. However, for most women the "change of life" tends to be accompanied by some physiological disturbances, and occasionally by temporarily depressed states, due to both physiological and psychological factors. Heredity has been credited with a part in the time of onset and the nature of the menopause for individual women. (It also has been reported, though not entirely authenticated, that earlier puberty usually is a forecast of a later menopause.)

In men, any so-called "change of life" is quite different from the woman's experience. It is not, as in her case, a fairly abrupt cessation of reproductive capacity during middle life or before, for in most men this capacity may continue well into the older years and sometimes into very old age. The change that does take place in men is no more than a gradual diminution in sexual capacities (and perhaps also, psychologically, in sexual interests) as an accompaniment of the general aging process. Heredity, insofar as it influences this process, may also, then, affect the relative "sex-life spans" of individual men.

Taken together, all of the biological differences between males and females in sex organs, mechanisms, functions and stimuli clearly make it difficult, if not impossible, to compare their sexual responses on the same scale. When there are added the innumerable psychological effects of conditioning, and of the different meanings sex has for the human male and female, one could very well reach the following conclusion (paraphrasing what was said about sex differences in intelligence):

> *No means or tests have yet been devised, and probably never can be devised, which can accurately measure the relative sexual capacities, performances, and satisfactions of men as a group, and of women as a group. And that being so, no one can say now, or may ever be able to say, whether the sexes are inherently equal or unequal in their sexual feelings and responses, or whether either is superior or inferior to the other in these respects.*

Turning now to *individuals* within each sex group, we may find less difficulty in comparing the sex lives of one person and another. That any two men, or any two women, may differ enormously in the extent of their sexual activities need hardly be said. Between the extremes of the male "sexual athlete" and the wholly impotent man, and the highly amorous female and the wholly frigid one, there is every possible gradation. The sexual range among men, however, is undoubtedly much narrower than among women. As already noted, the capacity for sexual *activity* is very much greater in women than in men; on the other hand, sexual restraint, or absence of sexual activity, is far more common among women. That much of the difference in sexuality between one individual and another of either sex relates to conditioning, opportunities or incentives, and states of mind is unquestioned. But there is every likelihood that inherent biological factors also contribute to this difference. For instance, the Kinsey investigators challenged the validity of the assumption that all sexually unresponsive individuals are "inhibited," holding that this "amounts to asserting that all people are more or less equal in their sexual endowments, and ignores the existence of individual variation." Further, the Kinsey report said, "No one who knows how remarkably different individuals may be in morphology, in physiological reactions and in other psychologic capacities, could conceive of erotic capacities (of all things) that were basically uniform throughout a population."

If heredity does account for some of the individual variability in sexual capacities and responses, the hypothetical "sexual" genes might work through producing variations in sex organs and mechanisms, sensory structures, and hormonal or other chemical components concerned with sexual activity. The best evidence relates to the hormonal influences. In men, for example, raising or lowering the androgen (male hormone) level can increase or decrease the frequency of sexual activity; and castration, which eliminates the sex glands and hence their production of androgen (although some sex hormones can still be produced by other glands), leads either to marked diminution in sexual activity or to inhibition or cessation, depending on individuals and on when the operation is performed.* Among aged people—women as well as men—sexual interest has been stimulated and sexual ardor increased or restored by the administration of sex hormones, but here again the effects may depend on individuals.

The pituitary gland is also a factor in sexual activity, since it helps to regulate the workings of the sex glands and is important in triggering sexual development (and, in women, the onsets of menstruation and the menopause). The ardenal glands, too, are involved in the production of sex hormones, and the thyroid glands may also affect sexual activity.

* Most important with respect to the effects of castration is the man's age. When castration is performed after a man is mature and has been having a regular sex life (as when the operation is a medical necessity), sexual activity may go on with only gradual diminution. But if the operation takes place before puberty, it may inhibit sexual functioning and performance thereafter. However, psychological factors, added to the individual's general biological makeup, may affect the development or continuance of a male's sexual desires and activities after castration. In a woman, *ovariectomy* (removal of the ovaries) tends to cause much less change in sex life, both for biological and psychological reasons, than does the analogous castration in males.

Thus, everything else being equal, genetic variability in the combined workings of the glands and their hormones may lead to individual differences in sexual tendencies and activities. Finally, the known effects on sexual responses of drugs, such as alcohol or narcotics (more often as repressants than as stimuli), together with evidence (see Chapter 38) that individuals may differ genetically in reaction to these drugs, suggests that a person's entire hereditary makeup with respect to biochemical functioning may bear importantly on his or her sex life.

However, we must not forget that what individuals actually *do* sexually—the specific forms their sex lives take—may be as much if not more a matter of conditioned behavior and circumstances than of hereditary tendencies. Nor need the relative sexual performances of two given individuals be any clear indication of how they compare in genetic capacity. Previously we emphasized that the sexual differences between a male and a female could not be weighed by the same standards. Similarly, though to a lesser extent, one might doubt that any system of rating sexual performance by "units"—one "outlet," however derived, being held equivalent to any other "outlet"—could fairly or scientifically measure the true quantity, quality and *satisfaction* of one person's sex life as compared with another's. Whatever sexual behavior may be in other animals, in human beings it unquestionably is an expression of the individual's whole personality; and more than a merely individual function, it also is a social function, involving some of the most vital and sensitive aspects of the relationships among people.

It is for these reasons that so much importance is attached to whether an individual's sexual behavior is average and normal, or nonaverage and abnormal. The full implications of these terms will be discussed in our next chapter.

CHAPTER FORTY

Sex Roles and Sex Life: II

Abnormal

EVERY SUMMER a strange phenomenon occurs in several New England seacoast towns. The resident males are chiefly rugged fishermen, of the most masculine type. But as the vacation season opens, large numbers of a *special kind* of male flock in from the big cities. They tend to be more than ordinarily good looking; they dress colorfully, though to extremes; they are many of them gifted in various artistic and intellectual ways.

To the unsophisticated girl vacationing there for the first time, the sight of so many attractive young single men may stir high hopes of romance. Alas, very soon she will discover that they have no interest in her as a woman. The romances of these males are reserved solely for members of their own sex.

Luckily, the young lady will find other male vacationers who, from her standpoint, are entirely normal. But the question will linger: What causes the "queer" men—the homosexuals—to be the way they are? Were they *born* to be like that? Are they really "feminized" males—a biological cross between male and female? Or are they merely *acting* queerly as the result of some peculiar conditioning?

These same questions have been asked over the ages by human beings everywhere, for there is no place or time where males of this type have not been known. But there are also other kinds of individuals, female as well as male, whose extreme deviations from customary sex roles and sexual behavior have caused wonder and concern: the Lesbian women whose sexual interests are only in other women; the males or females, who, whether or not normal in sexual tendencies, have a craving to dress and be like members of the opposite sex; the persons of both sexes who seem insatiable in their sexual appetites; and the ones who, while organically normal, evince no interest in sexual relationships.

Though much is still to be explained about deviant sex and sexual behavior, many misconceptions have been cleared up and important new insights provided as to the possible sources of such behavior (whether inherent or conditioned) by recent findings in human genetics, biochemistry, psychiatry, psychoanalysis, psychology and other fields. These findings are best approached by recalling how sex and sexual interests ordinarily develop in boys and girls.

As set forth in the preceding chapter, there is first, in infancy, preoccupation with oneself; then, as childhood advances, a neutral period of undifferentiated interest in children of both sexes; then attraction in play and companionship only with members of one's own sex; then social interest and sexual interest in the opposite sex in general; and finally, mature concentration on individuals of the opposite sex, proceeding from courtship to marriage. These are not merely average stages in male-female sexual development. They also are thought of as *normal* stages. For, when a boy or girl deviates too sharply from the pattern for each stage, or fails to pass from one stage to another, the beginnings of *sexual maladjustment* in later life may be present. Here are some examples:

> The individuals who remain in the infantile stage of preoccupation with their own bodies and who cannot transfer sexual interests to others. Classed with these are the chronic masturbators and the narcissists—persons "in love with themselves."
>
> Those who remain in the "neutral" stage of sexual indifference, not gravitating toward persons of the opposite sex any more than toward those of their own sex. They are apt to be individuals very low in sex drives or feelings.
>
> Those who continue in their strong attachment to their own sex, or feel aversion to sexual contact with the opposite sex. This may lead to homosexuality. On the other hand, those who repudiate their own sex and identify themselves instead with the opposite sex (in actions and interest) may also become homosexuals, of the invert type (femalelike males, or malelike females) or transvestites (dressing and acting like the opposite sex, but not necessarily homosexual).
>
> Those who never pair up or form any strong attachment with an *individual* of the opposite sex. They include the chronic "Don Juans," or their female counterparts, the "Doña Juanas" (the habitual flirts). While seemingly too strongly sexed, they may actually be persons with no deep sexual feeling, who never derive full satisfaction from their experiences.

The foregoing are, in brief, the main types of sexual deviation. But what is it that leads individuals of each type to fall out of line at the different stages of sexual development? With respect to normal sexual behavior, we noted in the last chapter that the results for any individual are governed by a complex interrelationship between biological factors (to a great extent genetic) and conditioned psychological and social factors. These factors, however, are related chiefly to the *quantity* of a person's sex drive and to the degrees of sexual response. Quite a different matter is the *direction* taken by the sexual impulses, especially with regard to the deviations previously mentioned. Are the persons so described biologically different in any way from those who do conform to the average or normal sexual pattern?

Among the sexual deviates, the homosexuals* have received the most attention. The popular concept of the homosexual is of an individual who not only prefers to engage consistently in sexual relations with members of his (or her) own sex, but even looks and behaves considerably like those of the opposite sex. This isn't necessarily so. There are many types, grades and intergrades of homosexuals, and those who conform to the popular picture make up only a minority group. For instance, there are male homosexuals who look and act as thoroughly masculine as possible, and women homosexuals who are utterly feminine; there are individuals who are both homosexual and heterosexual, at varying periods or simultaneously; there are some who start out as homosexuals and later become normal, or vice versa; some who believe that they are homosexual and live as such, although their true tendencies may be otherwise, and a great many who lead presumably normal lives without ever realizing that they are actually homosexual in tendency. Moreover, the classification of homosexuals as "active" and "passive," in terms of some playing the "husband" and others the "wife" role, does not apply to a great many, and perhaps most homosexuals, who engage with each other in both types of activity.

Thus, differences in the standards of diagnosing homosexuality, in methods of conducting sex studies, and in the samples of groups and individuals investigated may largely explain discrepancies in the reported

* The term "homosexual" is derived not from the Latin word *homo* ("man") but from the Greek word *homo*, meaning "same," and is thus used to describe a sexual relationship between individuals of the same sex. While such behavior in women is also called homosexuality, the popular term in their case is "Lesbianism," referring to the female homosexual practices described in the poetry of Sappho as having taken place on the Greek island of Lesbos.

incidences of various types and degrees of this behavior. Here we come back to the Kinsey studies, which have gone beyond all others in the attention given to homosexuality and in the estimates of its presumed incidence. Previous studies of male homosexuality in the United States (and in Europe, as well) had placed the incidence at from 2 to 5 per cent, although official figures of the Armed Forces in the United States had given it as no more than 1 per cent. But the staggering Kinsey estimates—based on the histories of the 5,300 males who had *volunteered* for questioning—were these:

> That about 4 per cent of the American males are exclusively homosexual throughout their lives; that about 13 per cent are predominantly homosexual, though engaging in opposite-sex relationships; and that if homosexuality is extended to include any kind of active homosexual experience, at any time, nearly half of the American males have had such experience.

Alarming figures, indeed. But the perspective of time and critical analysis reveals them in a more restrained light. The "Kinsey males" were in all probability a nonaverage group, there being indications or suspicions that those who volunteered comprised unduly large proportions of sexually deviated or maladjusted individuals, as well as considerable numbers of inmates of prisons and reformatories.* At the same time, the Kinsey standard of rating homosexuality went much beyond what had generally been understood by the use of the term. For example, it listed a man as having engaged in "homosexual activity" if this had been no more than a single, dimly recalled childhood incident of mutual sex play with another little boy at the age of six or seven. While the Kinsey report did class this as the mildest of six grades of homosexual behavior, when taken together with the apparently slanted

* While the Kinsey report on males refers at various points to volunteers in prisons and reformatories, it does not give their numbers. However, a clue that the criminal proportion was considerable appeared in the later report on females, which revealed that of 6,855 sex histories of White females originally gathered, 915 (or about 13 per cent) were of females who had served prison sentences, and that these histories had been omitted from the report because they would have "seriously distorted the calculations on the total sample." Whether or not the proportion of prison cases among the Kinsey males was as high as in the female histories gathered, or higher, and how far they may have affected the calculations on male sexual behavior must be left to conjecture. But the fact that the Kinsey statistics on males did include prison cases, while those on females did not, should call for caution in making comparisons of male and female sexual behavior on the basis of these two sets of Kinsey data.

sampling of males, the picture which as a whole was conveyed to the public may well have been of a far greater prevalence of homosexuality than actually exists.

What of the "Kinsey females"? The facts and implications presented in the report about them might be subjected to reservations similar to those expressed regarding the Kinsey males. For the women, too, there were startling statistics as to the incidences of various types of aberrant and socially disapproved sexual behavior. As one example, the incidence of female homosexual experience in one degree or another and at one time or another was placed at 20 per cent; and between 1 and 3 per cent of the unmarried females of all ages (though less than three in a thousand of the married ones) were labeled as exclusively homosexual for continued periods. However, whether or not the Kinsey figures for female homosexuality reflected the general situation, previous investigators had stressed the great difficulty of arriving at the actual incidence of female homosexuality because this behavior could be much more easily concealed or suppressed by women than could the corresponding behavior in men; and many more female than male homosexuals could voluntarily or by necessity submerge themselves in marriage, or could otherwise more readily carry on both homosexual and opposite-sex relations at the same time.

Extremely high incidences were also reported in the Kinsey study for other unconventional or aberrant types of female sexual behavior, including masturbation and adultery. But since women generally would be more reluctant than men to disclose their intimate sex histories, there was reason to suspect that the females who did volunteer for Kinsey interviews were an even more unrepresentative group than the males. Tending to confirm this suspicion were the findings of Professor A. H. Maslow, who had been at Brooklyn College at the time and had made studies of the personalities of girl students in his psychology classes. Subsequently many of these girls volunteered for Kinsey interviews, and a check then showed that they had been predominantly the more "sexually sophisticated and sexually unconventional ones." Going further, Dr. Maslow ventured the guess that the Kinsey women volunteers as a whole "were probably even more slanted toward those with unconventional or disapproved sexual behavior."

Regardless of the weight given to the Kinsey statistics for either sex, the most serious implications of the reports lay in the claim that homosexuality and other sexual deviations must be accepted as natural and normal forms of sexual behavior: in fact, that because of the incidences cited, *no* kind of human sexual behavior could any longer be termed abnormal. This contention has been violently disputed. For many critics have maintained that no matter what the incidences, they no more prove homosexuality and other sexual deviations to be "normal" than the very high incidences of mental diseases or neurological disorders prove *them to* be "normal"; and that, particularly, if the chronic, exclusive and freakish homosexuals cannot be called "abnormal," the very word would lose all meaning. The Kinsey group claims that these arguments misinterpret the scientific issues involved, confusing moral values with biological facts. Their critics answer that it is the Kinsey reports which cite biological facts to negate moral values. Is it then all a matter of semantics?

Let us recall here that the primary purpose of sexual behavior as designed by Nature is to insure the reproduction and continuance of the species—human or any other. Morals or religious codes aside, one clear fact is that homosexuality defeats completely the functions of reproduction. True, we now recognize that sexual relationships among humans serve other important purposes (emotional, psychological and social) and that even biologically, sex and reproduction do not necessarily, or even most of the time, go together. Women are fertile only one or two days in each month; and in many persons of both sexes, who are consistently sterile (or who have become so through disease, operations, or, in women, the menopause), the sex drive may be as strong as in those who are fertile. But whatever the objectives of sexual relationships, there is the further fact that the sexual structures of males and females are specifically designed to complement each other, and that those of the same sex are not. No amount of argument can therefore obscure the fact that homosexuality, as a *preferred* practice where opposite-sex relationships are available, goes contrary to Nature's design, and that socially and psychologically it sets the individual apart from the great majority of other persons.

Nor can we directly apply to human beings the evidence that homosexual activity occurs widely among lower animals and that it may be natural among them. For there are no groups among lower animals which deliberately *choose* homosexuality in preference to opposite-sex relations. As to a trait's being normal because it is natural, in the chapters on wayward genes we have discussed numerous conditions, behavioral as well as physical, which are entirely natural but yet must be regarded as abnormal.

If we agree, then, that *chronic homosexuality* and certain other types of extreme deviate sexual behavior are abnormal, we come to the question: *Is heredity in any way responsible?*

Since the conditions themselves are so varied in nature, there may be many causes for homosexuality and kindred deviations; and, as with many diseases, even the same condition may result from a number of influences. Those who regard homosexuality as primarily a state of mind point to many psychological factors which could help to bring it on. Among these, for a boy, might be a neurotic, possessive, sexually frustrated mother and/or an unsympathetic, autocratic or absent father, which would lead to "feminine" conditioning; an overly strong childhood fixation on some older person of the same sex; rigid training to regard sexual intercourse as indecent or repulsive; the development of an aversion to or dread of girls and women, or inability to adjust to them; seduction by older male homosexuals; and fear of fatherhood. In a girl, homosexuality might be fostered by corresponding adverse influences leading to "masculine" identification, aversion to sex relationships with males, fear of motherhood, and so on.

Again, in either sex, some of the factors noted may lead to two other related types of behavior: *transvestism* ("cross-dressing"), the urge to dress and behave like the opposite sex; and *transversion*, the wish to be transformed as completely as possible into one of the opposite sex. While these two conditions had long been considered as invariably linked with homosexuality, psychiatrists now recognize that either or both are sometimes independent of it, or of each other. Thus, many male transvestites are or wish to be "female" only in dress and behavior, but have normal sexual relationships with women and adjust fairly well to marriage and fatherhood. The *transvert*, however, may go to the extreme of combining transvestism with the desire to become *anatomically* like a female (at least outwardly) and may seek—or actually undergo as some have done—an operation to achieve this. Yet even such drastic behavior need not always be correlated with homosexuality. Likewise, some male homosexuals who play the role of a woman may be both transvestite and transvert, whereas others, who take only the male role with homosexual partners, may be neither.

As for *female transvestism* (even extreme and persistent dressing like a male) or *female transversion* (the urge to act and be as much like a male as possible), such behavior, also, may often be unrelated to homosexuality—the chance that it is unrelated, in fact, being much greater than with the corresponding behavior in a male. This is mainly because the urge for females to emulate males has a more valid psychological basis in our world and can be carried out with less social disapproval or sexual complications than can a male urge to emulate females.

Yet granted that psychological and social influences are major factors in the development of homosexuality or related aberrations, this need not be the whole story. To quote veteran psychiatric expert on homosexuality, Dr. George W. Henry, "Whatever the external influences may be, the majority of persons do not succumb to them and the minority who do succumb appear to be fundamentally predisposed to homosexuality. Some of this predisposition may be inherited." Is there evidence to support this theory?

The fact that homosexuality has appeared with unusual frequency in particular families might be ascribed as easily to environmental conditioning as to genetic tendencies. Much more significant have been studies of homosexuality in *twins*. In most cases, where one of a pair of identical twins was homosexual, so was the other, whereas this was not true of fraternal pairs. For example, Dr. Franz J. Kallmann and his associates studied 40 pairs of male identicals and 45 pairs of male fraternals among whom homosexuality had occurred and found that if the twins were identical, in almost every instance where one was overtly homosexual, so was the other; and usually the twins were not only fully concordant as to the type of homosexuality, but very similar in the visible extent of their "feminized" appearance and behavior. Further, most of these twins claimed to have developed their homosexual tendencies independently and often far apart from each other. Among the fraternals there was much less concordance, varying from only 11.5 per cent, for exclusive homosexual behavior, to about 42 per cent for occasional homosexual behavior. But since this concordance of homosexuality between fraternal twins was still very much higher than that between two nontwin siblings, we can gather that the chance of becoming homosexual is much increased if one is closely exposed to the behavior.

The foregoing facts suggest that the explanation for homosexuality may be much as we found it for various other complex human traits, physical or behavioral, normal or pathological: individual degrees of predisposition interoperating with degrees of environmental pressures. In other words, the two schools of thought on the subject—the biological and the all-out psychological—need not be entirely in conflict. We have dealt with many conditions in this book, involving behavioral as well as physical abnormalities, where a given condition may be either hereditary or environmental (abnormal intelligence, many physical deformities, many diseases) or may result from an interoperation of both forces. Thus, theoretically, some cases of homosexuality might be chiefly (or entirely) either biological or psychological; and in between there might be other cases where degrees of predisposition are dependent upon degrees of "homosexual" environment. In fact, with more accurate classification it may be found that different types of homosexuality are really different conditions, rather than gradations of the same behavior pattern. Only when we can more clearly diagnose homosexuality can we expect to identify the type or types in which genetic factors may be important.

If heredity does help to foster a homosexual tendency, by what conceivable means could it do so? The immediate thought here would be of the completely invert homosexuals—those in whom exclusive homosexual relations are coupled with extreme simulated femininity. Many special characteristics identify these individuals: eccentric movements of the hips, shoulders, hands, feet and head, certain expressions of the eyes and mouth, lisping or high-pitched voices, and so on. One might argue that all such mannerisms could be acquired, as by an actor learning them for a part. What makes it difficult to believe that they are purely conditioned is, first, that the beginnings of the patterns often are found in young boys, sometimes at early ages; and second—perhaps most important—that these behavior peculiarities appear in almost precisely the same form in individuals everywhere in the world, even among widely separated primitive peoples, from Eskimos to African cannibals. Thus, distracted American parents who now consult psychiatrists about their boys of this type have their counterparts among equally baffled primitive parents, although the latter sometimes regard the "feminine" boys as divine products and dedicate them to the priesthood (the shamans). In certain other tribes these boys are inducted at puberty into the ranks of women (to be known then as "berdaches") and are "married off" to men.

Looking for evidence of biological femaleness in male homosexuals, some authorities have reported that these individuals are more than ordinarily apt to have such physical characteristics as the feminine "carrying angle of the arm" (illustrated on page 120), deficient facial and body hair, and excess of fat on the shoulders. Among female homosexuals undue proportions are reported as having such masculine physical traits as muscularity, excess hair on face and body, underdeveloped breasts, and low-pitched voices. But these opposite-sex characteristics are far from general in homosexuals, for a great many of the males among them are more masculine in build than average men, and many of the women homosexuals are extremely feminine in appearance (sometimes strikingly so, and beautiful). However, the failure to conform physically or in behavior to the type appropriate for one's sex may be a conditioning influence toward homosexuality. That is, if a boy merely *looks* feminine—if he is regarded as pretty and delicate or behaves in what is thought of as a feminine way—he will have less appeal for girls and more appeal for older homosexuals, and so have a greater chance of being inducted into homosexuality. Similarly, a girl who looks or behaves in too masculine a way might be swerved more easily toward homosexuality because of her decreased appeal for men and increased appeal for certain women.

Thus, regardless of inherent sexual tendencies, the ranks of homosexuals would tend to have more than average numbers of individuals resembling their opposite sex. By the same token, any hereditary factors which could make an individual look or act more like a member of the opposite sex would strengthen the push toward homosexuality. This, however, relates only to external appearances and outside influences. Could it then not also be possible for *inner* biological traits and influences—genetically governed—to swerve the sexual feelings and drives of some individuals away from the opposite sex and toward their own sex? And if so, how?

Since hormones play so large a part in sexual behavior, one might look first to their possible part in misdirected sexuality. As noted in the last chapter, experiments with lower animals have shown that by changing the male-female hormonal balance, males can be made not only to behave like females but to be attracted to and to attract other males sexually. On the other hand, with added male hormones, male animals can be made more masculine in behavior and be stimulated to greater sexual activity with females. Corresponding results can be obtained with females, who, through hormonal treatments, can be made to imitate male sexual behavior and to be attracted to and attract other females. But in human beings, hormone treatments have not yet been found to produce such results. Male hormone administration has strengthened sex drives of feebly sexed males, and also, in some without previous interest in females, has strongly fostered such interests. However, lowering the male hormone level, and/or raising the female hormone level, has not diverted the sexual interests of human males to other males.

Most significantly it has not been found that male homosexuals themselves as a group are deficient in male sex hormones, or female homosexuals in female hormones. Nor have treatments of homosexuals with added hormones of their own sex proved effective in changing the direction of their sexual drives or interests. In consequence, hope of curing homosexuals through hormonal treatment has met with disappointment.

Another widely held earlier theory was that homosexuals of the invert type might be sexual "in-betweens" genetically—that is, *hermaphrodites* (as described in Chapter 27). The new chromosome-study techniques have blasted this theory by showing that cells of scores of male homosexuals studied had the normal XY male chromosome makeup. Similarly, cells of male transvestites or transverts, whether or not they were homosexuals, have proved to be entirely normal in their sex chromosomes. Thus, any *biological* explanation for homosexuality and transversion must be found not in mistakes in sex determination, but, at worst, in some biochemical malfunction or aberration which could swerve sexual feelings and interests from their normal course.

But a reverse possibility was suggested: What of the sexual behavior of individuals whose sex chromosomes *were* abnormal—for instance, the Klinefelter and Turner types (also discussed in Chapter 27)? A check on this point showed that the typical Klinefelter individuals (mainly those with two Xs plus a Y), while being biologically feminized males, nonetheless rarely showed any evidence of homosexual tendency. Likewise, in the Turner individuals ("incomplete" females, with only one X), when sexual responses do manifest themselves, they ordinarily are in the opposite-sex direction, without evidence of Lesbian tendency. However, both Klinefelter and Turner individuals, who usually are mentally retarded as well as physically abnormal, might run above-average risks of being inducted into sexually deviant behavior.

Particularly interesting are the studies of individuals who, while genetically of one sex were reared—because of outward sex-organ abnormalities at birth—as members of the opposite sex. (In some cases operations had been performed or treatments given to make these persons conform as much as possible to the sex chosen for them.) One group comprising more than 100 of these individuals—adults as well as children—was studied at the Johns Hopkins Institute by Drs. John Money, Joan G. Hampson and John L. Hampson. The sexual behavior and interests of all of these hermaphroditic individuals was found to have been determined mainly not by their sex chromosomes, sex glands, sex-hormone balance or internal sex mechanism, but by the sex assigned to each and the sex role for which each had been trained. However, questions about these conclusions have been raised by findings with regard to the sex behavior of pseudohermaphroditic monkeys—animals in which various abnormalities of sex development, equivalent to those of human beings—have been produced artificially through early sex-hormone treatment or through operations. Observations of such animals by Dr. William C. Young and his associates at the Oregon Primate Research Center have led to these conclusions: (1) The sex behavior of an individual, while influenced by psychological factors, is also dependent on the genetic makeup and the sex chromosomes with which he is started off; (2) the initial actions of the individual's sex chromosomes affect both the structure and functioning of the body cells and the types of sex hormones produced; and (3) in turn, the early activity of the sex hormones has a broad role in the determination of behavior, influencing not only behavior related to the reproductive processes, but the behavior which is part of the individual's masculinity or femininity. Thus, the foregoing evidence, pointing strongly to "predetermined psychosexuality" (as Dr. Young phrased it), would contradict the theory that individuals are started off at birth with "psychological sexual neutrality," and that only conditioning is responsible for the contrasting "maleness" and "femaleness" psychological behavior patterns.

One might further conclude that, regardless of being reared as of one sex or the other, a person who is genetically of one sex (whether normally so or pseudohermaphroditically) can never fully develop the behavior of a normal person of the opposite sex. Other investigators also have claimed that where sex rearing is contrary to a person's biological sex, it produces the greatest psychological disturbances in those who are intersexed. In any event, many authorities currently suggest that the individual circumstances should decide what is to be done about children in whom outward sex seems to be in conflict with genetic sex.

On the whole, one should be cautioned not to apply sex facts observed in hermaphroditic individuals too closely to normal males and females. In the abnormal persons mentioned, the biological makeup, internal and external, is so lightly poised that it can be diverted from one sex direction to another, although often only with the additional aid of surgery and treatment. Even then these individuals can never be made *fully* masculine or feminine. At best they can be only "reasonable facsimiles" of their adopted sex. The genetic male converted into a "female" lacks true female sex organs, does not have full feminine contours, cannot menstruate, cannot fully function sexually as a female, and, obviously, can never bear children. The genetic female converted into a "male" is not fully masculine in sex organs or in build, can only simulate masculine sex relationships (inadequately so) and cannot, of course, father children. In either case, since the "sexually converted" individuals are denied the full biological experiences and feelings of their adopted sex, *psychologically* they cannot be fully male or female. So, too, while it is now known that there are marriages where the "husband" is actually a genetic female or where the "wife" is a genetic male, such marriages can hardly be completely normal; not merely so far as the hermaphroditic mate is concerned, but with respect to the adjustment of the other partner.

The problems posed by the hermaphroditic individuals, which relate to clearly discernible and often remediable biological factors, are less serious and puzzling than those with respect to the homosexuals, whose numbers are far greater and whose conditions are much less subject to accurate diagnosis and treatment. What to do about homosexuality is a question which here can be touched on only lightly, for we have been dealing primarily with the possible biological factors and not with the moral, religious or legal aspects. Since the development of homosexuality —regardless of predisposition—is so largely dependent on conditioning, the major solution to the problem must lie in preventive measures. But once the behavior pattern has become ingrained—particularly in an overt homosexual adult—there is doubt at the present time whether it can be cured. Hormone treatments or other biological treatments, as was said before, have as yet proved of little effect. Psychoanalytic treatment has been reported successful to some extent where the individual was in the borderline areas of homosexuality; but with respect to overt homosexuality, many analysts believe the best that can be done at present is to adjust the individual to his pattern of life.

That something should be done, if possible, about homosexuality is obvious, both for social reasons and because the homosexual in our culture is hardly likely to be a well-adjusted or happy person. But if it can be proved that this form of sexual behavior is indeed unavoidable in many individuals—that they have inherited the tendency or acquired it without any volition, and that nothing can be done to change them— the question arises as to whether the attitude of our society should not be changed. A trend in this direction is already evident, especially in Great Britain, where the report of a committee of fifteen prominent Britons headed by Sir John Wolfenden recommended in 1957 that homosexuality no longer should be classed legally as a crime if it did not go beyond a private, consenting relationship between two adults, and, specifically, if it did not involve seduction of a minor by an older homosexual. Viewed from any standpoint, homosexuality calls for more sympathy and understanding than the public now gives it.

Yet it should not be ignored that while homosexuality may be tolerated, as other ills which cannot yet be cured or eradicated must be tolerated, it need not be accepted as desirable and inevitable. For there is no evading the fact that we live in a world where people everywhere are divided into two main groups, male and female, governed by prescribed norms of conduct for each sex, any marked departures from which will make social existence more than ordinarily difficult. We have seen that these norms derive to a considerable extent from biological and largely genetic sex differences. But, superimposed on these, we also know that there are many socially conditioned sex differences which have been subject to change—much more so in our time than in the past, when the biological and genetic sex differences, sex roles and divisions of labor were meshed into one big barrier which kept the sexes rigidly apart. A major problem of our time, therefore, is to determine whether and how far the behavior and activity patterns prescribed for the two sexes as groups can and should be modified, and also how far individuals of either sex can depart from the prevailing norms for their sex group to achieve greater personal fulfillment and happiness. All of this has had particular meaning for women.

Formerly, the emphasis on sex differences worked largely against women, for in a world governed by men's standards and ideals, *not* being male tended to subject women to an inferior status and excluded them from many activities and opportunities for advancement. So, understandably, the goal for many women in recent times has been to eradicate or minimize sex differentiation as much as possible. Innumerable modern developments have helped in this direction: mechanization in industry, which lessened the importance of the male strength factor and broke down previous work divisions; increased education and the opening of new career opportunities for women; effective means of limiting motherhood where desired; eased physical problems of homemaking; and granting women the vote, leading to greater participation in public and civic life. Accompanying and contributing to these changes—which were of far-reaching benefit to women and to human society as a whole—the feminist movement encouraged women to seek to be as much like men as possible in behavior, activities and thought. This, however, failed to reckon with the capacity of women—at least, the new generation of women—to change their minds.

In a Broadway musical in the late fifties there was a hit song, "I *enjoy* being a girl." It may have marked an end to the era of feminism (which actually was directed against *femininity*) and to the movement of women toward maleness. For, unmistakably, a *pro-femininity* trend had set in. Where in the past girls nearing puberty had been ashamed —or had been made ashamed—of the developmental changes and functions marking oncoming womanhood, American girls of the new generation can hardly wait for these changes to occur, view them with pride, and begin wearing brassieres even before there is need for them. With earlier puberty, too, earlier "going steady" and competition for males has led to earlier development of feminine wiles, feminine patterns and earlier marriage. New fashions have repudiated the flat-chested, boyish-bobbed, mannish-suited female model of a generation ago, with the accent being now on the female form and its exposure instead of concealment; and elaborate coiffures and eye and facial makeup have carried women back to the ancients. Above all, where childbearing had been depicted for a time as an infliction, to be spurned by the emancipated woman or, if duty demanded, to be reduced to a minimum, motherhood is definitely back in style, and most noticeably so among the younger, more educated and more socially advanced daughters of the former feminists. In countless instances, where the mother had chosen to have no more than one child, or at most, two, the daughter is having three or four.

Another change in women's viewpoint has been a re-evaluation of what men in the past had told them about their "biological inferiority." The growing evidence of how the female in many respects actually is a much more resistant and durable organism than the male—as noted in previous chapters—has raised the self-esteem of women regarding their physical selves. (In fact, there's now been much talk of their "natural superiority," which, scientifically, is no more valid than the theory of the "natural superiority" of males.) Also, there has been a questioning of the psychoanalytical concept of "penis envy"—that a feeling of inferiority develops in females, beginning in early childhood, when awareness comes that they lack the sex organ possessed by the male. But one of the leading women psychoanalysts, the late Dr. Clara Thompson, maintained that what women envy is not the male organ, but the male position of privilege and alleged superiority for which the organ had been made a symbol. With women accepting and esteeming their sexuality in its own right, it is held that only the one who is maladjusted for other reasons will, as an excuse, ascribe her failures to her lack of maleness.

Nevertheless, for the woman who wishes to and/or must pursue a career, as is the case with vast numbers today, there can still be serious conflict with her biological sexuality. During her working hours, while functioning as a colleague on an almost neutral plane with her male associates, she can largely erase the traditional sex-role differences. But returning to her home at night, she must switch on her sex role again as a woman, wife and sexual partner; and with pregnancy she enters a world completely apart from that of the male, where, as a *mother*, her parental role then and thereafter will always be uniquely distinguished from his as a father.

As for men, during the period that women were seemingly becoming more "masculinized," many authorities felt or feared that men were being "feminized" by the growing influence of women on their training and thinking, and by the softening effects of modern life. (There also has been the belief that homosexuality is on the increase, although what has occurred may be only a more open expression of the behavior.) However, if men are indeed tending to be less hard, tough and rough, and more aesthetic and sentimental, this may mean only that they have become not more "feminized" but more civilized. And if women are showing more enterprise, initiative, courage and capacity, this also may mean only that they, too, are becoming more civilized rather than more "masculinized." In a new and advanced world the old standards of masculinity and femininity may well be revised wherever they stand in the way of individual fulfillment.

Yet no matter how far the sexes may approach each other, we come back to the basic fact that their inherent biological differences must continue to set them apart in one degree or another in their sex, sexual and reproductive roles. That this biological sex differentiation—stemming from Nature's invention of two kinds of sex chromosomes—has been a highly important factor in human and other animal evolution will be told in our next chapter. It cannot be denied that the sex differences have added interest and zest to life.

In fact, if one places the large X chromosome above the small Y chromosome , one might say that Nature herself has provided the exclamation mark for that notable comment by a French deputy regarding the human male and female, "*Vive la difference* !"

PART IV

Human Evolution and Group Differentiation

CHAPTER FORTY ONE

Evolution

THERE IS ONE QUESTION which overshadows any of the others dealt with in our book:

How did all the different kinds of people and other living things in this world originate?

Whether you start far back before Adam and Eve, or with them—or long after with Noah and the animals in his Ark—the question is essentially the same.

Back in 1859, Charles Darwin set off an explosion throughout the civilized world with the answers embodied in his startling new theories of *evolution*. The resulting avalanche of controversy has never quite stopped. Echoes reverberated during the famous Scopes trial* in Tennessee, in 1925. Years later another uproar over evolutionary doctrines was precipitated by action in the Soviet Union (to be discussed further on).

Originally, the most violent of the reactions to Darwin's theories were

* In this famous test case, a high school biology teacher, John T. Scopes, was arraigned and convicted for having taught the Darwinian theory to his pupils, in violation of a Tennessee state law passed shortly before which prohibited the teaching in public schools of theories considered contrary to the Biblical concept of the creation of man. The trial, which stirred the world, was given special importance because the chief prosecutor was William Jennings Bryan, three-times Democratic candidate for the Presidency, and the chief defense counsel was Clarence S. Darrow, the nation's outstanding criminal lawyer. The conviction of Scopes was later set aside on a technicality, and he was never retried. The furore over the trial, however, discouraged any further attempts to enforce anti-evolution laws in other states, although teachers in some communities are still cautious about discussing Darwinism in classes.

those occasioned, as everyone knows, by his implied assault on the Biblical story of Creation. For Darwin dismissed the idea that man was dropped into the world ready-made. On the contrary, he advanced the belief that there had been a step-by-step development from the most elemental living things through fish, reptiles, lower mammals, up to apes, and then, by some "missing link," to man himself.

Despite continuing dissent, this aspect of the Darwinian theories eventually won acceptance from scientists and has been steadily reinforced by a wealth of evidence to which modern genetics has greatly contributed. But in some conclusions Darwin went along with common errors of his time. One involved the hoary old enigma:

Which came first, the chicken or the egg?

Or, from the standpoint of evolution, and thinking of chickens as symbolical of various species: Did a new kind of chicken arise, which then produced a new kind of egg? Or did a *new kind of egg* originate, which then produced a new kind of chicken?

Darwin held that the chicken came first. However, it was decades before the science of genetics was born, and Darwin—while brilliantly right in other respects—was merely taking for granted what almost every other contemporary scientist did: that the changes or improvements that each generation made in itself or which were made by environment could modify previously existing traits and then be passed on to the next generation. Thus:

> Giraffes had gotten their long necks by stretching higher and higher for choice top leaves on trees, each generation benefiting by the stretching done by their parents.
>
> Apes had developed their brains and muscles by the effort of keeping up with their respective Joneses and had passed on their accomplishments to their offspring.
>
> And conversely, in various species, certain organs had become atrophied or lost through disuse, the classic (supposed) example being that of the fish who swam into dark caves and by staying there generation after generation eventually gave rise to a race of blind fish.*

All this is in line with the theory of the *inheritance of acquired characteristics*. It is a very old theory. Years before Darwin's time it had been widely promulgated by the French biologist, Lamarck, who maintained that the transmission of acquired traits had been an important cause of evolutionary changes. But he could give no good explanation—

* The theory was also applied to other eyeless creatures—crayfish, beetles, crickets and bats—such as are found, together with blind fish, in Mammoth Cave, Kentucky.

WHICH CAME FIRST . . .?

ACCORDING TO DARWIN

1. *First there was a certain kind of bird.*

2. *It laid and hatched eggs,*

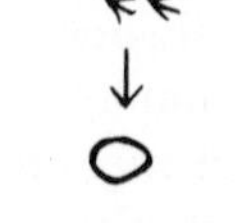

3. *Which produced offspring similar to itself.* But . . .

4. *As they developed, different environments, habits, etc., produced changes in the descendants,*

5. *Which were communicated and passed on* THROUGH THEIR EGGS

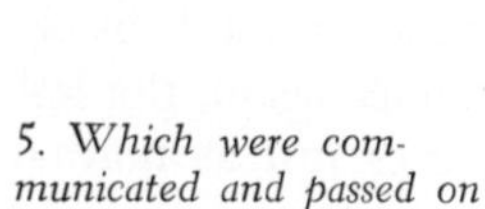

6. *Until, with many such changes in successive generations added together, eventually there resulted* THE CHICKEN,

7. *Which then produced the characteristic* CHICKEN EGG.

So by this reasoning THE CHICKEN CAME FIRST.

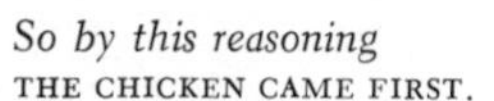

. . . BUT ACCORDING TO MODERN GENETICS

1. *First there was a certain kind of egg,*

2. *Which produced a characteristic kind of bird.*

3. *But in some of these birds something happened to produce* MUTATIONS

4. *Which resulted in their laying eggs with certain* CHANGED GENES,

5. *Which produced birds differing from their parents. And as mutations continued,*

6. *In the course of ages there resulted* A NEW KIND OF EGG WITH NEW GENES,

7. *Which produced* THE CHICKEN.

Thus, as science now indicates, THE EGG CAME FIRST.

and what he suggested was all wrong—as to how a bird who wanted or needed a longer bill and fancy feathers suddenly got them from some evolutionary Santa Claus. While Darwin considered the inheritance of acquired characteristics as of only secondary importance in the evolutionary process (placing his principal stress on "natural selection," as we'll tell later), he nonetheless went along with the then prevailing acquired-characteristic concept. To explain how acquired traits could be transmitted, he offered his gemmule theory—that various cells of the body produce minute units called gemmules, which concentrate in the germ cells and transmit the acquired changes to offspring. This assumption was shattered by later findings; and also, in time, the whole theory of the inheritance of acquired characteristics became discredited among reputable scientists (as noted at the very beginning of this book).

The process of disproof got under way even before Darwin's death (1882) when biologists began asking certain questions. Why hadn't the principle worked with regard to the binding of feet by the Chinese, circumcision by the Jews, tattooing by the savages, and the many other changes people had made in their bodies, through customs or habits, for generation after generation—with *no effect on their offspring?* Skeptical biologists tried some experiments. August Weismann (in 1880) gravely cut off the tails of mice for twenty successive generations, but in the last litter—just as he expected—the mice showed not the slightest shortening of their tails as compared with their ancestors.

The cave-blindness theory was tested by keeping flies in pitch blackness for sixty successive generations; and at the end, once again, the last batch of offspring, when born into the light, had eyes just as normal as ordinary flies.

The theory that acquired changes in thinking and behavior could be passed along by heredity was tested by training successive generations of experimental animals to do certain things or act in certain ways. And yet no effect of this training showed in their offspring (although for a time some experimenters thought it did, by confusing their own improved skill in training the animals, or in unconsciously selecting them, with changes in the capacities of the animals themselves).

The experiments made along these lines run into many hundreds. But when the science of genetics came into being, one could also point to these facts: Recessive genes, paired with dominants, may be carried hidden for generations in bodies of persons with characteristics entirely different from those which the recessives would tend to produce, and yet these genes are never affected. The "blue-eye" gene isn't changed if coupled for generations with a "brown-eye" gene; a recessive gene for idiocy isn't in any sense "brightened up" by being housed for a lifetime in a brilliant person. No "normal" gene of any kind is affected by living in an abnormal body, and no "abnormal" gene is normalized by living in a normal body. This in itself would show that characteristics developed by an individual cannot affect his genes and make them conform to these characteristics. But most important, as we know now, to become hereditary, any changes in the body cells would require automatic pin-pointed changes in the DNA coding of specific genes in the individual's germ cells—which would be almost as impossible as for you, by crossing out and rewriting a phrase in this sentence you are reading, to cause the type to change similarly in the publisher's plates from which this book has been and may continue to be printed.

One other fact should be apparent to every one. During the most important period of human life—from conception to birth—the individual lives in an environment exclusively conditioned by the mother. The only contribution from the father is through the genes brought into the egg by the sperm cell, matching the mother's genes. And yet, despite all the environmental influences that might come from the mother, children tend in no whit more to inherit—or acquire—their mother's general characteristics than their father's. (In a Negro-White mating, for instance, the results are the same whether it is the mother or the father who is Negro.)

In short, it was believed that the theory of the inheritance of acquired characteristics had virtually died with the tragic end of one of its principal exponents, Paul Kammerer, a Viennese zoologist, in 1926. It was in that year that a visiting American scientist, Dr. G. K. Noble, was amazed to find that some of Kammerer's experimental frogs had been doctored and some of his records faked to support the acquired-inheritance theory. Following this exposure, Kammerer went to Soviet Russia and, shortly after, committed suicide.

Why Kammerer, posthumously, then became a hero in the Soviet Union; why the theory of the inheritance of acquired characteristics was fervently revived and exalted there; and why, at the same time, Soviet authorities vetoed the science of modern genetics, is a long and sad story. But it involved not so much science as *politics*. For what the genetics storm in the Russian teapot boiled down to was the attempt to prove that any science which did not fit in with Marxist-Soviet political philosophy would have to be wrong.

Briefly, certain Soviet plant breeders, led by one Trofim Lysenko, claimed to have scored considerable success in growing and improving grains in new environments, and through using grafting and crossbreeding techniques with various plants. (In this work they relied heavily on the methods of the old practical plant breeders, notably the American, Luther Burbank, and the Russian, Michurin—the inspiration of the latter, especially, leading to their calling themselves Michurinists.) But Lysenko went on to claim that the environmental improvements had become *fixed and transmissible through inheritance;* that this constituted disproof of most of the theories of modern genetics; and—what most impressed the Soviet leaders—that his theories *were in complete accord with Marxist political principles,* whereas the theories of the "bourgeois-capitalist, ideological, American-Morgan-Mendelian geneticists" were anti-Marxist and "counter-revolutionary." He also convinced the practical-minded Soviet authorities that he was a man who got results quickly, whereas the classical geneticists were merely poking along with "useless fruit-fly experiments." Lysenko's triumph was climaxed in 1948 when the Soviet Central Committee formally pronounced his theories to be the only "correct" and "acceptable" ones and virtually made him overlord of the Soviet biological sciences.*

Yet the fact remained that Lysenko's theories of acquired-characteristic inheritance were scientifically invalid and, on examination by various authorities, proved to be largely a mass of jumbled fallacies, evasions and misstatements. (One of Lysenko's contentions, that *wheat had been turned into rye* by altering the environment, was considered just about as possible as causing a cat to give birth to puppies.) Most surprising was the failure of Soviet leaders to see the grave implications in the Lysenko thesis as applied to human beings: If the acquired effects of a good environment could be inherited, *the effects of a bad environment also could be inherited.* Were this true, Prof. Hermann Muller pointed out, "individuals or populations which have lived under unfavorable conditions and have therefore been physically or mentally stunted in their development would tend, through inheritance of these acquired characteristics, to pass on to successive generations an even poorer hereditary endowment; on the other hand, those living under favorable conditions would produce progressively better germ cells and

* Even before this a purge of geneticists who had challenged Lysenko's theories had already been under way. His chief opponent, Nicolai I. Vavilov, an outstanding geneticist, was kept from attending the World Genetic Congress at Edinburgh in 1939, which he had been chosen to head, and, after disappearing, was reported to have died in a concentration camp in Siberia a few years later. Several other Soviet scientists who did not recant their "anti-Lysenko heresies" also disappeared, and additional numbers were ousted from their positions. (See also footnote on Soviet twin studies, Chapter 37, page 485.)

so become innately superior. In a word, we should have innate master and subject races and classes, as the Nazis blatantly insisted." This erroneous concept, it should be noted, would hit particularly hard at the great mass of Russian people, who, for many centuries, lived under physical and intellectual conditions markedly inferior to those of most other Europeans. In short, the Lysenkoists had been rejecting one of the most important and socially constructive conclusions of modern geneticists, that peoples who have been continuously exposed to inferior environments are not *ipso facto* inherently less capable. No better proof of this could be offered than the amazing accomplishments of the Russians themselves within a few decades.

With successive changes in the Soviet political and ideological climate, Lysenko's shifts in power and prestige have been reminiscent of the "off again, on again, in again, Finnegan" of American lore. Fortunately, a complete ideological turn against Lysenkoism began toward the end of 1964, following the ouster of Premier Khrushchev. Articles in the leading Soviet newspapers and scientific journals now condemned Lysenko's theories and indicated that Soviet genetics would soon fall in step with modern genetics elsewhere.*

So we come back to this question: If the material for evolution was not provided by environmentally acquired characteristics, whence *did* it come? One way in which it undoubtedly came was through the spontaneous changes in genes and chromosomes which we have referred

* Certain findings which, it was thought, might point to acquired-characteristic inheritance, fall far short of upholding Lysenko's theories. For example, there have been experiments involving non-Mendelian, or "cellular" inheritance—transmission of traits in exceptional instances not through the chromosomes but through other factors in the cell outside the nucleus, such as viruses, plastids, plasmons, and plasmagenes. But any genetic influences of these cellular factors have been shown to be governed by characteristics of the genes which dictate their actions. In short, as Dr. Tracy M. Sonneborn (a leading "plasmagene" experimenter) has stressed, cytoplasmic (nonchromosomal) transmission involves only an addition to established genetic principles and in no sense supports the Lysenko theories. Similarly, there was no support to Lysenkoism in the reported findings by French scientists that when ducklings of the Pekin breed were injected with DNA from the germ cells of ducks of another breed (Khaki Campbell), the ducklings not only developed new traits of coloring different from the genetic characteristics of their breed, but, on maturity, proceeded to transmit the new traits to their progeny as far as the third generation. The reported phenomena, it may be noted, comprised the passing on not of environmentally acquired traits, but of new hereditary material acquired through injection, and which produced both the changed traits in the ducks and, presumably, changes in their germ plasm.

to as mutations. The whole concept of evolution now centers about these mutations in individuals and groups.

As geneticists look back, they see that mutated genes must have been responsible for innumerable new characteristics in all kinds of animals and plants within the last few hundred years. In 1791 a Massachusetts farmer found in his flock a peculiar lamb with very short legs. It offered him a distinct advantage: It couldn't jump the fences and get away. So from this single mutant (of course he didn't think of it as such) he bred the strain of sheep known as Ancon which was so popular for many years and is still extant.

In 1889 a hornless Hereford calf appeared in a Kansas herd and from this have been bred the present polled (hornless) Hereford cattle, valued because they suffer fewer injuries than horned cattle.

Also, in the late eighties, trappers or breeders on Prince Edward Island, Canada, found among red foxes a few whose coats had a silvery shade (black and white hairs, tipped with white) and from these were bred the silver foxes in numerous shadings which became enormously popular. Later "mutation" mink appeared and were bred in various light shades, including "platinum."

Among all sorts of animals, the list of comparatively recent mutations which were utilized in one way or another could be greatly extended. Vastly more numerous are those in the plant kingdom, many of which have given rise to new types of flowers or of highly desirable fruits, vegetables and grains.

But in human beings, also, mutations of various kinds have continued to take place. When discussing *hemophilia,* we noted that the gene for it which Queen Victoria passed along probably arose in one of her parents by mutation and that many of the other cases of hemophilia in each generation also arise in this way. So, too, innumerable other hereditary defects and peculiarities which appear spontaneously with no previous family history to explain them are often due to mutations. In fact, geneticists can even provide fairly close estimates of the specific rates at which mutations for many given conditions occur.*

* As examples, estimated mutation rates per 1 million sperms or ova in individuals from families without a previous history of the given condition are reported as, approximately: albinism, 28; achondroplasia, 40; Huntington's chorea, 5; micropthalmos, 5; retinoblastoma, 15 to 30; juvenile amaurotic idiocy, 38; neurofibromatosis, 100; hemophilia, 20 to 30; and Duchenne type muscular dystrophy, 40 to 60. (In the last two conditions, which are sex-linked, the mutation rates as applied to sperms are for those carrying Xs.) All estimates are as given in the 1962 report of the United Nations Scientific Committee on the Effects of Atomic Radiation.

(Varying mutation rates for designated conditions can be explained by the fact that genes differ greatly in "mutability," depending on their locations on chromosomes, their chemical structure, and other factors.) Nor are the mutations confined to defects and abnormalities. It is certain that mutations in genes concerned with all other human traits—coloring, features and body form, and mental and behavioral characteristics—are also continually arising. Altogether, it is estimated that any given human gene mutates about once in every 30,000 to 50,000 times that it is reproduced, and that at least one in every eight human sperms and ova, on the average, carries one such newly mutated gene. However, the great majority of mutations are of very slight degree, with their effects scarcely discernible and—especially in the case of the complex human traits—almost impossible to identify as due to genetic change.

But what causes gene mutations? The easily recognizable ones—involving changes in surface traits—appear so infrequently under natural conditions that geneticists in the first decades of the science had to wait interminably for a mutation to show itself. Even among the carefully watched Drosophilae, thousands of flies had to be studied for every mutation that was found. Then, in 1927, came the epochal discovery by Professor Muller that *if flies were exposed to X rays, the mutations would occur about 150 times as often.* Immediately, under X-ray bombardment, mutations in the Drosophilae began coming thick and fast. This discovery (which brought Muller the Nobel prize) was followed by findings that, under natural conditions as well, mutations were sometimes induced by radiation in the atmosphere caused by cosmic rays or emanating from certain radioactive rocks (such radiation accounts for about 3 per cent of natural human mutations); and further research showed that mutations could also be induced by certain chemicals and gases. Equally important was the finding that virtually every type of mutation which was or had been produced naturally could be produced artificially; and vice versa. Thus, the laboratory experiments were now providing clues to what actually had been taking place throughout evolutionary history.

How a given gene is caused to mutate was roughly set forth in Chapters 9 and 20. It has also been noted (Chapter 19) that every gene is a segment of the life stuff *DNA*, carrying its own specific coded instructions for certain chemical processes to be carried out. If the gene is hit or shaken up in such a way as to alter the arrangement of its atoms, and hence its DNA code, its workings and effects will be changed accordingly. As we also saw, there are chromosome mutations, in which chromosomes may be altered in their structures, or broken up, recombined and reassembled in new ways. These gene and chromosome changes unquestionably occurred repeatedly among all animals during their evolution, as we shall presently explain.

But this is another important point: *In the great majority of cases, mutations, whether of genes or of whole chromosomes, are harmful in their effects.* Only once in many times does a mutation produce a change which could be considered beneficial to individuals in a given time and environment. Nonetheless, it is mainly through the occasional and adaptable gene changes, eventually of innumerable kinds and in countless numbers, as selected, combined and perpetuated in groups over very extended periods, that the basic process of evolution may now be explained.

Tracing back even further, many scientists believe they can deduce *how life itself began* through changes from inorganic to organic substances. In fact, it is reported that some of the "building blocks" of DNA—certain nucleic acids and *adenine*, one of the four key components of DNA—have already been synthetically produced in laboratories. This by no means implies a challenge to the belief in divine creation. For there is the ultimate question of *how* the marvelous ingredients of life and its environment came to be present. To say that a single bit of living substance, however and whenever it arose, should have been endowed with the potentiality for evolving into all the forms of life on earth certainly implies no lesser miracle than that all living things were created directly without any preliminary steps.

So, hypothetically, in applying the theory of evolution, we might go back to a remote past when the earth's surface held pools of a steaming primordial "soup," containing a large variety of chemical ingredients. Indeed, you might think of it as an "alphabet soup," with the chemicals represented by their technical letters (C, Ca, H, N, Na, P, etc.). As the soup steamed and bubbled, the different chemicals (or "letters") combined to form compounds (or "words") and combinations of compounds (or "sentences"). Finally, after eons of such events, the gap was bridged between the nonliving and the living with the appearance of a bit of substance that in some way could reproduce itself. It may have resembled a segment of DNA—that is, consisting of a few links of adenine and thymine, cytosine and guanine, forming a simple AT, TA, CG, GC . . . coded arrangement. Here, in effect, we had a "gene," able to transmit a coded message. When a number of such genes combined and worked together—forming a chromosome—they were able eventually to surround themselves with a globule of the life stuff protoplasm within which they could operate and have their instructions carried out. The result now was a *living* cell. And at once this cell began to exercise the unique power given it by its genes to reproduce itself endlessly, until there were countless other one-cell organisms.

But all of these first unitary organisms, carrying only replicas of the same few original genes, were identical in their workings and effects. This situation did not long continue. First, changes occurred in the DNA of some of the genes here and there, resulting in varied types of cells. Then, every so often, several newly forming cells would stick and hold together, making a multicellular organism. This organism could work more efficiently and reproduce itself more effectively than the single-celled organisms. Presently, too, as more mutations began changing the DNA codes among the multicelled organisms, important differences began developing among them with respect to their genetic traits and capacities.

Continuing changes gave rise to organisms with still more cells and still more varied genes. As the genes within cells increased in number, the chromosomes in which they were linked together, like necklaces with new beads added, grew longer and longer. Now, twisting and turning as they multiplied, individual chromosomes would break up and re-form into two or more separate chromosomes. In time there would be many different kinds of chromosomes with many different types and arrangements of genes. Moreover, with cell clusters reaching into the hundreds, then thousands, then millions and billions, the same set of genes in different locations of the organism could do many different kinds of work, resulting in many different kinds of specialized cells (as explained in Chapter 9).

So here we had the means whereby, first, an infinite variety of genetic mechanisms, with endless capacities for specialization, could be produced; and second, through these, an infinite variety of living things could have evolved, ranging, in the animal kingdom, from the simple one-celled creatures still existing (such as amoeba and paramecia) to the most complex organisms—fish, birds and mammals, up to man himself—with billions of cells, thousands of genes, and every conceivable type of body construction and functioning.* In certain limited respects you yourself, like every other human being, went through these evolutionary steps: from a one-celled egg at conception—although already far more

* Evidence of the evolutionary affinity of almost all living things—not only animals, but plants—can be found in the fact that their genes and chromosomes, where present, are basically very similar in appearance, chemical makeup and functioning, with the major differences being in the DNA coding and the resultant makeup of the cells. (See also footnote, page 172.)

complex than a one-celled amoeba—to an individual with billions of cells at birth, and with the in-between stages having much in common with the development of embryo fish, frogs and other lower creatures.

But also, as animals grew in complexity and variety, there were several other important "biological inventions" which greatly speeded up evolutionary processes. Among these were the following:

MOBILITY. As animals developed various appendages for moving about, they greatly increased their capacity for seeking out new environments and using the new conditions and materials for evolving into more efficient, complex and varied forms.

SEXUAL REPRODUCTION. With the evolution of sexual mechanisms, including the production of sperm and egg cells, it became possible for two different individuals of a species to pool their genetic resources for transmission to their offspring and to combine in them their genes in many new ways. Where, in self-perpetuating, single individuals, important evolutionary changes by way of mutations might require many generations, in the union of two parents, an offspring with many new features could result immediately merely through a new combination of genes. (Just think how it would be if there were only mothers and no fathers—if all children were replicas of their mothers, with only slight modifications, as compared with the highly varied offspring when there are two parents.)

PARENTAL CARE. With sexual reproduction and the pairing of the sexes came the parental instincts which led animals of higher forms to take care of their young for longer and longer periods, permitting ever slower and more detailed development. Other instincts, such as those for *group behavior* and *cooperation* (discussed in Chapter 36, on behavior), likewise contributed greatly to the evolutionary processes.

ISOLATION (or "reproductive isolation"). Groups of animals who became isolated from others of their kind for very long periods developed special characteristics; and, in time, changes in their chromosomes rendered them infertile with the others, resulting in new species. (Lions, tigers, leopards, jaguars and domestic cats, or the varied species in the horse, ape, bird and other families are examples of segregation and evolution from common stocks.) Where segregated types developed different traits but remained fertile with one another, later crossbreeding produced further new types.

"KEY" AND "LARGE" MUTATIONS. As the biological mechanisms of animals grew more complex, a mutation in a single important key gene could lead to very great bodily changes. In the wayward gene chapters, we saw how one defective gene (singly or in a pair), or some chromosome change, could produce a whole series of defects. It follows just as easily that a single favorable major mutation—for example, in the brain or in the pituitary gland—might have led to some very big constructive change. Environment, too, could play a part in this, for often the potentialities of a gene or genes for doing important things may have been lurking for generations, awaiting the proper conditions. However, by and large, geneticists believe that the beneficial changes evolved in animals not by leaps and bounds (*macroevolution*), but by very small steps (*microevolution*), with the development from one form to another coming through a slow and continuous process.

To accomplish the myriads of evolutionary changes and developments and to account for all the types and species of living things that there are in the world would have required an enormous stretch of time. But that was no problem. The earth is, according to present estimates, from four to five billion years old, and life began well before two billion years ago. If it is borne in mind that what we think of as the long span back to the birth of Christ is less than two thousand years, it should not be hard to see how in *a million times* that span of twenty centuries all of the complex creatures now on earth—including man—could have evolved from the primordial one-celled organisms.

The processes described provided the *genetic raw materials* for the evolution of different species. But these raw materials were not sufficient in themselves. They had to be used properly, and, first, they had to be tested and *selected*. For they weren't ordered in response to the needs of any individuals. Most often, in fact, the contrary was true. Throughout its long history, each species found new materials for its evolution continually dumped on its doorstep, so to speak, the good mixed up with a great deal that was bad, the usable with far more that was bizarre and unusable (somewhat like most of the gifts showered on winners of radio give-away programs!).

In this process of selection there also was the constant requirement of *adapting* to the environment. Hazards or peculiarities of climate, or natural enemies, or limitations of food supply, made it essential for animals of different species to have special kinds of traits and equipment—particular types of fur, feathers or skin, toes or limbs, digestive apparatus, sensory mechanisms—in order to go on living successfully. The required assets didn't come all at once. Only by successive gene mutations and recombinations, here and there, did individuals and then groups appear, each with just a little more of what was needed under existing conditions. For environment, ever changing, posed, as Prof. Theodosius Dobzhansky has termed it, a continuous challenge. The individuals who at a given time were better able to meet the challenge and adapt to their environments were more likely to survive and *reproduce* themselves at higher rates, and thus were selected by circumstances to lead the evolutionary procession, while the less favored ones dropped out of the ranks.

In consequence, then, drawing always from the innumerable mutations perpetuated by the more favored individuals, each species and subspecies tended to develop the special kinds of *adaptive* characteristics which would add to its efficiency and safety under existing conditions. So came the hooves of horses, equipping them to run speedily over plains, and their teeth, adapted for grazing; the camouflage coloration of many birds and the stripes of the zebras and tigers; the antlers of deer, the horns of cattle, and the countless other biological devices and gadgets found in the animal kingdom.

Darwin's brilliant theories of natural selection and adaptation, which we have summarized, have in many respects been greatly strengthened by the findings of modern genetics. But when we go further into his theories of the *struggle for existence* and the *survival of the fittest*—or the way they were interpreted—we can see why certain modifications were called for. For instance, it is clear that without restrictive forces, infinitely more individuals of each species would be produced than there could possibly be room for. So Darwin held that there always was and had to be a constant battle not only between individuals and their environment but among members of each species, to determine which ones would survive and reproduce themselves.* Well aware of the touchy implications, Darwin himself had skirted the application of

* Darwin acknowledged that in formulating these theories he was initially influenced by the earlier doctrines of Thomas Malthus, who, in the famed *Essay on the Principles of Population* (1798), held that population increases in a much higher ratio (geometric) than the means of subsistence (arithmetic) and, therefore, that there must be a constant "struggle for existence in which the weaker perpetually perished." Adopting this view, Darwin (in *On the Origin of Species*, first edition, 1859) went on to conclude that in the competitive struggle the individual with any special inherent advantages would tend to be "naturally selected" and to "propagate its new and modified form." However, in a later work, *The Descent of Man* (1871), Darwin strongly emphasized the principle of *cooperation* as an important element in evolutionary advancement.

these principles to human beings, although there was no question that they might have been considered as applying to individuals and groups of the human species as well as to those of other species. But whatever Darwin may or may not have meant, before long the theories of the "survival of the fittest" were taken up and amplified by contemporary or succeeding intellectual leaders, such as Herbert Spencer. Biological Darwinism then was blended with social Darwinism to promulgate the belief that, in general, survival among human beings had always been determined heretofore—as it would continue to be—by "biological fitness."

But what was and is meant by "fitness"? In ordinary usage, "fitness" refers to the qualifications and capacities—physical and/or mental—for best meeting the challenges of one's life and environment. Darwin and others who followed gave the word the additional meaning that the "fittest" individuals, groups or species would be those most likely to reproduce themselves plentifully and to project relatively more descendants into future populations. But of late many scientists (notably population geneticists) have been using "fitness" in a technical sense only with respect to the reproductive aspect—that is, gauging the fitness of groups not necessarily by their own existing qualities and immediate capacities for survival, but solely by how many of their progeny and descendants survive. On this basis the individuals or groups popularly regarded as most fit and dominant during their lifetime might sometimes prove later to have been less fit than some inferior contemporaries who left relatively more descendants.

The foregoing interpretation of fitness is opposed by a number of authorities, among them Sir Julian Huxley. He holds that to equate fitness merely with reproductive results is "unscientific and misleading" and at variance with what was intended or implied by the phrase "survival of the fittest." True "evolutionary fitness," he maintains, is quite different from "net or differential reproductive advantage" and must carry with it the meaning of improvement achieved through natural selection in relation to the conditions of life. (How "reproductive fitness" might actually be going counter now to "evolutionary fitness"—and/or "social fitness"—in human beings will be discussed in Chapter 46.)

So in talking about the previous evolutionary selective processes in relation to genetic fitness and survival, we always must ask, *Fitness for what?*—and *when,* and *where?* Unquestionably, as we have seen, certain favorable mutations made some individuals exceptionally well-fitted for surviving and gaining ascendancy; and certain defective mutations rendered other individuals unfit under almost any conditions and destined to be weeded out. But there also were many mutations which produced merely minor adverse reactions to some special aspect of the immediate environment—its temperature, or a disease, or an enemy—and which could be overcome if the individual moved to another place (or, in the case of man, if he *changed* the conditions). Often, what in one environment might have been a handicap was in a different environment turned into an advantage, as with a heavy fur coat in a hot climate, and the same fur coat in a cold climate. In fact, with a mere change in the outer environment, many lowly creatures whose genetic potentialities had previously been suppressed could become dominant.*

On the other hand, what made some species or individuals supreme under certain conditions may have made it impossible for them to survive when conditions changed. The *dinosaurs* are the classic examples. Long before man appeared, these and innumerable other species of animals, once highly "fit,"—some the biggest and most powerful that ever lived—had become "unfit" and had been wiped out by changes in their environment. Even within our own time dozens of species of birds and mammals have become extinct. And a great many others, including the once lordly lions, tigers, and bison, exist now largely by the grace of man. Altogether, it is estimated that the animals living today represent only a fraction of 1 per cent of those that have appeared since life began on earth, and that of the tens of thousands of vertebrate species present a hundred million years ago (in the Mesozoic era), little more than a score have left any descendants today.†

* How this type of *selection* and *adaptation* is currently operating within given species is illustrated by the manner in which new strains of bacteria and insects are arising and supplanting others to meet the challenges of various antibiotics and insecticides. Dr. M. Demerec showed some years ago that strains of bacteria present in the colons of human beings and other animals had been killed off by *streptomycin* but that new types of colon bacteria, arising here and there through mutation, which would have died in the previous environment, were able not only to resist streptomycin but to thrive on it. These bacteria could then take over and by rapid propagation quickly create a new population of streptomycin-resistant organisms. A different type of antibiotic had to be devised if the new hardy bacteria were to be eliminated. Similarly, it has been found that in combating flies, mosquitoes and other pests, specific forms of DDT and other exterminating compounds often ceased to be effective when new types of resistant insects arose through mutation, requiring the preparation of new insecticides.

† A few animals, classed as "living fossils," have managed to remain almost exactly as their ancestors were millions or even hundreds of millions of years ago. Among these are the *coelacanth*, the unique, 5-foot fish found on rare occasions in Madagascan waters; the *tuatara* (or "beakhead"), a

Thus, the "survival of the fittest" must be weighed not only in relation to the *inherent* capacities of groups and species to reproduce and perpetuate themselves but in relation to their environments at given times—prescribed kinds of climate, terrain, bodies of water, vegetation, food facilities, the presence or absence of certain other animals, opportunities for expressing and developing their endowments, and so on. This means that the proper environmental stage always had to be set for whatever type of "fitness" was to make its bow and succeed; and even more, that as the stage settings—or environments—changed, groups in the roles of the "fittest" at one time could be ousted and replaced by groups which previously were in lesser roles.

We must therefore stop and consider whether or not the whole evolutionary process was indeed always working according to some plan as a constant movement *upward,* from lower to ever higher forms, with Man as the ultimate goal. For this would imply that from the very inception of the universe, some great Force was directing the production of mutations and environments for the needs of an infinitely far-off future; and that when the most elementary creatures had appeared, back in the dim Proterozoic era, the changes taking place in them were designed for the millions of years hence when they would evolve into all the present living things on earth and, as the crowning achievement, into Man himself. All this might be called the theory of predestination (*orthogenesis*), or *purposive* evolution.

The theory is open to considerable question. Viewed from the end results, it might appear that the evolutionary process has generally pursued an upward plan and that this plan might have in it the elements of purpose. But it need not follow that the specific stages and details fall into the pattern of predestination. In fact, with the knowledge that mutations have apparently always been haphazard, that the vast majority of them are harmful and very few can be regarded as advantageous, we might have to conclude that man was no more foreordained among other animals than Pikes Peak, the Sahara Desert or the Thames River were foreordained among the earth's physical features.

So it may be maintained that man as he now exists was a biological accident—or the fortuitous end result of many coincidences—and, happily, able to survive in the environment in which he found himself. But

---

lizardlike New Zealand reptile; and—familiar to us—our own little opossum and the horseshoe crab. Other animals have changed greatly from their original forms, the horse, for instance, having evolved from a remote fox-sized ancestor.

what if there had been a different environment, if the water had still covered the earth, with only marshes sticking out here and there? Or what if today, through some cosmic cataclysm, there should be a radical change in the earth's surface, or in its temperature, or in the chemical composition of the atmosphere, making it impossible for man and other higher vertebrates to survive, while some lowly creatures might find themselves quite at home? Do we not, in contemplating such a dire possibility, have to revise any lingering notion that the fitness for survival of a group or species is a stamp of fixed excellence and superiority? And do we not have to question whether man is inherently the "fittest" of all animals and predestined to always be the reigning monarch of the earth?

But we may not have to wait for the chance factors of evolution to alter our destinies. Until now man has been unique among all other animals in his power purposely to change his environment. Today we stand on the threshold of an ominous future, in which we also have the power *to change our heredity*.

Earlier in the chapter we spoke of the experiments in inducing mutations in fruit flies through X rays. Almost from the beginning, geneticists foresaw that ordinary X-ray treatment might have similar effects on human germ cells if special precautions were not taken. There was evidence that some persons (among them early radiologists) who had been carelessly exposed to radiation had been rendered sterile for varying periods; and there was a strong presumption that in other cases gene mutations had been induced in the sperm cells of men or the ova of women.

But no one dreamed that within a few decades *the whole world might yet be turned into a vast laboratory in which human beings would willfully subject themselves to mutation experiments.* Yet that is precisely what began to happen on that fateful day in 1945 when—for military reasons considered imperative—atomic bombs were dropped on Japan. Immediately it became apparent to geneticists that the effects of atomic radiation could be far more menacing in the induction of mutations than could ordinary X rays.

True, the immediate *reproductive* casualties among survivors in the bombed areas were no more than high incidences of sterility, miscarriages and the births of many defective babies who, as fetuses at the time the explosions occurred, had been damaged by the radiation. If there was any increase in *genetically* abnormal children, it was barely perceptible, nor has it been much more apparent to date. But this offers small comfort. For scientists have pointed out that any marked and large-scale genetic effects of atomic radiation may not be expected to assert themselves until many decades have elasped.

Specifically, as with mutations in general, most of the human mutations induced by atomic explosions would be minor ones, with the great majority of them *recessive*, or of the multiple-recessive type, in their workings. This means that it might take a number of generations for the mutated genes to be combined in given descendants of the persons exposed to atomic radiation and for the actual hereditary defects to crop out. (In Japan the process of combining the radiation-induced recessive mutations might be speeded up to some extent because of the relatively high incidence of cousin marriages in that country.) In any case, geneticists do not doubt that in the generations to come, descendants of persons in the bombed areas will show a higher than average number of genetic defects and abnormalities of the same types as were discussed in our wayward gene chapters. (It is a mistaken notion, however, that special kinds of "monsters" would be created by atomic mutation.)

But the effects of atomic radiation have not been confined to localities in which bombs have been exploded, including those in which the United States, the Soviet Union and other countries have conducted tests of atomic bombs and weapons. For the radioactive substances directly emanating from the nuclear blasts or from irradiated material stirred up by the explosions—all included under the term "fallout"—can be carried by the wind, air and water for great distances; or they might find their way into fish, animals, fruits and plants, and thence into people. Once in the body, some of these radioactive elements could remain for a considerable time, bombarding the individual's germ cells (as well as causing harmful changes in bodily cells). So apart from the immediate destruction wrought, any wide-scale atomic bombings or testings, wherever conducted, might contaminate the whole earth by radiation and thus affect the genes of mankind for the generations thereafter. This overwhelming genetic importance of radiation, coupled with all the other threats, has made it a matter of worldwide concern. For any conscientious person, it has become imperative to understand precisely what radiation is and what it can do.

First, then, radiation (as we are speaking of it) consists of ultra-small particles that move rapidly through space with a very high amount of energy attached to them—not unlike bullets discharged from a gun. Like ultra-tiny bullets, too, the rays (material particles) can crash or pass through bodies, or lodge in them, with various effects. Radiation causes ionization when it brings about a change in the substance of the object or tissues it hits—*ionization* being the process by which molecules of certain compounds are broken down into electrically charged particles.* If the ionizing radiation enters a person's body, it may strike some chromosomes, weakening or breaking them, or otherwise disrupting their normal workings; and if the person's germ cells—the ovaries of a woman or the testes of a man—are hit directly, mutations might be caused in some of the genes. In addition or alternatively, the radiation may directly injure many cells in the body, producing organic damage or disease (such as skin cancer, leukemia, cataracts and blindness), increased incidences of all of which followed the bombings in Japan. However, it should be stressed here that when cells which already are disrupted—for instance, cancer cells—are deliberately pin pointed as targets of medical X rays, radiation can be a curative and often lifesaving measure.†

It is particularly important to keep in mind that radiation is not merely man-made, or its genetic effects something recent. For *natural radiation* always has existed, coming from cosmic rays entering the atmosphere or emanating from certain types of radioactive soil or rocks (containing uranium, thorium or potassium 40). Radioactive substances, found in the air, water and food, and in various chemicals and other materials in constant use always have been part of the stuff of which our universe is composed. It is thus the new man-made radiation which has created our present problem. For decades (but with the genetic threats not recognized at first) there has been the radiation from medical and dental X-ray machines and fluoroscopes. (A very limited amount of radiation also has come from luminous watch and clock dials, and from high-energy radio and TV transmitters.)‡ But overshadowing all else in the threatened genetic consequences—although not yet in actual

* The main forms of ionizing radiation are the electromagnetic waves set loose by medical or dental X-ray machines; the closely allied gamma rays; and the particles, such as can come from atomic explosions (in addition to the gamma rays they produce) which include *alpha* particles, *beta* particles (high energy electrons), deuterons, neutrons and protons.

† The possible life-shortening effects of radiation are currently still under investigation. Experiments with rats and a few other lower animals have indicated that each dose of radiation may have the general effect of diminishing life expectancy by a certain small amount, but whether and how far this applies to human beings is not yet proved.

‡ There is also the possibility that some of the many chemicals added artificially to foods, although not actually radioactive, may perhaps turn out to produce mutations and thus mimic the effects of radiation. Increasing numbers of organic chemicals known to have such properties have been identified in recent years.

effects—has been the radiation from atomic devices, military and industrial.

To translate the possible genetic threats from all sources into specific terms, radiation dosage is measured by units called roentgens (after Wilhelm C. Roentgen, who discovered X rays in 1895). The more roentgens (r's) a person's *gonads* (ovaries or testes) have received, the more chance there will be of mutations having occurred among the genes transmissible to offspring. But under ordinary conditions to date only a very small part of any person's radiation dosage is apt to have struck at his or her germ cells: perhaps no more than 3 roentgens from natural radiation and only a little more, about 4 roentgens, from medical X rays during the average thirty years from birth (or before) until the completion of the reproductive period. So far, fallout from atomic explosions has actually added only a trifle more to the radiation received by the average person's germ cells—possibly no more than one-tenth to one-third of a roentgen. Nonetheless, while in terms of any given individual the total germ-cell radiation, with its mutating effects, has been small, when all mankind is considered it has been sufficient to cause large numbers of new gene mutations.

We have already noted that almost all mutations (whether arising naturally or through radiation) are likely to be harmful. When they are *dominant* and extreme in their effects, they may kill the embryo at an early stage, or, if it survives, usually will render it so defective that it has only a limited chance of passing the mutated gene along. The real danger comes when the mutated gene is *recessive*, for while in single form it may have no perceptible effect on the individual receiving it, when and if he or she, or a descendant, mates with a person carrying a similarly mutated form of the same gene (either one likewise produced in that particular part of the chromosome by recent radiation, or one which had previously existed), the result may be an offspring with a double dose of the gene, and thus seriously defective.

As matters stand now, it is estimated that every person, however normal, is carrying an average of about eight potentially dangerous recessive mutations. In double doses, four of them would probably be "killer" genes, bringing death before maturity to the offspring receiving them, or otherwise preventing him or her from reproducing; and four would be mutated genes which could produce serious abnormalities or defects. But since these recessives are of numerous kinds, each of limited incidence, the odds are much against any given two parents' carrying the same dangerous gene and a child's receiving a matching pair. On the other hand, if stepped-up radiation greatly increases the incidence of mutated genes of all kinds, not only will every person in the years to come carry more of the harmful recessives, but the chances of two parents' having matching mutated genes will be that much greater. In future generations, then, one could expect a sharp increase in the proportions of genetically defective children. (By the very roughest estimates, an additional 8,000 children are already being born annually the world over with genetic abnormalities and serious defects caused by atomic radiation-induced mutations.) In an all-out nuclear war, the future ranks of *genetically* deformed, mentally handicapped and diseased children entering the world would be swelled enormously.

One might ask, "But isn't it also possible that some of the 'atomic mutations' might be unusually 'good ones,' which could result in superior individuals—perhaps of higher levels than any heretofore existing?" True enough, among the mutations produced one in a hundred might be a good one and one in a thousand of a truly superior kind. But since any of the superior human genetic traits, such as high intelligence or talents of one kind or another, must be due to complex combinations of genes, long before any such good mutated genes could assemble and assert themselves, mankind could be overwhelmed and dragged down by the avalanche of bad mutations.

To lessen the fears of atomic radiation, suggestions have been offered for producing "clean" bombs, with a minimum of fallout; affording protection through fallout shelters; curing "radiation sickness"; and disinfecting radioactive environments. These measures, it is true, could reduce some of the threats of radiation, but nowhere near all of them. Likewise, there is much that is being or can be done to reduce some—but not all—of the genetic threats of medical radiation by using greater care in X-raying children and young adults whose reproductive years lie ahead of them.* And much also can be done to insure safer disposal of the radioactive waste products from atomic energy plants.

It must be emphasized that no matter what precautions are taken,

* The greatest genetic risk in X-raying lies, obviously, in direct radiation of the genital region and the gonads—rather more so in the case of males, whose testes are not covered up as are the ovaries of females. With increasing distance from the genital area—moving up to the abdomen, chest and head—the genetic risks of X-raying diminish greatly, but there still is a chance that some radiation will reach the gonads by scattering, or by traveling through the body. (In head or teeth X rays only a fraction of 1 per cent of the radiation dosage is apt to reach the germ cells.) But how and when to use X rays must be left up to the doctor. On the whole, the benefits of X-raying greatly outweigh the risks, especially in the case of therapeutic X-raying, which involves mostly persons past the reproductive age.

there can be no really "permissible" dose of radiation. The term has been used mainly with reference to the maximum radiation dosage that can be received in a given period by a person's whole body without showing any physical ill effects, particularly in the case of technicians and workers who are in regular contact with radiation devices or atomic-energy operations. (Current standards have set this "permissible" dosage at no more than 3/10 of a roentgen in any single week, but only for individuals over age eighteen, and with further restrictions for young adults to no more than a total of 5 roentgens a year.) But genetically there is no "permissible" dose. For no matter how small the quantity, if it gets to the germ cells, radiation can cause a mutation. And once a gene is hit, a small dose of radiation will not cause any milder mutation than will a massive dose. The only difference is that the larger the dose, the more chance there is that a gene, or several genes, will be mutated. Furthermore, since the genetic effects of radiation are cumulative, radiation exposure spread out over several moderate-sized doses will usually produce as many mutations as will the same amount of radiation delivered in a single large dose.

As for the incalculable toll of death and injury which could be levied by a large-scale atomic war, the die-hard optimist might say, "Well, there'll still be a lot of people left with whom to start afresh." Such an argument might have been tenable in the past, when devastating wars or plagues carried off large portions of the world's population. But this time there would be no "fresh" start—only a worse start. For the germ plasm of the survivors with which to replenish mankind would have been gravely damaged and would no longer be capable of producing human beings even as good as their predecessors. There also would be an almost ineradicable radiation poisoning of the environment, persisting for a very long time; and this, too, could upset the whole economy of nature in many ways not now fully foreseen and potentially just as dangerous for man as the extensive induction of human mutations.

The prospect is a staggering one. The first stages of human evolution were controlled largely by natural selection. This may not always have favored the greater propagation of the genetic traits today considered most desirable (as noted earlier). But, at least, nature and chance governed the results, and the assumption might be that the general trend for a long time was toward the genetic improvement and upward evolution of man. Later, with wars of wholesale extermination, and then with mass killings on solely religious, political or ethnic grounds (as with the genocidal holocausts of the Nazis and others), human beings deliberately took a hand in deciding which of whole groups of people should survive and which should not—regardless of their genetic value—and it began to be more and more dubious whether the quality of our genes was continuing to be improved. And what can we say now about upward evolution when our fumbling hands hold the overwhelming power not only to destroy countless numbers of persons without discrimination but also to warp our genes irremediably?

To quote biologist and paleontologist George G. Simpson: "Man has caused the extinction of numerous other organisms and is probably quite capable of wiping himself out, too. If he has not yet achieved the possibility, he is making rapid progress in that direction. What his future will be depends largely on him. In this matter he cannot place responsibility for rightness or wrongness on God or on nature."

But this, too, is important: We must not look at radiation only in its direst light, for it has a glowingly bright side. To date medical X rays alone have undoubtedly saved vastly more persons from death than have been killed by atomic bombings, and they have cured, eased or prevented far more disease and crippling than have been caused by atomic radiation. In scientific research in many vital areas, radiation has been an invaluable aid. In industry, radiation has made possible countless advances. What the further peaceful uses of atomic energy can accomplish in the years to come beggars our imagination.

Whether radiation is to become an ever greater and more beneficent good genie or a colossal Frankenstein's monster is for man to decide. Until we know the answers, our evolutionary road ahead must remain uncertain.

Yet if we are not sure where we're going as a species, we at least have been learning more and more about where we've been and how we evolved into all the varieties of people there are on earth today. This is what we'll talk about next.

CHAPTER FORTY TWO

The Human Races: I

Origins and Physical Differences

The most meaningful explosions of our time have not been those of atomic bombs: They have been the bursting forth of the energies of vast masses of people which long had been relatively dormant. By the millions whole groups of these people—in Africa, Asia and elsewhere—have emerged within a few decades from backward stages where other human groups had been hundreds and even thousands of years before, into far more advanced states of living, thinking and performance.

All of these have been *cultural* explosions, resulting only from changed environments and new opportunities. Other such explosions have taken place throughout mankind's history, now in one group, now in another, without any necessary relationship to inherent capacities. But also, and much more important, in the early evolutionary history of man as a whole, there were repeated *genetic* explosions which did involve changes in hereditary equipment. Like multistaged rockets, these new gene mechanisms had propelled man higher and higher into the biological stratosphere, ultimately to orbit as a species far above all other animals.

In the animal kingdom man's identification card (and *yours*) reads like this: (1) *vertebrate* (having a spinal column with segments, or vertebrae); (2) *mammal* (the class of animals whose young suckle from mothers' *mammae,* or breasts); (3) *primate* (the order of "first-ranking" animals, which includes monkeys, apes and man); (4) *hominid* (a two-legged primate); (5) *Homo* (the genus of man); and (6) *Homo sapiens* (the "wise" species of man as now existing, differing from extinct "less wise" species which preceded him).

The foregoing classification, as it happens, also roughly represents the steps by which human beings evolved from the most elementary creatures. As noted in our preceding chapter, this process may have required two or more *billions* of years. So it was very late on that vast time scale that our first higher primate ancestor appeared—by present estimates—some thirty million years ago. Only guesses can be made as to what it was and what it looked like. One theory has been that this creature, progenitor of both the ape and human families, was the monkeylike little animal referred to as *Proconsul.* But only a few fragments of Proconsul have been found, which indicate he may have been closer in ancestry to the early apes than to the forebears of man. Also speculative as to its place in man's remote ancestry is a much later creature called *Oreopithecus* (the "mountain ape"), which lived about ten million years ago and fragments of which have been found in soft coal beds in Italy.

In any case, anthropologists deduce that the most remote granddaddy of the primates was distinguished from other mammals chiefly by its slightly more specialized brain, its teeth, and the possession of fingered hands instead of inflexible forepaws. But this animal was still not yet fully an ape, any more than it was a man; it had a very long way to go to develop into either. In other words, whether or not apes as we see them today did come first and men later, as was once thought, both these apes and men evolved from the same primate ancestor—which, if he was an ape, differed very greatly from any we now know. The subtle distinction is that instead of thinking of ourselves as linear descendants of some chimpanzee or gorilla, we may consider these lower animals and ourselves as merely having had some very remote ancestor in common and being, at best or worst, only exceedingly distant relatives.

As to how humans and nonhumans arose from the same original ancestor, the prevalent theory is that there was a long, broad evolutionary primate highway from which there branched off, first, the *prosimians*—animals close to, but not fully related to the simian family of monkeys and apes, and which included tarsiers, lemurs, lorises and tree shrews. Then came the simian primates of two groups, the New World monkeys of South and Central America (such as the spider, squirrel, capuchin, howler monkeys) and also the Old World monkeys of Africa and Asia (macaques, baboons, langurs). Finally, from one forked branch, came the anthropoid apes (gibbons, chimpanzees, gorillas, orangutans) and, at about the same time, man.

Evidence for the foregoing evolutionary relationships is offered, on the one hand, by the many similarities which both apes and men share with monkeys in anatomy, blood components, color vision, "stereoscopic"

THE EVOLUTION OF APES AND MAN

HOMO SAPIENS SAPIENS

Caucasoids Negroids Mongoloids

(and Australoids, not shown here)

All existing human racial groups are of the same species, descended from the same original ancestral group.

vision,* control of facial muscles, menstruation in females, etc.; and on the other hand, by the many distinctive physical traits which are common only to man and the higher apes. Among the latter are broad and shallow chests (whereas monkeys have narrow, deep chests, like other mammals); erect or semi-erect posture; and the absence of tails. Man also is much more like apes than like monkeys in many details of anatomy and bodily proportion, and, also, in blood components. In fact, the hemoglobin (red blood pigment) of gorillas and chimpanzees is almost indistinguishable from man's.

But all of these similarities are overshadowed by the many biological differences between man and both apes and monkeys. First is the difference in the structure and relative size of man's brain—the mechanism chiefly responsible for his leaving all the other primates far behind him in development and achievement. As one example, the brain of a full-sized gorilla (the primate coming closest to man in intelligence) weighs only about a sixth that of the human brain in relation to body size; moreover, in the ape brain the frontal area, center of associative and symbolic thinking, is relatively much smaller than the corresponding area in the human brain. Another major difference is in the skeletal structure—the shape of the pelvis, and the longer, straighter legs—which enables man to walk and run fully erect. In turn, this has freed his arms and hands for innumerable uses not so easily permitted to apes, whose arms and hands also help out in walking. Not least, the hands and fingers of man are more flexible, and the thumbs longer and more useful, making the human hand a more pliable and efficient tool than the hand of the ape.

But precisely how did human beings evolve from their crude primate beginnings? For a long time the search, set in motion by Darwin's theories, centered on finding the "missing link" between man and apes—some fossil which represented the precise point where men and apes branched off and went separate ways. This quest has been abandoned, for it is clear now that instead of any one specific link there must have been a tangle of many links, stretching far back into the past through innumerable species of apes, man apes, ape men, pre-men and true men. Nor, is it believed, was there any direct, step-by-step progression whereby, as one higher form came into sight, a preceding lower form passed into the limbo. Rather, there is much evidence that at given times

* "Stereoscopic" vision is the kind that is possible when both eyes are in the same plane, so that they can focus together and see objects in depth, instead of being separated at an angle, placed on opposite sides of the head and focusing individually, as in most animals.

ape men and men of different types and levels existed in different parts of the world, and often close together.

It is in Africa that anthropologists now set the stage for the first appearance of the creatures who came closest to being men. Why Africa? The answer is that the primates generally, and the precursors of man particularly, were ill-equipped by nature to combat cold; and not until man forms had evolved sufficiently to be able to make clothing, kindle fire and find or provide adequate shelter could they have managed to survive and thrive in the colder regions. In the lush, tropical, food-abundant environment of Africa, however, things were happily different. And so man's story is now seen as really beginning in East and South Africa in the period between two million and one million years ago. Living then, fossil relics have shown, were species of creatures called *Australopithecines* ("Southern apes") which were in between ape and human types. (In view of the previously given estimate that the first primate may have appeared about thirty million years ago, there would have been adequate time in the long interval for these more advanced creatures to have evolved.)

Discovered first in South Africa in the mid-1920's by Dr. Raymond Dart and then by Dr. Robert Broom were fossil remains of individuals apelike in brain size and skull structure, but manlike in teeth and many of their skeletal bones. They were only about 4 feet tall, but walked erect. (A related individual, the Swartkrans man, had a huge, apelike jaw and big teeth, but while some thought he may have been a giant, others noted that he could have had great big teeth and still have been a runt.)

Farther north and east in Africa, a treasure-trove of fossils lay embedded in the Olduvain gorge of Tanganyika, many miles long and at places several hundred feet deep. Here, in exposed layers deep down, fossil hunters Louis S. B. Leakey and his wife found the bones of the being they called *Zinjanthropus* (the "East African" man, also referred to as the "Nutcracker" man, because of his huge molars). An analysis (by the potassium-argon radioactivity-dating process) of the rocks in which the remains were embedded showed them to be close to two million years old. With "Zinj" were found the remains of fantastically big animals—pigs the size of rhinos, sheep 6 feet tall, birds 12 feet tall. But "Zinj" himself was no more than 5 feet tall, despite his big jaw and short legs. He walked erect, though, and most important, he had made simple primitive cutting tools from pieces of quartz. He may also have used some crude weapons to kill animals (but only the infants of the giants, or other small animals). On this basis—that no primate other than man both makes and uses tools—Dr. Leakey felt that "Zinj" could be defined as a man, and was close to the line, if not actually in the line, of human evolution.

(A later find by Dr. Leakey, in Kenya, north of Olduvai, was the remains of *Kenyapithecus,* a creature dating back perhaps 14 million years ago, and apparently midway on the evolutionary scale between Proconsul and "Zinj." Judged by its jaw and tooth structures, this creature, while "emphatically not a man [to quote Dr. Leakey] was heading straight in man's direction." But unlike "Zinj," there was no evidence that he made or used stone tools.)

A long gap in time and space brings us to the next figures discovered on man's evolutionary highway—the beings who lived in eastern Asia between a million and a half-million years ago. Presumably they were descendants or offshoots of the South African men who long before, during a period when the ice had receded, had been able to trek by degrees through northeast Africa and across to east Asia. Named according to the places where their remains were located were the Java man (*Homo erectus erectus*) and the Peking man (*Homo erectus pekinensis*). While closely related, Peking man had a bigger brain than Java man (about 1075 c.c. versus 860 c.c.) which may perhaps account for the fact that crude stone implements were found with him, but not with his distant cousin. However, even Java man's brain was much bigger than any ape's (although far smaller than that of modern man's), and while both had low skulls with little forehead, massive brow ridges and almost no chin, they did have limbs and posture of a modern type and have therefore been definitely classed as *men.*

Now entering the procession, but further and further along, were the early men of Europe—Heidelberg man, Swanscombe man (whose skull was found in England), Florisbad man and Steinheim man.* Whether their forebears also came up from Africa or across from Asia is still to be learned, along with other facts about them.

The mists finally begin to clear with the emergence of the Neanderthal man, who lived from about 120,000 years up to about 40,000 years ago or later. The fact that his were the first of fossil man's remains to be discovered (in the Rhineland Neander Valley in 1856) and that many skeletons of his species subsequently turned up elsewhere in Europe and in Asia and Africa have made him the best known of the fossil men. It

* The Piltdown man, for years included among the fossil notables, was proved in 1952 to have been a hoax—a set of fossil bones so doctored up and buried as to have fooled many authorities into thinking they were 300,000 years old.

is he, also, with whom the traditional "cave man" picture was associated —a stooped, bandy-legged, heavy-browed, flat-headed brute clutching a club in one hand and with the other dragging a female by her hair to his lair. But for so portraying him we owe the Neanderthaler an apology. For, in 1958, two scientists, William Straus and A. J. E. Cave, examined the Neanderthal skeleton on which the original classic description had been based long before. To their surprise, they found that this particular Neanderthaler, a forty-year-old man, had been wracked by arthritis, which had given him his wry neck and warped, stooped, stunted figure. In his youth, it appeared, he may have been almost as tall and as well built as most Europeans. Further, the revised picture of the Neanderthaler shows that while low-browed, he had a cranial capacity of about 1,450 c.c.—equal to modern man's in size, though apparently not in quality. But he could *think*, act and behave in a human way. He made tools of flint, fashioned wooden spears with which he hunted skillfully, used fire, respectfully laid out his dead, and was solicitous of his sick and aged fellows. This latter trait was seen in the fact that the arthritic Neanderthaler previously mentioned had been enabled to survive until middle age; also, by evidence that another Neanderthaler, whose skeleton was found in the Shanidar Cave in Iran in 1957, had been congenitally crippled and one-eyed, but despite this had been kept alive by his kinsmen until age forty to fifty, when he was killed by a falling rock.

As also shown by the widely distributed remains of scores of Neanderthalers found to date, there were various types of these men, living or roving at successive stages over wide areas in Europe, Asia and North Africa. Those in Europe tended toward stockiness, with long trunks and big chests, but short arms and legs—much like many of today's Laplanders, Greenlanders and other Arctic dwellers. This body structure is especially well adapted to tolerate cold, which may explain why Neanderthalers were able to survive so well in regions which formerly were much colder than they now are. The Asiatic Neanderthalers tended to be taller and sparer. Their foreheads were higher, their chins more prominent and their looks as a whole closer to modern man's. Some anthropologists therefore believe that the modern Europeans may have been descended from men resembling the Asiatic Neanderthalers, if not in part from these Neanderthalers themselves. (Dr. Carleton Coon has noted that while no living person is quite like a Neanderthaler in all details, certain aspects of Neanderthal features may be seen among many people today.)

But even while Neanderthalers were still on the scene, the new type of man—*Homo sapiens*—was already being fashioned in the crucible of evolution. Earlier estimates were that he made his appearance only about 50,000 B.C. But many experts now feel that this was much too recent to have permitted the racial diversification which followed, and that long before—perhaps 100,000 B.C. or earlier—either the modern type of Homo sapiens, or one closely resembling him, was in existence. In fact, the theory has been advanced that some early variety of Homo sapiens may even have preceded Neanderthal Man, particularly in Europe. Further it is believed that the Neanderthalers themselves were a race of Homo sapiens. The assumption here is that both types of men had had a common origin, probably in North Africa; and after branching off, the Neanderthalers had evolved in Asia, the other early members of Homo sapiens in Europe. In time—to extend the theory—groups of the two types of men came together in crossroads between the two continents and interbred. A more "progressive" type of Neanderthaler may have resulted, and it is this one which would have moved into Europe.

In any event, the days of the Neanderthaler were numbered, because the new Homo sapiens who had evolved was *smarter* and abler.* At first, he may have been more of a developer than an initiator, for there is reason to believe that it was from the Neanderthalers that this Homo sapiens learned to make tools, weapons, fire and clothing. Nonetheless, once our Homo sapiens began to capitalize on his superior abilities, and the warming up of continents freed his movements, he was able to set out on a march of conquest which swept all other types of men before him. Whether he killed off or absorbed the Neanderthalers and any other species of men who had lingered on, or whether these other men died off through disease or inability to cope with changed environments, we know that before long the Homo sapiens from which we descended had the world to himself.

To go back now a little, precisely where did modern Homo sapiens arise? There is considerable evidence that it was in the region of Mesopotamia. This was then truly one of the most favored habitats of the world. It had the best of climates, an abundance of game, fruits, nuts and foods of every kind. In contrast to other regions where the man creatures battled for existence against great odds, there was every opportunity here for a species to thrive, multiply and develop. It was

* To distinguish the later type of Homo sapiens from his predecessors, anthropologists now sometimes use the term *Homo sapiens sapiens*. But for brevity (and also because of a certain hesitancy about giving double emphasis to the "wiseness" of the modern human species), we will continue in this chapter to use just one "sapiens" after Homo.

indeed a veritable Paradise. So, if you wish, you may think of it as the Garden of Eden, and of the first *new man* that arose in its midst as Adam, progenitor of Homo sapiens.

But whether one thinks of this "Adam" (the Hebrew for "man") as a newly created individual, or as symbolizing the end result of a long process of evolution which gave rise to a new type of human being, the essential fact is the same:

Every person on earth today, civilized or primitive, is descended from the same original human stock and belongs to the same species.

The first important point we made some pages back is that mankind descended not *from* apes, but *with* them, from some remote common ancestor. Oddly enough, those who have cited the Bible to challenge this point are most apt to ignore the Bible in challenging the second point—that *all* human beings had and have the same ancestor. For not only in the references to Adam, but at various other places in both the Old and New Testaments, the common origin and brotherhood of all races of men has been stressed. The scientific evidence to support this is based not only on the biological similarity of all existing human beings in all major respects, but on the fact that wherever tested, people of all races and subgroups have proved fertile with one another. This is not true of apes or monkeys, which include many varied species differing in their genes and chromosomes (some radically so), and all of which are infertile with one another. (The situation is similar among monkeys.)*

In particular, then, one can discard earlier theories that different divisions of men stemmed from different types of apes or of ape men. For to suppose that starting from quite different species, and within a comparatively short evolutionary period, the various existing races of man could have achieved the biological unity they possess—so vastly outweighing any of their differences—is quite inconceivable.

We can therefore safely assume that the history of all modern mankind begins with the same single group of the species Homo sapiens, clustered, as we believe, at the crossroads of Asia, Europe and Africa.

* Chromosome numbers identified for various species of apes and monkeys (to be compared with man's 46) include these: Chimpanzee and gorilla, 48; baboon, 42; gibbon (*hylobates hoolock*), 44; Rhesus monkey, macaque and mandril, 42; black lemur, 44; spider monkey, 34; ringtail monkey, 54; African green or white monkey, 60; langur, 50; guenon 66–72. Further, among these species characteristic differences appear in the shapes and sizes of their chromosomes and in the way in which these are "knotted" (where smaller chromosomes or parts of chromosomes had in times past been joined to form larger chromosomes).

Favored by environment as this group apparently was, it could have multiplied rapidly. Within five or six hundred years, it is not impossible that there could have developed a population of one million. But long before such numbers were reached, dispersal inevitably would have begun. For in those early days very little was required to set people moving. There was no strong sense of fixity in habitat, no dwellings, no cultivated fields, no domesticated cattle, no extensive social or tribal organization. Following the hunt or good weather, or dispersed by quarrels or natural forces, or often simply lost, little bands or families of Homo sapiens broke away and became established in separate units ever farther and farther from their former fellows.

But how, once isolated, did various groups become changed into *races*, markedly distinguished from one another in many biological traits? There are several possible explanations: among them, *random genetic drift, mutations*, and *natural selection.*

Random genetic drift is now known to be one factor in racial differentiation. It is the process whereby a small group, breaking away from a large group, may carry with it an assortment of genes differing on the average from the proportions in the parent population. We can start with the assumption that in the original core group of Homo sapiens there were already many genetic differences among individuals. Varied genes for given traits must have been derived from the different stocks from which Homo sapiens had arisen, and other gene variations would have resulted from mutations. Let us suppose that 10 per cent of the original groups were blue-eyed, 90 per cent dark-eyed; 10 per cent wavy-haired, 90 per cent straight-haired; and that in skin-color variations, there were such and such proportions. Now imagine that a very small band, or a few families, broke away from the main body of people and took up life elsewhere. Quite probably the proportions of those with given traits in this group would not have been the same as in the main group. Instead of one in ten having blue eyes, and one in ten wavy hair, there might have been one in five of each type; and there may have been relatively more lighter-skinned persons than in the parent population. Or, in another small break-away group, there may have been no blue-eyed persons, and no wavy-haired or lighter-skinned ones. Also, if in the parent group there were certain rarer genes present in only one in hundreds or thousands of individuals, the odds would be that many of these genes were lacking in one of the small offshoot groups and that different genes were missing in another small group. Thus, through genetic drift alone, two newly formed isolated groups could start off at once differing in their gene assortments not only from their parent group, but even more, from each other.

Mutations provide a second means by which two isolated groups could become differentiated; and the longer they were apart, the more likely it would be that certain mutations would occur in one group but not in the other, or in different proportions in the two groups.

Adaptation by natural selection would be another major evolutionary influence in creating racial distinctions. Suppose that one group wandered off and remained in a region that was or became much colder than their previous habitat, and another group into a region that was or became much warmer. In the first case there would be individuals—to begin with, or born later—who, for genetic reasons, couldn't adapt to the cold or to other aspects of the new environment and could not survive to reproduce; in the other case, individuals who couldn't adapt easily to hot and glaring sun (for instance, blue-eyed blonds) or to other new environmental conditions might be eliminated. Further, as new mutations arose within each group, the genes that offered advantages in one group might be disadvantageous in the other group, and so, by continuing differences in adaptation and selection, the two isolated groups would become more and more unlike each other genetically.

These, then, are the probable principal means by which isolated small groups of people, growing into larger groups, could have developed into the divisions of mankind referred to as *races* and *subraces*. But precisely what is meant by these terms is a matter of continuing debate among anthropologists and geneticists. Some argue that there are a few major races, and a limited number of subraces. Others maintain that mankind can now be classified into scores of groups each of which could properly be called a "race." At the opposite extreme are those who insist that there are *no* races at all today—only what they call *ethnic groups*—which is met by the counterargument that the word "ethnic" should be applied only to human cultural divisions and not to genetic divisions. And there are still other authorities who reject both "race" and "ethnic" as proper terms for classifying human beings biologically and speak instead of "breeding populations."*

But the foregoing may be chiefly a matter of semantics, and so to

* All of these divergent opinions can be heard expressed among the author's fellow members, or the guest speakers, of the Columbia University Seminar for Genetics and Evolution of Man. For example, Dr. Theodosius Dobzhansky, who holds strongly to the use of "race," may also stress that "major races" and "ethnic groups" are not two distinct categories, but rather, that ethnic groups may diverge and become major races, race being not a frozen unit but a stage in progress. Another colleague, Dr. Ashley Montagu, spurning altogether the term "races" as applied to human divisions, insists that there are only ethnic groups and that even these represent only vaguely defined categories.

avoid confusion so far as possible, we will here use the term "race" in the following sense (acceptable, we believe, to most authorities):

> A race is a large group of people, separated environmentally from other large groups for a long enough time and under sufficiently different conditions to have retained and/or developed an assortment of genetic traits which can distinguish its members, on the average, from those of the other groups.

It is on this basis, then, that we can differentiate among four major races and their geographic origins:

> The *Mongoloid*, or "Yellow-Brown" race, which developed in the region of eastern Asia.
>
> The *Negroid*, or "Black" race, which developed in Africa.
>
> The *Caucasoid*, or "White" race, which developed mainly in Europe and western Asia, and in the far north of Africa, bordering the Mediterranean.
>
> The aboriginal *Australoid* race, the derivation of which is uncertain, but whose members have some traits akin to the Negro, others like the Ainus of Japan.

There may still be doubt as to where and when the branching off of the major racial divisions—and then the subdivisions—began. It was long thought that all the modern human races sprang from one type of Homo sapiens, Cro-Magnon man—which appeared about 40,000 years ago and flourished in Europe for a long time thereafter. Among the reasons for doubting this now is that Cro-Magnon man resembled European Whites much more than other racial types; that Mongoloids, and presumably Negroids, now appear to have dated as far back as he did; and that the species of Homo sapiens may have been in existence for tens of thousands of years before any of these racial types had taken form. Thus, it seems more likely that Cro-Magnon man, while perhaps the (or one) progenitor of the White race, was merely one of the descendants of the same distant Homo-sapiens ancestor from whom, by various evolutionary routes, also came the Mongoloids, Negroids and Australoids.*

* A controversial alternate theory as to the origin of human races has been proposed by anthropologist Carleton S. Coon: that the main racial groups, instead of all deriving *from* a common *Homo sapiens* ancestry, had originally been distinct groups and then had all evolved *into* the *Homo sapiens* species by separate routes at different times. By Dr. Coon's theory, there had arisen long before Homo sapiens (about 700,000 years ago, and probably in Africa) a Man-type, *Homo erectus*, whose descendants, spreading to different regions, evolved under different conditions into the forerunners of five main

But however and whenever the major racial groups came into being, they never remained (if they ever were) intact units, completely confined to areas genetically fenced off from one another. For always the process of isolation and integration—of coalescing into distinctive large groups—was accompanied or followed by another process: the breaking away of small units to form new groups, either independently or by blending with offshoots from other racial stocks.

Among the clearest pathways we can follow is that of the Mongols. First, with regard to the populating of the Americas, it is now certain that at one time there was a wide land bridge between Siberia and what is now Alaska. That is to say, in place of the Bering Strait there was an isthmus, varying from hundreds to a thousand miles or more across, during periods when the water had been sucked up by huge glacial formations. Over this broad pathway connecting the two continents there had at first come hordes of Asiatic animals—mammoths, mastodons, elephants, bisons, horses and innumerable smaller species. Long after them, in pursuit of game, came men who had meanwhile evolved in eastern and northern Asia sufficiently to have become skillful hunters, to have mastered the means of coping with rigorous environments, and to have organized into tribes. Beginning about 30,000 B.C. and continuing to perhaps a few thousand years ago, successive waves of immigration brought in groups of these men from many different Asiatic regions. The one thing they had in common, as evidence now shows, was that all were of *Mongoloid* origin. However, not only were they widely separated by the times at which they came to the Americas, but

---

racial groups: the Caucasoids, from a pre-Neanderthal type, in western Asia and Europe; the Mongoloids, from Peking man, in China; the Australoids, from Java man, in the South Pacific area; and two Negroid groups, (a) *Congoids,* comprising Negroes and Pygmies, and stemming perhaps from Rhodesian man in North Africa and (b) *Capoids,* comprising Bushmen and Hottentots of the South African Cape area, whose origins are unclear. To continue with Dr. Coon's theory, each of these five groups, in turn, by slow evolutionary changes and some mingling with other groups, eventually crossed the bridge to Homo sapiens: The Caucasoids first, possibly 250,000 years ago; the Mongoloids about 100,000 years ago; the Australoids at varying points thereafter; and the Negroids—the Congoids in the forefront—not until about 50,000 years ago, with the Capoids trailing still later. It is the latter part of the Coon theory which has been most strongly disputed, especially by geneticists who question that evolution into the same species could have been achieved five different times by five different groups on separate routes. In any case, Dr. Coon's theory offers no support to certain erroneous "racist" implications read into it. (See footnote in Chapter 43, "The Human Races, II," p. 614.)

they also were in various degrees differentiated in their genetic makeup—more so than the Europeans who came to the Americas hundreds to tens of centuries later.

Thus, by different routes, at different periods and under different climatic conditions, came some Mongoloid groups which fanned out into North America and down into the plains of what are now Canada and the United States, settling in specific areas, to become identified as such and such Indian tribes. Other groups traveled down the West Coast and into what is now Mexico and Central America; and still others went on into South America—the tip of that continent being reached and settled by about 8,000 B.C. Only after that date, many facts indicate, did the Eskimos begin to come to North America, some settling in the Alaskan region, others moving farther and farther across Canada. Among the reasons for regarding the Eskimos as latecomers is that in their genetic traits they are much closer to North Asiatic Mongoloids than are the North American Indians, and in their techniques and equipment they had revealed a much more recent development. But from the Eskimos through the Algonquins, Apaches, Cheyennes, Chippewas, Choctaws, Iroquois, Navahos, Pueblos and other North American Indians, down through the Mayas, Aztecs, Incas, Toltecs and the farthermost Indians of Tierra del Fuego, greatly varied as these were or are in physical types and in ways of living, they all still could be identified as Mongoloids by such prevailing common features as straight, coarse black hair, dark eyes, sparse body hair, skin colors from yellowish-brown to reddish brown, and other traits to be mentioned later.

In Asia, before or while some Mongols were populating the Americas, others had been moving in many directions, but on paths not quite as clear. For additional races also had been on the march into and within Asia, and blending and recombining processes had been going on everywhere.* Whites, Blacks and Australoids joined with Mongols in populating India. Mongoloid stocks, Malayans and perhaps Polynesians combined with Ainus (an apparently archaic partly European group) to form the Japanese. And from Siberia to Asia Minor, from Mongolia through southeast Asia and spread through the islands around and beyond, the mixing of racial groups to form new subgroups went on.

So, too, in Africa, south of the Sahara, while the Negroes were branching off into various distinctive groups, some of the Eurasiatics from the North and East were blending with them. A mixed stream compounded

* While there still is much uncertainty as to details regarding the origins and formations of prehistoric populations, the general statements presented here conform to prevailing authoritative beliefs.

Table IX

The Melting Pots of Europe

If you are of European descent, you may see here the ethnic elements of which your ancestry was compounded. Listed for each country are, in approximate order and dating, the successive groups which went into its composition.

ENGLISH: *Primary native stock, ancient Celts. Beaker Folk, 1500* B.C. *Belgae, 1000* B.C. *Romans (Julius Caesar), 55* B.C., *and 43-400* A.D. *Germanic peoples, Anglo-Saxons, Jutes, 450* A.D. *Romans and Christianity (St. Augustine), 597* A.D. *Danish vikings, 793* A.D., *835-875* A.D. *Normans, 1066* A.D.

IRISH: *Primary natives, Picts. Arrivals before 2000* B.C., *Asturians (from coastal Spain); Campignians, from France or Denmark. Later* B.C., *other Spaniards, Celts from Rhine region, Anglo-Saxons, Scandinavians. Norsemen, 750* A.D. *Anglo-Normans, 1172* A.D.

FRENCH: *Before 2000* B.C., *Iberians from North Africa; Ligurians from Italy. Later* B.C., *Celts (Gauls), Romans, first and second centuries* B.C. *Germanic tribes (Visigoths, Burgundians, Franks), fifth century* A.D. *Huns, sixth and seventh centuries* A.D. *Norsemen, 841* A.D. *English, 1200-1300.*

GERMANS: *Primary, ancient Celts; later, unknown German peoples. Romans, first century,* B.C. *to fifth century* A.D. *Huns, 400* A.D. *Slavic peoples, Franks, Saxons, 500* A.D. *Norsemen, Slavs, Magyars, ninth and tenth centuries. Huguenots, 1650.*

ITALIANS: *Primary, ancient Latins, Ligurians, Iberians. Terranova, Villanova, Sabines, 2000-1000* B.C. *Etruscans and Phoenicians, 800* B.C. *Greeks, 700* B.C. *Celts (Gauls), 400* B.C. *Goths, 450* A.D. *Lombards, 550* A.D. *Normans from France, 1017* A.D. *German Swabians (Fred'k. Barbarossa), 1154. French, 1266.*

SPANISH: *Primary, ancient Basques, Iberians from North Africa, 3000* B.C. *or before, and Celtic tribes. Later* B.C., *Ionian Greeks, Phoenicians, Carthaginians, Romans. Germanic peoples (Suevi, Vandals, Visigoths), 400-500* A.D. *Mohammedans, 711-1212* A.D. *French (Charlemagne), 778* A.D.

RUSSIANS: *Primary,* B.C., *Scythians, Finns, Sarmatians. Germanic Goths, 200* A.D. *Huns, Slavs, 300* A.D. *Turk-Tartaric peoples, 500-750* A.D. *Scandinavians (Varangians), 800* A.D. *Mongols (Tartars), 1237-1450* A.D. *vikings, 1300. Germans, 1700.*

NORWEGIANS AND SWEDISH: *Primary, early Scandinavian and Germanic stocks by way of Denmark, from 10,000* B.C. *Later, Finns, Teutons, 1700* B.C. *Lapps, 100* A.D., *and 900-1000* A.D. *(Swedish forebears, the Svear,* B.C. *to sixth century* A.D., *merging with the Gotar.)*

HUNGARIANS: *Primary, native Celtic tribes,* B.C. *Romans, 2* A.D. *Vandals, 300* A.D. *Germanic Lombards and Goths, 500* A.D. *Huns, Avars, Slavs, 600-800* A.D. *Magyars (Finno-Ugrians), 800-950* A.D. *Germans, French, Italians, 1000* A.D. *and on. Tartars, 1240* A.D. *Turk-Tartaric tribes, 1300-1400.*

of Whites, Yellow-Browns and even some Blacks poured into Europe and gave rise, as human pools collected at various points, to the Nordic, the Alpine and the Mediterranean peoples. These, in turn, mingled with offshoots of each other and of invading hordes from Africa to form still further subdivisions of peoples. (See Table IX, "The Melting Pots of Europe.")

In short, however distinct the broad primary races may have been from one another at one time—the Mongoloids, Caucasoids, Negroids and Australoids—each had also been breaking up into subgroups; and while these offshoots might have been called subraces, there is much doubt as to whether the term can be applied to the many further subdivisions which developed thereafter, within comparatively short periods, as mixtures of different stocks, and which continued (and still continue) their mixing process. For example, although some still speak of different White European "races"—the "Nordic race," the "Alpine race," the "Mediterranean race"—it is apparent that these are groups differentiated from one another as much, if not more, by cultural and environmentally conditioned traits as by biological factors.

Nonetheless, if human beings are not neatly boxed off genetically into clearly separated racial divisions, and if no genetic traits are exclusive to specific racial groups, there are important *average* differences in the way that many particular genes occur or are distributed on a racial or subracial basis. What is further apparent now is that these average genetic differences go well beyond the surface traits formerly regarded as the principal racial distinctions. As researches have progressed, more and more significant differences, among both the main racial groups and the subgroups, have been found in the relative incidences of given blood groups and inherited diseases and defects, and in various aspects of internal anatomy and chemical makeup. We have touched on some of these physical differences among racial groups in previous chapters. Here we will summarize them and add additional details.

Do race differences in *coloring* have any practical significance? It was formerly doubted that they did, but there is growing evidence that many if not most of the surface racial traits do have what is called "adaptive" value—that is, in a given environment a particular trait confers some special benefit on individuals having it and may lessen the chances of survival for individuals without it. For instance, the heavier skin pigmentation of Negroes serves as a natural protection from the damaging ultra-violet rays of the hot sun and adapts them better to living in the tropics, whereas Whites are at a disadvantage (being, for one thing, more likely to develop skin cancer). On the other hand, in colder and cloudier areas heavily pigmented skins tend to block out the sun rays which are essential for vitamin-D production (important in bone growth); and thus, in the temperate and northern regions, light-complected persons are favored and darker-skinned persons are at a disadvantage (with their children more prone to develop rickets). So, too, heavily-pigmented eyes can be a shield against intense sun or against the glare of snow and ice in the Arctic; and heavily pigmented, thicker and more naturally oily hair might be more advantageous both in tropical environments and in Arctic regions.

What the coloring of the first men may have been is not known, but many anthropologists believe that if Africa was where man's evolutionary history began, dark pigmentation must have prevailed. In any case, if variations in coloring were already present in the early stages of human race formation, there would have been a tendency for the more heavily pigmented individuals to remain in or gravitate toward the hotter, tropical regions, and for those with lighter skin, hair and eyes to move off into the temperate regions. As subraces or ethnic groups were formed, this adaptive geographic separation with respect to coloring would have been likely to continue. (Among Caucasians, as anthropologist William W. Howells has noted, the general skin-color predominance still grades from blond around the Baltic to swarthy bordering the Mediterranean, then brunet in Africa and Arabia, and dark brown in India.) However, once racial and subracial groups took root, selection on the basis of coloring might in time have been cancelled out or become subordinate to other factors, as we shall see.

Features, as they differ among races, may also have some degree of adaptive significance. The wider nostrils of Negroes may possibly have evolved because they permitted better breathing in the hot, moist climates, whereas the smaller, narrower nostrils of Whites were better adapted to warming up cold air before it entered the lungs. (Eskimos, it is to be noted, have among the narrowest noses.) Among the Mongoloids the "almond" eyes with narrow slits, deeply set in protective fat-lined lids, as well as the flattish nose and the broad, fat-padded cheeks, are (to quote Dr. Howells) "in every way an ideal mask to protect eyes, noses and sinuses against the bitterly cold weather" which for the most part prevailed where the Mongoloids developed, and which still is encountered by those inhabiting northeast Asia and North America. The many distinctive racial aspects of *teeth* (such as the "shovel-shaped" incisors most marked in some American Indians, the high frequency of "chisel-shaped" incisors in Europeans, and the extremely large teeth of Australian aborigines, or other distinctive differ-

A RACIAL DIFFERENCE IN TEETH

(Upper central incisors)

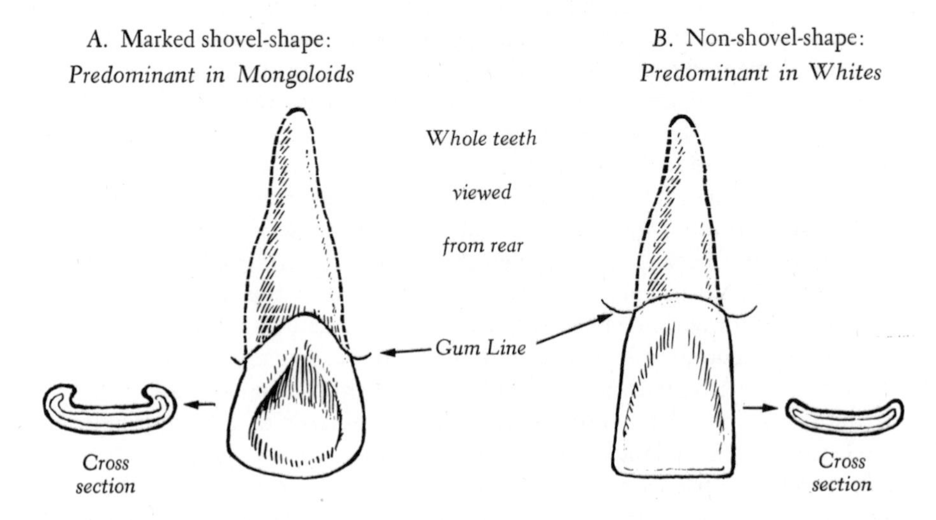

The markedly shovel-shaped incisor (A) occurs in a majority of Mongoloids (up to 100 per cent in some Eskimo groups), but in only about 15 to 25 per cent of Whites. Among many other racial differences, teeth of Negroes generally are larger, with relatively wider inner pulp chambers, than those of Whites; wrinkling of molar surfaces is much greater in Mongoloids than in Whites; and rates and sequences of eruption of teeth differ considerably among racial groups. (Prepared with aid of Dr. Melvin Moss, College of Physicians and Surgeons; Columbia University.)

ences among subgroups of Negroes, Whites and Mongoloids in tooth shapes and sizes) may possibly have evolved originally through particular chewing demands, although conditioned eating habits from childhood on may also contribute to existing differences. The woolly or tightly spiraled *hair* of some Africans, it is believed, may either have special adaptive value in their native habitats, or the gene for it may be involved in producing another trait which has such value. But evolutionary reasons for the thick lips of Negroes, as also for various characteristic Caucasian features, are yet to be found.

Race differences in *body size and form* may well have evolutionary meaning, many authorities believe. Among the theories, as summarized by anthropologist Dr. Stanley Garn, are these: Larger, longer and speedier bodies were advantageous in hunting and bringing in big game and thus may have evolved most pronouncedly among the original hunting peoples of Europe and North Africa. But such bodies require more food and are less efficient in hot weather, so the smaller peoples may have been more favored in the tropics. Similarly, in very cold regions thick-set, fat-padded bodies with short limbs can conserve heat better, whereas thinner bodies and long, thin extremities are more suited to desert and tropical regions. In a very general way, these body-form differences do distinguish the far northern dwellers from those who dwell in the tropicals, the most typical extremes being the squat Eskimos and the 7-foot, spindly-legged Nilotic Negroes. An odd anatomical feature, the heavy, fat rumps (*steatopygia*) of African Bushmen, Hottentots and certain other Negroes, is considered as having had value both as a natural "cooling device" (promoting heat loss) and for storing nourishment during periods of food shortage.

Many of the lesser physical differences among racial and subracial groups—as in head shapes, ear and nose shapes, and hairiness—may well have arisen and been retained without having any evolutionary importance. But where the differences are very marked, and have persisted for very long periods, it is possible that they originally had or still have some adaptive value.*

What is most important is that we should not assume that certain racial features are evidence of a more primitive or animal nature than others. Anthropologists point out that many of the peoples formerly considered most primitive in their looks are farther removed from the elementary primate types than are modern Europeans. The Negro's kinky hair and sparse body hair put him much further from the apes in these respects than the White Europeans who have straight or wavy hair and considerable body hair. The Negro's skin color is also further away from that of the brown ape than is the European's; the full, fleshy lips of the Negro are a highly specialized and advanced human feature, for it is the *thin* lips that characterizes the ape; and even the protuberant buttocks of the Hottentots and various other African Negroes are decidedly human, for it is the flat and skimpy buttocks which characterizes apes (and that is one reason why they can't easily walk erect). To point out, on the other hand, that Europeans may possess certain features less apelike than those of the Negroes would have just as little meaning.

Turning from the outsides to the insides of people, we find that the

* Not even such minor details as the racial variations found in *fingerprints* (the relative incidents of whorls, loops and arches) can be dismissed as without evolutionary importance, for (as noted in Chapter 25), a relationship has been reported between fingerprint patterns and susceptibility to mental illness.

ability to adapt to radical changes in temperature also shows considerable racial variability, although experience and training may contribute to the unusual capacities for "self-weather-conditioning" found among certain groups. For instance, in desert regions, where the temperature shifts rapidly from scorching days to freezing nights, extreme flexibility in metabolism is an essential; and over long periods individuals naturally so endowed would have been selected for survival. Thus, the Australian Pitjandjara tribesmen and desert-dwelling Patagonians and Bushmen can sleep nude or partly nude at very low temperatures, apparently because of an ability to raise their metabolic levels and generate more internal heat. Also, certain Indian tribes living at the tip of South America can sleep nude on the ground through bitter-cold nights because, it seems, they can allow only their skins to drop in temperature and then act as a protective shell for their bodies inside. Again, at very high altitudes a special adaptation of the breathing equipment and the blood is required, and this appears to have developed among certain isolated tribes living high in the Andes. Among Americans, Negroes have been reported as showing greater ability than Whites to adapt to humid (but not dry) heat, but poorer ability to adapt to cold.

With advancing civilization the increased means of adjusting artificially to extremes of climate may have lessened the importance of such inherent differences in adaptability. Nevertheless, some selection is bound to continue. Within the United States there is a continuous movement of people who "can't stand the cold," from northern states to southern states or California; of people who "can't stand the heat," from the south to the north; and of people who can't stand "damp" weather to dry regions. Probably, then, over long enough periods, the populations in areas or countries with marked differences in climate would tend to become genetically differentiated to some extent merely on the basis of climatic selection and adaptation. Other geographic factors could bear on individuals' abilities to get about and to cope with various physical forces, while types of vegetation and food in given regions would have meaning to those with nutritional eccentricities. So, just as in other animals varieties of each species are native to specific regions, it would be surprising if, in the course of time, different geographic areas of the world had also not become populated by human beings differing in some genetic respects from those elsewhere.

Some of the most significant internal distinctions among the racial divisions of mankind—but related only in part to geographic factors—are found in their blood groups and blood elements (discussed in Chapter 28). While the main blood groups, such as O, A, B and AB, and M and N, are present in individuals of all races and racial subdivisions, their relative frequencies vary considerably. Further, certain of the less common blood types may be entirely confined to one race or subrace. Data on the racial aspects of blood-group and blood-type frequencies have been worked out in great detail. We will present no more than a summary of the more pertinent facts.

Of the major blood groups, type O (generally commonest) has an incidence among most Europeans of about 35 to 45 per cent; among Asiatic Mongoloids about 30 per cent; and among African Negroes from 30 to 60 per cent. But it is almost 100 per cent in some of the pure American Indian tribes (a marked exception being the Blackfeet Indians, with only 23 per cent O).

Blood group A has an incidence generally of about 30 to 45 per cent in Europeans, 20 to 25 per cent in African Negroes, and 25 to 30 per cent in Asiatic Mongoloids. Among American Indians the A range is enormous, many South American tribes being entirely lacking in A, while in a few of the northern tribes the incidence reaches up to 70 or 75 per cent. (Even within given European groups the A incidence varies, as in England, where it is considerably higher in the south than in the north of the country.) However, the special "A_2" gene is limited largely to Europeans and Africans (10 to 15 per cent) and is wholly absent in Asiatic Mongoloids.

Blood group B is most common in Asiatic Mongoloids (25 to 35 per cent), is somewhat less common among African Negroes, and ranges among Europeans from about 23 per cent in the Russians to 7 per cent in the English and 5 per cent in the Spaniards (except for the Basques, among whom the incidence is less than 1 per cent). But B is almost entirely absent among American Indians.

The M-N blood types also have markedly different incidences among races, subraces and ethnic groups. Among most Europeans M occurs with only slightly higher frequency than N. But among American Indians, some tribes have up to 75 per cent or more M, with very little N, whereas it is N which greatly predominates among Pacific islanders, including original Hawaiians, Australian aborigines, and Fijians.

Some of the most striking blood-group differences relate to the Rh factor and its subtypes. As noted in Chapter 28, the incidence of Rh-positive is about 85 per cent in Europeans and White Americans, 92 per cent in American Negroes, and 99 per cent or over among most Mongoloids, the remainder in each case being Rh-negative. There are also differences in Rh distribution among subdivisions of each race, including Whites in different countries of Europe, but these are slight except for the uniqueness of the Spanish Basques of the Pyrenees region, who are almost 35 per cent Rh-negative. The subtype R^o is scarce among Europeans but is widespread among African Negroes, and also (though to a reduced extent) among American Negroes. The rare "R^z" gene has its highest frequency among the Mongoloids, is seldom found among Whites, and is apparently not present in Negroes.

Further race differences are found in several other of the most recently identified blood-group factors. Diego-positive is very rare among Whites, but very common among some Indian tribes of Brazil, the Caribbean and Peru. It is also quite frequent in Asiatic Mongoloids. The Duffy-positive blood type appears in a majority of White Europeans and a great many Chinese and Koreans, but in only a small minority of Negroes. On the other hand, the V blood antigen is present in 25 to 30 per cent of African Negroes, but in only a fraction of 1 per cent of the Europeans.

What has been given here is, to repeat, only a partial summary of the racial and ethnic differences in blood elements—all clearly hereditary—which have been found to date (with the likelihood that more will yet be identified). So we may ask again—as with the physical racial traits: How account for these blood differences?

We can assume, first, that by the time human beings began differentiating into racial groups, there already were among them considerably varied "blood-type" genes and that by chance or genetic drift the groups were started off with different assortments of these genes.* We also know now that the blood genes were not neutral in their action, there being mother-child conflicts with respect to the Rh's, the A-B-O types and other blood factors. Thus, dissimilar blood genes would have engaged in an evolutionary contest for survival, and where two racial groups differed in the relative proportions of given genes, the selection of one type over another would have gone on differently. For example, in a group which began with only a small proportion of "Rh-negative" genes, these might have been wholly weeded out in time (as among the Mongoloids); or, in a population preponderantly of the O type,

* Variations in A-B-O and other blood-type frequencies also appear in the lower primates, within each species and among the species. (These blood elements are not identical with those in human beings, but come closer to being so in the higher apes, and are least humanlike in the lower monkeys.) Java monkeys are the only ones (other than man) known to have A, B and O. Chimpanzees lack B, gorillas, orangutans, baboons and gibbons lack O. The rhesus monkey (after which the Rh factor was named) lacks O and A. The M and N antigens, quite similar to those in man, are found in chimpanzees, but the N is absent from other lower primates studied so far.

the A and B might have been eliminated or much reduced in incidence.

Contributing also to blood-type differentiation would be *mutations*, varying in amount and nature over long periods. They could account for some of the rarer blood types found exclusively in given racial groups, especially in smaller racial subdivisions which developed much later and which were so isolated as to prevent unique mutated genes from spreading to other groups. Still another selective influence could have been the possible relationship between given blood types and certain diseases, as between O and gastric or duodenal ulcers, A and stomach cancer (not fully proved), and others. Any such relationship would tend to reduce the frequencies of the respective blood types according to the incidences of the diseases.

By one means or another, then, human populations which started off with some blood-type differences would tend to have developed greater and greater differences the longer they were isolated; and so too, the more chance there was for populations to come together and exchange genes, the more similar they would become in their blood-type frequencies. On this principle it is possible to make much better deductions than formerly with respect to the origins of various subraces and ethnic groups, and the relationships among them. For instance, one mystery was why the Spanish Basques had more than double the incidence of Rh negative (35 per cent) and only a dab of B (under 1 per cent) as compared with other European peoples. In explanation, Dr. Alexander Wiener and others believe that originally the whole of Western Europe had much the same high incidence of Rh negative and low incidence of B as the Basques. Then came the invasion of the Asians—Mongols and Tartars—who were entirely Rh positive and much more B than Europeans. As they introduced their genes through matings, the ratio of Rh-positive to Rh-negative, and of B to A went up among all Europeans except one group isolated from them by the Pyrenees mountains—the Basques. (The Russians, it might be added, who were most directly in the Tartar or Mongol path and were then and since closer to the Asiatics, are higher in B and lower in Rh-negative frequencies than the more western European peoples.)

Similarly, the general relationships in most blood types among American Indians and Eskimos have shown their common Mongoloid origins, but sharp differences in certain blood-type frequencies among various tribes—as also between the Indians and the Eskimos—suggest either descent from initially different stocks, or results of later genetic drift and selection (explained previously). With regard to American Negroes, analyses of blood-type incidences have thrown light on the extent of White admixture among them (to be discussed in the next chapter) as well as the extent of Negro admixture among peoples of the West Indies and of South America. So, also, genetic origins and relationships of various European, Asian and African groups have been made clearer by following the blood-type trails. (An interesting example: The Hungarian gypsies, culturally long isolated, reveal by their blood types that their ancestors were probably immigrants from India who came to Europe in the fifteenth century.)

Hereditary diseases involving blood elements provide the most important evidence of how certain conditions became concentrated in particular racial groups. As noted in Chapter 28, two of these diseases, Cooley's anemia (confined largely to Whites of Mediterranean stock) and *sicklemia* (found largely in African Negroes and their descendants) are recessives which follow the same unusual pattern: A serious disease results only if a person gets two of the genes, whereas a single gene not only is *not* harmful (producing merely a mild blood-cell peculiarity) but is actually advantageous in that it endows the carrier with protection against *malaria*. This, then, could explain why the genes for these two diseases—each usually bringing early death—had managed to survive and even multiply in the racial groups mentioned: If, through mutation, Whites had been started off with the "Cooley" gene, the gene would have been perpetuated among those in the malarial regions around the Mediterranean; but elsewhere, in areas relatively free of malaria, the advantages of the "Cooley" gene would have been negligible, and the fatalities from the double-gene disease would in time have all but eliminated the gene in the stocks carrying it. Likewise, among African Negroes in malarial environments the advantage offered by the single-state "sicklemia" gene must have enabled it to persist. Many American Negroes still carry this gene, just as many American Whites of Mediterranean descent carry the "Cooley" gene, the one disease or the other developing in individuals who carry double genes of either type. But with malaria all but wiped out in the United States, the frequencies of these genes can be expected to diminish steadily; and if malaria is entirely wiped out elsewhere, sicklemia should eventually all but disappear as a Negro racial disease, and Cooley's anemia as a Mediterranean-White disease.

Also described in our blood chapter were the race-linked blood disorders resulting from a red-cell enzyme defect (G-6-P-D deficiency), which, as with the anemias previously discussed, involved an interplay between genetic and environmental factors: *favism* developing among the genetically susceptible Whites (mainly males) when they eat the fava bean popular in Mediterranean countries; and a corresponding blood-cell destruction occurring among the genetically predisposed Negroes when they are given certain antimalarial drugs, notably *primaquin*. While there may be some racial differences in the genes for G-6-P-D sensitivity, there is evidence that for those having this defect, whether Negroes or Whites, the effects either of eating the fava bean or of taking antimalarial drugs are much the same.

Several other hereditary diseases, and a few defects, are quite heavily race-slanted. One odd and serious degenerative disease, *kuru* (previously mentioned in Chapter 22 because of its discrimination against females), is highly prevalent and apparently exclusive in a primitive group in New Guinea. The tendency to the condition seems to be inherited, although some microbial infection may promote its appearance. Another serious degenerative disease, *amyotrophic lateral sclerosis*, is about one hundred times as common in some tribes of Guam and other Marianas islands as it is elsewhere in the world. *Porphyria*, a dominant metabolic disorder, is unusually frequent (in its acute "variegata" form) among White Afrikaners of South Africa, with an incidence of about 1 in 100, whereas among other Whites it is extremely rare. A family-history study of Afrikaner porphyria victims, by Dr. Geoffrey Dean, indicated that the gene for the condition had been introduced by one of the pioneer couples, the Gerrits, who settled in the area of South Africa in 1688, and that from these an estimated 10,000 descendants to date had inherited the gene. The facts as given would strikingly illustrate how, in the early periods of race or subrace formation, the chance appearance of a gene in one individual of an offshoot group could lead to the development of a racial trait as the group multiplied.

But for the most part the genetic racial differences in diseases or defects involve only variations in the frequency of given conditions. For example, *color-blindness* has an incidence among White Americans and Europeans of about 8 per cent; in Eskimos, less than 1/10 of 1 per cent. Selective factors over long previous periods may partly account for these differences, inasmuch as keen color vision could be essential for survival in primitive groups—in some more than others—and, depending on how much of a handicap color-blindness was, the incidence of the condition would diminish as afflicted individuals—and their genes—were weeded out.

Elaborating on the foregoing point, Dr. Richard H. Post has noted that the incidence of vision defects generally (including low grades of myopia) is much less among primitive peoples (various African Negro groups, as well as Eskimos) who depend on hunting and other activities demanding keen vision, than among nonhunting peoples, particularly those civilized groups which for the longest time have not engaged in hunting as a major activity (for instance, Jews, Egyptians, Chinese, and modern European peoples). It is also conjectured that as skilled crafts and activities developed which did not demand distance vision, more nearsighted people would have been enabled to survive; and, further, that as civilization progressed, the persons with color-blindness or other vision defects who made poor hunters and food gatherers turned to the less hazardous occupations of agriculture and domestic industries, and therefore even had an advantage in survival and propagation.

While the common congenital malformations are found in all racial groups, the relative incidences of given conditions may differ considerably. For example, extra fingers (six or more to a hand) occur about seven times as often among Negro babies as among White babies, whereas among Whites there is a considerably higher incidence of mongoloid idiocy, harelip and *spina bifida* (cleft spine). So, too, among Japanese infants congenital dislocation of the hip is far more common than among either Whites or Negroes.

Mental diseases and defects also show varying incidences among racial and ethnic groups, but while genes for these conditions (to the extent they are hereditary) ought to be highly prone to selective processes, it is as yet hard to explain the much greater prevalence or scarcity of this or that condition in one group as compared with another. (To illustrate, schizophrenia is reported as unusually prevalent and manic-depressive insanity as rare in some Swedish groups, whereas the incidences of these diseases are reversed among the Hutterites of North America.) Tay-Sachs disease has been said to occur predominantly among Jews, but it is found with varying incidences in many other peoples, ranging from the Swiss to the Japanese.

The major afflictions, such as heart diseases, cancer, diabetes and kidney ailments, likewise have different frequencies among racial, subracial and ethnic groups. But, whether or not hereditary factors are involved to any extent, in every instance there are complicating environmental influences which could contribute heavily toward producing these differences. For example, in a study of Hawaii's various racial groups, Dr. Walter B. Quisenberry listed these varying incidences of specific types of cancer and possible environmental influences: breast cancer, five times as high among White women as among Japanese (less breast-nursing by Whites?); stomach cancer, highest among Japanese men (diet?); cancer of the prostate, nine times as frequent among White males as among Japanese (sex-habit differences?); and lung cancer, much more common among Whites than among Japanese or Filipinos (cigarette smoking?).* One difference attributed mainly to hereditary factors—skin pigmentation—was the much higher incidence of skin cancer among Whites.

In heart and related diseases, while there are wide differences in incidences among racial groups, it must be noted that some of the biggest differences are found within the White race itself, as between the people of one European country and another, or between any of these people in their native environment and their relatives in the United States. Similarly, existing racial and subracial variations in the relative frequencies of diabetes, kidney disease, ulcers or other organic conditions—as well as in mental diseases and defects—must be weighed in the light of possible environmental causes before hereditary differences can be held responsible.

Nevertheless, since biochemical differences among individuals can account for varying degrees of susceptibility to almost any disease (as was brought out in the Wayward Gene chapters), one cannot rule out the likelihood that some such degrees of genetic susceptibility to certain conditions also exist among racial groups. This becomes more plausible as one realizes that genes for apparently minor surface racial traits may also be participating in important internal processes: for instance, that differences in pigmentation may be not unrelated to other aspects of body chemistry and that the development and shaping of various structural details of the body may actually be by-products or attendant effects of some overall chemical functioning. The blood elements have given us our most direct evidence of chemical differences among racial groups and subgroups and of how these might relate to diseases. Moreover, several studies have indicated that genetic differences also exist between racial groups in specific types of metabolic activity: for instance, as between Chinese and Whites in amino-acid metabolism (*beta-amino isobutyric acid* [BAIB] excretion) and between Negroes and Whites in serum albuminin values. *Testing* differences with respect to PTC among racial and ethnic groups (as noted in Chapter 36) may also involve body-chemical reactions or responses. In all, as Drs. H. Eldon Sutton and Philip J. Clark have suggested, there may be different "racial chemical factories" whose products may resemble one another and yet be distinctive. How the differential outputs of these "factories" might affect

* In other studies, cancer of the cervix has been reported as much less frequent among Jewish than non-Jewish women and, as with the much below average incidence of cancer of the prostate in Jewish men, is thought to be due in considerable measure to some presumed protective effects of circumcision in the males.

not merely disease incidences, but other functional traits, is only beginning to be explored.

Two *reproductive* phenomena in which there are striking racial differences were discussed in earlier chapters. One relates to *twinning*. As was noted in Chapter 18, marked differences in the incidence of twinning—mainly of fraternal, or two-egg twins—occur among the three main racial groups: Negroes (highest), Mongoloids (lowest), Caucasoids (in between). These differences appear to be largely genetic. A second reproductive difference, discussed in Chapter 7, is in the *sex ratio* at birth—the ratio of boys born being consistently higher among Whites (about 106:100 girls) than among Negroes (103:100), at least in the United States. While the reasons remain unexplained, the fact that this racial difference has persisted despite great environmental changes would suggest that genetic factors are involved to a considerable degree.

Still another race difference which gives evidence of being genetically influenced relates to birth weights. Negro infants, on the average, weigh significantly less at birth than White infants, with the percentage of those classed as premature (on the arbitrary basis of weights under 5½ pounds, or under 2,500 grams) being about 14 per cent among Negro infants—twice the incidence among Whites. While the difference may perhaps be ascribed in some measure to inferior prenatal conditions for Negroes, it has been shown that even among Negro mothers of the more advanced groups who receive the same hospital care as White women, the incidence of babies listed as premature remains markedly higher than the White rate. A question raised is whether low birth weights have the same significance in the two racial groups, for some studies have indicated that, weight-for-weight, Negro infants in their first weeks appear to be more advanced in development and behavior than White infants. Thus, investigators at the Leopoldville Maternity Hospital, Congo, Africa, have concluded that to take a birth weight of under 2,500 grams as indicating "prematurity" does not seem warranted so far as Africans are concerned, and that a birth weight of 2,250 grams for their babies would be a more realistic standard.

To conclude this chapter, it must be recognized that no matter which human groups are compared, so long as there were initial differences in the stocks from which they derived—genetically, geographically, culturally—one can expect some differences in their biological traits. These differences might be smaller or greater, depending on how long the ancestral stocks had been separated and to what extent their environments had favored the selection of some genes over others. But one must always bear in mind these facts: that whatever the *average* genetic differences may be among races, they are minor compared to the great similarities which all races share as members of the same human species; that with respect to given traits, there is much overlapping among all groups; and that within any racial or ethnic group there are many individuals who, in certain respects, are much more like persons of another racial group than they are like most members of their own race. Many Hindus, of the White race, have darker skins than many Negroes; many Englishmen are darker and swarthier than many Italians; there are blond, fair-skinned Spaniards and Arabs, and red-headed Egyptians; and there are many very tall Japanese and very short and stocky Scandinavians.

Yet the fact remains that when combinations of traits are taken together or when we look at peoples in the mass, it becomes fairly easy to classify human beings as of one racial group or another. And when more and more specific traits are identified as occurring in different proportions in various groups, the chances of assigning individuals to even smaller subdivisions of mankind increase correspondingly.

Which leads to our most vital questions: What *genetic* importance is there in being a member of any particular racial or ethnic group? And how might this bear on one's chances of inheriting specific mental and behavioral qualities? Such answers as there are will be given in our next chapter.

CHAPTER FORTY THREE

The Human Races: II

Cultural Evolution and Differentiation

MANKIND'S TURBULENT HISTORY has no darker and bloodier pages than those which record the mass killings, persecutions and repressions of some peoples by others in the name of *Race:* acts motivated or justified by the belief that there are genetically "superior" races endowed with the right to dominate or even exterminate those they designate as "inferior" ones.

Because these acts of man's inhumanity to man have reached extremes in modern times, and still continue, nothing in human relationships demands more careful scrutiny than the theories underlying the whole concept of racial "superiority" and "inferiority."

We have seen that among races and racial subdivisions there are a great many *physical* differences which are indeed hereditary; also, that in given environments this or that physical trait of one racial group might be more advantageous than another. But the theories of racial superiority do not relate to physical differences (for, in fact, some groups regarded as "inferior" could be considered physically "superior"). They refer to intelligence, character and capacities for achievement. Granted, then, that marked differences in ways of thinking and behavior and in degrees of civilization do exist among human groups, this is the question: Are the mental and cultural racial differences, like differences in physical traits, also in large measure *genetic?*

In previous chapters on intelligence, talents, behavior, and personality, we pointed out how difficult it is to link inheritance with complex psychological traits in *individuals* when we can't be sure of the role played by their environments. This applies in even greater measure to whole groups of people. Not until we dig deep below all the layers of environmental influences, often piled up through centuries or tens of centuries of living under distinctive conditions, can it be certain that what is being measured in comparing diverse peoples is the nature and quality of their genes. Thus, to evaluate the existing cultural differences among human racial groups properly, we must go back a very long way.

The preceding chapter traced man's biological evolution, from ape man to human, and then the development within the single species Homo sapiens of all the races and racial subdivisions which are found today. Now we will look into man's *cultural evolution*, first as it carried the entire species from the rudest beginnings of mental performance and social behavior to successively higher stages, and second, as this advancement proceeded differently in various groups.

A most important new point is this: Where formerly it was thought that man's cultural evolution came only after he had been equipped by biological evolution with a superior mental and physical mechanism, authorities now hold that both evolutionary processes proceeded together, step by step. In other words, it no longer is believed that Homo sapiens sprang into existence with a high-quality brain, upright posture, efficient hands, the gift of speech, and other qualities which then enabled him to do marvelous new things, but that these endowments had evolved by very slow stages only as his predecessors—from ape men through various fossil men—were impelled to and/or learned to do new things. Better equipment might have led to better performance; but better performance also led to better equipment, an advance in one part of the equipment often promoting advances in other parts.

To illustrate, there were the interacting effects of the human legs, hands and brain. What chiefly distinguished the ape man of close to two million years ago from the ape was not only a somewhat greater thinking capacity, but the fact that he could *walk erect*, though perhaps still awkwardly, for his legs were short and bowed. Thus, with his hands freed—not being needed to assist in walking, as were the ape's—he could grasp and use rocks as tools in a crude way. The ape man with slightly better brains could do more with these elementary stone tools and therefore had a better chance of surviving and reproducing. But also, those with slightly longer legs and a more upright posture could use their hands better, and they, too, would leave more survivors. So posture continued to interact with hands, hands with brains, brains with hands; legs grew longer and the pelvis more adapted to upright posture; hands became freer and more flexible, the brain more efficient in directing hands, and so on.

Another interaction was in the development of the human jaws, teeth and mouth. For all lower animals, these parts of the body had to be used as tools—in killing, tearing, breaking, carrying. Once hands became available as tools for these tasks, the jaws, teeth, mouth and facial muscles could develop for other uses; and in man this led to the perfecting of the vocal mechanism, then to speech, language and facial gestures, making possible communication among human individuals far beyond the range in any lower animal.

These changes were accompanied by successively increased powers to fashion and/or utilize artificial tools. First there were only the "pebble" tools—smoothed or jagged stones found along river beds or on shores which could be used as weapons or for pounding, digging or scraping. Then came chipped and fashioned stones, including the hand ax in various forms adapted for different purposes. And then, by stages of many thousands of years—and perhaps only with the arrival of the much-advanced Neanderthal man—came the shaping of sturdy wooden spears and the fixing to them of sharpened stone points; and with Homo sapiens came sharp flint blades and other hard stone tools and the carving with these of various implements, including needles, from bones, horns, antlers and tusks.*

By this time, too, man had learned to use and make *fire*, perhaps first found when set by lightning or by brush fires, or coming from volcanoes, or produced accidentally while rubbing stones and sticks together. With fire came cooking and warmth in caves which, supplemented by clothing, made it possible for men to live in colder regions. Other inventions, such as rafts, boats and canoes, were enabling men to fare out on the water; and nets and harpoons brought them large fish and sea animals to add to the game which was being hunted more and more effectively as bows and arrows were added to the growing list of weapons and equipment.

In one other way man had been progressing: in *social organization*. He had started with the nuclear unit of a family—father, mother, children. This, too, is considered as having been both the cause and effect of biological evolution. Specifically, to paraphrase anthropologist Weston La Barre, the human family arose from the male's sexual drive

* How difficult it was or is to produce these tools and implements becomes apparent when modern men try to imitate the techniques or when they see how primitives still using Stone-Age tools fashion them. Chipping flints or other very hard, grained stones into specific and detailed forms is a painstaking process much like that used in diamond cutting—finding the grain, splitting and chipping off so much here and there, finishing, and polishing, and so on. Without benefit of modern tools and equipment, the task of the Stone-Age toolmaker required the utmost skill and patience.

THE EVOLUTION OF MAN AND HIS TOOLS

PERIOD (Approx. Span of Yrs. Ago)	SPECIES	TYPE OF TOOLS	BIOLOGICAL TRAITS
1,750,000 to 1,000,000	MAN-APES Zinjanthropus, Small Man-Ape	PEBBLE TOOLS; STICKS ? BONES ? USED AS TOOLS	*Very small brain, but hands close to human. Posture approaching erect.*
600,000 to 400,000	PEKING MAN, JAVA MAN Sinanthropus, Pithecanthropus	CRUDELY-FASHIONED STONE, WOOD, BONE TOOLS	*Increasing brain size. Hands more human. Posture probably erect.*
250,000 to 75,000	EARLY EUROPEAN Swanscombe, Steinheim, Fontechevade, etc.	STONE KNIFE; STONE SCRAPER; CHOPPER; HAND AXE; FIRE	*Brain approx. 1,300 cc. Physique, hands, posture approaching later man's.*
120,000 to 40,000(?)	NEANDERTHAL MAN	FLINT BLADE KNIFE; CHISEL (BURIN); CHIPPED HAND AXE; WOOD SPEAR	*Brain in size at least equal to modern man's. Hands, physique close to modern.*
75,000(?) to 30,000	PRE-ADVANCED MAN Homo Sapiens	NEEDLES; STONE-TIPPED SPEAR; BONE HARPOON; STONE CHISEL; BONE SAW; SHARP FLINT KNIFE; STONE AXE	*Advanced human development in all genetic aspects.*
30,000 to 10,000 (and later)	ADVANCED MAN Homo Sapiens Sapiens	STONE SCYTHE; BOW AND ARROW; IRON INSTRUMENTS; THE WHEEL	*Fully developed human genetic traits of all kinds like those of man today. Changes from here on preponderantly environmental.*

and his desire to "go steady" with the female: from the interrelated facts that the female became sexually receptive to the male, and he sexually interested in her, continuously, on a year round basis, and not merely seasonally as in subprimate animals; that the male's need for the female extended to his remaining with and accepting her offspring, and protecting them together; that, as the trend proceeded, longer female pregnancies and more prolonged child development and dependency led to and required still more family unity and also caused the sexes to become more differentiated in their makeup and activities—the female tending more toward motherhood and domesticity, the male more toward being the strong, resourceful and protective *pater familias.* Following the single family unit, the logical next step was for an organized group of several or many families to band together, thus furthering both biological and social human evolution.

Yet all this time—up to 10,000 years ago or even later—all mankind was still in the *Stone Age,* so called because there were no tools or weapons other than those fashioned of stones or with their help. Moreover, without the equipment or knowledge for doing anything else, no human beings anywhere had progressed beyond the *hunting* and *gathering stage,* in which they were dependent for their food solely on what they could kill in the hunt or gather in their immediate surroundings. As the game was always on the move, and as food supplies changed with the seasons, men, too, were on the move; and since even a small band of people would have to range over a wide area for their food, no large population could gather and stay together for long in any one place. Man, up to this point, was at the mercy of his environment. He could not control it; it controlled him.

This situation began to change in one spot: the region of Mesopotamia in southwestern Asia known as the Fertile Crescent—"fertile" because of its natural richness, "crescent" because it curved up from what are now Israel and Jordan through Lebanon, Syria, Turkey, and down through Iran to the Persian Gulf. At some time between 8,000 and 7,000 B.C., bands of people, realizing that here was truly a "Garden of Eden," with an unusual abundance of food as well as an equable climate, stopped roving and began to settle down. Staying put, they could look closer at their environment and see its possibilities. They noted how food plants grew—wild grains, vegetables, fruits—and began cultivating them. They also found that certain animals there, such as sheep, goats, cattle and pigs, could be easily domesticated.* Thus came

* Many authorities believe that long before food animals were domesticated, *dogs,* because of their unusual natural affinity for people, had be-

the *agricultural revolution* and from it, rapidly, one civilizing advance after another.

People now moved out of caves and into dwellings—tents of goatskins or hides, or structures of stone and mud—clustered together into little settlements. Irrigation was devised to insure regular water supplies for the increasingly diversified crops. With steady food supplies more easily assured, time was left for other activities. Specialized occupations and crafts—basket and textile weaving, pottery making—developed, and one improved technique followed another. The smelting of metals—copper, tin, gold, silver—and their uses were discovered, and the fusion of copper and tin produced *bronze*. Now tools and weapons of stone or wood could be supplemented or replaced by far more malleable, durable and efficient ones of metal, and the Stone Age gave way to the *Bronze Age* (to be followed much later by the *Iron Age*). The *Wheel*, invented probably long before and in various places independently, could here be used in many new ways; and when domesticated horses were harnessed to sturdy wheeled vehicles, much more rapid land transportation became possible, which, with new and larger types of water craft also in production, could carry men far beyond their home territories for trade or conquest. Tribal settlements had grown into communities, and these in turn into towns, cities, domains, nations, with complex political, religious and military organizations. Man was at last fully on the march.

From the Fertile Crescent the sparks of civilization spread and ignited human energies in all directions. Cultural explosions were touched off during succeeding centuries throughout the surrounding Mediterranean area, in Egypt, Crete, Greece, Rome and many other places—ultimately in all parts of Europe—and eastward to India and China. Spoken languages became written languages as well, and intricate communication extended the transmission of ideas not only between existing groups of people living far apart, but from one generation to another. The advancement of any given group in the civilized world, therefore, was never solely of its own making, but at every stage was the end result of a series of chain reactions in which many other groups of people had participated. This suggests at least one answer to a familiar question:

Why did some racial groups lag so far behind others—many (notably Negroes, North American Indians, Eskimos and Australoids)—remaining at presumably Stone-Age levels up to recent times, and some into

---

come domestic pets and/or hunting companions of Stone-Age men in many places and that the success and experience with dogs eventually led to the raising of other animals.

the present? Before suspecting inherent failings as the causes, let us remember that similar lags had occurred in all other human groups regardless of their mental capacities. We know that fifty thousand or more years ago, human beings as a whole had the same mental capacities as we have today, and yet all of them, whatever their race, remained at much the same stages of Stone-Age backwardness for tens of thousands of years; that, mainly by chance, civilization began to take root and sprout in a particular group in a particular place; and that—again by the chance of being in the direct paths of advancement—this or that other group of people was drawn into the forward movement. We know, too, that tens of centuries after civilization was flourishing elsewhere—until 2,000 years ago or later—ancestors of the French, British, Germans, Scandinavians and Russians were in most respects still at Stone-Age levels perhaps not as high as those of various primitive groups surviving today.

Thus, the first reason one can give for the relative backwardness of some racial groups is that *these peoples were never in the original pathways of advancing civilization.* All had been cut off by geographic barriers: the Negroes of Africa by the Sahara and by walls of impenetrable jungles; the American Indians and Eskimos by the rising oceans which had flooded the Bering land bridge to Asia; and the primitive Australoids of the South Pacific, also, by water barriers, as well as jungles, which left them isolated in their habitats until recent times. Far removed from where the sparks that were igniting other people could reach them, and denied access to the stores of knowledge and inventions building up elsewhere, these peoples had to make do only with what their ancestors had brought along, and what they could devise thereafter through their own efforts. But their improvements were severely limited by these additional factors:

> The environments of most of the remaining primitives were unfavorable, in climate, terrain, or natural resources, for the initiation and development of large-scale cultural advances. (We are here reminded of the Arctic explorer's reply when a well-meaning but not too bright old lady expressed surprise that the Eskimos didn't drink milk. "Madam," said the explorer, "have you ever tried to milk a seal?" There are many things, important to the development of a culture, which one cannot "milk" out of an environment where they do not exist.)*

* Although this anecdote appeared in the author's first book, *You and Heredity*, in 1939, he had read of no actual attempt to milk a seal until an article in *Life* magazine, March 27, 1964, made clear how really difficult

The stage of advancement of an offshoot's parental group when it broke away and the amount of "know-how" an isolated group had to start out with in a new environment may have had much to do with the degree of its further progress. To state this another way, some primitive groups came from less educated backgrounds, or "left school" much earlier than others. For example, among African Negroes and among the Australoid peoples, it is likely that the relative stages of advancement in the past were much influenced by when and where different groups had been isolated and by what their environments and opportunities then had been. As for Whites, if no group among them remained primitive beyond one thousand five hundred or so years ago, it may well be because, after that, with all Whites much more compactly settled than those of other primary races, no group was isolated from the others or deprived of the chance to share in the general advancement. "Compulsory education," in fact, became the rule among the White groups.

(Regarding the various Mongoloid groups in the Americas, there is still mystery as to why some were or became so much more advanced culturally than others: for instance, the Mayas, Incas and Aztecs, compared with the North American Indians and the Eskimos. There are several theories. One has been that these groups—or their forebears—were already culturally dissimilar when they came from Asia over the Bering land bridge at successive and widely separated times, and thus were started off on different levels of achievement from which they went on to diverge still more in relation to their environments and opportunities. A second theory—favored now by many anthropologists—is that the forebears of the American Mongoloid groups had all come from Asiatic stocks which had been culturally much alike, and that the subsequent cultural differences among their descendants were produced environmentally. Further, in much the same way that Asian and European civilizations had a common nucleus in one group and region [i.e., Mesopotamia], it is thought that the major cultural advances among American Indians may have stemmed from some nuclear agricultural settlement in the Mexican highlands—possibly as early as 7,000 to 5,000 B.C.—and then spread to other groups in Central and South America. [It is worth noting that the North American Indian tribes which advanced most had been those closest to Mexico, whereas the least advanced were the

---

the procedure could be. The article told how a 1,300-pound female elephant seal on Guadalupe Island (off the Mexican coast) had to be sneaked up on from behind and jabbed in the tail with a hypodermic injection to put her under sedation before she could be milked. (The milk was wanted for scientific study.)

ones farthest off in the north and northeast.] Another theory is that the Central and/or South American Indians may have benefitted by renewed later contact at some point after 2,000 B.C. with Asiatics—Japanese or Chinese—over Pacific sea routes. In any case, it is not believed that the much higher levels of civilization reached by the Aztecs, Incas and Mayas [political, technological, artistic, linguistic, military] were due to any genetic superiority over other American Mongoloid groups.)

Extensive agriculture and animal domestication—the starting points of civilization—were extremely difficult to carry out in many places, whether tropical jungles, deserts or Arctic areas. The plants and fruits available to many primitives did not lend themselves easily to cultivation, and their food animals, such as the buffaloes, pronghorns, wild goats and deer of the Americas, or the wild cattle species and most other animals of Africa, were untameable.

While for many primitives the environments were so poor as to prevent them from doing much else than hunt and gather food, for other primitives the opposite may have been true: Their environments offered them all the essentials of comfortable living, and there seemed little incentive to devise anything new.

Conservatism—a reluctance to change, promoted by religions, customs and traditions ("what was good enough for our revered ancestors is good enough for us," an attitude not uncommon among some modern peoples)—further impeded progress in many if not most primitive groups. Tribal rites and routines, ceremonials, hunts and wars, often provided enough diversified excitement and activity to repress the desire for change.

Hostility, fear and taboos kept many primitive groups from contact with other groups, greatly limiting opportunities for advancing through the interchange of products, ideas and inventions.*

* A questionable new reason for modern racial differences in advances has been found by some in the controversial Coon theory (discussed in a footnote in the preceding chapter, p. 588) that the different major racial groups evolved into Homo sapiens at different times—the Negroids, specifically, much later than the Caucasoids or Mongoloids. Even assuming this theory to be correct (which most authorities doubt), it would not change the conclusions we have offered: that the differences in achievement which developed among human races after they had all evolved into the same Homo sapiens species, and which are now found among the existing major racial groups, may be attributed overwhelmingly to environmental rather than to genetic factors. Dr. Coon himself has stressed the point that Caucasoids and Mongoloids rose to their "positions of cultural dominance . . . because their ancestors occupied the most favorable of the earth's zoological regions. . . . Any other subspecies that had evolved in these regions would probably have been just as successful." (*The Origin of Races*, by Dr. Carleton S. Coon, Knopf, 1962, p. 663.)

But: Is it not possible that, apart from the environmental disadvantages mentioned, there were inherent mental differences which kept or still keep some racial groups from making the same advances as others? *Possible,* yes; *proved,* no. In our IQ chapter, we stressed the fact that, while long separated human groups may differ in *ways* of thinking, there is no evidence that they differ in relative *amounts* of intelligence. We also noted that no intelligence tests currently available have been able to measure fairly or accurately the IQs of groups living and conditioned under different circumstances, and that this holds for Negroes and other racial groups (whether in the United States or elsewhere) in comparison with Whites. What must always be kept in mind is that different peoples must train and exercise their intelligence in different ways to meet different needs. The person rated as "smart" among American intellectuals might easily be rated stupid if he tried to do many tasks and meet many situations which are commonplace among less advanced groups.

Anthropologists and others who have carefully and objectively studied primitives popularly thought of as still living in the Stone Age—various Negro tribes in Africa, Indians in South America, aborigines in New Guinea and Australia—have come away with the highest respect for their accomplishments in relation to their opportunities. There is no primitive group which is devoid of discipline or moral standards and which does not have stringent rules of conduct, sexual restraints, a well-organized social system, a religion and a complex language. Nor is any group lacking in inventiveness and creativeness. It would take a book in itself to list the devices and techniques invented by primitive peoples—hunting and fishing equipment, food-cultivating, gathering and preserving methods, dwellings to meet every requirement, pottery, woven materials, and so on. Innumerable such inventions have been taken over and utilized in our modern world. In another important category are the numerous drugs and medicinal herbs which came from primitives, among these some of the new drugs such as reserpine and mescaline which have revolutionized the treatment of mental diseases.

Technological or scientific achievements are far from being the only criteria of human mental capacity, for repeatedly—as during the Renaissance in Europe—peoples have made brilliant intellectual advances while remaining technologically almost as they had been for centuries before. Thus it is significant that Negroid, Mongoloid and Australoid primitives are now recognized as having long produced works of art of the highest order. In primitive sculpture and design many of our leading modern artists have been finding the inspiration for their own works. (See accompanying plate.) No less, modern serious music and modern ballet owe a great deal to the creations of African Negroes and other primitives, as do popular music—jazz and the blues particularly—and popular dances. Further, Negroes themselves are among today's foremost interpreters of music and the dance.*

How careful one must be not to mistake any people's crudity or backwardness at a given time for inherent incapacity has been proved repeatedly by misjudgments made in the past. Two thousand years ago, following the Roman invasion of Britain, the eminent Cicero wrote advising his friend Atticus not to buy any Britons as slaves "because they are so utterly stupid and incapable of learning." Later, a savant of the Black Moors—who were highly cultured whilst the Nordics were still close to barbarism—wrote of the northern peoples, "They are of cold temperament and never reach maturity. . . . They lack all sharpness of wit and penetration of intellect." Concretely, the underestimation of one group by another group has all too often led to disaster and humiliation. Early in this century the lesson was learned by the Russians when the presumably inferior Japanese, who only a few decades before had emerged from mediaevalism, routed the Czar's navies in the Pacific. Barely four decades later, in World War II, underestimation of Japanese aerial ability by a British admiral, Sir Thomas Phillips, brought disaster to a British naval squadron in the Gulf of Siam.† In the same war the Nazi Germans learned how grievously they had misjudged the Russian military capacity, which had undergone an amazing transformation in a few decades; and no less, old notions that Russians were inherently lacking in technological ability rapidly changed and were forever blasted with the launching of the first sputnik.

* We might wonder what the slave dealers who packed African Negroes into steaming holds of ships, with less concern than if they'd been cattle, would have thought if a crystal ball had revealed descendants of these slaves among the greatest future stars of grand opera, the concert stage and the theater, as well as among distinguished authors, scientists and statesmen. The story of what may come from Negroes in the new countries of Africa and elsewhere is still to be told.

† The incident is reported in the official British history of World War II (Vol. 1, *The War Against Japan*). In telling how and why the British admiral took the risk which led to the disaster (including the loss of the battleship Prince of Wales, the battle cruiser Singapore, and his own life), it said that he was influenced "by the generally held belief that the Japanese made poor airmen and that they would be incapable of delivering an attack with the boldness and skill which they did display."

PRIMITIVE OR MODERN?

Here are sculptures by four noted modernists and four African Negro primitives. Can you tell which are which? (For answers, turn over.)

Photograph by M. Lasser

CULTURAL TURNABOUTS

Paul Fung, Jr. (left), racially Chinese but completely American in culture, is shown here with Joseph Rhinehart (Fung Kwok-Keung), White American by birth, but Chinese in rearing and basic culture. Photographed on the evening when they met at the author's studio (February, 1950), Paul, a professional cartoonist, has just completed a sketch of Joseph, and the latter has written below it his name in Chinese. (For further description of the two men, see Text.)

Works shown on preceding page are: 1. Primitive: Man and owl. Ife (Yoruba), British Nigeria. 2. Primitive: Woman and child. Belgian Congo (Wa Regga). 3. Modern: Joan Miró, "Personage" (1956). 4. Primitive: Helmet mask. Ivory Coast (Senufo). 5. Modern: David V. Hayes, "Armored Animal" (1964). 6. Modern: Modigliani, Amedeo, "Head" (1915). 7. Modern: Pablo Picasso, "Head of a Woman" (1951). 8. Primitive: Rice spoon, Ivory Coast (Yakuba).

Credits: 1. Collection, Forschungsinstitute, Frankfort on the Main; photo, Walker Evans. 2. Collection, Mme. Bela Hein, Paris; photo, Walker Evans. 3. Pierre Matisse Gallery, New York. 4. Courtesy of The Museum of Primitive Art, New York; photo, Charles Uht. 5. Willard Gallery, New York. 6. Collection, The Museum of Modern Art, New York; gift of Abby Aldrich Rockefeller in memory of Mrs. Cornelius J. Sullivan. 7. Collection, The Museum of Modern Art, New York, Benjamin and David Scharps Fund. 8. Collection, Charles Ratton, Paris; photo, Walker Evans.

Caution might therefore well be exercised in appraising any existing deficiencies of this or that racial group. Particularly, with regard to Negroes in the United States, no better example could be given of how lack of performance may be confused with lack of capacity than in sports and athletics. Early in the century there was scarcely a Negro outstanding in any sport other than prizefighting—and Negro achievement in this area was popularly ascribed to the supposition that Negroes were more "animal-like," "thicker-skinned," "harder-skulled" and more insensitive to punishment. Until 1947 there were no Negroes in big-league baseball, and they were rarities in football, basketball and track. How vastly all this has changed need hardly be told. Not only are American Negroes now in the top ranks of baseball and almost every field of sports and athletics, but they are there far in excess of their proportion in the population—actually dominating basketball and many of the track events. Especially striking have been their victories in the Olympics, the title of "world's leading male athlete" (the decathlon winner) going in the 1956 games to one Negro, Milt Campbell, and in 1960 to another, Rafer Johnson. Also in the 1960 Olympics, an American Negro girl, Wilma Rudolph, was proclaimed "world's leading woman athlete." In fact, these and many other Negro athletic stars have been the mainstays of the United States teams in the recent Olympic games.

The possibility has been suggested, and must be considered, that some physical differences among Negroes—in leg structure, musculature, organic functioning—may confer an advantage in track and certain other sports. However, since Negro athletes include individuals of various physiques, from short and stocky to long and rangy, their amazingly sudden rise to athletic eminence may be more readily explained in other ways. For one thing, there was the previous denial of opportunity. Big league baseball had been completely barred to Negroes (until Jackie Robinson's debut in 1947), and track athletics demanded special training which relatively few Negroes could get. Once barriers were let down, "over-compensation" followed, for with other paths to achievement limited, Negroes tended to go in more for athletic activities and to pursue them more intensely and longer than would White athletes. In prizefighting—formerly the high road to fame and fortune for many Negroes—there was a reverse change. When Negro baseball, basketball and athletic stars became the new idols among their people, prizefighting fell off in esteem, and fewer Negroes went in for it. The same thing happened in certain other racial and ethnic groups. As with Negroes, White prizefighters had come largely from underprivileged and immigrant groups in city slum areas, where training usually began with street fighting. Irish, Jews, Poles and Italians were therefore prominent among the earlier prizefight champions. But as these groups advanced socially and economically, prizefighters dropped in status among them, and progressively fewer went in for ring careers. Repeating this trend, prizefighters now are coming increasingly from the new underprivileged American ranks—chiefly Puerto Ricans and other West Indians—and from underprivileged groups in other countries, with Africa being one of the likeliest sources of future ring champions. (A Nigerian Negro, Dick Tiger, won the middleweight championship in 1962.)

We've dwelt so much on the racial ups and downs in sports achievement because of the parallels which can be found throughout history in mental and cultural performances. In any area of accomplishment one people after another which was on top at one period was on the bottom later: Egyptians, Persians, Greeks, Romans, and so on. Likewise, both cultural and power balances have shifted enormously among European nations in modern times. It is conceivable that at different periods unusual combinations of "talent" genes might have become concentrated first in one people, then in another. But the rapidity with which transitions have occurred from dominant nations to subordinate ones leaves much doubt on this score. So we may question now whether the reign of art among the Athenians, the era of conquests among the Romans, the exploits of the Vikings, the Renaissance among the Italians or the literary flowering among the British had any more meaning from the standpoint of genetics than the American reign in business and industry or the French reign in fashions and cuisine.

Actually, while at some period almost every people has had the feeling of being superior to others, it has not usually been in a biological sense but only in the sense of its superiority in military might, enterprise, cleverness, courage, leadership and divine support. But the theory of *racial* superiority as we now know it—in a *genetic* sense—is essentially a product of modern times. It did not exist among the ancients. Kings would often take as their brides princesses of different races. The Greeks and Romans were ready to accept as equals individuals of any race who were up to them culturally, and to regard as inferiors members of their own race who were not. Coming closer to our own times, in the court of Peter the Great of Russia a Negro was such a favorite that he was vied for by the aristocratic ladies, and from his marriage to one of them descended the famous poet, Alexander Pushkin. Where taboos or restrictions against racial intermarriages formerly existed, they were almost always on religious or social grounds. Even *slavery* was not on a racial basis until a few centuries ago. Long before Whites began enslaving Negroes (in the fifteenth century), they had been enslaving other Whites (as among the Romans, Greeks and Arabs), and African Negroes had been enslaving Negroes, while Mongoloid peoples had been enslaving persons of their kindred stocks.

The attempts to classify different racial or ethnic groups as "superior" and "inferior" in varying degrees began only a century ago. First a Frenchman, Count Arthur de Gobineau, later a fanatical Englishman, Houston Chamberlain (son-in-law of Richard Wagner), sought to show that Whites in general and "Nordics" in particular were biologically superior to other peoples. The theories were perverted to fantastic limits by the Nazis, who sought to prove that the Germans constituted a special and exalted "race" (*herrenvolk*), authorized by nature to subjugate or exterminate "inferior" people and to become overlords of the world. This, fortunately, is for the most part history. But one fallacy fostered by the Nazis persists: It involves the terms "Aryan" and "non-Aryan."

In the 1880's, a German philologist, Professor Friedrich Max Mueller, coined the term "Aryan" to apply only to a large group of *languages*, European and Asiatic, including the Celtic, Teutonic, Italic, Hellenic, Albanian, Armenian, Balto-Slavic and Indo-Iranian. Used in this sense, "Aryan" would have to take in a wide sweep of the most diverse peoples, ranging from the Irish to the Veddas of Ceylon (one of the most primitive tribes in existence). To make his meaning unmistakably clear, Professor Mueller wrote: "I have declared again and again if I say 'Aryan,' I mean neither blood nor bones, nor hair, nor skull. I mean simply those who speak an Aryan language."

It is also worth noting that just a few years before Hitler came to power, the leading German treatise on human heredity (by Baur, Fischer and Lenz) stated: "It cannot be too emphatically insisted on that that which is common to the people of one nation, such as the German, the British or the French, and that which unites them as a nation, is not, properly speaking, their 'race,' but first and foremost, a common speech and culture." Other German authorities also had pointed out that the Germans were among the most highly mixed peoples of Europe, a blend of many ethnic groups—Nordics, Alpines, Mediterraneans and Slavs, with an admixture of Mongoloid genes as well. (See Table, p. 591.)

So it was in the face of all scientific findings that the Nazi leaders proclaimed that all persons of German *nationality* (excluding those of Jewish or Polish stock) were "blood brothers of the super-Aryan race," to which persons of all other races were biologically inferior. The height of this absurdity was reached when the Japanese joined forces with the Germans, and Nazi pseudo-scientists obligingly found evidence that their Oriental allies were also, in some way, "Aryans."

No more than there is an "Aryan race" is there a "Semitic" or "non-Aryan race," for *Semitic* also refers to a large family of languages, among them the Aramaic, Arabic and Hebrew. However, to classify the Jewish people—the chief Nazi victims—as *non-Aryan* on the basis of their language is to compound the fallacy. For the great majority of Jews in the world (except the Israelis) use Hebrew only in their religious services, as Roman Catholics use Latin. The characteristic spoken and written alternative language of most European Jews, and of those who emigrated to other countries, was long Yiddish, compounded largely with an Old German base but much modified everywhere with words of the countries in which they resided. Altogether, if judged only by their language, most Jews would have to be properly included among the *Aryans.*

Do the Jews in any sense constitute a "race"? One could certainly say "No."* While undoubtedly more homogeneous than many other large groups, the Jews are nonetheless compounded of many diverse peoples. The Bible reveals that, to begin with, they were of highly mixed stocks of Eurasiatics, little different from other groups then settled in the Fertile Crescent of Mesopotamia. Their "Chosen People" concept, when it arose, related chiefly to their One-God religion and their culture, not to any thought of being biologically unique and set apart from other groups. This was shown by the numerous recorded intermarriages and matings of Biblical notables with persons of alien stocks: Abraham, their founder (himself a Chaldean from Ur), with Hagar, an Egyptian; Esau, one of Isaac's sons, with a Hittite woman; Joseph with the daughter of a Pharaoh; Moses with a Cushite (possibly Ethiopian) woman, and then with the daughter of a Midianite (Zipporah). Again, Ruth, great-grandmother of King David, was a Moabite, and the sympathetic treatment given her in the book which honors her name further reveals a lack of antagonism to matings with persons of other stocks so long as they joined with the Hebrews in religion and spirit ("thy people shall be my people")—an attitude which has prevailed ever since.

After the Jews formally became an identifiable group, genes from

* An excellent discussion of this point, and of other aspects of Jewish biological makeup, appears in anthropologist Harry L. Shapiro's chapter, "The Jewish People: A Biological History," in the UNESCO book, *Race and Science,* Columbia University Press, 1961.

many diverse stocks continued to be brought in during their successive dispersals and emigrations—forced or voluntary—starting long before the Diaspora of A.D. 70 under the Romans and never ceasing thereafter. Conversions to Judaism of alien individuals and sometimes whole outside groups (including many Romans before and after the Christian era, the Khazars of Southern Russia in the eighth century, and groups in India, the Near East and North Africa at different times) added to the mixture. Today Jews of many varied types are found throughout the world, as is most strikingly shown among those collected in Israel—Yemenite, North African, Central and Eastern European, Spanish and Italian Jews—whose features, head shapes, body forms and sizes, blood-group incidences, diseases and other traits reveal not merely the effects of centuries of living in different environments but of differences in *genetic* makeup.* While it is generally true that certain characteristics are more common among Jews than among non-Jews, it also is obvious that large numbers of them do not have these characteristics, and that in every country a large proportion of Jewish individuals cannot be distinguished physically from their non-Jewish compatriots.

Of further importance, when Israeli Jews are classified by places of origin, many of the genetic traits of each group are found to be correlated with those of the non-Jewish population in its original habitat rather than with those of Jews from other regions. For instance, blood-group frequencies of Jews from Central Europe (Ashkenazis) are much like those generally prevailing in their native countries; and blood-groups of Spanish and North African Jews (Sephardics), as well as of those from long established settlements in the Near and Middle East, tend toward the averages for other peoples in their previous habitats. Likewise, in hereditary diseases, Israeli medical geneticists have reported that the *fava bean* condition has a high incidence among Yemenite and other Oriental Jews, and that *thalassemia* is common among Medi-

* A fallacy given considerable circulation is that unexpectedly large proportions of blond, blue-eyed children have cropped up in Israel among dark-haired, dark-eyed parents, suggesting some genetic change. This report, which the author found on a recent trip to Israel to be without basis, might be attributed to (1) underestimating how many light-haired, light-eyed individuals there always have been among Jews—up to 20 or 25 per cent in their major European groups, and some even in those from Asia and North Africa; (2) the bleaching effects of the sun in Israel, causing many Jewish children to be lighter-haired than they would be elsewhere; and (3) the frequent tendency of eyes and hair in any stock to be lighter in childhood than later, which, together with the action of recessive genes, may make many children lighter-haired and lighter-eyed than their parents.

terranean Jews (as it is among other peoples in the same area), whereas both conditions are almost nonexistent among Jews (and non-Jews) from central Europe. Similarly, a number of hereditary metabolic disorders (phenylketonuria, lipoidosis, and pentosuria) appear with markedly different frequencies among Oriental and European Jews.

Yet despite these genetic variations among themselves, there is evidence that Jews, when taken as a group, represent a population not quite the same in its pool of genes, and in combinations of genes for various traits, as other human populations. This is not surprising in view of the distinctive biological and social history of the Jewish people, who, while always undergoing some admixture, have managed through outer and inner pressures against assimilation to retain an inner core of genetic identity. But such limited homogeneity as the Jews may have, as compared with the great genetic diversity among them, certainly makes it impossible to refer to them as a "race." On the other hand, the fact that various biological differences, *on the average*, distinguish Jews as a group from other populations wherever they may be, makes it equally questionable to regard them solely and entirely as a cultural unit.

As to what, then, constitutes a Jew, there is much disagreement—as much within Israel as elsewhere.* Those who contend that Jews are merely a religious group must face the fact that there are wide differences in religious belief among them, from the strictest Orthodox on through Conservative to highly Reformed, with the extremes being further apart in faith than are ultra-Reformed Jews and Unitarians or certain other

* One example of the difficulty of establishing Jewish identity has been offered in Israel by the dispute over the B'nai Israel sect, a group numbering about 7,000 whose origins were in India. Early in this century a Rabbinical council ruled that they were not Jews and that Jews were forbidden to intermarry with them. But in 1961 the Supreme Rabbinical council reversed this decision and accepted the members of the sect as Jews. Another question involves the child of a mixed marriage. The rule is that if the mother is Jewish (or a convert to Judaism), the child is also Jewish, regardless of the father's faith; but if the mother is not Jewish, even though the father is, the child is not considered Jewish. Apart from the religious aspect, this may decide in Israel whether an immigrant of mixed parentage will or will not immediately qualify for Israeli citizenship, which goes automatically to anyone held to be a Jew. Bearing on the foregoing point, again, is the celebrated case of Brother Daniel, a Roman Catholic monk of Jewish birth who came to Israel to live. The courts there decided in 1962 that because of his conversion—and hence his presumptive disavowal of Jewish identity—he was not entitled to automatic Israeli citizenship under the "Law of Return." It was added, however, that he would be welcomed as an Israeli citizen if he went through the same naturalization process as is required of any other non-Jew settling in the country.

Christian sects. Many Jews who are nonbelievers still consider themselves Jews; and even many of those converted to Christianity have been regarded or may regard themselves as Jews (as did Disraeli, Heinrich Heine and Felix Mendelssohn, and later many victims of Nazi persecution). Are Jews a *nationality?* Again, quite clearly, no. Their nationality is that of the country of their birth or legal adoption: American Jews, British Jews, Canadian, French, Italian, Brazilian Jews, and so on. Nor did the establishment of the Israeli State (whose citizens, including now about 10 per cent of Arabs and other non-Jews, are identified as all "Israelis") make any change in the nationality of Jews elsewhere.

Perhaps the best way to describe the Jewish people is as a "kith." This technical term is applied to certain large human family groups with a more or less common ancestry for a long period, and therefore some biological unity, but distinguished mostly by the effects of common cultural conditioning—religion, traditions, training, habits (including diet) and social experiences (to which, in the case of the Jews, might be added psychological reactions to oppression and discrimination). Examples of other "kiths" would be the Irish, Scottish Highlanders, French Canadians and Roumanians. If we allow for combinations of many traits—biological, social and psychological—as determinants, it would be possible for individuals to deviate from the whole in some respects, and yet, with regard to other traits, to be recognized, or to recognize themselves, as belonging to their particular "kith."

Thus, many factors must have combined to produce any distinctiveness which Jews may have as a group. From the surface evidence, their achievements in Biblical times and in later history, extending to the present with their remarkably high representation in the sciences, arts, professions and other areas, would suggest some unusual hereditary endowment; but in view of facts previously dwelt on, it would appear much more likely that environmental influences, working through their culture and conditioning, have been the major spurs to their accomplishments as a people. Analogies could be found here to cultural heights reached by other groups at various stages (the Greeks, the Romans, and the Renaissance Italians, for example), the main difference in the case of the Jews being that the factors responsible, instead of being concentrated in one period, have persisted over a long stretch of time.

But whether one is speaking of Jews or other human groups, it is a common fallacy to attribute a set of specific traits to any group as a whole (and hence to all of its individual members). Psychologists designate as *stereotypes* such sweeping generalizations as that the Irish are inherently "pugnacious," the Japanese "sly," the Chinese "inscrutable," the Negroes "childlike," the Germans "regimented," the Jews "business-minded," the Swedes "stolid," the Russians "stubborn," the French "artistic but unstable." But we have seen clearly in the United States that in a little space of time persons descended from every race and nationality not only shed traits ascribed to their forebears, but take on common and presumably "American" characteristics to such an extent that when they go abroad—particularly as do our soldiers and sailors, dispersed throughout the world—they are thought of by peoples elsewhere as all typically American, regardless of their ancestral origins.

No experiment could be more convincing than the actual stories of two men now in their forties, living in New York and long known to the author: one, an American-Chinese; the other, a Chinese-American. To explain the distinction, the American-Chinese man was born Joseph Rhinehart, of German-American stock, and became Fung Kwok Keung when, as a baby (this was in 1920), he was adopted by a Chinese restaurateur on Long Island and taken to China to be reared in a small town near Canton with the family of his foster father. Not until he was twenty did he return to the United States, but by then was so completely Chinese in all but appearance that he had to be given "Americanization" as well as English lessons to adapt him to his new life.

Now for the completely opposite case of the Chinese-American man—racially Chinese but culturally American—Paul Fung, Jr., a talented comic artist with one of the largest syndicates (as had been his late father, a good friend of mine). With American-born parents and Chinese-American grandparents, Paul was educated and had lived among White Americans all his life. His thinking, behavior, speech, outlook and sense of humor were completely like that of any other American (and in nothing is this required more than in doing American comic strips!). But when World War II broke out and he enlisted for service, he found himself with the Chinese-American unit (composed mainly of immigrant Chinese youths or sons of immigrants) that had been organized. Knowing little of the Chinese language and customs, he was virtually a stranger among all the others of Chinese stock; and—no small matter to a GI—he couldn't adjust to having Chinese food served at almost every meal. Because he was regarded by the others as "putting on an act," conflicts ensued, and he had to seek companionship among White GIs from a near-by camp. Chance eventually brought him a transfer to a regular army unit with which he served in the Pacific area as an air-force photographer. (Paul participated in the atomic-bomb flight over Hiroshima.)

What happened during the war with our other man, who looked American but was culturally Chinese? As might be expected, he was placed not in the American-Chinese unit, but with regular army White troops with whom he had little in common save his racial origins. One special and congenial army service he could perform was as an interpreter in Italy when ships with Chinese sailors came into port. At the war's end Rhinehart-Fung worked for a while as compositor on a Chinese newspaper (an intricate job almost impossible for anyone not Chinese), but could never quite find himself and subsequently worked for White firms and institutions as a night watchman.

Linking up these stories, I arranged a meeting some years ago between the two men, Paul Fung, Jr., and Joseph Rhinehart-Fung Kwok Keung, as shown in the plate facing page 617. Almost at once the same thought occurred to both: They were *culturally transposed*. Said Joseph, spontaneously, "He should be like me and I should be like him." As a climax, the fact that both men bore the name "Fung" (in Joseph's case given first, in the traditional Chinese way) proved to be more than a coincidence. For—despite there being 600 million people in China and tens of thousands of towns—it developed that Paul Fung's grandfather had come from the very same town in China where Joseph had been reared and was a relative of Joseph's foster father!

Think carefully about this story. If ever you've believed, or still believe, that Chinese and Whites are born to be radically different in behavior, temperament, and speech mannerisms—that the sing-song voice, or the stoicism, or the eating habits of Chinese are traits produced in some way by special genes—let your mind dwell on these two men whose "natural" environments were completely reversed. (Or merely imagine yourself in Joseph Rhinehart's place.) There may be just as much reason to believe that a Negro child, reared among Chinese, would in his behavior be more Chinese than what we think of as "Negro," or that a child of aristocratic European parents, lost and reared among jungle Africans, might grow up with all the traits of a typical primitive Negro. We could expect corresponding results in the case of an Eskimo child reared among Danes in Copenhagen, or an Italian reared by Hindus in India, or a Turkish child reared by Boston Cabots.

Why offspring of one racial or subracial group are not so readily reared or accepted by people of a different racial group brings us to the crucial matter of *race-mixing*. There are two aspects of this, the biological and the social. To be realistic, we must consider both together. But what mainly concerns us here is the biological aspect: Is race-crossing, or interbreeding between peoples of different racial groups, *genetically* desirable? Undesirable? Or of little or no importance? While the major problem currently centers on Negro-White mixing, it also involves Mongoloid and White mixing (most strongly in Hawaii and other South Pacific areas), and to a lesser extent mixing between any subracial and otherwise biologically differentiated groups in this and other countries. Whatever the groups under consideration, and whatever the extent of their biological differences, the basic factors in race-mixing are essentially the same.

Most important is the question of whether one or both racial groups participating in a crossbreeding will lose or gain genetically by the mixture of genes. If there is no real proof of the inherent superiority of one of the racial groups, it cannot be said that its genes will be polluted or depreciated by the mixture. It may be argued, though, that there is nonetheless an advantage in keeping races and groups "pure" and in retaining their distinctive traits (in the way that various breeds of livestock and domestic animals are kept "pure"). But we have seen that there are no genetically "pure" races or groups of people today, and that all human beings are of highly mixed stocks—or, in a breeder's sense, "mongrels." The relatively "purest" peoples were those of thousands of years ago, and today the "purest" racial groups are the most primitive, isolated tribes of Africa, Australia, the South Pacific and South America. How mixed the European peoples are can be gathered from a look back at the table of origins on page 591. If one stops to think, further, that Americans today are a mixture of these mixtures, there would appear to be no substance to the theory that the "purest" people have made the greatest advances.

It has also been claimed that crossbreeding between peoples of markedly different genetic makeup would result in misshapen offspring—individuals with teeth too big for their jaws, or with other skeletal parts and organs that wouldn't fit or work together properly. This, too, has been disproved. The most obvious examples are the Mulattos in the United States, some of whom are among the most physically attractive—as well as gifted—of our theatrical stars. In Hawaii, crosses and intercrosses involving White, Chinese, Japanese, Polynesian, Puerto Rican and other racial groups not only have not resulted in structural abnormalities but have generally led to the production of unusually good-looking and well-built individuals. So, too, among the Pitcairn Islanders, the cross between English sailors (of *Mutiny on the Bounty* fame) and Polynesian women resulted in an attractive and sturdy group of descendants.

To further support the theory that racial intermixture, instead of being biologically harmful, may often be beneficial, are the facts cited with respect to crossbreeding, or *hybridization*, in domestic animals and plants. "Hybrid vigor"—greater size or strength, or greater resistance to certain diseases—comes frequently through the formation of new combinations of genes or the breaking up of previous undesirable combinations in the stocks which are crossed. By this means many improved strains of cattle, sheep, pigs and poultry, and many kinds of fruits, vegetables and plants, have been produced. While no clearly proved examples of "hybrid vigor" through race mixture in human beings can be offered as yet, here are some possibilities.

Taking first a simple genetic situation, we may recall that the "Rh disease" results from prenatal blood clashes between an Rh-positive fetus and an Rh-negative mother; and that since Mongoloids are almost entirely Rh-positive their babies escape this danger, whereas among Whites, with 15 per cent Rh-negative, many babies are menaced. Accordingly, Chinese-White matings on any broad scale in Oriental populations would be detrimental to the Chinese, for the introduction of "Rh-negative" genes would lead to many cases of the "Rh disease" which otherwise would not have occurred. But in a White population, the introduction of more "Rh-positive" genes through matings with Chinese would be beneficial, because it would diminish the incidence of Rh blood clashes. Similarly, but to a lesser extent, since Negroes are higher in Rh-positive and much lower in Rh-negative than Whites, Negro-White mixing would increase the risk of "Rh disease" among Negroes, and lower the risk among Whites. In the case of clashes or diseases involving the A-B-O blood groups and other blood elements, which have different frequencies among racial groups, changes in the proportions of given genes through race-mixing could also in some instances be beneficial, in others detrimental to one or another group. Again, where racially slanted hereditary diseases are due to different types of recessive genes (as in Mediterranean anemia, sickle-cell anemia, the "fava-bean" condition), mixed matings among the racial groups involved would decrease the chances for matching genes to come together and cause the diseases in question.

With regard to more complex genetic traits, a hypothetical situation would be this: Suppose that some desirable trait—a talent of a special kind, or improved physique—were dependent for its production on a four-gene combination, "A-B-C-D." If only the first two genes were common in one racial stock, and the second two in the other, crossbreeding would bring together the required four genes and prove advantageous to all concerned. On the other hand, where a given complex combination of genes causes harmful effects, if one racial group carries mostly one part of the combination, and a second group the other part, mixing of the two groups might lead to a more frequent occurrence of the full combination and its bad effects. For the most part, however, gene combinations for complex traits appear to be distributed not too differently among all human groups, so that the hypothetical situations cited, to whatever extent they do or could exist, would probably be exceptions.

If, then, we take all the possibilities together, there might be just as much *genetically* in favor of mixed matings as against them. (The social question, quite different, will be discussed later.) In any event, much of the argument for keeping people "racially pure" (to the relative degree they could be so at a given stage) is academic. For everything points to the fact that genetic fences between human groups cannot be long maintained. Moreover, regardless of what external barriers may be set up, *mutations* alone would constantly introduce into any supposedly "pure" strains many of the same genes which are found in other racial groups.

How can the foregoing facts be applied specifically to Negro and White mixing in the United States? An initial point is that the American Negro population from the beginning has been highly mixed in its genes, for it is a misconception that the Negroes brought in as slaves were all pretty much of the same stocks. That they were not is clearly shown on our accompanying map. In fact, there were more genetic differences among groups of Negroes taken from different regions than there were among White immigrants. (How enormously varied even today's African Negroes are in physical traits, language and culture —far more so than Europeans—can be seen by comparing representatives of the many African countries at the United Nations.) The present American Negro population is therefore made up of mixtures of many kinds of "African-Negro" genes, mixed in turn with perhaps 20 to 25 per cent "White" genes and additional genes from West Indian and other racial stocks. In the aggregate, American Negroes now comprise a new Negro racial group genetically unlike any African group.

But the Negro-White mixing has by no means been all in one direction. Many Negro genes also have found their way into the White stream. Moreover, the greatest amount of White-Negro mixing occurred not in recent generations, but in the Colonial days and up to the end of the Civil War. In the earlier periods the flow was mostly of "White" genes into the Negro population, with Mulattoes continuing to stay in the Negro ranks and adding their genes to make the Negroes lighter

ORIGINS OF AMERICAN NEGROES

Showing ports and regions in Africa from which Negroes were taken during the slave trade period, and their destinations in the Americas. Dates and lines are mainly of documented slave transports, but there were many others.

(Map also shows slave trade to the Arab countries.) *

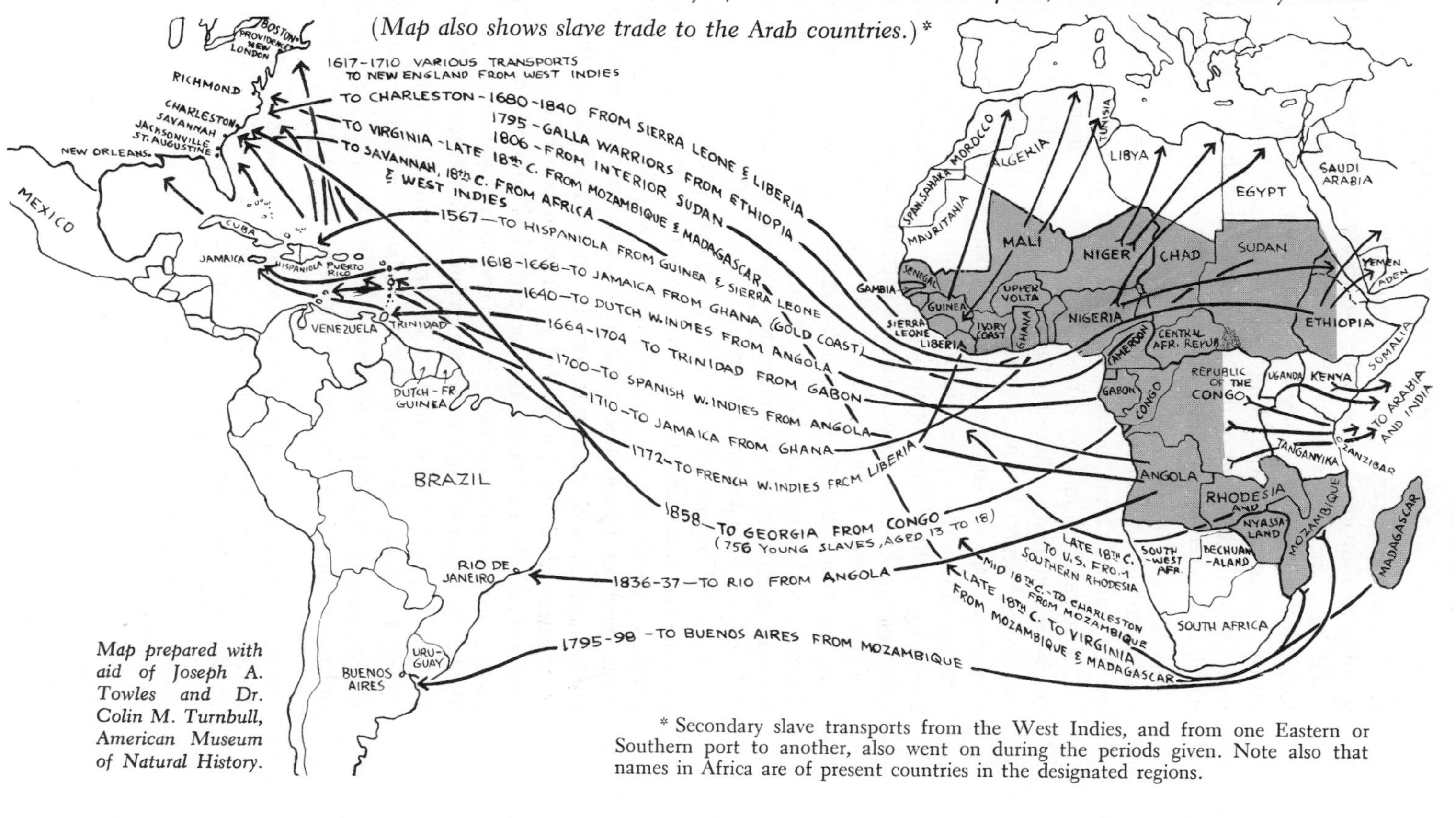

Map prepared with aid of Joseph A. Towles and Dr. Colin M. Turnbull, American Museum of Natural History.

* Secondary slave transports from the West Indies, and from one Eastern or Southern port to another, also went on during the periods given. Note also that names in Africa are of present countries in the designated regions.

in color, on the average, and less Negroid than the original stocks. Subsequently, as individuals with some Negro ancestry were produced who could "pass" as Whites, many of these married Whites. This had a twofold effect: It brought "Negro" genes (mostly for effects not visible on the surface) into the White population, and at the same time, it took "White" genes out of the Negro population. By this latter process, which is destined to continue, and with fewer reverse matings of Whites with Negroes than occurred in the days of slavery, it is very likely that American Negroes will for some time to come (so long as racial barriers are maintained) tend to *darken* again and to become more Negroid.*

But again, it should be stressed that neither "Negro" nor "White" genes are confined to those for easily recognizable surface traits, such as skin color, features and hair form. Many persons who can pass as Whites may carry more "Negro" genes for other traits than do some individuals who are distinctively Negroid in appearance. The selection of those remaining in the White population, and those in the Negro population, will therefore be carried out, as in the past, on the basis of the comparatively few surface traits. The end result may be that in outer characteristics the Negro and White populations will continue to be about as different as they now are for a long time to come, but that in the less obvious, or internal genetic traits (for instance, blood types), the genetic differences will be narrowed down.

The social aspects of interracial marriages are, of course, another matter. The most serious argument against race-crossing is that when people of widely diverse types are mated, their differences in temperament, behavior, background and family connections may make for social conflict which will react unfavorably on their offspring. But this is something outside of the field of genetics, to be weighed only on the basis of individual attitudes and the prevailing cultural environment in given groups. Marked contrasts in this respect are apparent in the Negro-White situations in the United States and in two other countries where the races have gathered together in large numbers, Brazil and South Africa. In Brazil, the Negro, like the Indian there before him, may be disappearing as a racial unit. Slavery in Brazil was never of the same order as in the United States and ended not through forced action but

* Much uncertainty exists as to the number of Americans of part Negro ancestry who have "passed over" or are continuing to do so. Some estimates are that up to two million such persons are in the present population of the United States and that thousands more—but vaguely placed at anywhere from a few thousand to as many as thirty thousand—cross over to the White side annually.

through gradual change. Likewise, barriers between Whites and Negroes were never very great, in large measure because they were not encouraged by the Catholic Church. While prejudice still exists in Brazil, it is more on the basis of *class* rather than race: to quote a commonly heard saying there, "A rich Negro is a White man, and a poor White man is a Negro." In White South Africa, on the other hand, Negroes have not even been recognized as citizens and are kept under many severe civil and social restraints which are relaxed only to some extent for those of mixed bloods called "Coloreds," as distinguished from the full Negroes or "Blacks." (Included also among the Coloreds are persons of Asian stock—preponderantly Malaysians—even if they have no Negro admixture.)

So we return to the point that racial discrimination, or the domination of one racial group by another, finds no justification in any hereditary differentiation among human beings but is an outgrowth of social conditioning, chance circumstances, expediency and—undoubtedly also—of selfishness: the desire of one group to further its interests at the expense of another. Yet if one group of human beings, on the presumption of real or fancied inborn racial differences, were content merely to stay apart from another group, this might be accepted as their right. That kind of racial separation could be illustrated simply in this way:

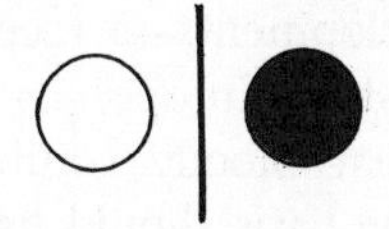

But racial segregation in intent and practice usually means this:

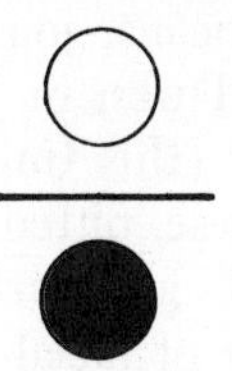

That is, the one people, which dictates the separation, considers itself "above" the other group, endowed by Nature or by divine intent with the right and mission of exercising dominance. Worst of all, when this theory is carried to its ultimate—as it was among the Nazis—it spawns the doctrine that the self-anointed "superior" people is justified in exterminating designated "inferior" groups for the presumed advancement of the human species.

No less insidious is the correlated doctrine: that periodic blood conflicts between racial or ethnic groups are vital to human progress because they supposedly lead to the purging of "inferior" stocks and the ascendance of "superior" ones. This is a gross perversion of the Darwinian theory of the "survival of the fittest." In the earlier evolution of mankind, it might be granted that competitive struggles among *individuals*, as well as natural selective processes, did help to weed out some of the least "fit" genetically and promote the survival of the more "fit," in relation to their existing environments. But Darwin himself pointed out that what became most important in human survival and advancement was not bitter competitiveness, but social organization and *cooperation*, and the development of the capacity among people to live and work together peacefully.

This policy of cooperation—coming not through instinct, as in bees, beavers, or various herd animals, but as a result of thought, deduction and learned experience—was the outstanding invention of human beings. It led, first, to cooperation of individuals within groups, a mutual respect for their differences, and the utilization of the special capacities of each for the good of all. Second, it led to growing cooperation among whole groups on the same basis. It is true that there were continual wars and conflicts. But far outweighing any gains which might have come through blood clashes were the advantages that accrued to different groups of people from peaceful contacts and the exchange of products, inventions and cultural developments—a recent notable example being the Common Market in Western Europe.

Any lingering thought that bloody conflicts among human racial groups have any valid genetic basis should have been dissipated by the hodge-podge of alignments in recent wars. In World War I, some White Europeans, plus Americans, joined with the Japanese against other White Europeans and Turks; in World War II, there was another arrangement of Whites (this time with Italians and Turks exchanging sides) plus the Chinese, pitted against other Whites and the Japanese. In subsequent "cold" and smaller "hot" wars, we have seen every conceivable combination of racial enmity or alliance throughout the world. Nor in any past or recent wars can it be said that the struggle was between superior and inferior genetic stocks. Victory, and survival, came to the sides which had the greatest resources, the best equipment, the best allies—and the most luck. (What if the Nazis had been first with atomic bombs?)

Not least insidious of the racial theories is that there are *inborn* aversions between racial groups, designed by Nature to keep them apart. This theory, too, is wholly without foundation. *Xenophobia*—the fear or dislike of strangers—has existed and is found as often as not between groups of the same race as between groups of different races, and in either case can be ascribed solely to conditioning. For example, without training, and in a prejudice-free environment, a White child would undoubtedly feel no more aversion or hostility to a Chinese or Negro child than a white animal of any species feels toward a brown or black one, or that an animal with such and such features feels toward one of its species with different features.* Nor should one forget the well-known saying (which may be equally wrong) that "opposites attract each other."

Thus, virtually every theory about racial or ethnic differences which has caused conflict, hatred and prejudice among human groups can be refuted by both science and experience. But how can people be made to accept and apply the *facts*—insead of the myths—in their everyday relationships? Those eager to promote tolerance have so far dwelt largely on the great *similarities* among all human groups. There was good reason for this, inasmuch as racial differences always have been too easily interpreted in terms of "superiority" and "inferiority"—with the currently dominant group ascribing the "superiority" to itself, and each oppressed group therefore resenting any stress on their differences from the others. Nevertheless, there is hope that growing public enlightenment and maturity will bring greater readiness to accept racial differences and to view them in a more intelligent and constructive light.

Let us assume—though proof is yet lacking—that genes for certain complex mental traits and capacities may be distributed in somewhat different proportions among various racial groups. This would in no sense add up to the overall "superiority" of one group over other groups. It might only mean that if one group has a slight advantage in some special gene combinations, another group has the advantage in other gene combinations. Nor would it mean that the specific traits and achievements in which a given group—or a few of its individuals—excelled at the moment were necessarily those for which its genes were superior. Some backward group might even have the better genes. For, as we've noted, where environments and opportunities have differed radically, the comparative performances of human groups at given times have often been unrelated to their genetic capacities.

* It might be added that antagonism on the basis of race in recent centuries has been largely on the part of Whites toward those of other races. Neither American Indians, nor African Negroes, generally showed any aversion to Whites when they first saw them. In fact, the friendliness of many primitives to Whites who came among them proved their undoing.

Remember, too, that we are always speaking only in terms of *averages*. Whatever the kind or degree of average differences among racial groups in given traits, it hardly entitles an individual to boast that because the group of which he is a member is somewhat better in a certain respect—whether genetically or environmentally—he himself excels in that respect. This is as silly as for a Swedish midget to draw himself up to his full 3-foot height and say, "Swedes are the tallest people in Europe. *I*, being a Swede, am therefore taller than any Frenchman or Italian." No matter, then, how given racial groups may differ on the average in genetic qualities, it is apparent that in groups now rated as highest in certain mental and cultural respects, there are innumerable members who are excelled in these respects by individuals in other groups; and that no group, however backward, is without outstanding individuals who are well above the average found in the so-called "superior" groups. The question to be asked of any person who boasts of racial superiority is "How do *you yourself* rate?"

In sum, not until human beings of all racial groups everywhere have had the same opportunities for education, training and advancement can we draw fair conclusions as to their relative genetic capacities. As matters stand, there is enough evidence to suggest that the genes for all types of achievement are spread among all racial groups, and that each group abounds in far more gifted individuals than have had a chance to assert themselves. If, in terms of the averages we've mentioned, there actually are special racial or subracial differences in relative capacities for designated kinds of thinking, temperament, performance and achievement, we should welcome the possibility that every group can make special contributions in addition to the general ones. Life would not be merely retarded, but very dull indeed, if all peoples everywhere and forever worked, behaved and performed in exactly the same way.

This idea has been summed up in a saying which you have heard, or have repeated yourself, hundreds of times: "*It takes all kinds of people to make a world.*" And only when there is a world that ensures equal freedom, justice, opportunity and tolerance for all kinds of people will it be possible to say that the fully civilized stage has been reached in human evolution.

CHAPTER FORTY FOUR

Ancestors and Relatives

MR. REGINALD TWOMBLEY DUNN-TWERPP—who is not very bright, weighs 110 pounds and is the first to climb on a chair at sight of a mouse—likes to boast that he is descended from William the Conqueror and that the steel-blue blood of ancient warriors flows in his veins. To prove it he will show you his family tree and a beautiful hand-painted crest, prepared by a genealogist in Boston for fifty dollars.

Even in these United States there are still a lot of people like Reginald, who point with pride to some remote ancestor; and, no doubt, others who feel humbled because they haven't any to point to. You yourself may have been among them. We say "may have been" because by now you must have gathered that the whole ancestry business has been shaken pretty badly by our genetic findings.

The importance previously attached to ancestry rested on a number of fallacies. First was the pre-genetic concept that heredity was a process of passing on "blood"—the blood of the parents being blended together to form that of the child. No matter how far one traced back, therefore, there was always a little of the blood of any ancestor flowing in one's veins. Also, as blood was thought to carry factors that influenced character, the greater the percentage of the "blue" or noble blood one carried, the more superior one would be; and the more "common" blood in one's veins, the more inferior one would be. Likewise, touches of genius, of great courage, of brilliance—or taints of criminality, shiftlessness and depravity—were thought to be carried in the blood. All that, of course, has been shattered by our knowledge that blood is merely a *product* of each individual's body and that not even a mother and her child have a single drop of blood in common.

ERRONEOUS CONCEPTS OF ANCESTRY

THE "BLOOD" THEORY

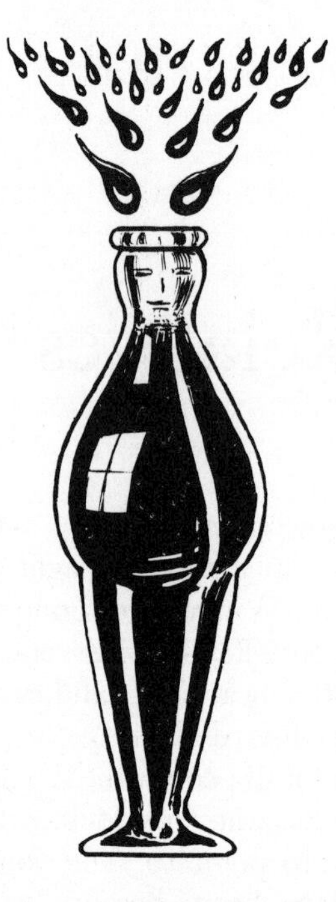

"A person is a mixture of the blood of all of his ancestors. No matter how far back an ancestor is, some of his 'blood' flows in one's veins."
Fact: *No one else's blood, but only what you yourself produce, flows in your veins. Nor does blood carry traits of character.*

THE "JIG-SAW" THEORY

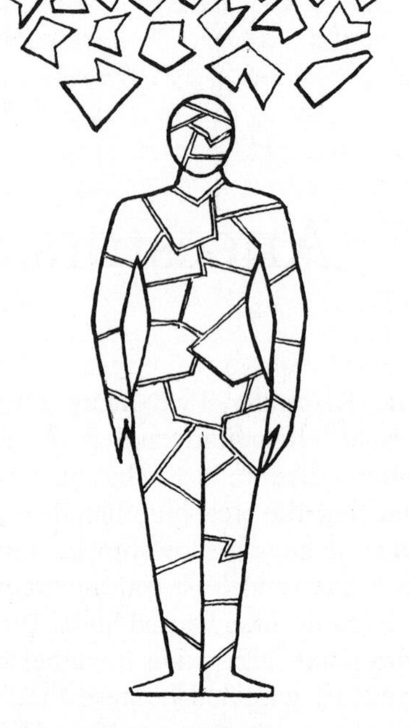

"A person is made up of parts of such-and-such ancestral stocks in given amounts. For example: 'One-quarter English, one-fourth Irish, one-eighth German, one-sixteenth Italian,' etc."
Fact: *Genetic ancestry cannot be broken down into identifiable ethnic fractions; nor, beyond one's parents, can anything more than guesses be made as to the derivation of one's ancestry in terms of chromosomes.*

Because we know now that all we inherit are *twenty-three chromosomes* from each parent, ancestry has been reduced to a simple mathematical formula: With each generation further back the *average* number of chromosomes you may have received from any ancestor is diminished by half. Note the qualification "average number." You can be quite certain you received twenty-three chromosomes from each of your parents, but in the combination from your father, for instance, any number of these twenty-three may have been derived from *his* father, with the rest from *his* mother. On an average, however, you can assume that you received eleven or twelve chromosomes from each of your grandparents, that an average of five or six of these came from each great-grandparent, an average of two or three from each great-great-grandparent, and so on, the number being halved with each generation backward.

Thus, as shown in Table X, "You and Your Ancestors," when we get to the fifth generation back, you may have received from any given ancestor one or two chromosomes, on the average. From that point, the more remote the ancestor, the greater would be the odds that you did not receive even a single one of his chromosomes. In other words, if you were a young adult today claiming descent from the Puritan Miles Standish, the chances would be great that you had no actual genetic link with him and that your relationship existed only on paper.

There are several qualifications to the foregoing. One, in the case of men, involves the small Y chromosome which is passed along from fathers to sons. If a man can trace his descent in a *straight male line* from a given male ancestor, he can be quite sure that he is carrying at least the Y chromosome from that ancestor. However, most "Y" lineages are broken somewhere along the way when there is no male heir and the ancestral Y drops out. This is so in the present British royal line, where, with King George VI having had no son, his grandsons—children of Queen Elizabeth and Princess Margaret—do not carry his Y but the Ys of their fathers. But even if the Y is carried along, it is uncertain whether it has any or more than a few specific genes, or whether it is important in determining any trait other than masculinity.

Another qualification relates to the fact that chromosomes may sometimes break and exchange parts in the process of "crossing over" when sperms or eggs are being formed (as was told on page 69, and illustrated on page 70). How often this happens among human beings is not yet known, but to the extent that it does it would further decrease the odds of anyone's having received whole chromosomes, or any one chromosome *intact,* from a given distant ancestor. On the other hand, the more that fractions of chromosomes are transmitted, the more chance there would be that a few genes from some remote ancestor would reach a descendant (though there would be no certainty that they were in their original form and had not mutated along the way).

Further increasing the possible genetic relationship to a given ancestor would be any marriage along the line between two of his or her descendants. For instance, if *both* your parents claimed descent from

TABLE X

You and Your Ancestors

Any hereditary link with a given ancestor could only be through some of his chromosomes and/or genes. How many of these, on the average, *you might have received from any specified ancestor in any generation back, is shown here.* (See Text for other details.)*

Generation Back	Approximate Period	Potential Ancestors	EXPECTED HEREDITARY LINK Chromosomes from Each (Avg.)	Genes from Each (Avg.) Proportion
First (Parents)	1930— (F.D.R.; World War II)	2	23	1/2
Second (Grand-parents)	1910— (World War I)	4	11-12	1/4
Third (Great-grandparents)	1890— (Cleveland)	8	5-6	1/8
Fourth (Great-great-grand-parents)	1860— (Lincoln; Queen Victoria)	16	2-3	1/16
Fifth (Grt.-grt.-grt.-grand-parents)	1830— (Andrew Jackson)	32	1-2	1/32
			Odds Against Having Even One of Ancestor's Chromosomes	
Sixth	1800— (Napoleon)	64	1.4 to 1	1/64
Seventh	1770— (Washington)	128	2.8 to 1	1/128
Eighth	1740— (Franklin)	256	5.6 to 1	1/256
Ninth	1700— (George I)	512	11 to 1	1/512
Tenth	1670— (Louis XIV)	1,024	22 to 1	1/1,024
Eleventh	1635— (The Pilgrims)	2,048	44.5 to 1	1/2,048
Twelfth	1600— (Capt. John Smith)	4,096	89 to 1	1/4,096
Thirteenth	1570— (Queen Elizabeth I)	8,192	178 to 1	1/8,192
Fourteenth	1535— (Michelangelo)	16,384	356 to 1	1/16,384
Fifteenth	1500— (Henry VIII)	32,768	712 to 1	1/32,768
Sixteenth	1470— (Columbus)	65,536	1,425 to 1	1/65,536

* (1) The "generations" are viewed in relation to the average young adult at this writing. (2) See Text for qualifications to the above Table, as related to inheritance of the Y chromosome in males, the passing on of fractions of chromosomes, and the effects of marriages between ancestral relatives.

Miles Standish, the chance of your carrying one of his chromosomes would be doubled; and if there were other marriages between Pilgrim descendants in your ancestry, you might very well be carrying a sizable quota of "Mayflower" genes. Thus, too, in families with considerable inbreeding, such as the European royalty, or in any groups or stocks which have tended to hold together for long periods (the Irish, Scotch and Jews, or certain small religious sects), there is a greater than average chance of carrying some chromosomes, or at least a number of genes, of a particular vaunted ancestor.

But in every person's pedigree, if one goes back any distance, there must have been many marriages between related persons. Were it not for this, it can readily be seen that if one's potential ancestors continually doubled with each generation back (as shown theoretically in our Table), the number would reach astronomical and impossible figures. Further, the longer any families and their ancestors have lived in a given country or area, the more the number of their potential ancestors is reduced, and the greater is the chance that they and others in the country had some common ancestry. Thus, it has been estimated that all persons of English descent, from the humblest to the most royal, are at least thirtieth cousins, and that the higher up they are on the social ladder, the closer must be their relationship by common ancestry to one another and to the royal family. For example, Queen Elizabeth shares a rather obscure sixteenth-century ancestor, Sir George Villiers, with such diverse figures as (or descendants of) Sir Winston Churchill, Sir Anthony Eden, Bertrand Russell and Arthur James Balfour, as well as any descendants of Henry Fielding (novelist), William Pitt (prime minister), the Earl of Chesterfield (essayist), and the Duke of Marlborough (military leader). Again, when Princess Margaret married the commoner, Antony Armstrong-Jones (later given the title Earl of Snowdon), avid genealogists dug up evidence that the "Jones Boy" was a direct descendant of Edward I, King of England from 1272 to 1307. Had the searchers burrowed further, to the ninth century, not only Margaret's husband, but *all* Englishmen today, except recent immigrants, could be shown to be related to her and the Queen as descendants of King Arthur (according to genealogist Sir Anthony Wagner's calculations).

As to other aspects of genealogies, most—if not all—are flawed by being compiled from dubious (and sometimes imaginary) sources; by the practice of disregarding the unimportant or undesirable ancestors; and by a general tendency to ignore the females unless their families had distinction.* Thus, when descent is claimed from William the Conqueror, the fact is conveniently overlooked that while half of his heredity was "royal," through his father, the other half came from the humble tanner's daughter who bore him illegitimately. There is little doubt that in every boasted lineage there were innumerable forgotten ancestors who, if they were alive today, wouldn't be allowed to enter their descendants' homes.

Still another fallacy in the ancestry field is that of comparing human families to strains of domestic animals—aristocrats to thoroughbreds and ordinary folks to mongrels. True enough, there are genetic aristocracies among horses, dogs, cows and cats; but bear in mind that they were derived only by the closest inbreeding—fathers with daughters, mothers with sons, brothers with sisters—and also by controlling every mating and by discarding those not wanted from every generation or litter. Human breeding, on the other hand, has been a haphazard process, and even the bluest of our blue-blooded families are a hodgepodge of unidentifiable genes. (Literally, no human pedigree would be "fit for a dog.") What we can point to, only, are certain families which by the achievements of their members in unusual proportions may give evidence of carrying more than average quotas of "superior" genes. Such families have included the Darwins and the Huxleys in England, and the Adamses, Edwardses, Jameses, Roosevelts and—most recently, the Kennedys—in the United States. We have also noted that a continuation of musical virtuosity and certain other types of performance in given families might suggest an unusual concentration of special genes, for a time, at least.

Yet it is the exception for outstanding persons or for geniuses to come from distinguished families; and even in the greatest families most mem-

* A contrary view—that ancestry should be traced only through females—was expressed in 1790 by the historian Edward Gibbon. Motivated by the erroneous notions of his day regarding the process of conception, he wrote: ". . . the future animal exists in the female parent; and the male is no more than an accidental cause which stimulates the first motion and energy of life. The genealogist who embraces this system should confine his researches to the female line—the series of mothers." In opposition, James Boswell, writing at the same time, argued on behalf of the male ancestry, citing "the opinion of some distinguished naturalists, that our species is transmitted through males, only, the female being all along no more than a nidus or nurse, as Mother Earth is to plants of every sort; which notion seems to be confirmed by that text of Scripture: 'He was yet in the loin of his Father when Melchisedec met him.' Heb. VII, 10." (Both quotations as given in *A Short History of Women*, by John Langdon-Davies, Viking Press, 1927, page 54.)

bers tend to be mediocrities who ride along, as in a trailer, pulled by their family influence, opportunity or wealth, and who, left to their own power (or genes) would get nowhere. One would have to conclude, therefore, that "distinguished pedigree" or "family" in human beings has genetic significance with regard to individuals only when they show clear evidence of continuing superiority through their own capacities, and irrespective of environmental advantages. As for the importance by itself of having such and such an ancestor, even if one could be certain of carrying a little driblet of his genetic material, in all probability it would have nothing to do with his particular merits—nor is there any guarantee that it wouldn't contain the very *worst* of his genes.

Not least of the common ancestry fallacies is that of ascribing to individuals inherited qualities of a whole group of ancestors—the *stereotypes* mentioned in our preceding chapter on race (page 623): that a person has ". . . the fighting spirit of his Irish ancestors," or ". . . the thriftiness of his Scottish ancestors," or ". . . the stubbornness of his Dutch ancestors"; or sometimes a combination of qualities from several ancestral groups. Even in the estimable journal *Science* an obituary some years ago referred to a scientist as "endowed with stamina derived from pioneer New England ancestors." Traits like this might well be achieved in race horses, fighting bulls or gamecocks through intensive inbreeding and selection over many generations, but it is hardly conceivable that in the highly mixed and random-bred human stocks they would be passed along from ancestors to descendants through genetic inheritance.

When we turn to "inferior" ancestry we find that the same fallacies underlie the compilation of "bad" pedigrees as of "superior" ones. The motives, however, are quite different. As we might express it in verse—

There was a Bostonese
Who searched out pedigrees
Which she stored in the middle of her forehead;
And when they were good, they were very, very good,
But when they were bad—they were horrid!

Which is by way of saying that compilers of pedigrees may be motivated by the very human urge to prove extremes. Those who compile genealogies of persons of "superior" stock are out to show how very good all these people are. In the compilation of pedigrees of "inferior" stock, investigators may unconsciously yield to the opposite impulse.

No clearer instance of gilding the lily and tar-brushing the weed could be found than in the classic study of the presumably two-branched

Kallikak family, which was published a half-century ago. Some decades previously another American clan, the *Jukes,* had been subject to scrutiny, and became linked in the popular mind with the degenerate branch of the Kallikaks as a glaring proof of what bad heredity could produce. Although geneticists soon showed that the methods and conclusions embodied in the studies of these families were wholly unscientific, the ghosts of the Jukes and Kallikaks continue to march on through many books still in use in some schools and colleges, or housed in public library shelves. So let us look at the facts.

First, the Kallikaks. In the early 1900's a pioneer psychologist and authority on mental defects, Dr. H. H. Goddard, was struck by the presence in an area of New Jersey of two family groups, distantly related to each other, but as different in character as the proverbial night and day. The one branch (as depicted) comprised almost uniformly upright, intelligent, prosperous citizens; the other abounded in mental defectives, degenerates, drunks, paupers, prostitutes and criminals. To emphasize the contrast, Dr. Goddard coined for the two-sided family the name "Kallikak" (compounded of Greek words meaning "good" and "bad") and after some research into the pedigrees published this explanation:

> Both clans had stemmed from the same remote ancestor—Martin Kallikak, a Revolutionary War soldier—*but through two different matings.* Martin Kallikak himself, it appeared, was of good stock and after the war had married a proper young Quakeress by whom he had seven children—progenitors of all the "good" Kallikaks. *However,* before his marriage, and while a-soldiering, Martin had dallied briefly with a feebleminded tavern maid, who, after he went his way, bore an illegitimate son *to whom she gave the name of Martin Kallikak, Jr.* This individual grew up to be so wicked and odious that he was known as "Old Horror," and unhappily also, sired ten worthless offspring. It was from these that Dr. Goddard traced all the many hundreds of "bad" Kallikaks. What is more, he concluded that all the differences between the very, very good Kallikaks and the very, very bad Kallikaks had been caused by "superior" and "inferior" heredity coming through the radically different female ancestresses—the prim Quakeress and the feebleminded slattern.

We must remember that the Kallikak study was made when both genetics and the social sciences were in their infancy, and when techniques of investigation and analysis in these areas were rudimentary. Today we can view the situation (if it was as reported) quite differently. For one thing, we can see that a crucial element in the Kallikak study was the *assumption* that the illegitimate child whom the feebleminded mother named after Martin Kallikak was indeed his son—which no court would accept as fact unless proved (and which no court did pass upon at that time). If we reject the premise—that all the "bad" Kallikaks descended from the same man as the "good" ones, but were genetically distinguished by the great difference in their female ancestress—then the whole double-pedigree aspect of the Kallikak study, balanced precariously on this one point, topples of its own weight.

Suppose we do accept the claim as to Martin Kallikak, Jr.'s, paternity. There is still this question: if he ("Old Horror") was so odious—a drunkard, criminal, degenerate, nitwit—because of *bad heredity*, by what gene mechanism did he become that way? No *single dominant* gene could possibly produce his complex state, nor is there any known gene that can singly produce even feeblemindedness. *Recessive* genes, and quite a number, would have to have been involved. Which means that as such genes must come from *both parents* for the effects to assert themselves, no matter how chock-full of wayward genes the feebleminded mother was, the worthy *Martin Kallikak, Sr., himself had to be carrying duplicates of the very same genes* that were involved in producing the obnoxious traits of his presumptive son. Further, it would mean that some of the "good" Kallikaks in all probability also received these genes. By what beneficent fortune, then, did all the "good" Kallikaks—without any exception noted in the study—turn out to be such genetic paragons?

There were other dubious aspects of the Kallikak study, among them the superficial manner in which conclusions were drawn about mental levels and the dogmatic classification of individuals as "prostitutes," "criminals" or "degenerates"—often when they had been dead for many years and when the ratings could be based only on the *recollections* of aged relatives or neighbors. Another important shortcoming was that the individuals traced or reported on in the study represented only *a part* of the Kallikak descendants. In all, the study could hardly be called a scientific one by present standards.

As for the Jukes (also a coined name), this was an unsavory clan clustered in a locality in New York state when first investigated in the 1870's by a prison inspector, R. L. Dugdale. His search revealed that all the members, abounding in every type of riffraff, and heavily inbred, were descended from two eighteenth-century brothers who married a pair of disreputable sisters. A follow-up study of the Jukes, made in 1916 by other investigators, showed that while degeneracy and defectiveness were still rampant among them, many of the newer members were honest, hard-working citizens, some even "superior." This improve-

THE GOOD AND BAD KALLIKAKS: WHAT'S WRONG HERE?

In Revolutionary War times there was a soldier named
"MARTIN KALLIKAK"

ment was ascribed to reduced inbreeding and the infusion of "good outside blood." Needless to say, the Juke studies were as flawed by questionable diagnoses and classifications as the Kallikak study; and the original Juke study—the one most often cited—went further with the conclusion that not only were pauperism and the tendency to have illegitimate children hereditary, but that the effects of an inferior environment were also hereditary and could help to account for the degeneracy of a human group.

Yet whatever the reservations, there could be no gainsaying that the Jukes who were observed, like the bad Kallikaks, were an unusually undesirable lot. But *why* were they so? Let us try to picture one of the girls from these families at the time the studies were made.

Mamie Juke (or call her Mamie Kallikak) lived in a squalid, festering nest of hovels, with a drunken, thieving father, a slut of a mother, and a swarm of untidy, unhealthy, neglected brothers and sisters. Nobody cared whether Mamie did or didn't go to school, and when she did, briefly, she hated it, for the children of decent folks had been warned to shun her and others of her brood. She was pretty lonely until, when she got to be about fourteen or fifteen, some men who hung around a pool hall took an interest in her. Mamie was a little bewildered and pathetically flattered by this sudden attention. She didn't quite understand what was happening, or how it happened, but . . . one day, in a dark corner where Mamie cowered like a sick animal, another illegitimate, potentially degenerate child, was added to her clan.

While this is a synthetic picture, no one who has studied the records would deny that it is a typical one. Knowing the conditions, then, under which such a girl as Mamie, or her child, had started out, can we still argue that they were predestined to inferiority by "inferior" genes? Or could we not equally predict a bad end, regardless of their genes, because of their inferior environments?

It is because earlier investigators did not see this distinction that the studies of the Jukes, Kallikaks and other such notorious families are now so greatly discounted. We are not ruling out the possibility that some or even a good part of the defectiveness of these clans might have been due to hereditary factors. Like often tends to mate with like, and in the course of time, with inbreeding, degenerate strains could develop in certain small areas. But there would have to be a lot better evidence than is now available to prove that the "horrible-example" clans cited were primarily products of wayward genes. In the words of one of the greatest geneticists, Thomas Hunt Morgan, "When groups of persons have lived under demoralizing social conditions that might swamp a family of average persons . . . it is not surprising that, once begun, from whatever cause, the effects may be to a large extent communicated rather than inherited."

The foregoing could easily be applied to many contemporary Kallikak-type groups clustered in city slums or other benighted areas and teeming with criminals, alcoholics, prostitutes, illegitimate children, degenerates and unwholesome characters of every kind. But we need only recall that at one time the whole human race, with genes no different than ours, could have been considered degenerate by modern standards, and that going back not too many centuries the ancestors of many good citizens of today were little different from the Jukes and Kallikaks. So, also, when even in our own time and in our own communities we see the enormous jumps that have been made in various groups from grossly inferior levels to ever higher stages of respectability, we cannot be too sure that some of our present "worst" families may not in time turn into some of the "best" families by changes in environment, education and opportunities.

Thus, the importance attached to either bad or good ancestry, as drawn on paper by genealogists or as deduced from the surface traits of selected individuals, has steadily lessened for geneticists. "Good family" still counts, for it has much to do with how the individual is started and carried along in life. Also—generally speaking—it may very well be an indication of better than average heredity. But for the individual, ancestry and family have genetic significance only insofar as he himself reveals the qualities ascribed to his family and its distinguished ancestors. In short, a family tree must be viewed not in regard to the fruit it once bore, but to the fruit it now is bearing—or can be made to bear with proper care.

The observations about ancestors can be further applied to the importance or unimportance of *immediate relatives* with respect to one's own genetic qualities. You might boast, understandably, that your father is the eminent Who Is Who, or that your brother is Such and Such, or that you're a first cousin of the famous So and So. You likewise might try to cover up kinship to a certain black sheep in the family, with the fear that his or her bad traits reflect on your own hereditary makeup. So let us see what *genetic* connection there actually is between you and given types of relatives in terms of chromosomes and genes shared with them, and what bearing their traits may have on your own.

The closest possible genetic relative—a 100 per cent one—would be an *identical twin,* as discussed in detail in previous chapters. With identical twinship it can be shown that almost invariably the chromosomes and genes of the one twin are fully matched by those of the other.* In the case of all other relatives, however, genetic relationships can be estimated only in terms of averages. We already have observed how the averages work with regard to ancestors. Here now are facts as they may be applied to various of one's close and existing relatives:

PARENTS AND CHILDREN. We know that a child receives from each parent 23 of his or her 46 chromosomes, which would imply that each parent is a *50 per cent* genetic relative of the child. But this isn't precisely so. For one thing, neither the mother's nor the father's genetic relationship to a son is the same as that to a daughter. The reasons lie in the two chromosomes, the X and Y, which determine sex (all the other chromosomes from among the parental pairs, being

* In rare instances, noted on page 194, identical twins may be genetically differentiated in some respect (1) because of an abnormality in the chromosome-formation mechanism during or after conception, before the two became separated, or (2) because of some gene mutations occurring in the one and not the other after their separation.

received impartially by sons and daughters). Since the small Y which the son gets from his father has very few genes at best, whereas the X from the mother is one of the largest chromosomes, it is estimated that a son has about 5 per cent more of the mother's genes than of the father's, and is to that extent genetically more closely related to her. (This also accounts for the many sex-linked diseases and defects which sons inherit solely through their mothers.) In a daughter's case, since she receives an X from her father (his only one) as well as an X from her mother (either of the two Xs), her genetic relationship is the same to each parent in terms of amounts of genes. A special point is that if the parents are first cousins, or otherwise are quite closely related, they could be carrying a number of matching chromosomes; and if so, in addition to the individual parent's own chromosomes, replicas of his or her chromosomes coming from the other parent would increase the genetic relationship to the child beyond 50 per cent. (In highly inbred lower animals a parent's relationship to an offspring may almost approach the identical-twinship degree.)

BROTHERS OR SISTERS. Theoretically, any two brothers or any two sisters are (like the parents) "50 per cent" genetic relatives, since each has received half of the father's chromosomes, and half of the mother's so, *on the average*, half of the chromosomes received by each child might be expected to be replicas of those received by another sibling. But, actually, it would be the exception for two brothers, or two sisters, to have received sets of chromosomes percisely half of which were the same. Apart from the father's Y going to every son, and his X going to every daughter, any proportion of the other 45 chromosomes received by given brothers or sisters from the two parents might or might not be matching ones. (As explained in Chapter 5, in the process of conception the parental chromosomes—one from each pair—are dealt out to a child at random.) Thus, if there were three brothers in a family, "Tom" might have 34 or 37 or more of his 46 chromosomes in common with "Dick," but only 18 or 16 or fewer chromosomes in common with "Harry." On this basis, one pair of brothers or sisters might be "80 per cent" genetically related, another pair only "30 per cent"—with the hypothetical extremes or relationship ranging up to complete genetic identity (as if they were one-egg twins) and down to zero. (The odds against either of these extremes are one in many millions.)

BROTHER-SISTER RELATIONSHIP. What distinguishes a brother-sister genetic relationship from that between two brothers, or two sisters in a family, is this: The brother has the father's Y chromosome, the sister the father's X, in addition to an X each received from the mother. Since the little Y is deficient in a great many genes carried by an X, two sisters, with their two Xs each, may be genetically more related

HOW MUCH ARE YOU RELATED?

Here are the average degrees of relationship between yourself and relatives of different kinds, in terms of average numbers of chromosomes you might expect to have in common.

You and:	GENETIC RELATIONSHIP (AVG.)	CHROMOSOMES SHARED (AVG. IN BLACK) *You* — *Relative*	QUALIFICATIONS
Identical Twin	100%	46	*In rare abnormal situations, identical twins may differ in a chromosome or two, or in some genes. (See Text.)*
Father or Mother	1/2	23	*A son inherits slightly more than half from mother, slightly less than half from father, because the X he gets from mother carries more genes than father's Y.*
Sister or Brother	1/2	23	*Two siblings may have more or fewer than half chromosomes matching. But two brothers always have same Y, two sisters same X from father.*
Uncle, Aunt, Nephew, Niece, Grandparent, Grandchild	1/4	11 — 12	*You and any of these relatives may have more or fewer than a fourth of your chromosomes in common.*
1st Cousins, Great Grand-parent, Great Uncle,-Aunt, Great Nephew, Great Niece	1/8	5 — 6	*You and any of these relatives might have more or fewer than the average of one-eighth of your chromosomes in common, or none.*
2nd Cousins, Great-Great Grand-parent,-Uncle, -Aunt,-Nephew, -Niece	1/16	2 — 3	*You and any of these relatives might have an average of two or three chromosomes in common, or none.*

chromosomes not shared

chromosomes from father

chromosomes from mother

to each other than are a brother and sister—that is, in the one-in-two cases where the two sisters carry the same X from their mother in addition to the X from their father. Other than this, as noted previously regarding siblings of the same sex, a brother and sister could be genetically related in any of various degrees, depending on how many or how few matching chromosomes they received from their parents.

UNCLE OR AUNT, NEPHEW OR NIECE, GRANDPARENT OR GRANDCHILD, A HALF-BROTHER OR HALF-SISTER. Each, theoretically, is a "one-fourth" relative of the person in question, with the average possibility of carrying replicas of 11 or 12 of that person's 46 chromosomes. Your uncle, for instance, if he is your father's brother, has an average of half of his chromosomes in common with your father; and since you have half of your father's chromosomes—or half of the uncle's half—½ × ½ equals ¼. The genetic ratio would be the same as between you and an uncle on your mother's side or an aunt on either parent's side, or between you and a niece or nephew of yours. Similarly, in the case of a grandparent—your father's father, for instance—if half of his and your father's chromosomes are the same, and you have half of your father's, again ½ × ½ equals ¼. Likewise, if you have a half-brother or half-sister, since he or she would carry ½ of only one parent's chromosomes, your ½ times his or her ½ would also equal ¼. But remember, once more, that as with brother and sister relationships, we are speaking of averages, and that in actuality any of these theoretical "one-fourth" relatives might have many more or many fewer than a fourth of their chromosomes (theoretically up to half or down to none) in common with you.

A FIRST COUSIN, OR A GREAT-UNCLE OR AUNT, OR A GREAT-NEPHEW OR GREAT-NIECE, OR A GREAT-GRANDPARENT OR GREAT-GRANDCHILD. Each is technically a *one-eighth* relative, with the average possibility of carrying replicas of either 6 or 5 of your 46 chromosomes. But, as previously observed, in individual instances the genetic relationship may be greater or less—up to one-fourth or down to none.

RELATIVES FURTHER REMOVED. Second cousins, or *great-great*-uncles, aunts, nephews, nieces, grandparents, grandchildren are technically *one-sixteenth* relatives, with the average possibility of carrying replicas of 2 or 3 of one's 46 chromosomes—but in actuality up to 6 or down to none. With each step further removed in relationship, the number of possible chromosomes in common is halved (as shown in Table X, page 638).

Altogether, it should be clear now that *genetic* relationships may be quite different from degrees of kinship as viewed legally and popularly. For example, in a court case involving the estate of a person who died without making a will, and who had an identical twin, that twin would not be regarded as having any closer relationship to the deceased, or any more claim to the estate, than an ordinary sibling—even though the twin was genetically a 100 per cent relative, whereas another sibling was no more than a 50 per cent one, or even less. The law also would disregard a daughter's somewhat greater genetic relationship than a son's to a father or mother (in the extra amounts of her "X" genes). Nor, obviously, could the law know or recognize the varying degrees of genetic relationship which different brothers or sisters, or relatives in the "one-quarter" or "one-eighth" groups, might have had to the deceased. Generally when there is no acceptable will and estates are in dispute among potential heirs, claims of relatives would be decided in the order of the categories shown on our accompanying Chart: First, the "half" relatives, sharing equally; if none of these, the "quarter" relatives; if none of these, the "eighth" relatives, and so on.

Degrees of "blood ties" as viewed popularly may also be different from what they are genetically. A grandparent, for example, might be thought of as "closer" than an aunt or an uncle, whereas genetically the relationship is the same. Or a first cousin may be thought of as "closer" to a given person than a great-aunt, great-uncle or great-grandparent, although all are of the same genetic rank (one-eighth). In these cases the degrees of social contact and familiarity between the relatives in question might sway the notions as to their genetic closeness or distance.

In any event, exaggerated importance is apt to be attached to similarities in traits as seen or believed to exist among relatives: "He's got his grandfather's head for business"; "She inherited her grandmother's deviltry"; "That lad has his Uncle Bill's mean streak." But as noted in previous chapters, if heredity does influence such mental or behavioral traits, it would be only through complex gene combinations, and the odds are heavily against any such combinations' being passed on intact from one individual to another. The further removed the relative, the less the chance that similar mental or behavioral traits recur through heredity. Thus, when an uncle, aunt or grandparent—not to mention someone at a greater distance—thinks a particular young relative is a reincarnation of himself in mind and character, that is pretty certain to be either an illusion or the result of strongly similar conditioning.

In your own case, it may well be true that one brother or sister, or some other individual relative, is much closer to you in thinking and behavior than another. But unlike similarities in coloring and features, which may be largely due to heredity, mental similarities cannot be so easily ascribed to siblings' having the same genes for the traits. Quite possibly the brother or sister who looks most like you may have fewer replicas of your "behavior" genes than the sibling who least resembles you. Uncertainties on these points may well be much reduced or entirely cleared up when geneticists, in time, succeed in mapping human chromosomes and identifying which genes are on which, as has been done with the chromosomes of fruit flies and several other experimental creatures. In human beings, exact identification has been approached only with respect to the X chromosome. But once all the human chromosomes have been tagged, it may be possible, by studying known inherited physical and pathological traits of given members of a family or pedigree, to estimate how many and which specific chromosomes they have in common, and thereby, also, to determine what their precise genetic relationships are, and to what extent similarities in their functional and behavioral traits may indeed be due to heredity.

All of which leads to this broader question: If we think not only of individual family trees, but of Mankind as a huge orchard filled with countless family trees, how well can we distinguish the groups of trees which may be destined to bear good fruit and those which are not? And if we wish to have a better human orchard in the future, what can be done about it? This will be the subject of the next chapters.

CHAPTER FORTY FIVE

Personal Problems of Heredity

A GIRL WHO'S JUST BECOME engaged learns that insanity runs in her fiance's family. A first child has been born with a serious genetic defect and the parents are fearful about having another. A couple eager for parentage finds that the stork may never come to them. A man wonders, "Is that baby really mine?"

These are among the many situations about which experts on human heredity are constantly consulted. You yourself may want or have wanted advice in similar situations. The answers that can be given have importance not only for individuals. In their broader meanings most of the personal problems of heredity and parenthood also concern the community, the nation, and ultimately, mankind as a whole; for the world of tomorrow and of the future stretching beyond can be no better than the potential quality of the children born into it. Thus, the genetic problems of individuals, with which we'll deal now, must be viewed as part of the larger problems of human reproduction to be discussed in the succeeding chapters.

We'll begin with this question:

Suppose you are on the threshold of marriage and are worried about the children you might have with a particular chosen person: How much can be foretold from the traits revealed in you, your prospective mate, and your two families?

Facts previously brought out in this book should provide some general forecasts regarding your prospective children's looks, physical fitness, defects, and mental capacities—always allowing for the interacting effects of the genes they receive and the environment provided for them. In the matter of your children's heredity, the best clues lie in your own genetic traits and your mate's. Any traits shown by immediate relatives on either side can have significance for your offspring only insofar as they may indicate what hidden genes you each may be carrying. (Here is where the Wayward Gene Forecast Tables in our Appendix, or, where required, medical-genetics counseling, may be helpful.) With respect to environment, however, your relatives may also be important, to the extent that they will affect your children's upbringing, directly, or through their influence—past or continuing—on your own behavior. In other words, prospective children must be thought of as products not only of a combination of their parents' genes, but of the combined influences of the two parental families.

Suppose there is clear evidence that a serious hereditary defect or disease is present in the person you're in love with, or in yourself, or in either or both of your families: How should this affect your plans for marriage?

If having children is a primary objective, you must weigh carefully the nature of the condition and its manner of transmission. Should the condition be a *dominant* one (the exception), so that a single gene from one parent can cause the damage, you know there is a straight fifty-fifty threat of its reappearing in a child. With the risk so great, your decision must be based on the degree to which the condition has affected the life of the one afflicted, how you would feel about having a child similarly afflicted, and whether you could compensate it for the handicap. Even if medical science has developed, or may develop, a means of easing the effects of the condition, you must keep in mind that a child inheriting it will continue to have a fifty-fifty chance of passing it on to any of his or her own children. Such a risk might call either for giving up the thought of having children, or—if it is the other person who has the affliction—for marrying someone else.

In the recessive diseases and defects, produced when at least two of the same genes come together, the problems posed are usually less serious, but they may also be more complex. That is to say, while the probability of transmitting a recessive defect is in most cases much less than with the dominant conditions, the odds and circumstances to be considered are much more varied. This applies particularly to the *recessive-plus* or *recessive-qualified* conditions, where, in addition to one pair of recessives, some other gene or genes, and/or something adverse in the environment, are required for the condition to manifest itself. In the sex-linked recessive conditions (of which the most serious ones, such as hemophilia and certain types of blindness or muscular defects, are rare), there is usually a threat to children only if the wife's father or brother is afflicted. If the husband is afflicted the threat is only to his daughter's children.

Suppose, then, you are worried because among your near relatives—or those of your mate, present or prospective—there is someone with inherited idiocy, congenital blindness or deafness, albinism, or some other condition listed as recessive. Regardless of which family has members so afflicted, *no child of yours will inherit that condition* unless both you and your mate are carrying a gene for it and *both* transmit the same gene to the child.

Here is one of the most significant and fear-easing findings of genetics. How many a hapless husband or wife has been scourged by an embittered spouse with the words, "You know the child inherited it from you!—it runs in your family!" What cannot be stressed enough is that wherever there is a child with a recessive defect, the inheritance has stemmed from both parents; and where there is hesitancy about a marriage or conception because of fear that such a child may result, the family histories on both sides should be considered with equal thoroughness. The same principle applies in large measure to our next question:

Is there hereditary insanity in a family?

Not unless the person marrying into the family *also* carries genes or supplementary genes for the same type of mental illness will a child of that person be likely to inherit the condition. And only in such an extremely rare condition as Huntington's chorea—which is more a neurological than a mental affliction—can transmission come through one parent alone. The most common mental ailment, *schizophrenia*, requires for its appearance at least a pair of recessive genes, plus, possibly, some other gene, and perhaps also an unfavorable environment. In manic-depressive insanity the genetic factors producing susceptibility are still not clearly identified, but even if (as some authorities think) *dominant-plus* genes are responsible—a dominant gene and one or more other genes—the probability is that one parent alone could not contribute the necessary genes. Thus, the situation in the common mental diseases would be the same as with the recessive conditions: Whether or not there is a mental disease on one person's side, it is unlikely an inherited predisposition to that disease will occur in a child unless the other parent carries and transmits some hidden matching gene or genes for the same condition.

In any case, wherever the inheritance of a disease or defect must come through both sides, the immediate threat lies in matching gene "carriers." Often the facts are quite clear. If either mate has or had a parent afflicted with a simple recessive condition, that mate must be a carrier. If both mates are carriers, the risk that any child of theirs will be afflicted with the condition is one in four; and if either mate has the condition and the other is a carrier, the risk to a child is fifty-fifty.

COUSIN-MARRIAGE RISKS

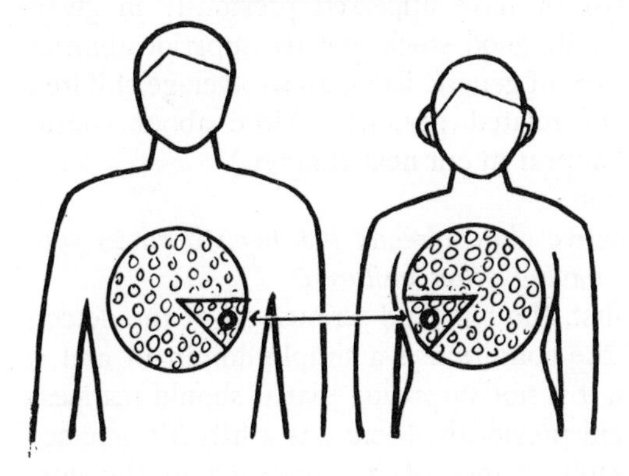

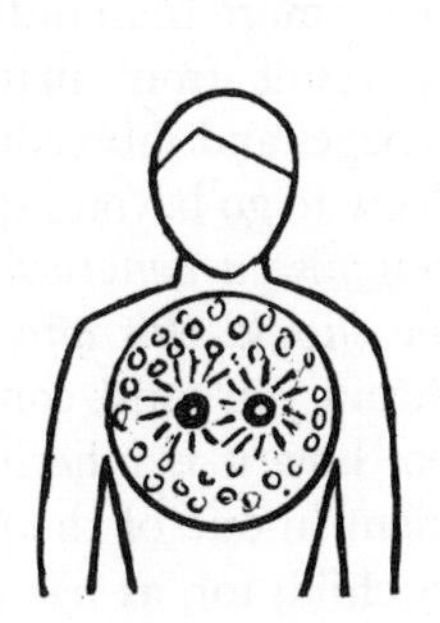

One-eighth of the genes of first cousins are the same. If one cousin carries a hidden (recessive) wayward gene, the chance is one in eight the other has the same gene.

If first cousins marry, the risk is thus much above average that a child will receive two of the same wayward genes and will develop the recessive defect.

If the condition has not appeared before in the immediate family of either mate, but appears in one of their children, the parents can know that both of them are carriers, and that a one-in-four threat faces each of their succeeding children. Most serious of all, if both parents themselves have the same recessive condition (for instance, common diabetes or albinism), *every one of their children* will inherit it.

(For other risks to children from hereditary conditions present in parents or prospective parents or their near relatives, see the Wayward Gene Forecast Tables in the Appendix.)

Suppose the two persons contemplating marriage are cousins: What special hereditary threats would there be to their children?

Popular notions about the grave dangers in cousin marriages may be exaggerated, but special genetic risks to the offspring do exist. As was explained in the preceding chapter, familially related persons can be expected to have a certain proportion of their genes in common—the closer the genetic relationship, the higher the average proportion of matching genes. In first cousins, for instance, one-eighth of their genes—sometimes more, sometimes less—might be the same. Accordingly, there is a far-above-average risk that when cousins marry, matching wayward genes will be passed along to their children. With second cousins (cousins once removed) who may share one-sixteenth of their genes, the genetic risks to progeny are correspondingly reduced, but are still much above average. However, all of this is dependent on the extent to which genetic defects exist or have appeared previously in given families. If these are of unusually good stock or have superior abilities and are more than ordinarily free of genetic flaws, above-average children may result from marriages of related persons. (More about cousin marriages and inbreeding will appear in our next chapter.)

Now to go beyond speculation:

Suppose a genetically defective child already has been born to you. How should this affect your having other children?

Again you must consider, first, the type and seriousness of the defect; then, how it is inherited. If the condition is a simple dominant and is present in one of the parents, it is not surprising that it should reappear in a child; for, as has been said previously, there was a fifty-fifty chance it would, and there is the same chance of its appearing in the next child, and the next. But if the condition, ordinarily dominant, is not present in either mate, (1) the gene for it may be "qualified" in its workings and have "skipped over" from a previous forebear without showing its effects in the parent carrying it; or (2) the gene may have arisen in one of the parents through a mutation in the germ cells. In the first case, if the condition has appeared in one child there is a good chance it will reappear in another. In the second case—of a mutated dominant gene—the odds are overwhelmingly against another child's receiving a similar gene. What the facts are may require expert opinion.

Situations very much as in dominant conditions may be presented by *sex-linked* diseases or defects, transmitted by mothers to sons. Once a son with one of these conditions has appeared—for instance, a *hemophiliac* or a child with a sex-linked type of muscular dystrophy—there is a fifty-fifty chance that any succeeding son will be similarly afflicted. But in actual experience two or more afflicted sons may follow one another, as has happened in families on record with four or five hemophiliac sons (including triplets in one family) or with four muscular-dystrophic sons.

Where the condition is recessive, the fact that the statistical averages call for a one-in-four risk to children should likewise not be misinterpreted. Some parents, having already had the "one-in-four" defective child, may hopefully think that there is no risk for the next three. But all too many cases can be cited where two or more children with some recessively inherited type of idiocy, or some crippling disease or other serious affliction, were born successively to the same parents. Thus, while

TABLE XI

Risks of a Repeat Defective Child

*Referring mainly to congenital abnormalities and mental defects apparent in early childhood, whether hereditary, environmental, or a combination of both.**

If a couple has had one child with this defect:	The chances of having another child with the same defect are:
Albinism	1 in 4
Cleft palate (alone)	1 in 7
Clubfoot	1 in 30
Dwarfism, achondroplastic (when hereditary)	Up to 1 in 2
Extra fingers and/or toes	1 in 2
Hands or feet malformed (hereditary types only)	Up to 1 in 2
Harelip (with or without cleft palate)	1 in 7
Heart, malformed	1 in 50
Idiocy	
Mongoloid	
(a) If mother has normal chromosomes: (Chances slightly increased over average for mother of same age. But still low risk)	1 in 50
(b) If mother has a chromosomal translocation (an extra No. 21 attached to another chromosome)	Up to 1 in 3
Amaurotic	1 in 4
Phenylpyruvic amentia	1 in 4
Cretinism	Uncertain
Intestines, pyloric stenosis	1 in 17
Kidney, polycystic congenital	1 in 4
Spina bifida	1 in 25

* For other conditions not listed above, see the Index, and then note facts in the Text and/or the Wayward Gene Tables. As a rule, for any simple recessive condition the repeat risk is 1 in 4; for a simple dominant, 1 in 2. But in no case make any final judgment without consulting a qualified doctor or a medical geneticist.

in lucky instances the genetic odds may be defied, in other less fortunate cases genetic lightning can indeed strike twice—or more often.* So once parents produce a child with a serious dominant or recessive hereditary defect, they must be prepared for the possibility that others may follow.

Suppose that all of the children you expect to have are alive and thriving, and no serious hereditary defect has yet shown up in any one of them. Is there any need to worry about defects appearing later?

The foreknowledge of hereditary threats offered by family histories may well be important, as indicated in our Wayward Gene chapters. In various serious conditions—heart, cancer, diabetes, severe allergies, certain functional disorders, mental illnesses—awareness by your doctor of any menaces to your children may help greatly in early detection, diagnosis, and treatment, often leading to prevention, reduction of the worst threats, or cure. To cite specific situations:

If the *heart and arterial ailments* are known to run in families, tests frequently may identify the vulnerable members, and special precautions may help to reduce the dangers.

In *cancer*, the previous appearance in parents or near relatives of breast, intestinal, rectal or stomach cancers should be a special reason for the periodic examination of the children when they reach maturity, particularly of those in the age range where the specific condition usually asserts itself. In *rectal polyps*, which may occur in childhood, detection in one member of the family should immediately put others on their guard.

If *diabetic* inheritance is suspected, tests can confirm the presence of the tendency well in advance of the serious manifestations, and control of diet and eating habits may thwart the development of the disease or impede its progress.

In the *mental diseases*, the foreknowledge of a familial predisposition may make possible various precautionary steps which a psychiatrist can recommend. Abnormal behavior in a child whose family has some history of mental disease should immediately call for psychiatric attention.

* Among the sad examples: In the case of *cystic fibrosis*—a recessive condition with a one-in-four repeat risk after an afflicted child has appeared—a family seen at the University of Wisconsin clinic in Milwaukee in 1960 had six out of eleven children with the disease; and in Chicago, the same year, a family was reported which had lost four out of five children through the disease. In muscular dystrophy and in hemophilia there also are many instances where the proportion of afflicted children (sons in the sex-linked conditions) has gone much beyond expectations. (For a discussion of whether society should interfere to prevent such repeat cases, see Chapter 47, pp. 709 ff.)

Where special kinds of *allergies* run in families, parents who promptly take note of any unusual reactions in their young children to specific foods or substances, and who obtain proper medical advice, may forestall serious disorders.

All of these facts are documented in medical literature by innumerable instances of children and adults who were saved from death or serious disease by prompt action based on knowledge of inherited tendencies. So, too (and unhappily more often), failure to be alert to hereditary threats or to act against them has brought tragedy. Nevertheless, there can be such a thing as worrying too much about heredity. Sometimes the very fear of having inherited a disease—especially, a mental illness—may actually produce some of its symptoms through *psychogenic* (mentally induced) effects; and constant concern or talk about some hereditary doom hanging over one's children can hardly produce an emotionally healthy home environment.

Several other points brought out in earlier chapters may again be stressed here. One is that inheritance of a serious condition, or the susceptibility to it, by no means implies that there is no hope of successful treatment or cure. Many serious hereditary ailments due to "errors of metabolism"—including diabetes, phenylketonuria, and galactosemia—have had their effects minimized or lessened by chemical treatments and diets. Some of the serious hereditary eye and ear defects are now being successfully corrected by operation or aided by various devices. Mental diseases are in many cases being ameliorated by drugs. Genetic sex abnormalities are often being modified by surgery and hormonal treatment. In fact, there is every reason to expect that as medical science advances and the causes and nature of more and more hereditary diseases and defects are clarified, their threats to *individuals* will be greatly reduced. (What the effects on society at large might be of helping genetically defective persons to survive and propagate is another matter, to be discussed in our next chapter.)

It is also worth emphasizing that a host of conditions formerly considered hereditary now have been proved otherwise, or, even if there is some hereditary element, have been shown to be largely dependent for their expression on environment. This applies to many congenital abnormalities and infectious diseases, some types of mental deficiency, and various forms of blindness and deafness, among other afflictions. On the whole, so much has been done to clear up the confusion as to what diseases and defects may be blamed on heredity, and if so, how, that an average couple today can have children with less worry regarding the genetic threats than ever before.

But being free in one's mind about having children does not always mean that the stork will obligingly deliver them. So to our next problem:

Suppose, after trying to have a child, you find you apparently can't have one?

In about one in ten couples this is the case; and it may be most upsetting to a man and woman who are unusually healthy, virile and ardent. But, alas, fertility and outward physique do not necessarily go together. Dr. Alan Guttmacher, Planned Parenthood president, tells of a strapping, all-American football star who was found to be sadly deficient in his sperms and incapable of fathering a child.* Similarly, "a woman who looked as though she were Mother Earth herself was totally barren." On the other hand, as he notes, an anemic little wisp of a man and his scrawny, frail little wife were producing a child every year.

When or how soon, then, can a couple suspect or conclude that they are barren? The time lapse is important. Ordinarily, if there is no contraception, about 60 per cent of the wives will conceive at the end of six months, 75 per cent in one year, 80 per cent in two years. (In any single sexual union without preventive measures, the probability of conception may be between one in 25 to one in 50.) When at the end of one year of trying, no pregnancy has occurred, there is only a fifty-fifty chance it will happen thereafter, with the chance diminishing the longer infertility continues. These are averages, however. The possibilities for conception, whether soon or at a later time, are best for women in their twenties, less for the very young, and least for women in their late thirties and thereafter.

Once it seems clear that something is keeping the stork away, aid can be sought from an expert on fertility (whom one's doctor can recom-

* Parallel situations in prize horses and cattle have brought grief to breeders. One of the greatest stallions of all time, the Triple-Crown winner, "Twenty Grand," proved to be totally sterile, his semen samples showing all the sperms to be dead. Moreover, according to Dr. John MacLeod of Cornell University, an expert on such cases, sterility or very low fertility ran extensively in the stallion's line and may have been hereditary—both his grandsire, Swinford (from whom the line got its name), and his sire, St. Germans, having been "shy breeders" who begot few progeny, as was also true of various descendants. Among bulls, the famed Hereford, T. Royal Rupert 99th, turned out after purchase for breeding purposes by Texas cattlemen to have underdeveloped sex organs. Again, in 1963, the prize Scottish bull Lindertis Evulse, bought by New York State breeders for $176,400, also appeared to be infertile. (In both instances the purchasers had their losses made up by repayment or insurance.) These experiences have now made all big sales of prize stallions and bulls conditional on proof of fertility.

mend) or at one of the Planned Parenthood clinics. The medical history of the husband and wife might offer initial clues. A next step may be an examination of the husband, for where formerly barrenness was almost always blamed on the wife, it is now known that in about one-third of the cases the deficiency lies in the husband. His sperms may be too few per ejaculation, or too sluggish, or preponderantly abnormal; or no sperms at all may be present, or, if they are produced, a blockage in the ducts may prevent their discharge.* These deficiencies or failures may be due to diseases or infections, sex-organ abnormalities or injuries, or other biological defects, while psychic factors, too, may also be involved.

In a woman the causes of infertility may be much more varied, and often more difficult to identify, than those in a man. Granted that she is producing eggs, one cannot examine them (as one can a sperm sample) to see if they are fertile. However, doctors may find obstacles in the Fallopian tubes which prevent eggs from descending, or malformations or defects in the reproductive organs. Or it may be that the vaginal and/or cervical secretions are hostile to sperms, or that the uterine environment is unfavorable for carrying through a pregnancy. As with a man, also, various diseases, sex-organ abnormalities, glandular defects or even psychic inhibitions and disturbances affecting the body processes may account for infertility in a woman. Again, in either a male or a female, infertility may be traceable to *genetic* factors—sex-chromosome abnormalities or gene defects (as explained in Chapter 27, page 312). Sometimes where each mate is reproductively normal as an individual (or with another mate), the two together might have their fertility impaired through a combination of their genetic factors. For example, there may be incompatibility between their blood groups (A-B-O, or Rh), so that blood elements of the mother may clash with those of the fetal child—because of the father's contribution—and result in prenatal death; or each mate might be carrying some wayward gene which, in a double dose, could prevent conception or induce an early miscarriage.

Happily, in at least a third of the cases of previously sterile couples, medical treatment and procedures can now overcome the difficulties and bring them children. Men may be helped to produce and release the required quotas of normal sperms; blocked Fallopian tubes can be opened;

* A possibility being explored is that in some cases a wife may be "allergic" to her husband's sperms: that is, her vaginal secretions may carry antibodies which are hostile to the sperms and could render them infertile. The reactions would be similar to the incompatibilities between the blood-group antibodies and antigens of two given individuals.

defects of the uterus may be corrected; hormonal or other drug treatments may stimulate ovulation, induce conception and make wombs more hospitable to fetuses; and if there are "psychic" hindrances, these, too, may be eased or removed by therapy. Through one means or another, conceptions for most couples who can respond to treatment usually follow within a year.

This still means that a majority of those who have sought sterility cures will be disappointed. When the wife has been proved irremediably sterile, nothing more can be done. But what if she is fertile, and it is the husband who is infertile—as is true, we noted, in a third of the cases? Here there is another possibility—but one involving a procedure which is highly controversial. We refer to *artificial insemination* of the wife with sperms from some unknown donor. The process has been in use for many years—the first attempts in the United States dating back to about 1870—and many thousands of individuals now alive, from children to middle-aged adults, owe their existence to its efficacy.* (Exactly how many Americans have been born through artificial donor insemination is hard to say because so much secrecy has surrounded use of the process. But conservative estimates place the number at from about 15,000 to 20,000 to date, with perhaps 1,000 to 1,500 now being added yearly.)

There may be any of several reasons for a couple to seek a child through artificial insemination. First, it should be noted that the pro-

* In the breeding of prize cattle and race horses artificial insemination has been standard procedure for decades. (It also has been used with pedigreed dogs, one notable case involving the famed male Scotch terrier pet of President Franklin D. Roosevelt, Fala, whose sperms were inseminated in a female after normal mating proved difficult. Twins resulted.) The principal advantage of the method in livestock breeding is that the services of a prize male can be vastly multiplied and carried over wide areas. Thus, in bulls, the sperms from one ejaculation (above five billion) can be divided into a great many effective portions, each of which will be enough to impregnate one female. Transferred to vials and refrigerated, the sperms can be kept fertile and shipped by airplane to almost any point—as when sperms from South American prize bulls are flown to the United States. Females can then be impregnated at the precise time most favorable for conception. Another important factor is that the breeding life of a highly desirable male can be much extended. For instance, the historic Hereford bull Hazford Rupert LXXXI—sire of the sterile Rupert mentioned in a previous footnote, as well as of vast numbers of fertile Herefords—was having his sperms used almost until his death at the age of twelve in 1947, and even in his last arthritic and enfeebled year was able through artificial insemination to sire 190 more calves. Another famed bull, the Holstein Red Apple, reportedly sired more than 15,000 offspring through artificial insemination during a period of three and a half years.

cedure can be of two kinds. In one, the most widely publicized, the sperms used are those of an anonymous donor. This is technically referred to as A.I.D.—the first two letters for "artificial insemination," the third letter for "donor." In the other procedure, to be discussed later, the husband's sperms are used, this being referred to as A.I.H.—the H standing for "homologous" ("the same," which might also mean "the husband's"). The A.I.D., or donor method, is most commonly resorted to when the husband is hopelessly sterile or infertile. But in certain cases where the husband is fully fertile there is the fear that he may transmit some serious hereditary defect to a child, or that he and his wife in combination, through matching recessive wayward genes they are known to carry, may produce a defective child. Or there may be the Rh situation—the husband Rh-positive, the wife Rh-negative—with previous conceptions having ended in miscarriages because of this. In any of the foregoing cases the donor-insemination method could be looked to as an alternative.

The procedure in the donor method, once both husband and wife have given formal, written consent, is fairly simple. The doctor engaging in the practice (there now are many throughout the country) has on file a number of prospective sperm donors—mostly young college men, and often medical students or internes, who receive modest fees—each of whom has been thoroughly investigated for his health, mentality, character, high-fertility sperms and genetic background. From this list the doctor selects a donor who is as close as possible to the husband in outward genetic traits, as well as in basic blood groups, to forestall any question of paternity which might come up later. (However, where the Rh problem threatens, an exception is made in favor of an Rh-negative donor.) The doctor then determines the nearly exact time of the month when the woman is likely to be ovulating and introduces sperms obtained from the donor. The process may have to be repeated for several successive months; but as a rule, unless the woman turns out to be infertile, conception results. The expectant mother is then turned over to an obstetrician who delivers the baby, and unaware of the circumstances, files a certificate naming the husband as the father.

Here, then, is a means whereby a wife who might otherwise be doomed to complete barrenness can have the experience of motherhood; where she and her husband can have children— as many as an average couple—each of whom would be at least "half theirs," but which the outside world would consider completely their own; and where no child need ever know there had been anything unusual about its conception. All of this is possible for hundreds of thousands of barren couples who are anxious to have children. Why, then, do so relatively few avail themselves of the opportunity?

The major reason for rejecting artificial insemination with donor sperms is obvious: The opposition of many husbands (and perhaps wives, too) on psychological grounds. Bound up with this is the antagonism to the method fostered by religious, moral and social attitudes. And there are also legal problems.

The religious opposition to donor insemination has come chiefly from the Roman Catholic Church and from the Anglican Church. But there has been no formal opposition from many Protestant groups, and at least one group—the United Presbyterian Church in the United States—has endorsed the procedure. Among Jews there has been little if any formal opposition. Thus, in most cases, it would appear that the question of whether or not to employ donor insemination is left for individual couples to decide on the basis of their personal feelings and attitudes.

The legal issues involved are extremely complex and yet to be settled. If a wife permits herself to be inseminated with the sperms of a man not her husband—but without any sexual contact with him (not even knowing his identity), and with the full consent of her husband—does this make her guilty of adultery? The few judicial decisions on the point have been in conflict, but most legal experts would say "No," and would support the verdict of a Scottish court in 1958 that "there cannot be adultery without actual sexual intercourse." Again, while some church authorities (Catholic and Anglican) have also construed the procedure as adulterous, other religious leaders have disputed this.* Another question: Is the child resulting from a donor insemination illegitimate (since the biological father is not married to the mother), or at least not legally the husband's child? To date one Chicago judge and a British judge have handed down the "illegitimate" verdict, whereas a New York State judge offered the *dictum* that the "insemination" baby in a case in dispute was not illegitimate and was as legally the husband's as if it had been formally adopted by the couple.

Another New York State judge held, in 1963, that a man whose marriage had been annulled on the grounds of his impotency was still legally responsible for the support of a child born to his wife through

* The assembly of the United Presbyterian Church previously mentioned, when it sanctioned donor insemination in 1962, also endorsed the statement, "To discover in artificial insemination by an anonymous donor an act of adultery is certainly to give the word a meaning that it does not have in the New Testament." The assembly (representing three million members) urged Presbyterians to work for uniform state laws that would protect the legal rights of "test-tube" babies.

artificial insemination to which he had formally consented while they were together. Also, in Israel, Chief Rabbi Nissim ruled in 1958 that children born to parents as the result of artificial insemination will be recognized by the Jewish religion as legitimate.

A further moot point relates to inheritance rights of a "test-tube" child. If the stipulation is that the inheritor must be the "issue" or "blood-relative" of the deceased, a donor-insemination offspring would be ruled out. A will, however, could be so phrased as to eliminate discrimination against an "A.I.D." child. But in the matter of hereditary titles, the British College of Arms, which passes upon these, flatly ruled in 1960 that no infant born through anonymous-donor insemination could be heir "of the body of the grantee" (i.e., the title) and that "if he is allowed to succeed . . . there is fraud on the sovereign . . . and on the nation."

Entirely different in its implications is the A.I.H. insemination procedure, using the husband's sperms. This has met with no religious or moral opposition and presents no legal obstacles. The method is actually employed far more often than is donor insemination and is of great value in the following circumstances: (1) Where impregnation of the wife is difficult or impossible by ordinary means; or (2) where she has proved to be of low fertility, and it is essential that she be impregnated at precisely the right time and under the most favorable conditions; or (3) where the husband is impotent or unable to impregnate his wife in a normal way, but yet produces fertile sperms. The A.I.H. method, too, has been bringing children to innumerable couples who otherwise might have remained barren.

The controversy over artificial insemination, then, relates almost entirely to the donor procedure. But if a couple has no religious or psychological restraints, and care is taken to insure secrecy and to forestall legal complications, the aftermath of a successful A.I.D. insemination is usually a very happy one. Added to the likelihood that the baby resulting will be more than ordinarily healthy, in almost every instance there not only is no marital conflict, but couples are brought closer together. For one thing, a husband and wife are not apt to undertake the procedure unless they already are strongly bound to one another and are eager to have a child. Further, there is every reason why a man should be at least as fond of an "insemination" baby borne by his wife as he would be of a child not his who comes to him through marriage to a widow or a divorced woman.*

* Artificial insemination has received such extended discussion here because the procedure seems almost certain to come into increasing use in the

But we must still think of the big majority of barren couples who, because donor insemination has been ruled out, or because the wife has been proved definitely unable to bear children, or for other reasons, face a life which they might feel empty without the satisfactions of parentage. To these couples one might say:

Adopt a child if possible, for an adopted child may often be reared to be quite as much like you as a child of your own. In fact, what is meant by a child "of one's own"?

We have seen how, with regard to its hereditary factors, every child is a gamble. No one can predict to what extent a child will be genetically like or unlike its parents. True enough, we often can make some definite forecasts about the child's physical traits and about its chances of having certain diseases or defects. But no one can know what the character, disposition, mentality or behavior of any given child of normal parentage will be. These traits are determined or influenced by such a multitude of genes, interoperating with so many environmental factors, that we cannot possibly expect to reproduce ourselves as individuals. An actual "chip off the old block" in humans is a genetic myth. On the other hand, while the genetic makeup of any potential child of yours is to a great extent unpredictable, you do have considerable power to control the environment which you will provide for it. Thus, if you adopted any healthy child of genetically normal stock and raised it carefully, it might in many respects turn out to be as much like you in character and social traits as many a potential child of "your own."

When adopted children do not turn out well—and there is some evidence of higher than average incidences of emotional disturbances among them—one should hesitate about ascribing this to "bad heredity." Obviously, emotional disorders in adopted children (as in children generally) may sometimes be due to genetic defects not apparent in infancy; but the environmental factors connected with adoption may be especially unfavorable. It should not be ignored that some foster parents do not treat an adopted child as they would their own; and often an adoption has resulted from the desire of an incompatible couple to hold a shaky marriage together, or from other purely selfish or neurotic motives. Under such conditions the home atmosphere is hardy likely to produce a well-balanced child, whether adopted or not. For this reason many agencies believe that the emotional stability of potential adoptive

---

future and will have particular importance if some of the proposals for breeding "superior" human beings are adopted. These will be discussed in our concluding chapter.

parents may be more important to the child than their financial status, education or religion.

But wanting and getting a baby for adoption are not the same thing. In the United States the supply of adoptable infants is so far short of the demand—scarcely ten or twelve White babies being available for every hundred couples applying—that a "Black Market" in babies for adoption, with enormous prices offered sometimes, has flourished for years. This shortage stems mainly from a combination of circumstances: As parental mortality rates (including mothers' deaths in childbirth) have been greatly reduced, many fewer infants are being orphaned, and if they are, their surviving parents or relatives, through better economic conditions and/or public or private subsidy, are able to take care of them. (Among the well over one hundred thousand legal adoptions annually in the United States now, more than half involve relatives—most often a step-parent of the child.) Also, while illegitimate births—at least a half-million a year—provide most of the babies for adoption, two-thirds of these are of non-White or mixed parentage and are not desired by the overwhelming majority of applicants, who are White. Further, of the illegitimately born White babies, a large proportion are quietly retained by their mothers or other relatives, or placed by doctors, lawyers or ministers, and so are not available to most applicants at agencies. Finally, religious stipulations augment the baby shortage in those groups, such as the Jewish, where the illegitimate birth rate is very low, and adoptable infants of their own faith are scarce.

Partly offsetting the adoptable-baby shortage have been several developments. For one thing, the effective treatments for sterility, and artificial-insemination procedures, have helped a great many couples—and could help many more—to have babies of their own. Moreover, numerous adoptable infants and children are being brought in from other countries. In addition, many couples are taking partially defective children who were formerly rejected. Also, as stigmas attached to illegitimacy are reduced and the burdens of unmarried mothers are eased, a large number of the babies now being lost through abortion can be brought into the world with happy futures awaiting them.

The question of illegitimacy brings us to another genetic problem of parenthood—that involving *disputed parentage*. Occasionally this relates to real or suspected *changelings*—babies given to the wrong parents after birth. But in the overwhelming majority of doubtful parentage cases, the question is whether or not a particular man fathered a given child. In former years this often was (and in many places still is) decided on the flimsiest circumstantial evidence, or perhaps merely on the woman's accusation. For instance, the author remembers a case in Milwaukee where a young unmarried mother testified that on the night of her dalliance with a boy friend there had been a full moon, and the presiding judge blandly said, "All we've got to decide now is whether the father of the child was this young fellow here, or *the man in the moon!*" But in more and more paternity cases justice and science are being served better by evidence regarding known hereditary traits common to or absent in a child and a putative father. Most often the decision hinges on the findings in *blood tests* for various blood groups and elements described in Chapter 28. First legally sanctioned in 1935 in New York State and in Wisconsin (through efforts of Drs. Alexander Wiener and Philip Levine, who later made the Rh discoveries), these tests have since been used in thousands of disputed paternity cases.

Primarily, the blood tests may often demonstrate that a particular man is not the father of a certain child. The tests cannot prove that any given man *is* the father. What they can only prove is that this or that man is not carrying the inherited blood-group factors which would be required for the child to be his. For example, if a baby has AB blood and the mother has A blood, the B would have had to come from the father. But if the accused man in the case is of the O type, or of the A type like the mother, he could not have been the child's father. Assuming he did carry the required B, one could test for other blood elements. An M man could not be the father of an N baby (or, if the mother is M, of an MN baby); or an Rh-negative man could not be the father of an Rh-positive child whose mother also was Rh-negative. Similarly, tests of child, mother and man could be made for various other inherited blood groups and elements, or their subcategories.

With the evidence available through blood tests, it now may be possible to clear a man wrongly accused of paternity in about 60 per cent of the cases. (If the mother refuses to have her blood tested—as some courts permit her to do—and only the child's blood and the accused man's are available, his exclusion chances may be halved.) But the 60 per cent exclusion represents the average in a mixed population, as in the United States, where people tend to be highly varied in their blood. In more closely knit or inbred groups, where persons are of similar stock, the same blood-type combinations may occur repeatedly, and a man's chances of parentage exclusion would be much reduced. (For example, in some American Indian tribes where all are of O blood, A-B-O testing is useless, and one could indeed say of an accused Brave, "Lo, the poor Indian!") But if a man is genetically quite different from other persons in a community, and/or his blood types tend to be uncommon, his chances of exoneration from a false paternity charge might be very high—possibly close to 100 per cent.

It must be added that the importance given to blood tests in paternity cases depends upon existing laws, practices and attitudes, which vary in different states. In some instances blood tests are required before a paternity case is brought to trial; in others the blood-test findings may be considered later, and then may or may not be accepted as decisive. A notable example was in the California case years ago in which Charlie Chaplin defended charges of paternity. Although the blood tests showed that with his O blood he could not have been the father of the B child in question, whose mother had A blood, a jury nonetheless convicted him, and an appeals court sustained the verdict.

Claims of persons that they *are* the parents of a given child may also be at issue in some parentage cases. In occasional instances this has been so where it is suspected that a child claimed by a wife or mistress as her own is actually one which she secured from another woman after feigning pregnancy. (Readers may also recall here the case mentioned in Chapter 17, page 149, of the two men who each claimed to be the father of their boarding-house lady's twins.) Much more common in the United States in recent years have been cases where naturalized citizens —often Orientals, sometimes Europeans—have asserted that some person from abroad seeking entrance to the country is a child of theirs. After blood tests revealed many of these claims to have been fraudulent, immigration officials now have made such tests mandatory in all these situations.

In changeling cases (increasingly rare because of precautions taken by hospitals to prevent baby mixups) the blood tests may be especially effective. This is because the tests in such instances involve two sets of putative parents—mothers as well as fathers, in addition to two children—providing much greater opportunity for establishing true parentage. While the chances of identifying the parents through the blood tests alone may be 90 per cent, important reinforcing evidence may come through evidence of other hereditary traits. This applies equally to any disputed parentage cases. We refer here to the fact that many surface physical traits, while not so clear-cut in inheritance or as easily revealed in a child as blood elements, may nonetheless prove of great significance when parentage is in question. As far back as 1910, long before the blood tests, a case in Norway involving a child with the dominant hereditary hand abnormality, *brachyphalangy,* was decided on the basis of the fact that the accused man was the only one in the community with that condition. Other dominant physical traits—normal or abnormal—if present in a child, could similarly provide evidence in disputed parentage cases where blood tests were not conclusive. These might include various recognizable inherited structural defects or anomalies in features and body form, as well as some of the diseases and defects which are dominant or partly dominant. Moreover, many conditions classed as recessive, but in which even the effects of single genes may now be detected (or in time, will be), may add to the evidence of parentage.

Normal traits of coloring, hair form and features may also prove useful in clearing up questions of parentage, as shown in Table XII, "Is This Man the Father?" While genetic similarity or dissimilarity between any child and a putative parent in one or two of these characteristics might have no meaning, concordance in a series of them, or lack of concordance, could prove significant where the choice was between two individuals or two sets of putative parents. But caution must be observed here, for as may be recalled from our chapters on coloring and features, it is possible for many parents to produce children who bear very little resemblance to them, or for children in some details to resemble unrelated persons more than their true parents.

However, as more and more known *combinations* of genetic traits in individuals are analyzed by geneticists, it may become possible to ascribe given children to given parents with a high degree of accuracy; and whatever the present inadequacies may be in establishing parentage, we have every reason to anticipate that as the hereditary mechanisms of more and more traits are pin pointed, there may be few cases where a clear-cut verdict cannot be given. In fact, in these days when experts can tell that a certain bullet was fired from a specified revolver, that a letter was typed on a specific typewriter, or—usually—that a brush stroke on a painting was made by a certain Old Master dead four hundred years, it would be strange if we should remain unable to determine whether any child—distinct in so many ways from other children—did or did not derive from such and such a parent.

Several other points might be made regarding doubts about parentage as they extend beyond a child's physical traits to other traits. "Is that child *really* mine (or ours)?" is often asked or pondered by puzzled parents when some behavioral quirk or mental trait inexplicably crops out—cruelty, dishonesty, sexual abnormality, alcoholism—or, in the desirable direction, a sky-high IQ, scientific brilliance, or musical or artistic genius. Devoid of any such traits themselves, parents may indeed wonder how they have appeared in their children. So it is necessary to distinguish first between "parentage" and "parenthood."

Table XII

Is This Man the Father?

In each of these doubtful paternity cases, the evidence may or may not rule out a man as the father of the child in question (although, if there is no exclusion, it does not prove he is the father). Helping you to decide will be the facts given on blood types in Chapter 28 and on feature inheritance in Chapter 15. (Answers at bottom of page.)

I. CASES INVOLVING ONLY BLOOD TESTS

Case	Child's Blood	Mother's	Man's
1.	O, M, *Rh-Pos.*	O, MN, *Rh-Neg.*	AB, M, *Rh-Pos.*
2.	A, MN, *Rh-Pos.*	A, N, *Rh-Pos.*	O, M, *Rh-Neg.*
3.	AB, MN, *Rh-Neg.*	*(Refused testing)*	O, MN, *Rh-Pos.*, Pos.
4.	O, N, *Rh-Neg.*	O, N, *Rh-Pos.*	A, M, *Rh-Pos.*
5.	AB, M, *Rh-Pos.*	A, M, *Rh-Neg.*	A, N, *Rh-Neg.*

II. CASES WHERE BLOOD TESTS ARE NOT CONCLUSIVE, AND OTHER INHERITED TRAITS (DOMINANT AND RECESSIVE) ARE CONSIDERED

Case	Child's Traits	Mother's	Man's
6.	*Eyes blue (Rec.)* *Skin fair (Rec.)* *Hair blond (Rec.)* *and straight (Rec.)* *Ear lobes affixed (Rec.)*	*Eyes blue* *Skin fair* *Hair blond,* *straight* *Ear lobes affixed*	*Eyes brown (Dom.)* *Skin darkish (Dom.)* *Hair black (Dom.)* *and curly (Dom.)* *Ear lobes free (Dom.)*
7.	*Eyes brown (Dom.)* *Hair black (Dom.)* *curly (Dom.)* *Skin darkish (Dom.)* *Lips full (Dom.)*	*Eyes green (Rec.)* *Hair blond (Rec.)* *straight (Rec.)* *Skin fair (Rec.)* *Lips thin (Rec.)*	*Eyes blue (Rec.)* *Hair blond (Rec.)* *straight (Rec.)* *Skin fair (Rec.)* *Lips thin (Rec.)*

JUDGMENTS—I. BLOOD TESTS: (1) Man excluded by his "AB." (2) Man not excluded; child could carry recessive "O" and "Rh-Neg." genes. (3) Double exclusion for man (even without testing mother) because of his "O" and "Rh-Pos." (4) Man excluded by his "M," lacking in child. (5) Man excluded by all three tests.
II. FEATURES: (6) Man neither excluded nor proved father. All traits given for baby are recessive, so man might be carrying hidden genes for them. (7) Man probably not father, on basis of all dominant traits in the child—eye color, hair color, hair form, complexion and lips—which, because not present in mother, could have come only from father.

Where psychological traits or tendencies are strongly influenced by heredity, they can as easily crop out in a child through combinations of genes carried but unexpressed in the two parents as can genetic physical traits. In this case *parentage*—the biological contribution of parents to their offspring—may be of major importance. But also contributing, and often the principal factor in the development of a child's mental and behavioral characteristics, are the effects of its environment. And here one must think of *parenthood*—the state of being a parent—which becomes effective after a child is born.

Thus, where parentage exerts its effects before a child comes into the world (and is not altered even if both parents die immediately thereafter or are no longer in the picture), parenthood continues to be an influence as long as the parents are in contact with the child. This distinction must always be kept in mind whenever we talk about what any children—or families, or groups—are, as compared with what they could have been or can be. And it is a fact which has its greatest implications when we consider, as we will next, any plans for improving the quality of human populations as a whole.

CHAPTER FORTY SIX

The Stork and Eugenics: I

Quantity and Quality Quandaries

THE MOST IMPORTANT CROP in the world is people.

As in agricultural production, there always has been concern about the human crop on two counts: *quantity* and *quality*.

The chief worry in given human populations usually has been about their *quantity*—most often, too many people to feed and care for with the available resources. But sometimes the anxiety in a particular group has been about not having enough members for proper defense and advancement.

Worry about the *quality* of populations was in the past confined mainly, and sporadically, to limited groups which sought to maintain or achieve high levels of "fitness," excellence and prestige: for instance, the ancient Greeks and Romans, who sanctioned the killing of defective offspring (as many primitive groups also have done or still do). With the advent of genetics and its first findings, the new idea arose of improving human quality scientifically through selective breeding, so as (1) to reduce the chances that defectives would be born, and (2) to increase the production of more than ordinarily desirable individuals. It further came to be recognized that the quantity and quality of human populations were related: that, as with crops planted too thickly, too many people crowded together without adequate resources could cause a deterioration of their overall quality, regardless of their genetic makeup. So if the objective in human breeding was to improve human quality—to produce the healthiest, ablest, happiest human beings possible—the problem of quantity would be an integral part of any program for its achievement.

Let us take up first the quantity aspect of human populations. Fears regarding *too many* people were growing ominously back in 1800, when it appeared that the world population had doubled in a century and a half to reach what was then regarded as the staggering total of almost one billion souls. (Previously it had taken mankind more than sixteen centuries to double itself from an estimated 250 million population in the year A.D. 1.) Many foresaw an imminent future when there wouldn't be enough food for everyone. So Thomas Malthus, a scholarly British clergyman, came out with this theory: Human population growth is biologically self-regulating. Whenever the numbers of people get too big for the available resources, then wars, plagues or famines will come along to trim mankind down to the required size. (Unfortunately, many persons interpreted this as meaning that wars and other depopulating calamities were not merely natural, but were necessary and desirable.)

The Malthusian doctrines became questionable when the world population went on to even more phenomenal growth, almost tripling in the next century and a half; and yet, through great advances in agriculture, technology and medicine, people in general were better fed and better off than ever before. Moreover, it was in the very countries in which food was most abundant that population growth was being slowed down—not by catastrophe, but by prosperity and the accompanying social advances which were leading to voluntary restraints on reproduction. On the other hand, the more backward countries, while being helped to greatly reduce their death rates, were doing little to limit their birth rates, and their populations were growing steadily.

Thus, by about 1930, the fears in the more advanced countries, including the United States, were of not having *enough* population to maintain their positions and to compete in the future with the prolific backward countries. (Inasmuch as there also were racial and ethnic differences between the advanced and the backward peoples, there was concern, too, on this score, as we'll see later.) In some European countries (Germany, Italy, the Soviet Union, Norway and Sweden) drives were under way to increase birth rates through bonuses and other incentives. Further worry about population deficiencies came when the World War II holocausts wiped out tens of millions of lives. But, almost immediately, unprecedented baby booms everywhere soon made up for the losses. Not only this, but increasing life expectancies were carrying along more and more individuals and adding them to the swelling populations. Once again there was the fear of *too many* people. And this time it could not be eased by any Malthusian notions that the population growth would check itself automatically in keeping with the needs.

Today the "population explosion"—more precisely, a series of continuing and diversified explosions all over the world—looms up as one of humanity's most ominous and urgent problems. Warning already had come that mushrooming human hordes were exhausting the earth's natural resources—tillable soil, forests, streams, minerals—faster than technological advances could provide adequate substitutes. Now it could be argued that the old specter of humanity choking itself to death, with groups battling each other for survival, would indeed become a reality unless there were drastic curbs on reproduction. Some optimists have taken another view. They maintain that the earth's present resources, if fully exploited by science and properly utilized, could keep pace with population growth for a long time to come. But for how long?

Between 1940 and 1960 the world's population jumped from about two billion to three billion. At this rate by the end of the century it could reach six or seven billion, in another generation twelve billion (theoretically), then twenty-four billion, then . . . ? Not even with the greatest achievements of science could this little old earth support more than certain numbers. (In six centuries at the current rate there would be only 1 square yard on earth for every inhabitant.) To suggest that atom bombs might take care of the problem is hardly an acceptable answer. And it would be going beyond the limits of science fiction—truly "out of this world"—to speculate that our excess population could be siphoned off to other planets.

If the problem were only one of human quantity it would not demand discussion in this book. What must interest us here is how the changes in the quantities of populations might affect their *biological qualities* or characteristics, and, specifically, the genetic makeup of future generations in given groups and in the world as a whole.

A first point is that birth rates and their relative effects—at the core of the population explosions—vary greatly in different parts of the world. Latin America is among the most fertile of all regions, with the annual birth rate in Central America averaging about 45 per 1,000 population (reaching 55 in Costa Rica), and in South America about 42. In Africa the birth-rate average is about 45 in the northern countries (including Egypt) and 48 in the tropical and southern countries. The birth rates in most of the Asian countries (China, India, Pakistan, Ceylon, Malaya, and so on) are close to or well into the 40's, and in the Arab countries reach 48 per 1,000. But in Japan the birth rate has been halved phenomenally, from a 1947 rate of 34 to the current 17, as a result of a rigorous program of state-sponsored and voluntary birth control. This brings Japan's birth rate well below the level of that in the United

States—now about 23 per 1,000, which, in turn is higher than the rates in most European countries, where the range is from 16 to 22.

The overall picture, then, is of the highest birth rates in the poorest and most underdeveloped countries. Moreover, the death rates in these countries no longer keep pace with the birth rates, as they once came close to doing. For instance, in India the death rate a half-century ago was about 47 per 1,000, the birth rate 48, leaving an annual population increment of only 1 per 1,000, or one-tenth of one per cent. By 1960, however, deaths had dropped to 18 per 1,000, but births were still about 40 per 1,000, which swelled the annual increment to 2.2 per cent. At this rate, the population in India would double in about thirty-two years. Doubling within a similar period, or less, would occur in China and various other Asian, as well as African, countries, on the basis of their current birth and death rates; and in such countries as Costa Rica and Brazil, population doubling might occur in a little over twenty years. Obviously, the most advanced countries, with their low birth rates—even though supplemented to some extent by increased life expectancies—would take far longer to double their population. In part, too, population brought into some countries through immigration or lost to others through emigration would affect the results. With these points taken into consideration, the United States at its current annual increase of 1.7 per cent might double its population in forty years (and at the same rate, in 200 years would have 50 per cent more people than the entire world has today). France, growing by less than 1 per cent annually, would achieve doubling in seventy years; the United Kingdom and Italy, at their current growth rates of only 0.5 (half of one per cent) would take 140 years to double; and the Scandinavian countries, at varying growth rates, might double in eighty to one hundred years or so.

The drastic social and economic consequences of populations bursting beyond resources can be foreseen. So can the attendant detrimental effects on health, mental development and opportunities for achievement and advancement. Nor is the problem only one for the now backward countries. Eventually, deterioration in living conditions, and hence in human quality, also would confront the United States, as one example. This may be hard for Americans to realize when they look at the vast open spaces that can be inhabited and made productive, the overflowing granaries, the lands being plowed under, the subsidies to keep down production. True, food is still overabundant, material goods plentiful. But yet there is unemployment—not enough jobs for the people already here, and threats of still fewer jobs through increasing automation. There are not enough hospitals, schools, good housing. What will it be if our population is doubled, then redoubled, in two generations?

The purely environmental effects on human physical makeup and mental functioning would not be the only results of the population explosions in separate countries and of their varying intensities in some countries as compared with others. There also would be *genetic* consequences, at least with respect to changing proportions of genes which are distributed differently among racial and ethnic groups (as listed in Chapter 42). On the major racial levels, with the Mongoloid and Negroid peoples multiplying at the faster rates, there would be relatively more of their genes in mankind's future gene pool, and relatively fewer of the "Caucasoid" (White) genes. Among Whites, again, any genes which have higher frequencies among the more rapidly reproducing Asian-Caucasoid and Latin-American stocks would be represented hereafter in increasing proportion, while genes more characteristic of North and Central Europeans would decrease in relative incidence.

Clearly, then, one could expect a shift toward proportionately more people in the world with inherited *physical* traits characteristic of Mongoloids or Negroids (features, coloring, certain hereditary defects and diseases, etc.) and relatively fewer people with Caucasoid physical traits; and among the Whites, more with Asian and Latin physical traits and fewer with "Nordic" traits. (The environmentally conditioned thinking and values of the world, and human behavior, would undoubtedly also be affected by any radical changes in the proportions of people from different racial and ethnic stocks, insofar as they carried along differences in habits, traditions, religion and political views.) But this by no means implies that there likewise would be changes in the future genetic *quality* of mankind's mental traits and performance capacities. For, as was stressed in Chapter 43, all human racial and subracial groups carry genes for every type of ability, talent and achievement; and while there may well be racial and ethnic differences in the frequency of genes for specific mental traits, there is no evidence that in the sum total any group is inherently superior in capacity to others. Nor could we tell what the genetic mental differences among human racial and subracial groups might be hereafter until all had the same opportunities for development and advancement.

The foregoing conclusions apply further to the population changes taking place *within* any country, as regards the relative proportions of its racial and ethnic stocks. In the United States, for instance, there have been radical ethnic and racial shifts, from the early preponderance of British and other North Europeans, to increasingly more Central,

Eastern and South European peoples, and, recently, more of the Puerto Rican and other West Indian groups (with mixtures of Latins, Mongoloids and Negroids). The latter influx has added still more to the Negroid ratio, which has been steadily increased by the higher birth rates among Negroes generally—32 per 1,000 as compared with 22 per 1,000 among Whites.* In another way, the makeup of the American population has been altered by differences in the birth rates among religious groups, with Catholics on the average outproducing Protestants, and Jews having the lowest birth rates. These changes too, have genetic meaning only insofar as the different religious groups vary in racial or ethnic composition and the variations involve differences in gene frequencies. But, as with the world picture, there is no evidence that shifts in the proportions of racial or ethnic components in the American population would alone have altered its quality with regard to mental traits and potentials for achievement.

It is quite another matter when we classify any country's population—or each of its racial, ethnic or religious groups separately—according to social and economic levels, and then analyze the variations in birth rates. Here we may find that the reproductive differences might well be producing changes in overall genetic *quality*. And this brings us to the next phase of our discussion.

Our Declaration of Independence states, "*We hold these truths to be self-evident: that all men are created equal. . . .*"

But everything we've learned about human heredity challenges this statement. Nor could our Founding Fathers have meant it literally. What they undoubtedly had in mind is that spiritually and ideologically, every human being, as one of God's creatures, is born with the right to be treated equally and fairly under the laws of man. Under the laws of Nature, unfortunately, there is no such justice. The further our knowledge of genetics proceeds, the more apparent it becomes that from the moment of conception, individuals are started off with every type of inequality in body and mind and in their chances for life, liberty and the pursuit of happiness. One can grant, also, that this inequality extends to the usefulness of people to society, and that by almost any

* Despite the recent increase in the Negroid ratio, however, the proportion of Negroes in the United States is still very much less than it was in Revolutionary days (about 1790), when it was over 19 per cent. White immigration mounted during the nineteenth century so that by the time of the Civil War the Negro proportion had dropped to 14 per cent, and by 1940 had come down to about 10 per cent. The gradual rise thereafter brought it to about 12 per cent in 1960, with the current ratio perhaps being a point or two higher.

standard, some types of people who are born into the world are more desirable than others. Why, then, should not efforts be made to improve human quality by increasing the production of the desirable individuals while curbing the production of the undesirable ones?

This was the thought behind the Eugenics movement launched by Sir Francis Galton of England (a cousin of Darwin) in 1883. (The name he coined was derived from the Greek *eugenes*—"well born.") However, eugenics should not be confused with genetics, which did not come into being until 1900. Thus, without knowledge of most of the important facts about heredity which we now have, Galton and his followers were necessarily motivated by many of the old fallacies concerning ancestry. They believed quite strongly that the people on top at the time—certain classes, nations, and races included—were there largely because of "superior" heredity, while those at the bottom were there because of "inferior" heredity. Nor was this belief much altered when the first findings regarding the workings of heredity did come out. Almost immediately the facts about genes as trait-determiners were seized upon as proof that "bad" genes were responsible for most of the major human defects and deficiences, whereas almost all human virtues and superior capacities could be credited to "good" genes. The challenge was simple: Out with the "bad" genes, in with more "good" ones, as quickly as possible.

In planning how this should be done, a twofold program was envisaged, (1) *negative eugenics* and (2) *positive eugenics*. The primary aim of negative eugenics was to stop or slow down the transmission of serious wayward genes by preventing reproduction of persons who were genetically defective, and—so far as possible—of persons themselves normal who gave evidence of carrying menacing recessive genes. With regard to gradations of genes and those carrying them, the population was seen as roughly divisible into the "less fit" or "unfit" groups, carrying most of the least wanted genes, and the "fit" groups, carrying, relatively, most of the best genes. ("Fitness" was used here in the sense of genetic desirability in health, physique, mental qualities and good-character traits.) Accordingly, if those in the "less fit" and "unfit" groups were reproducing at relatively high rates, an important additional objective of negative eugenics was to bring about a general reduction in births among them. As for the positive-eugenics program, this, of course, would include all ways of stimulating birth rates among the most "fit" elements and any other means of improving the genetic quality of the population.

Almost at once the eugenic concepts—particularly the negative-eugenics planks outlined—began running head on into a number of new developments. There was the burgeoning environmentalist movement, fostered by new studies in the psychological and social sciences which placed sweeping emphasis on the effects of environment and minimized hereditary influences. Politically and socially, too, in one nation after another, there were battering assaults on hereditary class structures, and everywhere groups and individuals once rated inferior were bursting forth with unexpected powers and achievements. And throughout the world, the tremendous advances in health, longevity, education and achievement, resulting solely from rapid changes in environment, seemed to be dwarfing further the importance of heredity. In fact, to many persons, any stressing of inherited human differences was regarded as "undemocratic" and an interference with social progress.

But opposition to the original eugenics concepts also came from most of the leading geneticists, who saw in them perversions of the scientific data. Moreover, as findings in human genetics became more precise and comprehensive, it was apparent that any eugenics program would face many more difficulties and complications than had been suspected. It wasn't all simply a matter of "bad" and "good" genes: With respect to most of the important human traits there was a complex interaction between environmental and genetic influences. While certain individuals in every group could be stamped as genetically defective, the terms "genetically superior" or "genetically inferior" could not be applied to whole segments of the population. And any program for human betterment would have to take into account innumerable factors—social, economic, political, religious—that went outside of the field of heredity.

Thus, as the facts have become clearer, the old eugenics program has given way to a much revised one. This "modern eugenics" can be defined as a movement to improve human quality *through both environmental and genetic measures*, in whichever proportions the situations and needs may dictate.* For instance—to start with the genetic factors—there are the many serious defects and diseases traceable almost directly to wayward genes. If given free rein to eliminate these genes, how would we go about it?

The first to be attacked among the wayward genes might be the *simple dominants*. These show their effects in any person carrying one of them. Preventing that person from reproducing, therefore, would halt the transmission of his or her "bad gene"; and if *all* persons with simple dominant defects were kept from reproducing, those defects might be almost wiped out in one generation. The "almost" refers to the likelihood that

* Reference here is to the general aims set forth in recent statements of the American Eugenics Society and the Eugenics Society of England.

some new genes of the same type would continue to crop up through mutations. Also, certain dominant conditions with late onset (Huntington's chorea, common adult glaucoma, etc.) may not reveal themselves until after the individual has had children and has passed the gene along. (Some of these situations may be met by tests which reveal carriers of the genes in their pre-adult years, as will be noted later.) Most elusive would be the incomplete or qualified dominant genes which often do not assert their effects in carriers. For these various reasons even the dominant wayward genes could never be wiped out entirely, even though stringent restrictive measures could greatly reduce their incidence. However, the really serious one-gene dominant defects are relatively few and/or uncommon, and have never been the major cause of eugenic concern.

The principal eugenic predicament is presented by the *recessive*, or other multiple-gene, conditions, which are responsible for most of the common serious hereditary mental diseases and defects, and the worst of the inherited neurological, ocular, muscular, skeletal and blood afflictions. The biggest obstacle to eliminating the genes for these conditions is that most, by far, are carried and transmitted by normal persons. For instance, consider the genetically produced idiots and imbeciles. These defectives pass on only a tiny percentage of their own genes, because the big majority are sterile or are institutionalized. The problem is principally with the *morons*—those with IQs of from 50 to 70. Assuming that there now are in the United States 600,000 low-grade morons produced by recessive genes (simple or multiple), geneticists estimate that there would be at least ten times as many, or 6 million normal or close to normal persons, each carrying a hidden one of these genes. Matings of these persons would continue to replenish the supply of morons, and even if not a single one of the existing morons had offspring, their counterparts in the next generation might be reduced by no more than 10 per cent. At this rate it might take ten generations—three centuries—to cut the incidence of these defectives to one-fourth their present number, and twenty-two generations—about seven hundred years—to bring it down to one-tenth (and this is not allowing for possible new mutations).

Equally difficult to root out would be any of the other common recessive defects or diseases; and even more tenacious would be the rare conditions, for the lower their incidence, the smaller the likelihood that the individual men and women carrying the genes would meet, mate, and bring their genes to the surface. Also to be considered are the qualifications wherever environmental influences are essential for genes to show their defects; or where gene mechanisms have not been clearly identi-

ELIMINATING WAYWARD GENES

How much could be accomplished if all genetically defective persons were kept from—or voluntarily ceased—reproducing

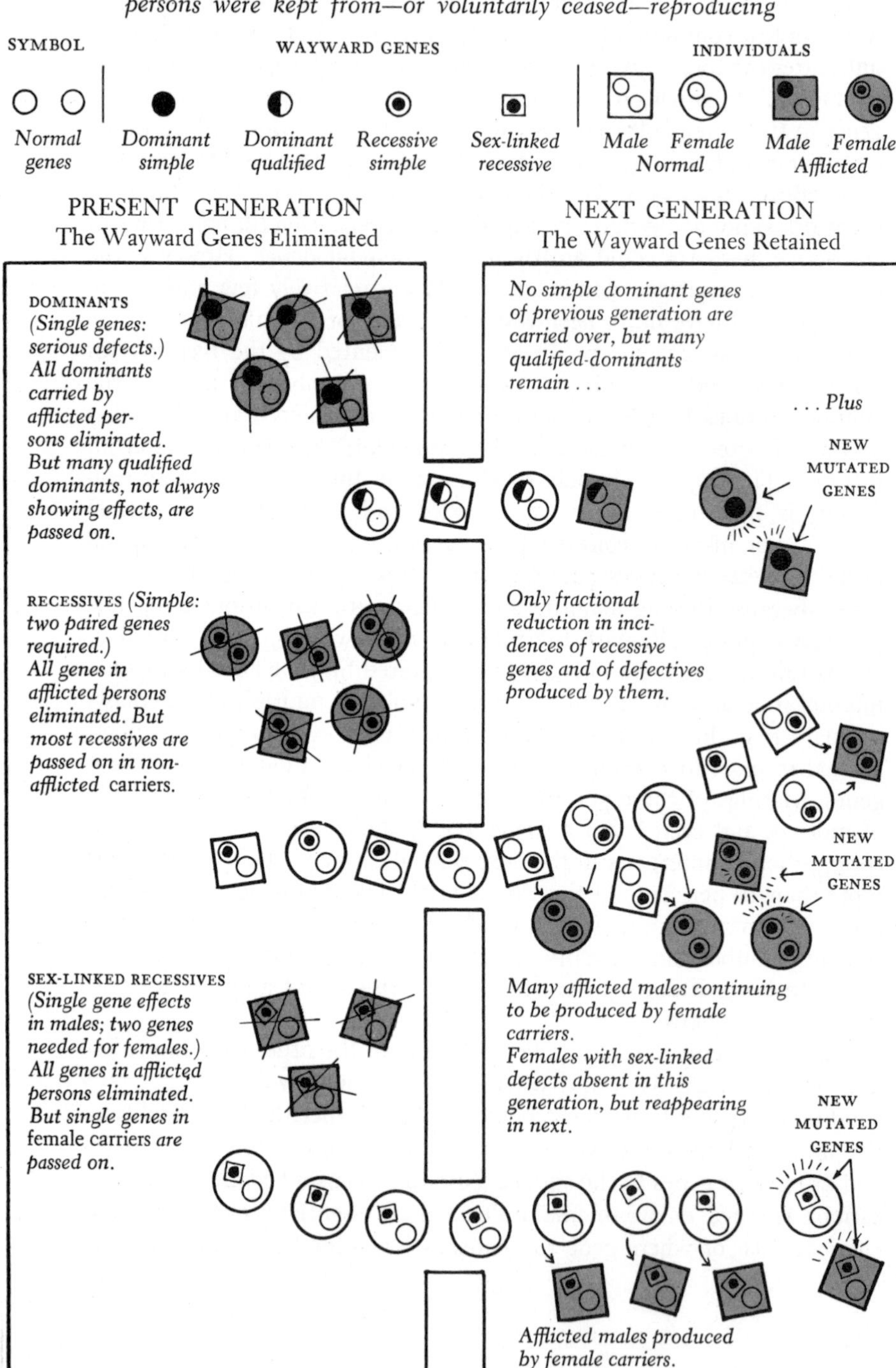

fied; or where multiple-gene mechanisms are involved, in which known recessives or dominants must combine with other, unknown genes to produce their effects. Any or all of these qualifications apply particularly to the major mental diseases, such as schizophrenia and manic-depressive insanity, any genes for which are carried by many normal persons.

Special problems are posed by the recessive sex-linked diseases and defects—those caused by "X-chromosome" genes which singly produce their effects only in males but require two genes to make a female defective. If all of the males afflicted with any of these conditions were kept from reproducing, there still would be great numbers of nondefective females, usually wholly normal themselves, who could pass on the genes and cause the conditions to appear in offspring and/or descendants. In addition, as with other wayward genes, mutations would replace many of the sex-linked genes which might be eliminated.

But for the most part the problem of trying to reduce the incidence of human hereditary defects must center on the *carriers*. These, technically speaking, are the fully or apparently normal individuals who carry wayward genes which produce no serious effects in them. (The genes may be either recessives in the single state, or "incomplete" or "qualified" dominants, or other genes largely inactive in the carriers but which could act seriously in their offspring.) Any thought of curbing the reproduction of carriers wholesale becomes preposterous with the knowledge that every single one of us carries some unrevealed wayward genes—an average of about eight to a person, Professor Hermann Muller has estimated.

What could be attempted is trying to keep the more menacing of the genes in carriers from combining or otherwise expressing themselves, so far as possible. Two aspects of this—discouraging matings of unrelated individuals known to carry matching wayward genes, and of *cousins* with above-average chances of carrying matching recessives—were discussed in the preceding chapter, where specific risks in these matings were set forth. We might now go into more detail about marriages between cousins and other near relatives with a view to learning further why they have been given so much attention in eugenics proposals and in medical-genetic research.

As was brought out on page 655, cousins have a high proportion of matching genes (first cousins averaging one in eight, second cousins, one in sixteen). So if any of these matching genes are seriously wayward in action, there would be far above average chances of gene combinations occurring in their offspring which could produce serious hereditary defects or diseases. Corroborative evidence has come from many studies.

In the United States a pioneer investigator, Alexander Graham Bell (inventor of the telephone) showed early in this century that among the nation's blind and deaf persons a large percentage were products of cousin marriages. Later studies revealed much-above-average incidences of defectives of many other kinds, mental as well as physical, among inbred families on the island of Martha's Vineyard (off Massachusetts), in the Chesapeake Bay peninsula, in the hill-country sections of New England, and among the Amish folk of Lancaster County, Pennsylvania (with many dwarfs). The story has been the same for highly inbred groups of any country. Generally, the rarer the recessive condition, the greater the likelihood that where it appears the parents are related. To state this another way, the more closely parents are related, the greater the chances that rare genetic defects may crop up in their children.*

Undoubtedly, then, banning of cousin marriages would greatly reduce the numbers of genetically defective persons—how much depending on the extent to which these marriages, or other inbreeding, had been occurring and on the incidences of particular wayward genes. For example, where 1 per cent of the marriages have been of first cousins (about the average now in most European countries), the ban might reduce the frequency of albinism by 30 per cent, congenital deafness by 25 per cent, juvenile amaurotic idiocy by 15 per cent. But the reductions in genetic defects could be far greater in countries with very high inbreeding rates, such as Japan, where the average of cousin marriages has been 8 per cent, and in some of its localities, up to 15 per cent. The latter high rate also has prevailed in some groups in Brazil, India and Israel; and in various regional groups in Switzerland and the Scandinavian countries the cousin-marriage rate has been 5 per cent or more.

However, while sizable inroads on genetic defectiveness can be achieved in the afore-mentioned countries merely by reducing cousin-marriage rates, not too much more can be accomplished in the United States by these means. Cousin marriages among Americans already have declined to an average incidence of no more than 1 in 500, and perhaps down to 1 in 1,000 (although considerably higher rates still prevail in some groups and localities). This big drop, from an average rate of

* Some interesting new evidence on this point came from a study of 100 first-cousin marriages in the Chicago Catholic Archdiocese. (The Church requires a special dispensation for such marriages and keeps a careful record of them.) While most offspring of the cousin couples were fully normal, the death rates in infancy and childhood among children in the group as a whole was about three times that of children of nonrelated parents, while the incidence of abnormalities was about double the average. Also, above-average numbers of the cousin marriages were sterile.

about 1 in 100 a century ago, has come much less through any legal or religious restrictions (laws in many states ban first-cousin marriages, and the Catholic church, as stated, frowns on them) or through individual eugenic considerations than through social changes: mainly the breaking up of isolated groups, and the greater mobility of Americans which has caused related persons to move apart far more and to seek mates farther afield geographically and socially than they once did. Similar trends occurring or impending in other countries are certain to make cousin marriages of lessening importance in eugenic programs.

Nevertheless, Cupid may continue capriciously to hit at cousins here and there. So let it be noted that there is nothing inherently wrong, sinister or "unnatural" about cousin marriages and that not *all* of them need be eugenically undesirable. For one thing, the laws, taboos or scruples against cousin marriages date far back to times when nothing was known or perhaps even suspected about the possible genetic threats. Nor was there any "instinctive" aversion to inbreeding. In Biblical times Jacob wed his first cousins Rachel and Leah; Abraham married a half-sister; and Moses sprang from a mating between an aunt (Jochebed) and nephew (Amram).* The Egyptian Pharaohs and Ptolemies mated with sisters wherever possible (and occasionally with a daughter), Cleopatra having stemmed from six generations of brother-sister marriages. And the ancient Peruvian rulers likewise deemed a king's sister the only bride royal enough for him. Nothing in the records of these ancient lineages suggests that the intense inbreeding had any disastrous results.

This leads to another point: Where wayward genes are minimal and a family is of unusually good stock, cousin marriages or other inbreeding may actually result in superior offspring. The physically superb Spartans were highly inbred; there were many cousin marriages among our Puritans; and the Hutterites, an American farming sect, distributed through the Midwest and central Canada, which abounds in cousin marriages, has the highest of all birth rates of any group (averaging ten children to a couple) and extremely low death rates. To take some notable individ-

* Exodus 6:20: "And Amram took him Jochebed his father's sister, to wife; and she bare him Aaron and Moses. . . ." (To readers who may have wondered about the author's rather unusual given name, Amram, this will explain its source.) Incidentally, there is an interesting–and perhaps Freudian angle–to the fact that Moses later put through a ban among the Hebrews on aunt-nephew marriages (like that of his own mother and father), but not on uncle-niece marriages. The distinction still holds among orthodox Jews. The reasons must be considered purely social and psychological, for, as the reader will recognize, there are no genetic differences in the two types of mating.

ual families, Charles Darwin, married to his first cousin, Emma Wedgewood, produced many distinguished descendants. And in the great Rothschild family almost half of the marriages of descendants of the family's founder, Mayer, in the past five generations have been between first cousins. (In domestic animals, of course, constant and intense inbreeding has produced our most valuable strains, but only by rigorously selecting for a few special traits and discarding individuals not possessed of them.) For the most part, however, unless a family is outstanding in both health and mentality, and with no evidence of serious hereditary defects, such marriages might well be discouraged.

One other thing to be said about human inbreeding is that, whatever its harm to individuals, it has helped to bring to light rare genetic defects and to further our knowledge about their workings. But in the case of many recessive conditions it now is no longer necessary to wait for matings between persons with matching genes—cousins or others—in order to identify the carriers. Continuing research has shown that various recessive genes (which include some sex-linked genes if carried by women) are not really "hidden" when in the single state, as was formerly thought, but reveal themselves by one symptom or another. (In fact, there is reason to believe that most if not all single recessive genes, as well as "incomplete" dominants, will in time be found to thus express themselves.) Some of the recessives and other conditions already detectable in carriers by designated tests are listed in the Clues to Carriers table in the Appendices. The information can be most important for members of families in which these defects have appeared, and who are anxious to know in advance whether they do or do not carry a given wayward gene. Not only can this provide foreknowledge of threats to potential offspring, but in cases where serious genetic conditions have a late onset, the identification of the carriers may permit precautionary or preventive measures to be taken by the individuals themselves.

A new eugenic complication has been presented by the finding that some genes which in the double state produce serious defects or diseases may, in single form, prove *beneficial* to their carriers under special conditions. This is true of the semidominant genes for *Cooley's anemia* and *sickle-cell anemia*, which can act as defenses against malaria (as noted in Chapter 42, page 600). The possibility is opened that genes for various other conditions—recessives as well as semidominants—may in the single state now have (or had in the past, or will have in the future) some beneficial effects under certain environmental circumstances. All of this calls for extreme caution—or, at least, discrimination—in trying to eradicate wayward genes.

The eugenic problems of individuals—carriers, or suspected carriers, or persons themselves genetically defective—and their possible courses of action were set forth at some length in the preceding chapter. We come now to eugenics as related to *whole groups*. Let us look more closely at the theory that populations can be divided into genetically "more fit" and "less fit" categories.

There would be little argument if we used the terms "plus" and "minus" to designate two groups as found within any population, community, racial or ethnic stock. The "pluses" would be the more intelligent, capable, law-abiding, socially conscious couples, eager to have as many children as they could afford to support and care for properly, and to whom they would try to give every possible advantage. The "minuses" would be the less intelligent, more defective, indifferent, irresponsible, unsocially minded couples, living from hand to mouth in unwholesome surroundings, careless about the upbringing of their children and often viewing them as inflictions. But to what can we ascribe the differences between the two groups? In Chapter 44 and elsewhere, it has been stressed that illiteracy, poverty, backwardness in achievement, irresponsibility, immorality, criminality—almost any undesirable social trait—as well as physical defectiveness, can be environmentally produced and carried along in families and whole groups from one generation to another.

Yet it also can be argued that over long periods—particularly if given considerable freedom of movement—there would be a greater tendency for the genetically defective and incompetent persons in any population to sink into the lower level groups and for abler individuals from the "minus" groups to rise into the "plus" groups. Thus, in time some *average* differences in genetic quality between the "plus" and "minus" groups would result, without implying that all or even most individuals in the one group were genetically "plus," and in the other group genetically "minus."

The basic question posed by the "pluses" and "minuses," however, is whether or how their differences in reproductive rates have affected mankind's genetic quality and its overall "fitness." And this, too, requires further clarification of what is meant by "fitness." Previously we mentioned that "fitness" in the eugenic terminology—implying "social fitness"—is not necessarily synonymous with "fitness" as referred to in our chapter on evolution. There we noted that the term was used technically to imply two things: (1) the inherent capacities of individuals and of groups for survival within their environments; and (2) the relative abilities of the survivors to reproduce themselves and project progeny and descendants into future generations. These capacities and abilities are still the major elements in our evolutionary processes; for it is a misconception that human evolution is something which took place in primeval epochs and stopped after Homo sapiens came into being. The fact is that human evolution has gone on continuously, is going on today, and will continue in the future. Gene mutations have been occurring uninterruptedly, with every probability that some of your own genes now are not the same in their workings as they were when you got them from your parents. Further, the proportion of your genes and your family's that have been or will be poured into mankind's collective gene pool may be greater or less than that contributed by another family with different types of genes. In other words, the genes selected for continuance in human evolution may differ in types and combinations from one generation to the next.

The eugenic concern, then, is that the processes of *gene selection* in human beings may have been proceeding in an unfavorable direction, and that action may be called for to redirect the processes. To oppose this as interfering with "natural" selection is to ignore the fact that ever since man began modifying his environment his evolution has been swerved from the "natural." As the abilities of individuals to cope with new environments have become less dependent on their own genetic makeup, as medical and social advances have lessened the importance of various handicaps, and as reproduction has come under the control of individuals in the degree that they choose, the "natural" selective processes have been greatly modified by "artificial" selection.*

* As with the definition of "fitness," the meaning of "natural selection" has been variously interpreted. In the view of Professor Theodosius Dobzhansky, the process of natural selection implies the selection and transmission of genes in accordance with their adaptation to given environments —regardless of how the environments came about. So even though man has artificially modified his environment, the operation of natural selection has not been altered; it has simply been redirected into new channels. Human evolution, then, will continue to be the result of interaction between man's biological and cultural forces. However, Dr. Dobzhansky also notes that if human reproduction (or the number of offspring per person) should come to be determined by the individuals or by some outside authority on the basis of genetic desirability, this would replace "natural selection" with

And this, it is held, has led to increased survival of the weaker and less capable and has reduced the advantages of the stronger and more capable.

As theorized by Harvard evolutionist Ernst Mayr, evolutionary selection for better human brains had already slowed down ages ago when people began forming into large social groups; the average and below-average individual benefitted from the efforts of the superior members and so was helped to survive and reproduce. Consequently, the relative contributions of the superior members to the gene pool of the next generation grew smaller. Further, Dr. Mayr suggests, this trend continued increasingly as all the members of a community began to benefit from technological advances and other achievements of the superior individuals, which helped those below average to survive, make a living and reproduce as successfully as the above-average ones. Added to this, the new means of conquering or overcoming physical detriments of the immediate environment—cold, heat, disease, dietary shortages—reduced the importance of such biological factors as body build, strength, and resistance to heat, cold or infection, which formerly conferred selective advantages on given individuals.

In short, where in former periods the biologically "fittest" individuals and groups were also as a rule the ones with the most surviving offspring, this no longer need be true. Even more, many among the genetically defective and deficient who once would have died when young are now enabled to survive, reproduce themselves and leave surviving progeny—sometimes in greater proportion than those who are physically and mentally more "fit" (in the eugenic sense, as we will continue to use the word in this chapter). Accordingly it is claimed that the modern evolutionary trend is not merely *not* in the eugenic direction—toward the genetic improvement of the human stock—but in the opposite or *dysgenic* direction—toward genetic deterioration. What evidence is there that this actually has been happening?

One approach to this question has been by the IQ route. As will be recalled from Chapter 31, children's IQs correlate, on the average, with the occupational levels of their fathers, the ones whose fathers are in professions and in managerial positions having considerably higher IQs than children of the unskilled or other lower-level workers. On this basis

---

"artificial selection." Conceding that some day this may come to pass, the professor maintains that, meanwhile, natural selection is going on—always within the context of the existing environment.

it might be assumed that if relatively more children have been and are being born to parents in the low socioeconomic-educational groups, a gradual drop in the overall IQ level would be occurring. This, some studies in the United States and England indicate, actually is (or was once) the case. Even so, a decline in average IQ would not be proof of deterioration in the stream of "mental" genes. For, as was brought out in Chapter 31, there is much uncertainty as to how far *IQs within the normal ranges* are measures of inherent mental capacity, especially as regards differences in test performance between whole groups whose environments, training and opportunities have not been the same.

What of the individuals who are below the normal range—those in the dull-normal or moron class? Do not these defectives tend to have more children than normal persons and so cause some drop in the general IQ level? This, too, had been believed—and some studies seemed to bear it out—until geneticists at the Dight Institute reanalyzed the data. Here is what they reported: While *couples* with very low IQs (averaging under 70) did tend to have above-average-sized families, an unusually high proportion of their equally low- or lower-IQ siblings and relatives did not marry and/or have children. Taken collectively, then, the low-IQ group was shown to produce no more children than normal persons; and on this basis the mean IQ of the population as a whole could be expected to remain at about the same level from one generation to the next. Also found was a tendency of children in lower-IQ families to have higher IQs than their parents, and to move up more toward the average. (A reverse "regression toward the mean" usually occurs where parents have very high IQs, their children's IQs tending to *drop* somewhat closer toward the average.)

Further lessening eugenic worries has been the steady narrowing down of birth-rate differences between the "plus" groups and the "minus" groups in the advanced countries. One post-World War II American phenomenon has been the rise in birth rates in the upper-level groups. Marriage and motherhood, previously downgraded among sophisticates, is back in style. Where a few decades ago a third to half of the women college graduates never married, and women graduates as a group averaged less than one child each, the newer coeds have been rushing into marriage and motherhood even before graduation; and in innumerable cases where their college-graduate mothers had only one or two children, the daughters have been producing three, four and five. Among men college graduates there has been a similar upsurge in reproduction, with this added fact: The brightest and most successful of these men have tended to produce more children than have their more backward classmates.* So, too, within the occupational ranks, the more successful and prosperous professional and business men have been fathering larger families than their less successful or poorer colleagues, while men in the minor professions and smaller businesses tend to father more children than do clerical workers. But in the manual occupations, the skilled workers still have fewer children than the semiskilled, who, in turn, have fewer progeny than the unskilled.

Taking members of the "minus" group as a whole (as they are or were), the trend has been toward smaller family sizes, in keeping with economic, educational and social advancement. But whether birth limitation is most marked among the genetically more "fit" individuals of the "minus" group—either those remaining there, or those rising into the "plus" ranks—is uncertain. Quite possibly an above-average proportion of the least "fit" among the "minuses" may not marry or reproduce (as with the low-IQ individuals previously mentioned) and so would tend to keep the genetic level of the whole "minus" group in balance.

Leaving unresolved the genetic aspects of differential reproduction, one need not doubt that on an environmental basis alone it is unwise to have higher birth rates among the more backward than among the more advanced couples in any country, community or group. Nor can it be contended that the underprivileged and less fortunate parents *want* more children than they properly can care for. The chief reasons for the disparity in birth rates between the "minuses" and the "pluses" are simple ones: The more educated, prosperous, advanced couples in all groups (socioeconomic, religious, racial) are, as a rule, not only more aware of the importance and benefits of planning their family sizes, but have more knowledge of and access to the *means* of carrying through their plans. Making this possible are various methods of regulating or spacing conceptions and preventing births when no more are desired. Whether or not these methods and means of controlling reproduction should be made more widely accessible, and whether or to what extent society or the state should encourage parenthood planning and/or birth limitation, are questions which carry us beyond genetics and into the highly controversial realms of religion, customs, politics, and ethics. But

* Among British university graduates, Dr. Cedric O. Carter has reported, those who'd led their classes–earning first-class honors–have averaged the most children, with family sizes diminishing successively in moving down to those who'd had second-class honors, then third-class honors, and, finally, to the "pass-only" tail-enders (those who didn't actually fail to graduate but who ranked too low for an honors degree).

some discussion of these points is essential here, for no eugenics program, and indeed, no far-ranging program for human betterment, can today be undertaken without considering measures for population control. Moreover, all of the problems related to human quality we have dealt with —the "population explosions," the transmission of hereditary defects, the differences in birth rates between and within various groups—all have a common denominator in the answers to the question just raised: How, and how far, are restraints on human reproduction to be carried out?

Birth control means or devices (some used by females, some by males) have been employed as far back, and wherever, human beings have had knowledge of the process of procreation.* In the past these means or devices usually were only temporary preventives of fertilization, limited to a given instance. Recent developments, however, have been pills or other chemical agents which can prevent conception for extended periods by inhibiting ovulation and/or the implantation of fertilized eggs, or by inhibiting the production of fertilizing sperms in the male. All of these methods have met with obstacles or limitations to being put into practice. The financial costs, ignorance about their use, indifference or carelessness, or possible side effects have been the most frequent factors. But there also has been strong religious opposition—not to birth control, *per se*, but to the methods employed. For it should be understood that no major religion—the Catholic Church included—is now opposed to the principle of birth limitation if and where advisable. In fact, in recent decades, papal pronouncements (by both Pius XII and John XXIII) have stressed the wisdom of avoiding conceptions (1) when necessary for a woman's health, (2) where genetic defectiveness in offspring is threatened, (3) where couples already have as many children as they can provide for properly, and (4) where the population density of a country and its lack of facilities are such as to warrant consideration in determining the number of offspring.

The objection of the Catholic Church is to "artificial" contraceptive methods which it considers "contrary to nature." What it does approve, as entailing "no frustration of nature's laws," is two methods: (1) abstention—which, as a permanent procedure, may be unacceptable to most couples, for many reasons, and (2) the "rhythm method"—the

* The Bible cites an instance of prevention of conception in the case of Onan (Genesis 38:9) who "spilled it [his 'seed'] on the ground"—akin to one of the contraceptive methods still widely practiced. However, Onan was doing this to thwart the religious law which had required him to marry the widow of his brother and father children by her on his brother's behalf.

alternative most widely recommended by Church authorities. The latter method requires confining sex relations to the so-called "safe" periods each month when the woman is unlikely to conceive. These periods are those before and after the week or so mid-way in a woman's menstrual cycle during which ovulation usually takes place. Actually, it is only during about one day each month that there is an egg in a position to be fertilized, but just when this happens may be uncertain (unless medical tests are made frequently) because many women are highly irregular in their menstrual cycles. Thus, to provide leeway, a week or more within the month is set off as a period in which not to have sex relations. But while adherence to this principle can greatly reduce the chances of conception—and if practiced widely could bring a sizable drop in birth rates—in *individual* cases where there is an urgent need or desire to prevent conception the rhythm method must be considered unreliable.*

Most extreme, and the ultimate method in birth control, is *sterilization*. This requires a surgical operation that is intended to end the individual's capacity for procreation: in a woman, cutting and tying the Fallopian tubes which carry eggs from the ovaries to the womb; in a man, cutting and tying the tubes which carry sperms from the testes. But this is not an *unsexing* operation and in no way inhibits sex desires or interferes with normal sex-functioning. Its effects, merely of sterility, are much the same as in many normal women and men who cannot procreate.

Unlike other birth-control methods, sterilization operations on a wide scale were first carried out *officially*, and for *eugenic* reasons (actual or presumed). In the United States laws authorizing sterilization of institutionalized mental defectives when deemed advisable were adopted in California and Indiana in 1907. Twenty-eight other states subsequently passed sterilization laws, but some states suspended them

* The possibility has been considered that the Catholic Church might yet sanction some chemical or medical treatment with contraceptive results, such as the contraceptive pill. This hope has been held forth by a leading Catholic gynecologist, Dr. John Rock of Harvard University, who himself had helped to develop one type of pill which prevents ovulation. However, despite Dr. Rock's argument (made in his book, *The Time Has Come*, Knopf, 1963) that the pill had the same effect and purpose as the rhythm technique, Catholic authorities have so far ruled it unacceptable for ordinary use. The one qualification, made by Pope Pius XII in 1958, was that a contraceptive pill could be taken by a Catholic woman if prescribed medically not to prevent conception, but as "a necessary remedy because of a disorder of the uterus or of the organism." While this might cause an *indirect* sterilization, the Pope explained it would be permissible "according to the general principle concerning acts which have a double effect."

later, or ceased to enforce them. Many court cases meanwhile had challenged the legality of these laws, until, in 1927, they were upheld by the Supreme Court in a decision in which the late Justice Oliver Wendell Holmes made the now classic comment, "Three generations of imbeciles are enough."

By 1960 it was estimated that under official state authorizations over 62,000 Americans had been sterilized, about 52 per cent of these having been mental defectives (the majority females), 44 per cent mentally diseased persons, and the remainder including some institutionalized epileptics, and (wholly unrelated to the foregoing cases) a considerable number of criminals.* Many additional thousands of legal sterilizations, mainly for eugenic reasons, have been performed in England, the Scandinavian countries, Switzerland, Japan and Germany. But to this one must add that under the Nazis sterilizations were carried to such frightful extremes—being inflicted on vast numbers of normal persons marked as "undesirable" (and with plans under way to sterilize untold numbers of others among conquered peoples)—that the world was shocked into a "go-slow" attitude thereafter toward any officially enforced sterilizations for eugenic reasons, real or assumed. However, years before this the United States already had been moved to such a policy when it was found that in a Kansas reformatory scores of girls had been arbitrarily sterilized for everything from sex offenses to "misbehavior" and "bad temper."

As a consequence of the sad lessons of the past, and also of new genetic findings which threw doubt on the eugenic justification of many of the operations, the numbers of official sterilizations—particularly, compulsory sterilizations—have been decreasing in the United States and elsewhere. On the other hand, there has been an enormous increase in *voluntary sterilizations*, requested by persons of both sexes primarily as a birth-control measure, but also for health reasons, or because of fear of having genetically defective children, or on social and psychological grounds. The numbers of these voluntary sterilizations are believed to be reaching up to 75,000 a year in the United States—at least fifty times those performed officially under state laws. Much more extensive have

* Authority to sterilize epileptic inmates is currently given to administrators of institutions in eleven states (Arizona, Delaware, Indiana, Kansas, Mississippi, Montana, New Hampshire, South Carolina, Utah, Virginia, and West Virginia). The *genetic* basis for this, however, is highly doubtful in most instances (as noted in Chapter 24). Even more dubious is the sterilization of criminals on genetic grounds (authorized, through castration, in Kansas and North Carolina), although in the case of habitual and serious sex offenders it might be justified for social reasons.

been the sterilizations in Puerto Rico, where it is estimated that the operation has been asked for and performed on up to one in every seven or eight women who have borne three or more children. In India, strong government endorsement has resulted in many more sterilizations of husbands than of wives, mainly because the procedure is easier for men and because bonuses have been offered to those undergoing the operation. As a result, close to 200,000 males in India, mostly in rural areas, have been sterilized to date.

Needless to say, in all countries sterilization—except where medically urgent—has met with particular opposition from Catholic authorities, although most Protestant authorities have favored it when needed as a eugenic measure or for mass birth control. However, the fact that sterilization and other birth-control measures are so widely resorted to in predominantly Catholic Puerto Rico emphasizes the point that official religious policies are often at variance with popular practice. Birth control has also been extensively utilized in such other predominantly Catholic countries as France, Italy, Spain and Chile; and in the United States some studies have indicated that Catholics are using contraceptives almost as much as are non-Catholics. In general, both among Catholics and others, it appears that birth control tends to be correlated more with social, educational and economic levels than with religion—the more advanced the group, the more its resort to the practice. The exceptions now occurring among the less advanced groups in Puerto Rico and India are dictated mainly by unusual social pressures and official encouragement.

Most controversial of all has been one other birth-control expedient —*abortion.* Despite religious and legal bans, it has long been one of the most widely used methods of preventing unwanted births. But a recent development has been the official and social endorsement of abortion in Japan as a birth-control measure, making it one of the principal means whereby that country's birth rate has been cut to one-half in a short period. One might add that in the United States and various other countries abortion has legal sanction only in the limited instances where there is medical certification that the continued pregnancy would seriously endanger the mother's life (or, sometimes, her sanity).

Whether abortion is justified as a *eugenic* measure—if the birth of a seriously defective baby is threatened—is another moot question. Readers may recall newspaper stories about this problem with the thalidomide babies (discussed in Chapter 6). In one case a pregnant American mother, fearful that she was carrying one of these babies, sought but was refused legal sanction for an abortion in her state, and then went to

Sweden, where it was authorized by a medical board. (Her fears were borne out when the fetus proved to be badly malformed.) The question at issue would also arise if there were evidence of serious genetic defectiveness in a baby being carried, and an abortion were sought. An argument then might be that aborting a fetus was no different in principle from killing a child already born—in other words, *infanticide* (another birth-control practice, widespread among primitives and many lower-level European and Asian groups). This was precisely the point in a second thalidomide baby case—that in Belgium where a mother (abetted by family members) painlessly ended the life of her armless, legless infant, and was acquitted by a jury.

The problems raised in the foregoing cases are similar in principle to various others which might arise with respect to *negative-eugenics* proposals. Which is to say, if *conceptions* in given instances are likely to result in badly deformed, diseased or retarded offspring, destined to be burdens to their families and to society, are the decisions regarding them to be left solely to the parents concerned, or is society also entitled or required to exert its influences and authority? And if so, by what means or pressures, and to what extent?

No less important are the questions raised by *positive-eugenics* proposals for furthering reproduction among persons designated as "superior" or "most desirable," and the extent to which society should act to carry them out. What these proposals may be, and other aspects of both negative and positive eugenics which may demand consideration in the future, will be our next topics for discussion.

CHAPTER FORTY SEVEN

The Stork and Eugenics: II

Program for Tomorrow

"WHO WOULD BE YOUR 'IDEAL' MAN?"

Ask this question of each of a few dozen unbetrothed young women picked a random. Or ask an assortment of fancy-free young bachelors to describe the "ideal woman."

The answers, you may be sure, would be extremely varied; and this is fortunate. Otherwise, if there were only one ideal type for each sex, a large majority of girls might be perpetual wallflowers, and most men would find no female who would truly dote on them.

Equally certain, the popular notions about desirability for either sex would conflict considerably with the standards of human "fitness" that eugenists had or have in mind. For social desirability and eugenic—or genetic—desirability may be far from the same at any given time and place.

Thus, in *positive eugenics*—the program for increasing reproduction of the individuals rated most "desirable" and carrying the presumably "best" genes—the complexities and doubts may be even greater than those of the *negative-eugenics* program dealt with previously. For it generally is easier to determine what isn't wanted in human traits than what is wanted. Adding further to the difficulties of the positive-eugenics proposals, as originally conceived, was much confusion between the environmental and genetic factors which made for human desirability. Granted that "superior" or "plus" labels could currently be affixed to some individuals and groups, it was not fully recognized that environmental advantages might have contributed greatly—perhaps preponderantly—toward their excellence, and that innumerable persons in the "minus" categories might have been endowed with genetic capacities for achieving similar "superior" or "plus" labels if they had had equal opportunities. To step up the breeding among the existing "pluses" might therefore not have been as crucial from the genetic standpoint as was thought.

The fact is that most of the positive-eugenics proposals of several decades ago had more to do with sociology than with genetics. To encourage the "plus" individuals to marry earlier, have more offspring, and rear families with less worry, there were suggested measures such as these: scholarships and educational grants to gifted students and graduate students; maternity-cost insurance; better and cheaper prenatal and obstetrical care; subsidized housing to reduce rentals for worthy young couples and to provide room for more children; greater job security; unemployment insurance; aid to widows; removing restrictions against women teachers' marrying and granting them maternity leaves; and raising salaries in the academic, research and other upper-IQ fields. As can be seen now, most of these proposals have been turned into reality, not through any special action by eugenists, but as a consequence of social, economic and political changes. Moreover, these measures, extending through all levels, have increased the production of "fit" or "fitter" offspring not only within the previous "plus" ranks, but among innumerable persons from the hitherto "minus" groups.

Not lost sight of, nonetheless, have been the human-breeding proposals which center about this point: That in all groups certain individuals can be recognized as unique and superior in their mental endowments, talents or other desirable traits. If these traits are in considerable degree the results of unusually good genes, which mankind could profit by having in much larger proportions, why should not every effort be made to propagate these "superior" genes beyond the limits of ordinary matings? The idea has for years had the support of such outstanding scientists as Professor Hermann J. Muller and Professor Julian Huxley. Certain measures which they and others have proposed may have seemed preposterous to many—although less so now in the light of new scientific developments. But it is not the means set forth for breeding superior human beings so much as the objectives that have aroused most controversy.

One proposal is that of storing and preserving the sperms of superior men or geniuses and, through artificial insemination, using these to father very large numbers of children—even after the sperm donors have long been dead. Far from being speculative, the procedure involved has been thoroughly tested and is widely employed today in breeding pedigreed livestock. In fact, fully one-third of the present dairy cows of the United States, and a large proportion of those in other countries, have been bred by artificial insemination with sperm from prize bulls. The method has many advantages. A single ejaculate from a bull can be divided, or diluted, into a great many effective portions and used to inseminate scores of cows—perhaps several hundred. Further, the bull semen, when deep-frozen or freeze-dried, can be stored for a number of years (so far up to eight years, but in time probably much longer) and be as fertile when thawed out for use later as when first obtained. Not only has this been proved repeatedly, but there are several instances where numerous progeny have been sired through sperms of bulls dead for several years.

The application of this method to human breeding was suggested far back in the 1930's, when it was given the name of *eutelegenesis*. As Dr. Muller sees it now, the sperms of designated superior men might be collected and stored in sperm banks, which would preferably not be drawn upon until there had been ample time to judge the donor's qualities—possibly after his death. Even with average men, however, Dr. Muller believes the method could be used in various situations. For instance, a young husband going to war—or an astronaut venturing into space—could leave behind his stored sperms, and should he not return his wife could still bear children by him.* Or, as insurance against a husband's later sterility, or in the face of radiation dangers to genes, frozen sperms (and possibly frozen eggs of women, as well) could be safeguarded in storage to be used as uncontaminated seeds for producing future human generations.

(The practicality of storing and using frozen human sperm in artificial insemination has already been established. In 1964 Dr. William H. Perloff [Albert Einstein Medical Center, Philadelphia] reported that six women had conceived through artificial insemination with donor

* The legal snarls which might ensue from the siring of posthumous children with the sperms of a deceased man—going much beyond the complications presented by ordinary artificial-insemination babies (dealt with in Chapter 45)—have been set forth by Harvard law professor W. Burton Leach. Among other things noted, there might have to be revisions in the rulings which stamp as illegitimate, or as not a man's "lawful" issue, a child born beyond a certain time after a husband's death and/or beyond the designated maximum period of gestation attributable to his direct sexual relationships with his wife. No less in need of resolution would be the legal status of babies resulting from transplanted human ova, as set forth in our next paragraphs.

sperms preserved for up to 5½ months by freezing. Of these women, four delivered full-term babies who, when checked months later, were physically and mentally normal in development. [Two of the women had aborted, but one of these had done so three times previously when inseminated with nonfrozen sperms.] The sperm-freezing technique employed, developed by Dr. J. K. Sherman, involves the preservation of freshly collected semen in liquid nitrogen. The semen can then be quickly thawed out and used at any time thereafter.)

What of women who are rated superior, or of genius caliber? The possibility is held forth not only that they, too, may become parents of large numbers of children after their death, but that while alive they and many other *living women may be enabled to have children without bearing them.* Incredible as this may seem, the biological equivalent has already been achieved with females among lower animals. In cattle (and in rabbits, mice and dogs), fertilized eggs have been withdrawn from some females and introduced into others, who nurtured the fetuses through to birth. An important outcome of these experiments is that the offspring are identical in all hereditary traits with what they would have been if the same female who produced the eggs had also mothered them. For example, when an egg from a white rabbit is gestated in a black foster-mother rabbit, the offspring are all white and take on none of the genetic traits of the foster mother. (Another disproof of the acquired-characteristic-inheritance theory.)

In transplanting ova in cattle, the objective is to enable a prize cow to produce not one calf in a season, as she normally does—and for a period of perhaps only ten years—but to produce twenty or more calves per season, and eventually, up to several hundred in all. The procedure involves (1) causing the blue-ribbon bossy, through hormonal stimuli, to mature many eggs at one time; (2) having the eggs fertilized in her by sperms from a prize bull; (3) flushing the fertilized eggs from her tubes; and (4) implanting one each of the eggs in a different cow—even a "scrub" cow. While this already has been done successfully, the method is still too difficult and costly to have had any extensive use in cattle breeding. Among dogs, not merely eggs but entire ovaries have been transplanted from pedigreed into mongrel females. After being bred to pure-blooded males, the mongrel foster mothers have produced *pure-blooded* puppies, exactly like those which the donor of the ovaries herself would have produced.

That these findings may have meaning some day for human beings isn't doubted by the experimenters. First (if and when the techniques are so employed) innumerable women who have healthy ovaries but cannot or should not undergo pregnancy for various reasons (organic defects, hormonal deficiencies, disease, age, etc.) may yet be able to have children through paid "proxy" mothers. To effect this, a fertilized egg from the woman donor could be transferred to the womb of the one who had agreed to gestate and bear the child. By another method, complete ovaries might be transplanted from an older woman, or one who otherwise risked miscarriage, to a young woman who could effectively carry children through to a healthy birth. The donor of the ovaries could then be genetically the mother of children borne by another woman, and the one in whom the ovaries were transplanted might bear a succession of children who would be genetically not her own but those of the ovary donor. (And couldn't this lead to complications on Mother's Day!)

Ova transplanting might also be undertaken for eugenic reasons similar to those prompting artificial insemination with donor sperms (A.I.D.), as set forth in Chapter 45. If it was the wife instead of the husband whose germ cells were infertile or carried the threat of transmitting some serious hereditary condition, she could be implanted with eggs from a healthy donor. The results and the parentage problems would then be analogous to those in the artificial insemination cases, with this difference: Instead of the child of a couple not being genetically the husband's, the child in the ova-transplant cases would not be the wife's.

Continuing with the theoretical possibilities, just as a great man's sperms might be stored and used after his death, the eggs—or ovaries—of a woman who was outstanding, or a genius, might be stored for mothering many children after her death. (A granddaughter might then conceivably bear the child of her grandmother!) Going to the ultimate of posthumous eugenic combinations, the eggs of a deceased great woman could be fertilized in test tubes with the sperms of a deceased great man, and these eggs could then be transferred to and gestated in different proxy mothers.* Once more let it be stressed that this is not

* The thought of the progeny that might result from combining the genes of a great man with those of a notable woman brings to mind the classic exchange between two famous personages of the past—George Bernard Shaw and the dancer, Isadora Duncan. Miss Duncan was reported as saying to Shaw, "Just imagine what a child we two could have—one with my beauty and your brains." To which Shaw was said to have replied, "But suppose the child had *your* brains and *my* looks?"

thought of as pure science fiction, but is regarded by Professor Muller and others as a feasible and not improbable development of the future. However, Dr. Muller comments that the use of any of the breeding methods mentioned—artificial insemination with sperms of deceased geniuses, ova transplants, and others to be mentioned later—"will not spring into existence *full-fledged* overnight, but may be taken up at first by tiny groups of the most idealistic, humanistic and at the same time realistic persons . . . with especially well-developed values."

Not least startling among the possibilities is that of "*virgin birth*" where an exceptional woman has been denied a suitable mate and yet wants to bear a child. That this might be achieved is held out by findings and experiments involving *parthenogenesis*—the process whereby females in some of the species which normally require mating with males have produced offspring from unfertilized eggs. In these cases the eggs have been activated—spontaneously or experimentally—to double their chromosomes as would ordinarily happen following fertilization by a sperm; and then, with each pair of duplicated chromosomes thus formed acting as if they had come from two parents instead of one, cell multiplication and fetal development can proceed as in a fertilized egg. Such "virgin" conceptions and births occur spontaneously sometimes in poultry, notably turkeys, and have been induced experimentally in rabbits.

One important aspect of the parthenogenic births relates to the sex of the fatherless offspring. Since their sex chromosomes are only a duplicated pair of one of the mother's, the results depend on which type of chromosome this was. Here there is a difference between the sex chromosomes of poultry and of mammals (including man), as noted in Chapter 7. In turkeys and other birds, the female is the one-X or XO individual, the male having the XX combination. Thus, when the turkey hen's single X doubles in parthenogenesis, the resulting XX offspring must be a male. (A turkey female can be produced only when an egg without an X—one in two of those laid—has been fertilized by the tom.) But in the species where the female is the XX individual (as in man and other mammals), parthenogenic sex determination has reverse results: Since a male can be produced only by a Y from a father, any parthenogenic offspring resulting from a doubling of the mother's chromosomes would be an XX individual like her—hence, invariably a female. Obviously, then, should an induced "virgin birth" ever become possible for and be sought by a woman, she would have to content herself with having only daughters.

What of the future hypothetical man who might want to become a father without a woman having any part in the child's heredity? This, too, is theoretically not impossible. It would require these steps: The sperms of a man would be separated into the Y bearers and the X bearers (by some means postulated in Chapter 7). An egg would be taken from a female and the nucleus, containing her chromosomes, would be withdrawn from it. Replacing this nucleus would be that from one of the male's X sperms. There would now be an X egg carrying the father's chromosomes. Next, this egg would be implanted in some woman host, where it would be fertilized with another of the father's sperms, through artificial insemination. Or it might previously have been fertilized in a test tube, before implantation. Whatever the procedure, the child so gestated and carried through to birth would be *only the father's offspring*—carrying only his chromosomes.* As for the sex of a "motherless" child so produced, since the father's sperm used for fertilization could be either an X bearer or a Y bearer, the offspring could be of the sex desired.

Reference to the separation of X-bearing and Y-bearing sperms leads to the point that control of the sex-determination process is considered one of the likeliest developments in the not too distant future. One of its eugenic benefits might be to increase the chances of producing sons to carry on the skills and achievements of outstanding men in arts and professions. Similarly, society might gain by ensuring the birth of daughters in certain families where unique feminine contributions had been and could continue to be made.

Another area in which reproductive control is foreshadowed is in *twinning*. In some lower-animal species experimental induction of twin-

* An erroneous assumption is that a hypothetical son produced only by the father's germ cells would be his "identical twin," genetically, or a hypothetical "virgin-birth" daughter would be her mother's "identical twin," inasmuch as in either case the child's chromosomes would be duplicates of those in the single parent. The fallacy is that the offspring would not receive duplicates of *all* of the single parent's chromosomes. The parthenogenic daughter would have doubled duplicates of only half of the mother's chromosomes, those which had been in the activated egg. The son stemming from two of the father's cells would carry his X and Y chromosomes, but some of the many other chromosomes of the parent might not have been present in either of these cells (keeping in mind the random process by which any sperm receives only one or the other of every pair of the father's chromosomes). In any case, there could be no "identical twinship," genetically, with a parent unless *all* of the parent's chromosomes were duplicated in the child.

ning is now commonplace; and it may be only a matter of time before human parents can have twins—or supertwins—if and when desired and of the types preferred. (Hormonal treatments of women for fertility already seem to have brought an unusual number of twin births.) Induction of fraternal twinning would entail the stimulation of a woman's ovaries to produce two or more eggs at a time. Controlled identical twinning would involve the ability to make an egg split shortly after fertilization (or, if higher multiples were desired, to induce further splitting). The practical eugenic advantages of artificially induced twinning might be those of (1) enabling unusually "desirable" parents to produce more than an average quota of offspring, and (2) enabling "superior" women in important careers to have several children with a minimum loss of time.

Direct attacks on wayward genes and their effects before birth—or even before conception—are also on tomorrow's genetic agenda. To cope with hereditary diseases or defects, ways may be found to block or mitigate the worst consequences of various conditions, especially those in which the biochemical pathways from the genes to their effects have been or will be traced. Where the "errors" of certain genes are anticipated, it may be possible to begin corrective processes by altering the chemistry of the fertilized egg or by setting up safeguards in the prenatal environment in which the fetus develops; or ways may be found to remedy ailing or deficient genes directly, with chemical additives injected into the cell or fetus. Control of the sex-determination process could also bring a marked reduction in the incidence of serious sex-linked conditions; for, if a man had hemophilia, for instance, his X-bearing sperms, carrying the harmful gene, could be separated out so that only his Y sperms would be used in procreation. He then would sire only sons, free from the threat of the disease, and with no daughters, his wayward X-linked genes would be taken out of circulation.

Also anticipated is the overcoming of biochemical genetic blocks between unrelated individuals in organ transplanting, which would enable such transplants to be made almost as effectively as they now are between identical twins. This would make practical the preservation and storage of healthy organs of various kinds taken from deceased persons (as with eye banks) and their use to replace defective parts in individuals having vital need of them. Among other things, it could reduce the hazards of genetic defect in some organ or other part of the body and could greatly extend life expectancies of persons whose genes were otherwise geared toward a long life span. This development will probably be preceded (and facilitated) by the marked reduction (if not elimination) of the hazards of blood-group incompatibilities between one individual and another. If this can be achieved—by reducing a person's antagonism to foreign blood substances or by neutralizing their effects—many of the difficulties now experienced in blood transfusions and the failures in grafting operations will be minimized.

Among the far-out possibilities is that of producing changes in specific genes themselves; that is, inducing mutations in desired directions. This might come about (1) when or if human chromosomes are "mapped" so that the precise locations of given genes are pin pointed (as can be done now with many of the genes of fruit flies); (2) when the genetic coding (the DNA composition) of the designated gene has been unraveled; (3) *if* the designated gene can be "hit" by radiation or some other force in such a way as to change its code and its workings; and (4) *if* all this can be done in an egg or a sperm before fertilization. Hypothetically, then, a gene that was wayward might be reset on a proper course. But also hoped for is that an "ordinary" gene for some trait might be turned into a "superior" one. This is the most remote of the objectives, because it is known that the overwhelming proportion of mutations—perhaps 98 in 100—tend to be disruptive and unfavorable; and a great many sperms and eggs would have to be sacrificed before the sought-for change could be attained. The process might become practical in lower and prolifically breeding animals. It may always remain only a fantasy for human beings.

Having gone over the various "atomic-age-parenthood" proposals, one might guess that those which would evoke the most challenge or criticism deal with plans to breed "superior" human beings of specific types. Worthy as this objective might seem, a first reservation is that human breeding can in no sense be likened to the breeding of lower animals, nor can it be carried out in the same manner. In domesticated lower animals, "superiority" and "desirability" (from man's viewpoint) can easily be defined in each species and linked with a few genetic and breedable traits: in the cow, milk output; in the steer, meat yield and quality; in the race horse, speed; in hens, egg-laying; and in turkeys, tenderness and breast size. By intensified breeding and selection in these species, and with the advantage that generations follow at short intervals, whole new populations with designated desired traits have been produced in the space of decades.

Obviously, in human beings the situation is quite different. Even if it were possible to dictate and control human breeding as easily or quickly as that of lower animals, there are these questions: What, precisely, do we mean by "superiority" or "desirability" in human makeup? If we could agree on this, how sure are we that the traits rated highest are primarily products of heredity and not in large measure environmentally produced? If the traits are genetic, how successfully can we identify, transmit and propagate the gene combinations responsible? And—assuming that we would be breeding not for the present but for the future (knowing the process might take generations)—how sure are we that the traits we now would aim for in human breeding would be the ones considered most desirable by people centuries hence?

Consider women (always fascinating subjects): If we had the power to breed them to order, what type would be striven for—leaving it to both sexes to decide? As indicated at the outset of this chapter, the notions about the "ideal" woman would be exceedingly varied. Moreover, throughout history, preferred "styles" and shapes in women have been as changeable as fashions. By present popular standards in the United States the ideal might be close to the "Miss America" type, accompanied by other "glamour-girl" traits, as shown in our Chart. Yet, as further revealed, these traits would be quite opposite in many ways to those deemed eugenically best suited to motherhood or to high achievement in a useful career. This does not imply that a woman with glamour cannot also be intelligent and a fine mother. What is meant is that to make the prevailing standards of female desirability the basis for breeding would rule out a great many of the women eugenically most superior and would include a large number of the less worthy ones.

Most fallible, however, may be contemporary judgments of desirability or superiority in mental traits, talents and achievements. Left to the majority opinions of their times, we know that the works of many of our greatest scientists, thinkers, artists, musical composers, writers and dramatists would have been doomed to speedy oblivion. In almost every field of human endeavor many men who were far—or too far—in advance of their times, or who did not conform to existing dictates or notions, were ignored, scorned, sometimes persecuted, sometimes killed. Grimmest of all, we know what happened during the demoniacal purges and wholesale exterminations that went on in Europe before and during World War II. All pre-existing standards of human worth were horribly warped, and countless individuals who previously or elsewhere would have been regarded as of the most superior types were destroyed as if they were vermin, while many individuals whom we now think of as degenerate, inferior, or menaces to civilization were exalted and idolized

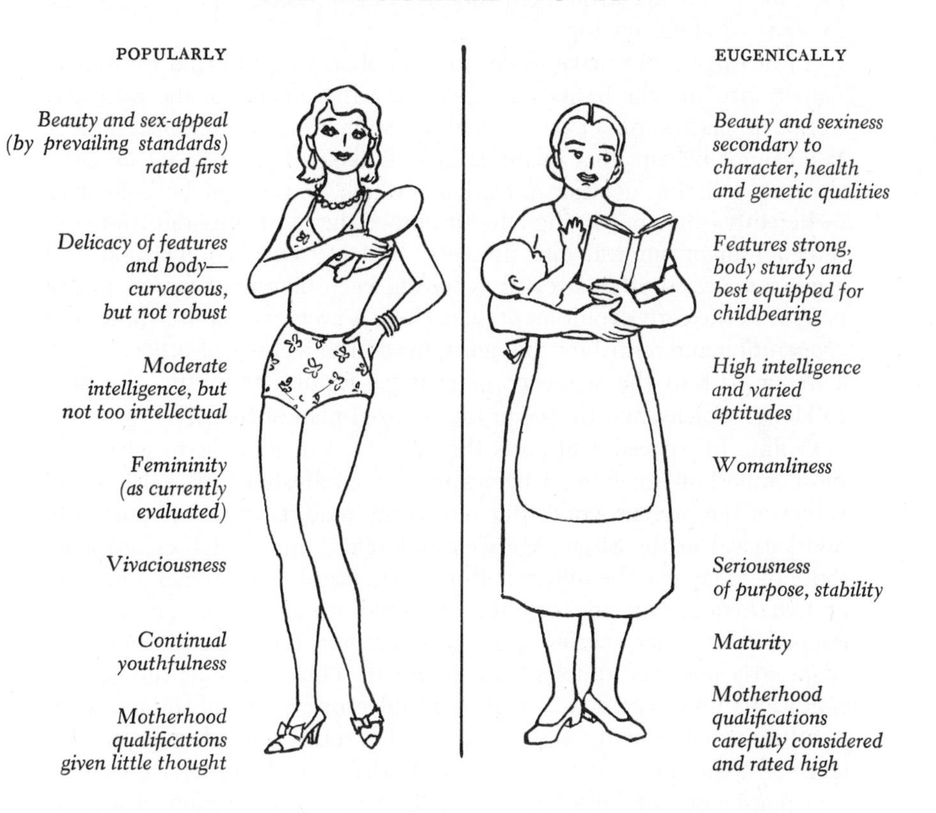

as paragons of human greatness. (Think of the many persons who now regret that their parents had proudly named them after Hitler, Mussolini, Stalin.)

Even at this moment, and among those close to you, it may be far from easy to evaluate qualities and rate individuals according to their intrinsic worth as human beings. The lying child with an overdeveloped imagination *may sometimes* be the forerunner of a gifted writer; the boy who balks at discipline or is indifferent to studies and marks *may sometimes* burst forth with creativity and become a blazer of important new trails. Extreme righteousness, on the other hand, may often be coupled with intolerance. Bravery may often mean callousness. The requirements for success in many fields may go with such traits as ruthlessness, insensitivity, unscrupulousness and selfishness. As human beings, many obscure little men who never get anywhere may be superior to some of those on top.

All of this should make us go slow about seeking to breed for human "superiority" on the basis of competitive achievements in the workaday world or classroom, or by prevailing evaluations of physical fitness. We cannot be sure that qualities such as sympathy, kindness, decency, altruism and consideration for others—which, too, might be influenced by heredity—may not in the long run prove the most important for continued human survival and progress. In fact, while competition and aggression are so often stressed as having been dominant factors in the past, or in the early evolution of man, there is every reason to believe that *cooperation* and regard for the rights, needs and feelings of others played a major part in the sequence of changes leading from the ape man to Homo sapiens, and thence to the civilized man of today.

Calling for special caution is the fact that our ideas as to what it is most important for human beings to be are adjusted to the needs and values of the present world. But just as the requirements for superiority and survival in the Stone Age were different from what they are now, those of today may be different from what may be demanded a century or two hence. Not to keep this in mind would be to ignore what happened to other species which had become too set and could not adapt to a new and different environment. Thus, some of the genetic conditions now regarded as defects could prove to be advantageous in a future environment; and some of the traits now deemed advantageous might prove detrimental in another era. As experts on evolution point out, the "flexibility" of a species—the possession of a great variety of genes, including some which might be rated "bad" at the moment—offers insurance for survival in the face of unforeseen future challenges: another way of saying that variety is not only the spice of life, but an essential for the continuing existence of our species.

At the same time, we cannot assume that certain of the positive-eugenic proposals now apt to be spurned as fantastic or ridiculous may not be regarded as practical and desirable in the social and psychological climates of the future. Conceivably, if children come to be thought of more as the concern of society than of the individual parents, controlled breeding may become more acceptable. If the importance of having children of "one's own blood" is de-emphasized, husbands may become more receptive to the idea of insemination of their wives with the sperms of geniuses or other selected donors. If women turn further toward careers and to equality with men, many might wish to relieve themselves of childbearing by having their eggs transplanted to proxy mothers. Perhaps, too, with more knowledge about the inheritance of complex traits, a future society may be readier to sponsor selective breeding for given talents and mental characteristics.

Yet it hardly seems likely that positive-eugenics proposals of the kinds discussed will become effective or be widely adopted within this century. More tangible and realistic for the moment are the various negative-eugenic measures aimed at preventing the deterioration of human quality, whatever the causes. Overshadowing all else in this respect are the effects of the "population explosions" discussed in the preceding chapter. While essentially environmental, they may also be interrelated with genetic factors. For example, various types of hereditary weaknesses and predispositions have much greater chance of expressing themselves under unfavorable conditions—inadequate hygiene, inferior diet, increased emotional stress—all of which could result from the deteriorated environments brought about by excessive population growth. Conversely, keeping populations within bounds would decrease the chances that hereditary weaknesses would assert themselves. Another favorable eugenic consequence of limiting family sizes would be a great reduction in childbearing by older women, who are most prone to producing children with congenital abnormalities, including mongoloid idiocy and other chromosomal aberrations.

However, whether it is a matter of curbing erratic reproductive trends in general, or of reducing the incidence of specific genetic or hereditary defects, there is always the question of how the objectives can be carried out. In Chapter 45 we presented certain eugenic problems as they might be viewed and considered by the individuals directly concerned. Now let us consider a number of actual cases from the standpoint of society. Think of yourself as sitting in judgment on them, with the power to decide what compulsory action, if any, should be taken:

1. Living in the Midwest is a man who, as a child, had the congenital eye cancer, *retinoblastoma,* usually inherited through a "dominant-qualified" gene. Unless operated on, the condition almost always brings early death, but thanks to surgery in which the affected eyes are removed, 70 per cent of those afflicted are enabled to survive and reproduce. This is what happened with the man in question. His cancerous eye was removed, he grew to maturity, married and fathered two children, *each* with retinoblastoma, and they, too, were saved—but only by the removal of both eyes in one child, and one

eye in the other. What would you suggest doing in situations such as these to prevent further transmission of the retinoblastoma gene: Not operate on the children, and let nature take its course? Operate on their eyes, but also sterilize them? Or would you say, operate without sterilization, leaving it entirely to those afflicted to decide in time whether they did or did not wish to have children with the disease?

2. A middle-income couple in an Eastern city has *four* hemophiliac sons. The special burdens have made the family dependent on their community for financial assistance, and so far, for donations of several thousand pints of blood needed in transfusions. Once the first hemophiliac son was born, it was certain there was a one-in-two risk of the disease's appearing in any succeeding son. Should pressure then have been brought on the mother not to have any more children? Should the four hemophiliac sons themselves be sterilized? Before answering, consider the next case.

3. The British Queen Victoria, as we know now, carried the gene for hemophilia which was transmitted to a number of her descendants (but not to any members of the present British royal family). The disease in one of Victoria's great-grandsons, the Czarevitch, caused his parents to be victimized by the evil Rasputin, and thus helped to bring on the Russian Revolution (explained in Chapter 22). If you had lived when Victoria was young and had known that she could transmit the hemophilia gene, would you have advocated her either not marrying, or taking steps not to have children? Or would you have held that her need to carry on the royal line made up for the danger?

4. Several thousand persons in the United States in the past three centuries have suffered the agonies of Huntington's chorea, most terrible of hereditary afflictions, as a result of receiving replicas of genes brought over and passed on by three English brothers in the 1700's. Many other descendants now face the same fate. If you had lived two centuries ago and could have looked into the future, would you have demanded sterilization of the three brothers? Would you today advocate sterilizing all identifiable carriers of Huntington's chorea genes before they reach maturity?

5. A delinquent girl in a New England state, where sterilization is authorized for mental defectives, was turned loose without an operation because she was just over the border of "legal feeblemindedness" (with an IQ of 72). In a few years she bore, illegitimately, three feebleminded children. What should have been done in such a case: keep the girl in an institution throughout her reproductive period unless she or her parents consented to her sterilization? Or release her without any operation, hoping she could be induced to abstain from sex relations or to use contraceptive measures?

6. Some years ago the daughter of a famous inventor (deceased) sued her mother for having had her sterilized on the grounds of feeblemindedness. The court upheld the mother. But there was this question: Might not the girl also have been carrying certain of her father's "superior" genes? And, lacking proof that the girl was *genetically* defective, should society have risked the passing on of some "feeblemindedness" genes in order to perpetuate the rare "superior" ones?

7. The *battered-child syndrome* is now the medical term applied to the cases—thousands yearly—of little children who have been badly beaten, often crippled, sometimes killed, by vicious parents who did not want them. In many instances the child was an addition to an already oversized and ill-cared-for family. Where parents have committed such acts, or have been proved capable of committing them, what would you suggest doing to prevent their having more children?

How difficult it is to answer these questions was pointed out by Professor Theodosius Dobzhansky in discussing the problem posed by *retinoblastoma* (Case No. 1). Noting that medical treatment now makes it possible for most of those afflicted with the disease to survive, reproduce and pass along the gene for it, Dr. Dobzhansky asks, "Is man thus not frustrating 'natural selection' and polluting his gene pool?" But he adds, "If our culture has an ideal, it is the sacredness of human life. A society that refused on eugenic grounds to cure children of retinoblastoma would, in our eyes, lose more by moral degradation than it gained genetically. Not so easy, however, is the question of whether a person who knows he carries a gene for retinoblastoma, or a similarly deleterious gene, has a right to have children." And, looking ahead, he concludes, "It may well be that . . . the social cost of maintaining some genetic variants will be so great that artificial selection against them is ethically, as well as economically, the most acceptable and wisest solution."

The preceding comments apply equally to our Case No. 2, dealing with hemophiliacs or hemophilia-gene carriers. But in No. 3 (Queen Victoria) we have the problem of a superior person capable of producing superior or much needed progeny, who yet might also produce seriously defective descendants as well. Here present benefits might be weighed against future hazards. The opposite dilemma—of present hazards versus possible future benefits—looms up in Case No. 6, that of a defective who might be carrying some "superior" genes. In Case No. 4, involving Huntington's chorea, the indicated action on eugenic grounds would seem clearest; while in Case No. 5, curbing of reproduction might be deemed advisable whether for genetic or social reasons. Case No. 7, of the "battered children," would have to be regarded as almost wholly a social problem.

However, all of the sample cases we have presented, and most others in the area of eugenics, are also social problems in one degree or another. In fact, the argument for restraining or limiting reproduction of any individuals because of their presumed unfitness for parentage may be strengthened if it is based not merely on possible genetic threats, but also on the social and psychological handicaps that might be suffered by their offspring. For with growing emphasis on the importance of early environments and conditioning, one could ask what chance children have of being "normal" if reared by parents who are psychotic, mentally retarded, habitually criminal, or afflicted with serious physical ailments or deformities? Yet even in these cases it is doubtful whether attempts at forcible or legal prevention of parentage could accomplish nearly as much as could guidance by doctors, competent social workers, marriage counselors and pastors, acting together where required with experts on genetics.

The hope for improving human genetic quality by some selective breeding may certainly continue in the background. For instance, even a limited reduction in the proportion of the seriously wayward genes could mean tens of thousands of fewer defective individuals in each generation, and the saving of innumerable heartaches and enormous expenses. More extensive elimination of wayward genes could vastly magnify these results. In the positive direction, also, when more is known about the genetic mechanisms of complex traits, efforts to improve human quality with respect to talents, mental capacities and behavioral tendencies may be more feasible and rewarding. For the immediate future, however, it is certain that whatever can be done to improve the human stock by making genetic changes is minimal compared to the possibilities offered by changes in environments. One has only to think of these facts:

All of the genes that produce mental retardation are not impeding human intellectual functioning a fraction as much as are lacks in education and opportunities for achievement. All the genes for mental disease aren't creating nearly as much havoc as can be ascribed to the warped conditioning of people's minds (past and present). The wayward genes responsible for the worst hereditary diseases, deformities and defects are not killing, crippling or handicapping a fraction as many people as are being victimized by preventable environmental catastrophes, wars, accidents or dietary deficiencies.

On the plus side, the advances made in human mental and physical quality by a few decades of environmental improvement have been amazing. Without any genetic changes, widened opportunities for training have vastly multiplied the number of exceptional and superior persons in the sciences, arts and professions, many of them from groups hitherto regarded as almost incapable of producing them. (And in this must be included the enormous numbers of women whose energies have been released by removal of barriers against their sex.)

With no change in genetic stamina or defense mechanisms, but only through improvements in living conditions, diet and medical care, people have been enabled to ward off or reduce the menace of one disease after another and to make greater gains in life expectancy than could have been achieved by centuries of breeding for longevity. With the same genes for "looks" as their ancestors had, girls are now far often more beautiful and shapely, and men handsomer; with the same "stature" genes as the shorties of old, our youths have shot up to phenomenal heights; and with no Spartan breeding, but purely by better nurture and training, athletes have been developed who run faster, jump higher and throw objects farther than would have seemed believable only a generation ago.

Nor can we assume that we have more than begun to exploit human capacities and control or improve our environments so that mankind's fullest genetic potentialities can express themselves. Such major advances as have been made, moreover, have been unevenly distributed and have barely touched some groups. As environmental improvements spread we are certain to see a shrinking in the disparities between and among peoples with respect to surface qualities and achievements. Already it is apparent that the economic and cultural worth of an average person in one country has been moving steadily closer to that of a person in another country; and as environmental differences grow smaller, one may expect that race, nationality or geography may have steadily less to do with individual production and attainments.

Our stressing of environment need not be taken as minimizing the significance of heredity. Environment, in the end, can do no more for human beings than their heredity permits it to do. Equally important, all of the environmental changes mentioned that have done so much to improve human quality, mentally and physically, are in themselves the product of man's unique hereditary capacities to alter his environment and adapt himself to the conditions around him, as his needs may dictate. But this is speaking of mankind as a whole. Individuals, it should always be remembered, differ in their genetic potentialities for coping with the hazards or utilizing the advantages of their environments, whether natural or created by other men.

Thus, as we move increasingly toward the equalization of environments for all people, the individual genetic differences will be asserting themselves more. This will be true with respect to disease, life expectancies, IQs, manifested talents, achievements and behavioral traits. It may then become easier to distinguish those with superior or more desirable *genetic* endowments from those less well endowed. And it is then, perhaps, that the need for planned eugenic measures will be more apparent and urgent. For to assume that no improvements can or need be made in man's genetic components is to conclude that human biological evolution has reached the ultimate of perfection; and to challenge the right to swerve the course of our evolution by exercising selection and control over our genes, is to forget that for a long time man has been deliberately doing just that—through wars, genocide and other artificially engineered catastrophes, or by making the environments of some groups more favorable or less favorable for survival than those of others.

A sensible eugenics program, therefore, would seek to replace the reckless or haphazard direction of human evolution with intelligent and carefully planned guidance. In this we must think not merely of ourselves, but of our descendants to come. For if it is recognized as unfair to inflate the national debt and impose a financial burden on future generations, it is hardly less fair to impose on them an unsupportable population or an excessive burden of wayward genes. The alternative, of controlled and discriminating human reproduction, is a worthy one. But whether there is or will be the wisdom and discretion to chart a proper eugenic course is another matter.

Certainly, until the facts are clearer, it would appear that whatever eugenic measures are to be carried out must depend on the wishes of individuals—those immediately concerned, or, if they are not mentally competent, those entrusted to act on their behalf. The thought that society or the state should try to impose involuntary restraints on propagation by persons rated undesirable, or, moving in the "positive-eugenic" direction, should exert pressure on selected individuals to propagate beyond their desires, is no longer tenable. Equally unacceptable is the notion that human beings should be bred to meet specified needs and with designated traits. For civilization, and democracy as an outgrowth of civilization, have been moving ever farther away from the stratification of human beings into rigid classes, castes or groups, with fixed functions and social roles. Rather, the trend has been ever more toward the elimination of group barriers and toward the consideration of each person as an individual, to be respected for his or her own intrinsic qualities and to be given every opportunity for development according to his or her capacities.

It is in this direction, of bringing out the uniqueness of every person, that human genetics has been making its greatest strides and can yet make its greatest contributions.

CHAPTER FORTY EIGHT

Your Heredity and Environment

"YOU," "YOUR," "YOURSELF," are words that have been used many times in this book. For, beginning with the very first line, "Stop and think about yourself," it has been our purpose to relate the facts presented to you, the reader, wherever and as much as possible.

But what is meant by "you . . . your . . . yourself"? Or, specifically, by *your* heredity and environment?

The question is a many-faceted one; and the answers, so far as they can be given, have much to do with the summing up of what might be considered the most important points in our book.

As you must have found in your own case, the concept of "self" undergoes many changes in the course of anyone's life. When you were an infant, "self" at first was only your own body and its functioning. Proceeding into childhood, you identified with yourself your parents, siblings, and others who were important to your welfare, while you also recognized the importance you had to them. With your further growth and development, the circle of people identified as part of your existence, and of whose lives you felt yourself a part as well, widened successively. Today, the measure of your maturity may be how much you realize and feel that who and what you are, or can be or will be, is bound up with what other human beings are, can be and will be, regardless of their distance from your immediate orbit.

To begin with, *your heredity* is not something unique to you, but is part of the human gene pool shared with hosts of other people, extending far back to the dimmest reaches of man's origins and spreading into every human group alive today. *Your environment*, in turn, is bound up not only with the environments of other people, past and present, but also with their heredity. For what is most important in human environment is not its physical aspects—geography, climate, natural resources and so on, which have been much the same throughout mankind's existence—but the changes that the genetic capacities of people have enabled them to make in their environments. And since all of us are products of a continuing interaction between our human hereditary endowments (or deficiencies) and our man-made environment, it is impossible to draw a line between the heredity and environment of any individual and the heredity and environment of other human beings in his world.

Viewed in this light, facts regarding hereditary traits which at first glance may seem to have little or nothing to do with you personally can take on new meaning. For instance, you and members of your family undoubtedly are free of most of the hereditary diseases and defects which we have discussed. Yet each of you might well be carrying hidden wayward genes for several or many of these conditions. Further, marriages can bring various additional wayward genes into your family group. If nothing else (and in callous material terms), any number of the serious wayward gene conditions in other persons—including mental diseases and defects, and many crippling disorders and major functional ailments—which now require vast public and private expenditures for hospitalization, treatment and research, are touching your life daily through taxation, charitable contributions or business payments.

Apart from the foregoing, the practical importance which many other facts about human heredity have had or can have for you and those close to you have been brought out in discussions of specific subjects: prenatal experiences and their effects, sex determination, features and "looks," twins and twinning, blood groups, longevity, IQs and talents, behavioral traits, sexual abnormalities, crime, racial origins and differences, ancestry, the "population explosion," eugenics, personal problems of parentage, and so on. We also have given much attention to "biochemical individuality," because of the great significance it has with respect to many aspects of human makeup and conduct. Understanding more about your own individuality, for instance, may strengthen your resolve not to try to be like everyone else in what you eat or drink, or in how you live and in what you do. You may find reason to say, "No, thank you," when someone insists that this or that is "good" for you, or, "it's worked for me." If you have children you may be more alert to their special needs, and, in general, you may be more tolerant of apparent idiosyncrasies or eccentricities in other people. As our knowledge of biochemical individuality grows, it will help greatly to insure better diet, medical care and training more closely adjusted to each person's makeup and requirements.

But of most value, perhaps, have been not the specific facts we have dealt with, but the insights which the facts, taken together, have afforded us into why people usually think, behave and perform as they do. What we now can see much more clearly is that, with all allowances for hereditary differences, the most important general distinctions among individuals and groups are due to environmental factors—to circumstances preceding their birth, and to their conditioning and their relative opportunities thereafter. Acceptance of this conclusion is having radical impacts in every social sphere, in education, psychology, industry and government, and may be the most potent force in changing the political complexion of the world.

Yet the emphasis on environmental influences, much as it has contributed to human understanding and progress, can be overdone if it ignores the many inherent differences among people. Unfortunately, despite all the new evidence of how strongly heredity is involved in the production of physical traits and biological functions, and the gradations of differences in these respects, there still is reluctance in many quarters to concede that genetic factors may also contribute importantly to the production of mental and behavioral traits, whether normal or abnormal. But it would seem obvious that the human mind is not disassociated from the body and its functions, and that inherited biochemical differences cannot fail to have an impact on the mental and neurological processes. The harm in not recognizing this fact becomes apparent in the many cases when a child's backwardness, emotional disturbance, aberrant behavior or what not, is immediately attributed to something connected with his weaning or rearing, the home environment, or the parental attitudes.

Ironically, the more sophisticated and presumably knowledgeable parents have been the ones most prone to accept personal blame for any and every problem or deficiency in a child—the "there are no problem children—only problem parents" approach. Overlooked all too often is the fact that children are always biological individuals, geared to react to the same environmental situations in innumerable ways, and that in many instances a child's inadequacies or aberrations might stem in considerable degree from genetic predispositions which cannot be countered easily by ordinary means. Awareness of this fact can lead to greater consideration for the genetically underprivileged child and increased efforts to compensate for his deficiencies. Equally important, more alertness to "plus" genetic capacities in a gifted child will stimulate development of his special talents.

Even where home influences were or are strongly accountable for a person's adverse traits, it may yet be difficult to decide where heredity and environment begin and end. Has a home been broken by the death of a parent? If the death was not purely accidental, heredity may have had something to do with it. If there was a divorce, heredity might conceivably have been involved in producing the parental traits which were in conflict. Likewise, whether a home is broken or not, its emotional climate (including financial and social factors) can be seriously affected by the presence in a parent or a child of some genetically caused or influenced mental condition, serious physical handicap or abnormality, or major disease. Again, "sibling rivalry" may arise from inherited advantages one child has over another. Further, the size of the family itself, which can much affect the home environment, often may be related to genetic factors affecting the parents' fertility potentials or desires.

Then there is the matter of a person's *sex*. In the interplay of hereditary and environmental determinants in your own life, we dare say that nothing has been more important than your being a male, or a female. For instance, can you conceive how different you might have been if you had gotten just one different sex chromosome—an X for a Y, or vice versa—and had developed as a member of the other sex? Or, rather, would it still be at all possible to think of the resulting individual as the person who now is *you?* At any rate, we have seen how the same environmental influences can interact very differently with the male and female genetic mechanisms and constitutions, beginning at conception and continuing throughout life, and how the biologic sex factor plays a vital role in every area of disease and defect, with the male, as a rule, at the disadvantageous end.

But it is with respect to the sex differences manifested in psychological traits—mental performance, achievements, behavior—that the relative influences of genetic factors and social conditioning are most in dispute. On the one hand, we might assume that the genetically produced differences between the sexes in bodily makeup, biochemical functioning and reproductive roles must certainly have marked effects on their behavioral traits, interests, drives and relative achievements—whatever the environment (granted it could be the same for both). On the other hand, it is obvious that social conditioning, strongly rooted in the past, has done much to accentuate the importance of the biological sex differences, causing males and females to become more differentiated than they would need to be by nature, and, perhaps, more than they should be.

The argument over sex differences can hardly be settled here. What can be said is that with the emphasis both in human genetics and in modern social trends on the importance of the individual, it might be well to do less generalizing about the sexes than has been the rule: instead of thinking so much of males and females only as members of the one group or the other, more attention might be given to the great range of differences existing within each sex group in individual capacities, personalities and inclinations, and to the means of enabling all persons, regardless of sex, to fulfill their potentialities to the greatest extent compatible with their personal desires and the best interests of society. Nor should this be interpreted as applying only or mainly to females and the discriminations against them. Modern environments also discriminate in many ways against males because they are males, and much could be done to ease their lot. On the physical side, moreover, there is hope that medical science may be able to reduce some of the effects of the genetic disadvantages suffered by males, as compared with females, with respect to diseases, defects, and life expectancy.

Having considered how identification with your sex group may have affected your life, let us think next of how your heredity and environment have interacted in your identification with other human groupings —racial or ethnic, and possibly even social and occupational. The nature and effects of these identifications are quite different from those going with your sex. For the sex differences provide a clear and constant biological division among human beings which inevitably must lead to significant environmental and social distinctions under all circumstances. In other human groupings, however, the boundaries have been set up arbitrarily by environmental and social conditions, and have been and are continually subject to removal or change. Biologically, as we have noted, there never have been any fixed fences among human beings; and such genetic differences (in terms of averages) as there are among existing groups derive their effects mainly from the way in which people regard them. For example, the chief effects of having Negro skin color result not from the nature of the trait itself, but from the prevailing attitudes toward it. The same observation applies to any surface features of your own which are linked with the particular racial or ethnic group with which you are identified.

Hardest for many to accept, quite probably, is the observation that even markedly different mental qualities and behavioral traits manifested by disparate racial and ethnic groups may be almost entirely, if not altogether, the result of environmental and social conditioning. Looking at documentary movies of painted primitives engaging in weird tribal dances, you might wonder, "Could *these* fantastically backward people really have hereditary capacities equal to mine?" Yet the weight of scientific evidence is that they could (for your own ancestors far back were little different) and that whatever distinctive psychological traits—superior or otherwise—you yourself now share with your own racial or ethnic group may derive far less from your genes than from your training and opportunities. (Remember, we are not speaking here of *individual* traits and capacities, for which there are many shades of genetic difference within every group. Nor are we referring to hereditary diseases and defects, in the average incidences of which there are many racial and ethnic differences, so that your being of this or that stock might significantly affect your chances of having or developing a given wayward-gene condition.)

Your identification with a particular socioeconomic class or occupational group might also conceivably involve some degree of hereditary and environmental interaction, but only to the extent that genetic qualities and tendencies in your forebears and/or yourself had something to do with belonging to that group. For instance, as noted in our chapters on eugenics, it is possible (although not proved) that certain genes for mental and behavioral traits might be distributed somewhat differently among the various socioeconomic groups. As for occupational groups, it is quite likely that one's membership in one or another of certain specialized professions—science, medicine, music, writing, art, acting and so on—in each of which given qualities of mind and temperament are required for success, may represent some community of predisposing hereditary and environmental factors with the others in that profession.

But, again, we are speaking only in terms of broad averages, and with regard to only a limited number of genetic traits. For the most part our findings have indicated that the genetic distinctions among existing social groups are relatively small, and too diffuse and unstable to be maintained for long. As a result, many of the old theories about the significance of ancestry have been nullified. So, too, there is the evidence now that innumerable persons have the genetic capacities to rise much above what members of their families previously had been, or what their backgrounds may seem to have destined them to be. The growing awareness of this fact has spurred the drive for widened opportunities for education and advancement; and it has bolstered the self-respect and confidence of many persons who might have believed that "inferiority" or "mediocrity," or "taints" or "evil streaks" were in their blood. Even more than in the case of sex differences, advancing civilization must demand that persons should be judged not as members of groups, however these may be classified, but as individuals, on the basis of their own traits, merits and capacities.

To state our general conclusions in another way, each of us is the product of two kinds of inheritance: *biologic* inheritance, transmitted through our chromosomal genes; and *environmental* inheritance, transmitted through what might be called our "*social genes*"—the elements of knowledge, techniques, language, customs, institutions and other cultural components, which have come down to us from previous generations, and which we of the present generation are continuing to modify and pass along to our successors. (These "social genes," it should be clear, are not transmitted in any way through biological reproduction, but only by means of written and spoken language, by teaching and conditioning, and by all the other means through which man records, preserves and passes on his accomplishments.) It is overwhelmingly in the capacity to select, utilize, change, improve and transmit "social genes" that human beings are so vastly different from other animals, and that one generation of man has been empowered to make itself different from preceding generations.

Our biologic genes, we know now, are virtually the same as those possessed by the men of tens of thousands of years ago—no better, no worse. But our "social genes" are immeasurably different. Were it not for these "social genes" of ours, developed, accumulated and increased enormously in diversity and complexity over the ages, you and every one of us would be denuded of all of our present knowledge, social attributes and practical accomplishments. Culturally, we would be behind even the crudest cave men, for even they already had a store of important "social genes" to draw upon. We would have no language, no technical skills, no science, no religion, no sense of morality. Stretching before us would be the same vast jungle of social evolution through which men had to hew their tortuous paths to reach the clearing of civilization. The simplest facts would again have to be learned by trial and error, or by accident. And thousands of generations would have to pass before we could achieve the cultural level of the most primitive men remaining today.

Thus you, like any one of us, is both a genetic heir and a social heir of countless generations and hosts of bygone men, of all races and strains. Nor can it be told from whom you derived the various social attributes you consider most significant in making you what you are. Some technique you now are using, some everyday bit of knowhow, some method of computation, some household remedy, some gesture, some snatch of a song, or any one of innumerable elements in your activities, thinking and behavior may have had its origins unbelievably long ago in an African jungle, a wilderness in Mongolia, a rain forest on a South Sea island, a cave in Europe. Nor can one assume that what came to you is the best of what men of the past had devised. As with many inheritances, there is little doubt that some of the most worthwhile "social genes" in the successive legacies from one generation to another were lost, destroyed or tossed away foolishly. When we think of the countless times in history when whole civilizations were demolished, when gifted leaders were slain, advanced ideas rooted out and books burned, we have every reason to believe that numerous "social genes" of inestimable value were not passed on, while some of the worst were.

Just as we have identified many biologic wayward genes, we can easily identify innumerable wayward "social genes" which are responsible for some of man's worst ills—poverty, ignorance, crime, corruption, tyranny, suppression of liberties, warping of thought, intolerance, fanaticism, bitter conflict among groups, and finally, war. In all that we have learned about heredity there is little to suggest that these blights on mankind are inherent in our germ plasm, or are inevitable components of human nature. As was suggested elsewhere in our book, if there is anything that is distinctive in *human* nature, it lies in the impulses of the vast majority of people everywhere toward sociability, kindness, peacefulness, tolerance, the desire to be and let be, to win the respect and love of others, and to be able to give to others. We need only think of the sacrifices parents make for their children, of the spontaneous impulses of so many persons to aid fellow beings in distress, of the legions of soldiers who have died for their countrymen, of the tens of millions praying each day to be made better and worthier. Nor should it be overlooked that one of the strongest of human desires is to *belong*, to be accepted and esteemed by other people, to feel oneself a part of the past, the present, the future.

So, again, if we can identify the wayward "social genes" which afflict humanity, we also can identify many superior and desirable "social genes." By the same token, if a program of biological eugenics is needed to improve human genetic quality, there is surely no less need for a concomitant program of "social eugenics," to breed out the undesirable "social genes" in our lives and to foster the development and growth of the desirable ones. Indeed, while much can be done to improve our collective germ plasm, or at least to keep it from deteriorating, much more can be accomplished, and more quickly, by making better use of the genetic resources already at our command. For scientists are certain that the biologic genes now in circulation offer all the potentialities for a race of men far superior physically and mentally to what we, as a species, now are: granted always that we can direct and control our environment so that the best in us can be brought to the surface and made to flourish.

What is equally clear is that true and lasting advancement for any group of human beings can be achieved not at the expense of others, but only with the accompanying advancement of mankind on a broad front. Within the last few decades, as never before, it has become startlingly apparent that whatever of importance happens to one segment of humanity anywhere must sooner or later affect all the rest of it. No human group can any longer expect to find peace and prosperity by sticking to its own knitting, staying in its own backyard, cultivating its own garden, or providing the best of everything only for itself. We have learned all too painfully that an epidemic, the fallout from an atomic bomb, a financial crash, a social upheaval, a brush war, starting no matter where, can soon spread its effects around the globe and ultimately strike at you and yours.

"Am I my brother's keeper?" "No man is an island . . . the bell tolls for thee." "We're all in the same boat." "It's a small world." These are phrases which have assumed ominous new significance. For this has truly become a very small world. In terms of time, distance and differences between the earth's inhabitants, the world has been shrinking fantastically under our very eyes. Through travel, commerce, reading, study, the radio and television, we are finding ourselves brought into ever closer contact with groups and peoples who had no meaning to our forebears, or of whose existence they did not even know. At the same time, other peoples the world over are becoming increasingly like us and like one another in their Main Streets, movies, supermarkets, shop-window displays, the attire of men and women, their attitudes and habits. In fact, many of the most colorful distinctions among present-day peoples exist only because they are artificially perpetuated. "The tourists are arriving" is a signal for African natives, South Sea islanders, American Indians and Eskimos to doff their business suits, overalls and houseclothes, put on their tribal outfits and go into the dances and ceremonials expected of them.

Not only have the distances been bridged among peoples living today, but our frontiers of knowledge have been pushed much further back, to the species of men and ape men who preceded us. The Zinjanthropus of almost two million years ago has taken his place in our ancestral portrait gallery with the Peking man, Java man, Neanderthal man and the Cro-Magnon man, among many others. We can see ourselves more and more as a continuation in an evolutionary process—social as well as biological—that has been going on for far longer than Darwin could have imagined, and that now may be in a more active and crucial stage than ever before.

The wish to keep on being part of the human stream is another aspect of "self" which our findings in genetics have revealed in a new light. Biologically, as we have emphasized, no one can perpetuate the unique living entity which is one's self. One's "blood" does not flow on with one's children. No child can be, genetically, a "chip off the old block." And if any significant combination of one's genes does persist in one's offspring, it will quickly be dissipated in later generations. In this regard, then, those who are parents will have little advantage genetically with respect to future generations over those who leave no children. For every person's genes, except for the rarest of mutations, have innumerable replicas in the existing gene pool and will continue to flow along in the human species.

In the transmission of "social genes," as well, all persons have an opportunity to actively insure themselves a stake in the future. Psychological and behavioral traits, inventions, new ideas and other cultural elements, which are developed in the present generation, can go on having their effects for a very long time. And even the humblest and most obscure person might yet make some contribution to the future through a deed or action which influences younger persons around him.

As conditions and opportunities among people come closer to being equalized, and more and more participate in decisions affecting their lives, human destinies will be subject less and less to the dictates of the greatest of all tyrants—*chance*. For in the past, neither despotic rulers nor the wisest of leaders had as much power over the broad course of human events or the fates of individuals as had the unknown, unpredictable and uncontrollable forces within human beings or the vagaries of nature and the turbulent currents of social evolution outside them. True, chance in many vital respects will continue to sway our collective and individual destinies. As a species, scientists hold, we are derived from a series of biological accidents—of countless haphazard mutations, combinations and recombinations of genes. Within this species, each of us has been a subsidiary product of chance: of a random combination of genes interacting with an initial fortuitous environment over which we had no choice.

Poised as we continue to be on the brink of statistical odds and unknown possibilities, those of us who are at the top of the social heap can hardly ascribe our success merely to our own intrinsic worth and superior efforts; and those of us at the bottom need not blame our failures all on inherent weaknesses or inadequacies. The findings in genetics and the social sciences must shake the old beliefs in predestination, and, equally, in the notion that the individual is truly the master of his fate and the captain of his soul. Aware now of how heavily the life of every person has been influenced by circumstances—and *luck*—you might well look at both the shining successes above you and at the misfits and unfortunates below you and say, "There but for the grace of my genes and my environment go I."

Yet we no longer are (or need not go on being) anywhere near as much at the mercy of chance as were human beings in the past. In the matter of our heredity, we are seeing how we can exercise some choice in selecting, recombining and transmitting our genes. Where wayward genes confront us, we are able in many instances to forestall or lessen their more serious effects; and in time even the changing of various genes in desired directions may become possible. But, as we've stressed before, our greatest hopes for improvement and our greatest control over our destinies will lie in the ability to identify and root out adverse factors in our environment. More and more the social sciences—psychology, psychoanalysis, sociology, anthropology—have been joining with human genetics to show us what is cause and effect in our lives. What formerly might have been thought of as mere accidents—applied to anything from individual misfortunes, maladjustments and everyday happenings, to cataclysmic world events—are revealed as the logical, probable or inevitable consequences of identifiable environmental situations. The World Wars of our time were not mere accidents. Hitler and the Nazis, and the horrors they set loose, were not mere accidents. The blood purges and wholesale atrocities in many countries have not been mere accidents. The flash of a gun which snuffed out the life of a virile young President was not a mere accident. The upsurges in crime and juvenile delinquency, the breakdowns in moral codes, the flaunting of human rights, the high incidences of alcoholism and emotional maladjustment, the race riots, the innumerable daily acts of dishonesty and callous cruelty, are not mere accidents. Nor are they products of our chromosomes and genes. All—whether expressed in individuals or in the mass—have been bred in environments of frustration, confused values, inequalities, intolerance, blind selfishness and, most of all, of ignorance.

But these malignant outgrowths must be recognized as the exceptions, no more representative of the orderly and essential processes of human society than are cancerous cells in our bodies. As was stressed earlier, the predispositions of the overwhelming majority of people are clearly toward peaceful, cooperative and constructive ends. Concrete evidences of these predominantly human impulses are seen all around us: in the hospitals, houses of worship, courts of justice; in the charitable institutions and agencies devoted to human welfare; in the greatly endowed foundations, universities, museums, art galleries, and temples of music. These, and the myriads of unrecorded, everyday acts of kindness, sympathy, friendship, loyalty, dedicated service, are the truer expressions of human nature.

We have said that the world has grown much smaller; and, we might add the famous Broadway line, "That's all there is; there isn't any more." It is within this world, an efflorescence on a ball 8,000 miles in diameter, which men can already encircle in a few hours—relatively, no more than a grain in the cosmic dust cloud of billions of other celestial bodies—that we will have to learn to live with all our fellow creatures. The fact that there now are no more new regions to be discovered on our little earth, and no peoples unknown to others, may be one reason why our explorers are turning their thoughts to the far outer spaces. Whether there are species of humans or near humans elsewhere in the vastness of the universe, and if so, whether any would be close enough for us ever to establish physical contact with them, is a matter of speculation. But whatever adventures and possibilities the outer spaces may hold for us, the most important imminent discoveries are likely to be those within our own *inner* spaces, in the microcosms of our cells and the secrets locked in the DNA codings of our genes. These are the new worlds already being explored by our scientific Columbuses.

In the years ahead, you and your children can expect exciting revelations which should make possible undreamed of expressions of human genetic potentialities. Among the developments one can envisage (some of which were touched on in preceding chapters) are startling improvements in human physique and body functioning, prevention or alleviation of many of our now most menacing diseases, replacements of defective organs, and much lengthened average life expectancies. Nor will the human brain and its functioning be neglected, for eager geneticists, biochemists and psychologists are certain to extend their combined researches ever farther into the nature and workings of the brain, and the means of directing its actions. If the possibilities are realized, intelligence levels for a great many persons will be raised, talents will be unloosed, impediments will be lifted from many sodden minds, various behavior disorders will be averted, and the whole level of human performance and creativity will be raised to new heights.

Yes, we have already come a long way from the primates in the treetops, from the ape men learning to walk erect, from the cave men with their implements of stone. We have moved on as a social species from the isolated little family huddles and the roving bands to the settled communities, the states, the nations, the alliances, and now to a family of nations in which voice is given to all peoples, and the humblest sit in council with the mightiest. And with this, too, the broadened concept of "self"—of the individual as interrelated with all other human beings—has extended to whole nations. For independence and interdependence have become one. No country, no matter how great and powerful, can any longer stand alone, thrive alone, fight alone, or fail to realize that its security and welfare are henceforth bound up with the desires and fates of other countries.

We have moved far, indeed, toward the One Human World. But let us not feel too smug about this progress, nor about our great advances in technology, science, education, material comforts, health, longevity. We have only to look through the album of our present era to see many pictures which should give us pause: of mass atrocities and inhumane acts worse than any in man's history; of vast numbers of people still without adequate food, housing, education or medical care; of countless millions of children denied any real opportunity for fulfillment and the pursuit of happiness; of nervous fingers poised to trigger off atomic missiles which could blast us all into nothingness; of widespread fear and distrust, violence, hate and bigotry.

So long as these conditions exist, we should be soberly aware that we have a much longer way to go before we have perfected a civilized environment befitting our heredity, and are fully worthy of the names we have given ourselves: Homo *sapiens* and *human* beings.

Appendices

GLOSSARY

*Terms not defined in this book but which the reader may encounter in other books or in technical articles dealing with human heredity.**

agglutination. Clumping of cells or other particles, as in the blood when there is a mixture or transfusion involving incompatible bloods.

agglutinin. The substance, or antibody, in the blood serum which causes the afore-mentioned clumping, or agglutination.

agglutinogen. The substance, or antigen, in the red blood cells which is affected by the conflicting agglutinin of another person.

alleles, allelomorphs. Genes which are assigned to the same function and are located in the same position in matching chromosomes, but which do not necessarily perform their function in the same way (as, a gene for blue eyes from one parent, which is the allele of the "eye-color" gene received from the other parent).

anaphase. The stage in cell division at which the newly formed duplicated chromosomes separate and go to opposite sides of the dividing cell.

antibody. A chemical substance in an individual which fights against or counteracts a specific foreign substance (antigen) that enters the body.

antigen. Any foreign substance entering an individual which induces the formation of antibodies.

atavism. The appearance in an individual of a trait like that of some distant ancestor, and which has not been seen in the family for some generations.

autosome. Any one of the chromosomes other than the sex chromosomes.

centromere. The region on a chromosome (shown by the "bend") where it becomes attached to the "spindle" or "spindle fiber" during the processes of chromosome-doubling and/or chromosome separation.

chromatid. Each half of a chromosome which is formed when it doubles and then splits before the cell divides.

chromatin. The substance in the chromosomes which is seen under the microscope after a cell is stained.

concordance. The likeness between members of matched groups or pairs (such as twins), in any one trait or in given traits.

* Adapted, with extensive revisions and additions, from the Glossary in the author's *Human Heredity Handbook.*

deletion. Loss of a portion of a chromosome at any time in its existence.

diploid ("*duplex*"). With two sets of chromosomes. In human beings a fertilized egg, normally with forty-six chromosomes, or twenty-three pairs, is in the diploid state. All the resulting body cells are also diploid. (See also *haploid.*)

discordance. Opposite of concordance: the dissimilarity between members of matched groups or pairs (such as twins) in any one trait, or in given traits.

disjunction. The process by which the two members of a chromosome pair, or the two parts of a newly doubled chromosome, separate and move apart during cell division.

dizygotic, dizygous. From two separate fertilized eggs, as in the case of fraternal twins. *Zygote* refers to the combination of a sperm and an unfertilized egg. (See also *monozygotic.*)

epistasis. The covering up, or "masking," of the effects of a gene by an "outsider" gene which is not one of its alleles. (See *alleles.*)

gamete. A mature germ cell, either a sperm or an egg.

genetic equilibrium. The maintaining of different forms of genes for given traits (see *alleles*) in the same relative proportions within a specific population.

genotype. The genetic makeup of individuals. Persons with the same genes have the same genotype. When persons merely show the same physical traits, they are referred to as having the same *phenotype.* Sometimes persons with the same genotype may differ in phenotype, as with identical twins who have developed differently because of environmental influences. Conversely, persons with the same phenotype, showing the same physical traits, may or may not be carrying the same genes, as with two dark-eyed persons, one of whom carries two "dark-eye" genes, the other one "dark-eye" gene coupled with a hidden "light-eye" gene.

germ plasm. The material of the germ cells, in the testes of the male or the ovaries of the female, from which sperms or eggs with their chromosomes are fashioned.

gonad. A reproductive gland: testicle in the male, ovary in the female.

haploid. With a single set of chromosomes. In human beings each sperm or egg is a haploid cell, containing only one set, or twenty-three single chromosomes, in contrast to a diploid cell, with two sets, as previously defined.

Hardy-Weinberg law. The principle that a certain balance is maintained in a population with respect to the proportions of persons carrying a single gene for any trait, those carrying two of the genes, and those free of the genes, used in estimating the expected ratios of all these individuals. Thus, if the number of persons with a recessive defect, such as albinism, is known, it is possible to estimate how many normal persons are carrying a hidden gene for the condition.

heterosis. The situation where two forms of a gene exist, and where a mixed pair in an individual is more beneficial than two of the same genes of either type. (See *heterozygote* below.)

heterozygote. A fertilized egg or individual in which paired genes differ, because of different contributions from the two parents. A person who is heterozygous for eye color would be carrying two different "eye-color" genes, as, for instance one "blue-eye" gene and one "brown-eye" gene. In a person who is *homozygous* the two "eye-color" genes would be the same.

holandric inheritance. Referring to the rare possibility of inheritance through a dominant gene in the Y chromosome, with no equivalent gene in the X chromosome. Such a gene and the trait it produces could be passed along by an affected father to all of his sons, but would not occur in females.

homologous chromosomes. Paired chromosomes with matching genes, one of the pair from the father, one from the mother. (See also illustrations, Chapter 2.)

homozygote. A fertilized egg or individual in which given paired genes are the same. (See also *heterozygote*.)

homozygous. Having both genes the same with respect to a given pair.

linkage. The connection of genes in the same chromosome. Linked genes are sometimes separated through chromosome "crossovers." (See illustration, Chapter 42.)

locus. The location of a gene on a particular chromosome.

maturation division. See *meiosis*, below.

meiosis. The process whereby a germ cell, with two of each chromosome, gives rise to sperms, or eggs, each with only one of every pair of chromosomes.

metaphase. The stage in the process of meiosis (above) or mitosis (see below) during which the chromosomes are doubled, lined up at the center of the cell, and are prepared to separate.

mitosis. The process whereby cells in the developing embryo divide and multiply, each cell forming two daughter cells with replicas of all the original chromosomes. (A simple aid in remembering the difference between meiosis and mitosis: Think of "mei-O-" as "mei-*One*"—forming a cell [sperm or egg] with *one* set of chromosomes; "mitosis" as "mi-*Two*"—forming *two* cells, with two sets of chromosomes in each.)

monosomic. The presence in a diploid cell of only one member of a given chromosome pair.

mutagen. Any force or influence—radiation, a chemical or other factor—which can cause mutations to occur in genes.

nondisjunction. Failure of the process of "disjunction" (see definition) to occur with respect to a given chromosome pair, so that both members, instead of just one, go into the same new body cell, or sperm or egg.

pangenesis. An erroneous early theory advanced by Darwin, that each cell of the body gives off little particles, or "pangenes," embodying the cell's characteristics (whether acquired or inherited), and that these "pangenes" then find their way into the eggs or sperms and determine an offspring's heredity.

pharmacogenetics. The study of genetically determined variations that are revealed by the effects of drugs.

phenotype. The outward expression of gene action in a given environment. (See *genotype*.)

polar body. A tiny cell containing a set of single chromosomes which is discarded during the process of egg formation so that the egg is left with just one of each pair of chromosomes.

polymorphism. The existence in a population of multiple genetic forms of a given characteristic.

position effect. The possible variable effects on a gene's behavior of being placed in different positions with respect to certain other genes on the same chromosome.

proband. The person whose condition or trait is the starting point for research into a family history.

prophase. The beginning of cell division when chromosomes are first visible in the nucleus. (See *meiosis* and *mitosis*.)

propositus. The person from whom a line of descent is reckoned. Sometime also used as a synonym for *proband* (see above).

reduction division. The process in the formation of eggs or sperms by which they receive only half of the individual's chromosomes—just one of each pair. (See *meiosis*.)

somatic. Referring to all cells in the body ("soma") outside of the germ cells.

spermatogenesis. The process by which sperms are produced.

spindle. A spindle-shaped structure of threads or fibers formed in the cell as it prepares to divide, and at the center of which the doubled pairs of chromosomes are arranged. When the two ends or halves of the spindle pull apart, each half carries one member of each pair of chromosomes with it.

synapsis. The stage (meiosis) in germ-cell development at which the matching chromosomes of each pair come together and pair up.

telophase. The stage in mitosis in which the doubled chromosomes have divided and moved apart into newly forming cells.

teratogenesis. (From the Greek word "teratos" meaning a "monster" and from "genesis": meaning "production of.") The prenatal development of a severely malformed baby, or "fetal monstrosity." Also applied now to congenital abnormalities and anomalies of many kinds, ranging from fairly mild to the most severe.

translocation. The attachment of part of one chromosome to another chromosome. It is called reciprocal when two chromosomes interchange parts.

zygosity. The derivation of twins or other multiples either from the same egg, or zygote (see below), or from two or more different eggs. (See also *heterozygote* and *homozygote*.)

zygote. The fertilized egg, or the individual, resulting from the union of a sperm with an unfertilized egg.

WAYWARD GENE TABLES

Hereditary Diseases, Defects and Abnormalities

Herewith are listed the principal diseases, defects and abnormal conditions (and rarer ones of special interest) in which there is proof of heredity or in which heredity is believed to play some part. Included are all the conditions discussed in the text and, in addition, a considerable number of other conditions which warrant attention. With each condition is given its type or types of inheritance as now established. Where there is uncertainty, a question mark follows. At the end of the tables will be found a forecast section giving the chances of inheritance, for yourself or your children, of any condition listed.*

Keep in mind these abbreviations and their meanings:

Dom. Dominant. Only *one* gene required to produce effect. (A parent with a dominant condition will pass it on to one in two children on the average.)

Dom.+ Dominant-*plus*. Condition is caused by a dominant gene *plus* some other gene or genes not identified.

Inc.-Dom. Incomplete dominant. Two of the same genes required to produce effect. One gene of this type by itself produces milder effect.

* *Note to physicians.* Some of the conditions listed as hereditary may be clinically similar to conditions caused by certain acquired diseases or environmental factors: for example, some cases of congenital deafness or blindness, and various eye defects and structural abnormalities. In many instances, careful diagnosis and study of the case history will reveal the distinction between the hereditary and nonhereditary types, as in eye-muscle paralysis, where the hereditary type is present from birth while the nonhereditary type is a sequel to injury or such diseases as meningitis. Not listed here are very rare conditions, of concern to only a few readers, but details regarding which may be found in the medical literature.

Dom.-q.—Dominant-qualified. A dominant gene which does not always produce its effects or may produce variable effects, depending on environment and sometimes on the workings of other genes.

Rec.—Recessive (simple). *Two* of the same genes required to produce effect. (For a child to have a recessive condition each parent must contribute exactly the same gene.)

Rec.+—Recessive-*plus*. Condition caused by a pair of the same recessives *plus* some additional different genes (another pair of recessives, or other genes not identified).

Rec.-q.—Recessive-qualified. A pair of recessives which do not always produce their effects or may produce variable effects, often depending on environment and sometimes on the workings of other genes.

Sex-L.—Sex-linked recessive. Gene carried on X chromosome. Since male has only one X chromosome, only *one* such gene needed to produce effect in male, but in female *two* are required (as with any other recessive).

Sex-L. Dom.—Sex-linked dominant. Rare type, in which a gene on X chromosome is dominant, producing, singly, same effect in females as in males.

Inc. Sex-L. Dom.—Severe in males, mild in females.

Sex-Lim.—Sex-limited. Conditions where genes, not on X chromosome, are inherited equally by both sexes, but have principal effects limited to one sex.

Chrom. Abn.—Chromosome abnormality. As in conditions due to one or more extra chromosomes; or the lack of a chromosome; or the presence of an abnormal chromosome, consisting of parts of two different chromosomes fused together (translocation).

?—Indicates heredity doubtful or uncertain. After gene mechanism, it means uncertainty as to exact manner of inheritance.

1. HEART AND BLOOD VESSELS

High blood pressure (hypertension). Predisposition possibly inherited. (?)

Arteriosclerosis. See below.

Atherosclerosis (a form of arteriosclerosis, or "hardening of the arteries"). Fatty deposits in blood-vessel walls, leading to blockage of blood flow. Hereditary when due to inborn errors of metabolism. For hereditary mechanisms, *see: hyperlipemia, hypercholesterolemia,* and *diabetes mellitus,* under Metabolic Disorders (p. 738).

Rheumatic fever (childhood rheumatism). Predisposition possibly inherited. (?)

Congenital heart disease. Many types, predisposition possibly inherited, in some cases as part of an inherited syndrome, such as Marfan's disease. (?)

Varicose veins. May involve hereditary predisposition. Dom.-q.

Multiple hereditary telangiectasia. Widespread dilations of small blood vessels with frequent bleeding. Inc.-Dom.

2. CANCER

COMMON TYPES

Possible predisposition or inherited susceptibility in some degree, but manner of inheritance uncertain. Present evidence suggests such predisposing factors in (a) breast cancer, (b) cancer of the intestines, (c) cancer of the stomach, (d) cancer of the prostate.

CANCER, RARE TYPES

(a) *Malignant freckles* (xeroderma pigmentosum). Multiple pigmented spots on skin which can undergo malignant change, especially when exposed to sunlight. Rec.

(b) *Epiloia* (tuberous sclerosis). Tumors of skin and brain, causing mental retardation, skin lesions (adenoma sebaceum). Dom.-q.

(c) *Multiple polyposis of the colon.* Multiple growths in the large intestine which frequently undergo malignant change. Dom.

(d) *Retinoblastoma* (glioma retina). Tumor of eye; fatal unless eye removed. Infancy or childhood. Dom.-q.

(e) *Retinal angioma.* Blood vessel tumor of eye. Sometimes familial, but manner of inheritance uncertain. (?)

(f) *Neurofibromatosis.* Lumpy growths on skin and in nerves, brain, etc., sometimes becoming cancerous. May also cause deafness or mental defect. Dom.-q.

(g) *Hodgkin's disease.* Malignant lymph-gland disease. Fatal. Heredity doubtful. (?)

(h) *Leukemia.* Several types. Heredity doubtful. One form, chronic myeloid leukemia, apparently associated with a chromosome defect, but not inherited. Chrom.-Abn. (?)

3. METABOLIC DISORDERS

CARBOHYDRATE METABOLISM DISORDERS

Diabetes mellitus (common type). "Sugar sickness" due to pancreas defect. Rec.

False diabetes, two types:

(a) *Diabetes insipidus.* Excessive urination. Usually not inherited, but when so: Dom., Sex-L., Rec.

(b) *Renal glycosuria.* Sugar in urine due to kidney defect. Not harmful. Dom.

Glycogen storage disease. Various types, each probably due to a different enzyme defect. Usually serious. Rec.

Galactosemia. Inability to metabolize the sugar galactose in milk because of enzyme defect. Birth on. Serious. Rec.

PROTEIN AND AMINO-ACID DISORDERS

Phenylketonuria (phenylpyruvic amentia). Inability to metabolize the amino acid phenylalanine in foods. Severe mental deficiency. Rec.

Histidinemia (histidinuria). Inability to metabolize histidine in foods. Mild mental retardation, speech impairment. Sometimes misdiagnosed as phenylketonuria. (*See* Text.) Rec.

Albinism. Blocked metabolism of amino acids preventing normal pigmentation. Rec. (rarely, Sex-L.)

Maple-syrup urine disease. Faulty metabolism of several amino acids, birth on. Serious. Rec. (?)

Alcaptonuria. Darkening of urine due to faulty amino-acid metabolism. Generally harmless, but may produce arthritis in maturity. Rec.

FAT (LIPID) METABOLISM DISORDERS

Essential familial hyperlipemia. Inability to clear dietary fat from blood. Produces atherosclerosis; also fatty skin deposits (xanthomas). Inc. Dom.

Essential hypercholesterolemia. Elevation of blood cholesterol, producing atherosclerosis; also fatty skin deposits. Dom. or Inc. Dom.

Amaurotic familial idiocy (Tay-Sachs disease). Fatty deposits in brain cells. Severe mental retardation; usually early fatality. Rec.

Niemann-Pick disease. Enlargement of liver and spleen, with neurological disturbances. Usually fatal early in life. Rec.

Gaucher's disease. Similar to Niemann-Pick, but more variable age of onset. Rec. (Dom. rarely ?)

OTHER METABOLIC DISORDERS

Gout. Disorder of uric-acid metabolism. Severe arthritic symptoms, more common in affected males than females. Dom.-q., Sex-Lim.

Wilson's disease. Abnormality of copper metabolism. Degeneration of liver and nervous system. Rec.

Periodic paralysis. *See under* NERVES AND MUSCLES (p. 743).

Porphyria. Abnormality in hemoglobin formation. Two types:

(a) Sensitivity to sunlight and reddish pigmentation of teeth. Rec.

(b) Intermittent abdominal pains and nervous-system abnormalities, with or without skin manifestations. Several subtypes. Dom. (?)

KIDNEY DISORDERS WITH URINE ABNORMALITIES

Vitamin-D resistant rickets (familial hypophosphatemia). Rickets due to defective kidney functioning not treatable with vitamin-D (as is ordinary rickets). Milder in females. Sex-L. Inc. Dom.

Fanconi syndrome. Loss of sugar, amino acids and phosphorus through urine, interfering with normal growth. Fatal early in life. Inherited and acquired types. When inherited: Rec.

Renal glycosuria. *See False diabetes* under *Carbohydrate Metabolism Disorders* (p. 737).

Cystinuria. Excessive urinary excretion of large amounts of cystine and other amino acids. May produce kidney stone. Rec. or Inc. Dom.

Hartnup's disease. Excessive loss of amino acids through urine. Skin and neurological abnormalities. Rec. (?)

4. GLAND AND ORGAN DISORDERS

Goiter (common). Thyroid-gland enlargement with deficient functioning. Generally environmental, but inherited susceptibility claimed in some cases. (?)

Exophthalmic goiter (Grave's disease). Overactivity of thyroid gland. Heredity claimed, but uncertain. (?)

Cretinism with goiter. Congenital thyroid deficiency causing mental retardation. Usually environmental, but several rare types hereditary, and if so: Rec.

Cystic fibrosis of the pancreas. Disturbed pancreatic and lung functioning, with infections. Usually early death. Rec.

Polycystic kidney disease. Multiple cysts in kidneys, with high blood pressure and kidney failure in adult life. Dom.-q. (?)

Hereditary kidney disease with deafness. Often fatal to young adults, especially males. Usually mild in females. Dom.-q.

5. SEX ABNORMALITIES

Klinefelter's syndrome. Sexual underdevelopment and feminization in males due to extra sex chromosome (XXY types). Chrom.-Abn.

Turner's syndrome. Sexual underdevelopment in females due to presence of only one X (XO type). Chrom.-Abn.

Testicular feminization. Female development in genetic males, due to abnormal sex-hormone production. Sex-Lim. (?) or Sex-L. (?)

Adrenal virilism (adrenal hyperplasia). Abnormal masculinization in children of either sex, due to adrenal-gland disorder. Rec.

Pubertas praecox. Abnormally early puberty. Several causes (*see* Text), but some cases hereditary. Dom. (?)

Hypospadias. In males, abnormal opening in penis, present at birth. Dom.-q.

Extra nipples (and/or breasts). In both sexes. (?)

6. BLEEDING DISEASES

Hemophilia A (classical type). Deficiency in a blood-clotting protein, with severe bleeding episodes throughout life. Sex-L.

Hemophilia B (Christmas disease). Distinguishable only by tests from classical hemophilia. (*See* Text.) Sex-L.

Von Willebrand's disease (vascular hemophilia, or pseudohemophilia). Sometimes confused with Hemophilia A. (*See* Text, Chapter 28.) *Not sex-linked.* Dom.-q. (?)

Other rare clotting defects. Mild bleeding tendencies, with deficiencies in various protein substances:

Plasma-thromboplastin-antecedent deficiency (PTA). Dom. (?)

Ac-globulin deficiency (parahemophilia). Rec. (?)

Stuart factor deficiency. Inc. Dom. (?)

Prothrombin-conversion-factor-"S" deficiency. Rec. (?)

Afibrinogenemia. Total absence of fibrinogen; bleeding, but not as severe as in hemophilia. Rec.

Thrombasthenia. Bleeding defect with multiple nosebleeds. More frequent in females. Heredity uncertain. (?)

7. ANEMIAS

Mediterranean anemia (Cooley's anemia, or thalassemia). Reduced ability to form normal hemoglobin. Single and double gene forms:

Mild anemia, not harmful. Inc. Dom. (one gene)

Severe anemia, usually fatal. Inc. Dom. (two genes)

Sickle-cell anemia. Sickle-shaped red cells, due to hemoglobin abnormality. Single and double gene forms:

Without anemia, but some abnormal hemoglobin and some sickled cells. Inc. Dom. (one gene)

Severe anemia and all hemoglobin and red cells abnormal. Usually fatal. Inc. Dom. (two genes)

Hemoglobin C disease. Single and double gene forms:

Some abnormal hemoglobin, but without anemia. Inc. Dom. (one gene)

All abnormal hemoglobin. With anemia and target-appearing red cells. Inc. Dom. (two genes)

Favism. Predisposition to serious red-cell destruction on eating the fava bean. (Due to inherited deficiency of glucose-6-phosphate dehydrogenase, which reduces resistance of red blood cells.) Sex-L.

Primaquine sensitivity. Same effects as in favism, on taking certain drugs. Sex-L.

Spherocytosis. Increased red blood cell fragility, with small round red cells (spherocytes), jaundice. Dom.

Nonspherocytic hemolytic anemia. Anemia and jaundice without spherocytosis. Dom.-q. (?)

Fanconi's anemia. Deficient blood-cell production, with multiple congenital defects. Rec.

Agammaglobulinemia. Absence of the blood protein gamma globulin and of plasma cells, predisposing to severe infections. Can be acquired. When hereditary, Sex-L., (Rarely Rec.)

Acholuric jaundice. Fragile blood cells, sometimes with anemia. Dom.-q.

Familial nonhemolytic jaundice. Milder form of above. Dom.-q.

Erythroblastosis fetalis. Jaundice in newborn, due to mother-child "Rh," "A-B-O," or other blood-factor conflict. (For mechanism, *see* Text.)

8. BLOOD-CELL ABNORMALITIES

Acanthrocytosis. Thorny-appearing red cells and absence of betalipoprotein from blood. Intestinal disturbances (resembling celiac disease) in childhood and nervous disorders. Rec.

Elliptocytosis. Oval red cells. Mild increase in red-cell destruction. Rec.

Pelger-Huet anomaly. Immature appearance of white blood cells. Not harmful. Dom.

9. ALLERGIC DISEASES

Common allergies. Often environmental, but heredity may also be involved in asthma, hay fever, and skin reactions to specific substances. If hereditary, usually Dom.-q.

Migraine headache. May have allergic background, although evidence is uncertain. Dom.-q. (?)

Collagen diseases. Related diseases involving allergies to foreign substances and body's own tissues, affecting various organs and areas. (*See* Text.) Heredity suspected, all appearing to be: Dom.-q. (?)

Rheumatoid arthritis.

Lupus erythematosus disseminata. Affecting small blood vessels and internal membranes.

Polyarteritis nodosa. Affecting larger blood vessels, nerves, muscles, organs.

Dermatomyositis. Affecting muscles and skin.

Scleroderma. Affecting skin and intestinal tract.

10. MENTAL DISEASES

Major psychoses (disordered thinking, or "insanity"). If and where heredity is a factor in these, it is believed to be largely as a predisposition.

Schizophrenia (dementia praecox). "Split personality." Rec.+ (?)

Manic-depressive insanity. Alternating cycles, excited and depressed states. Dom.-q. (?)

Involutional melancholia. Depression in middle life. Dom.-q. (?)

Psychopathic personality. Extreme unsocial behavior, but otherwise rational. Some inheritance suspected, but uncertain. (?)

Psychoneuroses. Emotional disorders with irrational fears, obsessions, anxieties, etc., but otherwise mentally lucid. Heredity suspected in some cases. (?)

11. DEGENERATIVE BRAIN DISEASES

Huntington's chorea. Progressive mental and physical deterioration, usually in middle age. Dom.

Pick's lobar atrophy. Similar to above, but rarer and differing in certain symptoms. (*See* Text.) Dom.

Cerebral sclerosis. Gradual failure of intelligence, vision, muscular power. Childhood or youth. Rec.

Other types, varying symptoms and ages. Dom., Sex-L.

12. MENTAL DEFECTS

Feeblemindedness (moron type). Subnormal mentality, IQ from 70 to 50. Where hereditary, may be due to a group of genes. Rec.+ (q) (?) or Dom.+ (q) (?)

Mongolian idiocy (Langdon-Downs syndrome). Congenital idiocy due to chromosomal aberration. (*See* Text.) Chrom.-Abn.

Cretinism. Mental defect due to prenatal thyroid deficiency. Only rarely any hereditary basis. (*See* Text.)

Amaurotic family idiocy (Tay-Sachs disease). *See* METABOLIC DISORDERS, under *Fat (Lipid) Metabolism Disorders* (p. 738).

Phenylketonuria (phenylpyruvic amentia). *See* METABOLIC DISORDERS, under *Protein and Amino-acid Disorders* (p. 737).

Gargoylism (Hurler's syndrome). Rare metabolic disorder with mental defect as one of many consequences (extreme dwarfism, etc.). Rec. or Sex-L.

Microcephaly. Underdeveloped brain and skull ("pinhead" type). Often environmental. Where hereditary Rec. (?)

Hydrocephaly ("water head"). Generally environmental, but heredity claimed in some cases. (?) or Dom.-q. (?)

Laurence-Moon-Biedl syndrome. Mental defect with obesity; often with retinitis pigmentosa, extra fingers. Rec.-q.

Wilson's disease. Retarded intelligence with peculiar jerky hand movements, facial distortions. Rec. (*See also* METABOLIC DISORDERS under *Other Metabolic Disorders*, p. 738.) Childhood.

Epiloia. See under CANCER, *Rare Types* (p. 737).

Sex abnormalities with mental defect: See Klinefelter's syndrome and *Turner's syndrome*, under SEX ABNORMALITIES (p. 739).

Epilepsy, with mental defect. *See* Text.

13. NERVES AND MUSCLES

Epilepsy. General name for a convulsive symptom of various types of nerve disorders. Hereditary predisposition present in some forms, but mechanisms uncertain. (?)

Myoclonus epilepsy. Rare type, with tremors and spasms between seizures. Childhood. Rec.

Hereditary ataxias. Varying types, depending on nerve centers affected.

Friedreich's spinal ataxia. With wobbly gait, speech defects. Childhood. Rec.

Marie's cerebellar ataxia. Jerky movements, speech defects. Adults. Dom.-q.

Hereditary tremor. Slight involuntary movements. Childhood. Dom.

Spastic paraplegia. Rigidity in lower limbs, spreading upward. Childhood. More common in males. Various types. Rec., Dom., Sex-L.

Muscular atrophies. Shriveling or degeneration of muscles.

Peroneal. Legs, feet, hands.

Childhood. Dom., Rec., Sex-L.

Spinal muscles. Infancy. Fatal. Rec.

Progressive spinal. Associated with cataract, sterility, etc. Maturity. Dom.-q.

Periodic paralysis. Episodes of muscle paralysis caused by disordered potassium metabolism. Dom.

Paralysis agitans (Parkinson's disease). Heredity suspected in senile type. Dom.-q. (?)

Myotonia atrophica. Slowed relaxation after muscular contraction. Early adult. Rec.

Thomsen's disease (myotonia congenita). Muscle stiffness, slow, delayed movements. Childhood. Dom.

Related types with varying ages of onset: *Dystrophica myotonica* and *Paramytonia* (rare). Dom.-q.

Amyotonia congenita. Flabby, immobile muscles. Infancy. Rec. (?)

Myasthenia gravis. Muscular weakness, without atrophy. Inheritance doubtful. (?)

Progressive muscular dystrophy. Serious muscular wasting, leading to weakness, invalidism. Several types with varying ages of onset, symptoms and heredity:

Childhood type, onset about age five. Often with declining mentality. Sex-L., Rec.

Pre-adolescent type, onset about age eleven. Rec.

Early adult type, onset about age eighteen. Dom.

Sporadic types. Heredity and/or mode of inheritance uncertain. (?)

Familial dysautonomia (Riley-Day syndrome). Abnormalities of tear and sweat glands, skin, nervous system. Rec.

14. EYES

MAJOR CONDITIONS

Cataract. Opaque lens, common cause of blindness. Onset and type varying in different families. (Old-age cataracts may not be hereditary.) Where inherited, several types: Dom.-q., Rec.

Glaucoma. Increase in pressure in eyeball, often leading to blindness. Types varying in onset, nature and importance of heredity:

Infantile, most serious. Rec.

Juvenile (rare), onset at puberty. Sex-L.

Adult type—common forms not usually hereditary but where so: Dom. or Sex-L.

Optic atrophy. Withering of optic nerve, leading to blindness.

Birth type. Occasionally associated with deafness. Dom.

Childhood type. Rec. (?)

Adult type (Leber's disease). Blindness only in center of eye. Sex-L.-q.

Associated with ataxia. Dom. (?)

Retinitis pigmentosa. Degeneration of retina due to deposit of pigment on sensitive regions. May result in night blindness or more serious vision loss, depending on severity. Various types and ages of onset. Dom., Rec., Sex-L.

Night blindness. Vision failing in dim light, with other visual defects. Birth. Various types:

With no other eye defect. Dom.

Oguchi's disease, Japanese type. Rec.

With nearsightedness. 99 per cent of cases males. Sex-L.

With extreme nearsightedness. Rec.

Day blindness. Inability to see in bright daylight, with total color-blindness. Dom., Sex-L.

Retinoblastoma (glioma retinae). Tumor of eye. *See* under CANCER, *Rare Types* (p. 737).

EYE-STRUCTURE DEFECTS

Defective cornea. (Defects in transparent covering of front of eyeball, iris and pupil.)

Opaque ring within cornea. Childhood. Dom.

Cone-shaped cornea (keratoconus). Extreme astigmatism. Childhood, progressing. Rec.

Enlarged cornea (megalocornea). Vision usually normal. Sex-L. or Dom.

Defective iris.

Segment of iris missing. Birth. Dom.-q.; also Rec. or Sex-L.

Complete absence of iris (aniridia). Birth. Dom.

"Pin-hole" pupil (persistent pupillary membrane). Iris almost closed; may cause blindness. Birth. Dom.

Displaced lens.

Due to atrophied suspensory ligament. Sometimes at birth. sometimes adult. Dom.-q.

Same as above, with displaced pupil. Rec. (?)

Small eyes (microphthalmia). Entire eye undersized, frequently with other eye defects. Sometimes environmental. Where hereditary: Dom., Rec., Sex-L., or Chrom. Abn.

Same as above, with teeth defects also. Rec.

Extreme form (anophthalmos), eyes completely absent, hence blindness from birth. Rec.+ (?) or Chrom. Abn.

Pink eye color (ocular albinism). Eyes unpigmented, but with no other albino effects. Birth. Confined to males. (Female carriers of gene show mottling of the retina.) Sex-L.

EYE-MOVEMENT DEFECTS

Cross-eyes (strabismus). Eyes not focusing together. Childhood, may disappear later. Not always hereditary. Rec.-q. or Dom.-q.

Quivering eyeball (nystagmus). Eye tremor, usually weak vision. Birth. Sometimes environmental, result of disease.

Common type, occurring by itself. Sex-L.

With head twitching. Dom.

Eye-muscle paralysis. Inability to move eyes. Birth or later, increasing in severity. (May result from injury, meningitis, etc.) Dom. or Sex-L. if hereditary.

Drooping eyelids (ptosis). Sometimes acquired. If hereditary: Dom.

OPTICAL DEFECTS

Astigmatism. Asymmetrical focusing of light rays. Birth. Heredity uncertain. Dom.-q. (?)

Nearsightedness (extreme).

Distant vision blurred. Birth, increasing with age. Probably several types. Rec. (?) or Sex-L. (?)

Associated with nystagmus and poor vision. Dom. or Sex-L.

Farsightedness (extreme). Inability to see clearly at far or near distance. Birth, may decrease with age.

Not pathological. Dom.-q.(?)

Serious, with other eye effects. Rec. (?)

Color-blindness (common). Confusion of red and green. Birth. Several types. Sex-L.

Total Color-blindness. See Day blindness (p. 744).

Mirror-reading (and writing). Seeing in reverse and upside down. Sometimes with stuttering. Birth. Dom.-q. (?)

Word blindness (dyslexia). Difficulty in reading and writing, unrelated to intelligence. Sometimes traceable to illness, but other cases possibly inherited. Dom.-q. (?)

15. EAR CONDITIONS

Deafness (excluding all environmental cases):

Deafness at birth, resulting in defective speech. Rec.+

Middle-ear deafness (otosclerosis). More frequent in women. Rec.+ or Dom.-q.

Inner-ear (or nerve) deafness. In middle age. Dom.

Rare types:

With kidney diseases. Dom.-q.

With goiter. Rec.

Word deafness. Hearing normal but with inability to interpret sounds. More common in males. Dom.-q.

Outer-ear deformities. Present at birth:

Cup-shaped ear (ear turned over at top). Dom.-q.

Imperfect double ear (one or both). Dom. (?)

Ear fistula. Gill-like opening near ear passage, external or internal, sometimes with discharge. Dom.-q.

Dense hair growth on ear. (No effect on hearing.) Direct father-to-son inheritance due to gene on Y chromosome suspected. Y-Dom. (?) (*See* Text, Chapter 22)

16. BODY STRUCTURE

Dwarfism

Midget type (ateleotic, or "Lilliputian"). Heredity believed through multiple genes. Rec.+ or Dom.+ (q)

Achondroplastic (chondrodystrophy). Stunted arms and legs. Where hereditary: Dom., (Rarely Rec.)

Gargoylism. Extreme dwarfism

as one effect of rare fat-metabolism disorder. *See* under MENTAL DEFECTS (p. 742).

Hand and foot abnormalities. Many different types, single genes producing variable effect within same family. (Sometimes environmental.) Most common hereditary types:

Stub fingers (brachyphalangy), middle finger joints missing; *extra fingers and toes* (polydactyly); *stiff fingers, joints fused; webbed fingers or toes, split foot or hand.* (*See* illustration, p. 238). Also *missing thumbs; clubbing of finger tips or toe tips.* All Dom. or Dom.-q.

Marfan's syndrome. Disproportionately long legs and arms, "spider fingers" (arachnodactyly), deformities of chest cage and of eye lenses, and frequently heart disease. Dom.-q.

Arm and/or leg abnormalities, congenital:

Phocomelia. Extreme shortening of one or more extremities. Usually environmental (as in "Thalidomide" babies). Very rarely hereditary, but if so: Rec.

Achiropodia. Almost total absence of arms and legs. Rec.

Clubfoot. Possible heredity in some cases. (Uncertain.) Rec. (?)

Brittle bones (osteogenesis imperfecta). Fragile bones, with bad teeth, deafness, bluish eyewhites. Birth. Dom.-q.

Deformed spine (spina bifida). With various other effects. Genetic predisposition, probably involving several genes. (?)

"Cobbler's (or "Funnel") chest." Abnormally hollow chest. Dom.-q.

"Tower skull" (Oxycephaly, Acrocephaly). High-pointed skull, several types, sometimes with eye, bone, or mild mental defects. Dom.-q.

Cranial soft spot (cleidocranial dysostosis). Infant "soft spot" in skull persisting to maturity, with other bone defects. Dom.-q.

Crouzon's disease (craniofacial dysostosis). Skull and jaw abnormalities. Dom.-q.

Forehead cleft (metopic suture). Incomplete fusion of frontal bones, shown by bluish line down forehead. Dom.-q. (?)

17. MOUTH AND TEETH

Cleft palate and harelip. Prenatal failure of palate and/or upper lip to fuse. Often hereditary predisposition, but with strong environmental influences.

Harelip with or without cleft palate. Involving either gene-induced or chromosome abnormalities. (?), or Chrom.-Abn.

Cleft palate without harelip. (?)

Missing teeth.

Upper lateral incisors absent or small. Dom., Rec. or Sex-L.

Molars or other teeth missing. Dom. (?)

"*Fang Mouth.*" All teeth missing except canines. Associated with inability to sweat. *See* under SKIN, *Sweat-Gland Defects* (p. 748).

Defective tooth enamel. (Referring only to conditions not environmental.)

Brownish discoloration ("honeycomb" teeth). Dom. or Sex-L. Dom.

Reddish enamel (porphyria). Rec.

Transparent (opalescent dentin). Dom.

Pitted teeth. (Sometimes with cataract.) Dom.

Extra teeth. Frequently associated with cleft palate. Dom. (?)

Teeth at birth. Incisors present at birth. Dom.-q.

Overgrowing gums (gingival hyperplasia). Gums covering nearly all of teeth. *Usually environmental* (infection, drugs, etc.) but in rare cases hereditary. Dom.

18. SKIN

Birthmarks (naevi). Common types—hairy, raised, warty; discolorations or pigmented spots of variable size. Also (one type) slight depression over eyebrows, extending to temple. *Rarely harmful.* Some hereditary. Where so: Dom.

Malignant types (rare, precancerous). *See* under CANCER, *Rare Types* (p. 737) and Text, p. 243.

Mongolian spot. Bluish patch near base of spine, at birth, later disappearing. Not harmful. Rec.+ or Dom.

Fatty skin growths. On eyelids or elsewhere (*xanthalesma,* or *xanthoma*). Inc. Dom. *See also Hypercholesterolemia* and *Hyperlipemia,* under METABOLIC DISORDERS, *Fat* (*Lipid*) *Metabolism Disorders* (p. 738).

Scaly or horny skin (ichthyosis).

Common type, with shedding. Infancy. Dom. (or Sex-L.)

Cracked skin, ears often defective. Birth, may disappear later. *Sometimes fatal.* Rec.

"*Elephant Skin,*" extreme form of above. Causes premature birth, death. Rec.

Psoriasis. Mottled scaly patches. If hereditary, only as tendency. Dom.-q.

Thick, or shedding skin (keratosis).

Skin flaking over entire body. Birth. Rec.

Same as above, but skin thicker. Several types, one with casts of palms and soles shedding. Birth or puberty. Dom.

Thick or discolored skin on limbs, but not shedding. Childhood, more among males. Dom.-q.

Abnormal scars (keloids). Unusually thick, overgrown scar tissue formed after operations or cuts. Dom. (?)

Blistering (epidermolysis).

Blisters easily raised. Childhood. Dom.-q.

Severe type, from birth, leaving scars with other defects. Rec.-q.

Rare extreme form, with bleeding. *Death in infancy.* Rec.

Sunlight blistering (porphyria congenita). With scarring. *See Porphyria,* under *Other Metabolic Disorders* (p. 738).

Darier's disease. Follicle defects, causing hair loss, goose flesh; often leads to mental deficiency and small stature. Several types. Birth or childhood. Dom.-q. (or Sex-L.)

Albinism. Lack of pigment in skin, hair.

Complete, skin and hair "dead white," also with pink eyes. Rec.; Sex-L. (rare)

Mild form (albinoidism), with some pigment. Dom.-q.

Partial, white stripes or patches on body. Dom. or Dom.-q.

White forelock, or "blaze." Unpigmented patch of skin on scalp, growing white hair. Dom.

Sweat-gland defects.

Complete inability to sweat, with missing teeth and other defects. (*Anidrotic ectodermal dysplasia.*) Mostly males. Sex-L.

Milder type of above. Dom.-q.

Excessive sweating (hyperhidrosis). Dom.

"Rubber" skin (cutis laxa, or Ehlers-Danlos syndrome). Highly elastic tissue, capable of freak stretching. Dom.-q.

Unelastic skin (pseudo-xanthoma elasticum). Lack of elasticity in skin and blood vessels (causing hemorrhages). Rec.

19. HAIR AND NAILS

Baldness (alopecia). Where not due to environmental factors:

Pattern baldness (common type). In maturity, almost exclusively in males. Sex-Limited Dom.

Patch baldness. Small bald area on scalp; may spread. Birth or puberty. Dom.-q.

Congenital (hypotrichosis). Hair defective or never developing. Various types, associated with teeth, nail or scalp defects. Rec., Dom., or Sex-L.

DEFECTIVE HAIR

Infantile down remaining through life. Dom.

Beaded hair (monilethrix). May lead to baldness. Infancy. Dom.-q.

Excessive long, soft hair on face and elsewhere. ("Dog face.") Other effects. Childhood. Dom.-q.

With abnormal nails. Congenital, worsens later. Puberty. Dom.

Premature grayness. Head hair only, beginning in adolescence. No relation to aging. Dom.

White forelock (or "Blaze"). *See* under SKIN, *Albinism* (p. 748).

DEFECTIVE NAILS

Nails absent, partially or wholly. Rec. or Dom.

Thick nails, protruding at angle. Dom.

Thickened nails with thickened skin on palms and soles. Dom.+

Spotted nails, bluish white. Dom.

Thin nails, several types, small and soft, or flat. Dom. or Dom.-q.

Milky-white nails. Dom-q.

20. MISCELLANEOUS

Sense of smell deficiency (anosmia). Total absence of sense of smell. Dom.

Transposed internal organs (situs inversus viscerum). Heart, liver, stomach, etc., on reverse side from normal location. (?)

Double (Multiple) eyelashes. Double row of lashes on each lid. Rec. (?)

Lymphedema (Millroy's disease). Abnormal swelling of legs and ankles. More common in women. Dom.-q. (?)

"Triple-E-group (chromosome 17 or 18)" syndrome. Multiple abnormalities at birth caused by extra one of chromosomes cited. Fatal. Chrom.-Abn.

Sturge-Weber syndrome. Defective blood vessels in brain and skin of head, frequently with mental retardation. Dom.-q. (?) or Chrom.-Abn. (?)

FORECAST TABLES:

Disease and Defect Inheritance

The following forecast tables should interest you if—

—You are still young, and there is some condition known in your family, generally appearing in later life, which you are worried may also appear in *you*.

—You are planning to marry and are worried that some condition in either you or your prospective mate or in one of your families may be passed on to your children.

—You are married and already have children and are worried that some condition may crop out in them later.

—You have a child with some condition and are worried about its appearing in future children.

When using these Forecast Tables, be as certain as possible that—

1. The condition you have in mind, as identified by your doctor, is the *one listed* in our tables.
2. That, where there are various methods of inheritance, you know which gene mechanism applies in your case.
3. That, where *environment* is a factor in the expression of a condition, you have ruled out the possibility that the condition has been covered up, or that, even if the genes are transmitted, the condition might be prevented from developing in a child. In *diabetes* (or some of the mental diseases), for example, the fact that parents or their families are affected does not positively indicate whether, or to what extent, their children might develop the condition if the environment is favorable. Conversely, the fact that parents themselves do not show the condition, when it has appeared in others in their families, is not conclusive proof that they are free of the genes involved. (This applies especially to all "qualified" conditions in the tables—where a "q." follows the genetic mechanisms given.)
4. That, where a condition can also be caused by environment, the one you have in mind is of the *hereditary type*.

5. That you have paid full attention to the question of "*onset.*" (For instance, where a condition appears late, you cannot be sure that the genes are, or are not, present, until the person reaches the required age.)

In no case consider your "forecast" conclusive, or take any action on the basis of it, without consulting competent medical authority. In all cases it is best to consult your family doctor first: then, if required, a medical geneticist.

I. RECESSIVE GENE FORECASTS

The most common form of wayward gene inheritance. Because of the vast number of persons carrying "hidden" recessive genes for various conditions, *complete* assurance never can be given that any common recessive condition may not crop out in some child. But the risk diminishes with the infrequency with which the condition appears in the parents' families. The more prevalent the condition in the general population, however, the more likely it is to turn up unexpectedly—especially if the parents are closely related. Also keep in mind points made above (in No. 3) regarding environmental factors in the expression or suppression of certain conditions, and the possibility of late onset.

	CHANCES CHILD WILL INHERIT:
1. If both parents are affected	Almost certainly
2. If one parent is affected, the other not, but if in the family of the "free" parent:	
a. The father or mother is affected, or a child with the defect already has appeared	Even chance
b. A brother, sister or grandparent is affected	Less than even chance
c. Some more remote relative is affected	Possible, but not probable
d. No one, near or far, has been known to have the condition	Very unlikely but still possible
3. If neither parent is affected, but:	
a. Both had an affected father or mother or already have an affected child	One-in-four
b. Each parent has an affected brother or sister	One-in-nine*
c. The condition occurs or has occurred in more distant relatives of the families of both	Extremely unlikely, but yet possible
d. The condition is wholly unknown in the family of either	Virtually nil

* The "one-in-nine" chance (which is contrary to the "one-in-four" estimate often but wrongly made for this situation) is arrived at as follows: If a person who is not affected by a recessive condition has a brother or sister

I-A. "QUALIFIED" RECESSIVE FORECASTS: Where conditions are "Rec.-q." the chances of inheriting the *genes* are exactly the same as shown in Table I, but the chances of the *condition's* actually appearing in a child may be altered by various circumstances. In general, however, the odds are somewhat lower than in simple recessives for each type of mating.

I-B. "RECESSIVE-PLUS" FORECASTS: Where listed as "Rec.+" the situations are about the same as in Table I, but with the *probability lessened* in most cases.

II. SEX-LINKED (RECESSIVE) FORECASTS

Where the wayward gene is carried in the X chromosome, and therefore acts as a dominant in the case of males, as a recessive in the case of females. Examples: Color-blindness, hemophilia, nystagmus, pink eye color (ocular albinoidism).

	CHANCES CHILD WILL INHERIT:
1. If both parents are affected	Almost certainly in all their children
2. If mother is affected, but father is free of it	Certain for every son, but no daughter
3. If father is affected, and mother is free of it, but:	
a. Her father, mother or a sister has or had the condition	Even chance in any child
b. One of her brothers is affected	One-in-four for any child

with the condition, one knows that *both of their parents* must have carried the recessive gene involved. These parents, then, could produce three types of offspring with these frequencies: (a) one in four carrying two recessive genes, and afflicted with the condition; (b) two in four each carrying a single gene, but not afflicted; and (c) one in four carrying no gene, and also not afflicted. An unafflicted brother or sister of an afflicted person must therefore be of type "b" or "c," but not knowing which, it can be said there is a *two-in-three* chance that he or she is carrying the hidden recessive gene matched with a normal gene (since of each three unafflicted members of the family two, on the average, would carry the gene, and one not); hence there is half the two-in-three chance, or a *one-in-three* chance, that a child of the unafflicted individual will receive the wayward recessive gene. If the same situation applies to both husband and wife, the chance for a child's getting *two* of the wayward genes (and so becoming afflicted), is one-third by one-third, or one in nine.

	CHANCES CHILD WILL INHERIT:
c. One of her more remote relatives is affected	Extremely unlikely, but yet possible
d. No known case in her family	Virtually nil for rare conditions; possible for common ones such as color-blindness
4. If neither parent is affected, but it occurs in the mother's family (in the same situations as noted above)	No chance for any daughter, but for sons same odds as in 3-*a.b.c.d.* above

III. DOMINANT GENE FORECASTS

For all conditions which can be produced in either sex by one gene acting singly, as in achondroplasia, drooping eyelids, various hand defects, etc.

	CHANCES CHILD WILL INHERIT:
1. If both parents are affected*	Very probable (3-in-4 chance)
2. If one parent is affected, the other "free"*	Even chance
3. Where neither parent is affected, but it appears in the family of one or the other:	
a. If the condition is always known to show itself when the gene is present	Nil
b. If the gene action is sometimes known to be suppressed by environment*	Some likelihood, but not great

* NOTE: In cases where the gene action is irregular, or influenced by environment, we have the situations which follow:

III-A. "QUALIFIED DOMINANT" FORECASTS: In all conditions given in the tables as "Dom.-q." (retinoblastoma, acholuric jaundice, albinoidism, etc.), the probabilities are modified downward from those shown in the preceding table, the forecasts depending upon the degree to which the gene expresses itself or is suppressed by environment.

III-B. "DOMINANT-PLUS" FORECASTS: ("Dom.+," very rare, as in thickened nails, etc.) Relative probabilities are about as shown in Table III above, but greatly reduced in most cases.

III-C. INCOMPLETE DOMINANT GENE FORECASTS: In the conditions where a single gene produces *mild* symptoms, but two of the same genes produce a *severe* effect (as in *Cooley's anemia* or *sickle-cell anemia*) the risks in the children for the severe or mild conditions are as follows:

	CHANCES CHILD WILL INHERIT:
1. Both parents with severe type	All children will probably have severe type
2. One parent severe, other mild condition	Chances fifty-fifty a child will have either the severe or mild type
3. One parent severe, other parent normal	All children will probably have the mild type of condition
4. Both parents with mild condition	One-in-four chance children will have the severe condition; one-in-two chance the mild condition; one-in-four chance child will be completely normal
5. One parent with mild condition, other parent normal	Chances fifty-fifty a child will either have the mild condition or else be completely normal

III-D. SEX-LINKED DOMINANT FORECASTS: Where the dominant gene is on the X chromosome (as in *Vitamin-D-resistant rickets* and *brownish-tooth* enamel): If the mother has the condition, the expected forecasts for *all* offspring, male and female, will be the same as for the regular dominants, Table III, 1 and 2. Where the father has the condition, *every daughter*, but not sons, will get it.

CLUES TO "CARRIERS"

I. AUTOSOMAL CONDITIONS, AFFECTING BOTH SEXES EQUALLY

How carriers of single genes are revealed in conditions which require two matching genes on any given chromosome (other than the sex chromosomes) to produce serious effects:

DOUBLE-GENE CONDITION AND EFFECTS	SINGLE-GENE EFFECTS IN CARRIERS
Sickle-cell anemia: Abnormal red cells, with severe anemia. Usually fatal.	Some abnormal hemoglobin on tests. No anemia.
Thalassemia (Cooley's anemia): Severe anemia.	Mild anemia, not harmful.
Congenital afibrinogenemia: Severe bleeding; absence of fibrinogen.	No bleeding, but reduced fibrinogen in some cases.
Elliptocytosis: Anemia due to blood destruction.	Abnormally shaped red blood cells.
Galactosemia: Damage to nervous system and liver; inability to metabolize galactose.	Reduced ability to metabolize galactose.
Wilson's disease: Damage to nervous system and liver; deficiency in the copper-carrying protein ceruloplasmin.	Altered metabolism of copper.
Glycogen storage (von Giercke's) disease: Liver and kidney disease; excessive accumulation of glycogen.	Abnormal levels of certain sugars in blood cells.
Hypophosphatasia: Bone disease resembling rickets.	Often, decreased alkaline phosphatase (an enzyme) in serum and abnormal substance in urine.
Cystic fibrosis of pancreas: Severe digestive disturbances; recurrent pulmonary infections.	Abnormally high salt excretion in sweat in many cases.

DOUBLE-GENE CONDITION AND EFFECTS	SINGLE-GENE EFFECTS IN CARRIERS
Cystinuria: Kidney stones, abnormal urine substances.	Some types have abnormal urine substances.
Phenylketonuria: Mental retardation; faulty metabolism of phenylalanine.	Reduced ability to metabolize phenylalanine.
Juvenile amaurotic family idiocy (*Tay-Sachs disease*): Blindness, neurological degeneration.	Abnormal cells in blood.
Gargoylism (one form): Grotesque appearance, abnormal formation of connective tissue.	Sometimes abnormal substances in urine.
Gaucher's disease: Bone abnormalities, anemia, bleeding; sometimes neurological disorder.	Unusual cells sometimes found in bone marrow.
Familial hyperlipemia: Markedly elevated blood lipids. Heart disease, abdominal pains.	Reduced ability to clear dietary fats from blood.
Familial hypercholesterolemia: Markedly elevated blood cholesterol; skin deposits; heart disease.	Elevated blood cholesterol levels.

II. SEX-LINKED CONDITIONS (*genes on X-chromosome*).

CONDITION AND SINGLE-GENE EFFECT IN MALES	SINGLE-GENE EFFECT IN FEMALES
Color-blindness: Inability to distinguish between red and green colors.	Reduced color discrimination, sometimes revealed by tests.
Ocular albinism: Absence of pigment in iris and retina of eye.	Patchy pigmentation of retina.
X-linked retinitis pigmentosa: Progressive loss of vision; pigmentation of retina.	Slightly abnormal appearance of retina.
Primaquin sensitivity (Favism): Absence of enzyme G-6-PD; massive destruction blood cells when ingesting certain substances.	Reduced enzyme levels in red cells.
Muscular dystrophy (Duchenne type): Progressive muscular weakness.	Sometimes certain enzymes increased in blood.
Vitamin D-resistant rickets: Bone disease (rickets) not treatable with vitamin D.	Less severe disease symptoms.
Nephrogenic diabetes insipidus: Excessive urination, severe thirst, dehydration.	Reduced ability to concentrate urine.
Anhidrotic ectodermal dysplasia: Absence of sweat glands, hair and teeth.	Patches of skin with absent sweat gland.

BIBLIOGRAPHY

Suggestions for Further Reading

In the following list of references, selected from the many thousands of publications drawn upon in preparing *Your Heredity and Environment*, the author has sought to include those which he felt would be most useful to readers in search of more specialized information on given subjects. Among other things, preference has been given to publications which are most recent, most comprehensive in treatment and/or which include bibliographies of previous works in their respective areas. Many of the basic references prior to 1950 will be found in the reading lists of the author's earlier books, *You and Heredity* (1939) and *The New You and Heredity* (1961), both available in many libraries. These older references, while sometimes mentioned in the present text, are with few exceptions omitted from the following list.

Most of the new literature on human genetics and the related sciences, and hence the references given here, may be too technical for the average nonprofessional or nonacademic reader. However, a special effort has been made to include many books and articles which are popular in nature, and these have been preceded by a bold paragraph symbol (¶) for easy reference. The medical and scientific journals listed are available in almost all major medical or scientific libraries. (The abbreviations—printed, like the book titles, in italics—are those in standard use. Full titles of the journals most frequently cited are given at the end of our bibliography, p. 806.) *Note* that in journal references connected by a *colon* (61:2), the first figure indicates the volume number, while the second figure indicates the issue number; elsewhere, the volume number and page reference are given, followed by the year of publication (32, p. 296, 1957).

For easier reference, the reading list has been arranged by topics which follow more or less their order in this book and in their respective chapters. References grouped under each heading are usually separated by a slant line. The word *also* indicates that the following item is on the same subject or by the same author as the reference that precedes it. In general, while book titles are given in full, the titles of articles are often paraphrased, condensed, or, occasionally, omitted (in such cases, the article may be looked up by author). Titles of books and articles in languages other than English are translated, with the original language indicated in parentheses. A place of origin given in parentheses after the name of a book indicates that no American edition has been published; otherwise, the books listed are available in United States editions.

PRENATAL DEVELOPMENT

(*Chapters* 1 *to* 6)

REPRODUCTIVE PROCESS (review, half-century research). Hartman, Carl G. *Fertil. & Steril.*, Jan.–Feb., 1961. / Fertilization in mammals. Austin, C. R., & Bishop, M. W. *Biol. Rev.*, 32, p. 296, 1957.

HUMAN SPERMS. *Morphology of the Human Sperm.* Schultz-Larsen, J. Munksgaard (Copenhagen), 1958. / Observations with electron microscope. Culp, O. S., & Best, J. W. *J. Urol.*, 61:2, Feb., 1949. ¶ Sperm maturescence. Bishop, D. W. *Sci. Monthly*, Feb., 1955.

HUMAN OVA. *Ovum Humanum.* (Text, photos, stages of development). Shettles, L. B. Hafner (N.Y.), Urban & Schwarzenberg (Munich, Berlin), 1960. / Studies on living ova: *Trans. N.Y. Acad. Sci.*, Dec., 1954, p. 99. / *The Mammalian Egg.* Austin, C. R. Chas. Thomas, 1961. / Description human ova in first 17 days. Hertig, A. T. et al. *Am. J. Anat.*, 98, p. 435, 1956.

MYTHS OF MATING. ¶ Congenital malformations in the past. Warkany, Josef. *J. Chron. Dis.*, 10:2, Aug., 1959. / Telegony retested. Daniel, J. C., Jr. *J. of Hered.*, Nov.–Dec., 1959.

HYBRIDS. Mules (horse-donkey), chromosomes in: Benirschke, Kurt, et al. *J. Reprod. Fertil.* 4:3, 1962. / Alleged fertile mare mules: *J. of Hered.*, 58:1, 1964. / "Zebronkey" (zebra-donkey): *Chromosoma,* 1964. / Turkey-chicken. Poole, H. K. *J. of Hered.*, May-June, 1963.

PLACENTA. *The Placenta and Fetal Membranes* (symposium). Viller, C. A., Ed. Williams & Wilkins, 1960.

HUMAN FETUS. From fetus to newborn baby. Patten, B. M. Chap. 13 in *Birth Defects.* Lippincott, 1963. ¶ *The First Nine Months of Life.* (Photos and text.) Flanagan, G. L. Simon & Schuster, 1962.

FETAL BEHAVIOR. Prenatal stress and later behavior. Lieberman, M. W. *Science,* 141, Aug. 30, 1963, p. 824. ¶ Fetal behavior and later personality. Sontag, L. W. *Vita Humana,* 6, 1963.

PRENATAL HAZARDS, GENERAL. Birth Defects. Symposium, Nat'l Found. Fishbein, Morris, Ed. Lippincott, 1963. / *Second International Conference on Congenital Malformations.* Proceedings. Lippincott, in press. / *Prenatal Influences.* Montagu, Ashley. Thomas, 1962. ¶ *Life Before Birth.* Montagu, Ashley, New American Libr., 1964.

DRUGS, EFFECTS ON FETUS. Thalidomide: Lenz, Widukind. In: Proceedings, *Second International Conference on Congenital Malformations.* Lippincott, in press. *Also:* Taussig, Helen B. *Sci. Amer.,*

Aug., 1962. / Conf. W. Berlin, 1963: Absts., Birth Defects, 1:1, Jan., 1964, Nos. 5, 6, 38. / Passage of drugs across placenta. Moya, Frank, Thorndike, Virginia. *Am. J. Obst. & Gyn.*, 84:2, 1963. / Drug-induced teratogenesis. Frazer, C. F. *Canad. Med. Ass. J.*, 87:3, Sept. 29, 1963. / Drug addiction in newborn: Vincow, Annabelle, & Hackel, Alvin. *GP*, 22, p. 90, 1963.

SMOKING BY MOTHER, EFFECTS ON FETUS. Herriot A., et al. *Lancet*, 1:7233, 1962. / Savel, L. E., & Roth, E. *Obst. & Gynecol.*, 20, Sept., 1962. / O'Lane, J. M., *Obst. & Gyn.*, 22:8, 1963.

GERMAN MEASLES (RUBELLA), DURING PREGNANCY. Lundstrom, R. *Acta Paediat.*, 51, Suppl. 133, 1962. / Blattner, R. J., *J. Pediat.*, 54, Feb., 1959.

DIABETIC MOTHERS, EFFECTS ON FETUSES. Driscoll, S. G., et al. *Am. J. Dis. Child.*, 100:6, Dec., 1960.

X RAYS, EFFECTS ON FETUS. Rugh, R. *J. Pediat.*, 52, May, 1958.

MOTHER'S AGE. Congenital malformations in children of mothers aged 42 and over. Böök, J. A., et al. *Nature*. 181, p. 1545, 1958.

BOY OR GIRL?

(*Chapter* 7)

PREDICTION OF SEX. ¶ Folklore and science. Forbes, T. R. *Proc. Am. Philosoph. Soc.*, 103:4, Aug., 1959.

GENETICS OF SEX DETERMINATION. Recent advances in human chromosome studies. Cooper, H. L., *J. Natl. Med. Ass.*, 54:4, July, 1962. / Historical review: Moore, K. L. Manitoba (Can.) *Med. Rev.*, 42:8, 1962. / On the sex ratio in man. Colombo, B. *Cold Spr. Harb. Symp. Quant. Biol.*, 22:193, Biologic concepts. Segal, S. J. *Eug. Qu., Sept.* 1960. / *Also*, on sex determ.: Stern, Curt. *Am. J. Med.*, 34:5, 1963.

SPERMS, MALE, FEMALE TYPES? Shettles, L. B. *Nature*, 186, May 21, 1960. *Also*: Human spermatozoan populations. *Int. J. Fertil.*, 7:2, 1962.

SEX RATIO AT BIRTH. U.S.: Vit. Statist., Specl. Reports, Natl. Summaries, 50:19, Nov. 27, 1959, p. 73. *Also*, 1961, Vol. 1, Sect. 1, Natality, general summaries. / Korea, sex ratio in: Kag, Y. S., & Cho, W. K. *Hum. Biol.*, 34:1, 1962. / Analyses of data: Makela, O.; Beilharz, R. G. (two articles). *Ann. Hum. Gen.*, 26:4, May, 1963.

HEREDITARY INFLUENCES ON SEX RATIO. Distribution, sequences of sexes, Swedish families. Edwards, A. W. F., & Fraccaro, M. *Ann. Hum. Gen.*, 24, 1960. / In mouse, male-transmitted sex-ratio factor. Weir, J. A. *Genetics* (Texas), 45:11, Nov., 1960. Single-sex sibships. Hewitt, D., et al. *Ann. Hum. Gen.*, 20–155, 1955.

BLOOD GROUPS (ABO) AND SEX RATIO. Allan, T. M. *Brit. Med. J.*, 1, 1959. / Cohen, Bernice H., Glass, Bentley. *Hum. Biol.*, 28:1, Feb., 1956.

MATERNAL AGE, AND SEX RATIO. Lowe, C. R., & McKeown, Thos. *Brit. J. Socl. Med.*, 4:2, Apr., 1960. / In Japan: Takahashi, Eiji. *Ann. N. Y. Acad. Sci.*, 57:5, Jan. 15, 1954. / Birth order, parental ages, sex of offspring: Novitski, E., Kimball, A. W. *Am. J. Hum. Gen.*, 10:3, Sept., 1958. / Parental factors, influences: Bernstein, Marianne E., & Martinez-Gustin, M. *J. of Hered.*, May-June, 1961.

WARTIME, SEX RATIO IN. U.S., World War II.: MacKahon, Brian, & Pugh, T. F. *Am. J. Hum. Gen.*, 6:2, June, 1954. / Various countries: Myers, R. J. *Hum. Biol.*, 21:4, Dec., 1949.

CONTROL OF SEX DETERMINATION. ¶ Gordon, M. J. *Sci. Amer.*, Nov., 1958. / Alkaline treatment, negative in rabbits: Emmens, C. W. *J. of Hered.*, 51:4, July-Aug., 1960. / Separation of bull X-Y sperm: Lindahl, P. E. *Nature*, 181, Mar. 15, 1958. / Control in rabbits by electrophoresis of spermatazoa. Gordon, M. J. *Proc. Nat. Acad. Sci.*, 43, p. 913, 1957.

RADIATION, AND SEX RATIO. Schull, W. J., & Neel, J. V. *Science*, Aug. 15, 1958.

PREFERENCE FOR BOY, GIRL. ¶ Size of family and: Freedman, Deborah S., et al. *Am. J. Sociol.*, Sept., 1960. / Differential treatment and effects on sex ratio (Guatemala): Cowgill, U. M., & Hutchinson, G. E. *Hum. Biol.*, 35:1, 1963. / Polygamy and infanticide, effect on sex ratio. Shaw, R. F. *Am. J. Phys. Anthrop.*, 19:1, Mar., 1961.

BOY-GIRL RATIO, GENERAL FACTORS. ¶ Scheinfeld, Amram. In *Women and Men.* Chapters 1–4). Harcourt, Brace, 1944.

GENES AND CHROMOSOMES

(*Chapters* 8, 9)

BIRTH OF GENETICS. Genetics: Suppl., Part 2, 35:5, Sept., 1950. / Glass, Bentley. Maupertuis and beginnings of genetics. *Qu. Rev. Biol.*, 22:3, Sept., 1947.

GREGOR MENDEL. ¶ Rediscovery of his work. Dodson, E. O. *Sci. Amer.* Oct., 1955. / Mendel's laws of inheritance. Glass, Bentley, in: *Studies in Intellectual History*, pp. 148, ff. Johns Hopkins Press, 1953. ¶ Facsimile, Mendel's original paper, 1965. *J. of Hered.*, 42:1, Jan.–Feb., 1961. ¶ Restoration of Mendel museum. Hutt, F. B., *J. of Hered.*, 53:1, Jan.-Feb., 1962. / ¶ Visit to Mendel's monastery. Boyes, J. W., & B. C. *Am. Inst. Biol. Sci. Bull.*, June, 1962. ¶ "I talked with Mendel." Eichling, C. W., Sr. *J. of Hered.*, July, 1942. ¶ Gregor Mendel and his work. Iltis, Hugo. *Sci. Monthly*, May, 1943.

THE GENE. (With background to Mendel.) ¶ Horowitz, N. H. *Sci. Amer.*, Oct., 1956. / In pursuit of a gene. Glass, Bentley. *Science*, 126, Oct. 11, 1957.

HUMAN CHROMOSOMES. Preparation, analysis, etc. Moore, K. L., & Hay, J. C. *Canad. Med., Ass. J.*, 88:20–21, 1963.

CHROMOSOME PUFFS. Wolfgang Beerman, Ulrich Clever. *Sci. Amer.*, April, 1964.

GENETICS, STUDIES IN. Muller, H. J. (Selected papers.) Indiana U. Press, 1962. / Human genetics, papers on. (Selection, dating back to 1900's.) Boyer, S. H., IV, Ed. Prentice-Hall, 1963.

THE CELL. ¶ Symposium. *Sci. Amer.*, 205:3, Sept., 1961.

DNA, GENE SPECIALIZATION. See refs. under Chemical Individuality, Chap. 19.

COLORING

(*Chapters* 10, 11, 12)

PIGMENTATION, GENERAL. Genes and the pigment cells of mammals. Silvers, W. K. *Science*, 134, Aug. 11, 1961. / The melanocytes of mammals. Billingham, R. E., & Silvers, W. K. *Qu. Rev. Biol.*, 35:1, Mar., 1960. / Symposium: The Pigment Cell. *Ann. N.Y. Acad. Sci.*, 100, Feb. 15, 1963. ¶ *The Nature of Animal Colours.* Fox, Munro, & Vevers, Gwynne. Macmillan (N.Y.), 1960.

EYE COLOR. Studies: Riddell, W. J. B. *Ann. Eugen.*, 1941, 1942, 1943. / Genetic analysis of human eye color. Brues, Alice M. *Am. J. Phys. Anthrop.*, 4:1, 1946. / Human mosaic, eye and hair color. Ahuja, Y. R. *Acta Genet. Med. Gemel.*, 9, Oct., 1960. ℂ "Heterochromatic girl." *Life* Mag., July 18, 1959. / Light eyes in a Brahman population. Rife, D. C., & Malhotra, K. C. *Acta Genet. Med.* (Rome), 12:2, 1963. / Iris-structure, inheritance. Hornbak, H. *Proc. 2nd. Intl. Congr. Hum. Gen.* Inst. Gregor Mendel, Rome, 1961.

RED HAIR. Reed, T. F. *Ann. Eugen.*, 17:2, 1952. / Barnicot, N. A. *Nature*, Mar. 3, 1956.

SKIN COLOR. Measurement and inheritance. Harrison, G. A. *Eug. Rev.*, 49:2, July, 1957. / Histology, skin pigmentation. Gates, R. R. *J. Royal Microscop. Soc.*, 80, 1961. / Studies. Edward, E. A., & Duntley, S. Q. *Am. J. Anat.*, 65, July, 1939. ℂ Hormones and skin color. Lerner, Aaron B. *Sci. Amer.*, July, 1961. / Menstrual cycle and skin pigmentation. McGuinness, B. W. *Brit. Med. J.*, 2, Aug. 26, 1961. / Sun-tanning. Lee, M. M. C., & Lasker, G. W. *Annls. Hum. Gen.*, 23:3, July, 1959. / Race differences, melanin pigmentation. Walsh, R. J. *J. Roy. Anthrop. Inst.*, G. B., 93:1, 1963.

RACE-CROSSING EFFECTS. *See* "Race crossing," under Races: II, Chap. 43.

NEGRO SKIN COLOR AND SOCIO-ECONOMIC STATUS. Horton, C. P., & Crump, E. P. *J. Pediat.* 53, p. 547, 1958.

RACE-CROSSING, PIGMENTATION RESULTS. Gates, R. R. *Hum. Biol.*, Feb., 1962. *Also*: New theory of skin color inheritance. *Intl. Anthrop. & Linguist.* Rev., 1:1, 1953. ℂ Skin color of children from White by near-White marriages ("Black baby" myth) Stern, Curt. *J. of Hered.*, Aug. 1947, p. 233. / Negro-White crosses, polygenic inheritance. Stern, Curt, In: *Principles of Human Genetics*. 2nd ed. (pp. 350–58, 692–3). W. H. Freeman, 1960. / *See also* refs. under "Races: I" (Chap. 42), on "Skin color," "Negro-White crossing."

FEATURES

(*Chapter* 13)

THE FACE AS A WHOLE. Brothers, similarities in: Howells, W. W. *Am. J. Phys. Anthrop.*, Mar., 1953. / Predetermination of adult face: Curtner, R. M. *Am. J. Orthodont.*, 29:3, 1953. / *Malformations of the Face*. Walker, D. G. Williams & Wilkins, 1961. / Facial depth, growth. Meredith, H. V. *Am. J. Phys. Anthrop.*, June, 1959. / Facial complex, family studies of. Hanna, B. L., et al. *J. Dental Res.*, 42:6, 1963.

EYE. Anthropology of: Krogman, W. M. *Ciba Symposia*, Nov., 1943.

EAR. Main types of ear shape (in German). Lundman, B. *Homo* (U. of Upsala), 3:2, 1952. / Ear shapes in twins, triplets. (Photos) Husén, Torsten. In: *Psychological Twin Research* (Appendix). Almqvist & Wiksell (Stockholm), 1959.

MOUTH, JAW. Morphology and heredity of the mouth and chin (in German). Pfannenstiel, Dora. *Arch. Julius Klaus-Stift,* 27:1–4, 1952.

TEETH. As basis for distinguishing MZ, DZ twins: Lundstorm, Anders. *Am. J. Hum. Gen.*, 15:1, Mar., 1963. / Genetic variations (twin study): Osborne, R. H., et al. *Am. J. Hum. Gen.*, 10:3, Sept., 1958. / Eruptive time. Hatton, Margaret. *J. Dental Res.*, 34:3, June, 1955. / Serial study of occlusion from birth to 12; Sillman, J. H. *Am. J. Orthodontics*, 37:7, July, 1951. *Also:* Clinical consideration of occlusion. *Am. J. Orthodontics*, 42:9, Sept., 1956.

Hair form. Symposium: Growth Replacement and Types of Hair. (Articles by Garn, Stanley M., et al.) *Annls. N.Y. Acad. Sci.*, 53:3, 1950. / Race differences in: Steggerda, Morris, & Seibert, H. C. *J. Hered.*, Sept. 1941. / Age changes: *J. Hered.*, Nov., 1941. / Chest-hair distribution. Setty, L. R. *Am. J. Phys. Anthrop.*, 19:3, Sept., 1961 / Curvature, transverse shape of human hair. Sergi, S. *Proc. 2nd Intl. Cong. Hum. Genet.*, Vol. 1, p. 451. / *See also* refs. on race differences in features, Races: I. (Chap. 42).

STATURE AND BODY FORM

(*Chapter* 14)

General. *Genetic Basis of Morphological Variation.* Osborne, R. H., & De George, F. V. Harvard U. Press, 1959. / Correlations of brothers. Howells, W. W. *Am. J. Phys. Anthrop.*, 11:1, Mar., 1953. *Also*, with Slowey, A. P.: Linkage studies. *Am. J. Hum. Gen.*, 8:3, Sept., 1956. / Twins, heritability of physical measurements. Clark, P. J. *Am. J. Hum. Gen.*, 8:1, Mar., 1956. / Familial correlations, height, weight, skeletal maturity. Hewitt, D. *Ann. Hum. Gen.*, 22:1, 1957. / Parent-child correlations, stature, body measurements: Tanner, J. M., Israelsohn, W. J. *Annls. Hum. Gen.*, 26:3, Feb., 1963. *Also:* Livson, Norman, et al. Science, 138, Nov. 16, 1962.

Changes in stature, growth. ❡ *Education and Physical Growth.* Tanner, J. M. U. of London Press, 1961. / Heights and Weights of Children and Youth in the U.S. Hathaway, Millicent L. Home Econ. Res. Report, 2, U.S. Dept. Agric., Oct., 1957. ❡ Growth trends in 'teen ages. *Statist. Bull.*, Metrop. Life Ins. Co., Oct., 1960. / Girls' body sizes, changes in U.S. Meredith, H. V., & Knott, V. B. *Growth*, 26:4, 1962. ❡ Our changing bodies. Scheinfeld, Amram & James, T. J. *Cosmopolitan* Mag., Apr., 1956. / Great Britain, changes in. Kemsley, W. F. F. *Ann. Eugen.*, 15:2, 1950. / Germany (Berlin youth), changes. Schröder, E. *Offentliche Gesundheitsdienst*, 22:3, June, 1960. / Japanese children, growth changes. Masao, Matsuo. *Paediat. Jap.*, 4:7, 1961. (*Hum. Gen. Abst.*, Apr., 1963, No. 1742.)

Influences on stature, body-size. Influence of heredity and environment on stature of school children. Quaade, F. *Acta Paediat.* (Upps.), 45:5, 1956. / Socioeconomic status and body size, boys. Meredith, H. V. A.M.A. *J. Dis. Child.*, Dec., 1951. / Japanese children, growth in different environments. Greulich, W. W. *Science*, 127, Mar. 7, 1958. / Adolescent boys, inconstancy of physique. Hunt, Edward E., Jr., & Barton, W. H. *Am. J. Phys. Anthrop.*, 17:1, Mar., 1959. / Aging effects on stature. Trotter, Mildred, & Gleser, Goldine. *Am. J. Phys. Anthrop.*, 9:3, Sept., 1951. / Body weight, race and climate. Roberts, D. F., *Am. J. Phys. Anthrop.*, 11:4, 1953.

Puberty onset, girls. Chinese, Hong Kong (socio-economic diffs.) Lee, M. M., et al. *Pediatrics*, 32:3, 1963. / In Nigeria: Tanner, J. M., & O'Keeffe, B. *Hum. Biol.*, 34:3, Sept. 1962. / In Israel: Hauser, G. A., et al. *Gynaecologia* (Basel), 155:1, 1963. / In West Bengal: Mukherji, P. S., & Sengupta, S. K. *Indian J. Child Hlth.*, 11:3, 1962.

(Hum. Gen. Abst., Sept., 1962, No. 75). / Twins and menarche. (In Spanish.) Dutrey, J. *Dia Medico*, 31.7, 1959.

PRECOCIOUS PUBERTY. *See* refs. under Sex Abnormalities, Chap. 27.

OBESITY. Genetic factors. Mayer, Jean. *Bull., N.Y. Acad. Med.*, 36:5–5, 1960. *Also*, on obesity in children: ¶ *Postgrad. Med.*, July, 1963. / Caloric intake in relation to children's physique. Peckos, Penelope S. *Science*, 117, June 5, 1953. / Abnormal obesity: *See* refs. under Structural Defects, Chap. 23.

SEX DIFFERENCES IN BODY FORM, GROWTH. Chaps. 5, 13 in: *Women and Men*. Scheinfeld, Amram. Harcourt, Brace, 1944. / Pelves, earliest indices for sexing, Hoyme, Lucille. *Am. J. Phys. Anthrop.*, 15:4, Dec., 1957.

HEAD FORM. Genetic model of the inheritance. Welon, Z. *Hum. Gen. Abst.*, Feb., 1963, No. 1287.

HANDS, FEET. Hand types in women, etc. Blincoe, Homer. *Am. J. Phys. Anthrop.*, 20:1, Mar., 1962. / Finger-length ratios, hereditary transmission. Rösler, H. D. *Acta Genet. Med. Gemel.*, 7:3, 1958. / Short thumbs, heredity. Stecher, R. M. *Acta Genet.*, Basel, 7:1, 1957. / Foot, digit-size inheritance. Kaplan, A. R. *J. Hered.*, 54:1, 1963.

TWINS AND SUPERTWINS

(*Chapters* 17, 18)

GENERAL. *Twins in History and Science*. Gedda, Luigi. Thomas, 1961. ¶ *Multiple Human Births*. Newman, H. H. Doubleday, 1940. ¶ *Twins and Supertwins*. Scheinfeld, Amram. Lippincott (in preparation).

CONCEPTION, FETAL DEVELOPMENT. The fetus of multiple gestations. Guttmacher, Alan F., & Kohl, S. G. *Obst. & Gynecol.*, 12:5, Nov., 1958. / The observed embryology of twins, other multiples. Corner, G. W. *Am. J. Obst. & Gynecol.*, 70:5, Nov., 1955. / Diagnosis of twin pregnancy. Novotny, C. A., et al. *J.A.M.A.*, Oct. 17, 1959. / Fetal electrocardiography. Larks, S. D., & Dasgupta, K. *Am. Heart J.*, 56:5, Nov., 1958. / Monoamniotic twin pregnancy. Timmons, J. D., & De Alvarez, R. R. *Am. J. Obst. Gynecol.*, 86:7, 1963.

TWIN-TYPE TESTS. Determination of zygosity of twins. Walker, Norma F. *Acta Genet.*, 7:1, 1957. / Smith, S. M., & Penrose, L. S. *Ann. Hum. Gen.*, 19–273, 1955. / Bulmer, M. G. *Ann. Hum. Gen.*, 22:4, 1958. / Diagnosis of zygosity by fingerprints. Slater, E. *Acta Psychiat. Scand.*, 39:1, 1963.

SITUS INVERSUS IN TWINS. Torgersen, Johan, *Am. J. Hum. Gen.*, 2:4, Dec., 1950. / Brandt, H., & Revaz, C. *J. Genet. Humaine* (Geneva), 7:1–2, 1958.

HANDEDNESS IN TWINS. See refs. on "Handedness," under "Personality," Chap. 37.

SIAMESE TWINS. ¶ Original pair: Daniels, W. B. *Med. Ann.*, D.C. (Washington), 32:7, 1963. ¶ —And their descendants: Taylor, S. S. *Trans. Med. Soc.* London, 79, 1963. (Same publication: Surgical separation of Siamese twins. O'Connell, J. J. A.) / Conjoined twins, zygosity, asymmetries. Walker, Norma F. *Acta Gen. Med. et Gemel.*, 1:2, May, 1962. Also, conjoined twins: Aird, L. *Brit. Med. J.*, 1, p. 1313, 1959.

DIONNE QUINTUPLETS. ¶ Birth of: Dafoe, Allan R. *J.A.M.A.*, 103, Sept. 1, 1934. ¶ *The Five Sisters*. Blatz, W. E. Wm. Morrow (N.Y.), 1938. / *Biological Study of the Dionne Quintuplets.* MacArthur, J. W., & Ford, Norma. U. of Toronto Press, 1937. / Genetics of quintuplets. MacArthur, J. W., & Dafoe, A. R. *J. Hered.*, Sept., 1939. / Taste reactions of Dionne quints. Ford, Norma, & Mason, A. D. *J. Hered.*, 32:10, Oct., 1941. / Dionnes in Maturity: *See* refs. under "Personality," Chap. 37.

OTHER QUINTUPLETS. ¶ Fischers, U.S.A., *Sat. Eve. Post,* May 2, Sept. 26, 1964. ¶ Prietos, Venezuela, *Sat. Eve. Post,* Oct. 3, 1964.

INCIDENCES, RATIOS OF TWINS AND TWIN-TYPES. ¶ Bio-social effects on twinning incidences. Scheinfeld, Amram. I (With Schachter, Joseph): Intergroup and generation differences, U.S. II. The world situation. *Proc. 2nd. Intl. Congr. Hum. Gen.*, Inst. Gregor Mendel (Rome), 1963. / Secular changes in rates of multiple births. U.S. Jeanneret, Olivier, & MacMahon, Brian. *Am. J. Hum. Gen.*, 14:4, Dec., 1962. / Twinning rate, Europe, Africa. Bulmner, M. G. *Ann. Hum. Gen.*, 24:2, May, 1960. *Also, Ann. Hum. Gen.* 22:2, 1958; *Brit. Med. J.*, Jan. 3, 1959; *Ann. Hum. Gen.*, 23:4, 1959. (latter on parental age and twinning). / Twin births, Italy. Karn, M. N. *Acta Gen. Med. et Gemel.*, Jan., 1954. / Twin Births, Italy. M. N. Karn, *Acta Gen. Med. et Gemel.*, May, 1953. / Japan, frequency multiple births. Inouye, Eiji. *Am. J. Hum. Gen.*, 9:4, Dec., 1957. / Variations in human twinning rate. Eriksson, A. *Acta Genet.*, 12:3–4, 1962. / Variations by race, socio-economic status. Lilienfeld, A. M., Pasamanick, Benj. *Am. J. Hum. Gen.*, 7:2, June, 1955.

MORTALITY OF TWINS. U. S. Vital Statist., 1960, Vol. II, Sec. 4, pp. 4–8. / Twin mortality during first year. Fiumara, A. *Acta Genet. Med.*, 12:3, 1963. / Influence of multiple births on perinatal loss. Donnelly, Madelene M. *Am. J. Obst. & Gyn.*, 72:5, Nov., 1956. / DZ twin survival, in relation to ABO groups. Osborne, R. H., & De George, F. V. *Am. J. Hum. Gen.*, 9:4, Dec., 1957.

INHERITANCE IN TWINNING. The familial incidence of twinning. Bulmer, M. G. *Ann. Hum. Gen.*, 24:1, Apr., 1960. *Also:* Repeat frequency, twinning. *Ann. Hum. Gen.*, 23:1, 1958. / Genetics of twinning. McArthur, Norma. Austral. Nat. U., Socl. Sci. Monogr., 1–1, 1953. / Twinning in twin pedigrees. Waterhouse, J. A. H. *Brit. J. Soc. Med.*, 4, 1950. / Hereditary and environmental factors in twinning. Torgersen, J. *Am. J. Phys. Anthrop.*, 9:4, Dec., 1951. / Tendency to twin birth (in German). Dahlberg, Gunnar. *Act. Gen. Med. et Gemel.*, Jan., 1952. / Parental age, sibship size, and twins. Stewart, A., & Barber, R. *Ann. Hum. Gen.*, 27:1, 1963. / Selection for twinning in Angus-Aberdeen cattle. Mechling, E. H. II, & Carter, R. C. *J. of Hered.*, 55:2, 1964.

"HELLIN'S LAW" AND SUPERTWIN INCIDENCES. Mathematical relations among plural births. Allen, Gordon, & Firschein, L. I. *Am. J. Hum. Gen.*, 9:3, Sept., 1957. / Method for estimation type frequencies in triplets, quadruplets. Gordon Allen. *Am. J. Hum. Gen.*, 12:2, June, 1960. / Quintuplet and sextuplet births in the U.S. Nichols, J. B. *Acta Genet. & Gemel.*, May, 1954.

TWIN STUDIES, METHODOLOGY. Waardenberg, P. J. Proc. 1st. Intl. Congr. Hum. Gen., 1957, Part 4,

p. 10. / Allen, Gordon. *Acta Gen. Med. et Gemel.*, 4:1, May, 1955. / Falkner, Frank. *Eug. Qu.*, June, 1957. / Waterhouse, J. A. H. Chap. 4 in: *Clinical Genetics*, Sorsby, Arnold, Ed. Butterworth (London), 1953. / Price, Bronson. *Am. J. Hum. Gen.*, 2, p. 293, 1950.

CHEMICAL INDIVIDUALITY

(*Chapter* 19)

GENERAL. *Human Biochemical Genetics*. Harris, H. Cambridge U. Press, 1959. / Biochemical variability. Cohen, Seymour S. *Science*, 139, Mar. 15, 1963. / Principles of biochemical genetics (in German). Kuhnau, J. *Internist* (Berlin), 4:9, 1963. / *Biochemical Individuality*. Williams, Roger J. Wiley, 1963. / Genetic variations in human biochemical traits. Sutton, H. E., *et al.* In: The hereditary abilities study. *Am. J. Hum. Gen.*, 14:1, Mar., 1962. / *The Chemistry of Heredity*. Zemenhof, Stephen. Columbia U. Press, 1959.

GENETIC CODE. Crick, F. H. C. *Science*, 139, Feb. 8, 1963. / Watson, J. D. *Science*, 140, Apr. 5, 1963. / Langridge, R., & Gomatos, P. J. *Science*, 141, Aug. 23, 1963. / *Molecular Genetics* (Part I). Taylor, H. H., Ed. Academic Press (N.Y.), 1963. ❡ *The Coil of Life*. Moore, Ruth. Knopf, 1961. ❡ The genetic code. Nirenberg, M. W. *Sci. Amer.*, Mar., 1963. ❡ Messenger RNA. Hurwitz, J., & Furth, J. J. *Sci. Amer.*, Feb., 1962.

TRANSPLANTS OF ORGANS. Genetics of transplantation in humans. Rogers, Blair O. *Dis. Nerv. Syst.*, Monogr. Suppl. 24:4, Apr., 1963. / Tissue homotransplantation. Sympos., Fifth Conf., N.Y. Acad. Sci., 99:3, 1962. ❡ Kidney transplants. Merrill, John P. *Sci. Amer.*, Oct., 1959. *Also*: Shackman, R., et al. *Brit. J. Urol.*, 35:3, 1963.

ANTIBODIES, ANTIGENS. ❡ How cells attack antigens. Speirs, Robert S. *Sci. Amer.*, Feb., 1964. ❡ The mechanism of immunity. Burnet, Macfarlane. *Sci. Amer.*, Jan., 1961. / Auto-immunity. MacKay, Ian R. *Postgrad. Med.*, Jan., 1964.

TWIN STUDIES, BIOCHEMICAL. Odours, discrimination of by dogs. Kalmus, H. *Brit. J. Anim. Behav.*, 3, p. 25, 1955. / Serum phosphateses variations, MZ-DZ twins. Arfors, K. E., et al. *Acta Genet.*, 13:2, 1963. / Serum lipids, adult twins. Osborne, R. H., & Adlersberg, David. *Science*, 127, May, 1958. / D-phenylalanine excretion, MZ-DZ twins. Gartler, S. M., & Tashian, R. E. *Science*, 126, July 12, 1957. / Very unequal identical twins. Sydow, G. von, & Rinne, A. *Acta Paediat.* (Uppsala), 47:2, 1958.

GLANDS, HORMONES, *Human Endocrinology*. *Kupperman*, Herbert S. (3 vols.) F. A. Davis, 1963. / *Sex and Internal Secretions*. Young, W. C., ed. 3rd. ed. (2 vols.) Williams & Wilkins, 1961. / Thyroid physiology. Sympos.: N.Y. Acad. Sci., 86:2, 1960. *Also*, Thyroid: Wilkins, Lawson. *Sci. Amer.*, Mar., 1960. / Hyperthyroidism: Engbring, N. H., et al. *Postgrad. Med.*, Jan., 1961. / Hypothyroidism: Lowrey, G. H., et al. *Postgrad. Med.*, Jan., 1960. / Parathyroid: Pasmussen, Howard. *Sci. Amer.*, Apr., 1961. / Thymus. Robert A. Good et al. *World Wide Abstr.*, 7:3, Mar. 1964. Miller, J. F. A. *Science*, 144, June 26, 1944. *Also:* Macfarlane Burnet. *Sci. Amer.*, Nov. 162. / Endocrine pharmacology. Robert Gaunt et al. *Science*, 133, Mar. 3, 1961.

DRUGS, INDIVIDUAL REACTIONS TO. Axelrod, Julius. Pharmacodynamics of human disease. *Postgrad. Med.*, Oct., 1963.

WAYWARD GENES

(*Chapter* 20)

General. *Introduction to Medical Genetics.* Roberts, J. A. Fraser. 3rd. ed. Oxford U. Press, 1963. / *Outline of Human Genetics.* 2nd. ed. Penrose, L. S. Heineman (London), 1963. / *Genetics for the Clinician.* Clarke, C. A. Oxford U. Press, 1962. / *Human Genetics.* Li, C. C. McGraw-Hill (Blakiston), 1961. / *Medical Genetics,* 1958–1960. *Also: Medical Genetics,* 1962. *J. Chron. Dis.*, 16, June, 1963. McKusick, V. A. C. V. Mosby, 1961. / *Human Genetics* (in German: *Genetik des Menschen*). Verschuer, O. Urban & Schwarzenberg (Munich; Berlin), 1959. / Genetic perspectives in disease resistance and susceptibility (Symposium). *Annls. N.Y. Acad. Sci.*, 91:3, June 7, 1961. / *Progress in Medical Genetics* (Symposium). Steinberg, A. G., & Bearn, I., Eds. Vol. I, 1961, Vol. II, 1962, and continuing. Grune & Stratton. / *Clinical Genetics* (Symposium). Sorsby, Arnold, Ed. Mosby, 1953. / Fifty years of medical genetics. Snyder, Laurence H. *Science,* Jan. 2, 1962. / Medical genetics and the practicing physician. Goodman, R. M. GP (Kansas), 27:2, 1963.

Methodology in human genetics. Sympos.: Burdette, W. J., Ed. Holden-Day (San Francisco), 1962. / Taking the family history. Fraser, F. C. *Am. J. Med.*, 34:5, 1963. / *Mathematical tables in human genetics.* Maynard-Smith, S., et al. Little, Brown, 1962. / Hardy-Weinberg law (and other methodology discussions) in: *Principles of Human Genetics.* Stern, Curt. 2nd. ed. W. H. Freeman, 1960. / The detection of linkage. Morton, N. E. *Am. J. Hum. Gen.*, 7–277, 1955; 8, p. 80, 1956; 9, p. 55, 1957.

Penetrance and expression. Sang, J. H. *J. Hered.*, 54:4, July–Aug., 1963. / Snyder, L. H., & David, P. R. In: *Clinical Genetics.* Sorsby, A., Ed. Mosby, 1953.

Syndromes. *Encyclopedia of Medical Syndromes.* Durham, Robert H. Hoeber (Harper), 1962. / Thirty important syndromes. World-Wide Abst., Nov.–Dec., 1962.

Chromosome abnormalities. In abortions, stillbirths. Carr, D. H. *Lancet,* 11, p. 7308, Sept. 21, 1963. / Recent developments in human cytogenetics. Miller, O. J., Cooper, H. L., & Hirschhorn, K. *Eug. Qu.*, Mar., 1961. / Trisomies, Nos. 18 & 21 groups. Heinrichs, E. H., et al. *Lancet,* 2, p. 7305, 1963. / Leukemia and the Philadelphia chromosomes. Benson, E. S. *Postgrad. Med.*, Dec., 1961. / Abnormal autosomes. Ford, C. E. In: Proceedings, *Second International Conference on Congenital Malformations.* Lippincott, in press. / *Genetic mosaics.* Hannah-Alava, Aloha. *Sci. Amer.*, May, 1960. / Opposite-sexed MZ twins. Turpin, R. et al. *Acad. Sci.* (Paris), 252, p. 2945, 1962. / *See also* refs, on sex-chromosome abnormalities, under "Sex Abnormalities," Chap. 27.

THE BIG KILLERS

(*Chapter* 21)

Heart and Blood Vessels

General. Genetics of cardiovascular diseases. McKusick, V. A. *Ann. Int. Med.*, 49:3, 1958. / Heredity and heart disease. Böök, Jan A. *Am. J. Pub. Health*, Mar., 1960. / Cardiovascular system, diseases. Herndon, C. Nash. Chap. 23 in: *Clinical Genetics*, Arnold Sorsby, Ed. Mosby, 1953. / Coronary artery disease, epidemiological aspects. Sympos.: N.Y. Acad. Sci., 97:4, 1962.

HYPERTENSION, INHERITANCE OF. Ostfeld, A. M., & Paul, O. *Lancet*, 1, p. 7281, 1963. / Essential hypertension, genetic factor. Pickering, G. W. *Ann. Int. Med.*, 43:3, Sept., 1955. Also: *Acta Med. Scand.*, 154, Suppl. 312, 1956. / Hypertension, twin study. Hines, E. A., Jr., et al. *Tr. Ass. Am. Phys.*, 70, p. 282, 1957.

ARTERIOSCLEROSIS, ATHEROSCLEROSIS. Familial hypercholesterolemia, inheritance. Hirschhorn, K., & Wilkinson, C. F., Jr. *Am. J. Med.*, 24, p. 60, 1959. / Atherosclerosis, problems in study. Spain, David M. *Annls. N.Y. Acad. Sci.*, 84:17, Dec., 1960. / Serum cholesterol, heredity, environment. Schaefer, L. E., et al. *Circulation*, 17, 1958. / Strain and sex differences, mice. Bruell, J. H., et al. *Science*, 135, Mar. 23, 1962.

RHEUMATIC FEVER. Advances in: Wilson, May G. Hoeber (Harper & Row), 1962. / Heredity in: Stevenson, A. C., & Cheeseman, E. A. *Ann. Hum. Gen.*, 21, p. 139, 1956. / Heredity and environmental factors in. Diamond, E. F. *Pediatrics*, 19, p. 908, 1957.

CONGENITAL HEART DISEASE. Polani, P. E., & Campbell, M. *Ann. Hum. Gen.*, 19, p. 209, 1955.

Cancer

GENERAL. Genetics of tumours. Von Verschuer, O. *Internist* (Berlin), 4:9, 1963. / *Genetics and Cancer* (Sympos.). U. of Texas, 1959. / Studies on human cancer families. Oliver, C. P. *Ann. N.Y. Acad. Sci.*, 71:6, 1958. / Viruses, Nucleic Acids, and Cancer (Sympos., U. of Texas). Cumley, R. W., Ed. Williams & Wilkins, 1963. / Cancer in infancy and childhood. Dargeon, Harold W. *Ann. N. Y. Acad. Sci.*, 114:2. Apr. 2, 1964.

BREAST CANCER. Macklin, Madge T. *J. Nat. Canc. Inst.*, May, 1959. / *Cancer in Families*. Murphy, D. P., & Abbey, Helen. Harvard U. Press, 1959. / *Variables Related to Human Breast Cancer*. Anderson, V. E., et al. U. of Minnesota Press, 1958.

STOMACH AND INTESTINES. Macklin, Madge T. *J. Natl. Canc. Inst.*, 24:551, 1960. / Woolf, C. M. *Am. J. Hum. Gen.*, 10:1, 1958. / Graham, S., & Lilienfeld, A. M. *Cancer* (Phila.), 11:5, 1958.

INTESTINAL POLYPOSIS. McKusick, V. A. *J.A.M.A.*, 182:3, Oct. 20, 1962. / Colon, intestines, polyposis of. Calabro, J. J. *Am. J. Med.*, 33:2, 1963.

LEUKEMIA. Chromosomes studies in. Hungerford, D. A. *J. Natl. Canc. Inst.*, 27, p. 983, 1961. / Chromosomes in adult. Fitzgerald, P. H., et al. *J. Nat. Canc. Inst.*, 32, p. 395, 1964.

SKIN CANCER. Sunlight and (as related to individual coloring). Winkler, Malcolm. Rhode I. Med. J., 46:7, 1963. / *See also* refs. under "Skin: Birthmarks, malignant," Structural Defects, Chap. 23.

TWINS, CANCER IN. Harvald, Bent, & Hauge, Mogens. *J.A.M.A.*, Nov. 23, 1963. / Aging twins. Jarvik, Lissy F., & Falek, Arthur. *Am. J. Hum. Gen.*, 13:4, Dec., 1961. *Also: Cancer*, 15:5, Sept.–Oct., 1962.

Ethnic differences, cancer. Newill, V. A. *J. Nat. Canc. Inst.*, 26: 405, 1961.

Lung cancer and smoking. (*Smoking and Health.* U.S. Dept. Health, Educ., Welf., U.S. Publ. Health Serv., Publ. 1103, 1964. / Report, Royal Collg. of Physicians. Pitman (London & N.Y.), 1962.

Rare cancers. Epiloia, neurofibromatosis, malignant birthmarks: *see* under "Skin," Structural Defects, Chap. 23. / Retinoblastoma, glioma: *See* under "Eyes," Functional, Chap. 24.

Diabetes

Diabetes mellitus. *Genetics and Constitutional Aspects of Diabetes Mellitus.* Nilson, Sven E. Almqvist & Wiksell (Stockholm), 1962. / The genetics of diabetes. Simpson, Nancy E. *Ann. Hum. Gen.*, 26:1, July, 1962. / Steinberg, A. G. *Ann. N.Y. Acad. Sci.*, 82, p. 197, 1959. *Also:* Diabetes (N.Y.), 7:3, 1958. / Diabetes mellitus—a "thrifty genotype"? Neel, J. V. *Am. J. Hum. Gen.*, 14:4, Dec., 1962.

Diabetes insipidus. Wenzl, J. E., et al. *Proc. Mayo Clin.*, 36, p. 543, 1961. / Forssman, H. *Acta Med. Scandinav.*, 121, Suppl. 159:1, 1956.

Other Major Conditions

Tuberculosis. Hereditary disposition. Diehl, K. (Review of papers.) *Internist* (Berlin), 4:9, 1963. / Current status. Sympos.: N.Y. Acad. Sci., 106:1, 1963.

Goiter. Fraser, G. R. *Ann. Hum. Gen.*, 26:4, May, 1963. / Hashimoto's disease (note on). *J. Hered.*, Mar.–Apr., 1963.

Kidney disease, hereditary chronic. J. B. Graham. *Am. J. Hum. Gen.*, 11:4, Dec., 1959. / Polycystic. P. M. Lundin, I. Olow. *Acta Paediat.* (Uppsala), 50:185, 1961. / Hereditary nephritis. P. J. Mulrow et al. *Am. J. Med.*, 35/6, p. 737, 1963. / Kidney stones. M. G. McGeown. *Irish J. Med. Sci.*, 6:451, 1963.

SEX DIFFERENCES IN DISEASE, DEFECT

(*Chapter* 22)

General. (The mortality of men and women. Scheinfeld, Amram. *Sci. Amer.*, Feb., 1958. (Scheinfeld. *Women and Men,* Chaps. 14–16. Harcourt, Brace, 1944. / Incidences, causes and significance of sex differences in inherited diseases. (In German: "Über haufigkert . . .") Franke, H., & Schröder, J. *Deutsche Medizinische Wochenscrift,* 84:14, Apr. 3, 1959.

Sex-linked diseases. *The X Chromosome of Man.* McKusick, V. A. Am. Inst. Biolog. Sci., 1964.

"Inactivated x," Lyon hypothesis. Sex chromatin and gene action in the mammalian X-chromosome. Lyon, Mary F. *Am. J. Hum. Gen.*, 14:2, June, 1962. / Discussions of: Reed, T. E., et al. *Lancet,* 2:7305, 7314, 7316, 7317, 1963. *Also:* De Mars, R. *Science,* 141:3581, 1963.

Y-linked inheritance. Gates, R. R., et al. *Am. J. Hum. Gen.*, 14:4, Dec., 1962. / In hairy ears, Israel. Slatis, H. M., & Apelbaum, Azay. *Am. J. Hum. Gen.*, Mar., 1963.

Hemophilia. ("The Royal Disease" (and British Royal family). Iltis, Hugo. *J. Hered.*, 39:4, Apr., 1948. / Principal refs. on Hemophilia: *See* under "Blood," Chap. 28.

Color-blindness. *Individual Differences in Colour Vision.* Pickford, R. W. Routledge & Kegan Paul (London), 1961. / Genetics of color vision. Kalmus, H. *Heredity,* 14, May, 1960. / Color defect and color theory. Graham, C. H., & Hsia, Yun. *Science,* 127, Mar. 28, 1958. / Natural selection and colour blindness. Pickford, R. W. *Eug. Rev.,* 55:2, July, 1963. / *See also* Richard H. Post, under Blood Groups, Races: I (Chap. 42).

Baldness. General aspects. Hamilton, James B. In *Hair Growth and Regeneration* (Symposium). N.Y. Acad. Sci., 83:3, Oct. 7, 1959. *Also,* same symposium: Histological and chemical changes in baldness, Ellis, R. A.; Alopecia areata, clinical aspects, Lubowe, Irwin I. / Age, sex and genetic factors in hair growth. In *The Biology of Hair Growth* (Symposium). Montagna, W. A., & Ellis, R. A., eds. Academic Press, 1958. / Women, baldness in. Maguire, H. C., & Kligman, A. M. *Geriatrics,* 184, 1963.

STRUCTURAL DEFECTS

(*Chapter* 23)

Congenital malformations, general. *Second International Conference on Congenital Malformations.* Proceedings. Lippincott, in press. / *Birth Defects.* (Sympos.), Fishbein, Morris, Ed. Nat'l. Foundation. Lippincott, 1963.

Dwarfism. Hereditary. Grebe, H. (In German.) *Med. Mschr.,* 17:11, 1963. (Summary, *Hum. Genet. Abstr.,* Mar., 1963, No. 644.) / Achondroplasia and heredity. Rothenberger, D. J. *J.A.M.A.,* 171:1: 2278, Dec. 19, 1959. / In Japan. Neel, J. V., *Jap. J. Hum. Gen.,* 4:165, 1959. / *Bird-Headed Dwarfs.* Seckel, Helmut P. Chas. Thomas, 1960. / Hormonal influences on skeletal growth. Wilkins, L. *Ann. N.Y. Acad. Sci.,* 60:763, 1955. / Adult midgets, size and proportion. Dupertuis, C. W. *Am. J. Phys. Anthrop.,* 3:2, 1945. / Dwarfism in cattle. Gregory, P. W., & Carroll, F. D. *J. Hered.,* May–June, 1956. Dwarfism in fowl. Hutt, F. B. *J. Hered.,* Sept.–Oct., 1959.

Pygmies. North Burma pygmies. Mya-Tu, M. *Nature,* 195, p. 4837, 1962. / "True pygmies found." (New Guinea.) *Sci. News Let.,* Oct. 13, 1956. ¶ *Madami: My Eight Years with the Congo Pygmies.* Putnam, Anne E., with Keller, Allan. Prentice-Hall, 1954.

Gigantism. ¶ Robert Wadlow, the "Alton giant." *Time* Mag., Mar. 1, 1937. ¶ Death of Robert Wadlow: N.Y. *Times* (Sun. science sect., Kaempffert, W.), July 21, 1940. ¶ "Young giant of Japan." *Time* Mag., July 14, 1958, p. 49.

Obesity. Familial aspects. Childs, B. *Pediatrics,* 20:3, 1957. / Etiology and pathogenesis of. Mayer, Jean. *Postgrad. Med.,* May, 1959. / Caloric intake in relation to physique in children. Peckos, Penelope S. *Science,* 117, June 5, 1953. / Obesity in children. Illingworth, R. S. *J. of Pediatr.,* 53, July, 1958. / Mice, obesity in. Falconer, D. S., & Isaacson, J. H. *J. Hered.,* July–Aug., 1963.

Connective tissue. *Heritable Disorders of Connective Tissue,* 2nd. ed. McKusick, Victor A. C. V. Mosby, 1960. *Also:* Heredity and diseases of connective tissue. *Annls. N.Y. Acad. Sci.,* 86:4, June 30, 1960. (Includes discussions of: Marfan's syndrome; Ehler's Danlos syndrome; osteogenesis imperfecta; pseudoxanthoma elasticum;

Hurler syndrome [gargoylism]; osteoarthritis.) / Marfan's syndrome. Bowers, D. *Canad. Med. Ass. J.*, 89:8, 1963. / In Abraham Lincoln? Schwartz, Harold. *J.A.-M.A.*, Feb. 15, 1964. / Osteogenesis imperfecta. Smars, Gunnar, with Berfenstam, Ragnar. *Svenska Bokforlaget* (Stockholm), 1961.

SKELETAL. Spina bifida and anencephalus. Milham, Samuel, Jr. *Science*, 138, Nov. 2, 1962. / Chondrodysplasias. Hobaek, Andreas. Oslo U. Press, 1961. / Chondrodystrophia congenita. Allansmith, M., & Senz, E. *A.M.A. J. Dis. Child.*, 100, p. 109, 1960. (*See also* Dwarfism, above.)

HANDS AND FEET. *Congenital Anomalies of the Hand.* Barsky, Arthur J. Chas. C. Thomas, 1958. / Polydactyly. Neel, J. V., & Rusk, M. L. *Am. J. Hum. Gen.*, 15:3, Sept., 1963. / Brachydactyly. Haws, D. V. *Ann. Hum. Gen.*, 26:3, Feb., 1963. / Syndactyly (in Indian kindred). Malhotra, K. C., Rife, D. C. *J. Hered.*, 54:5, Sept.–Oct., 1963. / Clubfoot, congenital. Orofino, C. F. *Acta Orthop. Scand.*, 29:1, 1959. / Clubbing, fingers and toes. Talbott, J. H., & Montgomery, W. R. *Arch. Int. Med.*, 92, p. 697, 1953. / Congenital amputations. Freire-Maia, N., et al. *Acta Gen. et Statist. Med.*, 9:1, 1959.

FACE, MOUTH. *Congenital Anomalies of the Face.* (Sympos.) Pruzansky, Samuel, Ed. Chas. C. Thomas, 1961. / Cleft lip, palate. Woolf, C. M., et al. *Am. J. Hum. Gen.*, 15:2, June, 1963. / Etiology of cleft lip and palate: Fraser, F. C. *Acta Genet. et Statist. Med.*, 5, p. 538, 1955./Harelip and cleft palate. Kobayasi, Y. *Jap. J. Hum. Gen.*, 3:73, 1958.

TEETH. *Genetics and Dental Health.* Witkop, Carl J., Jr., Ed. Blakiston, 1963. / Microdontia, hereditary. Steinberg, A. G., et al. *J. Den. Res.*, 40, 1961. / Heritability in dental caries. Goodman, H. O., et al. *Am. J. Hum. Gen.*, 11:3, Sept., 1959.

SKIN. *Clinical Genodermatology.* Butterworth, Th., & Strean, L. P. Williams & Wilkins, 1962. Uncommon congenital abnormalities of the skin. Newcomer, V. D., et al. *Pediat. Clin. N. Amer.*, 701, 1956. / Birthmarks, malignant: Epiloia. Stevenson, A. C., & Fisher, O. D. *Brit. J. Prev. Soc. Med.*, 10, p. 134, 1956. / Malignant freckles (xeroderma pigmentosum). Giller, H., & Kaufmann, W. C. *A.M.A. Arch. Ophthal.*, 62, p. 130, 1959. / Neurofibromatosis ("coffee-colored spots"). Crowe, F. W., & Schull, W. J. Chas. C. Thomas, 1956. / The pigmented mole. Pack, G. T., & Davis, Jeff. *Postgrad. Med.*, Mar., 1960. *Also*: Moles, melanomas, related to race, complexion. Fifth Intl. Pigment Cell Conf., Sloan-Kettering Canc. Cent., 1961. / Incidence of pigmented nevi. Allyn, B. *J.A.M.A.*, 186, p. 890, Dec. 7, 1963. / Cysts, hereditary. Witcop, C. J., & Gorlin, R. J. *Arch. Dermat.*, 84, p. 762, 1961. / Ehlers-Danlos syndrome ("rubber skin"). *See* McKusick above, under "Connective Tissue." / Ichthyosis. Curth, H. O., & Macklin, M. T. *Am. J. Hum. Gen.*, 6:4, Dec., 1954. Psoriasis. Ward, J. H. & Stephens, F. E. *Arch. Dermat.*, 84, p. 589, 1961. / Genetic aspects of Psoriasis: Aschner, B., et al. *Acta Genet. et Statist. Med.*, 7, p. 197, 1957. / Sympos., N.Y. Acad. Sci., 73:5, 1958. / Leprosy. Belknap, H. R., & Hayes, W. G.

Bull. Tulane Med. Fac., 19, p. 236, 1960. / Xanthomatosis. Jones, E. G. *Brit. J. Clin. Pract.*, 12:3, 1958.

ALBINISM. In: Congenital and hereditary disturbances of pigmentation. Lerner, A. B. & M. R. *Bibl. Paediatr.*, Basel, 66, 1958. / In Negroes: Massie, R. W., Hartmann, R. C. *Am. J. Hum. Gen.*, 9:2, June, 1957. / In Hopi Indians: Woolf, C. M., & Grant, R. B. *Am. J. Hum. Gen.*, 14:4, Dec., 1962. / In Caribe-Cuna Indians: Keeler, Clyde E. *J. Hered.*, May–June, 1964. / Partial albinism: Piebaldness. Jahr, H. M., & McIntire, M. S. *A.M.A. J. Dis. Child.*, 88:4, 1954. / White forelock. Campbell, B., & Swift, S. *J.A.M.A.*, 181:13, 1962.

MONILETHRIX ("beady hair"). Summerly, R., Donaldson, E. M. *Brit. J. Derm.*, 74:11, 1962. *Also*: Deraemaeker, R. *Am. J. Hum. Gen.*, 9:3, 1957.

FUNCTIONAL DEFECTS

(*Chapter* 24)

Eyes

GENERAL. Genetics and Ophthalmology. Waardenburg, P. J., Franceschetti, A., & Klein, D. Chas. C. Thomas, 1961, 1963. / Mutational origin of hereditary eye diseases. Penrose, L. S. *Bibl. Ophth.*, Basel, 47, 1957. / Heredity counseling in eye disease. Cuendet, Jean-François. *Eug. Qu.*, Sept., 1957. / Prognostication and counseling. Falls, H. F. *Tr. Am. Acad. Ophth. Otolar.*, 60:4, 1956.

CATARACT. Statistical, genetic aspects. Sorsby, Arnold. *Exp. Eye Res.*, 1:4, 1962. / Congenital. Wilson, W. A. *Arch. Ophthal.*, 67:2, 1962.

GLAUCOMA. Simple. Havener, W. H. *Ohio Med. J.*, 56, p. 824, 1960. *Also*: Matthew, W. B., et al. *J. Indiana State Med. Ass.*, 50, p. 1002, Aug., 1957. / Recessive juvenile. Beiguelman, B., & Prado, D. *J. Genet. Hum.*, 12:1–2, 1963. / Infantile. (In French.) Delmarcelle, Y. *J. Genet. Humaine*, 6, 1957. / Pseudoglaucoma, dominant. Sandvig, K. *Acta Ophth.* 39:30, 1961.

COLOR-BLINDNESS. *See* under refs. for Sex Differences, Chap. 22.

RETINOBLASTOMA. Macklin, Madge T. *Am. J. Hum. Gen.*, 12:1, Mar., 1960. *Also*, Manchester, P. T., Jr. *A.M.A. Arch. Ophth.*, 65, p. 546, 1961. / Management of: Reese, A. B., & Ellsworth, R. M. *Ann. N.Y. Acad. Sci.*, 114:2, Apr. 2, 1964.

RETINITIS PIGMENTOSA. And deafness. Hallgren, B. *Acta Psychiat. Neurol. Scand., Suppl.*, 138:34, 1959.

LINDAU'S DISEASE. Bird, A. V., & Mendelow, H. *Brit. J. Surg.*, 47, Sept., 1959. Nicol, A. A. *Ann. Hum. Gen.*, 22:1, 1957.

OCULAR ALBINISM. Gillespie, F. D. *Arch. Ophth.*, 66, p. 774, 1961.

NYSTAGMUS. Kestenbaum, A. *Bibl. Ophth.*, Basel, 49, 1957.

STRABISMUS. Parks, M. M. *A.M.A. Arch. Ophth.*, 60:1, 1958.

DYSLEXIA (reading disability). Sympos.: Money, John, Ed. Johns Hopkins Press, 1962. / Familial reading disability. Drew, A. L. *U. Michigan M. Bull.*, 21:8, 1955. / Specifix dyslexia in Helsinki. Gripdenberg, U. In: *Proc. 2nd. Intl. Congr. Hum. Gen.*, 1961. / Mirror reading (psychological aspects). Park G. E. *J. Pediatr.*, Jan., 1953.

ANIRIDIA. Shaw, Margery, W., et al. *Am. J. Hum. Gen.*, 12:4. Dec.. 1960.

MICROPHTHALMOS. Gill, E. G., & Harris, R. B. *Virginia Med. Monthl.*, 86, p. 33, 1959.

ANOPHTHALMIA, BILATERAL. Pinkerton, O. D. A.M.A. *Arch. Ophth.*, 63, May, 1960. / Inheritance of, in mice. Barber, A. M., et al. *Am. J. Ophth.*, 48, Dec., 1959.

MACULAR DEGENERATION, PROGRESSIVE FAMILIAL. Vancea, P., & Tudor, E. *Ann. Oculist* (Paris), 193, Nov., 1960.

Ears and Hearing

GENERAL. *Family and Mental Problems in a Deaf Population* (Sympos.) Rainer, J. D., et al., Eds. Dept. Med. Genet., N.Y. State Psychiatric Inst., Columbia U., 1963. (Includes: Genetic aspects of early total deafness. Sank, Diane, pp. 28–81.) / Multiple anomalies in congenitally deaf children. Danish, J. M., et al. *Eug. Qu.*, 10:1, Mar., 1963. / Sex differences, deafness. Corson, J. F. *Arch. Otalaryngol.*, 77, Apr., 1963.

DEAF-MUTISM, HEREDITARY. Stevenson, A. C., & Cheeseman, E. A. *Ann. Hum. Gen.*, 20, p. 177, 1956. / In N. Ireland: Slatis, H. M. *Ann. Hum. Gen.*, 22:2, 1958. / Deaf-mutism, sex-linked. Richards, B. W. *Ann. Hum. Gen.*, 26:3, 1963.

OTOSCLEROSIS AND MÉNIÈRE DISEASE. Damon, A., et al. *Tr. Am. Acad. Ophth. Otolar.*, 59:4, 1955.

NERVE DEAFNESS, HEREDITARY. Dolowitz, D. A., & Stephens, F. E. *Tr. Am. Otol. Soc.*, 49, p. 290, 1961.

Other Sensory Disorders

SPEECH DEFECTS. Language-formation disability (tachyphemia, studies). Arnold, G. E. *Logos* (Bull. Natl. Inst. Speech Disord.), 3:1, Apr., 1960, and 3:2, Oct. 1960. / Stuttering. Karlin, I. W. *N.Y. State J. Med.*, 56:23, Dec. 1, 1956. ¶ *Stuttering in Children and Adults* (Sympos.). Johnson, Wendell, Ed. U. of Minnesota Press, 1956. / *Stuttering* (Symposium mainly on psychological factors). Eisenson, Jon, Ed. Harper, 1958. / Cluttering. Bakwin, Ruth & Harry. *J. Pediatr.*, Mar., 1952.

TASTE AND SMELL. *The Chemical Senses in Health and Disease*. Kalmus, H., & Hubbard, S. J. Chas. C. Thomas, 1960. / Taste blindness (PTC-tasting, etc.) Fischer, Roland, & Griffin, Frances. *J. Hered.*, July–Aug., 1960. / *See also* refs. on Taste and Smell, under "Social and Sensory Behavior," Chap. 36.

ALLERGY. Heredity and allergy. Rapaport, H. G. N.Y. State J. Med., 58:3, 1958. / Allergic predisposition and heredity (in French). Cuendet, J. F. *Arch. Julius Klaus-Stif.*, 31:3–4, 1956. / Allergy susceptibility, heredity. Parrot, J. L., Saindelle, A. *Rev. Franc. Etud. Clin. Biol.*, 8:6, 1963. (*Hum. Gen. Abstr.*, Feb., 1964, No. 450). / Genetics, significance for classifying allergies. Klunker, W., & Schnyder, I. W. *Int. Arch. Allergy*, 15, p. 360, 1959. / *The Allergic Child*. Speer, Frederic, Ed. Hoeber (Harper & Row), 1963.

MIGRAINE. Familial. Goodell, Helen, et al. A.M.A. *Arch. Neurol. Psychiat.*, 72, 1954. / Migraine headache (physiology, biochemistry). Ostfeld, A. M. *J.A.M.A.*, 174:9, Oct. 29, 1960.

COLD-SENSITIVITY. Rothman, S., & Griem, S. F. *Ill. Med. J.*, 123, June, 1963.

COLLAGEN DISEASES. Chemistry and Therapy. Nustadt, D. H. Chas. C. Thomas, 1963.

Chemical Disorders

GENERAL. *Inborn Errors of Metabolism.* Hsia, D. Y. Yearbook Publ. (Chicago), 1959. / *Garrod's "Inborn Errors of Metabolism"* (with suppl.). Harris, H. Oxford U. Press, 1963. / Human Biochemical Genetics. H. Harris. Cambridge U. Press, 1959. / *The Metabolic Basis of Inherited Disease* (Sympos.). Stanbury, J. B., et al., Eds. McGraw-Hill, 1960. / Metabolic diseases, man and animals. (Sympos.): N.Y. Acad. Sci., 104:2, 1963.

GOUT. McGee, R. R. *Southern Med. J.*, 53:6, 1960. / Lockie, L. M., & Cooper, R. G. *Geriatrics*, 15:7, 1960. / Duncan, H., & Dixon, A. S. *Qu. J. Med.*, 29:129, 1960.

URINARY DISORDERS. Sutton, H. Eldon, & Tashian, R. E. In Sympos. on hereditary metabolic diseases. *Metabolism*, 9:3, Mar., 1960. / Pentosuria. Khachadurian, A. K. *Am. J. Hum. Gen.*, 14:3, Sept., 1962. / Porphyria. Granic, S. *Trans. N.Y. Acad. Sci.*, II, 25:1, Nov., 1962. / Porphyria in S. Africa: Dean, G. *Leech* (Johannesburg), 32:4, 1962. / Maple-urine disease. Norton, P. M., et al. *Lancet*, 1:26, 1962. *Also*: Lonsdale, D., et al. *Am. J. Dis. Child.*, 106:3, 1963. / Phenylketonuria, Histidinuria: *See refs.* under Mental Defects, Chap. 26.

FAT-METABOLISM DISORDERS. Lipomatosis. Stephens, F. E., & Isaacson, Arnold. *J. Hered.*, Mar.–Apr., 1959. / Lipid Metabolism, inborn errors of. Zöllner, N. *Gastroenterologia*, Basel, 97:4, 1962. / Lipidoses, genetics of. Herndon, C. Nash. In: *Proc. Assn. for Res. in Nerv. & Mentl. Dis.* Williams & Wilkins, 1954. / Xanthomatosis. Wolman, M., et al. *Pediatrics*, 28:742, 1961. / Hyperlipemia. Boggs, J. D., et al. *N. Eng. J. Med.*, 257, Dec. 5, 1957.

GALACTOSEMIA. Walker, F. A., et al. *Ann. Hum. Gen.*, 25–287, 1962.

VITAMIN-D RESISTANT RICKETS. Dancaster, C. P., & Jackson, W. P. *Arch. Dis. Childh.*, 34, 1959. / Rickets, hereditary forms. Dent, C. E., & Harris, H. *J. Bone Surg.*, (Brit. vol.), 38–B:1, 1956.

GLYCOGEN STORAGE DISEASE. Illingworth, B. *Am. J. Clin. Nutr.*, 9:683, 1961.

WILSON'S DISEASE. Symposium. Walshe, J. M., & Cumings, J. N., eds. Chas. C. Thomas, 1961. / Metabolic aspects of: Simpson, J. F. *U. Mich. Med. Bull.*, 29:3, 1963. / Fingerprints in: Hodges, R. E., & Simon, J. R. *J. Lab. & Clin. Med.*, 60, Oct., 1962.

PORPHYRIAS. *The Porphyrias.* Geoffrey Dean. Pitman (London), 1963.

CYSTIC FIBROSIS. Incidence of: Steinberg, A. G., Brown, D. C. *Am. J. Hum. Gen.*, 12:416, Dec., 1960. / Pilot survey of: Kramm, Eliz. R., et al. *Am. J. Pub. Health*, 52:12, Dec., 1962. / Cystic fibrosis of the pancreas. Di Sant' Agnese, P. A. *Am. J. Med.*, 21:3, 1956. / In animals as well as man? Warwick, W. J. *Canad. J. Comp. Med.*, 27:4, 1963.

JOINT DISEASES. Heredity of: Stecher, R. M. *Acta Genet.*, Basel, 7:1, 1957. / Rheumatoid arthritis, genetic studies: Lawrence, J. S., & Ball, J. *Ann. Rheumat. Dis.*, Lond., 17:2, 1958. / Familial occurrence: Ziff, M., et al. *Arthrit. & Rheum.*, 1:5, 1958. / In Sjögren's syndrome: Bloch, K. J., et al. *Trans. Ass. Am. Physicns.*, 73, p. 166, 1960.

Nerve and Muscle Disorders

GENERAL. *Hereditable Disorders of Connective Tissue*. McKusick, Victor A. 2nd. ed. Mosby, 1959. / *Myopathies, Chemistry of Hereditary*. Dreyfus, Jean-Claude, & Schapira, Georges. Chas. C. Thomas, 1962. / Demyelinating diseases (Sympos.): *N.Y. Acad. Sci.* (in press), 1964.

ATAXIA, HEREDITARY. Schut, J. W., Böök, J. A. *Arch. Neurol. Psychiat.*, 70, p. 169, 1953. / Friedreich's ataxia. Powell, E. D. *Brit. Med. J.*, 1, p. 868, 1961. / Marie's cerebellar ataxia. Matson, G. A. et al. *Ann. Hum. Gen.*, 25–7, 1961.

MYASTHENIA GRAVIS. Foldes, F. F., & McNall, P. G. *J.A.M.A.*, 174, p. 418, 1960.

MULTIPLE SCLEROSIS. A *Biochemical Basis of*. Swank, Roy L. Chas. C. Thomas, 1961. / Multiple sclerosis in twins, families. Myrianthopoulos, N. C., Mackay, R. P. *Acta Gen. et Stat. Med.*, 10, p. 33, 1960. / Familial and conjugal. Schapira, K., et al. *Brain*, 86:2, 1963.

MUSCULAR DYSTROPHY. In Man and Animals. Sympos.: Bourne, G. H., & Golarz, M. N., Eds. S. Karger (Basel & New York), 1963. Including, Chap. 8, Genetics of: Morton, N. E., et al. / Progressive muscular dystrophy. Walton, J. N. *Postgrad Med.*, Jan., 1964. / Duchenne type: Kloepfer, H. W., & Talley, C. *Ann. Hum. Gen.*, London, 22:2, 1958.

MYOTONIAS, GENETICS OF. Becker, P. E. *Internist* (Berlin), 4:9, 1963. (Summary: *Hum. Genet. Abst.*, 2:3, Mar., 1964, No. 645.)

PARKINSONISM. Epidemiologic and genetic characteristics. Kurland, L. T., & Darrel, R. W. *Int. J. Neurol.*, 2:1, 1961.

PERIODIC PARALYSIS. Poskanzer, D. C., & Kerr, D. N. *Am. J. Med.*, 31–328, 1961. / McArdle, B. *Brit. M. Bull.*, 12:3, 1956.

SPASTIC PARAPLEGIA. Schwarz, G. A., & Liu, C. N. *A.M.A. Arch. Neurol. Psychiat.* 75:2, 1956.

PROGRESSIVE SPINAL MUSCULAR ATROPHY. Myrianthopoulos, N. C., & Brown, I. A. *Am. J. Hum. Gen.*, 6:4, Dec., 1954.

DYSAUTONOMIA, FAMILIAL. Harris, J. R., et al. *Pediatrics*, 16:6, 1955. / De Hirsch, Katrina, & Jansky, Jeannette J. *J. Speech Disord.*, 21, 1956.

EPILEPSY. Hereditary factors: Harvald, B. *Med. Clin. N. Amer.*, 42:2, 1958. / Heredity of: Koch, G. *Psychiat. Neurol. Neurochir.* (Amsterdam), 66:3, 1963. (Summary: *Hum. Gen. Abst.*, 1:15, No. 3406, Nov., 1963.) / Familial epilepsy. Eisner, Victor, et al. *J. Pediat.*, 56, Mar., 1960. / Brain waves in epilepsy. Lennox, W. G. *J.A.M.A.*, 146, June 9, 1951. / Myoclonus epilepsy. Wada, T., et al. *Folia Psychiat. Neurol.* Jap., 14:3, 1960. / Carrier detection: Sarlin, M. B., et al. *Acta Gen. Med. et Gemel.*, 9:4, Oct., 1960.

CEREBRAL PALSY, genetic factors. Blumel, J. *Texas J. Med.*, 54:4, 1958.

MENTAL DISEASES

(*Chapter* 25)

INCIDENCES, RESEARCH. Hospitalized mental illness in the U.S. Progr. in Health Services, Bull. Health Inform. Found., 9:8, Oct., 1960. ¶ Highlights of progress in mental health research. U.S. Publ. Health

Serv., 1961. / Soviet Union, psychiatric care and research in. Kline, N. S. N.Y. Acad. Sci., 84:4, 1960. / Epidemiology of mental disorder. Sympos.: Pasamanick, Benj., Ed. *Am. Assn. Adv. Sci.*, Vol. 60, Dec., 1959. / Social class and mental illness. Hollingshead, A. B., & Rogler, L. H. *Sociol. & Socl. Res.*, July, 1962. / Race differences in mental disease—misconceptions. Pasamanick, Benj. *Am. J. Orthopsychiat.*, 33, Jan., 1963. / Twin research and psychiatry. Essen-Möller, E. *Acta Psychiat. Scand.*, 39:1, 1963.

GENETIC FACTORS, GENERAL. *Heredity in Health and Mental Disorder*. Kallmann, F. J. Norton, 1953. *Also*, Chap. 8 in: *American Handbook of Psychiatry*. Basic Books, 1959. / Psychiatric genetics. Cowie, V., & Slater, E. *Recent Progr. Psychiatry*, 3:1, 1959. / *Psychotic and Neurotic Illnesses in Twins*. Slater, Eliot. Spec. Rep., Ser. No. 278, Med. Res. Counc., London. Her Maj. Station. Off., 1953. / Mental illness in childhood and heredity. Bender, Lauretta. *Eug. Qu.*, 10:1, Mar., 1963. / EEG and Behavior. (Sympos.): Glaser, G. H., Ed. Basic Books, 1963.

HUNTINGTON'S CHOREA. Sympos.: *Proc. Mayo Clin.* 30:16, Aug. 10, 1955. / In Michigan: Demography, genetics. Reed, T. E., & Chandler, J. H. *Am. J. Hum. Gen.*, 10:2, June, 1958. / Selection and mutation. Reed, T. E., & Neel, J. V. *Am. J. Hum. Gen.*, 11:2, June, 1959. / Clinical observations. Chandler, J. H., et al. *Neurology*, 10:148, 1960. / In childhood: Jervis, G. A. *Arch. Neurol.* (Chicago), 9:3, 1963. / In India. Chhuttani, P. N. *J. Ind. M. Ass.*, 29:4, 1957. / In S. Africa: Klintworth, G. K. *S. Afr. Med. J.*, 36:43, Oct. 27, 1962. / In Trinidad: Beaubrun, M. H. W. *Indian Med. J.*, 12:1, 1963.

PICK'S DISEASE. Schenk, V. W. D. *Ann. Hum. Genet.*, 23:4, Dec., 1959.

SCHIZOPHRENIA: GENERAL. The genetics of schizophrenia. Sjögren, T. *Proc. 2nd. Intl. Cong. for Psychiatry*, Vol. 1, Zurich, 1957. / A clinical and genetico-statistical study of schizophrenia, etc. Hallgren, B., & Sjögren, T. *Acta. Psychiat. et Neurol. Scand.*, Suppl. 140, Vol. 35, 1959. / Genetic factors in schizophrenia. Gregory, Ian. *Am. J. Psychiat.*, 116:11, May, 1960. / *The Etiology of Schizophrenia.* (Symposium). Jackson, Don D., Ed. Basic Books, 1960. / Schizophrenic behavior, biological aspects. Sympos.: N.Y. Acad. Sci., 96:1, Jan. 13, 1962.

SCHIZOPHRENIA: SPECIAL ASPECTS. Twin studies of schizophrenia. Rosenthal, David. *J. Psychiat. Res.*, 1:2, 1962. / Schizophrenia in MZ quadruplet set: The Genain Quadruplets. Rosenthal, David, Ed. Basic Books, 1964. *Chemistry, Metabolism and Treatment of Schizophrenia.* Smythies, J. R. Chas. C. Thomas, 1963. / Chemical differences in schizophrenics: Friedhoof, A. J., & Van Winkle, Elnora. *Nature*, 199, p. 203, 1963. *Also*: Fujita, S., & Ging, N. S. *Science*, 134, p. 1687, 1961. Finger and palm prints in. Beckman, L., & Norring, A. *Acta Genet.*, Basel, 13.2, 1963. / Fingerprints in: Raphael, T. & L. *J.A.M.A.*, 180, Apr. 21, 1962. / "Nailfold capillary patterns in." Maricq, H. R. *J. Nerv. Ment. Dis.*, 136, Mar., 1963. *Childhood Schizophrenia.* Golfarb, William. Harvard U. Press, 1961. / Genetic aspects of childhood schizophrenia: Kallmann, F. J., & Roth, Bernard. *Am. J. Psychiat.*, 112:8, Feb. 1956. / When the childhood schizophrenic grows up. Bender, Lauretta, & Freedman, A. M. *Am. J. Orthopsychiat.*, 27, 1957. / Childhood intellectual development of adult schizophrenics. Lane, E. A., & Albee, G. W. *J. Abn. & Socl. Psychol.*, 67:2, 1963.

MANIC-DEPRESSIVE PSYCHOSIS. Genetics of. Reed, S. C. Chap. 26 in *Counseling in Medical Genetics*, 2nd. ed. Saunders, 1963. / Clinical, social, genetic aspects of: Stenstedt, A. *Acta Psychiat. et Neurol. Scand.*, Suppl. 29, 1952. / Lithium treatment of manic states. Strömgren, Erik, & Schou, Mogens. *Postgrad. Med.*, Jan., 1963.

INVOLUTIONAL MELANCHOLIA. Stenstedt, A. *Acta Psychiat. et Neurolog. Scand.*, Suppl. 127, Vol. 34, 1959.

SENILE DEMENTIA. (1) Sjögren, T., Jacobson, G.; (2) Larsson, T., et al. *Acta Psychiat. Scand.*, Suppl. 167, Vol. 39, 1963. / *Mental illness among older Americans.* U.S. Senate, Specl. Com. on Aging. U.S. Govt. Print. Off., Sept. 8, 1961. / Problems of the mind in later life. Sympos.: *Geriatrics*, 11:4, Apr., 1956.

PSYCHOPATHIC PERSONALITY. Psychopathic personality disorders in childhood, adolescence. Bender, Lauretta. In symposium, *Arch. Crim. Psychodyn.*, 4, 1961. / Sociopathic reactions, etiology. Thorne, F. A. Am. J. Psychotherapy, 13:2, Apr., 1959. / Psychopathic personality and organic behavior disorders. Weil, A. P. *Comprehensive Psychiat.*, 2–83, Apr., 1961. / Report on childhood of Lee Oswald (alleged assassin of Pres. Kennedy): Dr. Renatus Hartogs quoted. N.Y. *Times*, Dec. 4, 1963.

MENTAL DEFECTS

(*Chapter* 26)

GENERAL. *Biology of Mental Defect.* Penrose, Lionel S. 2nd. ed. Grune & Stratton, 1963. / Psychological Problems in Mental Deficiency. Sarason, Seymour B. Harper, 1959. / Genetics of mental deficiency, patterns of discovery in. Allen, Gordon, *Am. J. Ment. Defic.*, 62:5, Mar., 1958. / Biochemical and endocrinological causes. Rundle, A. T. *Am. J. Ment. Defic.*, 67:1, July, 1962. / Neuropathology of certain forms. Windle, William F. *Science*, June 14, 1963. / Low-grade mental deficiency (and schizophrenia). In: A clinical and statistical study. Hallgren, B. & Sjögren, T. *Acta Psychiatr. et Neurolog. Scand.*, Suppl. 140, Vol. 35, 1959. / Etiology of mental subnormality in twins. Allen, Gordon, & Kallmann, F. J. In: *Expanding Goals of Genetics in Psychiatry.* Grune & Stratton, 1962. *See also Heredity in Health and Mental Disorder.* Kallmann, F. J. Norton, 1953. / The mentally defective twin. Berg, J. M., & Kirman, B. H. *Brit. Med. J.*, 1:1911, 1960. / Sex-linked mental defect. Dunn, H. G., et al. *Am. J. Ment. Defic.*, 67:6, 1963. / Sex differences in mental deficiency. Malzberg, Benj. *Am. J. Ment. Defic.*, 58:2, 1953.

MONGOLISM. Chromosome studies. Dekaban, A. S., et al. *Cytogenetics*, 2:2–3, 1963. / 2nd. Sympos. on Mental Subnormality. *Brit. Med. J.*, 1963:5325, 1963. / Familial mongolism. Shaw, Margery W. *Cytogenetics*, 1:3–4,

1962. *Also,* Macintyre, M. N., et al. *Am. J. Hum. Gen.,* 14:4, Dec., 1962. / Triple-X syndrome and. Breg, W. R., et al. *Am. J. Dis. Child.,* 104:5, Nov., 1962. / Maternal age and miscarriage in mothers of mongoloids. Cowie, V., and Slater, E. *Act. Genet.,* Basel, 13:1, 1963. / In twin sibships. Allen. Gordon, Kallmann, F. J. *Acta Genet. et Statist. Med.,* 7:2, 1957. / In Orientals. Wagner, Henry R. *Am. J. Dis. Child.,* 103, 1962. / Dermal configurations in diagnosis of: Walker, N. F. *Pediat. Clin. N. Amer.,* May, 1958. *Also, J. of Pediatr.,* 50:19, 1957. / Intelligence, development in infants with mongolism: Dameron, L. E. *Child Develpm.,* 34, p. 733, 1963. / Intelligence in adult mongoloids: Sternlight, Manny, & Wanderer, Z. W. *Am. J. Ment. Def.,* 67, p. 301, 1962. / Heritable aspects, Mongolism: Jarvik, Lissy F., et al. *Psycholog. Bull.,* 61:5, 1964.

KLINEFELTER'S SYNDROME, TURNER'S SYNDROME. In: Chromosomal aberrations in human disease. Hirschhorn, Kurt, & Cooper, Herbert L. *Am. J. of Med.,* 31:3, Sept., 1961. *Also:* Recent developments in human cytogenetics. Miller, O. J., Cooper, H. L., Hirschhorn, K. *Eug. Qu.,* Mar., 1961. / Klinefelter's syndrome among mental patients. Forssman, H., & Hamberg, G. *Lancet,* 1, p. 7294, 1963. / Additional refs. under "Sex Abnormalities," Chap. 27.

CRETINISM (HYPOTHYROIDISM). Classification, diagnosis, etc. Goldsmith, Richard. *Postgrad. Med.,* Jan., 1961. / Defects of thyroid hormone synthesis and metabolism. Blizzard, R. M. *Metabolism,* 9, p. 232, 1960.

AMAUROTIC IDIOCY (TAY-SACHS DISEASE). Aronson, S. M., & Volk, B. W. *See* Chapter 27 in *Cerebral Sphingolipidosis.* Academic Press, 1962. / Aronson, et al: Identification of carriers in Tay-Sachs disease. *Proc. Soc. Exp. Biol.,* III:3, 1962. / A genetic approach. Kozinn, P. J., et al. *J. Pediat.,* 51:1, 1957.

GARGOYLISM (HURLER'S SYNDROME). Mittwoch, U. *Ann. Hum. Gen.,* 25, p. 424, 1962. / Gilbert, E. F., & Guin, G. H. *A.M.A.J. Dis. Child.,* 95:1, 1958. / Lamy, M., et al. *J. de Genet. Humaine,* 6, 1957.

PHENYLKETONURIA. Lyman, Frank L. Chas. C. Thomas, 1963. / Detection of in newborn. Guthrie, R., & Susi, A. *Pediatrics,* 32:3, 1963. / Diagnosis, treatment of. Gelinas, G. P. *Canad. J. Pub. Hlth.,* 53:12, 1962. / Carrier determination. Wang, H. L., et al. *Am. J. Hum. Gen.,* 13:2, June, 1961. / Iris color in. Berg, J. M., & Stern, J. *Ann. Hum. Gen.,* 22:4, 1958.

HISTIDINEMIA (HISTIDINURIA). Auerbach, V. H., et al. *J. Pediat.,* 60:4, 1962.

MICROCEPHALY. Cowie, V. *J. Ment. Defic. Res.,* 4:42, 1960. / A clinical and genetical study of. Böök, J. A., et al. *Am. J. Ment. Defic.,* 57:4, Apr., 1953. / In the Netherlands. Bosch, J. Van Den. *Acta Genet.* Basel, 7:2, 1957.

HYDROCEPHALUS. Banner, E. A. *J. Lancet,* 76:1, 1956. / Larson, Carl A. *Am. J. Hum. Gen.,* 6:1, Mar., 1954. / In cattle. Baker, Marvel L., et al. *J. of Hered.,* July–Aug., 1961.

CEREBRAL PALSY AND MENTAL RETARDATION. Incidence of in twins. Illingworth, R. S., & Woods, G. E. *Arch. Dis. Childh.,* 35, 1960. / Genetic factors pertaining to etiology. Blumel, J. *Texas J. M.*, 54:4, 1958.

STURGE-WEBER'S DISEASE. Patau, K., et al. *Am. J. Hum. Gen.*, 13, p. 287, 1961.

EPILEPSY (in relation to mental defect). *See* refs. "Functional Defects" (Chap. 24).

COUNSELING, PROGNOSIS, IN MENTAL DEFECTS. Counseling parents of mentally deficient children. Watson, E. H. *Pediatrics*, 22:2, Aug., 1958. / New hope for the retarded child. Jacob, Walter. Publ. Affairs Pamph. No. 210 (8th. ed.), 1958. / Mongolism, genetic counseling for. Reed, S. C. *Eug. Qu.*, 10:3, 1963.

SEX ABNORMALITIES

(*Chapter* 27)

GENERAL. Sex determination, recent chromosome studies on. Cooper, Herbert L. *J. Natl. Med. Ass.*, 54:4, July, 1962. / Chromosomal aberrations in human disease. Hirschhorn, Kurt, & Cooper, Herbert L. *Am. J. of Med.*, 31:3, Sept., 1961. / Biologic concepts of sex and reproduction. Segal, Sheldon J. *Eug. Qu.*, Sept., 1960. / *Sex and Internal Secretions*. 3rd. ed. Young, William C., Ed. Williams & Wilkins, 1961. / *Human Intersex*. Ashley, David, J. B. Williams & Wilkins, 1962.

SEX MOSAICISM. Hirschhorn, Kurt, et al. *Lancet*, 7145, 1960. / Harnden, D. G. In: Proceedings, *Second International Conference on Congenital Malformations*. Natl. Foundation. Lippincott, 1964. / In presumably MZ twins. Mikkelsen, M., et al. *Cytogenetics*, 2:2–3, 1963.

HERMAPHRODITISM. Lotfi, A. M., & Girgis, S. M. *J. Urol.* (Baltimore), 89:1, 1963. / True hermaphroditic siblings. Milner, W. A., et al. *J. Urol.*, 79:6, 1958. / Women with testes. Pimstone, B. *S. Afr. Med. J.*, 36:49, 1962. / Pseudohermaphroditism, diagnosis, treatment. Scott, William W. *Postgrad. Med.*, May, 1961.

KLINEFELTER SYNDROME. *See* Chromosomal aberrations in human disease, Hirschhorn, Cooper, above. / Developments in human cytogenetics. Miller, O. J., et al. *Eug. Qu.*, Mar., 1961. / An XXXXY male. Day, R. W., et al. *J. Pediat.*, 63:4, 1963. / For continuing reports see *Lancet* and *Brit. Med. J.*

TURNER'S SYNDROME (*gonadal dysgenesis*). See general refs., above. / XO syndrome. Lemli, L., Smith, D. W. *J. Pediat.*, 63:4, 1963.

LOWER ANIMALS, SEX ABNORMALITIES. Freemartins, sex chromatin in. Moore, Keith L., et al. *Anat. Rec.*, 121, p. 442, 1955. *Also:* Ewen, A. H., et al. *J. of Hered.*, May, 1947. / Sex reversal, poultry. Newcomber, E. H., & Donnelly, G. M. *J. of Hered.*, Sept.–Oct., 1961. / Sex reversal, mammalian embryos. Holyoke, E. A., & Beber, B. A. *Science*, 128, Oct. 31, 1958.

"SEX CHANGING," HUMAN. ¶ Case of "woman" doctor who became man. *Time* Mag., May 26, 1958.

EUNUCHISM, FAMILIAL HYPOGONADOTROPIC. Biben, R. L., & Gordan, G. S. *J. Clin. Endocr. Metab.*, 15:8, 1955.

TESTICULAR FEMINIZATION. Taillard, W., & Prader, A. *J. Genet. Hum.*, Vol. 6, 1957.

UNDESCENDED TESTES. Clatworthy, H. W., Jr., et al. *Postgrad. Med.*, Aug., 1957. *Also*, Anon: *Cancer Bull.*, July–Aug., 1960.

HYPOSPADIAS. Brown, J. B., & Fryer, M. P. *Postgrad. Med.*, Nov., 1957. *J. Obst. & Gyn.*, 74:4, Oct., 1957.

Also: Sörensen, H. R. *Munksgaard* (Copenhagen), 1953.

ADRENAL HYPERPLASIA. Wilkins, Lawson. *Postgrad. Med.*, Jan., 1961.

SEX-IDENTITY TESTS. Some properties of the sex chromosomes. Barr, Murray L. In: Proceedings, *Second International Conference on Congenital Malformations.* Lippincott, 1964. / Sex differences in cells. Mittwoch, Ursula. *Sci. Amer.*, July, 1963. / Lyon hypothesis, and sex chromatin patterns. De Mars, Robert. *Science*, Aug. 12, 1963. / *See also* refs. on "Inactivated X," Lyon hypothesis, under Sex Differences in Disease, Defect (Chap. 22). / The cytogenetics of sex in man. Ford, C. E. *Eug. Rev.*, 53:1, Apr., 1961. / The assignment of sex to an individual. Hamblen, E. C. *Am. J. Obst. & Gyn.*, 74:6, Dec., 1957.

PUBERTAS PRAECOX. *Precocious Sexual Development.* Thamdrup, Erik. Chas. C. Thomas, 1961. / Heredity of: Jacobsen, A. W., Macklin, Madge T. *Pediatrics*, 9–682, 1952. *Also:* Walker, Stuart H. *J. of Pediat.*, Sept., 1952. / Puberty onset, normal. *See* under Stature and Body Form, Chap. 14.

AGING CHANGES, HORMONAL, SEX EFFECTS. Masters, William H. *Am. J. Obst. & Gyn.*, 74:4, Oct., 1957.

BLOOD

(*Chapter* 28)

GENERAL. *Blood Groups and Transfusions*, 3rd ed. Wiener, Alexander S. Hafner, 1962. / Blood groups in man and lower primates. Alexander S. Wiener. *Transfusion* (Phila.), 3:3, 1963. / M-N types. Alexander S. Wiener, et al. *Exp. Med. Surg.*, 21:2–3, 1963. / *Blood Groups in Man*, 4th ed. Race, R. R., & Sanger, Ruth. Chas. C. Thomas, 1962. / ABO and Rh world distribution. Kirk, R. L. *Am. J. Hum. Gen.*, 13:2, June, 1961.

BLOOD GROUPS, SELECTIVE FACTORS. Polymorphism and natural selection in blood groups. Reed, T. Edward. In: *Proc. Conf. on Genetic Polymorphisms.* Blumberg, B. S., Ed. Grune & Stratton, 1961. / Infectious disease, ABO blood groups, and human evolution. Harris, R. *Eug. Rev.*, 54:4, Jan., 1963. / Prezygotic selection (ABO groups and disease). Matsunaga, Ei. *Eug. Qu.*, 9:1, Mar., 1962. / Effects of natural selection on human genotypes. Sympos.: Dunn, L. C., Levene, Howard & McConnell, R. B. *Annls. N.Y. Acad. Sci.*, 65, Art. 1, June 18, 1956. / Erythroblastosis. Diamond, Louis K. / *Birth Defects*, Chap. 21. Lippincott, 1963.

HEMOGLOBIN ABNORMALITIES. Ingram, Vernon M. Chas. C. Thomas, 1961. / Hemoglobin synthesis, genetic control. Nance, Walter E. *Science*, July 12, 1963.

HAPTOGLOBLINS, TRANSFERRINS. Genetic variations in serum proteins. Bearn, Alex. G. In: Proceedings, *Second International Conference on Congenital Malformations* (1963). Lippincott, 1964. / Haptoglobins. Nance, W. E., & Smithies, O. *Nature*, 198:4884, 163. / Genetic significance of haptoglobins: Bearn, A. G., & Franklin, E. C. *Science*, 128:3324, 1958. *Also:* Allison, A. C. *Proc. R. Soc. M.*, 51:8, 1958. / Haptoglobins and transferrin groups in S. and S. E. Asia. Kirk, R. L., & Lai, L. Y. C. *Acta Genet.*, 11:2, 1961. / In Pacific populations: Giblett, E. R. *Eug. Qu.*, 9:1, Mar., 1962. / Gamma globulins. Chodirker,

W. B., & Tomasi, T. B., Jr. *Science*, Nov. 22, 1963. / Antibodies to genetic types of gamma globulins. Allen, James C., & Kunkel, Henry G. *Science*, 139, Feb. 1, 1963.

HEMOPHILIA. Lewis, Jessica H., et al. *Am. J. Hum. Gen.*, 15:1, Mar., 1963. / Vascular: Graham, John B. *Am. J. Hum. Gen.*, 11:2, June, 1959. / Inheritance of, and related conditions. Bradlow, B. A. *Leech* (Johannesburg), 32:5, 1962. / Hemophilia A. Bitter, K., et al. *Internist* (Berlin), 4:9, 1963. / Christmas disease. Ikkala, E. *Scand. J. Clin. Lab. Invest.*, 12:46, 1960. / Hemophilia in a "girl" with male sex-chromatin pattern. Nilsson, I. M., et al. *Lancet*, 7097, p. 264, 1959. / Hemophilia in dogs: Brinkhous, K. M., & Graham, J. B. *Science*, June 30, 1950, p. 723.

TELANGIECTASIA. Saunders, W. H. *Arch. Otalaryng.*, 76:3, 1962. Ecker, J. A., et al. *Am. J. Gastroent.*, 33, p. 411, 1960.

ANEMIAS. Inherited: Allison, A. C. *Eug. Qu.*, 6:3, Sept., 1959. / Cooley's anemia (Mediterranean anemia, thalassemia). Symposium, N.Y. Acad. Sci., 1964. / Genetic basis of.: Ingram, V. M., & Stretton, A. O. *Nature*, 184, 1959. / Thalassemia, Survey: Bannerman, R. M. Grune & Stratton, 1962. / In English, Scots. (1) Roberts, P. D., (2) Buchanan, K. D., et al. *J. Clin. Path.*, 16:6, 1963. / Sickle-cell anemia. Studies: Porter, F. S., & Thurman, W. G. *Am. J. Dis. Child.*, 106, p. 35, 1963; In African populations, Allison, A. C. *Ann. Hum. Gen.*, 21, 1956; In Black Caribs, Brit. Honduras, Firschein, I. L. *Am. J. Hum. Gen.*, 13:2, June, 1961. / Hemoglobin C disease. Rice, H. M. *Brit. M. J.*, 5009, 1957. / Spherocytosis, genetics of. Morton, N. E., et al. *Am. J. Hum. Gen.*, 14:2, June, 1962. / Primaquin sensitivity. Zinkham, W. H. *Postgrad. Med.*, Oct., 1961. / Aggamaglobulinemia. Bridges, R. A., & Good, R. A. *Annls. N.Y. Acad. Sci.*, 86, Art. 40, June 30, 1960. *Also*: Gitlin, David, & Janeway, C. A. *Sci. Amer.*, July, 1957. *See also* discussion of anemias in Sympos.: Genetics of Migrant and Isolate Populations. (Conf. in Israel), Williams & Wilkins, 1963, and other refs. on race differences in disease, "Races: I" (Chap. 42).

LONGEVITY

(*Chapter* 29)

GENERAL. ¶ The life span of animals. Comfort, Alex. *Sci. Amer.*, Aug., 1961. ¶ Mortality estimates from Roman tombstones. Durand, J. D. *Amer. J. Sociol.*, Jan., 1960. / The change in mortality trends in the U.S. *Vital & Health Stat.*, Natl. Center, Ser., 3:1. U.S. Dept. Hlth., Ed., Welf., Mar., 1964. / Mortality and morbidity trends in the U.S. since 1900. Spiegelman, Mortimer. Am. Collg. Life Underwriters, 1964. *Also*, Mortality trends at older ages in countries of low mortality. Vol. 1, *Proc., Intl. Union for Sci. Study of Pop.*, Conf. N.Y., 1961. / Life expectancies, various countries. *U.N. Demographic Yearbook*, 1962 (also, succeeding issues). / Infant mortality, international trends. Shapiro, Sam, & Moriyama, I. M. *Am. J. Pub. Health*, 53:5, May, 1963. / Changing association between infant mortality and socio-economic status: Willie, Charles V. *Social Forces*, Mar., 1959. / Factors in human longevity. Yerushalmy, Jacob. *Am. J. Pub. Health,* 53:2, Feb., 1963.

LETHAL GENES. Gluecksohn-Waelsch, Salome. *Science,* 142, Dec. 6, 1963. / *Developmental Genetics and Lethal Factors.* Hadorn, Ernst. Wiley, 1961.

AGING PROCESS. Biological mechanisms underlying. Curtis, Howard J. *Science,* 141, Aug. 23, 1963. / Genetics of. Glass, Bentley. In Sympos.: Aging. Am. Assn. Adv. Sci., 1960. / Biochemistry of. Sinex, F. Marott. *Science,* 134, Nov. 3, 1961. ⁋ Physiology of. Shock, Nathan W. *Sci. Amer.,* Jan., 1962. / On nature of. Szilard, Leo. *Proc. Natl. Acad. Sci.,* 45:1, Jan., 1959. / Research Highlights in Aging. U.S. Natl. Inst. of Health, 1959. / Older Americans, population facts. U.S. Senate Com. U.S. Govt. Print. Off., 1961.

HEREDITY AND LONGEVITY. Van Zonneveld, R. J., Polman, A. *Acta Genet.,* Basel, 7:1, 1957. / Twins and longevity (MZ, DZ comparisons). Jarvik, Lissy F., et al. *Am. J. Hum. Gen.,* 12:2, June, 1960. *Also,* Kallmann, F. J. In: *Psychopathology of Aging.* Hoch, Paul H., Zubin, Joseph, Eds. Grune & Stratton, 1961. / Mice, life spans in inbred and hybrid. Chai, C. K. *J. of Hered.,* Sept.–Oct., 1959.

SEX DIFFERENCES IN LONGEVITY. ⁋ The mortality of men and women. Scheinfeld, Amram. *Sci. Amer.,* Feb., 1958. / Catholic nuns, brothers, comparisons. (Role satisfactions and length of life, etc.) Madigan, Francis G. *Am. J. Sociol.,* 67:6, May, 1962.

RACE, ETHNIC DIFFERENCES IN MORTALITY. Jacobson, P. H. Vol. 1, *Proc., Intl. Union for Sci. Study of Pop.,* N.Y., 1961.

INDIVIDUAL FACTORS IN LONGEVITY. *Handbook of Aging and the Individual.* Birren, J. E., Ed. U. of Chicago Press, 1959. / Prognostic value of life insurance mortality investigations. Bolt, William, & Lew, Edward A. *J.A.M.A.,* 160, Mar. 3, 1956. / Build and blood pressure study, Vol. I. Soc. of Actuaries (Chicago), 1959. / Body weight and longevity. Pauling, Linus. *Proc. Natl. Acad. Sci.,* 1958. / Marital status and mortality. Berkson, Joseph. *Am. J. Pub. Health,* 52:8, Aug., 1962. *Also:* Sheps, M. C. *Am. J. Pub. Health,* 51:4, Apr., 1961. / Athletes and longevity. Rock, Alan. *Brit. Med. J.,* Apr. 3, 1961. *Also:* ⁋ "Pigskin perils," *MD,* Nov., 1957, p. 64. ⁋ Smoking and longevity. Report, *Smoking and Health,* U.S. Health, Educ., Welf. Publ. 1103, 1964. / Weather and mortality. Kutschenreuter, P. H. *Trans. N.Y. Acad. Sci.,* Dec., 1961.

CONTINUING REFERENCES. Reports on mortality, U.S. Natl. Office of Vital Statistics. ⁋ *Statistical Bulletin,* Metropolitan Life Ins. Co. / *U.N. Demographic Yearbook* (annual). / *Geriatrics* (journal).

INTELLIGENCE

(*Chapter* 31)

LOWER-ANIMAL INTELLIGENCE. *Behavior Genetics* (Chapter 7). Fuller, John L., & Thompson, W. Robert. Wiley, 1960. ⁋ Elephants, intelligence of. Rensch, B. *Sci. Amer.,* 196:2, 1957. ⁋ Dolphin's IQ—and Man's. Bates, Marston. *N.Y. Times Sun. Mag.,* Oct. 16, 1960.

INTELLIGENCE TESTS. Background on Binet-Simon tests ("An individual who made a difference"). Wolf, Theta H. *Amer. Psychol.,* May, 1961. / *Measurement and Ap-*

praisal of Adult Intelligence, 4th ed. Wechsler, David. Williams & Wilkins, 1958. *Also:* Measuring the IQ test. N.Y. *Times Sun. Mag.*, Jan. 20, 1957. ❡ Understanding Testing. U.S. Dept. Health, Ed., Welf., OE-25003, 1960. / *Intelligence and Experience.* Hunt, J. McV. Ronald Press, 1961.

PREDICTIVE VALUE OF EARLY IQ TESTS. Prediction of school age intelligence from infant tests. Escalona, S. K., & Moriarity, Alice. *Child Develop.*, 32, 1961. ❡ Bayley, Nancy. Value and limitations of infant testing. *Children*, 5:4, July–Aug., 1958. *Also:* ❡ Can we predict a child's intelligence? *Natl. Parent-Teacher*, Dec., 1957.

HEREDITY AND INTELLIGENCE. Genetics and intelligence. Erlenmeyer-Kimling, L. Jarvik, Lissy F. *Science*, 142, Dec. 13, 1963. / Inheritance of human intelligence. Alstrom, Carl H. *Acta Psychiat. Neurol. Scand.*, 36, 1961. / Intellectual potential and heredity. Allen, Gordon (and reply, Knobloch, Hilda, & Pasamanick, Benj.). *Science*, 133, Feb. 10, 1961. / Inheritance of mental ability. Burt, Cyril. *Amer. Psychologist*, Jan., 1958. *Also: Nature*, 179, 1957; *Eug. Rev.*, 49, Sept., 1957. / School achievements, twin correlations. Husen, T. *Scand. J. Psychol.*, 4:2, 1963. *Also:* Sandon, Frank. *Brit. J. Statist. Psychol.*, 12:2, Nov., 1959. / Heredity, environment, and the question "How?" Anastasi, Anne. *Psychol. Rev.*, 65:4, 1958.

GROUP DIFFERENCES IN INTELLIGENCE. Class differences. Burt, Cyril. *Brit. J. Statist. Psychol.*, May, 1959. *Also:* Conway, J., exchange of views with Halsey, A. H., same publication. / Rural-urban differences. Lehmann, I. J. *J. Educ. Res.*, 53, 1959. / *See also*, "Social-class differences," in *Differential Psychology* (Chapter 15). Anastasi, Anne. Macmillan, 1958.

RACE DIFFERENCES. Comparative psychological studies of Negroes and Whites in the U.S. Dreger, Ralph M. *Psychol. Bull.*, 60:1, 1963. ❡ Negro-White differences in intelligence test performances. Klineberg, Otto. *Amer. Psychologist*, 18:4, Apr., 1963. / Personality and Negro-White intelligence. Roen, Sheldon R. *J. Abn. & Socl. Psych.*, 61:1, 1960. / African intelligence. Cryns, A. G. J. *J. Soc. Psychol.*, 57:2, 1962. / Racial differences and the future. Ingle, Dwight J., *Science*, Oct. 16, 1964, p. 375.

SEX DIFFERENCES. Brown, M. H., & Bryan, G. E. *J. Clinic. Psychol.*, 15–303, 1959. *Also* (comprehensive survey): *J. Educ. Psychol.*, 48–273, 1957. / Wechsler M-F Index. Levinson, Boris M. *J. Gen'l. Psychol.*, 69–217, 1963. / Male, female, kindergarten children on the WISC. Darley, F. L., Winitz, Harris. *J. Genet. Psychol.*, 99–41, 1961. / Should boys enter school later than girls? Pauly, Frank R. *Nat. Ed. Assn. J.*, 41, 1952. / *See also:* ❡ *Women and Men*, Scheinfeld, Amram. Chaps. 8, 24. Harcourt, Brace, 1944.

SPECIAL ABILITIES AND CREATIVITY. Three faces of intellect. Guilford, J. P. *Am. Psychologist*, Aug., 1959. *Also:* Primary abilities, nonverbal areas. *Proc. Natl. Acad. Sci.*, 1962. / Creativity ("Family environment and cognitive style"), Getzels, Jacob W., & Jackson, Philip W. *Am. Sociolog. Rev.*, 26:3, June, 1961. / The classification of abilities. Vernon, Philip E. *Educ. Res.*, 2, 1960. / General ability and special aptitudes. Burt, Cyril. *Educ. Res.*, 1:2, 1959.

MENTAL PROCESSES. Brain chemistry and learning capacity. Lund, F. H. *Education*, 80, 1959. / Memory (possible enzyme basis): Smith, C. E. *Science*, 138, p. 3343, 1962. / Cerebral localization of individual processes. Shure, G. H. Halstead, W. C. *Psychol. Monogr.*, 72:2, No. 465, 1958. Intelligence, quantum resonance and thinking machines. Wechsler, David. *Trans. N.Y. Acad. Sci.*, Feb., 1960. ¶ The evolution of mind. Munn, N. L. *Sci. Amer.*, June, 1957. / The psychology of thinking. Symposium: Fundamentals of psychology. Annls. N.Y. Acad. Sci., 91:1, Dec. 23, 1960.

MUSICAL TALENT

(*Chapter* 32)

TESTS FOR MUSICAL TALENT. *Psychology of Music*. Seashore, Carl E. McGraw-Hill, 1960. *See also* Seashore articles in *Science*, Apr. 24, 1942; and in *Sci. Monthly*, Feb., 1942, May, 1941, Apr. 1940, and (on vocal talent) Oct., 1939. / Validation of Seashore's measures. McLeish, John. *Brit. J. Psychol. Statist.*, Sect. 3, 1950. / Application of musical-ability tests. Wing, H. *Brit. J. Psychol.*, 24, Nov., 1954. / Tests, musical ability & appreciation, *Brit. J. Psychol.*, Monogr. Suppl. 27, 1948. / Assessment of musical ability in school children. Whittington, R. W. T. *J. Educ. Psychol.*, 48, 1957. / Measurement of music achievement at college entrance. Aliferis, James, & Stecklein, J. E. *J. Applied Psychol.*, 39:4, 1955. / Testing for musical talent. Farnsworth, Paul R. *Instrumentalist*, 16:3, 1961.

PSYCHOLOGY OF MUSIC, GENERAL. *Introduction to the Psychology of Music*. Révész, G. U. of Oklahoma Press, 1954. / Psychology of music, European studies. Jacobs, Camille. *Acta Psychol.* (Amst.), 17, 1960. / *The Social Psychology of Music*. Farnsworth, Paul R. Dryden, 1958. / *Exploring the Musical Mind*. Kwalwasser, Jacob. Coleman-Ross, 1955. / Foundations of musical aesthetics. Howes, Frank. *Proc. Roy. Mus. Assn.*, 83, 1957.

HEREDITY IN MUSICAL TALENT. Child singers of Sistine Chapel and their families. Gedda, Luigi, et al. *Proc. 2nd. Int'l Congr. Hum. Genet.*, 1961. (Rome), 1963. / Twin study, musical ability. Shuter, Rosamund P. G. (Ph.D. thesis.) U. of London, 1964. / Musical-talent clues (biological traits). Lehman, C. F. *J. Educ. Res.*, 45, 1952. / Inheritance of simple musical memory. Ashman, Richard. *J. of Hered.*, Jan.–Feb., 1953. / Sur la mémoire musicale. Guillaume, P. *Année Psychol.*, 50, p. 413, 1951. ¶ Tune deafness. Kalmus, H. (in article on inherited sense defects.) Sci. Amer., May 1952. ¶ On the Fishberg family ("The five generations"). Logan, Andy. *New Yorker*, Oct. 29, 1949.

ABSOLUTE PITCH. Ward, W. Dixon. *Sound*, 2:3, 1963.

IQ AND MUSICAL APTITUDE. Intellectual defect & musical talent. Anastasi, Anne, & Levee, R. F. *Am. J. Ment. Defic.*, 64, 1960 / Intelligence of music students. Wheeler, L. R. & V. D. *J. Educ. Psychol.*, 42, 1951.

RACIAL DIFFERENCES IN MUSIC. Bose, Fritz, *Homo*, 2:147, 1951.

MUSICIANS, INDIVIDUAL CASES. ¶ On Leonard Bernstein ("Wunderkind"). *Time* Mag., Feb. 4, 1957. ¶ On Nadia Boulanger.

Thomson, Virgil. N.Y. *Times Mag.*, Feb. 4, 1962. ¶ Slenczynaka, Ruth, with Lingg, Ann M. *Music at Your Fingertips.* (A prodigy's experience.) Doubleday, 1961. ¶ Richard Rodgers (profile of). Sargeant, Winthrop. *New Yorker,* Nov. 18, 1961. ¶ Rosina Lhevinne (profile of). Sargeant, Winthrop. *New Yorker,* Jan. 12, 1963. ¶ Szigeti, Joseph. *With Strings Attached.* (Autobiography.) Knopf, 1947.

SPECIAL TALENTS

(*Chapter* 33)

CREATIVITY AND TALENT IDENTIFICATION. In adolescents. Getzels, J. W., & Jackson, P. W. *Am. Sociol. Rev.,* 26:3, 1961. / Creativity and personality. Rees, M. E., & Goldman, Morton. *J. Genet. Psychol.,* 65, 1961. / Creativity, psychological study of. Golann, S. E. *Psychol. Bull.,* 60:3, 1963. ¶ What makes a person creative? MacKinnon, D. W. *Sat. Rev.,* Feb. 10, 1962. *Also:* Nature and nurture of creative talent. *Am. Psychol.,* July, 1962. / Interrelationship between certain abilities and certain traits. Guilford, J. P. *J. of Genl. Psychol.,* July, 1961. / Creativity in gifted adolescent artists. Hammer, E. F. Random House, 1961. / The hereditary abilities study. Vandenberg, S. G. *Am. J. Hum. Gen.,* 14:2, June, 1962.

SPECIAL APTITUDES. *The Psychology of Occupations.* Roe, Anne. John Wiley & Sons, 1956. / Scientists & non-scientists in group of 800 gifted men. Terman, Lewis M. *Psychol. Monogr.,* 378, 1954. / Diversity of talent. Wolfle, Dael. *Amer. Psychol.,* Aug., 1960. / Mathematical and related abilities, exceptional talent for. Hunter, I. M. L. *Brit. J. Psychol.,* 53.3, 1962. ¶ *The Gifted Group at Mid-Life.* (Terman "genius" study follow up. Terman, L. M., & Oden, Melita H. Stanford U. Press, 1959. / Overlap in characteristics in gifted children. Liddle, Gordon, *J. Educ. Psychol.,* 49, 1958.

PRODIGIES / *Mental Prodigies.* Barlow, F. Philosophical Libr., 1952. / Salo Finkelstein, memory prodigy. Weinland, J. D. *J. Genl. Psych.,* 39, 1948. ¶ An exceptionally gifted pupil. Witty, Paul, Blumenthal, Rochelle. Elementary Engl., Apr., 1957. ¶ Chess men and mentality. Lasker, E. N.Y. *Times* Mag., Apr. 24, 1949. ¶ *Personality of Chess Players.* Horowitz, I. A. & Rothenberg, P. L. Macmillan, 1963. ¶ Bobby Fischer (chess prodigy). O'Brien, Robert. *Sat. Rev.,* Apr. 27, 1963. ¶ *Ex-Prodigy: My Childhood and Youth.* Wiener, Norbert. Simon & Schuster, 1953. *Also:* Analysis of the child prodigy. N.Y. *Times* Mag., June 2, 1957. ¶ *Educating Gifted Children.* De Haan, R. F., & Havighurst, R. J. U. of Chicago Press, 1957.

SEX DIFFERENCES IN ACHIEVEMENT. ¶ *The Potential of Women.* (Symposium. Includes: "Women in the Field of Art," Howe, Thomas C.; "Problems of Creative Women," Mannes, Mary; "The Implication of Rivalry," Gunderson, Barbara B.; and "The Direction of Feminine Revolution," Hunt, Morton.) McGraw-Hill, 1963. / Woman as artist. Greenacre, Phyllis. *Psychoanal. Qu.,* 29:2, 1960. ¶ Achievement and genius (women in relation to men). In *Women and Men* (Chapter 25). Scheinfeld, Amram. Harcourt, Brace, 1944. / *See also* comparison of sexes in Terman "genius" study, listed under "Special aptitudes," above.

GENIUS

(*Chapter* 34)

Abnormality in genius. Creativity and mental illness. Herbert, P. S., Jr. *Psychiat. Qu.*, 33, 1959. / Problems of highly creative children (maladjustments). Torrence, E. P. *Gifted Child Qu.*, 5, 1961. / Survey of literature of artistic behavior in the abnormal. Anastasi, Anne, & Foley, J. P:, Jr. *Ann. N.Y. Acad. Sci.*, 42:1, Aug. 11, 1941. ¶ Illness and artistic creativity. Benda, C. E. *Atlantic Mo.*, July, 1961. / *Some Reflections on Genius.* (Biological aspects.) Brain, Russell. Lippincott, 1961. / Genius: Hereditary aspects and relationship to psychic anomalies. (In German: Hochstbegabung, etc.) Urban & Schwarzenberg (Munich), 1953. ¶ Ocular problems of some famous men. Ridgeway, W. G. *Missouri Med.*, Nov., 1961. / *See also:* ¶ Series on aberrations and pathology in artists (Modigliani, Goya, Van Gogh, Leonardo, Michelangelo), in *MD*, Jan., Apr., July, 1957; May, June, 1958. / Genius, alcohol, toxins and syphilis. Podolsky, Edward. *Med. Press*, Dec. 28, 1955. ¶ The childhood pattern of genius. McCurdy, H. G. Smithsonian Inst. Report, Publ. 4373, 1959.

Drugs as mental stimulants. The hallucinogenic drugs. Frank Barron et al. *Sci. Amer.*, Apr. 1964. / *The Doors of Perception.* Aldous Huxley. Harper, 1954. / Psychopharmacological drugs. H. E. Himwich. *Science*, 127, Jan. 10, 1958. / Psychopharmaceuticals: uses and abuses. Nathan S. Kline. *Postgrad. Med.*, May, 1960.

INSTINCTS AND HUMAN NATURE

(*Chapter* 35)

Behavior, general. *Behavior Genetics.* Fuller, J. L., Thompson, W. R. Wiley, 1960. ¶ *Human Behavior: Inventory of Scientific Findings.* Berelson, Bernard, Steiner, G. A. Harcourt Brace & World, 1964. / *Animal Behavior.* Scott, J. P. U. of Chicago Press, 1958. ¶ *The Science of Animal Behavior.* Broadhurst, P. L. Pelican, 1963. / *The Behaviour of Domestic Animals* (Symposium). Hafez, E. S. E., Ed. Williams & Wilkins, 1962. / Behavior genetics and individuality. Hirsch, Jerry. *Science*, 142, Dec. 13, 1963. / Mammalian behavior, inheritance. Broadhurst, P. L., Jinks, J. L. *J. of Hered.*, 54:4, July–Aug. 1963. / *Behavior and Evolution.* (Symposium). Simpson, George G., & Roe, Anne, Eds. Princeton U. Press, 1958. / *Brain and Behavior* (Symposium). Brazier, Mary A., Ed. Am. Inst. Biol. Sci., (Washington), 1962.

Instinct. *Learning and Instinct in Animals.* 2nd ed. Thorpe, W. H. Methuen, 1963. / *The Study of Instinct.* Tinbergen, N. Oxford U. Press, 1951. / Species-specific behavior. Beach, Frank A. *Am. Psychol.*, Jan., 1960. / (Sympos. on "Instinct" Princeton). *Psychol. Rev.*, 54:6, 1947.

Imprinting. Moltz, Howard. *Psychol. Bull.*, 57:4, 1960. / Hess, Eckhard H. *Science*, 130, July 17, 1959. *Also, Sci. Amer.*, Mar., 1958.

Conditioning and critical period. Misbehavior of organisms. Breland, Keller & Marian. *Am. Psychol.*, Nov., 1961. / Critical periods in behavioral development. Scott, J. P. *Science*, 138, Nov. 30, 1962. / Social development, critical period in dogs. Freedman, D. G., et al. *Science*, 133, Mar. 31, 1961. *Also:* Scott, J. P., *Psychosom. Med.*, 20:1, Jan.–Feb., 1958. / Affectional responses in infant monkey. Harlow, H. F., & Zimmerman, R. R. *Science*, 130, Aug. 21, 1959. *Also:* The nature of love. Harlow, H. F. *Am. Psychol.*, Dec., 1958.

Human infant behavior, development. Inheritance of behavior in infants. Freedman, D. G. & Keller, Barbara. *Science*, 140:196–198, 1963. ❡ *Birth to Maturity*. Kagan, Jerome, & Moss, H. A. Wiley, 1962. / Maternal deprivation: review of literature. Casler, Lawrence. Monog. Soc. Res. Child Develop., 26:2, 1961. / Imprinting in human infants. Gray, P. H. *J. Psychol.*, 46, 1958. / *The Evolution of Human Nature*. Herrick, C. Judson. U. of Texas Press, 1956. / Pattern vision in newborn infants. Fantz, Robert L. *Science*, 140:3564, 1963. / *See also* under "Personality," Chap. 37.

"Wolf children." ❡ The wolf boy of Agra. Ogburn, William F. *Am. J. Sociol.*, Mar., 1959. *Also:* same journal and date, Feral children and autistic children. Bettelheim, Bruno.

SOCIAL AND SENSORY BEHAVIOR

(*Chapter* 36)

Social behavior, general. Genetics and development of social behavior in mammals. Scott, J. P. *Am. J. Orthopsychiatry*, 32:5, Oct., 1962. / Basic social capacity of primates. Harlow, H. F. *Hum. Biol.*, 31:1, Feb., 1959. / Social organization of subhuman primates in natural habitat. Imanishi, K. *Current Anthrop.*, 1, 1960. / Social life of monkeys, apes, primitive man. Sahlins, M. D. *Hum. Biol.*, 31, 1959. / The biological basis of human sociality. Count, Earl W. *Amer. Anthrop.*, 60:6, Dec., 1958.

Genetic factors in behavior, general. *Biological and Biochemical Basis of Behavior* (Symposium). Harlow, Harry F., Woolsey, Clinton N., Eds. U. of Wisconsin Press, 1959. / Chimpanzees, individuality in behavior. Nissen, H. W. *Am. Anthrop.*, 58, 1956. ❡ Canine behavior patterns. Fox, M. W. *Mod. Vet. Pract.*, 44:9, 1963. / Individual differences in behavior and genetic basis. Hirsch, Jerry. In Sympos.: *Roots of Behavior*. Bliss, E. L., Ed. Hoeber (Harpers), 1962. / *See also* Behavior, General, above, under "Instincts and Human Nature," Chap. 35.

Parental behavior. *Maternal Behavior in Mammals* (symposium). Rheingold, Harriet L., Ed. Wiley, 1963. / *Behavioral Analysis*. (Maternal, etc.) Levy, David M. C. C. Thomas, 1958. / *Parental Care and Its Evolution in Birds*. Kendeigh, S. C. U. of Illinois Press, 1952.

Dominance, aggression. Hereditary differences in dominance development in puppies. Pawlowski, A. A., & Scott, J. P. *J. Compar. & Physiolog. Psych.*, 49:4, Aug., 1956. / Monkeys, influence of amygdalectomy on social behavior (dominance). Rosvold H. E., et al. *J. Comp. & Physiol. Psychol.*,

June, 1954. / Gamecocks, dominance-subordination. Fennell, R. A. *Am. Nat.*, 79 (781), 1945. ◖[*Aggression.* Scott, John P. U. of Chicago Press, 1958. / *Fear.* Psychopathology of fear and anger. Ax, Albert F. *Psychiat. Res. Rep.*, 12, 1960.

COMMUNICATION. Animal "languages" and human language. Hockett, Charles F. *Hum. Biol.*, 31, Feb., 1959. / Rhesus monkeys, development of communication. Mason, William A. *Science*, 130, Sept. 18, 1950. / Dolphins, vocal exchanges. Lilly, John C., & Miller, Alice M. *Science*, 134, Dec. 8, 1961. / Insects, communication. Dethier, V. G. *Science*, 125, Feb. 22, 1957.

SENSES, GENERAL. *The Senses of Animals.* L. Harrison Matthews, Maxwell Knight. Philosophical Library, 1963. ◖[*The Senses of Animals and Men.* Milne, Lorus, & Marjory. Atheneum, 1962. / *The Chemical Senses in Health and Disease.* Kalmus, H., & Hubbard, S. J. Chas. C. Thomas, 1960. *Also:* The chemical senses. Kalmus, H. *Sci. Amer.*, Apr., 1958. / Sensory evoked responses in man (symposium). N.Y. Acad. Sci., 112:1, 1964. / Pleasure centers in the brain. Olds, James. *Sci. Amer., Oct.*, 1956.

TASTE AND SMELL. *Physiological and Behavioral Aspects of Taste* (Symposium). Kare, M. R., & Halpern, B. P . Eds. U. of Chicago Press, 1961. / Complexities of human taste variation. Skude, Gunnar. *J. of Hered.*, 51:6, Nov.–Dec., 1960. / Taste thresholds and food dislikes. Fischer, R., et al. *Nature*, 191, Sept. 23, 1961. ◖[Taste interrelationships. Pangborn, Rose M. *Food Res.*, 25, 1960. / PTC tasting (study of 845 sibling pairs). Das, S. R. *Ann. Hum. Gen.*, 20:4, 1956. / Goitre, and taste sensitivity. Brand, N. *Ann. Hum. Gen.*, 26:4, 1963. / Cystic fibrosis, increased taste, smell sensitivity. Henkin, R. I., & Powell, G. F. *Science*, 138, p. 3545, 1962. / Cretins, increased incidence of PTC nontasters. Shepard, T. H. II, & Gartler, Stanley M. *Science*, 131, Mar., 1960. / Smoking and PTC sensitivity. Freire-Maia, Ademar. *Ann. Hum. Gen.*, 14, 1960. / *See also:* Refs. on Taste-blindness, under "Functional Defects" (Chap. 24), Other Sensory Disorders.

SOUND. Sound-precipitated convulsions (audiogenic seizures). Review of studies. Bevan, William. *Psychol. Bull.*, 52:6, Nov., 1955. / Audiogenic seizures. (Report on conf., *Science*, 136, Apr. 27, 1962, p. 334. / Loudness discrimination. Harris, J. Donald. *J. Speech, Hear. Disord.*, Monogr. Suppl. 11, 1963.

PAIN. ◖[The perception of pain. Melzack, Ronald. *Sci. Amer.*, 204:2, Feb., 1961. / Congenital insensitivity to pain. Sternbach, R. A. *Psychol. Bull.*, 60:3, 1963. *Also:* Magee, K. R., et al., *J. Nerv. Ment. Dis.*, 132, 1961; Phillips, L. I., *New Z. Med. J.*, 58, p. 327, Oct., 1959. / Cold sensitivity. Rothman, S., & Griem, S. F. *Ill. Med. J.*, 123, June, 1963. ◖[DIRECTION, SENSE OF, IN VERTEBRATES. Beecher, William J. *Sci. Monthly*, July, 1952. / Celestial navigation of birds. Sauer, E. G. F. *Sci. Amer.*, Aug., 1958.

SPATIAL VISUALIZATION. Genesis of visual space perception Epstein, William. *Psychol. Bull.*, 61:2, 1964. / Visual depth discrimination in animals. Shinkman, P. G. *Psychol. Bull.*, 59:6, 1962. / Visual depth perception in a ten-month-old infant. Walk, R. D., & Dodge, Sue H. *Science*, 137, Aug. 17, 1962. / Somatic sense of space and its threshold. Renfrew, S., & Melville, I. D. *Brain*, 83, 1960. ❡ "SEEING WITH FINGERS." R. K. Plumb. *New York Times*, Jan. 8 & 26, 1964.

EXTRASENSORY PERCEPTION. Report on (negative). Schmidt, H. E. *Percept. Mot. Skills*, 16:1, 1963. / ESP and flying saucers. Crumbaugh, James C. *Amer. Psychologist*, Sept., 1959. (Reply by Rhine, J. B., same journal and date.) / Review of Psychokinesis (PK). Girden, Edward. *Psychol. Bull.*, 59:5, Sept., 1962.

PERSONALITY

(*Chapter* 37)

FEATURES AND PERSONALITY. Preferences in feminine beauty Iliffe, A. H. *Brit. J. Psychol.*, 51, 1960. / Personalities in faces. Secord, Paul F., et al. *Genet. Psychol. Monogr.*, 49, 1954. ❡ Is there an executive face? Stryker, Perrin. *Fortune*, 1953. ❡ *Noses*. Holden, Harold M. World, 1950.

PLASTIC SURGERY. Motivational patterns in women seeking plastic surgery. Meyer, E., et al., *Psychosom. Med.*, 22, 1960. / Effects: The fate of one's face. Meerloo, Joost A. M. *Psychiat., Qu.*, Jan., 1956. ❡ Mongolian-fold operations. *Time* Mag. Nov. 4, 1957.

BODY FORM AND PERSONALITY. Somatyping. Sheldon, William H. In *Encyclopedia of Mental Health*. Deutsch, Albert, et al., eds. Franklin Watts, 1963. *Also: Atlas of Men. Sheldon, William H.*, Harper, 1963. / *Behaviour and Physique*. Parnell, R. W. Arnold (London), 1958. / Physique and behavior. In *Differential Psychology*, 3rd ed. (Chapters 5, 6). Anastasi, Anne. Macmillan, 1958. / Body build and behavior in children. Walker, R. N. *Child Develop.*, 34: 1, 1963. / Later careers of boys, early or late maturers. Jones, Mary C. *Child Develop.*, Mar., 1957. ❡ Tall girl, effects ("'Big bear' no more"). *Newsweek*, Apr. 24, 1961, p. 88. / Fears concerning physical changes of females, Angelino, Henry, & Mech, Edmund V. *J. of Physiol.*, 39, 1955. / Gynecomastia in adolescence. Schonfeld, W. A. *Arch. Gen. Psychiat.*, 5, July, 1961. / Somatopsychics of personality and body function. Sontag, L. W. *Vita Hum.*, Basel, 6:1–2, 1963. ❡ Obesity, anthropological approach. Powdermaker, Hortense. *Bull. N.Y. Acad. Sci.*, 36:5–6, 1960. / Physical disabilities, cultural reaction. Richardson, Stephen A., et al., *Am. Sociolog. Rev.*, Apr., 1961.

COLORING AND BEHAVIOR. Coat color, physique and temperament. Keeler, Clyd E. *J. of Hered.*, 38:9, Sept., 1948. / Hair and correlation with behavior. Clyd E. Keeler. *J. Tennessee Acad. Sci.*, 26:2, Apr., 1961. / Albinism and water-escape performance in mouse. H. D. Winston, Garden Lindzey. *Science*, 133, Apr. 10, 1964.

MANNERISMS AND MOVEMENTS. ❡ Posture, anthropology of. Hewes, G. W. *Sci. Amer.*, Feb., 1957. / Graphic movements and relationship to temperament. Talmadge, Max. *Psychol. Monogr.*, 72:16, 1958. / Tics, treatment in childhood. Zausmer, David M. *Arch. Dis. Childhd.*, 29, Dec., 1954.

HANDWRITING. Graphology, review of experimental research in. Fluckinger, F. A., et al. *Percept. Motor Skills*, 12, 1961. / Pathological aspects. Tholl, Joseph. *Postgrad. Med.*, Mar., 1962. / Psychodiagnostic value of handwriting analysis. Perl, W. R. *Am. J. Psychiat.*, 111, Feb., 1955. ¶ Graphodiagnosis. *MD*, Mar., 1958, p. 136. / Twins, handwriting differences. Husén, Torsten. In: *Psychological Twin Research* (Chapter 8). Acta U. Stockholm. Almqvist & Wiksell (Stockholm), 1959.

ASTROLOGY, PHRENOLOGY, PALMISTRY, ETC. ¶ Scientists look at astrology. Bok, B. J., & Mayall, M. W. *Sci. Month.*, Mar., 1941. ¶ The astrology racket. Meyer, A. E., *Am. Mercury*, Jan., 1945. ¶ *Phrenology: Fad and Science*. Davies, John D. Yale U. Press, 1955. ¶ Somatomancy. (Old notions, palmistry, phrenology, somatotypes, etc.) Lessa, W. A. *Sci. Monthly*, Dec., 1952.

HANDEDNESS. Neurological aspects of. Benton, A. L., et al. *Psychiat. et neurol.* 144:6, 1962. / Genetic aspects of. Falek, Arthur. *Am. J. Hum. Gen.*, 11:1, Mar., 1959. / In twins. Torsten, Husén. *See* under ref. for "Handwriting," above.

HAND-CLASPING, ARM-FOLDING: Freire-Maia, A. *Acta Gen. Med. et Gemel.*, 10:2, Apr., 1962; *Am. J. Phys. Anthrop.*, 19:3, Sept., 1960. *Also:* Pons, Jose. *Ann. Hum. Genet.*, 25, 1961.

TONGUE MANIPULATION: Lee, J. W. *J. of Hered.*, Nov.–Dec., 1955 and Jan.–Feb., 1956. / Whitney, David D. *J. of Hered.*, July, 1950 and Jan., 1949.

PERSONALITY CONCEPTS AND TESTS. *Personality Assessment*. P. E. Vernon. Methuen (London), 1964. / Problem of personality definition. McCreary, John K. *J. Gen. Psychol.*, 63, 1960. / *Theories of Personality*. Hall, C. S., & Lindzey. Gardner. Wiley, 1957. / Concept of the normal personality. Shoben, E. J., Jr. *Am. Psychol.*, Apr., 1957. / *Personality Tests and Assessments*. Vernon, P. E. Methuen (London). 1953. / Scaling of terms to describe personality. Buss, A. H., Gerjuoy, Herbert. *J. Consult. Psychol.*, 21:5, 1957.

CHILDREARING CONCEPTS AND EFFECTS. Impact of theories of child development. Caldwell, B. M., & Richmond, J. B. *Children* (U.S. Dept. Health, Educ., Welf.), 9:2, Mar.–Apr., 1962. / Infant care and personality. Orlansky, Harold. *Psychol. Bull.*, 46, 1949. / *Parental Attitudes and Child Behavior* (Symposium). Glidewell, J. C., Ed. Chas. C. Thomas, 1961. / Effects of early training on personality. Martin, W. E. *Marrg. & Fam. Living*, Feb., 1957. / Maternal deprivation (review of literature). Casler, Lawrence. *Monogr. Soc. Res. Child Develop.*, 26:2, 1961. / Earle, A. M. & B. V. *Am. J. Orthopsychiat.*, 31, 1961. / Yarrow, L. J. *Psychol. Bull.*, 58:6, 1961. / Pinneau, S. R., & reply by Spitz, René A. *Psychol. Bull.*, 52, 1955.

PERSONALITY DEVELOPMENT AND PREDICTION. Fetal behavior and later personality. Sontag, Lester W. *Vita Humana*, 6, 1963. / *Birth to Maturity*. (Report on Fels studies.) Kagan, Jerome & Moss, Howard A. Wiley, 1962. / Behavior in early childhood and adjustment of same persons as adults. Anderson, John E. Inst. of Child. Develop., U. of Minnesota, 1963. / "Shirley's babies" after fifteen years. Neilon, Patricia. *J. of Genet. Psychol.*, 73, 1948. / *Personality Development in Children* (Sympos.). Iscoe, Ira, & Stevenson, Harold, Eds. U. of Texas Press,

1960. / Infantile trauma, genetic factors and adult temperament. Lindzey, Gardner, et al. *J. Abn. & Socl. Psychol.*, 61:1, 1960.

SEX DIFFERENCES IN PERSONALITY. ¶ In: *Women and Men* (Chaps. 9, 17, 18). Scheinfeld, Amram. Harcourt, Brace, 1944.

HEREDITY AND ENVIRONMENT IN PERSONALITY. ¶ Determinants of personality. Murphy, Gardner. In: "*Man and Civilization.*" (U. of California.) McGraw-Hill, 1963. / *Determinants of Infant Behavior* (Symposium). Foss, Brian, Ed. Wiley, 1962. / Inheritance of personality. Cattell, Raymond B., et al. *Am. J. Hum. Gen.*, 7:2, June, 1955. *Also:* Nature-nurture ratios for primary personality factors, etc. *J. Abn. & Socl. Psychol.*, Mar., 1957. / Heredity and variations in human behavior patterns. Kallmann, F. J., Baroff, G. S. *Acta Gen. et Stat. Med.*, 7:2, 1957. / Inheritance of behavior in infants. Freedman, D. G., & Keller, Barbara. *Science*, 140, Apr. 12, 1963.

TWINS, PERSONALITY STUDIES. *Monozygotic Twins.* Shields, James. Oxford U. Press, 1962. / *Psychological Twin Research.* Husén, Torsten. (In English.) Almqvist & Wiksell (Stockholm), 1959. / Personality tests of identical twins. Magnusson, D. *Scand. J. Psychol.*, 1, 1960. / Personality development in identical twins. Lowinger, P., et al., *Arch. Gen. Psychiat.*, 8:5, 1963. ¶ *Twins: A Study of Heredity and Environment.* Newman, H., Freeman, F., & Holzinger, K. U. of Chicago Press, 1937. / Uniovular twins brought up apart. Juel-Nielsen, N., & Mogensen, A. *Acta Gen. et Statist. Med.*, 7:2, 1957. *Also,* EEG's in: Juel-Nielsen & Harvald, B. *Acta Gen. et Statist. Med.*, 8:1, 1958. / Studies of MZ twins reared apart. Burks, Barbara S. & Roe, Anne. *Psychol. Monogr.*, 63:5, 1949.

QUADRUPLETS, STUDIES OF: *J. of Hered.*, Apr., July, Oct., 1940; Sept., 1942; Sept., Oct., 1943; Mar., 1948. / *The Gerain Quadruplets.* (Schizophrenia in an MZ set.) Rosenthal, David, Ed. Basic Books, 1964.

DIONNE QUINTUPLETS. ¶ We were five. Annette, Yvonne, Marie and Cecile Dionne; James Brough, ed. *McCall's*, Oct., Nov., 1963.

CRIME

(*Chapter* 38)

CONSTITUTIONAL AND ENVIRONMENTAL FACTORS. ¶ Glueck, Sheldon & Eleanor. *Family Environment and Delinquency.* Routledge & Kegan Paul (London), 1962. *Also: Predicting Delinquency and Crime.* Harvard U. Press, 1959. *Also, articles: J. Crim. Law, Criminol. Police Sci.*, 48:6, Mar.–Apr., 1958, and 51:3, Sept.–Oct., 1960. / Assessments, follow-ups, of Glueck studies. Morris, Albert. *Sociol. & Socl. Res.*, Jan. 1962; Michael, C. M., & Coltharp, F. C. *Am. J. Orthopsychiat.*, Mar., 1962. / Application of prediction tables to study of delinquency. Briggs, P. F., et al. *J. of Consult. Psychol.*, 25:1, 1961. / Appearance and criminality. Corsini, R. J. *Am. J. Sociol.*, July, 1959. *Also:* Swedner, Harald. *Am. J. Sociol.*, p. 407, Jan., 1960.

HEREDITY AND CRIME. Genetics and the criminal. Penrose, L. S. *Brit. J. Delinq.*, 6:1, 1955. / Twins and crime, early studies (Lange, Rosanoff, Stumpfl, Kranz). Summaries in: Stern, C. *Principles of Human Genetics*, 2nd ed. (pp. 603–5). W. H. Freeman, 1960.

PATHOLOGICAL FACTORS IN CRIME. Crime and psychopathology. Gynther, Malcolm D. *J. Abn. & Socl. Psych.*, 64:5, 1962. / Psychopathy, criminality and mental deficit. Jacob, Walter. *Arch. Crim. Psychodyn.*, 4, p. 509, 1961. / Mental abnormalities in criminals. Hagopian, Peter B. *Am. J. Psychiat.*, Jan., 1953. / Crimes of violence and delinquency in schizophrenics. Kaufman, Irving. *J. Am. Acad. Child Psychiat.*, 1, Apr., 1962. / Organic factors in delinquency. Geiger, Sara G. *J. Soc. Therap.*, 6, 1960. / *Law and Psychiatry*. Glueck, Sheldon. Johns Hopkins Press, 1962. / Mental capacity and incompetency. Mezer, R. R., & Rheingold, P. D. *Am. J. Psychiat.*, 118:9, Mar., 1962. / Personality factors associated with criminal behavior. Siegman, Aron W. *J. Consult. Psychol.*, 26:2, 1962. / Personality factors related to juvenile delinquency: Peterson, Donald R., et al. *Child Develop.*, 32, 1961.

INCIDENCES, U.S., OTHER COUNTRIES. Crime and delinquency in the U.S. Sellin, Thorsten. *Annls. Am. Acad. Pol. & Socl. Sci.*, 339, Jan., 1962. ❡ Crime, Southern and non-Southern. Porterfield, A. L., & Talbert, R. H. *Socl. Forces*, Oct., 1952. / Homicide, U.S. and each state. See: Specl. Reports, U.S. Natl. Off. Vit. Statistics, Vol. 49, No. 62, Aug. 21, 1959, and succeeding issues. / International trends in juvenile delinquency. Neumeyer, Martin H. *Sociol. & Socl. Res.*, Nov.–Dec., 1956. ❡ Crime wave in England. Samuels, Gertrude. N.Y. *Times* Mag., May 13, 1962.

RACE, ETHNIC DIFFERENCES. The ecological structure of Negro homicide. Pettigrew, T. F., & Spier, R. B. *Am. J. Sociol.*, 67:6, May, 1962. / *See also*: "Crime, Southern and non-Southern," under Incidences, above. / Crime and the foreign born. Wickersham, G. W., et al. U.S. Natl. Com. on Law Observ. & Enforc., Rep. No. 10. Govt. Printg. Off., Wash., 1933.

SOCIO-ECONOMIC FACTORS AND DELINQUENT BEHAVIOR. Nye, F. Ivan, et al. *Am. J. Sociol.*, Jan., 1958. / Distribution of juvenile delinquency in the social class structure. Reiss, A. J., Jr., & Rhodes, A. L. *Am. Sociol. Rev.*, Oct., 1961. / Anti-social behavior of adolescents from higher socio-economic groups. Hershkovitz, H. H., et al. *J. Nerv. Ment. Dis.*, 129, Nov., 1959. ❡ *Troublemakers: Rebellious Youth in an Affluent Society*. Fryvel, T. R. Shocken (N.Y.), 1962.

WAR AND "FIGHTING INSTINCT." ❡ Science and human survival. AAAS Committee report on "War." *Science*, 134, Dec. 29, 1961. ❡ The fighting behavior of animals. Eibl-Eibesfeldt, Irenäus. *Sci. Amer.*, Dec., 1961. Comment on preceding: Gross, Seymour. *Sci. Amer.* (Letters), Mar., 1962. / The causes of fighting in mice, rats. Scott, J. P., & Frederickson, Emil. *Physiol. Zool.*, 24:4, Oct., 1951.

SEX DIFFERENCES IN CRIME. ❡ The female offender (Symposium). *Natl. Probation & Parole Assn. J.*, 3:1, Jan., 1957. / Sex differences among juvenile offenders. Gibbons. D. C., & Griswold, M. J. *Sociol. & Soc. Res.*, 42:2, Dec., 1957. *Also*: Wattenberg, W. W., & Saunders, Frank. *Sociol. & Socl. Res.*, Sept.–Oct., 1954. ❡ *The Criminality of Women*. Pollak, Otto. U. of Pennsylvania Press, 1950. / Menstruation and crime. Dalton, Katharina. *Brit. Med. J.*,

2, p. 1752, Dec. 30, 1961. / *See also:* ◖ *Women and Men* (Chap. 20). Scheinfeld, Amram. Harcourt, Brace, 1944.

SEX OFFENSES. ◖ Some myths about the sex offender. Tappan, Paul W. *Fed'l. Probation* (U.S.) 19:2, June, 1955. / *The Sexual Offender and His Offenses.* Karpman, Benjamin. Julian Press, 1954. / Psychodynamic patterns. Glueck, B. C., Jr. *Psychiat. Qu.*, 28, Jan., 1954. / Diagnosis and treatment. Pacht, Asher R., et al. *Am. J. Psychiat.*, 118, Mar., 1962. / Prostitution. Fairfield, L. *Brit. J. Delinq.*, 9:3, Jan., 1959. *Same issue:* Prostitution in Australia. Vincent, Barry J. ◖ Juvenile prostitution. Gibbons, T. C. D. *Brit. J. Delinq.*, 8:1, July, 1957. / *The Wolfenden Report* (report on homosexual offenses and prostitution, England). Stein & Day, 1963.

ALCOHOLISM. Genetic implications of. Stromgren, E. *Acta Genet. Med.*, 11:3, July, 1962. / Drinking habits in twins. Kaij, L. *Acta Genet., Basel*, 7:2, 1957. / *Disease Concept of Alcoholism.* Jellinek, E. M. Hillhouse Press, 1960. / Blood characteristics in alcoholics. Williams, Roger J., et al. *Proc. Natl. Acad. Sci.*, Nov., 1957. / Mice and alcohol. McClearn, G. E., & Rodgers, D. A. *Qu. J. Stud. Alc.*, 17, Dec., 1956. / Preadolescent sons of male alcoholics. Aronson, H., & Gilbert, Anita. *Fed'l. Probation* (U.S.) 19:2, *Am. Arch. Gen. Psychiat.*, 8:235, 1963. / The woman alcoholic. Lisansky, Edith S. *Ann. Am. Acad. Pol. Soc. Sci.*, 315, 1958.

DRUG ADDICTION. ◖ *The Road to H.* (Narcotics, delinquency, social policy.) Chein, Isidor. Basic Books, 1964. / *Drug Addiction.* Ausubel, D. P. Random House, 1959. / Narcotic addicts in U.S., statistics. Bureau of Narcotics (U.S. Treas. Dept.), Dec. 31, 1958.

SUICIDE. Causes and prevention. Robinson, Paul I. *Postgrad. Med.*, Aug., 1962. ◖ *The Cry for Help* (symposium). Farberow, N. L., & Shneidman, E. S., Eds. McGraw-Hill, 1961. ◖ *Suicide and Mass Suicide.* Meerloo, Joost A. M. Grune & Stratton, 1962. / Suicide in twins and only children. Kallman, F. J., et al. *Am. J. Hum. Gen.*, 1:2, Dec., 1949. / Negro vs. White suicide rates. (Suicide, homicide and socialization of aggression.) Gold, Martin. *Am. J. Sociol.*, May, 1958. / Occupation, status and suicide. Powell, E. H. *Am. Sociol. Rev.*, 23:2, Apr., 1958. / Suicide in children, adolescents. Bakwin, Harry. *J. Pediat.*, 50, 1957. / World suicide rates. World Health Org. (U.N.), *Epidem. & Vit. Stat. Rep.*, 14:5, 1961. ◖ U.S., frequency of suicide. *Metrop. Life Ins. Co. Stat. Bull.*, Dec., 1960. ◖ *Suicide and Scandinavia.* Hendin, Herbert. Grune & Stratton, 1964. / Japanese youth, suicides. Iga, Mamoru. *Sociol. & Soc. Res.*, Oct., 1961. / *African Homicide and Suicide.* Bohannan, Paul, Ed. Princeton U. Press, 1960.

SEX ROLES AND SEX LIFE: I. NORMAL

(*Chapter* 39)

GENERAL. *Encyclopedia of Sexual Behavior.* Ellis, Albert, & Abarbanel, Albert, Eds. Hawthorn, 1961. / *The Psychology of Sexual Emotion.* Grant, Vernon W. Longmans, Green, 1957. ◖ *Patterns of Sexual Behavior.* Ford, Clellan S., & Beach, Frank A.

Harper, 1951. ❙[*Sex and the Nature of Things*. Berrill, N. J. Dodd, Mead, 1953. (Paperback, Pocket Books, 1955.) ❙[*Women and Men*. Scheinfeld, Amram. Harcourt, Brace, 1944. (Chaps. 9, 12, 17, 18, 19, 26.)

LOWER ANIMALS, SEXUAL BEHAVIOR. ❙[The courtship of animals. Tinbergen, N. *Sci. Amer.*, Nov., 1954. / The heterosexual affectional system in monkeys. Harlow, Harry F. *Amer. Psychologist*, Jan., 1962. ❙[The female primate. Jay, Phyllis. In: *The Potential of Woman* (Sympos.). McGraw-Hill, 1963. / Sex differences in affective-social responses of rhesus monkeys. Mason, W. A., et al. *Behaviour*, 16, 1960. / Animal sexuality. Scott, J. P. *See: Encyclopedia of Sexual Behavior*, above. *See also: Behavior Genetics* (pp. 164–179). Fuller, J. L., & Thompson, W. R. Wiley, 1960. / Inheritance of sexual-behavior patterns in female guinea pigs. Goy, R. W., & Jakaway, J. S. *Anim. Behav.*, 7, 1959.

HORMONES AND SEXUAL BEHAVIOR. Sex hormones and human behavior. Bishop, P. M. F. *Brit. J. Anim. Behav.*, 1:1, 1953. / Testosterone-induced crowing in young domestic cockerels. Marler, P., et al. *Anim. Behav.*, 10:1–2, 1962. / Responses of male rats to androgen. Beach, Frank A., & Fowler, Harry. *J. Comp. Physiol. Psychol.*, 52, 1959. / Hypothalmic regulation of sexual behavior in male guinea pigs. Phoenix, C. H. *J. Comp. Physiol. Psychol.*, 54, 1961. / Castration, emotional changes following. Bowman, K. M., & Crook, G. H. *Psychiat. Res. Rep.*, 12, 1960. / Relation of castration, androgen therapy, to aggression in male mice. Bevan, William, et al. *Anim. Behav.*, 8, 1960.

MASCULINITY-FEMININITY DEVELOPMENT. Sex differences in identification of (note). Brown, Daniel G. *Psychol. Rev.*, 66, p. 126, 1959. *Also:* Sex-role preference in children. *Psychol. Reports*, 11, p. 477, 1962. / Measurement of, in children. Rosenberg, B. G., & Sutton-Smith, B. *Child Developmt.*, 30, p. 373, 1959. / Childhood experience and M-F scores. Steimel, Raymond J. *J. Consult. Psychol.*, 7, p. 212, 1960. / *See also:* Child Sexuality, Adolescent Sexuality. Reevy, W. R. In: *Encyclopedia of Sexual Behavior* (above).

MASCULINITY-FEMININITY TESTS AND EVALUATIONS. ❙[*Sex and Personality*. Terman, L. M. & Miles, C. C. McGraw-Hill, 1936. / Construct validity of three M-F tests. Barrows, G. A., & Zuckerman, Marvin. *J. Consult. Psychol.*, 24, 1960. / Stereotype measures of M-F. Nichols, Robert C. Ed. *Psychol. Measmt.*, 22, p. 449, 1962. / Rorschach, sex differences on. Baughman, E. E., & Guskin, Samuel. *J. Consult.Psychol.*, 22, p. 400, 1958. *Also:* Richards, T. W., & Murray, D. C. *J. Clin. Psychol.*, 14, p. 61, 1958. / Child Rorschach responses. Ames, Louise B. *Genet. Psychol. Monogr.*, 61, p. 229, 1960. / Opposite-sex scales, as measures of psychosexual deviancy. Rosenberg, B. G., et al. *J. Consult. Psychol.*, 25:3, 1961. / Men's and women's beliefs, ideals, self-concepts. McKee, J. P., Sherriffs, A. C. *Am. J. Sociol.*, 64, p. 356, 1959. / Men and women: personality patterns and contrasts. Bennett, E. M., & Cohen, L. R. *Genet. Psychol. Monogr.*, 59, p. 101, 1959.

FEMALE SEXUALITY. Symposium. Lehfeld, Hans, & Ellis, Albert, Eds. *Qu. Rev. Surg. Obst. Gyn.*, 16:4, Oct.–Dec., 1959. ❙[*The*

Cold Woman. Levine, Lena, & Loth, David. Messner, 1962. / *Psychosexual Functions in Women*. Benedek, Therese. Ronald Press, 1952. / Menstruation and accidents. Dalton, Katharina. *Brit. Med. J.*, 2, p. 1425, 1960. *Also:* Schoolgirls' misbehavior and accidents. *Brit. Med. J.*, 2, p. 1647, 1960. / The menstrual cycle and disorders in psychiatric patients. Gregory, B. A. *J. Psychosom. Res.*, 2, 1957. / Biological effects of female behavior. Overstreet, E. W. *In: The Potential of Woman* (symposium). McGraw-Hill, 1963.

MALE SEX DRIVE, CHARACTERISTICS OF. Beach, Frank. Nebraska symposium on motivation. Jones, M. R., Ed. 1956. (*Psych. Abstr.*, 31, p. 2316, 2372, 1957.) / Impotency, psychogenic and other factors. El Senoussi, Ahmed, et al *J. of Psychol.*, 48:1, 1959.

SEX ROLES, SEX LIFE: II. ABNORMAL

(*Chapter* 40).

SEXUAL DEVIATION, GENERAL. *California Sexual Deviation Research*. Reports, Jan., 1953, Mar., 1954. California Dept. of Mentl. Hygiene. / Experience of 500 children with adult sexual deviation. Landis, J. T. *Psychiat. Qu. Suppl.* 1, 1956. / Sexual disorders and behavior therapy. Rachman, S. *Am. J. Psychiat.*, Sept., 1961.

KINSEY STUDIES. *Sexual Behavior in the Human Male*. Kinsey, A. C. Pomeroy, W. B., & Martin, C. E. Saunders, 1948. / *Sexual Behavior in the Human Female*. Kinsey, A. C., Martin, C. E., & Gebhard, P. H. Saunders, 1953.

CRITICAL EVALUATIONS OF THE KINSEY REPORTS. Statistical problems of. Cochran, W. G., et al. *J. Am. Statist. Assn.*, 48, Dec., 1953. / Volunteer error in Maslow, A. H., & Sakoda, J. M. *J. Abn. & Socl. Psychol.*, 47:2, Apr., 1952. / Sexual behavior in American society (symposium). *Socl. Problems*, 1:4, Apr. 1954. ¶ How good are sex studies? Scheinfeld, Amram. *Cosmopolitan* Mag., Aug., 1953. *Also:* ¶ Sex behavior of women. *Cosmopolitan*, Sept., 1953; *Reader's Digest*, Oct., 1953. / *See also*, reviews of Kinsey reports by Hyman, H. & Marmack, J. E., *Psychol. Bull.*, 51:4, 1954; Hobbs, A. H., Kephart, W. M., *Am. J. Psychiat.*, 110:8, Feb., 1954; Kluckhohn, Clyde. *N.Y. Times Book Rev.*, Sept. 13, 1953, p. 3.

HOMOSEXUALITY, MALE AND FEMALE. *See* Kinsey reports, above. / *The Wolfenden Report (England): Homosexual Offenses and Prostitution*. Wolfenden, Sir John, Chairman. Stein & Day, 1963. / *See also: The Male Homosexual in Britain—A Minority Report*. Westwood, Gordon. Longmans, Green, 1960. / Genetic aspects of male homosexuality. (Twin study.) Kallmann, F. J. *J. Nerv. Ment. Dis.*, 115, 1952. *Also:* Twinship study, overt homosexuality. *Am. J. Hum. Gen.*, 4:2, 1952. / Homosexuality and heterosexuality in MZ twins (non-concordance). Rainer, J. D., et al. *Psychsom. Med.*, 22:4, July–Aug. 1960. / Male MZ twins discordant for homosexuality. Klintworth, G. K. *J. Nerve. Ment. Dis.*, 135:2, 1962. / Homosexuality and genetic sex. Pritchard, M. *J. Ment. Sci.*, 108, p. 456, 1962. *Also:* Sex chromosomes and the law. *Lancet*, 2, p. 7256, 1962. / Female homosexuality (symposium). *J. Am.*

Psychoanalyt. Ass., 10:3, July, 1962. *Also:* Socarides, C. W. *J. Am. Psychoanalyt. Ass.*, 11:2, Apr. 1963. / Homosexual behavior in children. Bakwin, Harry and Ruth. *J. Pediat.*, 43, 1953. / Development of sex-role inversion and homosexuality. Brown, D. G., *J. Pediat.*, 50:5, May, 1957. / Homosexuality and arts. Alexander, M., & Anatole, *J. Int. J. Sexol.*, 8, 1954. / Recovery from sexual deviations. Stevenson, Ian, & Wolfe, Joseph. *Am. J. Psychiat.*, 116:8, Feb. 1960. / The cure of homosexuality. Hadfield, J. A. *Brit. Med. J.*, 1, June 7, 1958. / Psychotherapeutic aspects. Rubinstein, L. H. *Brit. J. Med. Psychol.*, 31, 1958.

TRANSVESTISM AND SEX-ROLE INVERSION. Brown, D. G. *Marr. & Fam. Livg.*, Aug., 1960. / In children. Bakwin, Harry. *J. Pediatr.*, Feb., 1960. / In boys: Friend, Maurice R., et al. *Am. J. Orthopsychiat.*, 24, 1954. / Chromosomal sex in transvestites. Barr, M. L., & Hobbs, G. E. *Lancet*, May 29, 1954. / Homosexuality, transvestism and transsexualism. Prince, C. V. *Am. J. Psychotherapy*, 11, Jan., 1957. / Female transvestism and homosexuality. Barahal, H. S. *Psychiat. Qu.*, 27, 1953. / Winnebago Berdache. Lurie, Nancy O. *Am. Anthrop.*, Dec., 1953.

INTERSEXES AND SEXUAL BEHAVIOR. Imprinting and the establishment of gender role. Money, John, et al. *A.M.A. Arch. Neurol. & Psychiat.*, 77, Mar., 1957. / Hormones in relation to sexual morphology and sexual desire. Money, John. *J. Nerv. & Ment. Dis.*, 132, 1961. / Hormones and sexual behavior. (Experiments with ape intersexes; comments on Money studies, above.) Young, W. C. et al. *Science*, 143, Jan. 17, 1964. / Sexually deviant behavior in Klinefelter's syndrome. Mosier, H. D., et al. *J. Pediat.*, 57, 1960. / Psychological test findings in girls with ovarian dysgenesis. Cohen, Haskel. *Psychosom.* Med., 24, May–June, 1962.

MALE-FEMALE DIFFERENTIATION, SOCIAL, PSYCHOLOGICAL. ¶ Emotional differences of the sexes. Reik, Theodor. (Discussion by Scheinfeld, Amram.) *Psychoanalysis*, 2, 1953. / The psychopathology of egalitarianism in sexual relations. Schimel, John L. *Psychiatry*, May, 1962. / Some effects of the derogatory attitude towards female sexuality. Thompson, Clara. *Psychiatry*, 13:3, Aug., 1950. / A biosocial and developmental theory of male and female sexuality. Shuttleworth, F. K. *Marr. & Fam. Livg.* May, 1959. / Cross-cultural survey of sex differences in socialization. Barry, Herbert III, et al. *J. Abn. & Socl. Psychol.*, 55:3, Nov., 1957.

EVOLUTION

(*Chapter* 41)

SCOPES TRIAL. ¶ *Six Days or Forever? Tennessee vs. John T. Scopes.* Ginger, Ray. Beacon Press, 1958. ¶ A witness at the Scopes trial: Cole, Fay-Cooper. *Sci. Amer.*, Jan., 1959. / *"Chicken or Egg?"* Wiselogle, Frederick Y. *Trans. N.Y. Acad. Sci.*, Sec. II, 24:3, Jan. 1962.

ACQUIRED-CHARACTERISTIC INHERITANCE. Horsfall, Frank L., Jr. *Science*, 136: 472, 1962. *Also:* Simpson, George G. (comments). *Science*, 137, Aug. 10, 1962. / Reply to Lamarckians. Dodson, Edward O. *J. Hered.*, 54:1, Jan.–Feb., 1963. / DNA-modified ducks. Benoit, J., et al. *Trans. N.Y. Acad. Sci.* 22:7, 1960. / Experiments in acquired characteristics. Waddington, C. H. *Sci. Amer.*, Dec., 1953.

LYSENKO THEORIES: ❡ *Soviet Marxism and Natural Sciences.* Joravsky, David. Columbia U. Press, 1961. *Also:* Article in *Sci. Amer.*, Nov., 1962. / Lysenko at bay. Dobzhansky, T. *J. of Hered.*, Jan.–Feb., 1958. ❡ The Russian purge of genetics (symposium). *Bull. Atom. Sci.* 5:5, May, 1949.

ORIGIN OF LIFE. Sympos. Moscow Acad. of Sci. Pergamon Press, 1960. / *Life in the Universe.* Oparin, A., & Fesenkov, V. Twayne, 1962. / On the origin of life. Keosian, John. *Science*, 131, Feb. 19, 1960. / How did life begin? Fox, Sidney W. *Science*, 132, July 22, 1960. / Modern ideas on spontaneous generation (Sympos.). *Annls. N.Y. Acad. Sci.*, 69:2, Aug. 30, 1957.

DARWIN, DARWINIAN THEORIES. ❡ The Origin of Species (1859). The Descent of Man (1871). Darwin, Charles. Reprinted, combined ed. Modern Library. (Also many other editions.) / The world into which Darwin led us. Simpson, George G. *Science*, Apr. 1, 1960. / Pre-Darwinian theories. Glass Bentley, *Proc. Am. Philosoph. Soc., Forerunner of Darwin.* 104:2, Apr. 1960. *Also:* Glass, Bentley, Ed. Johns Hopkins Press, 1959. / The origin of Darwinism. Darlington, C. D. *Sci. Amer.*, May, 1959. ❡ Darwin's Century. Eiseley, Loren C. Doubleday, 1958. / Darwin discovers nature's plan. Huxley, Julian. *Life* Mag., June 30, 1958. ❡ Darwin's World of Nature. Barnett, Lincoln. *Life* Mag., series, 1958, 1959. Publ. as book: *Wonders of Life on Earth.* Life, 1960. / *Evolution after Darwin.* Tax, Sol, & Callender, Charles, Eds. U. of Chicago Press, 1960. / Evolution at work. Dobzhansky, T. *Science*, 127, p. 3306, May 9, 1958. / "Fitness," interpretations of. Dobzhansky, Th., & Huxley, Julian. *Eug. Rev.*, 55:2, July, 1963.

MUTATIONS. Alteration of the existing gene. Muller, H. J., et al. *Genetics,* 46:2, Feb., 1961. *Also* Muller: Our load of mutations, *Am. J. Hum. Gen.*, 2:2, June, 1950. Artificial transmutation of the gene. *Science*, 66, p. 84–87, 1927. / Mutations in the human population. Neel, J. V. In: Methodology in Human Genetics. Burdette, W. J., Ed. Holden-Day, 1962. / Genetic loads in natural populations. Dobzhansky, Th. *Science*, 126, Aug. 2, 1957.

EXTINCT ANIMALS: ❡ Crises in the history of life. Newell, Norman D. *Sci. Amer.*, Feb., 1963. ❡ *Vanishing Animals.* Street, Philip. Dutton, 1963.

RADIATION AND ATOMIC-BOMB EFFECTS. ❡ Radiation and heredity. Muller, H. J. *Am. J. Pub. Hlth.*, Jan., 1964. ❡ *Radiation, Genes and Man.* Wallace, Bruce & Dobzhansky. Th. Holt, 1959. / Genetics, radiation and people. Newcombe, Howard B. *Canad. J. Gen. & Cytol.*, 2:3, Sept., 1960. / Nuclear war, biology of. Glass, Bentley. *Am. Biol. Teach.*, 24:6, Oct., 1962. ❡ Symposium on Radiation. *Sci. Amer.*, 201:3, Sept., 1959.

THE HUMAN RACES: I. ORIGINS AND PHYSICAL DIFFERENCES

(*Chapter* 42)

HUMAN EVOLUTION: GENERAL. ❡ *Mankind Evolving.* Dobzhansky,

Theodosius. Yale U. Press, 1962. / *Ideas on Human Evolution* (sympos.) Howells, William, Ed. Harvard U. Press, 1962. ❡ Howells, William *Mankind in the Making*. Doubleday, 1959. / *Classification and Human Evolution* (Sympos.). Washburn, Sherwood L., Ed. Aldine, 1963. / *The Evolution of Man*. Lasker, Gabriel W. Holt, Rinehart & Winston, 1961. / *The Human Races*. Garn, Stanley M. Chas. C. Thomas, 1961. / *The Human Species*, Hulse, Frederick S. Random House, 1963. ❡ *The Antecedents of Man*. Clark, W. E. LeGros. Quadrangle Books (Chicago), 1960. ❡ *The Epic of Man*. Life, Editors of, & Barnett, Lincoln. Time, Inc., 1961.

PRIMATE EVOLUTION. Proconsul. Cole, S. M. *Discovery* (England), Nov., 1961. *Sci. Digest*, Mar., 1952. / Oreopithecus. Eiseley, L. C. *Sci. Amer.*, June, 1956. / New approach to the problem of man's origin. de Terra, Helmut. *Science*, 124, Dec. 28, 1956. / A new look at the evolution of the primates. Buettner-Janusch, John. *Trans. N.Y. Acad. Sci.*, Ser. II, 24:1, Nov., 1961. / Chromosome cytology and evolution in primates. Chu, E. H. Y., & Bender, M. A. *Science*, 133, May 5, 1961. / Chromosomes in apes. Hamerton, J. L., et al. *Cytogenetics*, 2:4–5, 1963. / Blood groups in apes, baboons. Wiener, A. S. & Moor-Jankowski, J. *Science*, 142, Oct. 4, 1963. / The relatives of man (symposium.) N.Y. Acad. Sci., 102:2, 1962–63. / *The Primates*. John Napier, N. A. Barnicot, eds. Symposium, Zoolog. Soc., London, No. 10. Academic Press, 1963.

APE-MEN, FOSSIL MAN. ❡ South African Ape-Man. Leakey, L. S. B. *Natl. Geogr.* 123:1, Jan., 1963. *Also: Natl. Geogr.* Oct., 1961; *Nature*, 188:1050, 1960. / Fossil man in China, new evidence on. Chang, Kwang-chih. *Science*, 136, June, 1962. ❡ Java Man. E. Dubois, Dunlop, Eugene W. *Science Digest*, Apr., 1960. / The great Piltdown hoax. Straus, William L., Jr. *Science*, 119, Feb. 26, 1954. See also: *Amer. Anthropol.*, 56, 1954.

PREHISTORIC MEN. Neanderthal Man. Brace, C. Loring. *Am. Anthropol.*, 64, p. 729 ff., 1962. / Neanderthal skeletons in Shanidar Cave, Iraq. Solecki, Ralph S. *Trans. N.Y. Acad. Sci.*, June, 1959. *See also* note, *Science*, 132, Oct. 14, 1960, p. 1002. / Saldanha man and his culture. Straus, W. L., Jr. *Science*, 125, May 17, 1957. / Swanscombe man. Straus, W. L., Jr. *Science*, 123, Mar. 9, 1956, p. 410. / The Villafranchian and human origins. Howell, F. Clark. *Science*, 130, Oct. 2, 1959.

HOMO SAPIENS, ORIGINS & DIFFERENTIATION. ❡ The distribution of man. Howells, William W. *Sci. Amer.*, Sept., 1960. / *The Origin of Races*. Coon, Carleton S. Knopf, 1962. See also Reviews and critiques by Mayr, Ernst, *Science*, 138, Oct. 19, 1962; Opler, Morris E., N.Y. *Herald Trib. Books*, Dec. 9, 1962; Howells, W. W., N.Y. *Times Book Rev.*, Dec. 9, 1962; Pollitzer, W. S. *Am. J. Hum. Gen.*, 15:2, June, 1963. ❡ *The Story of Man* Coon, Carleton S. (rev. ed.) Knopf, 1962. / Migration, isolation and ongoing human evolution. Lasker, Gabriel W. *Hum. Biol.*, 32:1, Feb., 1960. / Genetic drift. Glass, H. Bentley. *Am. J. Phys. Anthrop.*,

14:4, Dec., 1956. / *See also* refs. under "Human evolution: general," above.

RACIAL DISTRIBUTION. ¶ *Americas: Early Man in the New World* Macgowan, Kenneth, & Hester, J. A. Doubleday (Anchor paperback), 1962. ¶ The Bering Strait land bridge. Haag, W. G. *Sci. Amer.*, Jan. 1962. / Prehistoric connections between Siberia and America. Griffin, J. B. *Science*, 131, Mar. 18, 1960. / The Eskimos and Aleuts, origins and evolution. Laughlin, W. S. *Science*, 142, Nov. 8, 1963. / The Eskimos (rev. ed.) Birket-Smith, Kaj. Humanities Press. 1960. / Early man in the Andes. Mayer-Oakes, W. J. *Sci. Amer.* May, 1963. ¶ Pre-Columbian American Indians. Wauchope, Robert. *Lost Tribes and Sunken Continents*. U. of Chicago Press, 1962. / Northeast Asia, ethnic origins of peoples (Symposium). Levin, M. G. Michael, H. N., Eds. U. of Toronto Press, 1963. ¶ *The Prehistory of Africa*. Cole, Sonia. Macmillan, 1963. ¶ Early man in Africa. Desmond, C. J. *Sci. Amer.*, July, 1958. / *African Ecology and Human Evolution* (Symposium). F. Clark Howell, Ed. Aldine, 1963.

RACE DIFFERENCES, PHYSICAL. The genetic basis of human racial differences. Glass, Bentley, *Bios*, 31:1, Mar., 1960. / Anthropometry, genetics and racial history. Hunt, Edward E., Jr. *Am. Anthrop.*, 61:1, Feb. 1959. / *Dental Anthropology*. (Symposium on racial traits in teeth). Press, 1963. *Also:* Lasker, G. W., Brothwell, D. R., Ed. Pergamon & Lee, M. M. *J. Forensic Sci.*, 2:4, Oct., 1957. / Tooth anomalies, Japanese, in Brazil. Beiguelman, Bernardo. *Hum. Biol.*, 34:3, Sept., 1962. / Fingerprints, race, ethnic differences. Rife, D. C. *Am. J. Hum. Gen.*, 6:3, Sept., 1954. / Ear shapes, main types. (In German.) Lundman, B. *Homo*, 3:2, 1952. (*Biol. Abst.*, *Aug.*, 1953, No. 21287.) / Compact bone in Chinese, Japanese (cf. Caucasoids). Stanley M. Garn, et al. *Science*, 143, Mar. 27, 1964. / Skin resistance, racial differences. Laverne, C. J., & Corah, N. L. *Science*, 139, Feb. 22, 1963. / Skinfold thickness, in primitive peoples native to cold climates. Elsner, R. W. In Symposium, Body Composition. *Annls. N.Y. Acad., Sci.*, 110, Sept. 26, 1963. / Negro-White differences in thermal insulative aspects of body fat. Baker, P. T. *Hum. Biol.*, 31:4, Dec. 1959. / Adaptability to high altitudes. Verzar, F. *Ciba Sympos.*, 6:4, Oct. 1958. / Adaptation in the Andes. Newman, M. T. *Natl. Hist.*, Jan. 1958. ¶ Life at high altitudes. Gray, G. W. *Sci. Amer.*, Dec., 1955. / Visual acuity, racial differences. Karpinos, Bernard D. U.S. Publ. Health Reports, 75:11, Nov., 1960. *See also:* Color-blindness, under *Disease differences, racial*, below. / PTC tasting (Eskimos and other populations). Allison, A. C., & Blumberg, B. S. *Hum. Biol.*, 31:4, Dec., 1959. / BAIB excretion, racial variations. Gartler, S. M., et al. *Am. J. Hum. Gen.* 9:3, Sept., 1957. / Sex ratio; race differences. *See* under "Boy or Girl?" Chap. 7. ¶ Twinning, race differences (biosocial effects on). Scheinfeld, Amram. *Proc. 2nd Intl. Cong. Hum. Gen.*, Rome, 1961. / N. American Negro infants, size at birth and growth during first postnatal year. Meredith, H. V. *Hum. Biol.*, 24:4, 1952. / Birth weights, racial differences: natality characteristics. Vit. Stat. of U.S., 1960, Vol. 1, Sec. 2, Tables, 2, 6, 7, 8, 9. / Prematurity. Edward H. Bishop. *Postgrad. Med.*, Feb. 1964. / Low birth-weight of the premature African infant (Congo babies, in French). M. Vincent, J. Hugon. *Bull. World. Hlth., Org.*, 26–143, 1962. / African babies alert. *Sci. News Let.*, 85, p. 125, Feb. 22, 1964.

BLOOD GROUPS. Genetics and the human race. Boyd, William C. *Science*, 140, June 7, 1963. *Also:* ¶ Boyd, W. C. *Genetics and the Races of Man.* Little, Brown, 1960. / *Heredity of the Blood Groups.* (Chap. 12.) Wiener, A. S., Wexler, I. B. Grune & Stratton, 1958. / Effects of natural selection on human genotypes (Symposium). *Ann. N.Y. Acad. Sci.*, 65:1, June 18, 1956. / Negro-White, blood groups, biochemical differences. Cooper, A. J., et al. *Am. J. Hum. Gen.*, 15:4, Dec., 1963. / Gamma globulin, Negro-White differences. Steinberg, A. G., & Wilson, J. A. *Am. J. Hum. Gen.*, 15:1, Mar., 1963. / Blood groups, race and prehistory. Ottensooser, F. *Hum. Biol.*, 27:4, 1955.

DISEASE DIFFERENCES, RACIAL, ETHNIC. Discussion in symposium, Culture, Society and Health. N.Y. Acad. Sci., 84:17, 1960–61. / Hemoglobinopathies, Geography of. Neel, James V. In: *Proc. Conf. on Genet. Polymorphisms.* Blumberg, B. S., Ed. Grune & Stratton, 1961. / G-6-PD deficiency, Caucasians, Negroes. Oski, Frank A. et al. *Science*, 139, Feb. 1, 1963. / Favism, G-6-PD, anemias, etc. In: *The Genetics of Migrant Populations.* (Proc., Conf. on Population Genetics, Israel). Goldschmidt, Elisabeth, Ed. Williams & Wilkins, 1963. / Cardiovascular-renal mortality in Hawaii. Bennett, C. G., et al. *Am. J. Pub. Health*, 52:9, Sept., 1962. / Cancer of skin, race differences. Howell, J. B. *Arch. of Dermatol.*, 82–865, Dec., 1960. / Cancer in Jews. Margolis, Emmanuel. *Eug. Qu.*, 8:3, Sept., 1961. / Porphyria. (in S. Africa pedigree). Dean, Geoffrey. *Sci. Amer.*, Mar., 1957. P. 133. / Kuru. Bennett, J. H.; *also* Gajdusek, D. C. *Eug. Qu.*, 9:1, Mar., 1962. / Color-blindness, population differences in. Post, Richard H. *Eug. Qu.*, 9:3, Sept. 1962, 9:4, Dec. 1962, and 10:2, Sept., 1963. Comments on: Neel, J. V., Richard H. Post, *Eug. Qu.*, 10:1, Mar., 1963; Salzano, F. M., *Eug. Qu.*, 10:2, June, 1963. / Congenital abnormalities, race differences. Ivy, R. H. *Plastic Reconstr. Surg.*, 30:5, Nov., 1962. *Also:* Differences between Caucasians and Japanese in incidence of. Tanaka, Katumi. In: *The Genetics of Migrant and Isolate Populations*, above, under "Favism."

THE HUMAN RACES: II CULTURAL EVOLUTION

(*Chapter* 43)

MAN'S CULTURAL EVOLUTION, GENERAL. Coon, Carleton S. *The Story of Man*, 2nd ed. Knopf, 1962. / The study of race. Washburn, S. L. *Am. Anthrop.*, 65:3, 1963. ¶ Symposium: Evolution of tools (Washburn, S. L.); of cultures (Steward, J. H.); of agriculture (Braidwood, R. J.); of cities (Adams, R. M.); of society (Sahlins, M. D.); in speech (Hockett, C. F.); of science (Butterfield, Herbert). *Sci. Amer.*, 203:3, Sept.,

1960. / Culture and the Evolution of man (symposium). Montagu, Ashley, Ed. Oxford U. Press, 1962. ¶ *The Evolution of Culture.* White, L. A. McGraw-Hill, 1959. / *The Evolution of the Human Brain.* Von Bonin, Gerhardt. U. of Chicago Press, 1963. ¶ *Social Life of Early Man.* Washburn, S. L. Aldine, 1963. ¶ *Prehistory and the Beginnings of Civilization.* Hawkes, Jacquetta, Woolley, Leonard. Harper & Row, 1963. ¶ *The Epic of Man. Life* Editors, and Barnett, Lincoln. Time, Inc., 1961. ¶ *Vanished Civilizations of the Ancient World* (symposium). Bacon, Edward, Ed. McGraw-Hill, 1963. / Ancient Mesoamerican civilization. MacNeish, R. S. *Science,* 7, p. 143, Feb. 7, 1964. / Upper paleolithic cultures in Western Europe. Sonneville-Bordes, Denise de. *Science,* 142, Oct. 18, 1963. ¶ Graziosi, Paolo. Palaeolithic Art. McGraw-Hill, 1960. *See also:* ¶ Prehistoric art in the Alps. Anati, Emmanuel. *Sci. Amer.,* Jan., 1960. Rock art of Africa (review of books). Coon, C. S. *Science,* 142, Dec. 27, 1963.

PRIMITIVE MAN. ¶ *The World of Primitive Man.* Radin, Paul, Grove Press, 1960. Primitive races now dying out. Gusinde, M. *Int. Soc. Sci. Bull.,* 9:3, 1957. / Primitive Art. Wingert, Paul S. Oxford U. Press, 1962. / Primitive architecture and climate. Fitch, J. M., & Branch, D. P. *Sci. Amer.,* Dec., 1960.

AMERICAN NEGROES. ¶ *Black Cargoes: A History of the Atlantic Slave Trade.* Mannix, D. P., & Cowley, Malcolm. Viking, 1962. / How the Negro came to slavery in America. Wallace, Robert. *Life* Mag., Sept. 3, 1956. / Migration, mobility and the assimilation of the Negro. Taeuber, Irene B. *Population Bull.* (Pop. Ref. Bur.), 14:7, Nov., 1958. ¶ The American Negro, "his hope, his future" (symposium, Natl. Urban League). N.Y. *Times Sun.* Mag., Sect. 10, Jan. 17, 1960. ¶ Negro prizefighting. In: The occupational culture of the boxer. Weinberg, S. K., & Arond, Henry. *Am. J. Social.,* Mar., 1952. / The Negro in track athletics. Meade, George P. *Sci. Monthly, Dec.,* 1952. ¶ Our Negro aristocracy. Davidson, Bill. *Sat. Eve. Post.,* June, 1962.

AFRICAN NEGROES. *The Prehistory of East Africa* (rev. ed.). Sonia Cole. Macmillan, 1963. ¶ *The Human Factor in Changing Africa.* Herskovitz, Melville, J. Knopf, 1962. / African Negroes, types. (Ruanda-Urundi and Kivu: Photos and French text.) Hiernaux, J. *Ann. du Musée Royal du Congo Belge,* Tervuren (Belgium), 1956. / African material culture and technology. Bascom, William. *Annls. N.Y. Acad. Sci.,* 96:2, Jan. 10, 1962. / Anthropology and Africa today (symposium). N.Y. Acad. Sci., 96:2, 1961. / Bambuti pygmies, recent developments in sociology of. Turnbull, Colin M. *Trans. N.Y. Acad., Sci.,* Feb., 1960.

THE JEWISH PEOPLE. ¶ Shapiro, Harry L. In: *Race and Science.* (UNESCO Symposium.) Columbia U. Press, 1961. / Genetic differences among Jews. In: *The Genetics of Migrant and Isolate Populations.* (symposium, Proc., Conf. in Israel, 1961.) Goldschmidt, Elisabeth, Ed. Williams & Wilkins, 1963. / Surveys of deleterious genes in different population groups in Israel. Sheba, Chaim, et al. *Am. J. Pub. Health*, July, 1961. ❡ Legal definition of Jew. Israeli court ruling in case of Brother Daniel. N.Y. *Times*, Dec. 7, 1962.

RACE PREJUDICE, RACIST THEORIES. ❡ Science and the race problem. (Report, Committee Am. Assn. Adv. Sci.), *Science*, 142, Nov. 1, 1963. ❡ (UNESCO Symposium.) *Race and Science*. Columbia U. Press, 1961. / Racist theories, discussions (in re, Race and Reason, by Carleton Putnam). *Science*, 134, Dec. 8, 1961, pp. 1868–69, and Mar. 16, 1962, pp. 961, 982. ❡ *Changing Patterns of Prejudice*. Marrow, Alfred, Jr., Chilton, 1962. ❡ *National Character and National Stereotypes*. Duijker, H. C. J., & Fridja, N. H. Humanities Press, 1961. / Social structure and race relations. (Latin-America.) Bonilla, E. Seda. *Socl. Forces*, Dec., 1961. ❡ *Colour Prejudice in Britain*. Richmond, Anthony H. Grove Press, 1954. ❡ *White stereotypes of the Negro*. *Newsweek*, Oct. 21, 1963, p. 50. ❡ *Negro conceptions of White people*. *Am. J. Sociol.*, Mar., 1951. / Current misconceptions about Australian aborigines. Berndt, Ronald M. *Unesco Courier*, Apr., 1959. / Genetics of race equality. Theodosius Dobzhansky. *Eug. Qu.*, 10:4, Dec., 1963.

RACE CROSSING, HYBRIDIZATION. The dynamics of racial intermixture. Roberts, D. F., & Hiorns, R. W. *Am. J. Hum. Gen.*, 14:3, Sept., 1962. / Genetic aspects of race mixture. In: *Principles of Human Genetics*, 2nd ed. (Chapter 32). Stern, Curt. W. H. Freeman, 1960. / Brazil, gene flow, White into Negro. Saldanha, P. H. *Am. J. Hum. Gen.*, 9:4, Dec., 1957. / The mixed-blood in modern Africa. Smythe, Hugh H. *Sociol. & Socl. Res.*, Jan., 1961. / Hawaii, genetics of interracial crosses in. Morton, Newton E. *Eug. Qu.*, 9:1, Mar., 1962. / Australian aborigines and Whites. Gates, R. R. *Acta Gen. Med. Gemel.*, 9:1, Jan., 1960. / Indians and Europeans, race-crossing. Tiwari, S. C. *Annls. Hum. Gen.*, 26:3, Feb., 1963. / Hybrids and history. (Role of race-crossing in achievement.) Snell, George D. *Qu. Rev. Biol.*, 26:4, 1951.

ANCESTRY

(*Chapter* 44)

❡ Genealogy hunting. *Life* Mag., Jan. 18, 1963, p. 13. (Under "Life Guide.") *Also:* Coats of arms and family trees. *Life* Mag., May 3, 1963. ❡ *English Ancestry*. Wagner, Anthony A. Oxford U. Press (Paperback), 1961. ❡ "This special breed." (British royal family and kinships.) *Newsweek*, Aug. 26, 1957, p. 100. *The James Family*. Matthiessen, F. O. Knopf, 1947. ❡ Mayflower descendants. *Life* Mag., Nov. 29, 1948. p. 129. ❡ A *Family of Friends*. (Quaker lineage.) Allerton Parker, R. Museum Press (London), 1960. / *The Kallikak Family*. Goddard, Henry H. Macmillan, 1912. (Reprinted, 1939.) / In defense of the Kallikak study. Goddard, Henry H. *Science*, June 5, 1942. ❡ The Kallikaks after thirty years. Scheinfeld, Amram, *J. of Hered.*, 35:9, Sept., 1944. / *The Jukes*, Dugdale, R. L. 4th ed. G. P. Putnam, 1888. / Closeness to blood relatives (social times). Robins, Lee N., & Tomanec, Miroda. *Marrg. & Fam. Livg.*, Nov., 1962.

PERSONAL PROBLEMS OF HEREDITY

(*Chapter* 45)

GENETIC COUNSELING. *Heredity Counseling.* (Symposium, Amer. Eugenics Soc.), Hammons, H. G., Ed. Hoeber-Harper, 1959. *Also: Eug. Qu.*, 5:1, 1958. / *Counseling in Medical Genetics.* 2nd ed. Reed, Sheldon C. Saunders, 1963. / Genetic prognosis. Roberts, J. A. F. *Brit. Med. J.*, 1:587, Mar. 3, 1962. / Psychiatric aspects of genetic counseling. Kallmann, F. J. *Am. J. Hum. Gen.*, 8:2, June, 1956.

COUSIN MARRIAGES. ¶ Should cousins marry? Scheinfeld, Amram. *Today's Health*, 34:6, June, 1956. *See also:* refs. under "Eugenics: I" (Chapter 46.)

FERTILITY AND STERILITY. Sterility and impotence (Sympos.). *N.Y. Medicine.* 13:3, Feb. 5, 1957. / *Impotence and Frigidity.* Hastings, Donald. Little, Brown, 1963. ¶ Hereditary sterility. In female cattle: Kidwell, J. F., et al. *J. Hered.*, May–June, 1954; in male saanen goats: Soller, M., et al. *J. Hered.*, 54:4, Sept.–Oct., 1963. / Male infertility: sex chromatin, chromosomes in. Sohaal, A. R. *Fertil. & Steril.*, 14:2, 1963. / Endocrine approach to management and diagnosis of infertility. Kupperman, Herbert S. *Postgrad. Med.*, Nov., 1959. *Also: Am. J. Obst. & Gyn.*, 73, Feb., 1958. ¶ "To those denied a child" (with list of infertility clinics). Planned Parenthood Federation, Jan., 1955. (Write, 501 Madison Ave., N.Y. 22, N.Y., enclosing 10 cents.) *See also:* Continuing files on fertility and sterility, Am. Soc. for the Study of Sterility.

ARTIFICIAL INSEMINATION. British Home Office, Report: Dept. Com. on Hum. Artificial Insem. Lord Feversham, Ch. H. M. Stationery Off., 1960. / Therapeutic donor insemination. Kleegman, Sophia J. *Fertil. & Steril.*, 5:1, Feb., 1954. / Medicolegal aspects of. Lombard, John F. *Postgrad. Med.*, Oct., 1958. / Husband responsible for child's support (legal) decision). Gursky vs. Gursky. N.Y. Sup. Ct., Kings Co. (N.Y., July 26, 1963). / Sociolegal aspects of. Levisohn, A. A. *J. Forensic Med.*, 4, Oct.–Dec., 1957. / Catholic attitude toward. Leonard, J. T. *Am. Eccl. Rev.*, 140, 1959. ¶ United Presbyterian Church on. *Time* Mag., June 1, 1962, p. 83. / Artificial insemination not adultery (editorial). *Lancet*, 1:157, Jan. 8, 1958. ¶ Attitudes toward artificial insemination. Vernon, G. M., & Boadway, J. A. *Marrg. & Fam. Liv.*, Feb., 1959.

ADOPTION. ¶ *Adoption of Children.* (Booklet, with many refs.) *Am. Acad. Pediatrics.* (1801 Hinman Av., Evanston, Ill.), 1959. / *Independent Adoptions.* Witmer, Helen L., et al. Russell Sage Found., 1963. ¶ You and your adopted child (pamphlet). Public Affairs Committee. (22 E. 38th St., New York 16, N.Y.) ¶ Legislative guides in adoptions. U.S. Children's Bur., 1962. (Write, Supt. of Documents, U.S. Govt. Print. Off., Washington, 25, D.C., enclosing 30 cents.) ¶ Motives and conflicts in foster parenthood. McCoy, Jacqueline. *Children* (U.S. Children's Bur.), Nov.–Dec., 1962. ¶ Mixed racial ancestry, adopting children of. Genetic counseling. Reed, Sheldon C., & Nordlie, Esther B. *Eug. Qu.*, Sept., 1961.

DISPUTED PARENTAGE TESTS. Chances of disproof of false claims of parent-child relationship. Boyd, W. C. *Am. J. Hum. Gen.*, 9:3, Sept., 1957. / Non-paternity, estimation of frequency. MacCluer, J. W., & Schull, W. J. *Am. J. Hum. Gen.*, 15:2, June, 1963. / Medicolegal applications of blood grouping tests. Wiener, A. S., et al. *J.A.M.A.*, 164:18, 1957. *Also:* J. of Forens. Sci., 4:3; July, 1959. / *Disputed Paternity Proceedings*, 3rd. ed. Schatkin, Sidney. Bender & Co., 1953. *Also:* The scandal of our paternity courts. *Reader's Dig.*, May, 1960. / Review of 1000 disputed cases. Sussman, L. N. *Am. J. Clin. Path.*, 40:1, 1963. / Advances in anthropological paternity testing. Keiter, F. *Am. J. Phys. Anthrop.*, 21:1, 1963.

APPLICATION OF THE GC SYSTEM IN PATERNITY CASES. Hirschfeld, Jan, & Heiken, Aage. *Am. J. Hum. Gen.*, 15:1, Mar., 1963.

EUGENICS: I

(*Chapter* 46)

POPULATION GROWTH. ¶ *The Population Dilemma.* (Symposium.) Hauser, Philip M., Ed. Prentice-Hall, 1963. ¶ Population. Davis, Kingsley. *Sci. Amer.*, Sept., 1963. / World population growth, dilemma. Dorn, Harold F. *Science*, 135, Jan. 26, 1962. ¶ This crowded world. Osborn, Frederick. Publ. Affairs Pamphl., 306, 1960. / Malthusian theory. Petersen, William. *Population Rev.*, 5:2, July, 1961. / *See also: Population Bulletin*, 18:1, Feb., 1962 ("How many people have ever lived on earth?"); 18:6, Oct., 1962 (Latin America and population growth); also, continuing issues of *Pop. Bull.*

DIFFERENTIAL FERTILITY. Birth-rate changes, N. W. Europe and N. America. Carter, Cedric O. *Eug. Qu.* 9:1, Sept., 1962. / Negro-White birth-rate differences, U.S., *U.S. Health, Educ. Welf. Indicators* (U.S. Natl. Off. Vit. Statist.), March, 1963. / Jewish fertility in the U.S. Rosenthal, Erich. *Eug. Qu.*, Dec., 1961. / India, Japan, population control in. Freymann, Moye W., and Takeshita, J. Y. (two articles). *Marrg. & Fam. Livg.*, Feb., 1963. ¶ Irish population decline. *Sci. News Let.*, 83, p. 367, June 8, 1963.

DIFFERENTIAL FERTILITY, SOCIO-ECONOMIC. Differentiation in mating and fertility trends (symposium). *Eug. Qu.*, 6:2, June, 1959. / Social and psychological factors affecting fertility. *Milbank Mem. Fund Qu.*, 36:3, July, 1958. / Marriage and fertility patterns of college graduates. Lauriat, Patience. *Eug. Qu.*, Sept., 1959. / *The Fertility of American Women.* Grabill, W. H., Kiser, C. V., & Whelpton, P. K. Wiley, 1958.

EUGENICS. Galton's heritage. Slater, Eliot. *Eug. Rev.*, 52:2, July, 1960. ¶ Cook, R. C. *Eug. Rev.*, 55:3, Oct., 1963. / *Eugenics.* Haller, Mark H. Rutgers U. Press, 1963. / Cross currents in the history of human genetics. Dunn, L. C. *Am. J. Hum. Gen.*, 14:1, Mar., 1962. / Eugenics in evolutionary perspective. Huxley, Julian. *Eug. Rev.*, 54:3, Oct., 1962. / The Eugenic position, statement. American Eugenic Soc. *Eug. Qu.*, 8:4, Dec., 1961. ¶ Changing attitudes toward human genetics and eugenics. Scheinfeld, Amram. *Eug. Qu.*, 5:3, 1958. / Genetics and equality.

Dobzhkansky, Th. *Science*, 137, July 13, 1962.

CARRIERS. In: *Medical Genetics* (pp. 39, 65). Lenz, Widukin. U. of Chicago Press, 1963. *See also* Table, "Clues to Carriers," in our Appendix, p. 804.

COUSIN MARRIAGES, INBREEDING. Morbidity of children from consanguineous marriages. Morton, N. E. In: *Progress in Medical Genetics* (Chapter 7). Steinberg, A. G., Ed. Grune & Stratton, 1961. / Consanguineous marriage and genetic implications. Freire-Maia, Newton & A. *Ann. Hum. Genet.*, 25, 1961. *Also:* Inbreeding levels in different countries. *Eug. Qu.*, 4:3, Sept., 1957. / Cousin marriages. Darlington, C. D. *Eug. Rev.*, 51:4, Jan., 1960; 53:3, Oct., 1961. / Consanguineous marriages (among Catholics), Chicago region. Slatis, Herman M., *et al. Am. J. Hum. Genet.*, 10:4, Dec., 1958. ¶ Brother-sister, father-daughter marriage in ancient Egypt. Middleton, Russell. *Am. Sociol. Rev.*, 27:5, Oct., 1962.

BIRTH CONTROL, METHODS. ¶ *Planning Your Family; The complete guide to contraception.* Guttmacher, Alan F., Best, Winfield, & Jaffe, Frederick. Macmillan, 1964. *Also: The Complete Book of Birth Control.* Ballantine, 1961. *Manual of Contraceptive practice.* Calderone, Mary S., Ed., Williams & Wilkins, 1964. ¶ Pamphlets: Modern methods of birth control; The safe period. Planned Parenthood Federation, 1964. (Obtainable from any office of this organization.) ¶ *Science and the safe Period.* Hartman, Carl G. Williams & Wilkins, 1962. / Child spacing: mathematical probabilities. de Bethune, André J. *Science*, 142, Dec. 27, 1963. / Birth control (historical background). Pyke, Margaret. *Eug. Rev.*, 55:2, July, 1963.

BIRTH CONTROL, RELIGIOUS ATTITUDES. ¶ *The Time Has Come.* Rock, John. (Discussion of Catholic issue.) Knopf, 1963. ¶ Catholic theologians on birth control. N.Y. *Times*, July 8, 1963, p. 1. ¶ *The Population Explosion and Christian Responsibility.* Fagley, R. M. Oxford U. Press, 1960.

STERILIZATION. Voluntary sterilization. Blacker, C. P. *Eug. Rev.*, 54:1, Apr., 1962, and 54:3, Oct., 1962. *Also:* New Virginia law on. Reed, S. C. *Eug. Qu.*, 9:3, 1962. / Eugenic sterilization (in German). Stöckmann, F. Gesundheitsfursorge, 1962. / Sterilization in Japan. Koya, Yoshio. *Eug. Qu.*, Sept., 1961. / Reactions of men to vasectomies. Poffenberger, Thomas and Shirley. *Marrg. & Fam. Livg.*, Aug., 1963. *Also:* Rodgers, David, et al. *Marrg. & Fam. Livg.*, Nov., 1963. / The male operation. *Newsweek*, Sept. 16, 1963, p. 84. / Prognosis after sterilization on socio-psychiatric grounds. Eklad, Martin. *Acta Psychiatr. Scand., Suppl.* 161, Vol. 36, 1961. / Sterilized mental defectives look at eugenic sterilization. Sabagh, G., Edgerton, R. B. *Eug. Qu.*, 9:4, Dec., 1962. / For further data on sterilization, write: Human Betterment Assn. for Voluntary Sterilization, 515 Madison Ave., New York 10022.

EUGENICS: II

(*Chapter* 47)

EUTELEGENESIS (SELECTIVE HUMAN BREEDING). Human evolution by voluntary choice of germ plasm.

Muller, Hermann J. *Science,* 134, Sept. 8, 1961. *Comments on* (and rejoinder by Dr. Muller): Jackson, Don D., Kane, W. H., Classen, H. G., *Science,* 134, Dec. 8, 1961.

SPERM PRESERVATION, SELECTIVE UTILIZATION. Improved methods of preserving human spermatazoa by freezing. Sherman, J. K. *Fertil. & Steril.,* 14, p. 49, 1963. *Also:* Research on frozen human semen—past, present, future. *Fertil. & Steril.,* 15:5, 1964. ¶ Artificial breeding of livestock. *New York Times,* Mar. 16, 1958. ¶ Test-tube tempest (fertilization, human fetal development in vitro). *Newsweek,* Feb. 6, 1961, p. 78. ¶ Legal problems. Perpetuities in the Atomic Age: The sperm bank and the fertile decedent. Leach, W. Barton. *Am. Bar Assn.,* N.J., Oct., 1962. / *See also* references under "Artificial Insemination," Chapter 45 (Personal Problems of Heredity).

OVA TRANSPLANTATION. Storage of fertilized ova. Havez, E. S. E. *Int. J. Fertil.,* 8, p. 459, 1963. / Long distance transfer of sheep ova. Adams, C. E. et al. *Proceedings 4th Int. Cong, on Animal Reprod.,* The Hague, 1961, v. 2, p. 381. / Reciprocal transfer of cattle and rabbit embryos. Havez, E. S. E. & Sugie, T. *J. Animal Sci.,* 22, p. 30, 1963. / Asynchronous transfer of mouse ova. Doyle, L. L. et al. *Fertil. & Steril.,* 14:2, 1963. / *See,* for continuing references on sperm preservation, ova transplantation, etc., later files, *Fertil. & Steril.*

PARTHENOGENESIS (*virgin birth*). Performance record of a parthenogenetic turkey male. Olsen, M. W. *Science,* Dec. 2, 1960, p. 1661. / The mitotic chromosomes of parthenogenetic and normal turkeys. Poole, H. K. *J. of Hered.,* July–Aug., 1959. ¶ Fatherless gobbler (with photo). *Sci. News Let.,* Nov. 9, 1957. ¶ Life sans father. *MD,* Nov., 1957, p. 58.

EUGENIC PROBLEMS (*See* Text, pp. 709–711). ¶ The present evolution of man. Dobzhansky, Th. *Sci. Amer.,* Sept., 1960. ¶ Battered child syndrome. Elmer, Elizabeth. *Children* (U.S. Children's Bur.), Sept.–Oct., 1963, p. 180. *Also: Children,* Nov.–Dec., 1963, p. 237.

FUTURE DEVELOPMENTS, CONTINUING HUMAN EVOLUTION. ¶ *Mankind Evolving.* Dobzhansky, Th. Yale U. Press, 1962. / *Animal Species and Evolution.* Mayr, Ernst. Harvard U. Press, 1963. / The human future. Muller, H. J. In: *The Humanist Frame.* Huxley, Julian, ed. Harper. 1961. ¶ *The Torch of Life. Dubos, René.* Simon & Schuster, 1962. / Education and the human revolution. Huxley, Julian. *Eug. Rev.,* 55:2, July, 1963. / Science and the new humanism. Hoagland, Hudson. *Science,* Jan. 10, 1964. *Comments on: Science,* Mar. 13, 1964, pp. 1121–23. / Implications of the new genetics for biology and man. Sonneborn, T. M. *A.I.B.S.* (Am. Inst. Biolog. Sci.), Apr., 1963. / Biological discoveries and the future of the family. Nimkoff, M. F. *Socl. Forces,* 41:2, Dec., 1962. / The psychology of 1975. Murphy, Gardner. *Am. Psychologist,* Nov., 1963. ¶ The nonprevalence of humanoids (men on other planets). Simpson, George C. *Science,* 143, Feb. 21, 1964. / Exobiology: approaches to life beyond the earth. Lederberg, Joshua. *Science,* Aug. 12, 1960. / Life forms elsewhere than on earth. Muller, H. J. *Am. Biol. Teacher,* 23:6, Oct., 1961.

BASIC PERIODICALS DEALING WITH HUMAN HEREDITY

Published in the United States unless otherwise noted. The abbreviations by which these journals are cited in the bibliography are printed in **boldface** *type.*

Journals of Genetics

Archiv der **Julius Klaus Stiftung** für Vererbungsforchung. Bi-ann. (German or French.) ART, Inst. Orell-Füssli. Dietzinger Strasse 3, Zurich, Switzerland.

Acta Geneticae **Med**icae et **Gemel**lologiae. Qu. (English, other languages, stressing twin studies.) Piazza Galeno 5, Rome, Italy.

Acta Genetica et **Statist**ica **Med**ica. (Or, "**Acta Genet**ica.") Qu. (English, French or German.) Basel, 11, Switzerland. (Also, S. Karger, N.Y.)

American **J**ournal of **Hum**an **Gene**tics. Qu. Grune & Stratton, 381 Park Ave. S., New York 16, N.Y.

Annals of **Hum**an **Gen**etics (formerly Annals of Eugenics). Qu. Cambridge U. Press, 32 E. 57th St., New York 22, N.Y.

Cytogenetics. (Chromosome studies; English, other languages.) 6 times yearly. S. Karger, AG. Arnold-Bucklin-Strasse, 25, Basel, Switzerland.

Eugenics **Q**uarterly. American Eugenics Society, 230 Park Ave., New York 17, N.Y.

Eugenics **Rev**iew. Qu. Eugenics Society, England, 69 Eccleston Sq., London, S.W.1, England.

Human **Biol**ogy. Qu. Wayne State U. Press, Detroit 2, Mich.

Journal **de Génét**ique **Hum**aine. Qu. (French, German; English summaries.) 22 rue Micheli-du-Crest, Geneva, Switzerland.

Journal of **Hered**ity. Bi-monthly. Amer. Genetics Assn., 1507 M St., N.W., Washington 5, D.C.

Abstract Journals (see under desired topics)

Human **Genet**ics, **Abstr**acts. Monthly (from Sept., 1962). Excerpta Medica Found., N.Y. Academy Medicine Bldg., 2 E. 103rd St., New York 29, N.Y. (Also, London, Amsterdam, other cities.)

Abstracts of **Hum**an **Develop**mental **Biol**ogy. Monthly. Excerpta Medica Found. (Address above.)

Birth Defects. Monthly (from Jan., 1964). Natl. Foundation, 800 2nd Ave., New York, N.Y., 10017.

Psychological **Abstr**acts. Monthly. (Frequent reports on behavioral genetics.) Amer. Psychological Assn., 1333 Sixteenth St., N.W., Washington, D.C., 20036.

Medical and other journals (with frequent articles on human heredity)

Basic Periodicals

American Journal of Physical Anthropology. Qu. Wistar Inst., 36th & Spruce, Philadelphia 4, Pa.

British Medical Journal. Tri-ann. (Jan., May, Sept.) The British Council, 65 Davies St., London, W.1, England.

Journal of American Medical Assn. Weekly. 535 N. Dearborn St., Chicago 10, Ill.

Journal of Medical Genetics. Quarterly (from Sept., 1964). Lippincott, Philadelphia 5.

Journal of Pediatrics. Monthly. C. V. Mosby Co., 3207 Washington Blvd., St. Louis 3, Mo.

Lancet. Weekly. 7 Adam St., Adelphi, London, W.C.2.

Medical Tribune. Tri-weekly. Frequent late news items on human heredity. 624 Madison Ave., New York, N.Y.

Metabolism. Bi-monthly. Grune & Stratton, 381 Park Ave. S., New York 16, N.Y.

Nature. Weekly. Macmillan, St. Martin St., London, W.C.2.

Science. Weekly. Amer. Assn. Advancement of Science. 1515 Massachusetts Ave., N.W., Washington, D.C., 20005.

Miscellaneous

Annual Review of Medical Genetics. (From 1961.) C. V. Mosby, St. Louis 3, Mo.

Annual Review of Teratology. (Congenital malformations.) Excerpta Medica Found. (See address under "Human Genetics, Abstracts.")

HUMAN HEREDITY CLINICS

For counseling and information on problems of heredity.

UNITED STATES

ARIZONA: Department of Genetics, Arizona State University, Tempe.

CALIFORNIA: (1) Department of Genetics, University of California, Berkeley. (2) Los Angeles Medical Center.

DISTRICT OF COLUMBIA: Genetics Counseling Research Center, George Washington University Hospital.

ILLINOIS: Children's Memorial Hospital, Chicago.

LOUISIANA: Genetic Counseling Service, Tulane University, New Orleans.

MARYLAND: Department of Medical Genetics, Johns Hopkins Hospital, Baltimore.

MASSACHUSETTS: Department of Immunochemistry, Boston University Medical School.

MICHIGAN: (1) The Heredity Clinic, University of Michigan, Ann Arbor. (2) Department of Zoology, Michigan State University, East Lansing.

MINNESOTA: (1) Dight Institute, University of Minnesota, Minneapolis. (2) Human Genetics Unit, State Board of Health, Minneapolis. (3) Mayo Clinic, Rochester.

NEW YORK CITY: (1) Human Genetics Laboratory, New York University Medical Center. (2) Department of Genetics, Albert Einstein College of Medicine of Yeshiva University. (3) Department of Medical Genetics, New York State Psychiatric Institute. (4) Department of Medical Genetics, Rockefeller Institute.

NEW YORK STATE: Genetic Counseling Service, New York State Department of Health, at Albany Medical College, Albany.

NORTH CAROLINA: Department of Medical Genetics, Bowman Gray School of Medicine, Winston-Salem.

OHIO: Department of Biology, Western Reserve University, Cleveland.

OKLAHOMA: Department of Zoological Sciences, University of Oklahoma, Norman.

RHODE ISLAND: Department of Biology, Brown University, Providence.

TEXAS: The Genetics Foundation, University of Texas, Austin.

VIRGINIA: School of Medicine, University of Virginia, Charlottesville.

WASHINGTON STATE: Department of Medicine, University of Washington, Seattle.

WISCONSIN: Department of Medical Genetics, University of Wisconsin, Madison.

CANADA

ALBERTA: Heredity Counseling Service, University of Alberta, Edmonton.

MANITOBA: Hospital for Sick Children, Winnipeg.

ONTARIO: Hospital for Sick Children, Toronto.

QUEBEC: (1) Department of Medical Genetics, McGill University, Montreal. (2) Children's Memorial Hospital, Montreal.

Index

Note: The primary objective in this Index is to aid the reader in finding as quickly as possible the page or pages where a sought for discussion or illustration occurs. If there are several or many page references, the location of the principal discussion is indicated by bold (black) type. In the listing of *names*, space limitations have required that indexing be mainly of persons whose works or contributions are directly cited, or of others important in given discussions. The extensive Bibliography has not been indexed, but names of persons most likely to be sought for will usually be found under the subject classifications of areas with which they have been predominantly identified.

Index

Index

Index

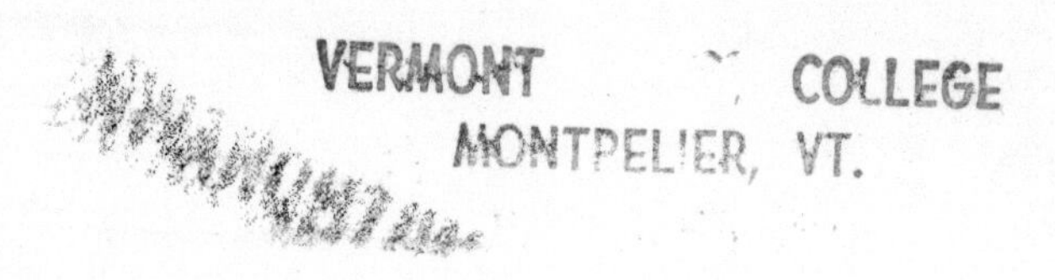